猪营养需要

（第十一次修订版）

NUTRIENT REQUIREMENTS OF SWINE

(ELEVENTH REVISED EDITION)

美国国家科学院科学研究委员会　著

印遇龙　阳成波　敖志刚　主译

科学出版社

北　京

图字：01-2013-5621号

内 容 简 介

美国科学研究委员会（NRC）（1998）第十版《猪营养需要》出版已经17年了，这期间全世界乃至我国的养猪业都发生了巨大的变化。随着中国养猪业规模化的迅猛发展、猪品种的改良、养殖模式的更新，对猪的营养需求也有了新的要求。在广大养殖界朋友的期待中，《猪营养需要》（第十一次修订版）于2012年7月正式出版。本书共17章，更新了上一版中9个章节：能量，蛋白质和氨基酸，猪营养需要量估测模型，矿物质，维生素，水，非营养性饲料添加剂，营养对养分排泄和环境的影响，营养需要量表和饲料原料组成；删除了上一版中的第9章"日粮配制"；同时增加了7个章节：脂类，碳水化合物，玉米和大豆加工副产物，饲料污染物，饲料加工，营养物质和能量的消化率，未来研究方向。

营养需要量标准和原料数据库是猪饲料配方的基石，精准的营养需要量和原料数据库的使用不仅能有效促进猪的生产性能，而且可以更加高效和经济地使用饲料原料，从而节约饲料配方成本。本书综述了猪营养和原料营养成分的最新信息，这些新知识将优化养猪生产，另外对生产乙醇所产生的新原料与饲料污染物及环保等相关问题都一并进行了描述，所以对维持高效而环保的养猪生产而言，本书是良好的指南。

本书适合于农业大学和科研院所的老师和学生、养猪业和饲料业的营养专家，以及从事开发和生产一线工作的技术人员阅读。

This is the translation of *Nutrient Requirements of Swine: Eleventh Revised Edition*, by Subcommittee on Swine Nutrition, Committee on Animal Nutrition and National Research Council. © 2012 National Academy of Sciences. First published in English by the National Academies Press. All rights reserved.

图书在版编目（CIP）数据

猪营养需要：第11次修订版/美国国家科学院科学研究委员会著；印遇龙等译. —北京：科学出版社，2014.8
书名原文：Nutrient Requirements of Swine (Eleventh Revised Edition)
ISBN 978-7-03-040955-3

Ⅰ.①猪… Ⅱ.①美… ②印… Ⅲ.①猪-家畜营养学 Ⅳ.①S828.5

中国版本图书馆CIP数据核字（2014）第121939号

责任编辑：李秀伟 刘 晶 王 静／责任校对：朱光兰
责任印制：苏铁锁／封面设计：北京茗轩堂广告设计有限公司

科学出版社 出版
北京东黄城根北街16号
邮政编码：100717
http://www.sciencep.com

北京凌奇印刷有限责任公司 印刷
科学出版社发行 各地新华书店经销

*

2014年8月第 一 版　开本：889×1194 1/16
2020年5月第七次印刷　印张：27 3/4
字数：700 000
POD定价：138.00元
（如有印装质量问题，我社负责调换）

译者名单

本书由 国家生猪产业技术创新战略联盟
　　　 亚热带农业生态研究所 组织翻译

主　译　印遇龙　阳成波　敖志刚

译　者（按姓氏汉语拼音排序）

敖志刚　鲍三高　陈代文　崔志英　方　俊　方热军　何庆华
贺　喜　黄克和　黄逸强　赖州文　李　彪　李　勇　李德发
李芳溢　李铁军　李兴国　廖益平　刘钧贻　卢德秋　穆玉云
欧迪军　彭险峰　谯仕彦　邱　卫　沈慧乐　史清河　苏　宁
谭碧娥　王　琪　王文策　王勇飞　吴　德　夏壮华　肖俊峰
阳成波　杨小军　杨晓建　易敢峰　印遇龙　张志博　郑根华
周　樱　周响艳　邹仕庚　左建军

主　审　李德发　沈慧乐

美 国 国 家 科 学 院
全国科学、工程和医药的顾问

美国国家科学院根据美国国会法令于1863年成立，是民间的、非营利的、科学家的荣誉性自治组织，旨在推动自然科学和技术的发展及应用，提高人们的福利。国家科学院必须为联邦政府就科学和技术发展方面提供建议。Ralph J. Cicerone 博士担任美国国家科学院主席。

美国国家工程学院于 1964 年成立，是隶属于美国国家科学院但与之平行的、由杰出工程学家组成的机构。国家工程学院是一个自治组织，可以独立选拔成员，与国家科学院一样为联邦政府提供建议。国家工程学院可以根据国家的需要进行资助工程技术研究项目，并推动教育和研究，奖励工程界杰出的成就。Charles M. Vest 博士担任美国国家工程学院主席。

美国国家医学院于1970年由美国国家科学院创建，旨在为参与民众健康的杰出医学专家提供服务，并管理医生资格考试等事宜。根据国家科学院的政策，国家医学院为联邦政府提供建议，同时根据需要可以提出医疗卫生、研究和教育方面的项目，并进行研究和解决。Harvey V. Fineberg 博士担任美国国家医学院主席。

美国科学研究委员会于 1916 年由美国国家科学院创建，服务自然科学和工程科学的研究，旨在推广自然科学和工程技术，并为联邦政府提供建议。根据国家科学院的政策，研究委员会已经成为服务国家科学院和国家工程学院的主要机构，为联邦政府、民众、自然科学和工程技术学术界提供服务。研究委员会由国家科学院和国家医学院共同管理，Ralph J. Cicerone 博士和 Charles M. Vest 博士分别为美国科学研究委员会主席和副主席。

www.national-academies.org

《猪营养需要》（第十一次修订版）编写委员会

L. LEE SOUTHERN，主席，路易斯安那州立大学农业研究中心，巴吞鲁日

OLAYIWOLA ADEOLA，普渡大学，西拉斐特，印第安纳

CORNELIS F.M. DE LANGE，圭尔夫大学，安大略

GRETCHEN M. HILL，密歇根州立大学，东兰辛

BRIAN J. KERR，美国农业部农业研究中心，埃姆斯，艾奥瓦

MERLIN D. LINDEMANN，肯塔基大学，列克星敦

PHILLIP S. MILLER，内布拉斯加大学，林肯

JACK ODLE，北卡罗来纳州立大学，罗利

HANS H. STEIN，伊利诺伊大学厄本那香槟分校

NATHALIE L. TROTTIER，密歇根州立大学，东兰辛

员工

AUSTIN J. LEWIS，首席科学家

RUTHIE S. ARIETI，研究助理

外部支持

DAVID BRUTON，计算机程序员

PAULA T. WHITACRE，编辑

农业和自然资源委员会

NORMAN R. SCOTT，主席，康奈尔大学，伊萨卡，纽约
PEGGY F. BARLETT，埃默里大学，亚特兰大，佐治亚
HAROLD L. BERGMAN，怀俄明大学，拉勒米
RICHARD A. DIXON，塞缪尔·诺贝基金会，阿德莫尔，俄克拉何马
DANIEL M. DOOLEY，加州大学，奥克兰
JOAN H. EISEMANN，北卡罗来纳州立大学，罗利
GARY F. HARTNELL，孟山都公司，圣路易斯，密苏里
GENE HUGOSON，明尼苏达农业部，圣保罗
MOLLY M. JAHN，威斯康星大学，麦迪逊
ROBBIN S. JOHNSON，嘉吉基金会，韦扎塔，明尼苏达
A.G. KAWAMURA，Solutions from the Land 公司，欧文，加利福尼亚
KIRK C. KLASING，加州大学，戴维斯
JULIA L. KORNEGAY，北卡罗来纳州立大学，罗利
VICTOR L. LECHTENBERG，普渡大学，西拉斐特，印第安纳
JUNE B. NASRALLAH，康奈尔大学，伊萨卡，纽约
PHILIP E. NELSON，普渡大学，西拉斐特，印第安纳
KEITH PITTS，Curragh Oaks 咨询公司，费尔奥克斯，加利福尼亚
CHARLES W. RICE，堪萨斯州立大学，曼哈顿
HAL SALWASSER，俄勒冈州立大学，利瓦利斯
ROGER A. SEDJO，未来资源组织，华盛顿特区
KATHLEEN SEGERSON，康涅狄格大学，斯托斯
MERCEDES VÁZQUEZ-AÑÓN，诺伟司国际公司，圣查尔斯，密苏里

员工

    ROBIN A. SCHOEN，首席科学家
    KAREN L. IMHOF，管理助理
    AUSTIN J. LEWIS，资深项目负责人
    EVONNE P.Y. TANG，资深项目负责人
    CAMILLA YANDOC ABLES，项目负责人
    KARA N. LANEY，项目负责人
    PEGGY TSAI，项目负责人
    RUTH S. ARIETI，研究助理
    JANET M. MULLIGAN，研究助理
    KATHLEEN A. REIMER，资深项目助理

译 者 序

养猪业是世界肉食工业的重要组成部分。饲料成本是养猪生产中最大的投入，占 70%左右。随着中国养猪业规模化的迅猛发展，品种的改良、养殖模式的更新、原料质量和价格的复杂多变，以及人们对生态环境的关注，对猪日粮配制提出了新要求。正确配合日粮是提高养猪生产效率和减少环境污染的基本途径，而营养需要量标准和原料数据库是猪饲料配方的基石。

美国科学研究委员会（NRC）于 1998 年出版的营养推荐量标准已经成为国内很多技术人员的必备参考书，对我国猪饲料配方和养殖水平的提高做出了卓越的贡献。但是，随着养猪业的发展，有些当时设定的营养需要参数和推荐量已不再适用。因此，美国 NRC 收集了新的研究成果，于 2012 年 7 月出版了新的猪营养需要量推荐标准。新 NRC 标准主要改动了猪的能量和营养需求、部分新原料信息，还新增了评估猪营养需求的最新计算机模型等。

由于该版书籍为原版英文书籍，价格昂贵，而且众多饲料工业和养殖业的技术与管理人员对英文存在一定的阅读困难，业界非常期待新版中译本的出版。应国家生猪产业技术创新战略联盟各成员单位的要求和委托，与 NRC 出版社沟通后达成协议，中译本由国家生猪产业技术创新战略联盟组织国内具备较高英文水平的专业人士进行翻译编写并由科学出版社出版。参与翻译人员组成搭配合理，有在生产一线经验丰富的饲料企业技术总监和饲料添加剂企业的技术总监，也有进行基础研究的高校和研究所教授。希望本书能为广大养猪生产者进行日粮配制提供指南，有效促进猪的生产性能，更加高效和经济地使用饲料原料，节约饲料配方成本，从而提高生产效率，减少环境污染。

印遇龙

国家生猪产业技术创新战略联盟

2013 年 12 月

前　言

《猪营养需要》（第十一次修订版）是由美国科学研究委员会（NRC）在已出版系列丛书基础上修订和颁布的，尤其是第十版，为本版本提供了重要的基础。在过去的15年期间，虽然科学家们发表了很多新的研究报告并且报道了大量的新信息，但是，对于许多营养素（如维生素等）的需要量而言，仍然很少或几乎没有找到任何新的研究数据。

《猪营养需要》（第十一次修订版）修订委员会确立了一个基本原则，即如果没有新的研究成果，就表明不需要修订该营养素的需要量，则保留第十版发表的相关数据。该原则也适用于本书文本，因此本修订版保留了第十版的部分文本。从这个意义上讲，本书是真正的修订本，并且可以避免读者再参考旧的版本。

相反，《猪营养需要》（第十一次修订版）修订委员会对现有饲料原料营养成分表进行了大幅度更新。正如第17章所释，本书对已发表的饲料营养价值数据作了详细的综述，并对饲料原料营养成分表的格式和内容进行了全面修订。

致　　谢

这篇报告以草稿的形式经由不同观点和专业知识的专家进行审阅，过程与美国科学研究委员会报告复核委员会的程序一致。独立审阅的目的在于提供公正的批评性意见，使得学会发表的文章更加具有合理性，并保证其满足客观、有依据和影响力的科研要求。审阅意见和原稿皆以保密形式进行，以保证审阅过程的公正性。我们对以下参与审阅的专家表示感谢：

Michael J. Azain，雅典，乔治亚大学；

R. Dean Boyd，肯塔基州，富兰克林，Hanor 公司；

Patrick C. H. Morel，新西兰，北帕莫斯顿，梅西大学；

Paul J. Moughan，新西兰，北帕莫斯顿，梅西大学；

Elizabeth (Betsy) A. Newton，俄亥俄州，刘易斯堡，Akey 公司；

C. M. (Martin) Nyachoti，加拿大，温尼伯，曼尼托巴大学；

John F. Patience，埃姆斯，艾奥瓦州立大学；

Gerald C. Shurson，圣保罗，明尼苏达大学。

尽管以上专家提出了很多建设性意见和建议，但并未在最后结论及推荐信上署名，在发表之前也未见过终稿。美国科学研究委员会任命康奈尔大学 **Dale E. Bauman** 监督完成整篇文稿的审阅工作，同时负责确保本报告独立审查的程序满足美国科学研究委员会要求，且仔细斟酌所有专家的审阅意见。作者委员会和学会对报告终稿负全部责任。

感谢伊利诺伊州玉米市场销售委员会（the Illinois Corn Marketing Board）、美国饲料教育和研究所（the Institute For Feed Education and Research）、美国国家猪肉委员会（the National Pork Board）、内布拉斯加州玉米委员会（the Nebraska Corn Board）、明尼苏达玉米种植者协会（the Minnesota Corn Growers Association）及美国食品和药物管理局（the U.S. Food and Drug Administration）对委员会工作的经济支持。

同时也要感谢资深项目官员 Austin Lewis 博士和研究协会 Ruthie Arieti 女士对本项目的竭力支持。Lewis 博士在报告的准备过程中进行了出色的指导、建议和鼓励，委员会对他的支持和情谊表示衷心的感谢。Arieti 女士在写作、校正、章节编辑和协调工作中做出了杰出贡献。Arieti 女士同时也是会议的组织和计划者。委员会亦对农业和自然资源董事会董事 Robin Schoen 对获得修订报告的顺利进行及其准备工作的支持和鼓励表示感谢。

此外，一些专家对委员会的工作也做出了重要的贡献。委员会成员感谢 Jason Schmidt 和 Stephen Treese（动物科学学院，路易斯安那州立大学农业中心）提供饲料原料表。对 Jean-Yves Dourmad 博士、Jaap van Milgen 博士和 Jean Noblet 博士（国家农业研究院，法国）以及 Allan Schinckel 博士（普渡大学）公开并指导建立生长营养需要模型的贡献表示感激。Dean Boyd 博士（Hanor 公司）、Mike Tokach 博士（堪萨斯州立大学）和 Soenke Moehn 博士（阿尔伯塔大学）提供了有关饲养管理和商品猪生产力水平的重要资料与反馈。委员会对母猪生殖组织氨基酸水平的评估得益于 Walter Hurley 博士（伊利诺伊大学）对妊娠母猪乳腺组织的慷慨贡献，以及 Robert Payne 博士和 John Thomson 博士（赢创德固赛公司，Evonik Degussa）对乳腺、胎盘、胎儿和子宫组织的氨基酸分析。

目 录

译者序
前言
致谢
总论 / 1
第1章 能量 / 4
 引言 / 4
 术语 / 4
 能量在动物体内的分配 / 4
 产热的组成 / 7
 生理状态 / 10
 能量利用模型——有效代谢能概念 / 12
 参考文献 / 13
第2章 蛋白质和氨基酸 / 17
 引言 / 17
 蛋白质 / 17
 必需、非必需以及条件性必需氨基酸 / 17
 氨基酸来源 / 18
 氨基酸分析测定 / 19
 氨基酸需要量的表示方法 / 20
 日粮氨基酸失衡 / 21
 氨基酸与赖氨酸之比 / 22
 氨基酸需要量的经验估计 / 23
 氨基酸需要量的测定——模型预测法 / 28
 氨基酸利用效率 / 37
 参考文献 / 44
第3章 脂类 / 51
 引言 / 51
 脂类的消化率及能值 / 51
 日粮脂肪和生长性能 / 52
 日粮必需及生物活性脂肪酸 / 53
 日粮脂肪、碘值与猪肉脂肪品质 / 54
 肉毒碱 / 56
 日粮脂肪的质量评价 / 56
 脂类的分析 / 59

 参考文献 / 59
第4章 碳水化合物 / 65
 引言 / 65
 单糖 / 66
 二糖 / 66
 寡糖（低聚糖）/ 66
 多糖 / 68
 碳水化合物的分析 / 71
 参考文献 / 72
第5章 水 / 74
 引言 / 74
 水的功能 / 74
 水的周转 / 74
 水的需要量 / 75
 水的品质 / 78
 参考文献 / 81
第6章 矿物质 / 84
 引言 / 84
 常量元素 / 84
 微量/痕量元素 / 92
 参考文献 / 100
第7章 维生素 / 116
 引言 / 116
 脂溶性维生素 / 117
 水溶性维生素 / 123
 参考文献 / 131
第8章 猪营养需要量估测模型 / 141
 引言 / 141
 生长育肥猪模型 / 142
 妊娠母猪模型 / 152
 泌乳母猪模型 / 156
 断奶仔猪 / 159
 矿物质与维生素需要 / 160
 氮、磷和碳沉积效率的估测 / 161

模型评估 / 162

参考文献 / 172

第9章 玉米和大豆加工副产物 / 174

引言 / 174

玉米副产物 / 174

大豆制品 / 177

粗甘油 / 179

参考文献 / 179

第10章 非营养性饲料添加剂 / 183

引言 / 183

抗菌剂 / 183

驱虫药 / 183

酸化剂 / 184

直接饲喂的微生物添加剂 / 184

寡糖 / 185

植物提取物 / 186

外源酶 / 186

调味剂 / 187

霉菌毒素吸附剂 / 188

抗氧化剂 / 189

颗粒黏合剂 / 189

流散剂 / 190

莱克多巴胺 / 190

肉毒碱和共轭亚油酸 / 191

臭味和氨气控制剂 / 191

参考文献 / 191

第11章 饲料污染物 / 197

引言 / 197

化学性污染物 / 197

生物性污染物 / 201

物理性污染物 / 202

将来潜在的问题 / 202

动物饲料安全体系 / 202

其他信息来源 / 203

参考文献 / 203

第12章 饲料加工 / 205

引言 / 205

饲料加工对养分利用率的影响 / 205

其他信息 / 206

参考文献 / 206

第13章 营养物质和能量的消化率 / 208

引言 / 208

粗蛋白和氨基酸 / 208

脂类 / 209

碳水化合物 / 210

磷 / 211

能量 / 212

参考文献 / 213

第14章 营养对养分排泄和环境的影响 / 215

引言 / 215

氮 / 216

钙和磷 / 216

铜、铁、锰、镁、钾和锌 / 217

硫 / 217

碳 / 217

日粮配方与气体排放 / 218

综合措施 / 219

参考文献 / 219

第15章 未来研究方向 / 225

引言 / 225

营养需要的评估方法 / 225

养分的利用和饲料采食量 / 225

能量 / 226

氨基酸 / 226

矿物质 / 227

脂类 / 227

维生素 / 227

饲料原料组成 / 227

其他领域和重点 / 228

营养需要、饲料组成及其他表格

第16章 营养需要列表 / 231

引言 / 231

表格 / 232

第17章 饲料原料成分 / 262

引言 / 262

常规成分和碳水化合物 /262
氨基酸 /262
矿物质 /263
维生素 /263
脂肪酸 /263
能量 /263
原料列表 /263
参考文献 /266
表格 /267

附录 A　模型使用指南 /392
　　概述 /392
　　使用模型 /392
附录 B　委员会的工作声明 /403
附录 C　缩略语和缩写词 /404
附录 D　委员会成员简介 /410
附录 E　美国国家科学院农业和自然资源委员会最近的出版物 /413
索引 /415

总　　论

自 1944 年以来，美国科学研究委员会（NRC）已出版了十版《猪营养需要》，这些书一直指导着养猪业和饲料业的营养专家，以及从事开发和生产一线的技术人员。自 1998 年出版第十版以来，养猪业已经发生了巨大变化，有些当时设定的营养需要参数和推荐量已不再适用，第十一版将之修订并体现了这些变化。

给予《猪营养需要》（第十一次修订版）修订委员会的任务呈现在附录 B 中。简言之，要求委员会提供一个关于猪各生长阶段能量和营养物质需要量的科学评述依据。其他的任务包括：关于来自生物燃料的饲料原料和其他新原料的信息，可消化磷的需要量和饲料原料中可消化磷的含量，饲料添加剂和饲料加工作用效果的评述，提高营养沉积从而降低粪尿排泄对环境污染的措施。

该研究工作得到了以下单位的资金支持：美国伊利诺伊州玉米市场销售委员会、美国饲料教育和研究所、美国国家猪肉委员会、内布拉斯加州玉米委员会、明尼苏达州玉米种植者协会、美国食品和药物管理局，同时也得到了 NRC 动物营养系列丛书销售收入的资金支持。

为达到本书写作目的，作者努力拓展原有主题并增加了新主题。新的研究成果都已收集在已修订和翻新的营养需要量表内。猪能量和营养需要量的计算机估测模型，经历了大的更新，并在深入细致综述信息的基础上，对饲料营养成分表进行了完全的修订。本书从论述能量和六大营养素开始，紧接着的是用计算机模型估测猪营养需要量。其余章节阐述了营养素利用效率的影响因素，而且列出了营养需要量和猪用饲料原料营养成分表。

第 1 章阐述了能量。首先描述从总能到净能的经典能量利用图谱及其在猪营养中的应用，讨论了用计算机模型来确定猪的能量需要。有关净能部分，已经大幅修订，以便从消化能、代谢能或用饲料化学组分推算净能值。本章还讨论了免疫去势和添加莱克多巴胺对能量利用效率的影响。

第 2 章是关于蛋白质和氨基酸，从日粮必需氨基酸和日粮非必需氨基酸的区别，以及那些随日粮其他组分和猪特定生理状态而定的条件性必需氨基酸开始讨论，而后又对来自整体蛋白质和晶体氨基酸两种来源的氨基酸作了评述。本章还讨论了确定和表示氨基酸需要量的不同方法（包括经验法、理想蛋白质概念法、析因计算法等），并综述了确定生长猪、母猪、种公猪氨基酸需要量的实验。

脂类在第十版中是放在"能量"章节中论述的，本版单独设置了章节（第 3 章）。第 3 章首先讨论了脂类作为能量来源及日粮脂肪对猪整个生命周期生产性能的影响，接着论述了具有生物活性的必需脂肪酸的特殊生理作用，然后讨论了日粮脂肪摄入量对猪体脂肪酸组成的影响，描述了碘价和碘价产物的计算方法。本章最后论述了脂肪质量的测定，如氧化程度和脂类分析方法。

上一版本，"碳水化合物"也含在"能量"章节中，本版把它单独列为第 4 章。虽然猪对日粮碳水化合物或纤维没有特殊的需要，但是猪日粮中大多数能量产自植物源的碳水化合物。本章描述了碳水化合物的大体分类、它们的消化及产能营养素的吸收等。

水，有时候描述为被遗忘的营养素，在第 5 章进行了论述。本章的主体是讨论不同阶段猪对水的需要量，但也涉及水的功能、水的周转与水的质量。

猪的矿物质营养研究依然是一个活跃的领域。第 6 章根据新的科研成果对常量和微量矿物质的相关内容都作了更新。其他方面，如某些矿物质的生物利用率及作为药物添加剂的应用效果也作了论述。

第 7 章更新了 1998 年版（第十版）对维生素需要量的论述。本章分为脂溶性和水溶性维生素两部分，涉及维生素的相对生物学利用率与在饲料中应用的稳定性。另外，在有数据的情况下讨论了一些维生素的毒性及其最大耐受量方面的试验数据。

上一版介绍了应用计算机模型估算能量和氨基酸需要量，那时开发的三个模型（生长育肥猪、怀孕母猪和泌乳母猪）现在已经更新并扩展了。正如第 8 章所描述的，在代表猪整体水平营养物质和能量利用的生物学理论方面，现在这三个模型是机械的、动态的和确定性的。除了能量和氨基酸，新模型还能估算钙和磷的需要量。生长育肥猪模型中包括了莱克多巴胺及针对公猪臭味进行免疫去势等因子的影响。本章描述了模型中出现的基本概念和用于计算的特定基本公式。

生物燃料产业的发展，尤其是用玉米生产乙醇，导致产生大量的辅产物（也叫副产物），现用于猪饲料中。第 9 章评述了这些产品对猪饲喂价值的相关信息，重点论述了来自玉米和大豆加工的副产物，同时也涉及了其他植物和动物的副产物。

第 10 章讨论了非营养性饲料添加剂，如抗菌剂和外源性酶。本章用多个不同类别添加剂的新信息对上一版作了修订。

由于 2007 年宠物饲料三聚氰胺掺假而引爆的重大新闻事件，动物饲料无论是意外还是故意的污染都成为人们日益关注的问题。第 11 章论述了饲料污染物并把它们分成三大类：化学的、生物学的和物理学的。在美国，动物饲料的安全性和合理性是受美国食品和药物管理局（FDA）监管的，本章引用了 FDA 的一些重要文件。

营养物质的利用率可能会受原料的加工方法和日粮配制方式的影响。这部分内容放在了第 12 章。本章论述了机械加工，如膨化、粉碎和制粒等对营养物质消化率及猪生产性能的影响。虽然大多数加工方式，尤其是富含复合碳水化合物原料的加工工艺，可以提高猪的生产性能，但必须权衡其收益与加工成本。

第 13 章论述了猪对营养物质和能量的消化率，内容涵盖蛋白质与氨基酸、脂类、碳水化合物、磷和能量。本章描述了测定消化率的理由及主要测定方法。猪饲料原料的消化率列入营养成分表中。

上一版介绍了使营养物质排泄最小化的相关饲喂技术，本书第 14 章扩展了该内容，新增了关于营养影响营养素排泄量与环境污染的相关信息，涉及的营养素有氮、钙、磷、微量元素、硫和碳。本章还论述了日粮组成对气体尤其是温室气体与氨排放的影响。

第 15 章确定了需要优先研究的课题，包括那些需要通过研究增加新信息或者数据来源不足需要通过进一步研究来肯定或否定这些数据的特定领域和课题。现已证明，有许多领域需要研究，但最重要的是各生理状态猪对氨基酸、钙和磷需要量的研究，重点是母猪。

第 16 章包含了各阶段猪营养需要的一系列表格。需要量以饲喂状态为基础表示。本委员会认真地综述了已发表的研究报告，得到了表中的评估值。因此，表中数据是本委员会评估得出的最佳值而非文献数据的平均值。与前几版一样，本书评估的营养需要量是没有任何安全阈值的最低标准，所以它们不应被认为是建议的供给量。在某种情况下，营养专家可以有选择地提高某些重要营养物质的水平从而起到安全阈值的作用（该建议在美国不适用于硒，因为 FDA 控制硒的用量）。另外一个重点是，矿物质和维生素需要量的评估值包括了天然饲料中含有的量，而不是日粮中应添加的量。

第 17 章由猪常用的 122 种饲料原料的表格组成，包括各种营养成分的平均值。在第十版基础上，我们对这些表格进行了全面修订并以每页一个原料的形式呈现。重点综述了过去 15 年间的文献，得出本书的原料组成数据。如果没有新的数据可用，就搜索更早的文献。有些情况下找不到数据，就拿其他已发表的营养成分表的参数作为数据的信息来源。

所有畜牧产业都需要聚焦在高效、盈利和环保型的产业中，也包括养猪业。营养在养猪生产链条中效率、利润和环保等每一个环节都起着重要的作用，饲料耗费占养猪生产成本的主要部分，营养物质消化率低会减少获利空间和利用效率，还会污染环境。本书综述了猪营养和原料营养成分的最新信息，这些新知识将优化养猪生产。生产乙醇所产生的新原料与饲料污染物及环保等相关问题在本书中也都一并作了描述。对维持高效而环保的养猪生产而言，本书是良好的指南。

第1章 能 量

引言

能量最初的定义是做功的潜力，而动物营养学家对能量的界定是有机化合物的氧化。能量虽有很多种形式，但营养学上涉及的主要是化学能和热能。虽然根据影响能量代谢的主要因素，目前已经建立了简单、准确预测猪能量需要的数学模型，但是由于动物能量代谢过程的复杂性、日粮组成和原料来源的多样性，猪能量评价体系仍然非常复杂。因此，本章描述的内容主要是针对饲料原料能量价值和猪的能量需要量，其中用于表示猪能量需要量的指标体系从最初的总可消化营养（NRC, 1971）发展为如今的代谢能（ME）和净能（NE）。此外，除进一步讨论上一版 NRC（1998）之前的重要研究发现、综述猪能量代谢利用和能量需要量计算机预测模型（第 8 章）中引用或发展的相关概念之外，本章重点阐述自上一版（NRC, 1998）之后的研究进展。

术语

饲料原料、排泄物及热损耗中的能量含量都可以用 cal（卡）、kcal（千卡）或者 Mcal（兆卡）来度量。同时，能量也经常用 J（焦耳）来表示，它与卡的转换关系是 4.184 J（焦耳）=1 cal（卡）。本版《猪营养需要》对饲料能量在猪体内分配和利用的讨论应用了大量的缩写，这些术语及其缩写的含义和定义可参考 NRC（1981）中"饲料能量含量和能量需要量"部分的有关描述。此外，本版描述能量需要和饲料能量含量所用的度量单位将用 kcal 来表示。

能量在动物体内的分配

图 1-1 表明了饲料总能（GE）在动物体内的经典分配，而猪能量需要评价体系正是根据这一分配图的描述得到发展的。根据图 1-1 的描述，能量在动物体内的分配是把采食的能量划分为三个部分，即产热、产品（组织）形成和排泄物。但是，其中的能量分配模式受饲料原料的理化特性和猪生理状态（生长、妊娠、泌乳）的影响，然后再影响饲料成分的能量价值和动物能量需要量。接下来的章节将会具体阐述饲料化学成分、动物生理状态和养殖环境对能量在动物体内分配的影响。虽然在这本书中的能量需要统一采用了 ME（有效代谢能，见"能量利用模式——有效代谢能概念"章节）来表示，但在饲料数据库中，饲料原料的能量含量有以下三种常用表示系统，即消化能（DE）、代谢能（ME）、净能（NE）。因此，日粮也用不同的能量基础（如 DE、ME 或 NE）来评估。此外，本章中所述饲料能量值的预测模型都是建立在对经验的总结基础之上，即这些回归方程是在特定情况（参数条件）下形成的，鼓励读者有必要参阅形成这些方程的原始文献。

总能

总能是物质完全氧化时所产生的能量总和。所有的有机化合物均含有一定量的 GE。通过测定饲料、粪便、尿液、气体和各种产品中的 GE 含量可以计算出饲料中 DE、ME、NE 的含量（具体见相应章节）。

而总能或燃烧热通常可直接用氧弹式测热器测定。另外，GE 也可以通过以下这些特定养分含量及其对应的单位能值来估计：碳水化合物 3.7（葡萄糖和单糖）~4.2 kcal/g（淀粉和纤维素），蛋白质 5.6 kcal/g，脂肪 9.4 kcal/g（Atwater and Bryant, 1900）。此外, Ewan（1989）提出如果知道一种饲料原料或饲料的化学组成，可以用如下的方程来预测 GE（kcal/kg）：

$$GE = 4143 + (56\% \times EE\%) + (15\% \times CP\%) - (44 \times 灰分\%) \quad (Ewan, 1989) \quad (式1\text{-}1)$$

式中，EE 是醚浸出物，CP 是粗蛋白。

由于以下原因：即使结构不同，但只要碳水化合物、脂肪或蛋白质元素组成相同，则其 GE 含量相近，所以测定总能对比较饲料原料或日粮能量价值的意义不大。

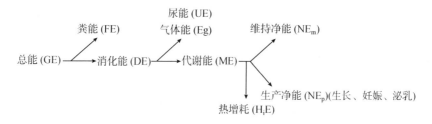

图 1-1　养分/饲粮能量在动物体内的分配

消化能

消化能是日粮总能与粪能之差（图 1-1）。由于无法直接把粪便中内源和来源于饲料的总能区分开来，所以未经特别说明，大多文献中 DE 值指的是表观 DE 值。DE 可以直接用动物试验来测定（Adeola, 2001），或通过把饲料化学组成代入预测方程来估算。通过日粮化学组成来估算 DE（kcal/kg 干物质）的预测方程有以下几种：

$$DE = 1161 + (0.749 \times GE) - (4.3 \times 灰分) - (4.1 \times NDF)（Noblet and Perez, 1993）$$
$$(式1\text{-}2)$$

$$DE = 4168 - (9.1 \times 灰分) + (1.9 \times CP) + (3.9 \times EE) - (3.6 \times NDF)（Noblet and Perez, 1993）$$
$$(式1\text{-}3)$$

上述式中 NDF 是指中性洗涤纤维（且所有化学组成均用 g/kg 干物质来表示）。通过方程预测饲料 DE 值时应注意评估方程的适用性（同样在应用 ME 和 NE 预测方程时需要注意到这点）。特别需要强调的是，做预测时输入值（式中独立变量）应该在方程的适用范围之内。另外，有些方程是用配合日粮拟合得到的，当应用于单一原料时要注意其可行性。

除化学组成之外，还有其他许多因素会影响饲料养分消化率和 DE 含量。Noblet 和 Shi（1993）、Le Goff 和 Noblet（2004）研究表明，随着生长猪逐步发育成熟（生长猪对比母猪），日粮脂肪和纤维消化率会得到逐渐提高，所以相对生长猪，母猪可获得对同一饲料能量更高的消化率（Noblet and Bach Knudsen, 1997）。正是因为生长猪和母猪对饲料能量的表观消化率存在差异，所以理论上推荐生长猪和母猪应该采用不同 DE、ME、NE 值（Noblet and van Milgen, 2004）。这种方法毫无疑问可以提高对饲料有效能含量评估的准确性，但由于受到现有研究结果和经验总结出来的数据量的限制，在本书及饲料原料数据库中的每一种饲料原料仅列出一套 DE、ME 和 NE 值，而且这些数据均来源于生长育肥猪。

尽管一些研究表明饲养方式，如群体饲喂相对单个饲喂会导致猪采食量的差异（Bakker and Jongbloed, 1994），但饲料采食量本身对能量消化率的影响很小（Haydon et al., 1984; Moter and Stein, 2004）。只不过在群养条件下，由于排空速率的加快还是会表现出能量消化率的降低（Bakker and

Jongbloed，1994）。此外，其他的因素，如饲料加工和热处理也会影响到饲料能量的消化率，具体参见第 12 章（饲料加工）中的论述。

虽然存在前述因素对猪消化率和 DE 值的影响，但是受现有数据不足等限制，本版饲料营养数据库和猪能量需要量还未能针对这些因素作相应调整。

代谢能

消化能减去尿能和消化道发酵气体能即等于 ME（图 1-1）。ME 占 DE 的比例很大，为 92%~98%（NRC，1981；1998）。气体能量损失是可变的，如生长育肥猪采食典型日粮时，产生的发酵气体通常很低，由此产生的能量损失仅占 DE 的约 0.5%（Noblet et al.，1994），但当母猪采食高纤维日粮时导致的气体能量损失可达 DE 的 3%（Ramonet et al.，1999）。猪产生的甲烷可直接根据可发酵纤维的含量进行估计（Rijnen，2003），据此也可预计到发酵气体能量损失占 DE 的比例是比较低的。这决定了影响 DE 转化成 ME 效率的主要因素是尿能。尿中能量的损失主要来源于排泄的氮（主要是尿素）。因而，在假定摄入的可消化粗蛋白有固定的比例用于尿氮排泄的条件下，ME/DE 可以通过可消化粗蛋白的含量来预测：

$$ME/DE = 100.3 - (0.021 \times CP) \quad \text{（Le Goff and Noblet，2001）} \quad \text{（式 1-4）}$$

式中，CP 用 g/kg 干物质表示。

当然，猪摄入可消化粗蛋白质中转变成尿氮的比例是可变的，这取决于日粮氨基酸平衡状况（蛋白质量）和猪对蛋白质的存留效率。

ME（kcal/kg）也可以直接用饲料营养成分来进行预测：

$$ME = 4194 - (9.2 \times 灰分) + (1.0 \times CP) + (4.1 \times EE) - (3.5 \times NDF)$$
$$\text{（Noblet and Perez，1993）} \quad \text{（式 1-5）}$$

或

$$ME = (1.00 \times DE) - (0.68 \times CP) \quad \text{（Noblet and Perez，1993）} \quad \text{（式 1-6）}$$

式中，营养成分用 g/kg 干物质表示；DE 用 kcal/kg 表示。

净能

代谢能减去热增耗（H_iE）（见"产热的组成"章节）等于净能，且净能分为用于维持 [NE_m] 和用于生产 [NE_p] 两部分。通常认为用 NE 来表示猪能量需要量是最理想的（Nobletand van Milgen，2004）。猪净能系统是采用生长育肥猪比较屠宰试验（Just，1982）或间接测热试验（Noblet et al.，1994）建立起来的。其中，间接测热促进了以可消化营养成分建立 NE 预测模型的发展（Noblet et al.，1994），而且这一预测结果也已经被实践应用于补充氨基酸的低蛋白日粮配方设计中（Le Bellego et al.，2001）。最近几年，采用比较屠宰试验测定大豆油和精选白油 NE 值的方法在北美得到了逐步发展和推广应用（Kil et al.，2011）。

在具体应用 NE 为日粮和饲料原料预测方程时需要注意一系列问题。例如，目前很少开展单一原料 NE 测定的研究，所以要特别注意现有的 NE 预测方程都是由全价日粮发展而来，当这些方程应用于单一原料时必须要注意其适用性和适当校正（DE 和 ME 也同样存在上述问题）。此外，在预测维持净能（NE_m）时的误差有时会非常大，这主要是因为 NE_m 通常是借助于绝食产热（FHP）法测定，而 FHP 的准确测定存在很多困难，这会直接影响到预测净能值的准确性（Birkettand de Lange，2001a）。以下四个方程式经过实践验证具有较好的饲料 NE（g/kg 干物质）预测效果：

其中的前三个方程来源于 Noblet 等（1994）的研究结果，且方程中所有的营养和可消化营养成分均以 g/kg 干物质来表示。

$$NE = (0.726 \times ME) + (1.33 \times EE) + (0.39 \times 淀粉) - (0.62 \times CP) - (0.83 \times ADF) \quad (式\ 1\text{-}7)$$

$$NE = (0.700 \times DE) + (1.61 \times EE) + (0.48 \times 淀粉) - (0.91 \times CP) - (0.87 \times ADF) \quad (式\ 1\text{-}8)$$

式中，ADF 是酸性洗涤纤维；ME 和 DE 用 kcal/kg 表示。

$$NE = (2.73 \times DCP) + (8.37 \times DEE) + (3.44 \times 淀粉) + (2.89 \times DRES) \quad (式\ 1\text{-}9)$$

式中，DRES=DOM−（DCP+DEE+淀粉+DADF）；DCP 为可消化 CP；DEE 为可消化 EE；DRES 为可消化残渣；DOM 为可消化有机物；DADF 为可消化 ADF。

第四个式来源于 Blok（2006）：

$$NE = (2.80 \times DCP) + (8.54 \times DEE_h) + (3.38 \times 淀粉_{am}) + (3.05 \times Sug_e) + (2.33 \times FCH) \quad (式\ 1\text{-}10)$$

式中，DEE_h 为酸水解后的可消化粗脂肪；淀粉$_{am}$ 为按淀粉葡萄糖酶方法测得的酶可消化淀粉；Sug_e 为总糖中酶降解部分；FCH（可发酵碳水化合物）=淀粉$_{am(ferm)}$ [可发酵淀粉$_{am}$，除马铃薯淀粉以外均假定为 0]+ Sug_{ferm}（可发酵糖）+DNSP（可消化非淀粉多糖）；DNSP=DOM−DCP−DEE_h −淀粉$_{am}$−（校正系数×Sug_e）；Sug_{total} = Sug_e +Sug_{ferm}；假定校正系数=0.95；所有的养分和可消化营养成分用 g/kg 干物质表示。

从现实可行性考虑，对 NE 的预测尽可能使用现有的饲料原料营养成分数据库。NRC 饲料原料数据库的建立和发展几乎完全依赖于已发表文献的数据，缺少针对 NE 预测方程建立的特殊参数值。所以，在不考虑 NE 预测准确性的条件下，相对其他三个方程，很显然 Blok（2006）的 NE 预测方程需要用到不同的数据库，这正好是本书所缺少的。此外，应用 Blok（2006）的 NE 预测方程还存在以下几个方面问题：从文献中得来的数值经过大量的求证筛选后经常仅能得到很少有效的淀粉和糖数值，以及 CP 和 EE 的消化率估计值，而所需其他更为复杂的参数值从现有文献中更难获得。所以，Blok（2006）NE 预测方程的应用需要引用其他专门饲料原料成分数据库中的相关参数（Sauvant et al., 2004; CVB, 2008）。

由于通过现有文献筛选到的有效数据有限，而糖和养分消化率的准确测定又相对困难，所以目前更趋向于采用常规营养成分来预测 NE 值，如 Noblet 等（1994）建立的式 1-8（见表 17-1）。

产热的组成

动物总产热（HE）由维持产热（H_eE）、热增耗（H_iE）、活动产热（H_jE）和维持体温（H_cE; 见 NRC[1981]）几部分组成。

$$HE = H_eE + H_iE + H_jE + H_cE \quad (式\ 1\text{-}11)$$

由 ME 转化成 NE 的效率（维持 + 生长、怀孕或泌乳）受 H_iE 影响显著：

$$ME = H_eE + H_iE + NE_p（生长、泌乳或妊娠） \quad (式\ 1\text{-}12)$$

因此，分配至生产特定产品（蛋白质、脂肪）、H_eE（通常只考虑 FHP）和 H_iE 的比例是影响 ME 用于维持和生产总效率的重要因素。此外，热增耗还可以进一步剖析如下：

$$H_iE = H_dE + H_rE + H_fE + H_wE \quad (式\ 1\text{-}13)$$

式中，H_dE 为消化和分泌产热；H_rE 为组织生产产热；H_fE 为发酵产热；H_wE 为废弃物生产产热。

H_iE 的组成可以通过试验测定或理论估计（Baldwin, 1995）。从数量上来看，H_dE 代表了较大比例的 H_iE（可占 ME_m 的 10%~20%; Baldwin and Smith, 1974）。营养和生理状态也是影响 H_iE 组成的重要因素，但是这些因素通常不适于当作猪的 ME 利用效率预测模型中独立或具体的预测因子。所以，我们发现尽管目前已开发出许多包括各种能量代谢机制元素的猪能量利用效率预测模型（Birkett and de

Lange, 2001a, b, c; van Milgen et al., 2001; van Milgen, 2002），而且这些模型在定义能量利用时也发挥了重要作用，但却得不到广泛的实际应用。现实中，被广泛用于分析 ME 分配的模型则来源于 Kielanowski（1965）：

$$MEI = ME_m + (1/k_p) PEG + (1/k_f) LEG \qquad (式1-14)$$

式中，MEI 为 ME 采食量；ME_m 为维持 ME；k_p 和 k_f 分别是 ME 用于蛋白质（PEG）和脂肪能量沉积（LEG）的分配系数。

接下来我们将在"生理状态"章节的"生长"部分对 k_p 和 k_f 做详细讨论。

维持

FHP（绝食产热）是维持能量需要（ME_m）中最大组成部分：

$$ME_m = FHP + H_iE（维持） \qquad (式1-15)$$

估计 FHP 的方法和设定条件在本章"净能"部分已有相应描述。通常，FHP 和 ME_m 用以代谢体重为基础的数学方程模型（aW^b）表示，而且已经有大量相关文献报道（Tess et al., 1984a; Noblet et al., 1994, 1999; de Lange et al., 2006）。但是，至今对于方程中的幂值（b）有很多争论和异议。指定和使用合适的幂对准确估计维持能量值、k_p 和 k_f 非常重要（Noblet et al., 1999; de Lange and Birkett, 2005）。以往多用 $BW^{0.75}$ 为基础来表示 ME_m，如 106 kcal ME/kg $BW^{0.75}$（NRC, 1998）、109 kcal ME/kg $BW^{0.75}$（ARC, 1981）。然而，也有很多数据表明以 $BW^{0.54~0.75}$ 更为合适（Tess, 1981），如 Noblet 等（1999）建议合适的幂应该是 0.60 左右。与此一致的报道所推荐的绝食产热估计方程有 137 kcal ME/kg $BW^{0.60}$（van Es, 1972）、179 kcal ME/kg $BW^{0.60}$（Noblet et al., 1994）、167 kcal ME/kg $BW^{0.60}$（van Milgen et al., 1998）等。一般认为 NE_m = FHP + 用于体力活动的能量（van Milgen et al., 2001）。

很多因素影响 FHP（ME_m; Baldwin, 1995; Birkett and de Lange, 2001b）。例如，测定试验之前的能量和营养（蛋白质）摄入会影响 FHP。其中，由于胃肠道或肝脏绝食产热均可占总 FHP 的 30%左右（Baldwin, 1995），因此增加能量和蛋白质摄入会增加肠道和肝脏的重量（Critser et al., 1995），进而会增加 FHP（Koong et al., 1983）。

目前，通常以 $BW^{0.75}$ 为基础来表示猪的 FHP 和 ME_m。其中，每千克代谢体重 ME 需要量估测值范围为 95~110 kcal/kg $BW^{0.75}$（Dourmad et al., 2008）。虽然没有证据表明初产母猪和经产母猪的 ME_m 存在差异，但是 Dourmad 等（2008）建议妊娠母猪和泌乳母猪的 ME_m 应分别为 105 kcal/kg $BW^{0.75}$ 和 110 kcal/kg $BW^{0.75}$，而本书推荐妊娠和泌乳模型的 ME_m 值分别是 100 kcal/kg $BW^{0.75}$ 和 110 kcal/kg $BW^{0.75}$（见第 8 章中"妊娠母猪模型和泌乳母猪模型"）。

同样，也没有数据表明阉公猪、小母猪和公猪的 FHP 或 ME_m 存在差异（NRC, 1998; Noblet et al., 1999）。然而，不同瘦肉生长率的群体之间却存在 FHP 和 ME_m 的差异（Noblet et al., 1999）。因而，如果基于瘦肉增重（潜力）来估计维持能量需要，由于小母猪和公猪更多的蛋白质沉积，可能会武断地认为它们比阉公猪的维持需要会更高一些。当然，无视群体、品系和性别差异而假定 FHP 或 ME_m 是一恒定值也是不合适的。所以，合理地针对性调整 FHP（估测 NE）或 MEI 分配比例非常重要。一般情况下，生长育肥猪的 ME_m 范围为 191~216 kcal/kg $BW^{0.60}$，平均 197 kcal/kg $BW^{0.60}$（Birkett and de Lange, 2001c）。

体温维持

以前的讨论主要集中于估计适宜温度环境条件下动物的维持能量消耗。然而，实际生产中同样存在低于下限临界温度（LCT）或高于上限临界温度（UCT）的异常环境状况，这些异常温度环境会影响猪的产热/损失和 MEI。例如，温度小于 LCT 时的动物平均日采食量（ADFI）会增加，大于 UCT 时会减少。其中，现有的报道大多数都集中在高于 UCT 情况的研究。同时，动物采食量对环境温度的反应还受到猪和环境的互作影响（如空气温度、风速、栏舍材料、饲养密度的影响；见 Curtis，1983 年的综述）。此外，能量浓度也影响猪的自由采食量（Stahly and Cromwell，1979；1986），而且能量浓度和冷热应激条件下动物采食量的互作与 H_iE 有关。尤其是高纤维日粮会产生更多的 H_iE，这有助于冷应激动物的产热；而补充油脂日粮产生较少的 H_iE，这有助于降低动物的热应激。

生长猪

体重（Holmes and Close，1977；Noblet et al.，2001；Meisinger，2010）和 MEI（Bruce and Clark，1979；Whittemore et al.，2001）会影响 LCT 和 UCT。采食量从维持增加到 3 倍维持时，60 kg 以上猪的 LCT 降低 6~10 ℃（Holmes and Close，1977）。Verstegen 等（1982）估计，猪在生长阶段，温度在 LCT 以下时，每降低 1 ℃，25~60 kg 生长猪需要增加采食 25 g 饲料/d（80 kcal 代谢能/d）；而 60~100 kg 育肥猪则需要增加采食 39 g 饲料/d（125 kcal 代谢能/d）。当环境温度低于 LCT 时，ME_m 中用于产热的 ME（kcal/d）=0.074 25 ×（LCT–T）× ME_m。

大量研究证实，当环境温度从 19 ℃ 左右上升至 31 ℃ 时，动物平均日采食量（MEI）会降低 10%~30%（Collin et al.，2001；Quiniou et al.，2001；Le Bellego et al.，2002；Renaudear et al.，2007）。Le Dividich 等（1998）估计采食量最高能降低 80g/(℃·d)。此外，环境温度对采食量的影响与体重存在交互作用（Close，1989；Quiniou et al.，2000）。Quiniou 等（2000）建立的由环境温度和体重预测动物自由采食量（VFI）的函数方程如下：

$$VFI(g/d) = -1264 + (73.6 \times BW) - (0.26 \times BW^2) + (117 \times T) - (2.40 \times T^2) - (0.95T \times BW)$$

（式 1-16）

式中，环境温度范围为 12~29 ℃；体重范围为 63~74 kg。

妊娠母猪

单栏饲养母猪的 LCT 为 20~23 ℃（Noblet et al.，1989），而群养母猪的 LCT 可能比单栏饲养的低 6 ℃（Verstegen and Curtis，1988）。当温度低于 LCT 时，增加 MEI 有利于促进产热维持体温，且额外增加的 ME 为 2.5~4.3 kcal /（$kg^{0.75}$·℃）（Noblet et al.，1997）。但是，由于妊娠母猪一般都采取限制饲喂，所以通常无需考虑环境温度高于 UCT 时对 ME_m 或 MEI 的影响。

泌乳母猪

泌乳母猪的 UCT 为 18~22 ℃（Black et al.，1993）。但是，通常情况下，由于产房多采取保温措施，因此泌乳母猪不存在温度低于 LCT 时的采食应激问题。在环境温度高于 UCT 时，泌乳母猪代谢能的摄入量会因为热应激而降低，而且 MEI 降低量会随着环境温度的升高幅度发生变化。Quiniou 和 Noblet（1999）研究表明 MEI 的降低量取决于温度：18~25 ℃ 时，每天为 0.33 Mcal 代谢能/℃；25~27 ℃ 时，每天为 0.76 Mcal 代谢能/℃。

活动

生理活动也影响产热。Petley 和 Bayley（1988）通过研究猪在跑步机上的产热，发现运动的猪产热比对照组猪高 20%。Close 和 Poorman（1993）计算出生长猪步行每千米需额外消耗 1.67 kcal 代谢能/BW。Noblet 等（1993）测得母猪每站立 100 min 增加的产热为 6.5 kcal /$BW^{0.75}$，该数字与 Hornicke（1970）的 7.2、

McDonald 等（1988）的 7.1、Susenbeth 和 Menke（1991）的 6.1 及 Cronin 等（1986）相近。此外，Noblet 等（1993）还研究发现猪每消费 1 kg 饲料需消耗 24~35 kcal 代谢能。

生理状态

尽管通常认为可以通过化学反应层面的能量转化机械地换算出总的能量利用效率，但是动物营养需要层面的能量转化复杂性决定了不可如此机械地估计可利用养分需要量。因此，在本书"营养和能量需要"相关章节中，考虑到猪不同的生理状态（生长、妊娠、泌乳），许多静态模型没有把参数设定为固定值。此外，本书引入了改进后的计算机动态模型来表示不同生理状态下猪能量摄入的反应，这就是最好的例证（见第 8 章）。

生长

测定生长的能量需要是（维持）体重和增重组织中蛋白质和脂肪的函数。因此，用于生长（维持以上）的能量利用效率（ME）是一个代谢能转化为蛋白质（k_p）和脂肪（k_f）沉积效率的函数（这在"产热的组成"章节中有过相关描述）。代谢能用于蛋白质沉积的效率为 0.36~0.57（Tess et al., 1984b），用于脂肪沉积的效率为 0.57~0.81（Tess et al., 1984b），或者每克蛋白质和脂肪沉积需要消耗的代谢能分别为 10.6 kcal/g 和 12.5 kcal/g（Tess et al., 1984b; NRC, 1998）。

Birkett 和 de Lange（2001c）用一个简化的营养通路模型预测出 k_p 和 k_f 分别为 0.47~0.51 和 0.66~0.72。k_p 和 k_f 的预测值受饲粮组成、生长模式或组成的影响。其中，k_p 受合成蛋白质的底物组成和蛋白质的沉积量及其速率的影响（Whittemore et al., 2001）；而 k_f 的整体效率则取决于沉积脂肪的组成、脂肪组织的周转、脂肪前体物质的组成（Birkett and de Lange, 2001c; Whittemore et al., 2001）。

代谢能的组成（即日粮蛋白、淀粉和脂类）影响代谢能利用的效率。Noblet 等（1994）估计蛋白质、淀粉和脂肪来源的代谢能转化为净能的效率（k）分别为 0.58、0.82 和 0.90，这与 van Milgen 等（2001）的测定结果（0.52、0.84 和 0.88）基本一致。总体而言，在有多种原料的混合日粮中，k 值的范围为 0.70~0.78（Noblet et al., 1994; van Milgen et al., 2001; Noblet and van Milgen, 2004）。

代谢能的摄入量是决定猪生长速度的一个关键因素。通过控制和调节采食量来调控猪生长速度的观点亦有过很多类似综述（NRC, 1987; Kyriazakis and Emmans, 1999; Ellis and Augspurger, 2001; Torrallardona and Roura, 2009）。Bridges 等（1986）提出了如下 MEI 预测公式：

$$MEI = a \times \{1 - \exp[-\exp(b) \times BW^c]\} \quad \text{（式 1-17）}$$

量化方程中 a、b 和 c 三个系数即可预测出不同性别和不同生长遗传潜力猪的 MEI（Schinckel et al., 2009）。

妊娠

怀孕期的营养摄入对于胎儿和相应组织（胎盘、子宫、泌乳组织）的生长和发育，以及母体脂肪和蛋白质的沉积都是十分关键的。妊娠母猪的营养和能量需要量已在几个重要的综述中总结（ARC, 1981; Aherne and Kirkwood, 1985; Dourmad et al., 1999, 2008; Boyd et al., 2000; Trottier and Johnson, 2001）。通常，由于妊娠母猪采取人为限饲，并不需要预测其采食量。

在妊娠后期增加能量的摄入对胎儿生长及母体增重都有正面的影响；但是，过量的能量摄入反而会对母猪接下来泌乳期的性能造成负面影响，妊娠期采食量增加会造成母猪在泌乳期能量采食量减少和失重加大（Williams et al., 1985; Weldon et al., 1994）。以前认为，妊娠期最大胎儿生长和母体增重所需要的 MEI 为 6.0 Mcal/d（ARC, 1981; Whittemore et al., 1984; NRC, 1998）。这相当于 1.6~2.4 kg/d 的采食量，具体数量取决于日粮 ME 的浓度。在最近的 NRC（1998）出版之后，随着养殖生产技术的发展，母猪产仔数和仔猪出生重都有一定的增加，因此，MEI 需要量可能需要上调至 6.5 Mcal/d，但是必须根据产仔数、平均初生重、泌乳阶段及母猪胎次进行校正。

妊娠期的体增重是母体蛋白、脂肪沉积和胚胎增重的结果，其中每个组成所需要的 ME 可以通过 ME 用于体增重（其中蛋白质为 k_p、脂肪为 k_f）和胚胎生长（k_c）的效率进行测定。同时，母体也可以动用自身蛋白质和脂肪用以支持胎儿和组织的发育（k_r）。举一个特殊的例子，如果在整个妊娠期实行固定的饲喂计划，母猪妊娠后期能量或营养摄入可能出现暂时的不足，估计此时动用母体自身蛋白质和脂肪的 k_p 和 k_f 分别为 0.6 和 0.8（Noblet et al., 1990），而 k_r 的估计值（0.8）正好与 k_f 相似，这提示母体动用自身储备去支持妊娠的能量大部分可能来自脂肪组织（Noblet et al., 1990; Dourmad et al., 2008）。

尽管胎儿生长相关的组织涉及胎儿、胎盘、羊水、子宫四个部分（Noblet et al., 1985），但是估测 k_c 时通常只考虑其与孕体（胎儿+胎盘+羊水）的关系。基于对孕体的剖析，计算出的 k_c 大约为 0.5（Close et al., 1985; Noblet and Etienne, 1987）；可是，如果子宫维持能量消耗没有划归到母猪的维持需要，则 k_c 的估测值会降低至 0.030（Dourmad et al., 1999）。此外，能量用于孕体生长的效率与妊娠阶段、预期的产仔数相关，计算公式如下：

$$\ln(ER_c) = 11.72 - 8.62\exp(-0.0138t + 0.0932LS) \quad (\text{Noblet et al., 1985}) \quad (\text{式 1-18})$$

式中，$\ln(ER_c)$ 是沉积到胎儿能量的自然对数；t 为妊娠期长度（d）；LS 是预期的产仔数（头）。

对于一个产仔数为 12 的母猪，沉积到胎儿的 ER_c 大约等于 15.2 Mcal，或者 1.3 Mcal/头。胎儿生长的代谢能需要量为 ER_c/k_c。

泌乳

改变泌乳期能量平衡可能会对母猪繁殖性能和使用寿命产生长期影响（Dourmad et al., 1994）。泌乳母猪能量需要的 MEI 分为维持（可能受温度和活动影响）与产奶。另外，因为能量摄入经常不足以支持产奶需要，导致母猪需要动用储存的体脂和蛋白质去支持泌乳，因此期望泌乳母猪能有最大的采食量。限制采食量对年轻泌乳母猪的代谢、繁殖以及相应组织动用的影响要比对老年母猪更大（Boyd et al., 2000）。

以前估计泌乳母猪 ME_m 为 106 kcal ME/$BW^{0.75}$（NRC, 1998），这与妊娠母猪的 ME_m 相同。但是，Noblet 和 Etienne（1986, 1987）的研究表明，泌乳母猪的 ME_m 要比妊娠母猪高 5%~10%。Noblet 等（1990）报道泌乳母猪的 ME_m=110 kcal/$BW^{0.75}$，该估计值比妊娠母猪的 ME_m 高 10%（100 Mcal ME/$BW^{0.75}$，见"妊娠"部分）。

泌乳母猪产奶的遗传潜力主要通过仔猪的生长速度来体现，这也是最先测定泌乳母猪产奶能量需要的方法。与产奶量相关的能量含量可以通过仔猪生长速度和窝产仔数进行估计（Noblet and Etienne, 1989; NRC, 1998）：

$$产奶能量（GE, kcal/d）= (4.92 \times ADG) - (90 \times LS) \quad (\text{式 1-19})$$

式中，ADG 为平均日增重（窝，g）；LS 为窝产仔数。因此，有可能用一个标准的泌乳生产曲线（Whittemore and Morgan, 1990）计算出每天产出的能量。

ME 转化为泌乳的能量效率（k_m）为 0.67~0.72（Verstegen et al., 1985; Noblet and Etienne, 1987）。上一版 NRC（1998）中的 k_m 假设为 0.72，这与 Dourmad 等（2008）描述的模型一致。而本书中泌乳母猪模型中的 k_m 为 0.70（见第 8 章中"摄入代谢能的分配"部分）。

Schinckel（2010）用一个非线性方程的方法描述了 MEI 对泌乳天数的响应。泌乳母猪日粮的 MEI

很少能满足泌乳的能量需要，母猪体组织因此被分解用于产奶的能量（营养）需要。和期望的一样，动员体组织满足产奶的能量效率 k_{mr} 高于 k_m，体组织能量转化为产奶的能量效率范围为 0.84（de Lange et al., 1980）~0.89（Noblet and Etienne, 1987; NRC, 1998）。

后备公猪和母猪

一般情况下，体重达到 100 kg 之前的后备公猪和后备母猪都采取自由采食的饲养模式，直到体重达到大约 100 kg 时，通过评估其潜在的生长速度和瘦肉率，选育出种用公猪和母猪。当被选入种猪群后，后备猪的能量摄入量就会限制，以便在用作种猪时达到预期的重量（Wahlstrom, 1991）。

公猪性活动

种公猪的能量需要是维持、配种、精液分泌及生长的总和。Kemp（1989）报道了与精液采集有关的产热，其中用假母猪支架采精时的产热为 4.3 kcal DE/ $BW^{0.75}$。Close 和 Roberts（1993）估计了精液生产的能量需要量，即平均每次射出精液的能量为 62 kcal DE，能量利用效率估计为 0.6，则每次射精的能量需要量为 103 kcal DE。

促性腺激素释放激素免疫拮抗剂的应用

近年来，为控制猪肉的风味，公猪使用促性腺激素释放激素免疫拮抗剂的化学去势技术已在多个国家批准应用。第二次免疫之前（上市前的 4~6 周），免疫公猪的生长性能和蛋白质沉积与未免疫公猪相似。第二次免疫之后，血液循环的激素水平及构象与阉公猪相似，7~10 天内转变成与阉公猪相似的生产性能。但是目前这种免疫的效果在不同试验间的变异较大，第二次免疫后 4~5 周，免疫公猪相对于未免疫且未去势公猪的日采食量和体增重改善效果显著，提高幅度最高可分别达到 18%和 13%，但结束时的背膘厚也提高了 17%（Bonneau et al., 1994; Dunshea et al., 2001; Metz et al., 2002; Turkstra et al., 2002; Zeng et al., 2002; Oliver et al., 2003; Pauly et al., 2009; Fabrega et al., 2010）。这种反应提示：当公猪被免疫时，其蛋白质沉积有一定的降低，大部分增加的能量摄入被用于脂肪的沉积。

莱克多巴胺的应用

日粮莱克多巴胺使用的效果被安排在第 10 章"非营养性饲料添加剂"部分论述。莱克多巴胺对蛋白质和脂肪的代谢有着其独特的影响，这种影响在生产中体现为对单位增重所需 MEI 的降低（NRC, 1994; Schinckel et al., 2006）。莱克多巴胺对 MEI/增重的改善效果与体增重和莱克多巴胺添加剂量表现为函数关系。莱克多巴胺可增加猪体蛋白的沉积，相应减少了一部分用于脂肪沉积的能量分配。莱克多巴胺对蛋白质和脂肪的沉积能量分配的影响取决于它的添加水平和使用持续时间（见第 8 章中"饲喂莱克多巴胺（RAC）和公猪进行针对促性腺激素释放激素（GnRH）的免疫注射对养分分配的影响"）。

能量利用模型——有效代谢能概念

目前已经开发出多种方法用于建立能反映生长猪和繁殖猪能量需要的数学模型（Black et al., 1986; Pomar et al., 1991; NRC, 1998; van Milgen et al., 2008）。本书第 8 章中详细解释了新模型中能量利用的计算原理。新模型涉及一个表示能量利用的关键概念——有效代谢能，在此将加以阐述。

概念上，当今的净能系统比消化能系统和代谢能系统更准确地反映了日粮能量来源（如淀粉、纤维、蛋白质和脂肪）对动物生产性能所需能量利用效率的影响（式 1-7~式 1-10）。但是，在净能体系中，猪利用能量的目的并没有被解释清楚。例如，当确定了生长猪日粮的净能含量时，就假定不同群体用于蛋白质、脂肪沉积及机体维持功能的能量是不变的，即使这些猪群机体的增重速度与组成都存在变化。目前已清楚，代谢能用于脂肪的沉积效率要高于代谢能用于蛋白质沉积或维持（式 1-14）。在更准确的能量系统中，日粮的能量来源和猪对能量的利用方式都应予以考虑，后者融入模型中能够清楚说明猪能量利用-营养产出之间的关系（Birkett and de Lange, 2001a, b, c; van Milgen et al., 2001）。但是，此类诸多模型中，一个重要的限制因素是原料和养分中的（净）能值并非恒定的，它随动物生产性能水平的变化而变，而这点在日粮配方中很难加以考虑。

作为目前净能体系和更机械的能量利用模型的补充，有效代谢能的概念在本书模型中被提出来。在该方法中，日粮有效代谢能的含量可通过日粮净能含量以固定的转化效率计算出来，其中 5~25 kg 小猪的转化效率为 1/0.72、25~135 kg 生长育肥猪的为 1/0.75、母猪的为 1/0.763。上述转化效率值是通过计算玉米-去皮浸提豆粕基础日粮的代谢能和净能而获得的，并假设这一基础日粮的应用可用于维持机体各项功能的代谢能利用效率最大化。这三个玉米-去皮浸提豆粕基础日粮的代谢能为 3300 kcal/kg（可灵活加入少量脂肪来满足能量）、0.1% 的赖氨酸盐酸盐、3%维生素和矿物质，用以满足三种不同阶段猪的赖氨酸需要量。在此模型中，有效代谢能用于反映摄入能量向维持、蛋白质和脂肪沉积、妊娠的能量沉积及泌乳能量输出的分配。使用有效代谢能的这一概念时会发现，热增耗低的日粮（如日粮中有大量的添加脂肪），它的有效代谢能含量比实际代谢能高，而高产热日粮（如日粮含有高水平的纤维原料）的有效代谢能比实际代谢能低。同理，小猪、生长育肥猪和母猪日粮（有效）消化能向有效代谢能转换的系数分别为 0.96、0.97、0.974。在本章描述的模型中，代谢能和有效代谢能可以交换使用。在本书饲料成分表中（见第 17 章），不同阶段猪对同种饲料原料的能量参数值一样（如每一种饲料提供给小猪、生长育肥猪和母猪的能值相同），原因是现已发表的数据不足以针对猪不同生长阶段的差异提供有效的校正。

预测猪能量摄入反应最准确的方法是：首先以日粮净能含量作为模型输入，然后用这个模型推算出预测猪能量摄入反应的有效代谢能值。当以日粮消化能或代谢能含量作为模型输入时，则忽略了个体营养素的能量生成差异对能量利用效率的影响。

参 考 文 献

Adeola, O. 2001. Digestion and balance techniques in pigs. Pp. 903-916 in *Swine Nutrition*, 2nd Edition, A. J. Lewis and L. L. Southern, eds. Boca Raton, FL: CRC Press.

Aherne, F. X., and R. N. Kirkwood. 1985. Nutrition and sow prolificacy. *Journal of Reproduction and Fertility* (Supplement) 33:169-183.

ARC (Agricultural Research Council). 1981. *The Nutrient Requirements of Pigs: Technical Review*. Slough, UK: Commonwealth Agricultural Bureaux.

Atwater, W. O., and A. P. Bryant. 1900. The availability and fuel value of food materials. P. 73 in *12th Annual Report of the Storrs, Connecticut Agricultural Experiment Station*.

Bakker, G. C. M., and A. W. Jongbloed. 1994. The effect of housing system on the apparent digestibility in pigs using the classical and marker (chromic oxide, acid-insoluble ash) techniques, in relation to dietary composition. *Journal of the Science of Food and Agriculture* 64:107-115.

Baldwin, R. L. 1995. *Modeling Ruminant Digestion and Metabolism*. New York: Chapman & Hall.

Baldwin, R. L., and N. E. Smith. 1974. Molecular control of energy metabolism. Pp. 17-34 in *The Control of Metabolism*, J. D. Sink, ed. University Park: Pennsylvania State University Press.

Birkett, S., and K. de Lange. 2001a. Limitations of conventional models and a conceptual framework for a nutrient flow representation of energy utilization by animals. *British Journal of Nutrition* 86:647-659.

Birkett, S., and K. de Lange. 2001b. A computational framework for a nutrient flow representation of energy utilization by growing monogastric animals. *British Journal of Nutrition* 86:661-674.

Birkett, S., and K. de Lange. 2001c. Calibration of a nutrient flow model of energy utilization by growing pigs. *British Journal of Nutrition* 86:675-689.

Black, J. L., R. G. Campbell, I. H. Williams, K. J. James, and G. T. Davies. 1986. Simulation of energy and amino acid utilisation in the pig. *Research and Development in Agriculture* 3:121-145.

Black, J. L., B. P. Mullan, M. L. Lorschy, and L. R. Giles. 1993. Lactation in the sow during heat stress. *Livestock Production Science* 35(1-2):153-170.

Blok, M. C. 2006. Development of a new NE formula by CVB using the database by INRA. Pre-symposium Workshop, Net energy systems for growing and fattening pigs, May 24, 2006, Munkebjerg Hotel, Vejle, Denmark.

Bonneau, M., R. Dunfor, C. Chouvet, C. Roulet, W. Meadus, and E. J. Squires. 1994. The effects of immunization against luteinizing hormone-releasing hormone on performance, sexual development, and levels of boar taint-related compounds in intact male pigs. *Journal of Animal Science* 72:14-20.

Boyd, R. D., K. J. Touchette, G. C. Castro, M. E. Johnston, K. U. Lee, and I. K. Han. 2000. Recent advances in amino acid and energy nutrition of prolific sows. *Asian-Australian Journal of Animal Science* 13:1638-1652.

Bridges, T. C., L. W. Turner, E. M. Smith, T. S. Stahly, and O. J. Loewer. 1986. A mathematical procedure for estimating animal growth and body composition. *Transactions of the American Society of Agricultural Engineers* 29:1342-1347.

Bruce, J. M., and J. J. Clark. 1979. Models of heat production and critical temperature for growing pigs. *Animal Production* 28:353-369.

Close, W. H. 1989. The influence of thermal environment on voluntary feed intake. *British Society of Animal Production* 13:87-96.

Close, W. H., and P. K. Poorman. 1993. Outdoor pigs—their nutrient requirements, appetite and environmental responses. Pp. 175-196 in *Recent Advances in Animal Nutrition*, P. C. Garnsworthy and D. J. A. Cole, eds. Loughborough, UK: Nottingham University Press.

Close, W. H., and F. G. Roberts. 1993. Nutrition of the working boar. Pp. 21-44 in *Recent Advances in Animal Nutrition*, W. Haresign and D. J. A. Cole, eds. Loughborough, UK: Nottingham University Press.

Close, W. H., J. Noblet, and R. P. Heavens. 1985. Studies on the energy metabolism of the pregnant sow. 2. The partition and utilization of metabolizable energy intake in pregnant and non-pregnant animals. *British Journal of Nutrition* 53:267-279.

Collin, A., J. van Milgen, S. Dubois, and J. Noblet. 2001. Effect of high temperature on feeding behaviour and heat production in group-housed young pigs. *British Journal of Nutrition* 86:63-70.

Critser, D. J., P. S. Miller, and A. J. Lewis. 1995. The effects of dietary protein concentration on compensatory growth in barrows and gilts. *Journal of Animal Science* 73:3376-3383.

Cronin, G. M., J. M. F. M. van Tartwijk, W. van Der Hel, and M. W. A. Verstegen. 1986. The influence of degree of adaptation to tether housing by sows in relation to behavior and energy metabolism. *Animal Production* 42:257-268.

CVB. 2008. CVB Feedstuff Database. Available online at http://www.pdv.nl/english/Voederwaardering/about_cvb/index.php. Accessed on June 9, 2011.

Curtis, S. E. 1983. *Environmental Management in Animal Agriculture*. Ames: Iowa State University Press.

de Lange, C. F. M., and S. H. Birkett. 2005. Characterization of useful energy content of swine and poultry feed ingredients. *Canadian Journal of Animal Science* 85:269-280.

de Lange, C. F. M., J. van Milgen, J. Noblet, S. Dubois, and S. Birkett. 2006. Previous feeding level influences plateau heat production following a 24 h fast in growing pigs. *British Journal of Nutrition* 95:1082-1087.

de Lange, P. G. B., G. J. M. van Kampen, J. Klaver, and M. W. A. Verstegen. 1980. Effect of condition of sows on energy balances during 7 days before and 7 days after parturition. *Journal of Animal Science* 50:886-891.

Dourmad, J-Y., M. Etienne, A. Prunier, and J. Noblet. 1994. The effect of energy and protein intake of sows on their longevity: A review. *Livestock Production Science* 40:87-97.

Dourmad, J-Y., J. Noblet, M. C. Pere, and M. Etienne. 1999. Mating, pregnancy and prenatal growth. Pp. 129-153 in *A Quantitative Biology of the Pig*, I. Kyriazakis, ed. New York: CABI.

Dourmad, J-Y., M. Etienne, A. Valancogne, S. Dubois, J. van Milgen, and J. Noblet. 2008. InraPorc: A model and decision support tool for the nutrition of sows. *Animal Feed Science and Technology* 143:372-386.

Dunshea, F. R., C. Colantoni, K. Howard, I. McCauley, P. Jackson, K. A. Long, S. Lopaticki, E. A. Nugent, J. A. Simons, J. Walker, and D. P. Hennessy. 2001. Vaccination of boars with a GnRH vaccine (IMPROVEST) eliminates boar taint and increases growth performance. *Journal of Animal Science* 79:2524-2535.

Ellis, M., and N. Augspurger. 2001. Feed intake in growing-finishing pigs. Pp. 447-467 in *Swine Nutrition*, 2nd Edition, A. J. Lewis and L. L. Southern, eds. Boca Raton, FL: CRC Press.

Ewan, R. C. 1989. Predicting the energy utilization of diets and feed ingredients by pigs. Pp. 271-274 in *Energy Metabolism*, EAAP Bulletin No. 43, Y. van der Honig and W. H. Close, eds. Wageningen, The Netherlands: Purdoc.

Fàbrega, E., A. Velarde, J. Cros, M. Gispert, P. Suarez, J. Tibau, and J. Soler. 2010. Effect of vaccination against gonadotropin-releasing hormone, using IMPROVEST, on growth performance, body composition, behaviour and acute phase proteins. *Livestock Science* 132:53-59.

Haydon, K. D., D. A. Knabe, and T. D. Tanksley, Jr. 1984. Effects of level of feed intake on nitrogen, amino acid and energy digestibilities measured at the end of the small intestine and over the total digestive tract of growing pigs. *Journal of Animal Science* 59:717-724.

Holmes, C. W., and W. H. Close. 1977. The influence of climatic variables on energy metabolism and associated aspects of productivity in pigs. Pp. 51-74 in *Nutrition and the Climatic Environment*, W. Haresign, H. Swan, and D. Lewis, eds. London: Butterworths.

Hornicke, H. 1970. Circadian activity rhythms and the energy cost of standing in growing pigs. Pp. 165-168 in *Energy Metabolism of Farm Animals*, EAAP No. 13, A. Suchurch and C. Wenk, eds. Zurich: Juris Druck Verlag.

Just, A. 1982. The net energy value of balanced diets for growing pigs. *Livestock Production Science* 8:541-555.

Kemp, B. 1989. Investigations on breeding boars to contribute to a functional feeding strategy. Ph.D. Dissertation, University of Wageningen, The Netherlands.

Kielanowski, J. 1965. Energy cost of protein deposition in growing animals. P. 13 in *Energy Metabolism*, EAAP Publication 11. New York: Academic Press.

Kil, D. F., F. Ji, L. L. Stewart, R. B. Hinson, A. D. Beaulieu, G. L. Allee, J. F. Patience, J. E. Pettigrew, and H. H. Stein. 2011. Net energy of soybean oil and choice white grease in diets fed to growing and finishing pigs. *Journal of Animal Science* 89:448-459.

Koong, L. J., J. A. Nienaber, and H. J. Mersmann. 1983. Effects of plane of nutrition on organ size and fasting heat production in genetically obese and lean pigs. *Journal of Nutrition* 113:1626-1631.

Kyriazakis, I., and G. C. Emmans. 1999. Voluntary food intake and diet selection. Pp. 229-248 in *A Quantitative Biology of the Pig*, I. Kyriazakis, ed. Wallingford, Oxon, UK: CAB.

Le Bellego, L., J. van Milgen, S. Dubois, and J. Noblet. 2001. Energy utilization of low-protein diets in growing pigs. *Journal of Animal Science* 79:1259-1271.

Le Bellego, L., J. van Milgen, and J. Noblet. 2002. Effect of high temperature and low-protein diets on the performance of growing-finishing pigs. *Journal of Animal Science* 80:691-701.

Le Dividich, J., J. Noblet, P. Herpin, J. van Milgen, and N. Quiniou. 1998. Thermoregulation. Pp. 229-263 in *Progress in Pig Science*, D. J. A. Cole, J. Wiseman, and M. A. Varley, eds. Nottingham, UK: Nottingham University Press.

Le Goff, G., and J. Noblet. 2001. Comparative total tract digestibility of dietary energy and nutrients in growing pigs and adult sows. *Journal of Animal Science* 79:2418-2427.

McDonald, T. P., D. D. Jones, J. R. Barret, J. L. Albright, G. E. Miles, J. A. Nienaber, and G. L. Hahn. 1988. Measuring the heat increment of activity of growing-finishing swine. *Transactions of the American Society of Agricultural Engineers* 31:1180-1186.

Meisinger, D. J., ed. 2010. *National Swine Nutrition Guide*. Des Moines, IA: U.S. Pork Center of Excellence.

Metz, C., K. Hohl, S. Waidelich, W. Drochner, and R. Claus. 2002. Active immunization of boars against GnRH at an early age: Consequences for testicular function, boar taint accumulation and N-retention. *Livestock Production Science* 74:147-157.

Moter, V., and H. H. Stein. 2004. Effect of feed intake on endogenous losses and amino acid and energy digestibility by growing pigs. *Journal of Animal Science* 82:3518-3525.

Noblet, J., and K. E. Bach Knudsen. 1997. Comparative digestibility of wheat, maize and sugar beet pulp non-starch polysaccharides in adult sows and growing pigs. Pp. 571-574 in *Digestive Physiology in Pigs*, EAAP Publ. No. 88, J. P. Laplace, C. Février, and A. Barbeau, eds. Saint-Malo, France: INRA.

Noblet, J., and M. Etienne. 1986. Effect of energy level in lactating sows on yield and composition of milk and nutrient balance of piglets. *Journal of Animal Science* 63:1888-1896.

Noblet, J., and M. Etienne. 1987. Metabolic utilization of energy and maintenance requirements in lactating sows. *Journal of Animal Science* 64:774-781.

Noblet, J., and M. Etienne. 1989. Estimation of sow milk nutrient output. *Journal of Animal Science* 67:3352-3359.

Noblet, J., and J. M. Perez. 1993. Prediction of digestibility of nutrients and energy values of pig diets from chemical analysis. *Journal of Animal Science* 71:3389-3398.

Noblet, J., and S. Shi. 1993. Comparative digestibility of energy and nutrients in growing pigs fed ad libitum and adult sows at maintenance. *Livestock Production Science* 34:137-152.

Noblet, J., and J. van Milgen. 2004. Energy value of pig feeds: Effect of pig body weight and energy evaluation system. *Journal of Animal Science* 82:E229-E238.

Noblet, J., W. H. Close, R. P. Heavens, and D. Brown. 1985. Studies on the energy metabolism of the pregnant sow. 1. Uterus and mammary tissue development. *British Journal of Nutrition* 53:251-265.

Noblet, J., J. Y. Dourmad, J. Le Dividich, and S. Dubois. 1989. Effect of ambient temperature and addition of straw or alfalfa in the diet on energy metabolism of pregnant sows. *Livestock Production Science* 21:309-324.

Noblet, J., J. Y. Dourmad, and M. Etienne. 1990. Energy utilization in pregnant and lactating sows: Modeling of energy requirements. *Journal of Animal Science* 68:562-572.

Noblet, J., X. S. Shi, and S. Dubois. 1993. Energy cost of standing activity in sows. *Livestock Production Science* 34:127-136.

Noblet, J., H. Fortune, X. S. Shi, and S. Dubois. 1994. Prediction of net energy value of feeds for growing pigs. *Journal of Animal Science* 72:344-354.

Noblet, J., J. Y. Dourmad, M. Etienne, and J. Le Dividich. 1997. Energy metabolism in pregnant sows and newborn pigs. *Journal of Animal Science* 75(10):2708-2714.

Noblet, J., C. Karege, S. Dubois, and J. van Milgen. 1999. Metabolic utilization of energy and maintenance requirements in growing pigs: Effects of sex and genotype. *Journal of Animal Science* 77:1208-1216.

Noblet, J., J. Le Dividich, and J. van Milgen. 2001. Thermal environment and swine nutrition. Pp. 519-544 in *Swine Nutrition*, A. J. Lewis and L. L. Southern, eds. Boca Raton, FL: CRC Press.

NRC (National Research Council). 1971. *Atlas of Nutritional Data on United States and Canadian Feeds*. Washington, DC: National Academy of Sciences.

NRC. 1981. *Nutritional Energetics of Domestic Animals and Glossary of Energy Terms*, 2nd Rev. Ed. Washington, DC: National Academy Press.

NRC. 1987. *Predicting Feed Intake of Food-Producing Animals*. Washington, DC: National Academy Press.

NRC. 1994. *Metabolic Modifiers: Effects on the Nutrient Requirements of Food-Producing Animals*. Washington, DC: National Academy Press.

NRC. 1998. *Nutrient Requirements of Swine*, 10th Ed. Washington, DC: National Academy Press.

Oliver, W. T., I. McCauley, R. J. Harrell, D. Suster, D. J. Kerton, and F. R. Dunshea. 2003. A gonadotropin-releasing factor vaccine (IMPROVEST) and porcine somatotropin have synergistic and additive effects on growth performance in group-housed boars and gilts. *Journal of Animal Science* 81:1959-1966.

Pauly, C., P. Spring, J. O'Dohetry, S. Ampuero Kragten, and G. Bee. 2009. Growth performance, carcass characteristics and meat quality of group-penned surgically castrated, immunocastrated (IMPROVEST) and entire male pigs and individually penned entire males. *Animal* 3:1057-1066.

Petley, M. P., and H. S. Bayley. 1988. Exercise and post-exercise energy expenditure in growing pigs. *Canadian Journal of Physiology and Pharmacology* 66:21-730.

Pomar, C., D. L. Harris, and F. Minvielle. 1991. Computer simulation model of swine production systems. I. Modeling the growth of young pigs. *Journal of Animal Science* 69:1468-1488.

Quiniou, N., and J. Noblet. 1999. Influence of high ambient temperatures on performance of multiparous lactating sows. *Journal of Animal Science* 77(8):2124-2134.

Quiniou, N., S. Dubois, and J. Noblet. 2000. Voluntary feed intake and feeding behaviour of group-housed growing pigs are affected by ambient temperature and body weight. *Livestock Production Science* 63:245-253.

Quiniou, N., J. Noblet, J. van Milgen, and S. Dubois. 2001. Modelling heat production and energy balance in group-housed growing pigs exposed to cold or hot ambient temperatures. *British Journal of Nutrition* 85:97-106.

Ramonet, Y., M. C. Meunier-Salaun, and J. Y. Dourmad. 1999. High-fiber diets in pregnant sows: Digestive utilization and effects on the behavior of the animals. *Journal of Animal Science* 77:591-599.

Renaudeau, D., E. Huc, and J. Noblet. 2007. Acclimation to high ambient temperature in Large White and Caribbean Creole growing pigs. *Journal of Animal Science* 85:779-790.

Rijnen, M. M. J. A. 2003. Energetic utilization of dietary fiber in pigs. Ph.D. Dissertation, Wageningen Institute of Animal Sciences, Wageningen University, Wageningen, The Netherlands.

Sauvant, D., J. M. Perez, and G. Tran, eds. 2004. *Tables of Composition and Nutritional Value of Feed Materials: Pig, Poultry, Sheep, Goats, Rabbits, Horses, Fish*. Paris: INRA, and Wageningen, The Netherlands: Wageningen Academic.

Schinckel, A. P., N. Li, B. T. Richert, P. V. Preckel, K. Foster, and M. E. Einstein. 2006. Development of a model to describe the compositional growth and dietary lysine requirements of pigs fed increasing dietary concentrations of ractopamine. *Professional Animal Scientist* 22:438-449.

Schinckel, A. P., M. E. Einstein, S. Jungst, C. Booher, and S. Newman. 2009. Evaluation of different mixed nonlinear functions to describe the feed intakes of pigs of different sire and dam lines. *Professional Animal Scientist* 25:345-359.

Schinckel, A. P., C. R. Schwab, V. M. Duttlinger, and M. E. Einstein. 2010. Analyses of feed and energy intakes during lactation for three breeds of sows. *The Professional Animal Scientist* 26:35-50.

Stahly, T. S., and G. L. Cromwell. 1979. Effect of environmental temperature and dietary fat supplementation on the performance and carcass characteristics of growing and finishing swine. *Journal of Animal Science* 49:1478-1488.

Stahly, T. S., and G. L. Cromwell. 1986. Responses to dietary additions of fiber (alfalfa meal) in growing pigs housed in cold, warm, or hot thermal environment. *Journal of Animal Science* 63:1870-1876.

Susenbeth, A., and K. H. Menke. 1991. Energy requirement for physical activity in pigs. Pp. 416-419 in *Energy Metabolism of Farm Animals*, C. Wenk and M. Boessinger, eds. Zurich: ETH.

Tess, M. W. 1981. Simulated effects of genetic change upon life-cycle production efficiency in swine and the effect of body composition upon energy utilization in the growing pig. Ph.D. Dissertation, University of Nebraska, Lincoln.

Tess, M. W., G. E. Dickerson, J. A. Nienaber, and C. L. Ferrell. 1984a. The effects of body composition on fasting heat production in pigs. *Journal of Animal Science* 58:99-110.

Tess, M. W., G. E. Dickerson, J. A. Nienaber, J. T. Yen, and C. L. Ferrell. 1984b. Energy costs of protein and fat deposition in pigs fed ad libitum. *Journal of Animal Science* 58:111-122.

Torrallardona, D., and E. Roura. 2009. *Voluntary Feed Intake in Pigs.* Wageningen, The Netherlands: Wageningen Academic.

Trottier, N. L., and L. J. Johnson. 2001. Feeding gilts during development and sows during gestation and lactation. Pp. 725-769 in *Swine Nutrition*, 2nd Ed., A. J. Lewis and L. L. Southern, eds. Boca Raton, FL: CRC Press.

Turkstra, J. A., X. Y. Zeng, J. T. M. van Diepen, A. W. Jongbloed, H. B. Oonk, D. F. M. van de Weil, and R. H. Meloen. 2002. Performance of male pigs immunized against GnRH is related to the time of onset of biological response. *Journal of Animal Science* 80:2953-2959.

van Es, A. J. H. 1972. Maintenance. Chapter in *Handbuch der Tierernahrung*, Band II. Hamburg, Berlin: Verlag Paul Parey.

van Milgen, J. 2002. Modeling biochemical aspects of energy metabolism in mammals. *Journal of Nutrition* 132:3195-3202.

van Milgen, J., J.-F. Bernier, Y. Le Cozler, S. Dubois, and J. Noblet. 1998. Major determinants of fasting heat production and energetic cost of activity in growing pigs of different body weight and breed/castration combination. *British Journal of Nutrition* 79:509-517.

van Milgen, J., J. Noblet, and S. Dubois. 2001. Energetic efficiency of starch, protein and lipid utilization in growing pigs. *Journal of Nutrition* 131:1309-1318.

van Milgen, J., A. Valancogne, S. Dubois, J-Y. Dourmad, B. Seve, and J. Noblet. 2008. InraPorc: A model and decision support tool for the nutrition of growing pigs. *Animal Feed Science and Technology* 143:387-405.

Verstegen, M. W., and S. E. Curtis. 1988. Energetics of sows and gilts in gestation crates in the cold. *Journal of Animal Science* 66(11):2865-2875.

Verstegen, M. W. A., H. A. Brandsma, and G. Mateman. 1982. Feed requirement of growing pigs at low environmental temperatures. *Journal of Animal Science* 55:88-94.

Verstegen, M. W. A., J. Mesu, G. J. M. van Kampen, and C. Geerce. 1985. Energy balances of lactating sows in relation to feeding level and stage of lactation. *Journal of Animal Science* 60:731-740.

Wahlstrom, R. C. 1991. Feeding developing gilts and boars. Pp. 517-526 in *Swine Nutrition*, E. R. Miller, D. E. Ullrey, and A. J. Lewis, eds. Stoneham, UK: Butterworth-Heinemann.

Weldon, W. C., A. J. Lewis, G. F. Louis, J. L. Kovar, M. A. Giesemann, and P. S. Miller. 1994. Postpartum hypophagia in primiparous sows: I. Effects of gestation feeding level on feed intake, feeding behavior, and plasma metabolite concentrations during lactation. *Journal of Animal Science* 72:387-394.

Whittemore, C. T., and C. A. Morgan. 1990. Model components for the determination of energy and protein requirements for breeding sows: A review. *Livestock Production Science* 26:1-37.

Whittemore, C. T., A. G. Taylor, G. M. Hillyer, D. Wilson, and C. Stamataris. 1984. Influence of body fat stores on reproductive performance. *Animal Production* 38:527(Abstr.).

Whittemore, C. T., D. M. Green, and P. W. Knap. 2001. Technical review of the energy and protein requirements of pigs: Energy. *Animal Science* 73:199-215.

Williams, I. H., W. H. Close, and D. J. A. Cole. 1985. Strategies for sow nutrition: Predicting the response of pregnant animals to protein and energy intake. Pp. 133-147 in *Recent Advances in Animal Nutrition*, W. Haresign and D. J. A. Cole, eds. London: Butterworth-Heinemann.

Zeng, X. Y., J. A. Turkstra, A. W. Jongbloed, J. Th. M. van Diepen, R. H. Meloen, H. B. Oonk, D. Z. Guo, and D. F. M. van de Wiel. 2002. Performance and hormone levels of immunocastrated, surgically castrated and intact male pigs fed ad libitum high- and low- energy diets. *Livestock Production Science* 77:1-11.

第 2 章 蛋白质和氨基酸

引言

本章的主要目的是阐述评定仔猪、生长育肥猪、母猪和公猪氨基酸需要量的方法。首先，简单描述了氨基酸的分类、来源和代谢；然后，根据已经发表的文献资料，综述了各个生长阶段猪的氨基酸需要量，且重点描述生长育肥猪、怀孕母猪、哺乳母猪的氨基酸需要量；最后，介绍了保育期仔猪和种公猪的氨基酸需要量。

蛋白质

蛋白质由氨基酸组成，但饲料中蛋白质（CP）的含量值通常是由饲料中含氮物质的含量换算而来的，即蛋白质的含量等于饲料中含氮量乘以 6.25。因此，饲料中的其他非蛋白含氮物质也被包括进蛋白质含量之中，这种方法测定出来的是粗蛋白（CP）含量。其中的系数 6.25 是基于每 100 g 蛋白质中平均含有 16 g 氮这一假设。但是，事实上每一种食品蛋白质中的含氮量是不同的，如下列食品中每 100 g 蛋白质含氮量分别为：大麦 17.2 g；玉米 16.1 g；小米 17.2 g；燕麦 17.2 g；大米 16.8 g；黑麦 17.2 g；高粱 16.1 g；小麦 17.2 g；花生 18.3 g；大豆 17.5 g；鸡蛋 16.0 g；肉 16.0 g；牛奶 15.7 g。从功能上来说，只有可消化的蛋白质才能提供机体所必需的氨基酸；从数量上来说，蛋白质是猪饲料中一种昂贵的营养成分，它转化为机体组织前需要经过消化、吸收以及代谢多个反应过程。所以，日粮是否能够提供适宜数量和比例的氨基酸决定着日粮蛋白质含量是否充足及其品质。

必需、非必需以及条件性必需氨基酸

组成蛋白质的 20 种基本氨基酸通常被分为必需氨基酸和非必需氨基酸两大类（表 2-1）。必需氨基酸是指猪在机体细胞内不能利用"通常可获得的"原料合成，或合成速率不能满足最佳生产性能（包括维持、正常生长和繁殖）所需的一类氨基酸。通常可获得的概念很重要，因为多个营养性氨基酸，如蛋氨酸、苯丙氨酸和支链氨基酸，能够通过各自相应的类似物——α-酮酸的转氨基作用合成，但是这些酮酸并不是猪日粮中的天然组成成分，所以在细胞中并不是通常可以获得的；以"一定速率"同样很重要，因为氨基酸的合成速率可能受到适宜数量代谢氮的可获得性的限制。在精氨酸、半胱氨酸、谷氨酸、甘氨酸、脯氨酸和酪氨酸利用的速率比合成的速率更快的情况下，这类氨基酸可以被称为条件性必需氨基酸（Reeds, 2000）。通常猪有足够的能力合成条件性必需氨基酸。因此，作为非必需氨基酸和条件性必需氨基酸合成的底物或前体物质，猪最重要的蛋白质营养需求取决于必需氨基酸和总氮含量。

利用一个限制性定义来区分一个氨基酸是否是必需的，Reeds（2000）明确提出，有几种必需氨基酸能够从结构类似的前体物质合成，如蛋氨酸（能够通过酮酸类似物转氨基作用和羟基丁氨酸再次甲基化合成）、亮氨酸、异亮氨酸、缬氨酸和苯丙氨酸（能够通过支链酮酸合成）。因此，使用这种代谢性定义真正的必需氨基酸只有苏氨酸和赖氨酸（可能还有色氨酸）。非必需氨基酸的代谢性定义是指这种氨基酸能够使用非氨基酸的含氮物质从头合成，如使用氨离子和合适的碳源（如 α-酮酸）合成。因此，严格来说，只有谷氨酸和丝氨酸才是真正意义上基于代谢性定义的非必需氨基酸。

表 2-1 必需、非必需和条件性必需氨基酸

必需的	非必需的	条件性必需的
组氨酸	丙氨酸	精氨酸
异亮氨酸	天冬酰胺	半胱氨酸
亮氨酸	天冬氨酸	谷氨酰胺
赖氨酸	谷氨酸	脯氨酸
蛋氨酸	甘氨酸	酪氨酸
苯丙氨酸	丝氨酸	
苏氨酸		
色氨酸		
缬氨酸		

在猪的生长早期，机体通过谷氨酸盐自身合成精氨酸的量不足以满足它的正常生长需要。因此，生长猪的日粮中必须含有一定量的精氨酸或其前体物质。而以玉米-豆粕为基础的饲料中，精氨酸的含量不足以满足仔猪最佳生长速度的需要（Kim et al., 2004; Wu, 2009）。Easter 和 Baker（1977）通过纯合日粮试验得出，精氨酸的合成量无法满足母猪妊娠和哺乳的需要。这一早期试验结果得到最近研究结果的进一步证实，即 Mateo 等（2007）研究发现，在玉米-豆粕型饲料中添加 0.83%的精氨酸能够提高新出生仔猪的数量和重量。

半胱氨酸能够满足约 50%的总含硫氨基酸的需要（Chung and Baker, 1992a; Lewis, 2003; Ball et al., 2006），由此降低对蛋氨酸的需要。半胱氨酸能够通过蛋氨酸合成，因此，在缺乏半胱氨酸的情况下，总含硫氨基酸的需要量可以通过蛋氨酸而满足，尽管当部分总含硫氨基酸由半胱氨酸来满足时，能够在一定程度上提高猪的生产性能（Lewis, 2003）。此外，由于半胱氨酸能够用于合成谷胱甘肽，因此它对猪的免疫系统也具有重要作用。

苯丙氨酸能够转换成酪氨酸，所以苯丙氨酸能够满足总的芳香族氨基酸的需要（苯丙氨酸和酪氨酸）；而酪氨酸最多仅能够满足这两种氨基酸需要量的 50%（Robbins and Bakers, 1977）。

因为大部分谷氨酰胺被肠道利用，所以进入门静脉血的谷氨酰胺不到日粮摄入总谷氨酰胺的 1/3（Boelens et al., 2003; Stoll and Burrin, 2006）。谷氨酰胺能够促进细胞增殖和发挥不同的细胞保护功能，所以它在应对营养缺乏、氧化损伤、应激和免疫挑战等特殊情况方面具有重要的意义（Rhoads and Wu, 2009）。

相对机体本身从头合成，脯氨酸的合成更多依赖于肠道代谢和日粮氨基酸前体物质的合成转化（Murphy et al., 1996; Stoll et al., 1998; Reeds, 2000）。肠道代谢改变会对机体脯氨酸的合成能力造成至关重要的影响。Wu（2009）指出，由日粮提供进入门静脉血的脯氨酸仅能满足生长育肥猪<60%的总脯氨酸需要，也就是说还有>40%的脯氨酸是体内合成的。

总之，必需氨基酸需要从猪饲料中提供，非必需氨基酸在有充足氮源的情况下不需要饲料提供，条件性氨基酸的需要则取决于日粮和机体生理状况。表 2-1 也据此将 20 种氨基酸分成上述三大类。

氨基酸来源

大部分猪饲料中的基本成分是谷物，如玉米、高粱、大麦或者小麦，它们通常能够提供 30%~60%的氨基酸需要量。由于谷物类原料通常会非常缺乏某些必需氨基酸，所以需要其他的一些蛋白质原料，如豆粕等添加到饲料配方中以确保充足的、平衡的必需氨基酸供应量，甚至发酵或者化学合成的氨基

酸单体也会添加到饲料中以提高某些特定氨基酸的摄入量和平衡日粮的氨基酸比例。

适宜的日粮必需氨基酸摄入量取决于组成饲料的各个原料成分。理想的蛋白质原料所含有的氨基酸组成模式应该接近猪只维持和生产的需要。与氨基酸组成模式不合适的原料相比，氨基酸组成模式互补的多个原料的混合能够在降低日粮总氮含量的情况下，更好地满足动物对必需氨基酸的需要。此外，合成氨基酸在饲料配方中合理使用有助于减少饲料总蛋白的含量，进而可减少养猪生产氮代谢物向环境中的排放量。另外，使用合成氨基酸还可以避免氨基酸不平衡，以及最小化由于脱氨基作用和尿素排泄导致的氨基酸代谢损失。

目前本书所提及的氨基酸都是指 L 型，植物和动物蛋白中的氨基酸大部分都是这种形式。当提供合成形式氨基酸的时候，DL-蛋氨酸能够替代 L-蛋氨酸以满足蛋氨酸需求（Reifsnyder et al., 1984; Chung and Baker, 1992c; Lewis, 2003）；但也有证据表明小猪对 D 型蛋氨酸的利用比 L 型蛋氨酸效果差（Kim and Bayley, 1983）。对于生长猪，D 型色氨酸的活性估计只有 L 型色氨酸的 60%~100%（Baker et al, 1971; Arentson and Zimmerman, 1985; Kirchgessner and Roth, 1985; Schutte et al., 1988）；D 型色氨酸的活性有可能取决于饲料中 D 型色氨酸和 L 型色氨酸的比例，或者取决于添加的合成氨基酸是 D 型的还是 DL 型的（消旋混合物）。D 型赖氨酸和 D 型苏氨酸不能被任何一种动物所利用，因为这两种氨基酸不能发生转氨基反应，而且它们的α-酮酸无法转换成 L 型，这也解释了为什么这两种氨基酸才是真正意义上的必需氨基酸。其他必需氨基酸的 D 型氨基酸形式对猪的效果目前还没有研究报道。

发酵生产的饲料级氨基酸包括 L 型赖氨酸盐酸盐（纯度 98.5%，相当于 78.8% 的赖氨酸活性）、L 型苏氨酸（纯度 98.5%）、L 型色氨酸（纯度 98.5%）；人工合成的饲料级氨基酸有 DL 型蛋氨酸（纯度 99%）和 DL 蛋氨酸羟基类似物（液体状，纯度 88%）。在家禽蛋氨酸营养方面，至今有超过 70 篇论文（大约涉及 500 个实验）和至少 3 篇综述发表；但是就不同形式蛋氨酸的生物活性，不同研究者的结论仍然存在差异。此外，目前市场上除了单一固体氨基酸之外，还有一些混合氨基酸产品（如赖氨酸和色氨酸混合物），以及液体氨基酸产品形式（如赖氨酸）。为了简化术语，不管是发酵生产的还是人工合成的氨基酸都统称为晶体氨基酸。

氨基酸分析测定

氨基酸分析测定是当前蛋白质营养学研究的基础。蛋白质营养学科的发展很大程度上取决于能否准确和精确地量化食物、饲料、组织、体液和食糜中的氮和氨基酸含量。氨基酸分析方法有可能导致目前已经发布的氨基酸需要量的变动。样品前处理技术（包括完整蛋白的水解和蛋白质沉淀以获得游离氨基酸）以及为定量而对氨基酸的分离技术是氨基酸测定的两个关键技术环节，这在 Williams（1994）和 Kaspar（2009）报道中均讨论过。建立在现有的技术方法上，含硫氨基酸和色氨酸含量的测定变异比较大，因此也普遍比较受关注。含硫氨基酸含量测定过程中，盐酸水解前，蛋氨酸和半胱氨酸经过甲酸氧化进行多重衍生的步骤非常重要，目的是控制蛋氨酸氧化成蛋氨酸砜、半胱氨酸氧化成磺基丙氨酸。色氨酸在大部分饲料原料中含量很低，而且它在使用标准盐酸水解时局部会遭到破坏，这些都为色氨酸的测定增加了相当的难度。考虑到这些因素，一些特殊的预防措施，如用氢氧化钡、氢氧化钠、氢氧化锂水解，或者避免其被酸氧化的保护剂等被应用于样品制备过程中，Fontaine（2003）就此提供了更为详细的讨论。另外，为了提高氨基酸水解效果，通常建议延长前处理过程的水解时间；但是，多肽中氨基酸组成不一样，水解时间也不一样，例如，含有异亮氨酸和缬氨酸的多肽比其他需要较长的水解时间，而延长水解时间可能导致苏氨酸和丝氨酸的破坏；因此，相比传统的 24h 水解模型，采用由多重水解时间模型得出的数学曲线模型进行前处理的方法能够得到更准确的氨基酸含量。

氨基酸需要量的表示方法

单位

猪的氨基酸需要量可以通过以下形式表示：日粮中氨基酸浓度、每日需要量、单位代谢体重（$BW^{0.75}$）需要量、单位蛋白沉积需要量、单位日粮能量需要量。当氨基酸的需要量以日粮中浓度表示时，它们随着饲料能量浓度的提高而增加。因此，相较于标准谷物-豆粕型日粮，氨基酸的需要量（在饲料的百分比含量）可能需要随着饲料能量浓度的变化而上调或下调。这就要求确定氨基酸需要量时，应该考虑能量实际摄入量的变化。对生长猪而言，当能量摄入量不同于推荐水平时，能量摄入量会限制体蛋白的沉积，这时氨基酸需要量应该根据固定的"氨基酸/能量比"来调整。而且，大多数情况下，尤其是实际生产中，能量的实际摄入量通常低于猪发挥最大遗传潜力所需要的能量，这也要求氨基酸需要量的确定应该建立在饲料氨基酸/能量比的基础上。

生物学利用率

大部分日粮蛋白质通常不能被完全消化，因而氨基酸也不能充分地吸收。而且，并不是所有被吸收的氨基酸都能够被代谢利用。设计氨基酸的比例必须考虑饲料氨基酸的生物学利用率。例如，一些来自奶制品的蛋白质的氨基酸几乎全部是能利用的，然而其他植物籽实类蛋白质中氨基酸有效利用率就要低很多（Lewis and Bayley, 1995; Moehn et al., 2007; Adeola, 2009）。因此，对猪来说，一种原料蛋白质的质量可以通过评价其蛋白质中的氨基酸生物学利用率来评估，同时生物学利用率也可以用来表示氨基酸需要量。使用生物可用氨基酸量来表达氨基酸需要量的方法具有很好的理论基础。必须了解猪日粮配方中可利用氨基酸含量的表示方法。斜率比值法是目前广泛应用于评价猪饲料氨基酸相对生物学利用率的重要方法之一（Batterham, 1992; Kovar et al., 1993; Adeola et al., 1994; Adeola, 2009），这种方法是建立在对猪随着待测原料中某一单一氨基酸浓度增加的生长反应与晶体氨基酸浓度增加导致猪生长反应的比较分析基础上的。

由于斜率比值法所需时间长、试验成本高，而且单一原料测定的氨基酸生物学利用率值在混合饲料中的可加性也存在质疑，所以目前更普遍地用氨基酸消化率法来评估氨基酸有效利用率。而且，斜率比值法在控制待测原料配制的日粮组成对测定结果影响而不是限制性氨基酸的影响方面存在很多不足，因此会对待测原料中氨基酸利用率的测定结果造成较大的偏差。通过收集回肠食糜，分析其中待测氨基酸含量，然后计算小肠消失的氨基酸占日粮氨基酸比例是测定氨基酸消化率最典型的方法。回肠食糜分析方法是对肛门氨基酸消化率法的改进，该方法可有效校正后肠对测定结果的影响，因为后肠氨基酸的消失主要是由于微生物发酵利用，而对动物没有起到实际意义。那些无法被消化吸收的蛋白质进入后肠后，一部分被后肠微生物发酵利用，其他的则随粪便排出体外。微生物氮占了总的粪氮的62%~76%，后肠微生物的活性依赖于可发酵利用碳水化合物的量。Zebrowska（1978）最初选用无氮日粮的研究表明，灌注到猪回肠末端的酪氨酸或水解酪氨酸可以被消化和吸收，然而被吸收的含氮物质（大部分是氨和部分的胺）几乎全部快速从尿中排出；Sauer 和 Ozimek（1986）的研究进一步说明后肠中的蛋白质和氨基酸对满足动物蛋白和氨基酸营养几乎没有作用。然而，在某些日粮条件下，当用于合成非必需氨基酸所需要的氮供给不足时，被后肠吸收的这些含氮物质能够用于合成非必需氨基酸，从而起到节约必需氨基酸的作用（Metges, 2000）。此外，后肠微生物合成的氨基酸能用以满足维持氨基酸量的等价需求，从而起到维持整个机体氨基酸平衡的作用，但是这些微生物氨基酸的合成和吸收过程可能发生的位点在回肠（Torrallardona et al., 2003a，b）。也有研究表明，后肠发酵，尤其是回肠末端的微生物发酵能够提供给猪氨基酸分解代谢的原料，从而可降低机体氨基酸需要总的供应量（Libao-Mercado et al., 2009）。这些研究指出，后肠微生物群落对猪净氨基酸需要量的影响有待于准确量化。

Sauer 和 Ozimek（1986）综述许多研究指出，回肠氨基酸消化率优于粪氨基酸消化率，回肠氮消化率能与日增重和饲料转化效率等生产指标有更好的相关性。通过这种方法测定的消化率之所以被称为回肠消化率而不是生物利用率，是因为吸收的氨基酸有时候并没有全部被用于机体代谢。氨基酸消化率的测定是建立在消化道氨基酸消失率的基础上，而不能反映氨基酸具体的吸收形式。以过度热处理的饲料原料为测定对象，相对于斜率比值法，回肠消化率法测定的赖氨酸、苏氨酸、蛋氨酸和色氨酸生物学价值通常会被高估（Batterham, 1994; Van Barneveld et al., 1994）。结合化学分析法和消化率法能够更好地估计氨基酸的生物学价值，如经过热处理饲料原料中有效赖氨酸的评定（Carpenter, 1973; Batterham, 1992; Rutherfurd and Moughan, 1997; Pahm et al., 2009）。这样的话，有必要对回肠食糜和饲料中有效氨基酸含量化学分析方法做进一步的改进。此外，越来越多的试验表明回肠消化率有可能因为一些因素而低估某些氨基酸的生物学价值，如投喂的饲料含有较多的可发酵纤维，或者投喂的饲料容易导致过高的肠道内源损失，又或者是发酵氨基酸的分解代谢（Zhu et al., 2005; Libao-Mercado et al., 2006, 2009）。

回肠表观消化率估测值不能将日粮中未被消化和吸收的氨基酸与回肠末端内源氨基酸加以区分。内源性的蛋白质和氨基酸包含了来自胃、胰液、胆汁、脱落的黏膜细胞的蛋白质和氨基酸，以及内源性的氨和尿素。氨基酸消化率的准确评定需要消除食糜中内源氨基酸的影响。内源氨基酸损失受到多种因素的影响，包括饲料中的抗营养因子（如胰蛋白酶抑制因子、丹宁酸）、脂肪、纤维和蛋白质含量等（Stein et al., 2007）。两种主要的回肠内源氨基酸组分分别是基本内源氨基酸损失和特异性内源氨基酸损失。基本内源氨基酸损失被称为与日粮无关的或非特异性的内源氨基酸损失，而特异性内源损失指的是与饲料有关的内源氨基酸损失。非特异性的和特异性的氨基酸损失构成了总的内源氨基酸损失。在回肠表观消化率的基础上，消除回肠总内源氨基酸损失后得到的消化率是氨基酸真消化率，而仅消除非特异性内源氨基酸损失后得到的消化率是回肠标准氨基酸消化率。现在普遍使用的回肠标准氨基酸消化率的定义和测定方法都是由 Stein 等（2007）提出的。在本书中，建立在非特异内源氨基酸损失分析的基础上，猪的氨基酸需要量和原料中氨基酸浓度都是采用回肠标准可消化氨基酸来表示。

Lewis 和 Bayley（1995）通过综合分析一些研究，发现晶体氨基酸能够完全被小肠内腔吸收，所以它们的生物利用率通常被设定为100%。但是，很多情况表明氨基酸的完全吸收并不等于100%的生物学效价。例如，热损害赖氨酸生成的衍生物（ε-N-deoxyketosyllysine，美拉德反应产生的 Amadori 化合物）能够被吸收但是无法被利用；饲喂频率低会导致晶体氨基酸吸收速度远高于蛋白质分解产生氨基酸。另外还存在其他涉及生物利用率方面的内容，特别是消化率方面的内容将会在第13章作详细的讨论。

日粮氨基酸失衡

摄入比例不当的氨基酸会带来很多不利的影响，如氨基酸缺乏、氨基酸中毒、氨基酸拮抗、氨基酸比例失调等（Harper et al., 1970; D'Mello, 2003）。氨基酸缺乏是指日粮中一种或几种氨基酸的供给量少于满足其他氨基酸或营养物质有效平衡利用的需要量。猪日粮中的蛋白质供应会完全缺乏某一种必需氨基酸，但是可能会缺乏其中的一种或几种。日粮中实际可提供的氨基酸与动物理论需要量比值最低的氨基酸称之为第一限制性氨基酸，比值其次低的氨基酸称为第二限制性氨基酸，其他的依此类推。猪氨基酸缺乏很少会表现出临床症状，最主要的表现是饲料采食量减少，以及饲料浪费和生长受阻。

猪能够耐受高蛋白的摄入量，除了偶尔的轻度腹泻外，很少会引起特殊的疾病反应。然而，饲喂高蛋白的饲料是浪费的（如生长育肥猪饲喂高于25%的蛋白质），由此还会加大对环境的污染，降低增重和饲料转化效率。过量摄入某些特定氨基酸会导致猪采食量减少、生长障碍、行为异常，甚至死亡。

氨基酸中毒带来很多不利影响（如病理症状）。大量摄取某种单一的氨基酸容易引起氨基酸的中毒，而且这种过量失衡引起的中毒无法通过添加其他一种或一类氨基酸的方式加以预防或控制。围绕蛋氨酸和半胱氨酸的毒性开展了大量动物试验，研究结果表明：在所有氨基酸中，这些含硫氨基酸的毒性最强（Baker, 2006; Dilger and Baker, 2008）。苏氨酸是毒性最低的必需氨基酸（Edmonds et al., 1987）。除丝氨酸之外，非必需氨基酸的毒性相对较低。氨基酸中毒导致的病理变化可能是由于个别氨基酸的结构和代谢特点造成的。

化学性质或结构上相似的氨基酸在发挥其合成蛋白质的功能过程中，可能存在相互间竞争和抑制的问题。氨基酸拮抗是发生在结构和化学性质相似的氨基酸之间的一种特殊的互作反应，是指当饲料中的某一种氨基酸远超过需要量时会引起机体对与它存在相互拮抗关系的另一种氨基酸需要量的增加。这种情况下，即使在问题日粮中添加第一限制性氨基酸也不能消除由此引起的不利影响，从而影响动物的生产性能。存在拮抗关系的常见氨基酸包括中性氨基酸和支链氨基酸（亮氨酸、异亮氨酸、缬氨酸），它们对生长猪（Langer and Fuller, 2000; Langer et al., 2000; Wiltafsky et al., 2010）和母猪（Guan et al., 2004; Perez-Laspiur et al., 2009）的营养价值非常重要。赖氨酸和精氨酸也存在相互间拮抗关系，只是它对猪实际生产的影响效应相对较小（Lewis, 2001）。支链氨基酸之间的拮抗有可能导致支链氨基酸分解代谢的增加，包括第一限制性支链氨基酸。通常，这种不利影响靠添加化学性质或结构相似的氨基酸来缓解。

氨基酸失衡与氨基酸结构无关，它可能是由于过多供应某种或某些非限制性氨基酸所致，这种情况通常会导致采食量的降低。氨基酸失衡可以通过添加少量某种或某一些限制性氨基酸来缓解。氨基酸拮抗和失衡可能会引起小肠对氨基酸的竞争性吸收和转运，导致小肠损伤、代谢混乱、大量释放毒性物质（如氨和同型半胱氨酸）。这些状况的共同特征是造成猪采食量的降低，消除不良氨基酸的影响可快速恢复机体的正常生产。Harper等（1970）、Benevenga和Steele（1984），以及Garlick（2004）先后综述了氨基酸过量摄食引起的猪生理和代谢反应。

氨基酸与赖氨酸之比

近年来，建立在对生长动物高品质蛋白质（接近动物组织中蛋白质的氨基酸组成）氨基酸组成分析的基础上，用于表示动物氨基酸需要的"理想氨基酸模型"的概念形成了。这种按理想氨基酸模型组成的蛋白质被称为"理想蛋白质"。理想氨基酸模型或理想蛋白质假定这种蛋白质的氨基酸组成和比例与特定生理条件下猪维持和生产所需要的氨基酸组成比例一致。第十版《猪营养需要》（NRC, 1998）中，理想必需氨基酸模型反映的是完成机体关键生理活动对氨基酸的需要量。因此，最佳日粮氨基酸模型会随动物生理状态和生产水平的变化而变化。在第十版中，理想氨基酸模式是建立在其他氨基酸与赖氨酸的相对比例基础上，同时考虑了其他可以获得的信息，如机体肠道基本内源氮损失的氨基酸组成、体表氮损失（皮肤和毛）、蛋白质沉积（生长猪的空腹条件下的整个机体、妊娠母猪的孕体和母体组织、泌乳母猪的奶和母体组织），本版本的理想氨基酸模型对第十版做了进一步扩展。有关理想氨基酸模式建立方法的内容将在本章接下来的部分中阐述。

氨基酸需要量的经验估计

传统上，营养物质需要量完全建立在经验性研究总结的基础上。然而，这种方法存在很多的限制因素，涉及经验性研究的特定试验条件，包括瘦肉和脂肪沉积速率、采食量、健康和环境条件等，这些特定的经验性研究存在时间阶段的依赖性。因此，人们越来越重视对氨基酸需要量的析因估测。针对经验性研究的全面综述对于按激素需要量模型的开发和测试非常必要。氨基酸需要量的经验性测定方法需要特别仔细地考虑合适的动物模型、适宜的环境条件和试验日粮，这种条件下的测定结果才能适用于实际生产条件的需要。虽然至今已有大量的相关试验研究积累，但是仍然存在一些研究不清楚的影响因素（如可加性和环境条件等），即使是一些实际生产日粮中经常缺乏的关键氨基酸，如赖氨酸、蛋氨酸、色氨酸和苏氨酸。至今关于第五~第八限制性氨基酸需要量的知晓甚少，但是随着越来越多的晶体氨基酸被更广泛地应用，对所有必需氨基酸需要量的准确评估也变得越来越急迫。开展氨基酸需要量评定试验的关键点在于：①使用缺乏待测氨基酸的原料设计出缺乏该种氨基酸的基础日粮（即把经添加其他晶体氨基酸调整之后的基础日粮中待测氨基酸设计成为第一限制性氨基酸）；②基础日粮中，除缺乏待测氨基酸之外，其他营养物质组成合理；③至少设计4个待测氨基酸的梯度水平（水平范围涵盖缺乏和过量；以估计需要量为参考，设置两个低剂量和两个高剂量）；④设计足够的试验持续时间，具体的试验周期取决于是否达到试验效果的评判标准；⑤为能客观分析试验效果和确定需要量选择合适的统计方法。此外，本书还总结了大量关于猪氨基酸需要量的文献资料，接下来将在后面做进一步阐述。

为了保持针对各种氨基酸和生长阶段的估测需要量一致性，氨基酸需要量的估计值采取分阶段表示的方法（Robbins et al., 2006）。对生长猪，氨基酸需要量的确定是建立在多个日粮氨基酸水平范围内"能获得最佳平均日增重生长成绩的氨基酸水平"这一判断标准基础上；而怀孕和哺乳母猪则需要考虑其他更多的参数。此外，在缺少日粮氨基酸和回肠标准可消化氨基酸含量参数的情况下，可用常规营养成分和回肠氨基酸消化率数据（NRC, 1998）来替代，以降低对比试验研究的变异。当然，有些特殊饲料原料的常规营养成分或消化率估计值在猪数据库里面没有，这时可参考其他的一些数据库（AmiPig, 2000; AminoDat, 2006）。

仔猪和生长育肥猪

在开展仔猪和生长育肥猪适宜氨基酸需要量的筛选试验时，多个标准条件需要满足（但这些条件的具体参数不受限定），包括可用于计算回肠标准氨基酸消化率和代谢能的日粮化学成分和/或营养成分、足够的重复数、缺乏待测氨基酸但是其他营养成分组成适宜的基础日粮、梯度范围涵盖了不足和过量情况的多个待测氨基酸水平、一个反应灵敏的重要生产指标（如平均日增重）。氨基酸需要量筛选试验过程中需要记录日粮代谢能浓度、猪的体重（平均体重、初始体重、结束体重）以及相关生长性能指标（平均日增重和平均日采食量）。通过这些大量筛选试验，我们可以获得氨基酸需要量的估计值，同时可通过日粮成分估计出回肠标准可消化氨基酸需要量，甚至每千克体增重所需要的回肠标准可消化氨基酸的克数也可以通过总结的数据计算得到。有关文献综述见表2-2。

表2-2 生长育肥猪氨基酸需要量估测值的总结及相关生长性能参数[a]

参考文献	体重/kg			生长性能		日粮	SID	
	平均值	初始重	末重	ADG	ADFI	ME	%	g/kg 增重
赖氨酸								
Lewis 等（1980）	10.0	5	15	397	710	3300	1.100	19.67
Martinez 和 Knabe（1990）	10.6	6	15	325	631	3400	1.060	20.58
Kendall 等（2008）	15.0	11	19	526	688	3421	1.350	17.66
Schneider 等（2010）	15.2	9	21	588	783	3667	1.350	17.98
Oresanya 等（2007）	15.5	8	23	554	840	3500	1.480	22.44
Schneider 等（2010）	16.0	10	22	584	900	3667	1.150	17.72
Williams 等（1997）	17.0	7	27	677	977	3452	1.218	17.58
Nam 和 Aherne（1994）	17.5	9	26	612	1035	3513	1.179	19.94
Kendall 等（2008）	18.0	11	25	625	865	3421	1.260	17.44
Yi 等（2006）	18.5	12	25	586	889	3420	1.280	19.42
Kendall 等（2008）	19.0	11	27	646	958	3421	1.300	19.28
Urynek 和 Buraczewska（2003）	21.9	13	31	634	1190	3346	1.148	21.55
O'Connell 等（2005）	30.5	21	40	789	1354	3166	1.153	19.78
Bikker 等（1994b）	32.5	20	45	768	1272	3671	0.827	13.69
Batterham 等（1990）	32.5	20	45	680	1288	3511	0.840	15.91
Batterham 等（1990）	32.5	20	45	625	1299	3511	0.713	14.82
Martinez 和 Knabe（1990）	34.8	21	49	786	1994	3264	0.820	20.80
Lawrence 等（1994）	35.0	20	50	968	1976	3362	0.880	17.96
Krick 等（1993）	39.5	20	59	921	2198	3350	0.942	22.47
Williams 等（1984）	40.0	25	55	875	2144	3348	0.757	18.54
Warnants 等（2003）	40.0	31	49	601	1260	3166	1.090	22.85
Warnants 等（2003）	40.0	31	49	649	1400	3166	1.140	24.59
O'Connell 等（2005）	51.0	40	62	833	1922	3166	0.994	22.94
O'Connell 等（2005）	55.0	42	68	968	1967	3166	1.118	22.71
Hahn 等（1995）	71.5	52	91	970	2798	3485	0.640	18.46
Hahn 等（1995）	71.5	52	91	1150	3497	3485	0.560	17.03
O'Connell 等（2006）	75.5	60	91	980	2427	3166	0.950	23.54
Williams 等（1984）	80.0	55	105	870	2540	3315	0.651	19.02
Ettle 等（2003）	83.5	56	111	1068	2890	3227	0.675	18.27
Cline 等（2000）	85.0	54	116	850	2730	3370	0.748	24.02
Friesen 等（1995）	88.0	72	104	890	2890	3462	0.710	23.06
O'Connell 等（2006）	89.5	80	99	905	2525	3166	0.818	22.83
O'Connell 等（2006）	91.5	81	102	880	2451	3166	0.871	24.26
Dourmad 等（1996b）	95.5	80	111	902	2832	3075	0.600	18.84
Dourmad 等（1996b）	95.5	80	111	896	2822	3075	0.602	18.96
Yen 等（2005）	98.5	84	113	790	2990	3400	0.440	16.65
Hahn 等（1995）	99.5	91	108	993	2796	3468	0.520	14.64
Hahn 等（1995）	99.5	91	108	1118	3945	3468	0.500	17.64
King 等（2000）	100.0	80	120	934	2479	3327	0.580	15.39
King 等（2000）	100.0	80	120	976	2390	3327	0.667	16.33
Loughmiller 等（1998a）	102.0	91	113	800	3000	3303	0.469	17.59
Friesen 等（1995）	120.0	104	136	830	3150	3462	0.650	24.67
精氨酸								
Southern 和 Baker（1983）	12.0	9.0	15.0	508	806	3582	0.480	7.62
组氨酸								
Izquierdo 等（1988）	14.8	10.0	19.5	453	594	3200	0.252	3.31

续表

参考文献	体重/kg			生长性能		日粮	SID	
	平均值	初始重	末重	ADG	ADFI	ME	%	g/kg 增重
异亮氨酸								
Becker 等（1963）	8.2	5.1	11.2	197	340	3799	0.616	10.63
Kerr 等（2004）	8.3	6.6	9.9	255	355	3440	0.654	9.11
Kerr 等（2004）	8.8	6.6	10.9	314	410	3440	0.690	9.01
Oestemer 等（1973）	11.6	5.8	17.4	385	648	3143	0.514	8.64
Wiltafsky 等（2009）	15.5	7.7	23.2	444	621	3251	0.601	8.41
Wiltafsky 等（2009）	17.1	8.0	26.2	433	616	3251	0.501	7.12
Becker 等（1957）	21.5	14.7	28.2	450	957	3152	0.350	7.44
Becker 等（1957）	21.5	14.2	28.7	484	848	3335	0.513	8.98
Parr 等（2003）	34.5	27.0	42.0	709	1464	3430	0.453	9.35
Taylor 等（1985）	40.0	25.0	55.0	630	1598	3590	0.381	9.68
Becker 等（1963）	53.0	44.6	61.3	595	1780	3533	0.291	8.71
亮氨酸								
Augspurger 和 Baker（2004）	13.4	9.2	17.5	480	797	3490	1.050	17.44
蛋氨酸								
Chung 和 Baker（1992b）	8.4	6	11	321	518	3476	0.315	5.08
Owen 等（1995）	8.9	5	13	372	413	3478	0.363	4.03
Matthews 等（2001）	10.2	6	14	367	546	3354	0.420	6.25
Owen 等（1995）	10.6	6	15	439	658	3326	0.319	4.78
Chung 和 Baker（1992b）	18.1	11	25	645	1174	3476	0.275	5.01
Yi 等（2006）	19.5	13	26	650	956	3420	0.440	6.47
Schutte 等（1991）	25.5	13	38	440	1010	3221	0.320	7.35
Schutte 等（1991）	26.0	14	38	628	1212	3221	0.290	5.60
Leibholz（1984）	28.0	21	35	505	1353	3465	0.180	4.82
Lenis 等（1990）	50.0	35	65	835	1990	3268	0.270	6.43
Lenis 等（1990）	50.0	35	65	847	2070	3268	0.230	5.62
Leibholz（1984）	53.0	35	71	618	2064	3465	0.157	5.23
Chung 等（1989）	66.4	53	80	946	2680	3512	0.175	4.96
Roth 等（2000）	79.0	53	105	769	2410	3083	0.180	5.64
Roth 等（2000）	80.5	54	107	837	2440	3083	0.220	6.41
Roth 等（2000）	80.5	54	107	869	2500	3083	0.210	6.04
Loughmiller 等（1998b）	82.5	54	111	890	3050	3203	0.230	7.88
Loughmiller 等（1998b）	89.0	74	104	880	2410	3474	0.125	3.42
Knowles 等（1998）	92.7	74	111	780	3320	3478	0.135	5.75
蛋氨酸+半胱氨酸								
Matthews 等（2001）	10.2	6	14	367	546	3354	0.801	11.92
Yi 等（2006）	19.5	13	26	650	956	3420	0.770	11.32
Schutte 等（1991）	25.5	13	38	440	1010	3221	0.520	11.94
Schutte 等（1991）	26.0	14	38	628	1212	3221	0.540	10.42
Lenis 等（1990）	50.0	35	65	835	1990	3268	0.460	10.96
Lenis 等（1990）	50.0	35	65	847	2070	3268	0.430	10.51
Chung 等（1989）	66.4	53	80	946	2680	3512	0.410	11.61
Roth 等（2000）	79.0	53	105	769	2410	3083	0.366	11.47
Roth 等（2000）	80.5	54	107	837	2440	3083	0.350	10.20
Roth 等（2000）	80.5	54	107	869	2500	3083	0.413	11.88
Loughmiller 等（1998b）	82.5	54	111	890	3050	3203	0.392	13.43
Loughmiller 等（1998b）	89.0	74	104	880	2410	3474	0.335	9.17
Knowles 等（1998）	92.7	74	111	780	3320	3478	0.250	10.64

参考文献	体重/kg			生长性能		日粮	SID	
	平均值	初始重	末重	ADG	ADFI	ME	%	g/kg 增重
苏氨酸								
Ragland 和 Adeola（1996）	15.1	9.8	20.3	405	1158	3456	0.398	11.38
Kovar 等（1993）	15.2	10.9	19.4	442	975	3388	0.455	10.03
Adeola 等（1994）	15.4	9.9	20.9	416	998	3936	0.454	10.90
Adeola 等（1994）	15.4	9.9	20.9	492	1068	3936	0.507	11.01
Bergstrom 等（1996）	17.1	11.4	22.7	497	1117	3314	0.475	10.67
Ferguson 等（2000）	19.0	12.9	25.0	621	1034	3327	0.622	10.36
Conway 等（1990）	33.5	17.0	50.0	486	1208	3180	0.514	12.77
Sève 等（1993）	37.5	25.0	50.0	635	1501	3072	0.503	11.90
de Lange 等（2001）	58.0	39.0	77.0	866	1620	3262	0.538	10.06
Cohen 和 Tanksley（1977）	74.0	58.9	89.1	756	2961	3064	0.298	11.67
Saldana 等（1994）	75.7	58.0	93.3	897	3020	3245	0.299	10.06
Rademacher 等（1997）	81.5	60.0	103.0	976	3243	3107	0.411	13.66
Johnston 等（2000）	103.9	92.0	115.8	873	2953	3373	0.338	11.44
色氨酸								
Guzik 等（2002）	6.3	5.2	7.3	190	300	3300	0.205	3.24
Guzik 等（2002）	8.3	6.3	10.2	322	511	3300	0.182	2.88
Burgoon 等（1992）	11.0	6.2	15.7	343	500	3446	0.168	2.46
Cadogan 等（1999）	11.4	6.1	16.6	498	526	3442	0.257	2.71
Guzik 等（2002）	13.0	10.3	15.7	440	765	3300	0.180	3.13
Sato 等（1987）	13.3	10.0	16.6	314	775	3226	0.153	3.78
Eder 等（2001）	13.4	7.5	19.3	344	600	3107	0.154	2.69
Boomgaardt 和 Baker（1973）	15.1	10.4	19.7	396	896	3182	0.111	2.52
Borg 等（1987）	15.9	9.7	22.0	437	943	3192	0.135	2.91
Russell 等（1983）	26.4	18.4	34.3	620	1500	3285	0.153	3.71
Schutte 等（1995）	30.0	20.0	40.0	734	1393	3212	0.188	3.57
Quant 等（2012）	34.1	25.7	42.5	801	1721	3349	0.112	2.40
Burgoon 等（1992）	36.2	21.9	50.5	815	1723	3600	0.127	2.68
Quant 等（2012）	37.3	28.5	46.2	844	1738	3325	0.114	2.34
Eder 等（2003）	37.5	25.0	50.0	774	1640	3344	0.131	2.77
Eder 等（2003）	65.0	50.0	80.0	876	2150	3331	0.147	3.61
Burgoon 等（1992）	76.4	55.4	97.3	998	3090	3456	0.075	2.34
Guzik 等（2005）	89.9	74.6	105.1	900	3400	3297	0.094	3.54
Eder 等（2003）	97.5	80.0	115.0	746	2752	3243	0.093	3.43
缬氨酸								
Mavromichalis 等（2001）	7.6	5.8	9.4	258	292	3445	0.863	9.77
Wiltafsky 等（2009）	14.8	7.9	21.6	409	573	3275	0.659	9.24
Mavromichalis 等（2001）	15.1	10.9	19.2	519	847	3487	0.674	11.00
Barea 等（2009）	17.8	12.8	22.7	473	843	3233	0.659	11.75
Wiltafsky 等（2009）	18.8	14.1	23.4	333	516	3275	0.614	9.51
Gaines 等（2011）	20.3	13.5	27.0	641	1100	3350	0.683	11.72
Gaines 等（2011）	27.0	21.4	32.6	805	1378	3350	0.724	12.38

a 每一篇参考文献中，日粮代谢能（ME）和回肠标准可消化率都是通过文章中描述的日粮组成计算得来的。

妊娠母猪

对于妊娠母猪，除少量试验特意仅设计 3 个待测氨基酸水平外，氨基酸需要量筛选标准基本与生长育肥猪的近似。除了生长育肥猪中可用到的有效参数外，还需要记录以下生产性能评估参数：母猪采食量、受孕（第 1 天）和妊娠结束（第 113 天）时母猪体重、出生仔猪数（包括活仔数和死胎数）、仔猪初生重、氮的沉积、血浆氨基酸浓度、氨基酸氧化指标等。同生长育肥猪一样，在每个试验日粮营养成分和回肠标准氨基酸消化率基础上，可计算获得回肠标准可消化氨基酸需要量值。不像生长育肥猪氨基酸需要量研究的数据资源非常丰富，关于妊娠母猪氨基酸需要量的研究，仅有 4 个赖氨酸、4 个苏氨酸、3 个色氨酸、1 个异亮氨酸、2 个蛋氨酸和半胱氨酸、1 个缬氨酸的研究报道。有关文献综述见表 2-3。

表 2-3 归纳怀孕母猪氨基酸量及相关生长性能参数

参考文献	胎次	体重（第1天）	体重（第113天）	窝产仔数	出生体重/kg	ADFI/kg	日粮ME/(kcal/kg)	日粮SID[a]/%	日粮SID[a]/(g/d)	氮沉积/(g/d)
赖氨酸										
Rippel 等（1965a）[b]	1	—	—	10.88	1.224	1.82	3340	0.358	6.51	13.95
Duée 和 Rérat（1975）[c]	1	109.4	156.7	8.00	1.250	2.00	3226	0.542	10.85	12.80
Woerman 和 Speer（1976）[d]	1	130.3	142.4	9.80	1.306	1.82	3263	0.547	9.95	9.40
Dourmad 和 Étienne（2002）[e]	>1	228.0	265.0	12.80	1.450	2.75	3278	0.430	11.84	14.70
苏氨酸										
Rippel 等（1965）[b]	1	—	—	8.90	1.476	1.82	3340	0.389	7.07	16.68
Leonard 和 Speer（1983）[f]	2,3	131.0	184.6	9.45	1.407	1.82	3360	0.299	5.44	7.10
Dourmad 和 Étienne（2002）[e]	—	219.0	259.0	12.10	1.540	2.75	3078	0.271	7.46	13.20
Levesque 等（2011）[g]	—	191.5	230.4	—	—	—	—	—	—	—
Phe AA 氧化	2~3	191.5	236.9	13.30	1.526	2.40	3442	0.247	8.5	ND
血浆 Thr	2~3	191.5	236.9	13.30	1.526	2.40	3442	0.218	7.5	ND
色氨酸										
Rippel 等（1965c）[b]	1	—	—	9.00	1.400	1.82	3340	0.083	1.505	16.51
Easter 和 Baker（1977）[h]	1	—	—	—	—	2.00	2960	0.070	1.400	9.80
Meisinger 和 Speer（1979）[i]	1	—	—	8.50	1.294	2.00	3355	0.086	1.729	5.00
异亮氨酸										
Rippel 等（1965a）[b]	1	—	—	9.57	1.237	1.82	3340	0.317	5.769	16.79
蛋氨酸+半胱氨酸										
Rippel 等（1965a）[b]	1	—	—	8.56	1.360	1.82	3340	0.200	3.642	17.31
Holden 等（1971）[j]	1	—	—	7.60	1.220	1.82	3466	0.217	3.958	9.38
缬氨酸										
Rippel 等（1965c）[b]	1	—	—	9.75	1.313	1.82	3340	0.517	9.416	16.88

a 每一篇参考文献中日粮代谢能（ME）和回肠标准可消化率都是通过文章中描述的日粮组成计算得来的。
b N 平衡试验在怀孕 100~110 天。
c N 平衡试验从怀孕 80 天开始。
d N 沉积值取自于从怀孕 0 天、30 天、60 天和 95 天开始的 N 平衡试验的平均值。
e N 平衡试验包括怀孕 20~104 天的 4 个阶段，作者只报道了平均值。
f N 平衡试验开始于怀孕 45 天和 90 天，作者只报道了平均值。
g 取自于怀孕 30~54 天和 87~111 天的 N 平衡试验中的平均值。
h N 平衡试验在怀孕 80~107 天，作者只报道了平均值。
i N 平衡试验在怀孕 45~70 天和 90~115 天，作者只报道了平均值。
j N 沉积值取自于从怀孕 0 天、30 天、68 天和 106 天开始的 N 平衡试验的平均值。
"—"表示没有测定。

哺乳母猪

哺乳母猪氨基酸需要量试验的筛选标准与之前所述基本相同，额外需要记录的参数包括：哺乳期长度、断奶仔猪数、母猪初始和结束体重及体重变化、窝增重（或者产奶量）。关于哺乳母猪氨基酸需要量的研究，仅收录到 10 篇赖氨酸（Lewis and Speer, 1973; O'Grady and Hanrahan, 1975; Chen et al., 1978; Johnston et al., 1993; King et al., 1993b; Knabe et al., 1996; Tritton et al., 1996; Sauber et al., 1998; Touchette et al., 1998; Yang et al., 2000）、3 篇苏氨酸（Lewis and Speer, 1975; Westermeier et al., 1998; Cooper et al., 2001）、2 篇蛋氨酸+半胱氨酸（Ganguli et al., 1971; Schneider et al., 1992b）、2 篇色氨酸（Lewis and Speer, 1974; Paulicks et al., 2006）和 2 篇缬氨酸（Rousselow and Speer, 1980; Paulicks et al., 2003）的报道。有关文献综述的概要见表 2-4。

表 2-4 归纳哺乳母猪氨基酸及相关生长性能参数[a]

参考文献	胎次	哺乳日龄/d	断奶仔猪数	母猪体重变化/(kg/d)	平均体重/kg	ADFI/kg	日粮ME/(kcal/kg)	日粮SID/%	日粮SID采食量/(g/d)	窝增重/(g/d)
赖氨酸										
Chen 等（1978）	1~2	21	9.5	−0.410	142	5.01	2888	0.535	26.80	1429
Johnston 等（1993）	1~9	24	9.9	−0.086	199	6.27	3270	0.687	43.07	2120
King 等（1993b）	1	29	9.0	−0.821	137	3.81	3456	0.910	34.67	1971
Knabe 等（1996）	1	21	9.7	−0.152	185	5.64	3378	0.590	33.28	1668
Lewis 和 Speer（1973）	2~6	21	9.0	−0.762	192	5.45	3224	0.490	26.71	1665
O'Grady 和 Hanrahan（1975）	1~4	21	8.6	−0.319	161	5.45	2880	0.470	25.61	1348
Sauber 等（1998）[b]	1	28	14	−1.224	144	4.74	3224	0.66	31.28	2286
Touchette 等（1996）	1	17	10.0	−0.539	178	3.96	3400	0.986	39.05	2015
Tritton 等（1996）	1	23	9.9	−1.139	162	4.45	3174	0.655	29.15	2000
Yang 等（2000）	1~3	18	9.9	0.122	186	6.10	3309	0.726	44.28	2277
苏氨酸										
Cooper 等（2001）	1~3	20	10.9	0.235	—	7.15	3173	0.491	35.09	2487
Lewis 和 Speer（1975）	3~7	21	9.0	−0.400	—	5.45	3269	0.384	20.95	1581
Westermeier 等（1998）	1	21	9.3	−0.050	—	4.37	3278	0.487	21.27	1804
蛋氨酸+半胱氨酸										
Ganguli 等（1971）	1~5	21	8.0	−0.819	—	5.00	3442	0.294	14.71	1400
Schneider 等（1992b）	2~8	21	9.5	−0.520	—	4.53	3096	0.646	29.25	1891
色氨酸										
Lewis 和 Speer（1974）	3~6	21	9.0	−0.562	—	5.45	3304	0.082	4.49	1360
Paulicks 等（2006）	>1[c]	28	10.3	−0.685	—	4.66	3158	0.148	6.88	1896
缬氨酸										
Paulicks 等（2003）	>1	21	11.0	−0.787	—	4.45	3206	0.570	25.36	1802
Rousselow 和 Speer（1980）	3~7	21	9.0	−0.238	—	5.50	3466	0.531	29.20	1022

a 赖氨酸数据用来估测氨基酸的利用效率，而其他氨基酸数据（苏氨酸和缬氨酸）则用测试模型。每一篇参考文献中日粮代谢能（ME）和回肠标准消化率是通过文章中描述的日粮组成计算得来的。
b 数值取自于低瘦肉率和高瘦肉率的猪的平均值，并用这些数据来估测赖氨酸的利用效率。
c 试验使用经产母猪，但具体母猪胎次情况作者没有报道。

氨基酸需要量的测定——模型预测法

猪生理过程需要的氨基酸都来自于胃肠道摄入蛋白消化、吸收的产物，然后通过代谢反应提供给机体分解和合成代谢需要，以及体蛋白沉积（生长和繁殖，包括乳蛋白生产）。因此，机体总氨基酸的需要包括了维持机体功能的氨基酸需要和体蛋白沉积的氨基酸需要二者之和。生产乳蛋白的氨基酸有可能来自于外界摄入的饲料蛋白，也可能来自自身体蛋白的分解。哺乳期间，尽可能降低母体自身蛋白的损失有利于提高以后的繁殖性能，尤其是初产母猪（Boyd et al., 2000）。如果哺乳母猪在哺乳期间氨基酸的供应和摄入充足的话，母体蛋白的动用情况将受能量摄入情况的驱动。因此，体蛋白的分解对猪日粮氨基酸需要量的影响可根据能量分配估计出来，这将在后面"随母猪体重变化表现出的体蛋白含量变化"部分中做进一步讨论。在生长育肥猪、怀孕母猪和哺乳母猪的维持氨基酸需要中，共性需要部分是为了弥补肠道内源损失、皮肤和毛损失的需要量。

维持需要量

Moughan（1999）介绍了维持氨基酸和氮需要量的主要决定因素在于内源性肠道的氨基酸损失（与饲料摄入有关）、皮肤和毛氨基酸损失（可表示为代谢体重 $BW^{0.75}$ 的函数）、最低的氨基酸分解代谢损失（主要包括基本的体蛋白更新、用于必需含氮化合物不可逆的合成以及尿素氮排泄）。现有的数据信息量不足以有效支持对单个氨基酸最小代谢损失的合理估计。因此，用于补充肠道、皮肤和毛氨基酸损失的回肠标准氨基酸摄入量中，吸收后无效的部分被假设为机体基础蛋白质周转的氨基酸损失。在能量和营养成分供给适宜且氮沉积为零的情况下，用于猪维持氨基酸的需要直接被用于前面所提到的这些生理过程。

分泌到肠道且不被重吸收的内源氨基酸（来自于肠道蛋白）最低损失量与干物质的摄入有关。Moughan（1999）推算了大肠对总的肠道基础氨基酸损失的贡献值接近于回肠基础内源性损失的10%，总的肠道基本内源性损失被认为是基本回肠内源性损失的110%。以安装有回瘘管的生长育肥猪为模型动物，Moughan（1999）通过57个试验获得一个内源性回肠氨基酸损失的平均值，以此为依据可推导出肠道各个氨基酸总的平均损失量（g 氨基酸/kg 干物质）及其相对赖氨酸比例模型（表2-5）。57个试验内源性回肠赖氨酸损失的平均值为0.417 g/kg 干物质摄入量。相对而言，有关怀孕母猪和哺乳母猪肠道氨基酸损失组成模型的数据之间报道很少，表2-5中的数据也被借用于怀孕和哺乳母猪，而怀孕母猪和哺乳母猪的肠道赖氨酸损失量分别是0.552 g/kg 干物质摄入量和0.292 g/kg 干物质摄入量（Stein et al., 1999）。

表2-5 肠道和皮毛中损失的蛋白质的氨基酸组成

氨基酸	肠道损失				皮毛损失	
	g/100 g 赖氨酸	g/kg DMI			g/100 g 赖氨酸	mg/kg $BW^{0.75}$
		生长-育肥阶段	妊娠阶段	哺乳阶段		
精氨酸	116.4	0.485	0.608	0.340	0	0
组氨酸	48.7	0.203	0.254	0.142	27.9	1.26
异亮氨酸	91.9	0.383	0.480	0.268	55.8	2.51
亮氨酸	125.9	0.525	0.657	0.368	116.3	5.23
赖氨酸	100	0.417	0.522	0.292	100	4.5
蛋氨酸	27.3	0.114	0.143	0.080	23.3	1.05
蛋氨酸+半胱氨酸	78.1	0.326	0.408	0.228	127.9	5.76
苯丙氨酸	82.2	0.343	0.429	0.240	67.4	3.03
苯丙氨酸+酪氨酸	150.4	0.627	0.785	0.439	109.3	4.92
苏氨酸	145.1	0.605	0.757	0.424	74.4	3.35
色氨酸	31.8	0.133	0.166	0.093	20.9	0.94
缬氨酸	129.8	0.541	0.678	0.379	83.7	3.77
N×6.25	3370.4	14.05	17.59	9.84	2325.6	104.7

皮肤和毛的氨基酸损失也占据了维持氨基酸需要量的一部分，作为$BW^{0.75}$的函数，van Milgen 等（2008）还用氨基酸比例（相对于赖氨酸）来生成氨基酸维持需要估测值，见表2-5。

基础肠道内源氨基酸损失没有考虑日粮抗营养因子和纤维对它的影响。表 2-6 列举了每日胃肠道基础内源氨基酸损失量。例如，假设生长猪每天采食 2 kg 干物质，基础肠道内源氨基酸损失量为每摄入 1 kg DM 导致基础回肠内源氨基酸损失量的 110%（例如，赖氨酸为 0.417×1.1，见表2-5；10%的校正值是反映后肠对整个肠道内源氨基酸损失的贡献）。表2-6列出的每日皮肤和毛氨基酸损失量是通过表 2-5 中的氨基酸损失值和代谢体重（$BW^{0.75}$）计算而来的。维持氨基酸的需要量可通过总的物理损失除以氨基酸利用效率计算得到（见表 2-12）；估计氨基酸利用效率的方法在接下来的章节中有详细描述。表 2-7 列出了 50 kg 生长猪、200 kg 怀孕母猪、200 kg 哺乳母猪的氨基酸维持需要量，并采用了 g/kg、$mg/BW^{0.75}$ 和相对于赖氨酸的氨基酸模型三种表示方法。不同体重和不同生长阶段猪的维持氨基酸需要模型都可以根据前面描述的方法获得。其中，50 kg 生长猪、200 kg 怀孕母猪、200 kg 哺乳母猪的维持赖氨酸需要量分别为 71 $mg/kg\ BW^{0.75}$、35 $mg/kg\ BW^{0.75}$ 和 46 $mg/kg\ BW^{0.75}$（表2-7），这不同于第十版 NRC（1998）的估计值，即 36 $mg/kg\ BW^{0.75}$。通过特别地明确用于补充皮肤、毛和肠道内源损失的氨基酸维持需要，内脏器官氨基酸代谢损失（可假定为饲料摄入导致的基础内源肠道氨基酸损失量）可解析得更为明白。

表 2-6　生长猪，怀孕母猪和哺乳母猪每天通过肠道和皮毛损失的氨基酸

氨基酸	50kg 的猪 (2 kg DMI/d)		200kg 的妊娠母猪 (2 kg DMI/d)		200kg 的泌乳母猪 (5 kg DMI/d)	
	肠道的损失 /(g/d)	皮肤和毛的损失 /(g/d)	肠道的损失 /(g/d)	皮肤和毛的损失 /(g/d)	肠道的损失 /(g/d)	皮肤和毛的损失 /(g/d)
精氨酸	0.726	0.000	0.909	0.000	2.045	0.000
组氨酸	0.447	0.024	0.574	0.069	0.967	0.083
异亮氨酸	1.110	0.062	1.406	0.178	1.890	0.171
亮氨酸	1.538	0.131	1.607	0.309	2.497	0.344
赖氨酸	1.223	0.113	1.531	0.319	2.141	0.319
蛋氨酸	0.343	0.027	0.414	0.074	0.480	0.061
蛋氨酸+半胱氨酸	1.189	0.179	1.459	0.498	1.553	0.379
苯丙氨酸	1.123	0.085	1.137	0.194	1.690	0.207
苯丙氨酸+酪氨酸	1.850	0.124	2.101	0.318	2.982	0.323
苏氨酸	1.748	0.083	2.140	0.229	2.805	0.214
色氨酸	0.478	0.029	0.512	0.070	0.676	0.066
缬氨酸	1.489	0.089	1.773	0.238	3.193	0.307
N × 6.25	36.376	2.315	45.536	6.548	63.681	6.548

表 2-7　用于维持的回肠标准可消化氨基酸需要量及最佳的氨基酸与赖氨酸比值

氨基酸	50kg 的猪 (2 kg DMI/d)			200kg 的妊娠母猪 (2 kg DMI/d)			200kg 的泌乳母猪 (5 kg DMI/d)		
	g/d	$mg/kg\ BW^{0.75}$	与赖氨酸的比值	g/d	$mg/kg\ BW^{0.75}$	与赖氨酸的比值	g/d	$mg/kg\ BW^{0.75}$	与赖氨酸的比值
精氨酸	0.73	38.62	54.4	0.91	17.09	49.1	2.04	38.45	83.1
组氨酸	0.47	25.00	35.2	0.64	12.09	34.8	1.05	19.74	42.7
异亮氨酸	1.17	62.32	87.7	1.58	29.78	85.6	2.06	38.76	83.8
亮氨酸	1.67	88.78	124.9	1.92	36.03	103.6	2.84	53.41	115.4
赖氨酸	1.34	71.05	100.0	1.85	34.79	100.0	2.46	46.26	100.0

续表

氨基酸	50kg 的猪 （2 kg DMI/d）			200kg 的妊娠母猪 （2 kg DMI/d）			200kg 的泌乳母猪 （5 kg DMI/d）		
	g/d	mg/kg BW$^{0.75}$	与赖氨酸的比值	g/d	mg/kg BW$^{0.75}$	与赖氨酸的比值	g/d	mg/kg BW$^{0.75}$	与赖氨酸的比值
蛋氨酸	0.37	19.68	27.7	0.49	9.17	26.4	0.54	10.16	22.0
蛋氨酸+半胱氨酸	1.37	72.77	102.4	1.96	36.80	105.8	1.93	36.33	78.5
苯丙氨酸	1.21	64.27	90.5	1.33	25.03	72.0	1.90	35.66	77.1
苯丙氨酸+酪氨酸	1.97	104.96	147.7	2.42	45.49	130.8	3.31	62.14	134.3
苏氨酸	1.83	97.33	137.0	2.37	44.53	128.0	3.02	56.78	122.7
色氨酸	0.51	26.98	38.0	0.58	10.94	31.4	0.74	13.97	30.2
缬氨酸	1.58	83.89	118.1	2.01	37.82	108.7	3.50	65.81	142.3
N×6.25	38.69	2057.73	2896.0	52.08	979.33	2814.9	70.23	1320.51	2854.3

蛋白质的沉积和存留及其氨基酸组成

生长猪

在生长猪中，日粮所提供的高于维持需要的氨基酸可用于体蛋白的沉积，直至猪的最大限度体蛋白沉积能力。猪生长期体蛋白的沉积和相应的蛋白增加反映了体蛋白合成代谢和分解代谢之间的差异。体重、性别、饲喂莱克多巴胺或者针对促性腺激素释放激素进行免疫接种对体蛋白沉积的影响见本书第 8 章。

综合分析 Batterham 等（1990）、Kyriazakis 和 Emmans（1993）、Bikker 等（1994a）、Mahan 和 Shields（1998）的研究报道，可获得整个体蛋白中氨基酸的含量和体蛋白增量的氨基酸组成。Batterham 等（1990）对体重 20~45 kg 生长猪进行了试验研究，试验期间猪只每天饲喂 3 次，日粮中不限制赖氨酸的摄入量，结果建立了整个体蛋白含量对其中氨基酸含量的线性回归方程，并推导出机体增重蛋白质中氨基酸的组成。Kyriazakis 和 Emmans（1993）试验获得了体重 12~32 kg 生长猪体蛋白和氨基酸之间的回归系数，通过回归系数进行回归分析获得了机体增重的蛋白质的氨基酸组成。Bikker（1994a）研究报道了采食 3 倍于维持需要营养水平的日粮条件下，20~45 kg 生长猪机体增重蛋白质的氨基酸组成。Mahan 和 Shields（1998）报道了 8~146 kg 体重猪、涉及 9 个屠宰体重点的一个健全的数据集，以及一个整个体蛋白对其氨基酸组成的线性回归方程，这个回归方程反映了 21~127 kg 期间 7 个屠宰点的猪体蛋白和氨基酸沉积之间的关系，并应用这个方程推导了生长-育肥猪机体增重蛋白质的氨基酸组成。取上述 4 个数据集的平均值可获得机体增重蛋白质的赖氨酸含量（7.1 g 赖氨酸/100 g 体蛋白）、机体增重蛋白质的氨基酸组成以及以赖氨酸为基础的各种氨基酸比例。其中，以赖氨酸为基础的各种氨基酸比例见表 2-8。

Schinckel（2003）和 Webster（2007）的研究表明，当日粮中提供 10 mg/kg 的莱克多巴胺时，莱克多巴胺诱导了整个体蛋白沉积的增加并改变了其中的氨基酸比例模型，其中对肌肉蛋白质的影响要大于非肌肉蛋白质（表 2-8）。上述体蛋白中氨基酸组成模型的改变值是基于肌肉（Lloyd，1978）和非肌肉（即整个体蛋白减去肌肉蛋白质）蛋白氨基酸组成，以及一个假设（即莱克多巴胺的添加使肌肉蛋白沉积占体蛋白的比例由 54% 上升为 81%）的基础之上。

表 2-8 生长育肥猪增重的整体蛋白中和莱克多巴胺诱导增重的整体蛋白中赖氨酸和其他氨基酸的含量

氨基酸	整体蛋白增重	莱克多巴胺诱导的整体蛋白增重
	赖氨酸, g/100g 整体蛋白增重	
	7.10	8.24
	g 氨基酸/100 g 赖氨酸	
精氨酸	90.2	79.4
组氨酸	45.2	37.5
异亮氨酸	50.8	56.6
亮氨酸	100.0	93.7
赖氨酸	100.0	100.0
蛋氨酸	27.9	30.2
蛋氨酸+半胱氨酸	41.8	44.1
苯丙氨酸	52.2	49.5
苯丙氨酸+酪氨酸	89.9	89.7
苏氨酸	53.1	54.4
色氨酸	12.8	14.3
缬氨酸	66.2	64.2

怀孕母猪

在 NRC（1998）中，怀孕母猪氨基酸的营养需要量包括母体和胎儿的需要，怀孕期间组织沉积的氨基酸组成是借用生长育肥猪数据。而在本书中，充分考虑 6 个蛋白库的蛋白质存留及其氨基酸模型：胎儿、乳腺组织、胎盘（包括与之相关的绒毛尿囊液）、子宫，以及能量摄入量和时间依赖性的母猪体蛋白沉积。其中，不同怀孕日龄母猪中 4 个蛋白库［胎儿、乳腺组织、胎盘（包括与之相关的绒毛尿囊液）、子宫］的粗蛋白质量可以通过本书引用文献中每个蛋白质库重量和粗蛋白质浓度计算出来。数据来源、样本日龄和各蛋白库参数等信息见表 2-9。随时间和能量摄入量的变化，母体蛋白质库中蛋白质质量的变化情况将在下文中进行评估。

表 2-9 选择的用来估测怀孕期间组织蛋白库中氮含量及相应对的取样日龄的研究的总结

参考文献	胎儿组织		乳腺组织		胎盘组织		子宫内液		子宫组织	
	体重	粗蛋白	体重	粗蛋白	体重	粗蛋白	体积	粗蛋白	体重	粗蛋白
Biensen 等（1998）	70~75, 90,110				70~75, 110		70~75, 90,110			
Freking 等（2007）	45,65, 80~85, 105				45,65, 80~85, 105					
Ji 等（2005）	45,60, 70~75, 90,102, 112~114						45,60, 70~75, 90,102, 112~114			
Ji 等（2006）			45,60, 70~75, 90,102, 110~114		45,60, 70~75, 90,102, 110~114					
Kensinger 等（1986）			110							

续表

参考文献	胎儿组织		乳腺组织		胎盘组织		子宫内液		子宫组织	
	体重	粗蛋白	体重	粗蛋白	体重	粗蛋白	体积	粗蛋白	体重	粗蛋白
Knight 等（1977）	45,60,70~75,90,102				45,60,70~75,90,102		45,60,70~75,90,102	45,60,70~75,90,102	45,60,70~75,90,102	
McPherson 等（2004）	45,60,70~75,90,102,112~114				45,60,70~75,90,102,112~114					
Noblet 等（1985）	50,70~75,102	50,70~75,102			70~75,102	50,70~75,102	50,70~75,102	50,70~75,102	50,70~75,102	50,70~75,102
Pike 和 Boaz（1972）			70~75							
Wu 等（1999）	45,60,90,110,112~114	45,60,90,110,112~114								
Wu 等（2005）	45,60,90,110				45,60,90,110		45,60,90,110			
目前的研究			80,100,110							

蛋白库（protein pools）

胎儿粗蛋白含量根据 Noblet 等（1985）、Wu 等（1999）、Mathews 等（2004）、Canario 等（2007）、Pastorelli 等（2009）、Charneca 等（2010）的研究结果统计而来。怀孕第 45 天、60 天、72.5 天、90 天、102 天、110 天和 113 天的胎儿粗蛋白含量见图 2-1A。乳腺组织粗蛋白含量的数据依 Kensinger 等（1986）和 Ji 等（2006）的研究结果计算而得到，怀孕第 0 天时乳腺组织的粗蛋白含量设定为 0 g，因为在未怀孕母猪中几乎没有乳腺薄壁组织。怀孕第 45 天、60 天、72.5 天、90 天、102 天、110 天和 113 天时乳腺粗蛋白含量见图 2-1B。胎盘粗蛋白含量根据 Noblet 等（1985）、McPherson 等（2004）的研究结果统计而来，怀孕第 45 天、60 天、72.5 天、90 天、102 天、110 天和 113 天时胎盘的粗蛋白含量见图 2-1C。子宫的粗蛋白含量根据 Knight（1977）和 Noblet（1985）的研究结果统计而来，怀孕第 0 天、50 天、72.5 天、102 天时子宫粗蛋白含量见图 2-1D。

Dourmad 等（1998）报道，在怀孕的不同阶段，随时间和能量摄入变化，母猪体蛋白库中蛋白存留量的变化可以根据整个机体不同妊娠阶段的氮存留量估计得到，其中"整个机体不同妊娠阶段的氮存留量"信息参见第 8 章中 Dourmad 等（2008）的概述。总之，假设在整个怀孕期，高于维持能量需要之上的能量摄入量和随能量摄入导致的母猪体蛋白沉积之间始终保持固定的线性关系，那么，整个机体的氮存留量就与能量摄入或者是生殖系统发育无关，并主要贡献为随妊娠时间变化的蛋白沉积。基于对表 2-10 中引用的研究结果的总结归纳，我们对随妊娠时间变化的母猪体蛋白沉积模式进行了微调。在这个总结归纳中，各个试验研究获得的氮存留量分别分配到 4 个怀孕阶段（分别是 10~40 天、40~65 天、65~90 天和 90~114 天），平均后，以 65~90 天阶段结果的相对值表示出来。由于 Dourmad 等（1998）报道的氮存留数据高于表 2-10 中列举的研究结果，这些数据通过表 2-10 中相对值，即 0.84、

0.75、1.00 和 1.36 调整后，构建的母体随着妊娠期天数发展的蛋白沉积模型见图 2-2。

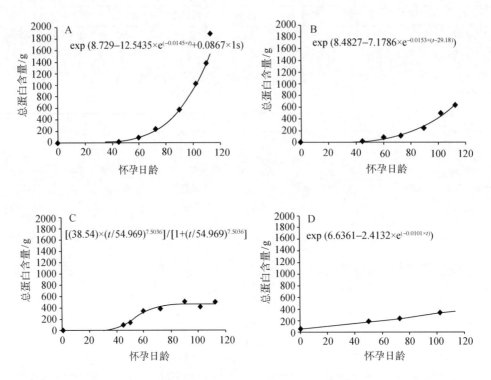

图 2-1　胎儿（$n=12$）（A）、乳房（B）、胎盘和羊水（C）及空子宫中总蛋白含量与怀孕日龄之间的关系。◆代表在表 2-9 中报道的试验值，线条则代表通过第 8 章中的模型公式预测得到的值，式 8-55、式 8-59、式 8-56 和式 8-58 分别用来计算胎儿、乳房、胎盘和羊水及空子宫中的总蛋白含量。"ls" 代表窝产仔数；t 代表时间（如怀孕天数）。

表 2-10　总结怀孕天数和仔猪生长性能对氮沉积（g/d）的影响

参考文献	胎次	代谢能/(kcal/d)	N摄入量/(g/d)	窝产仔数	出生体重/kg	怀孕天数 10~40	40~65	65~90	90~114
Rippel 等（1965b）	1	6078	34.94	10.4	1.365	—	—	13.67	16.88
Woerman 和 Speer（1976）	1	5939	25.50	10.2	1.245	7.90	6.80	8.50	—
Willis 和 Maxwell（1984）	1	6585	40.80	—	—	13.90	14.60	20.50	
King 和 Brown（1993）[a]	1	9499	23.31			10.00	12.10	16.50	
Everts 和 Dekker（1994）	1	7775	42.50			13.40		17.80	
Dourmad 等（1996a）[b]	>1[c]	8160	54.31			10.75	9.20	12.05	17.10
Clowes 等（2003）[d]	1	7120	52.73	9.3	1.450	17.70	—	14.80	21.20
相对于怀孕 65~90 天的平均值						0.84	0.75	1.00	1.36

a 怀孕 10~40 天、40~65 天和 65~90 天相对应的 N 摄入量 22.72 g/d、21.28 g/d 和 25.92 g/d 的平均值。
b 试验 1 和试验 2 中 N 摄入量和沉积量的平均值。
c 试验中使用经产母猪，但是没有注明具体胎次。
d N 摄入量和沉积量来自于对照组的数据。氮摄入量取自于怀孕 10~40 天、65~90 天和 90~114 天时的氮摄入量 52.1 g/d、51.8 g/d 和 54.3 g/d 的平均值。窝产仔数没有报道，表中数值是指断奶仔猪数。

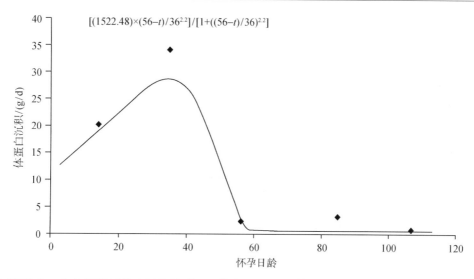

图 2-2 体蛋白沉积（g/d）与怀孕天数之间的关系。◆代表 Dourmad 等（1998）中报道的试验值；线条则代表通过模型公式预测得到的值，反映了表 2-10 中的所有数据。

怀孕母猪蛋白库的氨基酸组成

Everts 和 Dekker（1995）以怀孕 108 天的初产母猪为研究对象，研究报道了整个母猪体蛋白的氨基酸组成，其中包括乳腺组织但排除了子宫、胎儿和毛因素。胎儿增重蛋白质的氨基酸组成的数据来源于 Wu（1999）的研究。在怀孕第 40 天、60 天、90 天、108 天和 114 天时，把胎儿的每种氨基酸的量与胎儿体蛋白的量进行回归分析，可获得一个截距为 0 的线性回归曲线，氨基酸模式可用 100 与回归曲线斜率的乘积表示，即表示为 100 g 粗蛋白中所含的氨基酸的克数。

至今还没有关于母猪怀孕阶段乳腺组织氨基酸组成模型的公开发表资料。美国伊利诺伊大学的 Walter Hurlsy 教授提供了初产母猪分别在怀孕第 80 天、100 天和 110 天的乳腺组织样本，德固赛公司根据 Llames 和 Fontaine（1994）所述方法对这些样本进行氨基酸含量检测分析。Ji 等（2006）研究表明，在怀孕第 70 天、90 天、100 天和 110 天，乳腺个体干重分别为 74 g、81 g、101.1 g 和 118.4 g。第 80 天的乳腺重可以用第 70 天和第 90 天的平均重表示，即 77g。第 80 天、100 天和 110 天的乳腺组织粗蛋白含量分别为 23.44%、35.23%和 43.98%，据此可用以估测每个腺体粗蛋白质的量（即 18.05 g、35.61 g 和 52.07 g）。因此，每个腺体氨基酸质量的计算是建立在乳腺蛋白质的氨基酸组成和每个腺体粗蛋白含量的基础上。在怀孕的第 80 天、100 天和 110 天，把每一种氨基酸的质量［氨基酸（g）/乳腺］与每一个乳腺蛋白质量作回归分析，可推算出该妊娠阶段乳腺腺体增加蛋白质的氨基酸组成。Ji 等（2006）认为，由于乳腺个体蛋白质量在怀孕第 80 天时为 18.05 g，而在第 45 天时仅为 1.5 g，所以怀孕第 0 天乳腺蛋白质量可设定为 0 g。在整个怀孕期间，乳腺增重蛋白质的氨基酸组成是基于一个回归线的斜率推算得到的，这是一个胎儿增重蛋白质氨基酸组成的回归曲线。

至今尚无关于怀孕母猪胎盘氨基酸组成模型的公开发表资料。为此，从 22 头怀孕第 43 天、57~58 天、90~92 天和 100~109 天的初产母猪取得胎盘组织样本，像前面提到的乳腺组织一样，对这些样品的氨基酸含量进行检验，取所有样本的平均值构建出了一个妊娠母猪胎盘的氨基酸模型。Wu 等（1995）研究整个胎液（绒毛尿囊液）的氨基酸值，结果表明在怀孕第 45 天时尿囊液和羊膜囊液中仅有游离氨基酸（非结合蛋白）。绒毛尿囊液的氨基酸模型的建立是基于对尿囊液和羊膜囊液的数据的整合，即假定尿囊液和羊膜液分别占总绒毛尿囊液的 65%和 35%。由于胎盘的蛋白质占"胎盘和绒毛尿囊液"总蛋白质的 96%左右，所以，"胎盘+液体"的总氨基酸模型可用 96%的胎盘和 4%的绒毛尿囊液氨基酸来估测。

至今尚无关于母猪怀孕阶段子宫组织氨基酸组成模型的公开发表资料。子宫组织样品的获得与上文提到的胎盘组织样本的渠道一样，又额外增加了 8 头未怀孕的小母猪确定未怀孕子宫的氨基酸含量；组织样本的处理和分析同上文提到的胎盘组织一样，取所有妊娠期样本的平均值以代表一个妊娠母猪子宫组织的氨基酸模型。除了亮氨酸和苏氨酸外，胎盘和子宫的蛋白质氨基酸组成存在差异，这就为单独考虑这两个蛋白库提供了生物学基础。

对于上述提到的 5 个蛋白库，用于蛋白质沉积的赖氨酸含量及氨基酸模型（以赖氨酸为参照）见表 2-11。Cuaron 等（1984）认为，其他至今还没有详细研究的蛋白库可能也会对氨基酸需要量的估测产生一定影响，如黏蛋白和免疫球蛋白等。此外，Carlstedt 等（1983）发现，子宫分泌物中包括大量的黏液糖蛋白，且这种黏液糖蛋白中苏氨酸含量丰富，但是这些物质至今还难以定量。

哺乳母猪

随母猪体重变化表现出的体蛋白含量变化

总共 12 篇研究报道（Lewis and Speer, 1973; O'Grady and Hanrahan, 1975; King et al., 1993b, Dove and Haydon, 1994; Weeden et al., 1994; Coma et al., 1996; Knabe et al., 1996; Richert et al., 1997; Dourmad et al., 1998; Touchette et al., 1998; Guan et al., 2004; dos Santos et al., 2006）被用于分析哺乳期母猪体蛋白质量变化规律，在这 12 个研究中得到了哺乳期母体重量变化、背膘厚度变化，以及式 8-48~式 8-51 等丰富信息。这些信息随后被用来估测母体蛋白分解产生的赖氨酸用于补充泌乳赖氨酸需要的作用。被挑选的研究报道中提供了以下研究重要信息：母猪体重、母猪产后第 1 天和断奶时的 P2 点背膘厚及泌乳周期。根据这些信息，计算出下限临界氨基酸摄入量和上限临界氨基酸摄入量情况下的母体蛋白质损失分别为体蛋白的 9.9%和 10.1%，它们的平均值即 10%被用于预测哺乳期母猪随体重变化而可能发生的体蛋白质量变化（第 8 章）。

表 2-11 母体和胎儿蛋白质增重，以及胎盘、子宫、羊水、乳房和奶中赖氨酸和其他氨基酸组成（以赖氨酸的百分比来表示）

氨基酸	母体	胎儿	子宫	胎盘+羊水	乳房	奶
	赖氨酸, g/100 g 粗蛋白					
	6.74	4.99	6.92	6.39	6.55	7.01
	g 氨基酸/100g 赖氨酸					
精氨酸	105	113	103	101	84	69
组氨酸	47	36	35	42	35	43
异亮氨酸	54	50	52	52	24	56
亮氨酸	101	118	116	122	123	120
赖氨酸	100	100	100	100	100	100
蛋氨酸	29	32	25	25	23	27
蛋氨酸+半胱氨酸	45	54	50	50	51	50
苯丙氨酸	55	60	63	68	63	58
苯丙氨酸+酪氨酸	97	102	—	—	—	115
苏氨酸	55	56	61	66	80	61
色氨酸	13[a]	19	15	19	24	18
缬氨酸	69	73	75	83	88	71

a 值来自于整体蛋白增重中色氨酸与赖氨酸的比值（12.8；表 2-8）。

产奶

乳蛋白的产量可以通过窝产仔数和仔猪生长速度来预测（见第 8 章）。哺乳期第 5 天和第 26 天时，奶中粗蛋白和氨基酸含量以以下 9 个研究报道为基础进行估测：Elliott 等（1971）；Duée 和 Jung（1973）；Dourmad 等（1991）；Schneider 等（1992a）；King 等（1993a）；Csapó 等（1996）；Dourmad 等（1998）；Guan 等（2002）；Daza 等（2004）。选择这些研究结果的重要原因是因为以上每一个研究都提供了以下有效信息：乳中总乳蛋白氮（非蛋白氮+真蛋白氮）、乳中氨基酸含量或者是氨基酸占总乳蛋白的百分含量。另外，有些研究报道了氨基酸占乳中粗蛋白（氮×6.25）的百分含量，据此乳中氨基酸含量可通过"氮百分含量×6.38"计算出来。表 2-11 中归纳了常乳（哺乳期第 5~26 天）中赖氨酸含量和氨基酸模型（以赖氨酸为基础）。乳中 CP 含量的平均值为 5.16%（氮×6.38），相应的赖氨酸的含量为 7.01 g/100 g 乳粗蛋白。

氨基酸利用效率

概念

Moughan（1999）认为各种机体功能氨基酸利用效率不高反映了存在最低的和必然的氨基酸分解代谢，以及动物间生长潜力的差异（Pomar et al., 2003）。对具平均生长潜能的猪来说，如果赖氨酸最低+必然分解代谢假设为回肠标准可消化赖氨酸摄入量的 0.25，则摄入的回肠标准可利用赖氨酸用于各种器官功能运行的最大利用效率为 0.75。这个效率值基于 Bikker 等（1994b）和 Moehn 等（2000）对个体生长猪的观察和对体重 30~70 kg 猪只的连续屠宰试验分析；而且，这种效率似乎不受体重的影响（Dourmad et al., 1996b; Moehn et al., 2000），但随猪生长性能的提高而略有提高（Moehn et al., 2000）。0.25 的无效氨基酸代表基础肠道内源赖氨酸损失和体表赖氨酸损失，以此来估测赖氨酸分解代谢的最小损失和随尿氮排出的量，进而得到维持赖氨酸需要量。正如前面提到的，Moughan（1999）和 van Milgen 等（2008）认为氨基酸分解代谢的最低效率可以用来估计整个体蛋白的周转。然而，动物品种和日粮对整个体蛋白周转和最低氨基酸分解代谢的影响还没有充足的资料来定量估测。对其他氨基酸的最低+必然分解代谢的估测将在仔细筛选分析氨基酸需要量的相关研究资料基础上，在下文中介绍。

考虑到不同动物间的差异，超过维持需要以上的回肠标准可消化赖氨酸用于蛋白沉积的最大效率要在 0.75 的基础上降低，以保证模型预测的回肠标准赖氨酸需要量与实际经验性试验测定的结果相一致，且生长育肥猪（与体重有关）、哺乳母猪和怀孕母猪分别采用各自独立的调整比例；这些比例也适用于其他氨基酸，并且所有的氨基酸可采用同样的调整方法。因此，对于生长育肥猪、怀孕母猪和哺乳母猪中的任何一个对象来说，所有氨基酸的维持氨基酸利用效率/沉积蛋白质的氨基酸利用效率均保持一致。

生长育肥猪的氨基酸利用效率

对于生长育肥猪，收集了 35 篇研究赖氨酸需要量的数据，以估计赖氨酸用于体蛋白沉积的利用效率校正值。这些研究考虑到了每日采食量、体重和体蛋白沉积变化对氨基酸利用率的影响，具体内容将在第 8 章中介绍猪动态生长模型时说明。以采食量测定值（假设有 5%的饲料浪费）和标准维持代谢能量需要量为基础，通过改变体蛋白沉积的平均效率，保证体增重测定值与预测值一致，构建能量利用的模型。用于维持的回肠标准可消化赖氨酸需要量，可以通过肠道、皮肤、毛的损失以及维持赖氨酸利用效率几个方面估测。用于蛋白质沉积的回肠标准可消化赖氨酸需要量，可以通过沉积蛋白质的赖氨酸含量和蛋白质沉积的赖氨酸利用效率计算得到。那么，总的回肠标准可消化赖氨酸需要量等于维持的需要量和机体沉积蛋白质的需要量之和。最初，高于维持需要部分的回肠标准可利用赖氨酸摄入量被认为是仅反映氨基酸最低+必然分解代谢的需要量，现在认为其利用效率应该与用于维持的回肠标准可利用赖氨酸摄入的利用效率（0.75）相一致。对高于维持需要部分的回肠标准可利用赖氨酸摄入用于赖氨酸沉积的边际利用效率进行不断调整，直到预测值与通过经典试验实际测定赖氨酸需要量之间达到很好的一致性（图 2-3）。这些分析揭示了回肠标准可利用赖氨酸摄入中用于沉积蛋白质的边际效率随着体重的增长而下降。对体重 20 kg 的猪边际效率的下调幅度为 9.1%（从 0.75 下降到 0.682），体重 120 kg 的猪下调了 24.3%（从 0.75 下降到 0.568），据此可以推测出其他体重与这个体重存在线性相关。以 100 g 体蛋白沉积需 7.1 g 赖氨酸为基础，那么相对于某一特定生长性能的猪（最大体蛋白沉积为 145 g/d），当体重为 20 kg 和 120 kg 时，100 g 蛋白质沉积分别需要 10.4 g 和 12.5 g 回肠标准可消化赖氨酸。Moehn 等（2000）认为在最大体蛋白沉积的基础上每增加 1 g，则氨基酸最低+必然分解代谢的比率将下降 0.002。这与 NRC（1998）观点不一致，NRC（1998）报道每沉积 100 g 体蛋白需要 12.0 g 回肠标准可消化赖氨酸，且各个体重阶段猪的需要量保持不变。

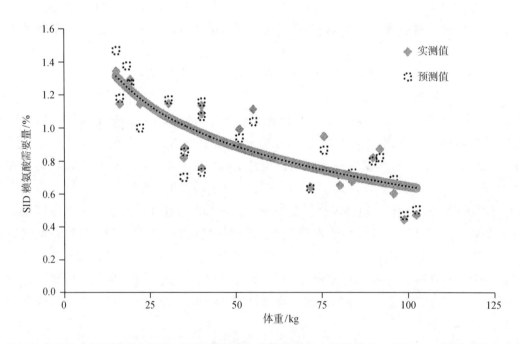

图 2-3A 通过经典试验实际测定赖氨酸需要量及预测值与猪体重增长之间的关系。来源：15 篇发表论文中的 24 个观察值，Martinez 和 Knabe（1990）；Lawrence 等（1994）；Williams 等（1998, 2 个观察值）；Hahn 等（1995）；Dourmad 等（1996b, 2 个观察值）；Loughmiller 等（1998a）；Ettle 等（2003）；Urynek 和 Buraczewska（2003）；Warnants 等（2003, 2 个观察值）；O'Connell 等（2005, 3 个观察值；2006, 3 个观察值）；Yen 等（2005）；Yi 等（2006）；Kendall 等（2008, 3 个观察值）；Schneider 等（2010）。

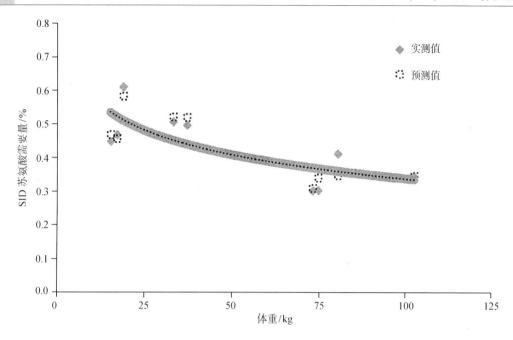

图 2-3B　通过经典试验实际测定苏氨酸需要量及预测值与猪体重增长之间的关系。来源：9 篇发表论文中的 9 个观察值，Cohen 等（1977）；Conway 等（1990）；Kovar 等（1993）；Sève 等（1993）；Saldana 等（1994）；Bergstrom 等（1996）；Rademacher 等（1997）；Ferguson 等（2000）；Johnston 等（2000）。

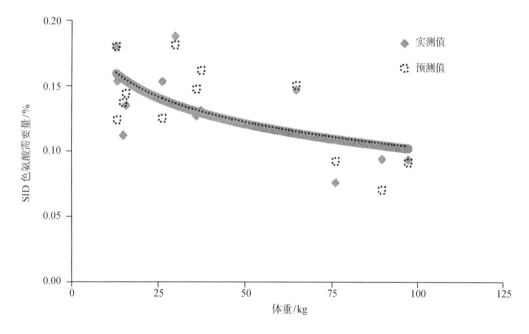

图 2-3C　通过经典试验实际测定色氨酸需要量及预测值与猪体重增长之间的关系。来源：9 篇发表论文中的 12 个观察值，Boomgaardt 和 Baker（1973）；Russell 等（1983）；Borg 等（1987）；Sato 等（1987）；Burgoon 等（1992, 2 个观察值）；Schutte 等（1995）；Eder 等（2001, 2003, 3 个观察值）；Guzik 等（2002, 2005）。

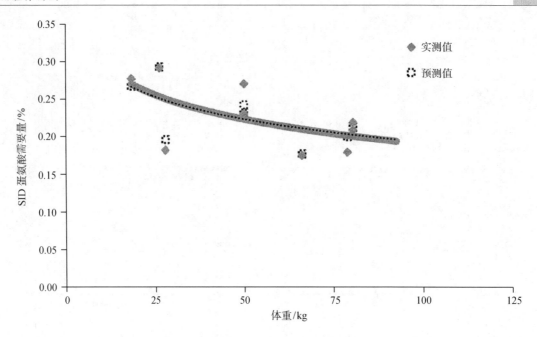

图 2-3D 通过经典试验实际测定蛋氨酸需要量及预测值与猪体重增长之间的关系。来源：6 篇发表论文中的 9 个观察值，Leibholz（1984）；Chung 等（1989）；Lenis 等（1990, 2 个观察值）；Schutte 等（1991）；Chung 和 Baker（1992b）；Roth 等（2000, 3 个观察值）。

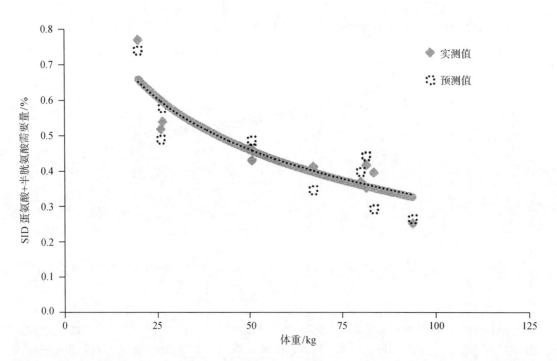

图 2-3E 通过经典试验实际测定蛋氨酸+半胱氨酸的需要量及预测值与猪体重增长之间的关系。来源：7 篇发表论文中的 11 个观察值，Chung 等（1989）；Lenis 等（1990, 2 个观察值）；Schutte 等（1991, 2 个观察值）；Knowles 等（1998）；Loughmiller 等（1998b）；Roth 等（2000, 3 个观察值）；Yi 等（2006）。

除了赖氨酸以外，其他氨基酸的最低+必然分解代谢需要量都将通过氨基酸需要量试验，并基于同赖氨酸一样的分析思路来估测。对于单个氨基酸，为了使在试验中得到的氨基酸需要量实际测定值和预测值相匹配，氨基酸最低+必然分解代谢值将被调整，而对反映体重、动物品种和用于体蛋白沉积的氨基酸最大利用效率的边际效率的调整值在所有氨基酸中应保持一致（例如，体重为 20 kg 和 120 kg 时，分别调整 9.1%和 24.3%）。图 2-3B~E 分别显示了随着猪体重变化，回肠标准可消化苏氨酸、色氨酸、蛋氨酸和蛋氨酸+半胱氨酸需要量的预测值和实测值的变化趋势。

由于亮氨酸、苯丙氨酸、苯丙氨酸+酪氨酸没有可靠的资料来源，NRC（1998）通过设置一个具有一定生长性能的动物模型来估测最低+必然分解代谢的氨基酸需要量。体重 50 kg 生长猪用于维持和生长的回肠标准可利用赖氨酸摄入的效率见表 2-12。

表 2-12 生长育肥猪、怀孕母猪和哺乳母猪用于维持、蛋白质沉积和奶蛋白产出的日粮回肠标准可消化氨基酸的利用效率

氨基酸	维持			沉积		
	生长育肥猪	怀孕母猪	哺乳母猪	生长育肥猪	怀孕母猪	哺乳母猪
精氨酸	1.470	1.470	0.914	1.270	0.960	0.816
组氨酸	1.000	0.973	0.808	0.864	0.636	0.722
异亮氨酸	0.760	0.751	0.781	0.657	0.491	0.698
亮氨酸	0.751	0.900	0.810	0.649	0.588	0.723
赖氨酸	0.750	0.750	0.750	0.648	0.490	0.670
蛋氨酸	0.730	0.757	0.755	0.631	0.495	0.675
蛋氨酸+半胱氨酸	0.603	0.615	0.741	0.521	0.402	0.662
苯丙氨酸	0.671	0.830	0.820	0.580	0.542	0.733
苯丙氨酸+酪氨酸	0.746	0.822	0.789	0.645	0.537	0.705
苏氨酸 [a]	0.780	0.807	0.855	0.671	0.527	0.764
色氨酸	0.610	0.714	0.755	0.527	0.467	0.674
缬氨酸	0.800	0.841	0.653	0.691	0.549	0.583
N ×6.25	0.850	0.850	0.850	0.735	0.555	0.759

a 苏氨酸的利用效率适用于可发酵纤维含量为 0%的日粮。随着日粮中可发酵纤维含量的增加，苏氨酸利用效率下降（式 8-46）。

怀孕母猪的氨基酸利用效率

目前，除了赖氨酸和苏氨酸以外，在怀孕母猪上还未见用于氨基酸存留的回肠标准可消化氨基酸利用效率的直接估测资料；而且，不同氨基酸、不同怀孕阶段条件下，氨基酸利用效率是否存在差异还未知。为了建立模型，假设用于蛋白质存留的氨基酸利用效率在所有的蛋白库和不同怀孕天数期间都相同。同生长育肥猪一样，用于蛋白质存留的赖氨酸利用效率根据赖氨酸经验性试验研究结果来估测。在三个研究报道中（表 2-3），为了使赖氨酸需要量的预测值与实测值相匹配，用于蛋白质沉积的赖氨酸最大利用效率（相当于赖氨酸用于维持的效率；0.75）下调 34.7%至 0.49。当观测需要量和预测需要量相匹配时，可以进行母猪怀孕 90~114 天期间的蛋白质存留和赖氨酸利用效率的估测，因为母猪怀孕后期的赖氨酸需要量最高且母猪生产性能对这期间的赖氨酸摄入量最为敏感。Everts 和 Dekker（1995）进行的代谢试验中证实，0.49 用于估测赖氨酸利用效率的合理性，其中当平均日采食 74.4 g 氮时的赖氨酸利用效率为 0.46，当平均日采食 50.8 g 氮时的赖氨酸的利用效

率为 0.59。基于以上这些分析，所有用于蛋白质存留的氨基酸利用效率可以假设为低于维持利用效率 34.7%。目前还没有可靠资料用来估测其他氨基酸的最低+必然分解代谢的比例，进而估测用于维持需要和蛋白沉积的氨基酸利用效率。因此，效率值的估测要通过模式预测需要量和NRC（1998）报道的怀孕母猪氨基酸需要量，以及第8章详尽解释的微调相匹配而得到。用这种方法，可以得到用于蛋白质存留的苏氨酸和总含硫氨基酸的利用效率值分别为 0.509 和 0.402。Everts 和 Dekker（1995）采用代谢试验估测了苏氨酸的边际利用效率为 0.44~0.67，含硫氨基酸的边际利用效率为 0.34~0.47；这些值与前面提及的数值相当一致。怀孕母猪的氨基酸利用效率的估测值见表 2-12。

哺乳母猪的氨基酸利用效率

根据表 2-4 中研究报道的经验性赖氨酸需要量估测值，可以估测出用于产乳的赖氨酸的利用效率。其中 5 篇研究（Lewis and Speer, 1973; Chen, 1978; King et al., 1993b; Tritton et al., 1996; Sauber et al., 1998; Yang et al., 2000；对高和低瘦肉型母猪的需要量分开估测）根据拐点分析的条件进行试验设计，然后进行拐点分析确定或者调整研究结果来估测每日赖氨酸需要量。而对其他的研究以及数据分析中没有确定一个拐点的研究，只是报道了赖氨酸相对窝增重的比值及其平均值（Lewis and Speer, 1973; O'Grady and Hanrahan, 1975; Johnston et al., 1993; Knabe et al., 1996; Tritton et al., 1996; Touchette et al., 1998）。Lewis 和 Speer（1973）、King 等（1993b）的研究中，在估量窝增长速率的基础上还考虑了其他指标，如血浆尿素氮、血浆氨基酸含量、产奶量或氮平衡，这些指标将连同窝重一起评价以确定或者调节需要量。在某些情况下，拐点分析得到的赖氨酸需要量可用于该研究的所有指标（如窝增长速率、血浆尿素氮和产奶量），取得的平均值可作为试验的最终结果。泌乳母猪氨基酸利用效率的估测基于最短 17 天，最长 29 天的泌乳期。在一些研究中的母猪哺乳期超过 28 天，但生产性能参数仅涉及 21 天，这种情况下的参数仅适用于前 21 天哺乳阶段。另外，对于一些特殊胎次的母猪，估测值常取多个试验结果的平均值（O'Grady and Hanrahan, 1975; Chen et al., 1978; Yang et al., 2000）。另一些研究选用多个胎次的母猪（Lewis and Speer, 1973），这种情况下需要在统计分析时把胎次作为一个固定因素（Johnston et al., 1993），或选用初产母猪为研究对象。

用于产乳部分的回肠标准可消化赖氨酸的利用效率可通过回归分析估测（图 2-4）。为了进行回归分析，所选择的每一篇关于机体赖氨酸需要量的研究报告在第 8 章都以概要形式作了单独表述。用于机体维持功能的赖氨酸需要量等于每日回肠标准可消化赖氨酸摄入量减去每日用于产乳的回肠标准可消化赖氨酸估计值。总的乳赖氨酸产量可通过窝仔数和哺乳仔猪平均体增重来计算。当观察到母猪体重下降时，为保证总乳赖氨酸产量的需要量，应与母猪体蛋白的动员联系起来。正如图 2-4 所示，用于产乳的日粮赖氨酸消耗与用于产乳的回肠标准可消化氨基酸摄入量之间高度相关，截距为 0；斜率可被作为用于产奶部分的回肠标准可消化氨基酸的利用效率。图 2-4 中显示的数据相关性好于图 2-5 中每日窝增重与回肠可利用赖氨酸的需要量之间的相关性。后者是 NRC（1998）用于估测哺乳母猪赖氨酸需要量的方法；前者方法上的改进有利于更详细地解释个体赖氨酸需要，这样我们可以更准确地估测赖氨酸的需要量。基于以上研究分析，所有用于产乳蛋白的回肠标准可消化氨基酸的各种氨基酸利用效率可以假设是 10.7%，这稍低于维持的利用效率。只有苏氨酸和色氨酸的需要量是根据基于试验研究获得的氨基酸利用效率调整值估测（表 2-3）。而对于其他氨基酸而言，哺乳母猪氨基酸利用效率值的估测是建立在需要量预测值与 NRC（1998）中哺乳母猪氨基酸需要量一致的基础上，具体内容见第 8 章。

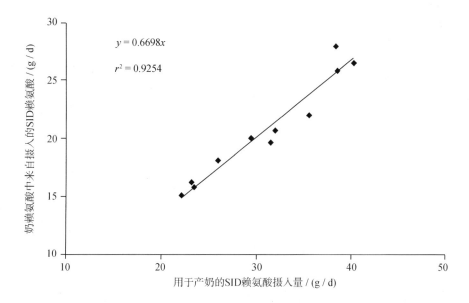

图 2-4　奶中来自 SID 赖氨酸摄入量的赖氨酸与用于产奶的 SID 赖氨酸摄入量的关系，线性关系公式为：$y=0.6698x$，截距为 0，$r^2=0.925$，斜率则代表日粮赖氨酸用于奶中赖氨酸生产的利用效率。来源：10 篇发表论文中的 11 个观察值，Lewis 和 Speer（1973）；O'Grady 和 Hanrahan（1975）；Chen 等（1978）；Johnston 等（1993）；King 等（1993 b）；Knabe 等（1996）；Tritton 等（1996）；Sauber 等（1998，2 个观察值）；Touchette 等（1998）；Yang 等（2000）。

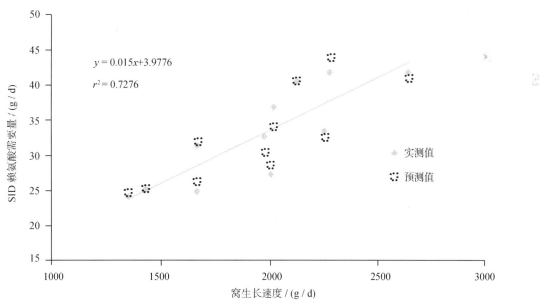

图 2-5　回肠标准可消化赖氨酸需要量（通过试验测定的回肠标准可消化赖氨酸）与窝仔猪生长速度之间的关系。线条代表两者之间的线性关系，线性关系公式为：$y=0.015x+3.9776$，$r^2=0.73$。来源：10 篇发表论文中的 11 个观察值，Lewis 和 Speer（1973）；O'Grady 和 Hanrahan（1975）；Chen 等（1978）；Johnston 等（1993）；King 等（1993 b）；Knabe 等（1996）；Tritton 等（1996）；Sauber 等（1998，2 个观察值）；Touchette 等（1998）；Yang 等（2000）。

保育猪氨基酸的需要量

我们对哺乳仔猪氨基酸利用率的了解远不足以通过模型估测氨基酸需要量（像生长育肥猪那样）。另外，也没有足够的数据来直接用经验方法，利用体重估测 5~11 kg 仔猪的氨基酸需要量。以这些原因，5~11 kg 保育猪的氨基酸需要量的估测建立在每千克体增重对回肠标准可消化赖氨酸需要量之上。到目前为止，只有 2 篇同行审阅的文摘研究了最初的 5 kg 或 6 kg 体重和最终体重 15 kg 或以下仔猪的赖氨酸需要量。研究结果表明，每千克体增重的平均回肠标准可消化赖氨酸需要量为 20.1 g（表 2-2）。在 12 篇研究报道中使用了大量数据列出了最初体重 5~13 kg 和最终体重 15~31 kg 的保育猪每千克体增重的回肠标准可消化赖氨酸需要量，得出平均值为 19.3 g（表 2-2）。对体重 5~11 kg 的猪，应用一个定值和推理值都存在局限性，但是，好在最近几年有研究者（Gaines et al., 2003; Dean et al., 2007; Nemechek et al., 2011）为此提供了一定的支撑数据，他们给出了每千克体增重的平均回肠标准可消化赖氨酸需要量约为 19g。然而，一些研究者认为，一些因素，如回肠标准可消化氨基酸（Eklund et al., 2008）、日粮蛋白质来源（Jones et al., 2011）、体重（Stein et al., 2001），或者是体蛋白沉积和体增重之间的关系，在小猪和中大猪之间均存在差异。由于越来越多的可用资料的报道，关于仔猪赖氨酸需要量的最新估测值将更为准确。

通过使用每千克体增重的平均回肠标准可消化赖氨酸摄入量为 19g 这个定值，5~7 kg 和 7~11 kg 猪只的平均日增重和平均日采食量见表 2-2。Meisinger（2010）研究结果表明，体重 5~11 kg 猪只的生长性能水平稍高于保育猪的平均生长性能水平。表 2-2 中列出了体重 11~25 kg 猪的回肠标准可利用赖氨酸需要量，这些数据表明从试验结果中可以得出最初体重 9~13 kg 到最终体重 19~31 kg 猪只的赖氨酸需要量平均值。随着体重分别是 5~7 kg、7~11 kg、11~25 kg 猪的回肠标准可消化赖氨酸需要量估测值的确定，就可以利用特定体重外推法估测维持氨基酸需要量和整个机体增重的最佳氨基酸比例（正如在第 8 章所描述）。其他氨基酸用于整个体蛋白沉积的需要量、维持氨基酸需要量和体蛋白沉积所需最佳氨基酸比例在第 8 章论述。

种公猪的氨基酸需要量

Kemp 和 Soede（2001）综述了生长期和成年公猪的能量、氨基酸、矿物质和维生素的需要量。与商品猪相比，成年公猪数量很少；相比生长育肥猪、怀孕母猪和哺乳母猪，成年公猪的氨基酸需要量研究报道很少。NRC（1998）建议性活跃期种公猪日粮中的赖氨酸需要量为 0.6%或者是 12 g/d（假定采食量为 2 kg/d）；这一需要量值是建立对精液量和精子质量分析测试基础之上的（Meding and Nielsen, 1977; Yen and Yu, 1985; Kemp et al., 1988; Louis et al., 1994a,b）。Rupanova（2006）的最新研究指出，在公猪精液质量方面，饲粮中赖氨酸为 1.03%处理组优于饲粮中赖氨酸为 0.86%的处理组，但射精量没有改变。但是，这个研究存在一定的局限性，因为试验动物仅有 10 头（5 头/组），并且试验只进行了 46 天。另一篇报道指出，公猪赖氨酸需要量为 0.92%（0.76%可消化赖氨酸），但是没有提供较具体的试验过程（Golushko et al., 2010）。因此，尽管饲粮中赖氨酸含量大于 0.60%可能对种公猪更有利，但是仍然没有充分的证据改变 NRC（1998）估测的赖氨酸需要量。另外，种公猪对其他必需氨基酸需要量是基于使用母猪体蛋白沉积的可利用氨基酸模型得到的估测值（表 2-11）。

参 考 文 献

Adeola, O. 2009. Bioavailability of threonine and tryptophan in peanut meal for starter pigs using slope-ratio assay. *Animal* 3:677-684.

Adeola, O., B. V. Lawrence, and T. R. Cline. 1994. Availability of amino acids for 10- to 20-kilogram pigs: Lysine and threonine in soybean meal. *Journal of Animal Science* 72:2061-2067.

AminoDat. 2006. AminoDat 3.0 Platinum Version, Amino Acid Composition of Feedstuffs. Hanau-Wolfgang, Germany: Degussa AG.

AmiPig. 2000. *Ileal Standardized Digestibility of Amino Acids in Feedstuffs for Pigs*. AFZ, Ajinomoto Eurolysine, Aventis Animal Nutrition, INRA, and ITCF.

Arentson, B. E., and D. R. Zimmerman. 1985. Nutritive value of D-tryptophan for the growing pig. *Journal of Animal Science* 60:474-479.

Augspurger, N. R, and D. H. Baker. 2004. An estimate of the leucine requirement for young pigs. *Animal Science* 79:149-153.

Baker, D. H. 2006. Comparative species utilization and toxicity of sulfur amino acids. *Journal of Nutrition* 36:1670S-1675S.

Baker, D. H., N. K. Allen, J. Boomgaardt, G. Graber, and H. W. Norton. 1971. Quantitative aspects of D- and L-tryptophan utilization by the young pig. *Journal of Animal Science* 33:42-46.

Ball, R. O., G. Courtney-Martin, and P. B. Pencharz. 2006. The in vivo sparing of methionine by cysteine in sulfur amino acid requirements in animal models and adult humans. *Journal of Nutrition* 136:1682S-1693S.

Barea, R., L. Brossard, N. Le Floc'h, Y. Primot, D. Melchior, and J. van Milgen. 2009. The standardized ileal digestible valine-to-lysine requirement ratio is at least seventy percent in postweaned piglets. *Journal of Animal Science* 87:935-947.

Batterham, E. S. 1992. Availability and utilization of amino acids for growing pigs. *Nutrition Research Reviews* 5:1-18.

Batterham, E. S. 1994. Ileal digestibilities of amino acids in feedstuffs for pigs. Pp. 113-131 in *Amino Acids in Farm Animal Nutrition*, J. P. F. D'Mello, ed. Wallingford, UK: CABI.

Batterham, E. S., L. M. Andersen, D. R. Baigent, and E. White. 1990. Utilization of ileal digestible lysine by growing pigs: Effect of dietary lysine concentration on efficiency of lysine retention. *British Journal of Nutrition* 64:81-94.

Becker, D. E., A. H. Jensen, S. W. Terrill, I. D. Smith, and H. W. Norton. 1957. The isoleucine requirement of weanling swine fed two protein levels. *Journal of Animal Science* 16:26-34.

Becker, D. E., I. D. Smith, S. W. Terrill, A. H. Jensen, and H. W. Norton. 1963. Isoleucine need of swine at two stages of development. *Journal of Animal Science* 22:1093-1096.

Benevenga, N. J., and R. D. Steele. 1984. Adverse effects of excessive consumption of amino acids. *Annual Review of Nutrition* 4:157-181.

Bergstrom, J. R., J. L. Nelssen, M. D. Tokach, R. D. Goodband, K. Q. Owen, B. T. Richert, W. B. Nessmith, Jr., J. A. Loughmiller, and S. S. Dritz. 1996. Determining the optimal threonine:lysine ratio in diets for the phase III nursery pig. *Journal of Animal Science* 74(Suppl. 1):56 (Abstr.).

Biensen, N. J., M. E. Wilson, and S. P. Ford. 1998. The impact of either a Meishan or Yorkshire uterus on Meishan or Yorkshire fetal and placental development to days 70, 90, and 110 of gestation. *Journal of Animal Science* 76:2169-2176.

Bikker, P., M. W. A. Verstegen, and M. W. Bosch. 1994a. Amino acid composition of growing pigs is affected by protein and energy intake. *Journal of Nutrition* 124:1961-1969.

Bikker, P., M. W. A. Verstegen, R. G. Campbell, and B. Kemp. 1994b. Digestible lysine requirement of gilts with high genetic potential for lean gain, in relation to the level of energy intake. *Journal of Animal Science* 72:1744-1753.

Boelens, P. G., R. J. Nijveldt, A. P. J. Houdijk, S. Meijer, and P. A. M. van Leeuwen. 2003. Glutamine alimentation in catabolic state. *Journal of Nutrition* 131:2569S-2577S.

Boomgaardt, J., and D. H. Baker. 1973. Tryptophan requirement of growing pigs at three levels of dietary protein. *Journal of Animal Science* 36:303-306.

Borg, B. S., G. W. Libal, and R. C. Wahlstrom. 1987. Tryptophan and threonine requirements of young pigs and their effects on serum calcium, phosphorus and zinc concentrations. *Journal of Animal Science* 64:1070-1078.

Boyd, R. D., K. J. Touchette, G. C. Castro, M. E. Johnston, K. U. Lee, and I. K. Han. 2000. Recent advances in amino acid and energy nutrition of prolific sows. *Asian-Australian Journal of Animal Science* 13:1638-1652.

Burgoon, K. G., D. A. Knabe, and E. J. Gregg. 1992. Digestible tryptophan requirements of starting, growing, and finishing pigs. *Journal of Animal Science* 70:2493-2500.

Cadogan, D. J., R. G. Campbell, and J. Less. 1999. Effects of dietary tryptophan on the growth performance of entire male, female, and castrated male pigs between 6 and 16 kg live weight. *Journal of Animal Science* 77(Suppl. 1):57 (Abstr.).

Canario, L., M. C. Père, T. Tribout, F. Thomas, C. David, J. Gogué, P. Herpin, J. P. Bidanel, and J. Le Dividich. 2007. Estimation of genetic trends from 1977 to 1998 of body composition and physiological state of Large White pigs at birth. *Animal* 1:1409-1413.

Carlstedt, I., H. Lindgren, J. K. Sheehan, U. Ulmsten, and L. Wingerupl. 1983. Isolation and characterization of human cervical-mucus glycoproteins. *Biochemical Journal* 211:13-22.

Carpenter, K. J. 1973. Damage to lysine in food processing: Its measurement and its significance. *Nutrition Abstracts and Reviews* 43:424-451.

Charneca, R., J. L. T. Nues, and J. Le Dividich. 2010. Body composition and blood parameters of newborn piglets from Alentejano and conventional (Large White × Landrace) genotype. Instituto Nacional de Investigación y Tecnología Agraria y Alimentaria (INIA) *Spanish Journal of Agricultural Research* 8:317-325.

Chen, S. Y., J. P. F. D'Mello, F. W. H. Elsley, and A. G. Taylor. 1978. Effect of dietary lysine levels on performance, nitrogen metabolism and plasma amino acid concentrations of lactating sows. *Animal Production* 27:331-344.

Chung, T. K., and D. H. Baker. 1992a. Maximal portion of the young pig's sulfur amino acid requirement that can be furnished by cystine. *Journal of Animal Science* 70:1182-1187.

Chung, T. K., and D. H. Baker. 1992b. Methionine requirement of pigs between 5 and 20 kilograms body weight. *Journal of Animal Science* 70:1857-1863.

Chung, T. K., and D. H. Baker. 1992c. Utilization of methionine isomers and analogs by pigs. *Canadian Journal of Animal Science* 72:185-188.

Chung, T. K., O. A. Izquierdo, K. Hashimoto, and D. H. Baker. 1989. Methionine requirement of the finishing pig. *Journal of Animal Science* 67:2677-2683.

Cline, T. R., G. L. Cromwell, T. D. Crenshaw, R. C. Ewan, C. R. Hamilton, A. J. Lewis, D. C. Mahan, and L. L. Southern. 2000. Further assessment of the dietary lysine requirement of finishing gilts. *Journal of Animal Science* 78:987-992.

Clowes, E. J., R. Kirkwood, A. Cegielski, and F. X. Aherne. 2003. Phase-feeding protein to gestating sows over three parities reduced nitrogen excretion without affecting sow performance. *Livestock Production Science* 81:235-246.

Cohen, R. S., and T. D. Tanksley, Jr. 1977. Threonine requirement of growing and finishing swine fed sorghum-soybean meal diets. *Journal of Animal Science* 45:1079-1083.

Coma, J., D. R. Zimmerman, and D. Carrion. 1996. Lysine requirement of the lactating sow determined by using plasma urea nitrogen as a rapid response criterion. *Journal of Animal Science* 74:1056-1062.

Conway, D., W. C. Sauer, L. A. den Hartog, and J. Huisman. 1990. Studies on the threonine requirements of growing pigs based on the total, ileal and faecal digestible contents. *Livestock Production Science* 25:105-120.

Cooper, D. R., J. F. Patience, R. T. Zijlstra, and M. Rademacher. 2001. Effect of nutrient intake in lactation on sow performance: Determining the threonine requirement of the high-producing lactating sow. *Journal of Animal Science* 79:2378-2387.

Csapó, J., T. G. Martin, Z. S. Csapó-Kiss, and Z. Házas. 1996. Protein, fats, vitamin and mineral concentrations in porcine colostrum and milk from parturition to 60 days. *International Dairy Journal* 6:881-902.

Cuaron, J. A., R. P. Chapple, and R. A. Easter. 1984. Effect of lysine and threonine supplementation of sorghum gestation diets on nitrogen balance and plasma constituents in first-litter gilts. *Journal of Animal Science* 58:631-637.

Daza, A., J. Riopérez, and C. Centeno. 2004. Changes in the composition of sows' milk between days 5 to 26 of lactation. *Spanish Journal of Agricultural Research* 2:333-336.

Dean, D. W., L. L. Southern, B. J. Kerr, and T. D. Bidner. 2007. The lysine and total sulfur amino acid requirements of six- to twelve-kilogram pigs. *Professional Animal Scientist* 23:527-535.

de Lange, C. F. M., A. M. Gillis, and G. J. Simpson. 2001. Influence of threonine intake on whole-body protein deposition and threonine utilization in growing pigs fed purified diets. *Journal of Animal Science* 79:3087-3095.

Dilger, R. N., and D. H. Baker. 2008. Excess dietary L-cysteine causes lethal metabolic acidosis in chicks. *Journal of Nutrition* 138:1628-1633.

D'Mello, J. P. F. 2003. Adverse effects of amino acids. Pp. 125-142 in *Amino Acids in Animal Nutrition*, J. P. F. D'Mello, ed. Wallingford, UK: CABI.

dos Santos, J. M. G., I. Moreira, and E. N. Martins. 2006. Lysine and metabolizable energy requirements of lactating sows for subsequent reproductive performance. *Brazilian Archives of Biology and Technology* 49:575-581.

Dourmad, J. Y. 1991. Effect of feeding level in the gilt during pregnancy on voluntary feed intake during lactation and changes in body composition during gestation and lactation. *Livestock Production Science* 27:309-319.

Dourmad, J. Y., and M. Étienne. 2002. Dietary lysine and threonine requirements of the pregnant sow estimated by nitrogen balance. *Journal of Animal Science* 80:2144-2150.

Dourmad, J. Y., M. Étienne, and J. Noblet. 1996a. Reconstitution of body reserves in multiparous sows during pregnancy: Effect of energy intake during pregnancy and mobilization during the previous lactation. *Journal of Animal Science* 74:2211-2219.

Dourmad, J. Y., D. Guillou, B. Sève, and Y. Henry. 1996b. Response to dietary lysine supply during the finishing period in pigs. *Livestock Production Science* 45:179-186.

Dourmad, J. Y., J. Noblet, and M. Étienne. 1998. Effect of protein and lysine supply on performance, nitrogen balance, and body composition changes of sows during lactation. *Journal of Animal Science* 76:542-550.

Dourmad, J. Y., M. Étienne, A. Valancogne, S. Dubois, J. van Milgen, and J. Noblet. 2008. InraPorc: A model and decision support tool for the nutrition of sows. *Animal Feed Science and Technology* 143:372-386.

Dove, C. R., and K. D. Haydon. 1994. The effect of various diet nutrient densities and electrolyte balances on sow and litter performance during two seasons of the year. *Journal of Animal Science* 72:1101-1106.

Duée, P. H., and J. Jung. 1973. Amino acid composition of sow milk. *Annales de Zootechnie* 22:243-247.

Duée, P. H., and A. Rérat. 1975. Etude du besoin en lysine de la truie gestante nullipare. *Annales de Zootechnie* 24:447-464.

Easter, R. A., and D. H. Baker. 1977. Nitrogen metabolism of gravid gilts fed purified diets deficient in either leucine or tryptophan. *Journal of Animal Science* 44:417-421.

Eder, K., S. Peganova, and H. Kluge. 2001. Studies on the tryptophan requirement of piglets. *Archives of Animal Nutrition* 55:281-297.

Eder, K., H. Nonn, H. Kluge, and S. Peganova. 2003. Tryptophan requirement of growing pigs at various body weights. *Journal of Animal Physiology and Animal Nutrition* 87:336-346.

Edmonds, M. S., H. W. Gonyou, and D. H. Baker. 1987. Effect of excess levels of methionine, tryptophan, arginine, lysine or threonine on growth and dietary choice in the pig. *Journal of Animal Science* 65:179-185.

Eklund, M., R. Mosenthin, H.-P. Piepho, and M. Rademacher. 2008. Effect of dietary crude protein level on basal ileal endogenous losses and standardized ileal digestibilities of crude protein and amino acids in newly weaned pigs. *Journal of Animal Physiology and Animal Nutrition* 92:578-590.

Elliott, R. F., G. W. Vander Noot, R. L. Gilbreath, and H. Fisher. 1971. Effect of dietary protein level on composition changes in sow colostrum and milk. *Journal of Animal Science* 32:1128-1137.

Ettle, T., D. A. Roth-Maier, and F. X. Roth. 2003. Effect of apparent ileal digestible lysine to energy ratio on performance of finishing pigs at different dietary metabolizable energy levels. *Journal of Animal Physiology and Animal Nutrition* 87:269-279.

Everts, H., and R. A. Dekker. 1994. Effect of nitrogen supply on the retention and excretion of nitrogen and on energy metabolism of pregnant sows. *Animal Production* 59:293-301.

Everts, H., and R. A. Dekker. 1995. Effects of protein supply during pregnancy on body composition of gilts and their products of conception. *Livestock Production Science* 43:27-36.

Ferguson, N. S., G. A. Arnold, G. Lavers, and R. M. Gous. 2000. The response of growing pigs to amino acids as influenced by environmental temperature: 1. Threonine. *Animal Science* 70:287-297.

Fontaine, J. 2003. Amino acid analysis of feeds. Pp. 15-40 in *Amino Acids in Animal Nutrition*, J. P. F. D'Mello, ed. Wallingford, UK: CABI.

Freking, B. A., K. A. Leymaster, J. L. Vallet, and R. K. Christenson. 2007. Number of fetuses and conceptus growth throughout gestation in lines of pigs selected for ovulation rate or uterine capacity. *Journal of Animal Science* 85:2093-2103.

Friesen, K. G., J. L. Nelssen, R. D. Goodband, M. D. Tokach, J. A. Unruh, D. H. Kropf, and B. J. Kerr. 1995. The effect of dietary lysine on growth, carcass composition, and lipid metabolism in high-lean growth gilts fed from 72 to 136 kilograms. *Journal of Animal Science* 73:3392-3401.

Gaines, A. M., D. C. Kendall, G. L. Allee, M. D. Tokach, S. S. Dritz, and J. L. Usry. 2003. Evaluation of the true ileal digestible (TID) lysine requirement for 7 to 14 kg pigs. *Journal of Animal Science* 81(Suppl. 1):139 (Abstr.).

Gaines, A. M., D. C. Kendall, G. L. Allee, J. L. Usry, and B. J. Kerr. 2011. Estimation of the standardized ileal digestible valine-to-lysine ratio in 13- to 32-kilogram pigs. *Journal of Animal Science* 89:736-742.

Ganguli, M. C., V. C. Speer, R. C. Ewan, and D. R. Zimmerman. 1971. Sulfur amino acid requirements of the lactating sow. *Journal of Animal Science* 33:394-399.

Garlick, P. J. 2004. The nature of human hazards associated with excessive intake of amino acids. *Journal of Nutrition* 134:1633S-1639S.

Golushko, V. M., V. A. Roschin, and S. A. Linkevich. 2010. Modern norms of energy and amino acid nutrition of breeding boars. *Proceedings of the National Academy of Sciences of Belarus* 2:84-88.

Guan, X., B. J. Bequette, G. Calder, P. K. Ku, K. N. Ames, and N. L. Trottier. 2002. Amino acid availability affects amino acid flux and protein metabolism in the porcine mammary gland. *Journal of Nutrition* 132:1224-1234.

Guan, X., J. E. Pettigrew, P. K. Ku, N. K. Ames, B. J. Bequette, and N. L. Trottier. 2004. Dietary protein concentration affects plasma arteriovenous difference of amino acids across the porcine mammary gland. *Journal of Animal Science* 82:2953-2963.

Guzik, A. C., L. L. Southern, T. D. Bidner, and B. J. Kerr. 2002. The tryptophan requirement of nursery pigs. *Journal of Animal Science* 80:2646-2655.

Guzik, A. C., J. L. Shelton, L. L. Southern, B. J. Kerr, and T. D. Bidner. 2005. The tryptophan requirement of growing and finishing barrows. *Journal of Animal Science* 83:1303-1311.

Hahn, J. D., R. R. Biehl, and D. H. Baker. 1995. Ideal digestible lysine level for early- and late-finishing swine. *Journal of Animal Science* 73:773-784.

Harper, A. E., N. J. Benevenga, and R. M. Wohlheuter. 1970. Effects of ingestion of disproportionate amounts of amino acids. *Physiological Reviews* 50:428-558.

Holden, P. J., R. C. Ewan, and V. C. Speer. 1971. Sulfur amino acid requirement of the pregnant gilt. *Journal of Animal Science* 32:900-904.

Izquierdo, O. A., K. J. Wedekind, and D. H. Baker. 1988. Histidine requirement of the young pig. *Journal of Animal Science* 66:2886-2892.

Ji, F., G. Wu, J. R. Blanton, Jr., and S. W. Kim. 2005. Changes in weight and composition in various tissues of pregnant gilts and their nutritional implications. *Journal of Animal Science* 83:366-375.

Ji, F., W. L. Hurley, and S. W. Kim. 2006. Characterization of mammary gland development in pregnant gilts. *Journal of Animal Science* 84:579-587.

Johnston, E., D. R. Cook, R. D. Boyd, K. D. Haydon, and J. L. Usry. 2000. Optimum threonine:lysine ratio in a corn-soybean meal diet for pigs in the late nursery phase (12-23 kg). *Journal of Animal Science* 78(Suppl. 2):57 (Abstr.).

Johnston, L. J., J. E. Pettigrew, and J. W. Rust. 1993. Response of maternal-line sows to dietary protein concentration during lactation. *Journal of Animal Science* 71:2151-2156.

Jones, C. K., J. A. Acosta, M. D. Tokach, J. L. Usry, C. R. Neill, and J. F. Patience. 2011. Feed efficiency of 7 to 16 kg pigs is maximized when additional lysine is supplied by L-Lys instead of intact protein, but is not affected when diets are supplemented with differing sources of nonessential amino acid nitrogen. *Journal of Animal Science* 89(E-Suppl. 1):439 (Abstr.).

Kaspar, H., K. Dettmer, W. Gronwald, and P. J. Oefner. 2009. Advances in amino acid analysis. *Analytical and Bioanalytical Chemistry* 393:445-452.

Kemp, B., and N. M. Soede. Feeding of developing and adult boars. 2001. Pp. 771-782 in *Swine Nutrition*, 2nd Ed., A. J. Lewis and L. L. Southern, eds. Boca Raton, FL: CRC Press.

Kemp, B., H. J. G. Grooten, L. A. Den Hartog, P. Luiting, and M. W. A. Verstegen. 1988. The effect of a high protein intake on sperm production in boars at two semen collection frequencies. *Animal Reproduction Science* 17:103-113.

Kendall, D. C., A. M. Gaines, G. L. Allee, and J. L. Usry. 2008. Commercial validation of the true ileal digestible lysine requirement for eleven- to twenty-seven-kilogram pigs. *Journal of Animal Science* 86:324-332.

Kensinger, R. S., R. J. Collier, and F. W. Bazer. 1986. Effect of number of conceptuses on maternal mammary development during pregnancy in the pig. *Domestic Animal Endocrinology* 3:237-245.

Kerr, B. J., M. T. Kidd, J. A. Cuaron, K. L. Bryant, T. M. Parr, C. V. Maxwell, and J. M. Campbell. 2004. Isoleucine requirements and ratios in starting (7 to 11 kg) pigs. *Journal of Animal Science* 82:2333-2342.

Kim, K. I., and H. S. Bayley. 1983. Amino acid oxidation by young pigs receiving diets with varying levels of sulphur amino acids. *British Journal of Nutrition* 50:383-390.

Kim, S. W., R. L. McPherson, and G. Wu. 2004. Dietary arginine supplementation enhances the growth of milk-fed young pigs. *Journal of Nutrition* 134:625-630.

King, R. H., and W. G. Brown. 1993. Interrelationships between dietary protein level, energy intake, and nitrogen retention in pregnant gilts. *Journal of Animal Science* 71:2450-2456.

King, R. H., C. J. Rayner, and M. Kerr. 1993a. A note on the amino acid composition of sow's milk. *Animal Production* 57:500-502.

King, R. H., M. S. Toner, H. Dove, C. S. Atwood, and W. G. Brown. 1993b. The response of first-litter sows to dietary protein level during lactation. *Journal of Animal Science* 71:2457-2463.

King, R. H., R. G. Campbell, R. J. Smits, W. C. Morley, K. Ronnfeldt, K. Butler, and F. R. Dunshea. 2000. Interrelationships between dietary lysine, sex, and porcine somatotropin administration on growth performance and protein deposition in pigs between 80 and 120 kg live weight. *Journal of Animal Science* 78:2639-2651.

Kirchgessner, V. M., and F. X. Roth. 1985. Biologische wirksamkeit von DL-tryptophan bei mastschweinen. *Zeitschrift für Tierphysiologie, Tierernährung und Futtermittelkunde* 54:135-141.

Knabe, D. A., J. H. Brendemuhl, L. I. Chiba, and C. R. Dove. 1996. Supplemental lysine for sows nursing large litters. *Journal of Animal Science* 74:1635-1640.

Knight, J. W., F. W. Bazer, W. W. Thatcher, D. E. Franke, and H. D. Wallace. 1977. Conceptus development in intact and unilaterally hysterectomized-ovariectomized gilts: Interrelations among hormonal status, placental development, fetal fluids and fetal growth. *Journal of Animal Science* 44:620-637.

Knowles, T. A., L. L. Southern, and T. D. Bidner. 1998. Ratio of total sulfur amino acids to lysine for finishing pigs. *Journal of Animal Science* 76:1081-1090.

Kovar, J. L., A. J. Lewis, T. R. Radke, and P. S. Miller. 1993. Bioavailability of threonine in soybean meal for young pigs. *Journal of Animal Science* 71:2133-2139.

Krick, B. J., R. D. Boyd, K. R. Roneker, D. H. Beermann, D. E. Bauman, D. A. Ross, and D. J. Meisinger. 1993. Porcine somatotropin affects the dietary lysine requirement and net lysine utilization for growing pigs. *Journal of Nutrition* 123:1913-1922.

Kyriazakis, I., and G. C. Emmans. 1993. Whole body amino acid composition of the growing pig. *Journal of the Science of Food and Agriculture* 62:29-33.

Langer, S., and M. F. Fuller. 2000. Interactions among the branched-chain amino acids and their effects on methionine utilization in growing pigs: Effects on nitrogen retention and amino acid utilization. *British Journal of Nutrition* 83:43-48.

Langer, S., P. W. D. Scislowski, D. S. Brown, P. Dewey, and M. F. Fuller. 2000. Interactions among the branched-chain amino acids and their effects on methionine utilization in growing pigs: Effects on plasma amino- and keto-acid concentrations and branched-chain keto-acid dehydrogenase activity. *British Journal of Nutrition* 83:49-58.

Lawrence, B. V., O. Adeola, and T. R. Cline. 1994. Nitrogen utilization and lean growth performance of 20- to 50-kilogram pigs fed diets balanced for lysine:energy ratio. *Journal of Animal Science* 72:2887-2895.

Leibholz, J. 1984. A note on methionine supplementation of pig grower diets containing lupin-seed meal. *Animal Production* 38:515-517.

Lenis, N. P., J. T. M. van Diepen, and P. W. Goedhart. 1990. Amino acid requirements of pigs. 1. Requirements for methionine + cystine, threonine and tryptophan of fast-growing boars and gilts, fed ad libitum. *Netherlands Journal of Agricultural Science* 38:577-595.

Leonard, R. P., and V. C. Speer. 1983. Threonine requirement for reproduction in swine. *Journal of Animal Science* 56:1345-1353.

Levesque, C. L., S. Moehn, P. B. Pencharz, and R. O. Ball. 2011. The threonine requirement of sows increases late in gestation. *Journal of Animal Science* 89:93-102.

Lewis, A. J. 2001. Amino acids in swine nutrition. Pp. 131-150 in *Swine Nutrition*, 2nd Ed., A. J. Lewis and L. L. Southern, eds. Boca Raton, FL: CRC Press.

Lewis, A. J. 2003. Methionine-cystine relationships in pig nutrition. Pp. 143-156 in *Amino Acids in Animal Nutrition*, 2nd Ed., J. P. F. D'Mello, ed. Wallingford, UK: CABI.

Lewis, A. J., and H. S. Bayley. 1995. Amino acid bioavailability. Pp. 35-65 in *Bioavailability of Nutrients for Animals: Amino Acids, Minerals, and Vitamins*, C. B. Ammerman, D. H. Baker, and A. J. Lewis, eds. New York: Academic Press.

Lewis, A. J., and V. C. Speer. 1973. Lysine requirement of the lactating sow. *Journal of Animal Science* 37:104-110.

Lewis, A. J., and V. C. Speer. 1974. Tryptophan requirement of the lactating sow. *Journal of Animal Science* 38:778-784.

Lewis, A. J., and V. C. Speer. 1975. Threonine requirement of the lactating sow. *Journal of Animal Science* 40:892-899.

Lewis, A. J., E. R. Peo, Jr., B. D. Moser, and T. D. Crenshaw. 1980. Lysine requirement of pigs weighing 5 to 15 kg fed practical diets with and without added fat. *Journal of Animal Science* 51:361-366.

Libao-Mercado, A. J., S. Leeson, S. Langer, B. J. Marty, and C. F. M. de Lange. 2006. Efficiency of utilizing ileal digestible lysine and threonine for whole body protein deposition in growing pigs is reduced when dietary casein is replaced by wheat shorts. *Journal of Animal Science* 84:1362-1374.

Libao-Mercado, A. J. O., C. L. Zhu, J. P. Cant, H. Lapierre, J. N. Thibault, B. Sève, M. F. Fuller, and C. F. M. de Lange. 2009. Dietary and endogenous amino acids are the main contributors to microbial protein in the upper gut of normally nourished pigs. *Journal of Nutrition* 139:1088-1094.

Llames, C. R., and J. Fontaine. 1994. Determination of amino acids in feeds: Collaborative study. *Journal of the Association of Official Analytical Chemists* 77:1362-1366.

Lloyd, L. E., B. E. McDonald, and E. W. Crampton. 1978. *Fundamentals of Nutrition*, 2nd Ed. San Francisco, CA: W. H. Freeman and Co.

Loughmiller, J. A., J. L. Nelssen, R. D. Goodband, M. D. Tokach, E. C. Titgemeyer, and I. H. Kim. 1998a. Influence of dietary lysine on growth performance and carcass characteristics of late-finishing gilts. *Journal of Animal Science* 76:1075-1080.

Loughmiller, J. A., J. L. Nelssen, R. D. Goodband, M. D. Tokach, E. C. Titgemeyer, and I. H. Kim. 1998b. Influence of dietary total sulfur amino acids and methionine on growth performance and carcass characteristics of finishing gilts. *Journal of Animal Science* 76:2129-2137.

Louis, G. F., A. J. Lewis, W. C. Weldon, P. M. Ermer, P. S. Miller, R. J. Kittok, and W. W. Stroup. 1994a. The effect of energy and protein intakes on boar libido, semen characteristics, and plasma hormone concentrations. *Journal of Animal Science* 72:2051-2060.

Louis, G. F., A. J. Lewis, W. C. Weldon, P. S. Miller, R. J. Kittok, and W. W. Stroup. 1994b. The effect of protein intake on boar libido, semen characteristics, and plasma hormone concentrations. *Journal of Animal Science* 72:2038-2050.

Mahan, D. C., and R. G. Shields, Jr. 1998. Essential and nonessential amino acid composition of pigs from birth to 145 kilograms of body weight, and comparison to other studies. *Journal of Animal Science* 76:513-521.

Martinez, G. M., and D. A. Knabe. 1990. Digestible lysine requirement of starter and grower pigs. *Journal of Animal Science* 68:2748-2755.

Mateo, R. D., G. Wu, F. W. Bazer, J. C. Park, I. Shinzato, and S. W. Kim. 2007. Dietary arginine supplementation enhances the reproductive performance of gilts. *Journal of Nutrition* 137:652-656.

Mathews, S. A. 2004. Investigating the effects of long chain polyunsaturated fatty acids on lipid metabolism and body composition in the neonatal pig. Ph.D. Dissertation, North Carolina State University, Raleigh, NC.

Matthews, J. O., L. L. Southern, and T. D. Bidner. 2001. Estimation of the total sulfur amino acid requirement and the effect of betaine in diets deficient in total sulfur amino acids for the weanling pig. *Journal of Animal Science* 79:1557-1565.

Mavromichalis, I., B. J. Kerr, T. M. Parr, D. M. Albin, V. M. Gabert, and D. H. Baker. 2001. Valine requirement of nursery pigs. *Journal of Animal Science* 79:1223-1229.

McPherson, R. L., F. Ji, G. Wu, J. R. Blanton, Jr., and S. W. Kim. 2004. Growth and compositional changes in fetal tissues in pigs. *Journal of Animal Science* 82:2534-2540.

Meding, A. J. H., and H. E. Nielsen. 1977. Fortskellige proteinnormers indflydelse pa frugbarheden hos orner, der anvendes til kunstig saerdoverforing. Copenhagen, Denmark: Statens Husdyrbrugsforog.

Meisinger, D. J., ed. 2010. National Swine Nutrition Guide: Collaboration Among Universities, Agri-businesses, and the U.S. Pork Center of Excellence. Iowa State University, Ames.

Meisinger, D. J., and V. C. Speer. 1979. Tryptophan requirement for reproduction in swine. *Journal of Animal Science* 48:559-569.

Metges, C. C. 2000. Contribution of microbial amino acids to amino acid homeostasis of the host. *Journal of Nutrition* 130:1857S-1864S.

Moehn, S., A. M. Gillis, P. J. Moughan, and C. F. M. De Lange. 2000. Influence of dietary lysine and energy intakes on body protein deposition and lysine utilization in the growing pig. *Journal of Animal Science* 78:1510-1519.

Moehn, S., E. Martinazzo-Dallagnol, R. F. P. Bertolo, P. B. Pencharz, and R. O. Ball. 2007. Metabolic availability of lysine in feedstuffs determined using oral isotope delivery. *Livestock Science* 109:24-26.

Moughan, P. J. 1999. Protein metabolism in the growing pig. Pp. 299-331 in *Quantitative Biology of the Pig*, I. Kyriazakis, ed. Wallingford, UK: CABI.

Murphy, J. M., S. J. Murch, and R. O. Ball. 1996. Proline is synthesized from glutamate during intragastric infusion but not during intravenous infusion in neonatal piglets. *Journal of Nutrition* 126:878-886.

Nam, D. S., and F. X. Aherne. 1994. The effects of lysine:energy ratio on the performance of weanling pigs. *Journal of Animal Science* 72:1247-1256.

Nemechek, J. E., M. D. Tokach, S. S. Dritz, R. D. Goodband, J. M. DeRouchey, J. L. Nelssen, and J. L. Ursy. 2011b. Evaluation of SID lysine level, the replacement of fish meal with crystalline amino acids, and lysine:CP ratio on growth performance of nursery pigs from 6.8 to 11.3 kg. *Journal of Animal Science* 89(Suppl. 3):80 (Abstr.).

Noblet, J., W. H. Close, and R. P. Heavens. 1985. Studies on the energy metabolism of the pregnant sow. 1. Uterus and mammary tissue development. *British Journal of Nutrition* 53:251-265.

NRC. 1998. *Nutrient Requirements of Swine*, 10th Ed. Washington, DC: National Academy Press.

O'Connell, M. K., P. B. Lynch, and J. V. O'Doherty. 2005. Determination of the optimum dietary lysine concentration for growing pigs housed in pairs and in groups. *Animal Science* 81:249-255.

O'Connell, M. K., P. B. Lynch, and J. V. O'Doherty. 2006. Determination of the optimum dietary lysine concentration for boars and gilts penned in pairs and in groups in the weight range 60 to 100 kg. *Animal Science* 82:65-73.

Oestemer, G. A., L. E. Hanson, and R. J. Meade. 1973. Reevaluation of the isoleucine requirement of the young pig. *Journal of Animal Science* 36:679-683.

O'Grady, J. F., and T. J. Hanrahan. 1975. Influence of protein level and amino-acid supplementation of diets fed in lactation on the performance of sow and their litters. *Irish Journal of Agricultural Research* 14:127-135.

Oresanya, T. F., A. D. Beaulieu, E. Beltranena and J. F. Patience. 2007. The effect of dietary energy concentration and total lysine/digestible energy ratio on the growth performance of weaned pigs. *Canadian Journal of Animal Science* 87:45-55.

Owen, K. Q., J. L. Nelssen, R. D. Goodband, M. D. Tokach, L. J. Kats, and K. G. Friesen. 1995. Added dietary methionine in starter diets containing spray-dried blood products. *Journal of Animal Science* 73:2647-2654.

Pahm, A. A., C. Pedersen, and H. H. Stein. 2009. Standardized ileal digestibility of reactive lysine in distillers dried grains with solubles fed to growing pigs. *Journal of Agricultural and Food Chemistry* 57:535-539.

Parr, T. M., B. J. Kerr, and D. H. Baker. 2003. Isoleucine requirement of growing (25 to 45 kg) pigs. *Journal of Animal Science* 81:745-752.

Pastorelli, G., M. Neil, and I. Wigren. 2009. Body composition and muscle glycogen contents of piglets of sows fed diets differing in fatty acids profile and contents. *Livestock Science* 123:329-334.

Paulicks, B. R., H. Ott, and D. A. Roth-Maier. 2003. Performance of lactating sows in response to the dietary valine supply. *Journal of Animal Physiology and Animal Nutrition* 87:389-396.

Paulicks, B. R., F. G. Pampuch, and D. A. Roth-Maier. 2006. Studies on the tryptophan requirement of lactating sows. 1. Estimation of the tryptophan requirement by performance. *Journal of Animal Physiology and Animal Nutrition* 90:474-481.

Pérez-Laspiur, J., J. L. Burton, P. S. D. Weber, J. Moore, R. N. Kirkwood, and N. L. Trottier. 2009. Dietary protein intake and stage of lactation differentially modulate amino acid transporter mRNS abundance in porcine mammary tissue. *Journal of Nutrition* 139:1677-1684.

Pike, I. H., and T. G. Boaz. 1972. The effect of condition at service and plane of nutrition in early pregnancy of the sow. 1. Uterine and extra-uterine changes. *Animal Production* 15:147-155.

Pomar, C., I. Kyriazakis, G. C. Emmans, and P. W. Knap. 2003. Modeling stochasticity: Dealing with populations rather than individual pigs. *Journal of Animal Science* 81:E178-E186.

Quant, A. D., M. D. Lindemann, B. J. Kerr, R. L. Payne, and G. L. Cromwell. 2012. Standardized ileal digestible tryptophan-to-lysine ratios in growing pigs fed corn-based and non-corn-based diets. *Journal of Animal Science* 90:1270-1279.

Rademacher, M., L. Babinsky, and J. Tossenberger. 1997. Digestible threonine requirement of growing and finishing pigs. *Journal of Animal Science* 75(Suppl. 1):183 (Abstr.).

Ragland, D., and O. Adeola. 1996. The response of 10-kg pigs to increasing dietary threonine levels. *Journal of Animal Science* 74(Suppl.1):55 (Abstr.).

Reeds, P. J. 2000. Dispensable and indispensable amino acids for humans. *Journal of Nutrition* 130:1850S-1840S.

Reifsnyder, D. H., C. T. Young, and E. E. Jones. 1984. The use of low protein liquid diets to determine the methionine requirement and the efficacy of methionine hydroxy analogue for the three-week-old pig. *Journal of Nutrition* 114:1705-1715.

Rhoads, J. M., and G. Wu. 2009. Glutamine, arginine, and leucine signaling in the intestine. *Amino Acids* 37:111-122.

Richert, B. T., M. D. Tokach, R. D. Goodband, J. L. Nelssen, R. G. Campbell, and S. Kershaw. 1997. The effect of dietary lysine and valine fed during lactation on sow and litter performance. *Journal of Animal Science* 75:1853-1860.

Rippel, R. H., B. G. Harmon, A. H. Jensen, H. W. Norton, and D. E. Becker. 1965a. Essential amino acid supplementation of intact proteins fed to the gravid gilt. *Journal of Animal Science* 24:373-377.

Rippel, R. H., B. G. Harmon, A. H. Jensen, H. W. Norton, and D. E. Becker. 1965b. Response of the gravid gilt to levels of protein as determined by nitrogen balance. *Journal of Animal Science* 24:209-215.

Rippel, R. H., B. G. Harmon, A. H. Jensen, H. W. Norton, and D. E. Becker. 1965c. Some amino acid requirements of the gravid gilt fed a purified diet. *Journal of Animal Science* 24:378-382.

Robbins, K. R., and D. H. Baker. 1977. Phenylalanine requirement of the weanling pig and its relationship to tyrosine. *Journal of Animal Science* 45:113-118.

Robbins, K. R., A. M. Saxton, and L. L. Southern. 2006. Estimation of nutrient requirements using broken-line regression analysis. *Journal of Animal Science* 84:E155-E165.

Roth, F. X., K. Eder, M. Rademacher, and M. Kirchgessner. 2000. Influence of the dietary ratio between sulphur containing amino acids and lysine on performance of growing-finishing pigs fed diets with various lysine concentrations. *Archives of Animal Nutrition* 53:141-155.

Rousselow, D. L., and V. C. Speer. 1980. Valine requirement of the lactating sow. *Journal of Animal Science* 50:472-475.

Rupanova, M. 2006. Influence of different lysine's levels in the compound feeds for boars on quantity and quality of the semen. *Zhivotnovdni Nauki* 4:45-50.

Russell, L. E., G. L. Cromwell, and T. S. Stahly. 1983. Tryptophan, threonine, isoleucine and methionine supplementation of a 12% protein, lysine-supplemented, corn-soybean meal diet for growing pigs. *Journal of Animal Science* 56:1115-1123.

Rutherfurd, S. M., and P. J. Moughan. 1997. Application of a new method for determining digestible reactive lysine to variably heated protein sources. *Journal of Agricultural Food Chemistry* 45:1582-1586.

Saldana, C. I., D. A. Knabe, K. Q. Owen, K. G. Burgoon, and E. J. Gregg. 1994. Digestible threonine requirements of starter and finisher pigs. *Journal of Animal Science* 72:144-150.

Sato, H., T. Kobayashi, R. W. Jones, and R. A. Easter. 1987. Tryptophan availability of some feedstuffs determined by pig growth assay. *Journal of Animal Science* 64:191-200.

Sauber, T. E., T. S. Stahly, N. H. Williams, and R. C. Ewan. 1998. Effect of lean growth genotype and dietary amino acid regimen on the lactational performance of sows. *Journal of Animal Science* 76:1098-1111.

Sauer, W., and L. Ozimek. 1986. Digestibility of amino acids in swine: Results and their practical applications. A review. *Livestock Production Science* 15:367-388.

Schinckel, A. P., N. Li, B. T. Richert, P. V. Preckel, and M. E. Einstein. 2003. Development of a model to describe the compositional growth and dietary lysine requirements of pigs fed ractopamine. *Journal of Animal Science* 81:1106-1119.

Schneider, J. D., M. D. Tokach, S. S. Dritz, J. L. Nelssen, J. M. DeRouchery, and R. D. Goodband. 2010. Determining the effect of lysine:calorie ratio on growth performance of ten- to twenty-kilogram of body weight nursery pigs of two different genotypes. *Journal of Animal Science* 88:137-146.

Schneider, R., M. Kirchgessner, B. R. Paulicks, and F. J. Schwarz. 1992a. Concentrations of protein and amino acids in the milk of sows in dependence of dietary methionine supply. 3. Contribution to the requirement of suckling sows for S-containing amino acids. *Journal of Animal Physiology and Animal Nutrition* 68:254-262.

Schneider, R., M. Kirchgessner, B. R. Paulicks, and F. J. Schwarz. 1992b. Feed intake and body weight of suckling sows in dependence of dietary methionine supplementation. 1. Contribution to the requirement of suckling sows for S-containing amino acids. *Journal of Animal Physiology and Animal Nutrition* 68:235-243.

Schutte, J. B., E. J. van Weerden, and F. Koch. 1988. Utilization of DL- and L-tryptophan in young pigs. *Animal Production* 46:447-452.

Schutte, J. B., M. W. Bosch, J. de Jong, E. J. van Weerden, and F. Koch. 1991. Factors affecting the requirement of dietary sulphur-containing amino acids of young pigs. *Netherlands Journal of Agricultural Science* 39:91-101.

Schutte, J. B., A. J. M. A. Verstraten, N. P. Lenis, J. De Jong, and J. T. M. Van Diepen. 1995. Requirement of young pigs for apparent ileal digestible tryptophan. *Netherlands Journal of Agricultural Science* 43:287-296.

Sève, B., P. Ganier, and Y. Henry. 1993. Response curve of growth performance to true digestible threonine measured at the ileal level. *Journées de la Recherche Porcine en France* 25:255-262.

Southern, L. L., and D. H. Baker. 1983. Arginine requirement of the young pig. *Journal of Animal Science* 57:402-412.

Stein, H. H., N. L. Trottier, C. Bellaver, and R. A. Easter. 1999. The effect of feeding level and physiological status on total flow and amino acid composition of endogenous protein at the distal ileum in swine. *Journal of Animal Science* 77:1180-1187.

Stein, H. H., S. W. Kim, T. T. Nielsen, and R. A. Easter. 2001. Standardized ileal protein and amino acid digestibility by growing pigs and sows. *Journal of Animal Science* 79:2113-2122.

Stein, H. H., B. Sève, M. F. Fuller, P. J. Moughan, and C. F. M. de Lange. 2007. Amino acid bioavailability and digestibility in pig feed ingredients: Terminology and application. *Journal of Animal Science* 85:172-180.

Stoll, B., and D. G. Burrin. 2006. Measuring splanchnic amino acid metabolism in vivo using stable isotopic tracers. *Journal of Animal Science* 84:E60-E72.

Stoll, B., J. Henry, P. J. Reeds, H. Yu, F. Jahoor, and D. G. Burrin. 1998. Catabolism dominates the first-pass intestinal metabolism of dietary essential amino acids in milk protein-fed piglets. *Journal of Nutrition* 128:606-614.

Taylor, S. J., D. J. A. Cole, and D. Lewis. 1985. Amino acid requirements of growing pigs. 6. Isoleucine. *Animal Production* 40:153-160.

Torrallardona, D., C. I. Harris, and M. F. Fuller. 2003a. Lysine synthesized by the gastrointestinal microflora of pigs is absorbed, mostly in the small intestine. *American Journal of Physiology* 284:E1177-E1180.

Torrallardona, D., C. I. Harris, and M. F. Fuller. 2003b. Pigs' gastrointestinal microflora provide them with essential amino acids. *Journal of Nutrition* 133:1127-1131.

Touchette, K. J., G. L. Allee, M. D. Newcomb, and R. D. Boyd. 1998. The lysine requirement of lactating primiparous sows. *Journal of Animal Science* 76:1091-1097.

Tritton, S. M., R. H. King, R. G. Campbell, A. C. Edwards, and P. E. Hughes. 1996. The effects of dietary protein and energy levels of diets offered during lactation on the lactational and subsequent reproductive performance of first-litter sows. *Animal Science* 62:573-579.

Urynek, W., and L. Buraczewska. 2003. Effect of dietary energy concentration and apparent ileal digestible lysine:metabolizable energy ratio on nitrogen balance and growth performance of young pigs. *Journal of Animal Science* 81:1227-1236.

Van Barneveld, R. J., E. S. Batterham, and B. W. Norton. 1994. The effect of heat on amino acids for growing pigs. 2. Utilization of ileal-digestible lysine from heat-treated field peas (*Pisum sativum* cultivar Dundale). *British Journal of Nutrition* 72:243-256.

van Milgen, J., J. Noblet, A. Valancogne, S. Dubois, and J. Y. Dourmad. 2008. InraPorc: A model and decision support tool for the nutrition of growing pigs. *Animal Feed Science and Technology* 143:387-405.

Warnants, N., M. J. Van Oeckel, and M. De Paepe. 2003. Response of growing pigs to different levels of ileal standardised digestible lysine using diets balanced in threonine, methionine and tryptophan. *Livestock Production Science* 82:201-209.

Webster, M. J., R. D. Goodband, M. D. Tokach, J. L. Nelssen, S. S. Dritz, J. A. Unruh, K. R. Brown, D. E. Real, J. M. DeRouchey, J. C. Woodworth, C. N. Groesbeck, and T. A. Marsteller. 2007. Interactive effects between ractopamine hydrochloride and dietary lysine on finishing pig growth performance, carcass characteristics, pork quality, and tissue accretion. *Professional Animal Scientist* 23:597-611.

Weeden, T. L., J. L. Nelssen, R. C. Thaler, G. E. Fitzner, and R. D. Goodband. 1994. Effect of dietary protein and supplemental soyabean oil fed during lactation on sow and litter performance through two parities. *Animal Feed Science and Technology* 45:211-226.

Westermeier, C., B. R. Paulicks, and M. Kirchgessner. 1998. Feed intake and body weights of suckling sows and piglets in dependence of dietary threonine supplementation. 1. Contribution about the threonine requirement of suckling sows. *Journal of Animal Physiology and Animal Nutrition* 79:33-45.

Williams, A. P. 1994. Recent developments in amino acid analysis. Pp. 11-36 in *Amino Acids in Farm Animal Nutrition*. J. P. F. D'Mello, ed. Wallingford, UK: CABI.

Williams, N. H., T. S. Stahly, and D. R. Zimmerman. 1997. Effect of level of chronic immune system activation on the growth and dietary lysine needs of pigs fed from 6 to 112 kg. *Journal of Animal Science* 75:2481-2496.

Williams, W. D., G. L. Cromwell, T. S. Stahly, and J. R. Overfield. 1984. The lysine requirement of the growing boar versus barrow. *Journal of Animal Science* 58:657-665.

Willis, G. M., and C. V. Maxwell. 1984. Influence of protein intake, energy intake and stage of gestation on growth, reproductive performance, nitrogen balance and carcass composition in gestating gilts. *Journal of Animal Science* 58:647-656.

Wiltafsky, M. K., B. Schmidtlein, and F. X. Roth. 2009. Estimates of the optimum dietary ratio of standardized ileal digestible valine to lysine for eight to twenty-five kilograms of body weight pigs. *Journal of Animal Science* 87:2544-2553.

Wiltafsky, M. K., M. W. Pfaffl, and F. X. Roth. 2010. The effects of branched-chain amino acid interactions on growth performance, blood metabolites, enzyme kinetics and transcriptomics in weaned pigs. *British Journal of Nutrition* 103:964-976.

Woerman, R. L., and V. C. Speer. 1976. Lysine requirement for reproduction in swine. *Journal of Animal Science* 42:114-120.

Wu, G. 2009. Amino acids: Metabolism, functions, and nutrition. *Amino Acids* 37:1-17.

Wu, G., F. W. Bazer, and W. Tou. 1995. Developmental changes of free amino acid concentrations in fetal fluids of pigs. *Journal of Nutrition* 125:2859-2868.

Wu, G., T. L. Ott, D. A. Knabe, and F. W. Bazer. 1999. Amino acid composition of the fetal pig. *Journal of Nutrition* 129:1031-1038.

Wu, G., F. W. Bazer, J. Hu, G. A. Johnson, and T. E. Spencer. 2005. Polyamine synthesis from proline in the developing porcine placenta. *Biology of Reproduction* 72(4):842-850.

Yang, H., J. E. Pettigrew, L. J. Johnston, G. C. Shurson, and R. D. Walker. 2000. Lactational and subsequent reproductive responses of lactating sows to dietary lysine (protein) concentration. *Journal of Animal Science* 78:348-357.

Yen, H. T., and I. T. Yu. 1985. Influence of digestible energy and protein feeding on semen characteristics of breeding boars. Pp. 610-612 in *Efficient Animal Production for Asian Welfare, Proceedings of the 3rd Asian-Australian Association of Animal Production Animal Science Congress, Seoul, South Korea*, Vol 2, J. K. Ha, ed. Seoul, Korea: AAAP Animal Science Congress.

Yen, J. T., J. Klindt, B. J. Kerr, and F. C. Buonomo. 2005. Lysine requirement of finishing pigs administered porcine somatotropin by sustained-release implant. *Journal of Animal Science* 83:2789-2797.

Yi, G. F., A. M. Gaines, B. W. Ratliff, P. Srichana, G. L. Allee, K. R. Perryman, and C. D. Knight. 2006. Estimation of the true ileal digestible lysine and sulfur amino acid requirement and comparison of the bioefficacy of 2-hydroxy-4-(methylthio)butanoic acid and DL-methionine in eleven- to twenty-six-kilogram nursery pigs. *Journal of Animal Science* 84:1709-1721.

Zebrowska, T. 1978. Digestion and absorption of nitrogenous compounds in the large intestine of pigs. *Roczniki Nauk Rolniczych* 95B:85-90.

Zhu, C. L., M. Rademacher, and C. F. M. de Lange. 2005. Increasing dietary pectin level reduces utilization of digestible threonine intake, but not lysine intake, for body protein deposition in growing pigs. *Journal of Animal Science* 83:1044-1053.

第3章 脂 类

引言

在描述脂质的术语中，"脂肪"（固态甘油三酯）和"油"（液态甘油三酯）常常被交换使用，但"脂质"一词一般是指可溶于脂溶性溶剂的所有物质的总称，包括甾醇、蜡酯、甘油一酯、甘油二酯、甘油三酯、磷酯、糖酯、游离脂肪酸、长链醛类、乙醇、脂溶维生素及其他非极性物质。脂肪及其组成成分脂肪酸在猪日粮配制中发挥着很多重要作用（Azain, 2001; Gu and Li, 2003; Rossi et al., 2010; Lin et al., 2012）。日粮中脂肪的作用包括：提供浓缩的能量来源、提供必需脂肪酸、减少热增耗、促进脂溶性维生素的吸收、润滑制粒、减少饲料粉尘，以及咀嚼和吞咽过程中的润滑作用。

脂肪是猪常用饲料原料的天然组成成分（表 17-1），也可以提炼浓缩后直接添加到日粮中（表 17-4）。当日粮脂肪能满足猪对必需脂肪酸的营养需要时，添加脂肪大多是出于经济方面的考虑，即所提供单位能量的成本。考虑到日粮操作的特点，典型日粮中脂肪的添加量不超过6%。但若采用制粒后喷涂工艺，脂肪的添加量可提高。日粮中添加脂肪在提高日粮能量的同时，可能会造成动物采食量的下降，但能提高饲料转化效率（Engel et al., 2001）。因此，在设计日粮配方时，应有一个适宜的养分与能量比例来满足猪的营养需要。此外，日粮脂肪的脂肪酸组成可直接影响肉的脂肪酸组成，进而影响肉的品质（Warnants et al., 2001; Wood et al., 2008）。日粮中添加的脂肪容易氧化酸败，进而降低日粮的营养价值，因此必须保证脂肪质量的稳定。这些影响因素将在下面的综述中作进一步的讨论，如第11章中提到的，环境中的某些脂溶性化合物（杀虫剂等）可残留在日粮脂肪中，增加了污染的潜在风险。

脂类的消化率及能值

通常认为脂肪和油是动物可高效利用的能量来源（Babatunde et al., 1968; Cera et al., 1988 a,b, 1989a, 1990; Li et al., 1990; Jones et al., 1992; Jorgensen et al., 1996; Jorgensen and Fernandez, 2000）。不论日粮中不饱和脂肪酸与饱和脂肪酸的比例如何，短链与中链脂肪酸（14个碳或更少）的消化率均为80%~95%（Stahly, 1984; Cera et al., 1990）。其中，脂类来源、日粮中的比例、分子间饱和脂肪酸与不饱合脂肪酸的分布均可影响脂类的消化和代谢（Allee et al., 1971, 1972; Mattson et al., 1979; Jorgensen et al., 1996; Averette Gatlin et al., 2005; Duran-Montgé et al., 2007），同时对氮的利用和氨基酸的吸收也有一定影响（Lowrey et al., 1962; Cera et al., 1988a, 1989a,b; Li et al., 1990; Li and Sauer, 1994; Jorgensen et al., 1996; Jorgensen and Fernandez, 2000; Cervantes-Pahm and Stein, 2008）。一般来说，仔猪对各种脂类的表观消化率随日龄的增长（Hamilton and McDonald, 1969; Frobish et al., 1970）和脂肪不饱和程度的提高（Powles et al., 1995）而提高，但动物源脂肪（猪油和牛油）的表观消化率随日龄增长提高的程度比植物油高（Cera et al., 1988a, b, 1989a, 1990）。在不同类型的脂肪中，动物更易于消化不饱和脂肪（Wiseman et al., 1990; Powles et al., 1994），但此结论尚不一致（Jorgensen and Fernandez, 2000, Kerr et al., 2009; Kil et al., 2010a）。游离脂肪酸对脂类消化率的负面影响尤其值得注意，Brambila 和 Hill（1966）、Jorgensen 和 Fernandez（2000）报道，游离脂肪酸的消化率低于甘油三酯。日粮中游离脂肪酸含量越高，有效能值越低（Wiseman and Salvador, 1991; Powles et al., 1994, 1995; Jorgensen and Fernandez, 2000）。

也有报道认为，精炼白脂（DeRouchey et al., 2004）和大豆皂角（Atteh and Leeson, 1985）的脂肪消化率与游离脂肪酸的含量无相关性。另外，日粮中每添加1%的粗纤维，脂肪消化率降低1.3%~1.5%（Just, 1982a, b, c; Dégen et al., 2007）。最近，Kil等（2010b）的研究发现，饲喂外加脂肪与内在脂肪比较，引起的内源脂肪损失更少，纯化中性洗涤纤维对脂肪真消化率和表观消化率无影响。

表17-4主要以Wiseman等（1990）和Powles等（1993, 1994, 1995）的研究工作为基础，通过下述公式估测不同来源脂肪的消化能值：

$$DE（kcal/kg）=\{36.898-[0.005×FFA, g/kg]-[7.330×exp（-0.906×U:S）]\}/4.184$$

（式3-1）

式中，FFA为游离脂肪酸；U:S为不饱和脂肪酸/饱和脂肪酸的比例。

在计算时，代谢能约为消化能的98%，代谢能转化为净能的效率为88%（van Milgen et al., 2001）。虽然近年来关于不同类型精炼油脂的消化能和代谢能的试验结果与NRC（1998）报道的十分相近（Jorgensen and Fernandez, 2000; Kerr et al., 2009; Silva et al., 2009; Anderson et al., 2012），但通过公式来预测各种类型脂类的能量含量和脂肪质量的准确性尚不清楚。另外，消化能与代谢能体系未真正说明代谢日粮中脂质的能量转化效率，有可能低估了净能值（Noblet et al., 1993; de Lange and Birkett, 2005）。Galloway和Ewan（1989）估测牛油的净能值为4180 kcal/kg，低于Halas等（2010）推测合成体脂肪的不饱和脂肪酸的转化效率。近来的研究表明，豆油（4679 kcal/kg）与精炼白脂（5900 kcal/kg）的净能估测值（Kil et al., 2010a）大幅低于Sauvant等（2004）的推荐值（7120 kcal/kg）。考虑到代谢能转化为净能的效率较高，这两个估测值也低于预期的结果（Just, 1982d; Noblet et al., 1993; Jorgensen et al., 1996）。这些差异存在的同时，脂肪酸组成、游离脂肪酸水平及氧化程度之间的互作对于消化能值、代谢能值及净能值的影响均不甚清楚，而这些指标对于更好地理解脂肪的净能值是必要的。

日粮脂肪和生长性能

断奶仔猪日粮中添加脂肪的价值尚不确定（Gu and Li, 2003）。Pettigrew和Moser（1991）对5~20 kg断奶仔猪日粮中添加脂肪的92个数据进行了总结，结果表明，在这一体重范围内，日粮中添加脂肪会降低采食量并显著改善饲料转化效率。同样地，通过喷雾干燥或乳化的方法只能中等程度地改善脂肪的利用（Xing et al., 2004）。日粮中添加脂肪效果的差异受一系列因素的影响，包括猪的日龄、脂肪添加数量、类型及添加方式，这些均会影响脂肪在动物日粮中的应用效果。Pettigrew和Moser（1991）报道，在蛋白能量比不变的情况下，添加脂肪对动物的生长速度无影响，会减少采食量，但对于饲料转化效率则有一定的改善作用。

对于生长育肥猪（20~100 kg），日粮中添加脂肪能提高生长速度，在减少采食量的同时提高饲料转化效率，但增加育肥猪的背膘厚（Coffey et al., 1982; Pettigrew and Moser, 1991; Øverland et al., 1999; Benz et al., 2011a）。Chiba等（1991）报道，当日粮赖氨酸消化能比为3 g/Mcal（或蛋白质能量比49 g/Mcal）时，才能使日粮中添加脂肪的效益最大化。脂肪作为一种能量饲料，日粮脂肪的消化率、日粮中代谢能和脂肪水平以及动物圈舍环境温度均能影响其营养价值（Stahly, 1984）。通常情况下，猪处于温度适中的环境中时，用脂肪代替碳水化合物作为能量来源有助于提高动物生长性能，同时降低单位体增重所需的代谢能。由于脂肪比碳水化合物的热增耗低，所以圈舍环境温度较高时，日粮中每添加1%的脂肪可使动物自由采食代谢能摄入量增加0.2%~0.6%（Stahly, 1984）。

许多证据表明，在母猪妊娠后期及哺乳期日粮中添加脂肪，能有效提高母猪产奶量，改善初乳及常乳品质，有效提高仔猪出生到断奶的增重和成活率，对于体重较小的初生仔猪作用更为显著（Moser and Lewis, 1980; Boyd et al., 1982; Coffey et al., 1982; Seerley, 1984; Pettigrew and Moser, 1991; Averette et al., 1999; Quiniou et al., 2008）。日粮添加脂肪对断奶成活率的改善取决于母猪产前摄入脂肪的数量（>1000 g）和对照组仔猪存活率（<80%）。仔猪出生体重较小时，口服中链脂肪酸也能提高其成活率（Lepine et al., 1989; Odle, 1997; Casellas et al., 2005; Dicklin et al., 2006）。日粮添加脂肪可减少母猪哺乳期间的体重损失，缩短发情间隔（Moser and Lewis, 1980; Pettigrew, 1981; Cox et al., 1983; Seerley, 1984; Moser et al., 1985; Shurson et al., 1986; Pettigrew and Moser, 1991; Averette Gatlin et al., 2002a）。最近，Rosero（2011）和 Rosero 等（2012）用现代高产母猪作为试验动物，以精炼白脂和动植物混合油为脂肪来源，设置 4 个脂肪添加水平（0%、2%、4%、6%），结果发现，随脂肪添加水平的提高，精炼白脂能减少母猪体重损失、提高窝增重，动植物混合油则无明显作用；两种脂肪均可缩短母猪发情间隔，提高分娩率。日粮添加脂肪改善母猪繁殖性能的原因可能与必需脂肪酸的供给有关（下面讨论）。

日粮必需及生物活性脂肪酸

脂肪酸不仅是高浓度能量来源，某些脂肪酸还是动物必需的、具有生物活性的成分，影响多种重要的生理过程，包括脂质代谢、细胞分裂和分化、免疫功能及炎症反应。最初认为亚油酸和花生四烯酸均为必需脂肪酸（EFA；Cunnane, 1984）。现在认为这两种脂肪酸属于 n-6 必需脂肪酸，亚油酸在体内分别通过Δ^6-脱氢酶、脂肪酸延长酶和Δ^5-脱氢酶的作用转化为花生四烯酸（图 3-1，Jacobi et al., 2011）。除了 n-6 必需脂肪酸，猪也需要 n-3 脂肪酸（α-亚麻酸、二十碳五烯酸、二十二碳六烯酸）（Palmquist, 2009，综述）。与 n-6 脂肪酸相似，长链 n-3 多不饱和脂肪酸可由 α-亚麻酸合成，猪日粮中可能含有足够的 α-亚麻酸，但仍然缺乏完整的数据。

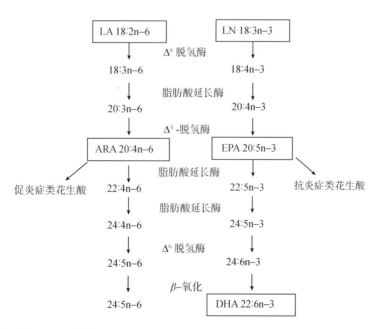

图 3-1 由 C18 前体物合成多不饱和脂肪酸。LA，亚油酸；ARA，花生四烯酸；LN，α-亚麻酸；EPA，二十碳五烯酸；DHA，二十二碳六烯酸。引自 Nelson（2000）。

猪日粮中特有的高比例 n-6/n-3 脂肪酸已引起关注。因为 C18 前体物脂肪酸竞争性地参与脂肪酸延长和脱氢代谢途径（图 3-1），这种不平衡会限制由二十碳五烯酸产生具有抗炎症作用的类花生酸（Wall et al., 2010，综述）。尽管有这种潜在的不平衡，但猪没有表现明显的必需脂肪酸缺乏。例如，Enser（1984）报道，从断奶到屠宰期间，仅含 0.1%亚油酸的日粮不影响猪的生长性能。英国农业研究委员会（ARC, 1981）建议，小于 30 kg 体重的猪，必需脂肪酸的需要占消化能的 3.0%，30~90 kg 体重的猪则占消化能的 1.5%，分别相当于日粮的 1.2% 和 0.6%。Christensen（1985）报道，5 周龄断奶到 100 kg 体重的猪，获得最佳生长性能和饲料转化效率的日粮亚油酸需要应占总能的 0.2%左右，相当于日粮的 0.1%。因此，通常情况下用谷物和蛋白质原料配制的日粮含有足够的亚油酸和 α-亚麻酸。一些证据表明，幼龄动物会出现脂肪酸延长和脱饱和途径受限的情况。因此，FDA（美国食品和药物管理局）2002 年批准婴儿食品中添加花生四烯酸和二十二碳六烯酸（最多占日粮脂肪含量的 1.25%），其部分依据来源于哺乳仔猪的试验结果（Huang et al., 2002; Mathews et al., 2002）。此外，一些研究还探讨了富含 n-3 脂肪酸的鱼油对公猪（Penny et al., 2000; Rooke et al., 2001a; Estienne et al., 2008; Castellano et al., 2010）和母猪（Perez Rigau et al., 1995; Rooke et al., 2001b; Laws et al., 2007; Brazle et al., 2009; Gabler et al., 2009; Mateo et al., 2009; Papadopoulos et al., 2009; de Quelen et al., 2010; Cools et al., 2011; Leonard et al., 2011; Smits et al., 2011）繁殖性能的影响。这些研究工作中，均能观察到组织中 n-3 脂肪酸的富集，但繁殖性能的结果不一致，而且以推荐值为基础配制的日粮缺乏剂量反应数据。一些研究报道了 n-3 脂肪酸对幼龄猪免疫反应的影响（Fritsche et al., 1993; Turek et al., 1996; Thies et al., 1999; Carroll et al., 2003; Liu et al., 2003; Jacobi et al., 2007; Lauridsen et al., 2007; Binter et al., 2008），但同样缺乏剂量反应的数据。

由于通过日粮方法改变猪肉脂肪酸组成相对比较容易，因此，大量工作研究了各种脂肪酸在组织中的富集，并以此为人的食品提供生物学活性的脂类，包括油酸（Miller et al., 1990）、共轭亚油酸（Averette Gatlin et al., 2002c, 2006; Dugan et al., 2004; Weber et al., 2006; Martin et al., 2007; Latour et al., 2008; Jiang et al., 2009; Larsen et al., 2009; White et al., 2009; Cordero et al., 2010）及 n-3 脂肪酸（Palmquist, 2009; Bryhni et al., 2002; Duran-Montgé et al., 2008; Flachowsky et al., 2008; Huang et al., 2008; Jaturasitha et al., 2009; Meadus et al., 2010; Realini et al., 2010; Wiecek et al., 2010）。但猪肉脂肪中 α-亚麻酸的半衰期预计超过 300 天（Anderson et al., 1972），而日粮改变至少两周才能检测出脂肪组织中脂肪酸组成的变化（Averette Gatlin et al., 2002b）。为此，一些研究工作探讨了用数学方法描述日粮脂肪酸组成与猪肉脂肪酸富集的关系（Lizardo et al., 2002; Nguyen et al., 2003）。

日粮脂肪、碘值与猪肉脂肪品质

许多年来一直认为日粮脂肪组成可直接影响猪肉的脂肪酸组成。1926 年，Ellis 和 Isbell 报道，饲喂各种不饱和油脂的猪，猪油中的不饱和脂肪酸含量增加。事实上，人们也在应用这种方法为注重健康的消费者开发富含生物活性脂肪酸的猪肉。但提高肉中不饱和脂肪酸后所含的"软脂"肉在猪肉加工中面临挑战（如制作培根时的腹部切割效率；脂肪污点），同时氧化酸败导致货架期缩短（Apple 综述，2013）。因此，当饲喂如 DDGS 等富含不饱和脂肪的饲料时，极容易导致上述问题（White et al., 2009; Xu et al., 2010）。

对于现代瘦肉基因型猪，不饱和脂肪酸含量增加导致的腹肉加工问题更加突出。研究者采用了多种日粮措施来消除这一问题，包括：① 饲喂天然饱和脂肪，如牛油（Averette Gatlin et al., 2002b; Apple et al., 2009）；②饲喂化学氢化脂肪（Averette Gatlin et al., 2005）；③更换谷物（Carr et al., 2005; Lampe et al., 2006）；④饲喂共轭亚油酸（Thiel-Cooper et al., 2001; Wiegand et al., 2001; Averette Gatlin et al., 2002c, 2006；Dugan et al., 2004; Weber et al., 2006; Martin et al., 2007; Latour et al., 2008; Jiang et al., 2009; Larsen et al., 2009; White et al., 2009; Cordero et al., 2010）。共轭亚油酸（CLA）能抑制硬脂酰辅酶A，进而减少C16:1和C18:1的从头合成，同时增加C16:0和C18:0的浓度（Demaree et al., 2002; Averette Gatlin et al., 2002c）。因此，共轭亚油酸可与日粮中不饱和脂肪结合，减少后者对猪肉脂肪品质的负面影响（Larsen et al., 2009）。几个研究表明，新生仔猪和生长育肥猪日粮中添加共轭亚油酸会减少肉中的脂肪沉积（Ostrowska et al., 1999, 2003; Thiel-Cooper et al., 2001; Corl et al., 2008）。

以脂肪碘值（IV）为依据配制日粮是解决"软脂"猪肉的实用方法。碘值是指每100g脂肪所吸收碘的克数，它能粗略显示脂肪中双键的相对含量。碘值越高，脂肪的不饱和程度越高，软脂程度越高。碘值可用AOAC（1997）的方法直接测定，也可通过气相色谱测定由脂肪衍生的脂肪酸甲酯（FAME）含量，并用下述公式计算：

$$IV = \sum 100 \times \frac{FAME_i \times 253.81 \times db_i}{MW_i} \quad (式3-2)$$

式中，$FAME_i$为混合物中脂肪酸甲酯在脂肪中的比例；253.81为碘的分子质量；db_i为脂肪中的双键数量；MW_i为$FAME_i$的分子质量（AOCS, 1998; Knothe, 2002; Pétursson, 2002; Meadus et al., 2010）。

以脂肪酸为基础的这种转化为：

总 IV $_{基于脂肪酸}$ = %C16:1（0.9976）+ %C18:1（0.8985）+ %C18:2（1.8099）
+ %C18:3（2.7345）+ %C20:1（0.8173）+ %C20:4（3.3343）+ %C20:5（4.1956）
+ %C22:1（0.7496）+ %C22:5（3.8395）+ %C22:6（4.6358） （式3-3）

以纯的甘油三酯为基础表示的碘值为：

总 IV $_{基于脂肪酸}$ = %C16:1（0.9502）+ %C18:1（0.8598）+ %C18:2（1.7315）
+ %C18:3（2.6152）+ %C20:1（0.7852）+ %C20:4（3.2008）
+ %C20:5（4.0265）+ %C22:1（0.7225）
+ %C22:5（3.6974）+ %C22:6（4.4632） （式3-4）

式中，%表示每个FAME在气相色谱分析中占全部总FAME的百分比。

表17-1和表17-4中所列碘值为基于这些饲料原料的脂肪酸组成，是用式3-2推算的，而脂肪酸含量则以占总乙醚浸出物的百分比来表示。值得注意的是，粗加工玉米油的碘值（107，表17-1）比精加工玉米油低（125，表17-4，USDA, 2011）。这主要是由于粗玉米油在经过加工纯化后，影响碘值含量的磷脂及成分被去除（www.corn.org）。一些组分明显影响粗油的碘值。表格中也包含了一些饲料原料的IVP值（Madsen et al., 1992），IVP值是饲料成分碘值乘以其脂肪含量再乘以换算系数0.1后得到的：

IVP = 原料中脂肪碘值 × %脂肪含量 × 0.1 （式3-5）

日粮配方中使用IVP可推测肉的碘值（Cast, 2010）。下面是从日粮IVP预测胴体碘值的回归方程：

胴体 IV = 47.1 + 0.14 × 日粮IVP；r^2=0.86（Madsen et al., 1992） （式3-6）
胴体 IV = 52.4 + 0.32 × 日粮IVP；r^2=0.99（Boyd et al., 1992） （式3-7）

上述两个公式的差异来源于IVP值的范围以及试验中猪自由采食的饲喂模式（Boyd et al., 1997）。由于预测方程间存在差异，可用数据又很少，无法建立日粮脂肪IVP与胴体脂肪IV间的有力的定量关系，因而没有应用到计算机模型中。Benz等（2011b）以IVP为基础配制日粮进行验证，发现日粮中C18:2n-6脂肪酸的含量比IVP更能有效地预测胴体碘值。

肉毒碱

肉碱是机体将长链脂肪酸跨膜转运到线粒体进行β氧化的条件性必需养分。猪和其他哺乳动物可以通过赖氨酸合成肉碱，但也有证据表明，幼龄猪未必总是能够合成足够量的肉碱（van Kempen and Odle, 1993; Owen et al., 1996; Heo et al., 2000a, b; Lyvers-Peffer et al., 2007）。因此可以在猪的日粮中添加一定量的L-肉碱。在断奶仔猪日粮中添加肉碱有可能提高其生长性能（Owen et al., 1996），但也并非总是如此（Hoffman et al., 1993; Owen et al., 2001）。肉碱并不能提高生长育肥猪的生长性能（Owen et al., 2001）。但母猪日粮中添加肉碱可改善胎儿的代谢（Xi et al., 2008）及大小（Brown et al., 2008），并增加出生的活仔数（Eder, 2010; Musser et al., 1999b; Ramanau et al., 2002），但也有研究表明并不总有这种效果（Musser et al., 1999a）。然而，母猪采食了添加肉碱的日粮后，仔猪断奶重得到改善（Ramanau et al., 2004）。

日粮脂肪的质量评价

脂质氧化会产生初级、次级和三级氧化产物，这些产物会产生与酸败相关的臭味和异味。因此，脂质氧化是决定某些饲料营养价值或保质期的重要因素。脂质可以被各种酶和氧自由基催化氧化，其过程包括：①脂质氧自由基的形成，启动氧化过程；②形成初级氧化产物过氧化氢；③形成次级氧化产物；④形成三级氧化产物（AOCS, 2005）。脂质的氧化速率主要取决于其饱和程度，多不饱和脂肪氧化速率较单不饱和脂肪快，饱和脂肪则基本稳定。氧化速率会随着温度、氧分压及光照强度的增大而提高，重金属离子与非游离的盐、水或各种非脂质成分也能影响这一过程（AOCS, 2005）。这些氧化产物不仅会产生酸败的臭味和异味，过氧化氢及其分解产物还会与其他养分或细胞组分（蛋白质、细胞膜和酶）进行互作，进而影响细胞功能（Comporti, 1993; Frankel, 2005）。

检测脂质的氧化程度是一项复杂的工作。氧化过程中，氧化反应产物的产生与转化同时进行（图3-2）。

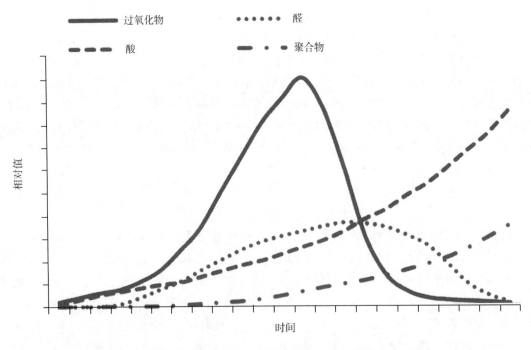

图3-2 脂质氧化过程中氧化产物组成的变化（改编自 Liu, 1997）

因此，实验室检测的相关氧化指标不一定能确切地反映脂质的氧化状态，也不一定能确切地预测脂质的保质期。下面描述的一些脂质氧化测定方法很常用，但没有一种被公认的最好方法。多数情况下，可能需要几种方法的结合才能对脂质的氧化程度做一个可信的估计。

传统的分析方法（氧化状态分析）

过氧化值（PV）用来估测氢过氧化物（包括二氢过氧化物与环状过氧化物），主要是对脂质氧化初级产物的形成进行评价。但是由于初级产物很快分解为二级氧化产物，因此 PV 值低估了真正的氧化程度（Ross and Smith, 2006）。影响 PV 值的因素很多，也有很多的表述方法，常用 1 kg 中的毫当量（milliequivalents）表示，也可用 1 kg 中的毫分子表示（millimoles，相当于毫当量的 1/2），或者以每千克脂肪中氧气的毫克数表示（相当于毫当量的 8 倍），但多种表述为数据的统一增加了难度。

羰基化合物，即醛和酮及其氧化产物或环氧化物（环氧乙烷衍生物）是一些活性最强的脂质氧化产物，都是脂质氢过氧化物分解产生的。羰基化合物一直作为脂肪氧化的重要指标。联苯胺值（BV）和对位甲氧基苯胺值（AV）在方法上十分接近，且产生的缩合产物的结构类似，但它们的共轭双键的长度不同，所以两种方法得到的绝对值不同。同样地，丙二醛（主要在多不饱和脂肪酸或不饱和醛氧化时产生）与二硫代巴比妥酸（TBA）反应时，因而会产生共轭双键化合物，因此可作为脂肪氧化的另一指标。但硫代巴比妥酸除与丙二醛反应外，还能与其他许多物质发生反应，因而会过高估计脂质氧化的程度，因此一些研究用硫代巴比妥反应物（TBARS）来报告结果（Ross and Smith, 2006）。虽然一直有人建议用气相色谱（GC）法或高效液相色谱法替代 TBARS 方法（Frankel, 2005; Ross and Smith, 2006），但 TBARS 仍是分析脂质氧化最常用的方法之一，具有简便、快速、成本相对较低等特点，样品量大时更为合适。由于 TBARS 方法的局限性，特定挥发性物质便成为常用的测定脂质氧化的指标。在氢过氧化物（烷烃、烯烃、醛类、酮类、醇类、酯类、酸类、烃类）的次级氧化产物中，醛类（辛醛、壬醛、戊醛和乙醛）是与正己醛反应产生的最主要的挥发性物质，是反映脂质氧化程度的最好指标之一（Ross and Smith, 2006）。羟基化醛也可作为醛类不同生物效应的中间产物，由于 4-羟基-2-壬醛（4-HNE）具有某些有害的生理作用，而被认为是具特点的羟基化醛（Seppanen and Sarri Csallay, 2002; Poli et al., 2008）。与其他许多化合物一样，可采用不同可靠性水平的方法来测定 4-HNE 的含量（Uchida et al., 2002; Zanardi et al., 2002）。上述分析方法皆可用以测定脂质氧化的程度，从而对脂质的质量进行粗略评价。但是，它们不能为样品的进一步氧化程度（如货架期等）提供信息。

加速稳定性测试（预测性分析）

加速试验的目的是为了估测产品的氧化稳定性，即货架期。最常用的加速试验方法是将样品放置在提高温度和氧分压的环境中。在史卡尔耐热试验（Schaal Oven test）中，将脂肪样品加热到 50~60℃，通过感官指标、PV 值或 TBA 值来确定氧化的终点。这种方法的测定结果与实际货架期的相关性较高，但需要耗费时间和劳动力。而活性氧法（AOM）主要是将纯化空气吹过加热至 97.8℃ 的脂质样品，通过绘制 PV 值与时间的关系来确定脂质 PV 值达到 100 mEq/kg 时所需的时间。AOM 法也需要耗费时间和劳动力，但存在自身缺陷，得出的数据变异较大。氧化稳定指数（OSI）是替代 AOM 的一种方法，该法基于脂质在温度和空气作用下氧化产生挥发性脂肪酸，被通过样品的空气携带进入含有去离子水的检测池内，然后通过自动软件连续检测其导电性。相对于 AOM 方法，OSI 方法的优点有：对氧化诱导点的测定更加准确，受气流影响小，主要基于稳定的三级氧化产物，重复性好，自动化程度高（Shahidi and Wanasundara, 1996）。

脂质氧化的调控

由于脂质氧化产物对其他营养物质的负面影响（如维生素 E；Mahan, 2001）并降低动物生产性能，必须慎重对待含有不饱和脂肪酸日粮的氧化稳定性。而对脂质氧化的控制则基于对脂质氧化过程的基本认识。因此，将脂质部分氢化、减少亚麻酸的含量和日粮在氧环境下的暴露（充氮气）、添加金属钝化剂（柠檬酸和磷酸）、保护其不受紫外线辐射（深色容器或限制叶绿素污染）、降低温度、添加抗氧化剂，这些方法均可降低脂质氧化速率（Frankel, 2007）。合成抗氧化剂（如乙氧基喹啉、丁基羟基茴香醚[BHA]、二丁基羟基甲苯[BHT]、没食子酸丙酯[PG]、叔丁基对苯二酚[TBHQ]）、天然抗氧化剂（维生素 E、类胡萝卜素）、植物提取物和螯合物（如抗坏血酸、柠檬酸、黄酮类化合物、磷酸，EDTA 以及 8-羟基喹啉）都已用于饲料及食品工业来抑制脂质氧化和食物酸败（Frankel, 2005, 2007; Wanasundara and Shahidi, 2005）。有关这些物质在畜禽日粮中使用价值的报道很少（Fernandez-Duenas, 2009），但最近在肉鸡上的研究（Tavarez et al., 2011）表明，肉鸡日粮中添加抗氧化剂防止脂质的进一步氧化，与未加抗氧化剂的对照组相比提高了肉鸡的生长性能。已经允许几种抗氧化剂（BHA、BHT 和 TBHQ）在人的食品中添加（单独添加或组合添加），但限量为 200 mg/kg（21CFR）。同样，已允许在畜禽及宠物饲粮中添加乙氧基喹啉，最高添加量为 150 mg/kg，在动物生肉中的最高残留量不得超过 0.5 mg/kg（21CFR）。

脂类质量对动物生理及生产性能的影响

饲喂氧化脂肪可提高生长猪小肠氧化应激指标（Ringseis et al., 2007）和血液中甘油三酯氧化指标（Suomela et al., 2005），减少小鸡小肠绒毛高度（Dibner et al., 1996a，b）。另外的研究表明，给肉鸡饲喂氧化的脂肪会减少体内对病原菌的初级抗体的产生（Takahashi and Akiba, 1999）。而饲喂某些特定的羟基化醛会产生生理反应，饲喂含 4-HNE 的日粮或用 4-HNE 处理细胞会产生共轭谷胱甘肽聚体（Uchida, 2003），增加应激通路的活性（Biasi et al., 2006; Yun et al., 2009）和巨噬细胞炎症因子的表达（Kumagai et al., 2004），降低 IgA 结合细菌抗原的能力（Kimura et al., 2006），同时阻断巨噬细胞信号转导机制（Kim et al., 2009）。

上述数据表明，氧化脂质对肠道功能有不良影响，畜禽似乎对氧化程度低的脂质有一定的耐受力。由于动植物蛋白饲料（如鱼粉、肉骨粉、DDGS）中含有高达 15%左右的脂肪，且大多经过热处理，因此这些产品中的脂类极易被氧化。但更应关注这些原料在饲料中的添加量、脂质含量和组成，以及该产品加工过程中的温度。迄今为止，有关各种脂类产品或蛋白原料中脂肪氧化程度的数据很少，氧化脂质对饲料的营养价值及畜禽生长的影响尚不甚清楚。在肉鸡上，仅有水分、不溶物、无皂化物及游离脂肪酸与其生长性能相关，而 AOM 稳定性与 PV 值则不在其中（Pesti et al., 2002）。与饲喂没有氧化脂肪的日粮相比，生长猪采食含 10%的肉骨粉（脂肪含量 17%；脂肪 PV 值为 210 mEq/kg，相当于日粮 3.6 mEq/kg 的 PV 值），或采食相同比例的肉骨粉但肉骨粉中脂肪含量和 PV 值不同（脂肪含量 17%；PV 值 210 mEq/kg 或相当于日粮 3.6 mEq/kg 的 PV 值）的日粮，其生长性能没有差异（Carpenter et al., 1966; L'Estrange et al., 1967）。相反，给仔猪饲喂含 6%精炼白脂（脂肪 PV 为 105 mEq/kg 或相当于日粮 6.3 mEq/kg 的 PV 值）的日粮，仔猪采食量及增重均有下降（DeRouchey et al., 2004）。

虽然增加日粮氧化脂肪及其相关氧化产物的含量影响血脂氧化、肠道屏障功能及炎症反应指标，但对脂类氧化指数与这些影响间的关系并不十分清楚。另外，对脂质氧化与营养利用、动物生产及动物胴体品质之间的关系也不甚了解。

脂类的分析

由于法规（营养标签）、经济（产品贸易）、健康（能量摄入）及质量控制（食品加工）等原因，准确测定饲料中脂类含量有重要意义。此外，测定肠道内容物和粪中的脂类含量也很重要，能帮助我们理解动物对脂肪的消化及能量利用。但脂质分析比较困难（Hammond，2001）。迄今为止，最常用的分析脂肪的方法有半连续提取法（索氏萃取）、连续溶剂提取法（Goldfisch）和兰德尔浸提法（Randal submersion）。随着分析技术的发展，快速溶剂萃取技术、过滤袋技术、超临界流体萃取技术、通过液相色谱测定脂肪酸总量、核磁共振技术以及近红外光谱技术均可作为高效、快捷、准确的脂类分析方法。不管采用哪种方法，样品干燥度、颗粒大小、溶剂类型（乙醚、己烷、氯仿）、提取时间、提取温度、压力以及设备校准程度均能影响提取脂肪的数量，从而造成不同实验室间分析结果的差异（Matthaus and Bruhl，2001；Palmquist and Jenkins，2003；Thiex et al.，2003a，b；Luthria，2004；Thiex，2009；Liu，2010）。

传统的萃取方法并不能将脂肪酸（如甘油三酯）或是脂类化合物完全提取出来，当它们以二价阳离子盐的形式存在或与各种碳水化合物和蛋白质结合在一起时尤其如此。在酸水解脂肪的过程中，盐酸将甘油三酯、糖脂和磷脂、胆固醇中的脂肪酸裂解，同时破坏脂质与碳水化合物和脂质与蛋白质之间的键结合以及细胞壁，使得脂质变为更加复杂的提取物（Palmquist and Jenkins，2003）。结果，酸水解脂肪含量要比相应的粗脂肪含量高，虽然不同饲料原料间的差异很大（Jongbloed and Smith，1994；Palmquist and Jenkins，2003；Karr-Lilienthal et al.，2005；Moller，2010）。分析技术的发展可有效减少分析方法学带来的差异（Schafer，1998；Toschi et al.，2003）。由于饲料中粗脂肪和酸水解脂肪之间存在差异，而且回肠内容物中可能存在阳离子结合脂肪，因此有必要采用常用的方法分析日粮和食糜中的脂肪含量，才能更公正地理解脂肪的消化。

参 考 文 献

Agricultural Research Council. 1981. *The Nutrient Requirements of Pigs: Technical Review*, Rev. Ed. Slough, UK: Commonwealth Agricultural Bureaux.

Allee, G. L., D. H. Baker, and G. A. Leveille. 1971. Influence of level of dietary fat on adipose tissue lipogenesis and enzymatic activity in the pig. *Journal of Animal Science* 33:1248-1254.

Allee, G. L., D. R. Romsos, G. A. Leveille, and D. H. Baker. 1972. Lipogenesis and enzymatic activity in pig adipose tissue as influenced by source of dietary fat. *Journal of Animal Science* 35:41-47.

Anderson, D. B., R. G. Kauffman, and N. J. Benevenga. 1972. Estimate of fatty acid turnover in porcine adipose tissue. *Lipids* 7:488-489.

Anderson, P. V., B. J. Kerr, T. E. Weber, C. J. Ziemer, and G. C. Shurson. 2012. Determination and prediction of energy from chemical analysis of corn coproducts fed to finishing pigs. *Journal of Animal Science* 90:1242-1254.

AOAC (AOAC International). 1997. Method 920.159. *Official Methods of Analysis*, 16th Ed., 3rd Rev. Gaithersburg, MD: AOAC International.

AOCS (American Oil Chemists Society). 1998. *Official Methods and Recommended Practices of the AOCS*, 5th Ed. Champaign, IL: AOCS.

AOCS. 2005. *Analysis of Lipid Oxidation*, A. Kamal-Eldin and J. Kororny, eds. Champaign, IL: AOCS Press.

Apple, J. 2013. Swine nutrition and pork quality. In *Sustainable Swine Nutrition*, L. I. Chiba, ed. Hoboken, NJ: Wiley-Blackwell.

Apple, J. K., C. V. Maxwell, D. L. Galloway, S. Hutchison, and C. R. Hamilton. 2009. Interactive effects of dietary fat source and slaughter weight in growing-finishing swine: I. Growth performance and longissimus muscle fatty acid composition. *Journal of Animal Science* 87:1407-1422.

Atteh, J. O., and S. Leeson. 1985. Effects of dietary soapstock on performance, nutrient digestibility and bone mineralization in weaner pigs fed two levels of calcium. *Canadian Journal of Animal Science* 65:945-952.

Averette, L. A., J. Odle, M. H. Monaco, and S. M. Donovan. 1999. Dietary fat during pregnancy and lactation increases milk fat and insulin-like growth factor I concentrations and improves neonatal growth rates in swine. *Journal of Nutrition* 129:2123-2129.

Averette Gatlin, L. A., J. Odle, J. Soede, and J. A. Hansen. 2002a. Dietary medium- or long-chain triglycerides improve body condition of lean-genotype sows and increase suckling pig growth. *Journal of Animal Science* 80:38-44.

Averette Gatlin, L. A., M. T. See, J. A. Hansen, D. Sutton, and J. Odle. 2002b. The effects of dietary fat sources, levels, and feeding intervals on pork fatty acid composition. *Journal of Animal Science* 80:1606-1615.

Averette Gatlin, L. A., M. T. See, D. K. Larick, X. Lin, and J. Odle. 2002c. Conjugated linoleic acid in combination with supplemental dietary fat alters pork fat quality. *Journal of Nutrition* 132:3105-3112.

Averette Gatlin, L. A., M. T. See, and J. Odle. 2005. Effects of chemical hydrogenation of supplemental fat on relative apparent lipid digestibility in finishing swine. *Journal of Animal Science* 83:1890-1898.

Averette Gatlin, L., M. T. See, D. K. Larick, and J. Odle. 2006. Descriptive flavor analysis of bacon and pork loin from lean-genotype gilts fed conjugated linoleic acid and supplemental fat. *Journal of Animal Science* 84:3381-3386.

Azain, M. J. 2001. Fat. Pp. 95-105 in *Swine Nutrition*, 2nd Ed., A. J. Lewis and L. L. Southern, eds. Boca Raton, FL: CRC Press.

Babatunde, G. M., W. G. Pond, E. F. Walker, Jr., P. Chapman, and R. J. Banis. 1968. Hematological changes, skin changes and apparent digestibility of lipids and protein in male and female growing pigs fed diets containing safflower oil, hydrogenated coconut oil, cholesterol or no fat. *Journal of Animal Science* 27:985-991.

Benz, J. M., M. D. Tokach, S. S. Dritz, J. L. Nelssen, J. M. DeRouchey, R. C. Sulabo, and R. D. Goodband. 2011a. Effects of choice white grease and soybean oil on growth performance, carcass characteristics, and carcass fat quality of growing-finishing pigs. *Journal of Animal Science* 89:404-413.

Benz, J. M., M. D. Tokach, S. S. Dritz, J. L. Nelssen, J. M. DeRouchey, R. C. Sulabo and R. D. Goodband. 2011b. Effects of dietary iodine value product on growth performance and carcass fat quality of finishing pigs. *Journal of Animal Science* 89:1419-1428.

Biasi, F., B. Vizio, C. Mascia, E. Gaia, N. Zarkovic, E. Chiarpotto, G. Leonarduzzi, and G. Poli. 2006. c-Jun N-terminal kinase upregulation as a key event in the proapoptotic interaction between transforming growth factor-β1 and 4-hydroxynonenal in colon mucosa. *Free Radical Biology and Medicine* 41:443-454.

Binter, C, A. Khol-Parisini, P. Hellweg, W. Gerner, K. Schafer, H. W. Hulan, A. Saalmuller, and J. Zentek. 2008. Phenotypic and functional aspects of the neonatal immune system as related to the maternal dietary fatty acid supply of sows. *Archives of Animal Nutrition* 62:439-453.

Boyd, R. D., B. D. Moser, E. R. Peo, Jr., A. J. Lewis, and R. K. Johnson. 1982. Effect of tallow and choline chloride addition to the diet of sows on milk composition, milk yield and preweaning pig performance. *Journal of Animal Science* 54:1-7.

Boyd, R. D., M. E. Johnston, K. Scheller, A. A. Sosnicki, and E. R. Wilson. 1997. Relationship between dietary fatty acid profile and body fat composition in growing pigs. PIC USA T&D Technical Memo 153. Franklin, KY: Pig Improvement Company.

Brambila, S., and F. W. Hill. 1966. Comparison of neutral fat and free fatty acids in high lipid-low carbohydrate diets for the growing chicken. *Journal of Nutrition* 88:84-92.

Brazle, A. E., B. J. Johnson, S. K. Webel, T. J. Rathbun, and D. L. Davis. 2009. Omega-3 fatty acids in the gravid pig uterus as affected by maternal supplementation with omega-3 fatty acids. *Journal of Animal Science* 87:994-1002.

Brown, K. R., R. D. Goodband, M. D. Tokach, S. S. Dritz, J. L. Nelssen, J. E. Minton, J. J. Higgins, X. Lin, J. Odle, J. C. Woodworth, and B. J. Johnson. 2008. Effects of feeding L-carnitine to gilts through day 70 of gestation on litter traits and the expression of insulin-like growth factor system components and L-carnitine concentration in foetal tissues. *Journal of Animal Physiology and Animal Nutrition* 92:660-667.

Bryhni, E. A., N. P. Kjos, R. Ofstad, and M. Hunt. 2002. Polyunsaturated fat and fish oil in diets for growing-finishing pig: Effects on fatty acid composition and meat, fat, and sausage quality. *Meat Science* 62:1-8.

Carpenter, K. J., J. L. L'Estrange, and C. H. Lea. 1966. Effects of moderate levels of oxidized fat in animal diets under controlled conditions. *Proceedings of the Nutrition Society* 25:25-31.

Carr, S. N., P. J. Rincker, J. Killifer, D. H. Baker, M. Ellis, and F. K. McKeith. 2005. Effects of different cereal grains and ractopamine hydrochloride on performance, carcass characteristics and fat quality in late-finishing pigs. *Journal of Animal Science* 83:223-230.

Carroll, J. A., A. M. Gaines, J. D. Spencer, G. L. Allee, H. G. Kattesh, M. P. Roberts, and M. E. Zannelli. 2003. Effect of menhaden fish oil supplementation and lipopolysaccharide exposure on nursery pigs. I. Effects on the immune axis when fed diets containing spray-dried plasma. *Domestic Animal Endocrinology* 24:341-351.

Casellas, J., X. Casasc, J. Piedrafita, and X. Manteca. 2005. Effect of medium- and long-chain triglyceride supplementation on small newborn-pig survival. *Preventive Veterinary Medicine* 67:213-221.

Cast, W. 2010. Formulating diets to iodine product specifications. Pp. 153-159 in *Proceedings of the 71st Minnesota Nutrition Conference, September 21-22, 2010, Owatonna, MN*. St. Paul: University of Minnesota.

Castellano, C.-A., I. Audet, J. L. Bailey, P. Y. Chouinard, J.-P. Laforest, and J. J. Matte. 2010. Effect of dietary n-3 fatty acids (fish oils) on boar reproduction and semen quality. *Journal of Animal Science* 88:2346-2355.

Cera, K. R., D. C. Mahan, and G. A. Reinhart. 1988a. Effects of dietary dried whey and corn oil on weanling pig performance, fat digestibility and nitrogen utilization. *Journal of Animal Science* 666:1438-1445.

Cera, K. R., D. C. Mahan, and G. A. Reinhart. 1988b. Weekly digestibilities of diets supplemented with corn oil, lard or tallow by weanling swine. *Journal of Animal Science* 66:1430-1437.

Cera, K. R., D. C. Mahan, and G. A. Reinhart. 1989a. Apparent fat digestibilities and performance responses of postweaning swine fed diets supplemented with coconut oil, corn oil or tallow. *Journal of Animal Science* 67:2040-2047.

Cera, K. R., D. C. Mahan, and G. A. Reinhart. 1989b. Postweaning swine performance and serum profile responses to supplemental medium-chain free fatty acids and tallow. *Journal of Animal Science* 67:2048-2055.

Cera, K. R., D. C. Mahan, and G. A. Reinhart. 1990. Evaluation of various extracted vegetable oils, roasted soybeans, medium-chain triglyceride and an animal-vegetable fat blend for postweaning swine. *Journal of Animal Science* 68:2756-2765.

Cervantes-Pahm, S. K., and H. H. Stein. 2008. Effect of dietary soybean oil and soybean protein concentration on the concentration of digestible amino acids in soybean products fed to growing pigs. *Journal of Animal Science* 86:1841-1849.

Chiba, L. I., A. J. Lewis, and E. R. Peo, Jr. 1991. Amino acid and energy interrelationships in pigs weighing 20 to 50 kilograms: I. Rate and efficiency of weight gain. *Journal of Animal Science* 69:694-707.

Christensen, K. 1985. *Determination of Linoleic Acid Requirements in Slaughter Pigs*, Research Report Number 577. Copenhagen, Denmark: Beret. Statens Husdyrbrugsforsøg.

Coffey, M. T., R. W. Seerley, D. W. Funderburke, and H. C. McCampbell. 1982. Effect of heat increment and level of dietary energy and environmental temperature on the performance of growing-finishing swine. *Journal of Animal Science* 54:95-105.

Comporti, M. 1993. Lipid peroxidation: Biopathological significance. *Free Radical Biology and Medicine* 7:333-349.

Cools, A., D. Maes, G. Papadopoulos, J.-A. Vandermeiren, E. Meyer, K. Demeyere, S. De Smet and G. P. J. Janssens. 2011. Dose-response effect of fish oil substitution in parturition feed on erythrocyte membrane characteristics and sow performance. *Journal of Animal Physiology and Animal Nutrition* 95:125-136.

Cordero, G., B. Isabel, D. Menoyo, A. Daza, J. Morales, C. Piñeiro, C. J. López-Bote. 2010. Dietary CLA alters intramuscular fat and fatty acid composition of pigs skeletal muscle and subcutaneous adipose tissue. *Meat Science* 85:235-239.

Corl, B. A., S. A. Mathews Oliver, X. Lin, W. T. Oliver, Y. Ma, R. J. Harrell, and J. Odle. 2008. Conjugated linoleic acid reduces body fat accretion and lipogenic gene expression in neonatal pigs fed low- or high-fat formulas. *Journal of Nutrition* 138:449-454.

Cox, N. M., J. H. Britt, W. D. Armstrong, and H. D. Alhusen. 1983. Effects of feeding fat and altering weaning schedule on rebreeding in primiparous sows. *Journal of Animal Science* 56:21-29.

Cunnane, S. C. 1984. Essential fatty-acid/mineral interactions with reference to the pig. Pp. 167-183 in *Fats in Animal Nutrition*, J. Wiseman, ed. London: Butterworths.

Dégen, L., V. Halas, and L. Babinszky. 2007. Effect of dietary fibre on protein and fat digestibility and its consequences on diet formulation for growing and fattening pigs: A review. *Acta Agriculturae Scandinavica* 57:1-9.

de Lange, C. F. M., and S. H. Birkett. 2005. Characterization of useful energy content in swine and poultry feed ingredients. *Canadian Journal of Animal Science* 85:269-280.

Demaree, S. R., C. D. Gilbert, H. J. Mersmann, and S. B. Smith. 2002. Conjugated linoleic acid differentially modifies fatty acid composition in subcellular fractions of muscle and adipose tissue but not adiposity of postweaning pigs. *Journal of Nutrition* 132:3272-3279.

de Quelen, F., G. Boudry, and J. Mourot. 2010. Linseed oil in the maternal diet increases long chain-PUFA status of the foetus and the newborn during the suckling period in pigs. *British Journal of Nutrition* 104:533-543.

DeRouchey, J. M., J. D. Hancock, R. D. Hines, C. A. Maloney, D. J. Lee, H. Cao, D. W. Dean, and J. S. Park. 2004. Effects of rancidity and free fatty acids in choice white grease on growth performance and nutrient digestibility in weanling pigs. *Journal of Animal Science* 82:2937-2944.

Dibner, J. J., C. A. Atwell, M. L. Kitchell, W. D. Shermer, and F. J. Ivey. 1996a. Feeding oxidized fats to broilers and swine: Effects on enterocyte turnover, hepatocyte proliferation and the gut associated lymphoid tissue. *Animal Feed Science and Technology* 62:1-13.

Dibner, J. J., M. L. Kitchell, C. A. Atwell, and F. J. Ivey. 1996b. The effect of dietary ingredients and age on the microscopic structure of the gastrointestinal tract in poultry. *Journal of Applied Poultry Research* 5:70-77.

Dicklin, M. E., J. L. Robinson, X. Lin, and J. Odle. 2006. Ontogeny and chain-length specificity of gastrointestinal lipases affect medium-chain triacylglycerol utilization by newborn pigs. *Journal of Animal Science* 84:818-825.

Dugan, M. E. R., J. L. Aalhus, and J. K. G. Kramer. 2004. Conjugated linoleic acid pork research. *American Journal of Clinical Nutrition* 79(Suppl):1212S-1216S.

Duran-Montgé, P., R. Lizardo, D. Torrallardona, and E. Esteve-Garcia. 2007. Fat and fatty acid digestibility of different fat sources in growing pigs. *Livestock Science* 109:66-69.

Duran-Montgé, P., C. E. Realini, A. C. Barroeta, R. Lizardo, and E. Esteve-Garcia. 2008. Tissue fatty acid composition of pigs fed different fat sources. *Animal* 2:1753-1762.

Eder, K. 2010. Influence of L-carnitine on metabolism and performance of sows. *British Journal of Nutrition* 102:645-654.

Ellis, N. R., and H. S. Isbell. 1926. Soft pork studies. II. The influence of the character of the ration upon the composition of the body fat of hogs. *Journal of Biological Chemistry* 69:219-248.

Engel, J. J., J. W. Smith, J. A. Unruh, R. D. Goodband, P. R. O'Quinn, M. D. Tokach, and J. L. Nelssen. 2001. Effects of choice white grease or poultry fat on growth performance, carcass leanness, and meat quality characteristics of growing-finishing pigs. *Journal of Animal Science* 79:1491-1501.

Enser, M. 1984. The chemistry, biochemistry and nutritional importance of animal fats. Pp. 23-51 in *Fats in Animal Nutrition*, J. Wiseman, ed. London: Butterworths.

Estienne, M. J., A. F. Harper, and R. J. Crawford. 2008. Dietary supplementation with a source of omega-3 fatty acids increases sperm number and the duration of ejaculation in boars. *Theriogenology* 70:70-76.

Fernandez-Duenas, D. M. 2009. Impact of oxidized corn oil and synthetic antioxidant on swine performance, antioxidant status of tissues, pork quality and shelf life evaluation. Ph.D. Dissertation. University of Illinois, Urbana-Champaign. Available online at http://www.ideals.illinois.edu/handle/2142/14588. Accessed November 15, 2011.

Flachowsky, G., E. Schulz, R. Kratz, and P. Glodek. 2008. Effects of different dietary fat sources on the fatty acid profile of backfat and intramuscular fat of pigs of various sire breeds. *Journal of Animal and Feed Sciences* 17:363-371.

Frankel, E. N. 2005. *Lipid oxidation*. Bridgwater, UK: The Oily Press.

Frankel, E. N. 2007. *Antioxidants in Food and Biology: Facts and Fiction*. Bridgwater, UK: The Oily Press.

Fritsche, K., D. W. Alexander, N. A. Cassity, and S. Huang. 1993. Maternally-supplied fish oil alters piglet immune cell fatty acid profile and eicosanoid production. *Lipids* 28:677-682.

Frobish, L. T., V. W. Hays, V. C. Speer, and R. C. Ewan. 1970. Effect of fat source and level on utilization of fat by young pigs. *Journal of Animal Science* 30:197-202.

Gabler, N. K., J. S. Radcliffe, J. D. Spencer, D. M. Webel, and M. E. Spurlock. 2009. Feeding long-chain n-3 polyunsaturated fatty acids during gestation increases intestinal glucose absorption potentially via the acute activation of AMPK. *Journal of Nutrition and Biochemistry* 20:17-25.

Galloway, S. T., and R. C. Ewan. 1989. Energy evaluation of tallow and oat groats for young swine. *Journal of Animal Science* 67:1744-1750.

Gu, X., and D. Li. 2003. Fat nutrition and metabolism in piglets: A review. *Animal Feed Science and Technology* 109:151-170.

Halas, V., L. Babinszky, J. Dijkstra, M. W. A. Verstegen, and W. J. J. Gerrits. 2010. Efficiency of fat deposition from non-starch polysaccharides, starch and unsaturated fat in pigs. *British Journal of Nutrition* 103:123-133.

Hamilton, R. M. G., and B. E. McDonald. 1969. Effect of dietary fat source on apparent digestibility of fat and the composition of fecal lipids of the young pig. *Journal of Nutrition* 97:33-41.

Hammond, E. W. 2001. Lipid analysis—A 20th century success? *Journal of the Science of Food and Agriculture* 82:5-11.

Heo, K., X. Lin, J. Odle, and I. K. Han. 2000a. Kinetics of carnitine palmitoyltransferase-I are altered by dietary variables and suggest a metabolic need for supplemental carnitine in young pigs. *Journal of Nutrition* 130:2467-2470.

Heo, K., J. Odle, I. K. Han, W. Cho, S. Seo, E. VanHeugten, D. H. Pilkington. 2000b. Dietary L-carnitine improves nitrogen utilization in growing pigs fed low energy, fat-containing diets. *Journal of Nutrition* 130:1809-1814.

Hoffman, L. A., D. J. Ivers, M. R. Ellersieck, and T. L. Veum. 1993. The effect of L-carnitine and soybean oil on performance and nitrogen and energy utilization by neonatal and young pigs. *Journal of Animal Science* 71(1):132-138.

Huang, M. C., A. Chao, R. Kirwan, C. Tschanz, J. M. Peralta, D. A. Diersen-Schade, S. Cha, and J. T. Brenna. 2002. Negligible changes in piglet serum clinical indicators or organ weights due to dietary single-cell long-chain polyunsaturated oils. *Food and Chemical Toxicology* 40:453-460.

Huang, F. R., Z. P. Zhan, J. Luo, Z. X. Liu, and J. Peng. 2008. Duration of dietary linseed feeding affects the intramuscular fat, muscle mass and fatty acid composition in pig muscle. *Livestock Science* 118:132-139.

Jacobi, S. K., A. J. Moeser, B. A. Corl, K. Ryan, A. T. Blikslager, R. J. Harrell, and J. Odle. 2007. Prophylactic enrichment of ileal enterocyte phospholipids with polyunsaturated fatty acids facilitates acute repair following ischemic injury in suckling piglets. *Gastroenterology* 132(4 Suppl. 2):A-242.

Jacobi, S. K., X. Lin, B. A. Corl, H. A. Hess, R. J. Harrell, and J. Odle. 2011. Dietary arachidonate differentially alters desaturase-elongase pathway flux and gene expression in liver and intestine of suckling pigs. *Journal of Nutrition* 141:548-553.

Jaturasitha, S., R. Khiaosa-ard, P. Pongpiachan, and M. Kreuzer. 2009. Early deposition of n-3 fatty acids from tuna oil in lean and adipose tissue of fattening pigs is mainly permanent. *Journal of Animal Science* 87:693-703.

Jiang, Z. Y., W. J. Zhong, C. T. Zheng, Y. C. Lin, L. Yang, and S. Q. Jiang. 2009. Conjugated linoleic acid differentially regulates fat deposition in backfat and longissimus muscle of finishing pigs. *Journal of Animal Science* 88:1694-1705.

Jones, D. B., J. D. Hancock, D. L. Harmon, and C. E. Walker. 1992. Effects of exogenous emulsifiers and fat sources on nutrient digestibility, serum lipids, and growth performance in weanling pigs. *Journal of Animal Science* 70:3473-3482.

Jongbloed, R., and B. Smits. 1994. *Effect of HCl-hydrolysis for Crude Fat Determination on Crude Fat Content, Digestibility of Crude Fat and NE_f of Feeds for Fattening Pigs*. Rapport IVVO-DLO, No. 263. 8200 AD. Lelystad, The Netherlands: Institute of Animal Science and Health.

Jorgensen, H., and J. A. Fernandez. 2000. Chemical composition and energy value of different fat sources for growing pigs. *Acta Agriculturae Scandinavica Section A—Animal Science* 50:129-136.

Jorgensen, H., S. K. Jensen, and B. O. Eggum. 1996. The influence of rapeseed oil on digestibility, energy metabolism and tissue fatty acid composition in pigs. *Acta Agriculturae Scandinavica Section A—Animal Science* 46:65-75.

Jorgensen, H., V. M. Gabert, M. S. Hedemann, and S. K. Jensen. 2000. Digestion of fat does not differ in growing pigs fed diets containing fish oil, rapeseed oil or coconut oil. *Journal of Nutrition* 130:852-857.

Just, A. 1982a. The influence of crude fiber from cereals on the net energy value of diets for growth in pigs. *Livestock Production Science* 9:569-580.

Just, A. 1982b. The influence of ground barley straw on the net energy value of diets for growth in pigs. *Livestock Production Science* 9:717-729.

Just, A. 1982c. The net energy value of balanced diets for growing pigs. *Livestock Production Science* 8:541-555.

Just, A. 1982d. The net energy value of crude fat for growth in pigs. *Livestock Production Science* 9:501-509.

Karr-Lilienthal, L. K., G. M. Grieshop, J. K. Spears, and G. C. Fahey, Jr. 2005. Amino acid, carbohydrate, and fat composition of soybean meals prepared at 55 commercial U.S. soybean processing plants. *Journal of Agricultural and Food Chemistry* 53:2146-2150.

Kerr, B. J., T. E. Weber, W. A. Dozier, III, and M. T. Kidd. 2009. Digestible and metabolizable energy content of crude glycerin originating from different sources in nursery pigs. *Journal of Animal Science* 87:4042-4049.

Kil, D. Y., F. Ji, L. L. Stewart, R. B. Hinson, A. D. Beaulieu, G. L. Allee, J. F. Patience, J. E. Pettigrew, and H. H. Stein. 2010a. Net energy of soybean oil and choice white grease in diets fed to growing and finishing pigs. *Journal of Animal Science* 89:448-459.

Kil, D. Y., T. E. Sauber, D. B. Jones and H. H. Stein. 2010b. Effect of the form of dietary fat and the concentration of dietary neutral detergent fiber on ileal and total tract endogenous losses and apparent and true digestibility of fat by growing pigs. *Journal of Animal Science* 88:2959-2967.

Kim, Y. S., Z. Y. Park, S. Y. Kim, E. Jeong, J. Y. Lee. 2009. Alteration of Toll-like receptor 4 activation by 4-hydroxy-2-nonenal mediated by the suppression of receptor homodimerization. *Chemico-Biological Interactions* 182:59-66.

Kimura, H., M. Mukaida, K. Kuwabara, T. Ito, K. Hashino, K. Uchida, K. Matsumoto, and K. Yoshida. 2006. 4-Hydroxynonenal modifies IgA in rat intestine after lipopolysaccharide injection. *Free Radical Biology and Medicine* 41:973-978.

Knothe, G. 2002. Structure indices in FA chemistry. How relevant is the iodine value? *Journal of the American Oil Chemists Society* 70:847-854.

Kumagai, T., N. Matsukawa, Y. Kaneko, Y. Kusumi, M. Mitsumata, and K. Uchida. 2004. A lipid peroxidation-derived inflammatory mediator. *Journal of Biological Chemistry* 279:48389-48396.

Lampe, J. F., T. J. Baas, and J. W. Mabry. 2006. Comparison of grain sources for swine diets and their effect on meat and fat quality. *Journal of Animal Science* 84:1022-1029.

Larsen, S. T., B. R. Wiegand, F. C. Parrish, Jr., J. E. Swan, and J. C. Sparks. 2009. Dietary conjugated linoleic acid changes belly and bacon quality from pigs fed varied lipid sources. *Journal of Animal Science* 87:285-295.

Latour, M. A., B. T. Richert, J. S. Radcliffe, A. P. Schinckel, and H. M. White. 2008. Effects of feeding restaurant grease with or without conjugated linoleic acid or phase-integrated beef tallow on finishing pig growth characteristics and carcass fat quality. *The Professional Animal Scientist* 24:156-160.

Lauridsen, C., J. Stagsted, and S. K. Jensen. 2007. n-6 and n-3 fatty acids ratio and vitamin E in porcine maternal diet influence the antioxidant status and immune cell eicosanoid response in the progeny. *Prostaglandins and Other Lipid Mediators* 84:66-78.

Laws, J., A. Laws, I. J. Lean, P. F. Dodds, and L. Clarke. 2007. Growth and development of offspring following supplementation of sow diets with oil during early to mid gestation. *Animal* 1:1482-1489.

Leonard, S. G., T. Sweeney, B. Bahar, B. P. Lynch, and J. V. O'Doherty. 2011. Effect of dietary seaweed extracts and fish oil supplementation in sows on performance, intestinal microflora, intestinal morphology, volatile fatty acid concentrations and immune status of weaned pigs. *British Journal of Nutrition* 105:549-560.

Lepine, A. J., R. D. Boyd, J. A. Welch, and K. R. Roneker. 1989. Effect of colostrum or medium-chain triglyceride supplementation on the pattern of plasma glucose, non-esterified fatty acids and survival of neonatal pigs. *Journal of Animal Science* 67:983-990.

L'Estrange, J. L., K. J. Carpenter, C. H. Lea, and L. J. Parr. 1967. Nutritional effects of autoxidized fats in animal diets. 4. Performance of young pigs on diets containing meat meals of high peroxide value. *British Journal of Nutrition* 21:377-390.

Li, D. F., R. C. Thaler, J. L. Nelssen, D. L. Harmon, G. L. Allee, and T. L. Weeden. 1990. Effect of fat sources and combinations on starter pig performance, nutrient digestibility and intestinal morphology. *Journal of Animal Science* 68:3694-3704.

Li, S., and W. C. Sauer. 1994. The effect of dietary fat content on amino acid digestibility in young pigs. *Journal of Animal Science* 72:1737-1743.

Lin, X., M. Azain, and J. Odle. 2012. Lipid nutrition and metabolism in swine. In *Sustainable Swine Nutrition*, L. I. Chiba, ed. Hoboken, NJ: Wiley-Blackwell.

Liu, K. 1997. Properties and edible applications of soybean oil. Pp. 347-378 in *Soybeans: Chemistry, Technology, and Utilization*. New York: Chapman & Hall.

Liu, K. S. 2010. Selected factors affecting crude oil analysis of distillers dried grains with solubles (DDGS) as compared with milled corn. *Cereal Chemistry* 87:243-249.

Liu, Y. L., L. M. Gong, G. F. Yi, A. M. Gaines, and J. A. Carroll. 2003. Effects of fish oil supplementation on the performance and the immunological, adrenal, and somatotropic responses of weaned pigs after an *Escherichia coli* lipopolysaccharide challenge. *Journal of Animal Science* 81:2758-2765.

Lizardo, R., J. van Milgen, J. Mourot, J. Noblet, and M. Bonneau. 2002. A nutritional model of fatty acid composition in the growing-finishing pig. *Livestock Production Science* 75:167-182.

Lowrey, R. S., W. G. Pond, J. K. Loosli, and J. H. Maner. 1962. Effect of dietary fat level on apparent nutrient digestibility by growing swine. *Journal of Animal Science* 21:746-750.

Luthria, D. L., ed. 2004. *Oil Extraction and Analysis: Critical Issues and Comparative Studies*. Champaign, IL: AOCA Press.

Lyvers-Peffer, P. A., X. Lin, S. Jacobi, L. A. Gatlin, J. Woodworth, and J. Odle. 2007. Ontogeny of carnitine palmitoyltransferase I activity, carnitine-Km, and mRNA abundance in pigs throughout growth and development. *Journal of Nutrition* 137:898-903.

Madsen, A., K. Jakobsen, and H. P. Mortensen. 1992. Influence of dietary fat on carcass fat quality in pigs: A review. *Acta Agriculturae Scandinavica Section A—Animal Science* 42:220-225.

Mahan, D. C. 2001. Selenium and vitamin E in swine nutrition. Pp. 281-314 in *Swine Nutrition*, 2nd Ed., A. J. Lewis and L. L. Southern, eds. Boca Raton, FL: CRC Press.

Martin, D., T. Antequera, E. Gonzalez, C. Lopez-Bote, and J. Ruiz. 2007. Changes in the fatty acid profile of the subcutaneous fat of swine throughout fattening as affected by dietary conjugated linoleic acid and monounsaturated fatty acids. *Journal of Agricultural and Food Chemistry* 55:10820-10826.

Mateo, R. D., J. A. Carroll, Y. Hyun, S. Smith, and S. W. Kim. 2009. Effect of dietary supplementation of n-3 fatty acids and elevated concentrations of dietary protein on the performance of sows. *Journal of Animal Science* 87:948-959.

Mathews, S. A., W. T. Oliver, O. T. Phillips, J. Odle, D. A. Diersen-Schade, and R. J. Harrell. 2002. Comparison of triglycerides and phospholipids as supplemental sources of dietary long-chain polyunsaturated fatty acids in piglets. *Journal of Nutrition* 132:3081-3089.

Matthaus, B., and L. Bruhl. 2001. Comparison of different methods for the determination of the oil content in oilseeds. *Journal of the American Oil Chemists Society* 78:95-102.

Mattson, F. H., G. A. Nolen, and M. R. Webb. 1979. The absorbability by rats of various triglycerides of stearic and oleic acid and the effect of dietary calcium and magnesium. *Journal of Nutrition* 109:1682-1687.

Meadus, W. J., P. Duff, B. Uttaro, J. L. Aalhus, D. C. Rolland, L. L. Gibson, M. E. R. Dugan. 2010. Production of docosahexaenoic acid (DHA) enriched bacon. *Journal of Agricultural and Food Chemistry* 58:465-472.

Miller, M. F., S. D. Shackelford, K. D. Hayden, and J. O. Reagan. 1990. Determination of the alteration in fatty acid profiles, sensory characteristics and carcass traits of swine fed elevated levels of monounsaturated fats in the diet. *Journal of Animal Science* 68:1624-1631.

Moller, J. 2010. Cereals, cereals-based products and animal feeding stuffs—determination of crude fat and total fat content by the Randall extraction method: A collaborative study. *Quality Assurance and Safety of Crops and Foods* 2010:1-6.

Moser, B. D., and A. J. Lewis. 1980. Adding fat to sow diets. *Feedstuffs* 52:36-37.

Moser, R. L., J. E. Pettigrew, S. G. Cornelius, and H. E. Hanke. 1985. *Feed and Energy Consumption by Lactating Sows as Affected by Supplemental Dietary Fat*, Minnesota Swine Research Report. St. Paul: University of Minnesota Press.

Musser, R. E., R. D. Goodband, M. D. Tokach, K. Q. Owen, J. L. Nelsen, S. A. Blum., R. G. Campbell, R. Smits, S. S. Dritz, and C. A. Civis. 1999a. Effects of L-carnitine fed during lactation on sow and litter performance. *Journal of Animal Science* 77:3296-3303.

Musser, R. E., R. D. Goodband, M. D. Tokach, K. Q. Owen, J. L. Nelsen, S. A. Blum., S. S. Dritz, and C. A. Civis. 1999b. Effects of L-carnitine fed during gestation and lactation on sow and litter performance. *Journal of Animal Science* 77:3289-3295.

Nelson, G. 2000. P. 489 in *Fatty Acids in Foods and Their Health Implications*, 2nd Ed., C. K. Chow, ed. New York: Marcel Dekker.

Nguyen, L. Q., M. C. G. A. Nuijens, H. Everts, N. Salden, and A. C. Beynen. 2003. Mathematical relationships between the intake of n-6 and n-3 polyunsaturated fatty acids and their contents in adipose tissue of growing pigs. *Meat Science* 65:1399-1406.

Noblet, J., H. Fortune, C. Dupire, and S. Dubois. 1993. Digestible, metabolizable and net energy values of 13 feedstuffs for growing pigs: Effect of energy system. *Animal Feed Science and Technology* 42:131-149.

NRC (National Research Council). 1998. *Nutrient Requirements of Swine*, 9th Rev. Ed. Washington, DC: National Academy Press.

Ostrowska, E., M. Muralitharan, R. F. Cross, D. E. Bauman, and F. R. Dunshea. 1999. Dietary conjugated linoleic acids increase lean tissue and decrease fat deposition in growing pigs. *Journal of Nutrition* 129:2037-2042.

Ostrowska, E., D. Suster, M. Muralitharan, R. F. Cross, B. J. Leury, D. E. Bauman, and F. R. Dunshea. 2003. Conjugated linoleic acid decreases fat accretion in pigs: Evaluation by dual-energy X-ray absorptiometry. *British Journal of Nutrition* 89:219-229.

Øverland, M., K.-A. Røvik, and A. Skrede. 1999. High-fat diets improve the performance of growing-finishing pigs. *Acta Agriculturae Scandinavica Section A—Animal Science* 49:83-88.

Odle, J. 1997. New insights into the utilization of medium-chain triglycerides by the neonate: Observations from a piglet model. *Pig News and Information* 127:1061-1067.

Owen, K. Q., J. L. Nelsen, R. D. Goodband, T. L. Weeden, and S. A. Blum. 1996. Effect of L-carnitine and soybean oil on growth performance and body composition of early weaned pigs. *Journal of Animal Science* 74:1612-1619.

Owen, K. Q., J. L. Nelsen, R. D. Goodband, M. D. Tokach, and K. G. Friesen. 2001. Effect of dietary L-carnitine on growth performance and body composition in nursery and growing-finishing pigs. *Journal of Animal Science* 79:1509-1515.

Palmquist, D. L. 2009. Omega-3 fatty acids in metabolism, health and nutrition and for modified animal product foods. *The Professional Animal Scientist* 25:207-249.

Palmquist, D. L., and T. C. Jenkins. 2003. Challenges with fats and fatty acid methods. *Journal of Animal Science* 81:3250-3254.

Papadopoulos, G. A., D. G. D. Mayes, S. Van Weyenberg, T. A. T. G. van Kempen, J. Buyse, and G. P. J. Janssens. 2009. Peripartal feeding strategy with different n-6:n-3 ratios in sows: Effects on sows' performance, inflammatory and periparturient metabolic parameters. *British Journal of Nutrition* 101:348-357.

Penny, P. C., R. C. Noble, A. Maldjian, and S. Cerolini. 2000. Potential role of lipids for the enhancement of boar fertility and fecundity. *Pig News and Information* 21:119N-126N.

Pesti, G. M., R. I. Bakalli, M. Qiao, and K. G. Sterling. 2002. A comparison of eight grades of fat as broiler feed ingredients. *Poultry Science* 81:382-390.

Perez Rigau, A., M. D. Lindemann, E. T. Kornegay, A. F. Harper, and B. A. Watkins. 1995. Role of dietary lipids on fetal tissue fatty acid composition and fetal survival in swine at 42 days of gestation. *Journal of Animal Science* 73:1372-1380.

Pettigrew, J. E., Jr. 1981. Supplemental dietary fat for peripartal sows: A review. *Journal of Animal Science* 53:107-117.

Pettigrew, J. E., Jr., and R. L. Moser. 1991. Fat in swine nutrition. Pp. 133-146 in *Swine Nutrition*, E. R. Miller, D. E. Ullrey, and A. J. Lewis, eds. Stoneham, UK: Butterworth-Heinemann.

Pétursson, S. 2002. Clarification and expansion of formulas in AOCS recommended practice Cd 1c-85 for the calculation of iodine value from FA composition. *Journal of the American Oil Chemists Society* 79:737-738.

Poli, G., R. J. Schaur, W. G. Siems, and G. Leonarduzzi. 2008. 4-Hydroxynonenal: A membrane lipid oxidation product of medicinal interest. *Medicinal Research Reviews* 28:569-631.

Powles, J., J. Wiseman, D. J. A. Cole, and B. Hardy. 1993. Effect of chemical structure of fats upon their apparent digestible energy value when given to growing/finishing pigs. *Animal Production* 57:137-146.

Powles, J., J. Wiseman, D. J. A. Cole, and B. Hardy. 1994. Effect of chemical structure of fats upon their apparent digestible energy value when given to young pigs. *Animal Production* 58:411-417.

Powles, J., J. Wiseman, D. J. A. Cole, and S. Jagger. 1995. Prediction of the apparent digestible energy value of fats given to pigs. *Animal Science* 61:149-154.

Quiniou, N., S. Richard, J. Mourot, and M. Etienne. 2008. Effect of dietary fat or starch supply during gestation and/or lactation on the performance of sows, piglets' survival and on performance of progeny after weaning. *Animal* 2:1633-1644.

Ramanau, A., H. Kluge, J. Spilke, and K. Eder. 2002. Reproductive performance of sows supplemented with L-carnitine over three reproductive cycles. *Archives of Animal Nutrition* 56:287-296.

Ramanau, A., H. Kluge, J. Spilke, and K. Eder. 2004. Supplementation of sows with L-carnitine during pregnancy and lactation improves growth of piglets during the suckling period through milk production. *Journal of Nutrition* 134:86-92.

Realini, C. E., P. Duran-Montgé, R. Lizardo, M. Gispert, M. A. Oliver, and E. Esteve-Garcia. 2010. Effect of source of dietary fat on pig performance, carcass characteristics and carcass fat content, distribution and fatty acid composition. *Meat Science* 85:606-612.

Ringseis, R., N. Piwek, and K. Eder. 2007. Oxidized fat induces oxidative stress but has no effect on NF-κB-mediated proinflammatory gene transcription in porcine intestinal epithelial cells. *Inflammation Research* 56:118-125.

Rooke, J., C.-C. Shao, and B. Speake. 2001a. Effects of feeding tuna oil on the lipid composition of pig spermatozoa and in vitro characteristics of semen. *Reproduction* 121:315-322.

Rooke, J. A., A. G. Sinclair, and M. Ewen. 2001b. Changes in piglet tissue composition at birth in response to increasing maternal intake of long-chain n-3 polyunsaturated fatty acids are non-linear. *British Journal of Nutrition* 86:461-470.

Rosero, D. S. 2011. Response of the modern lactating sow to source and level of supplemental dietary fat during high ambient temperatures. M.S. Thesis, North Carolina State University.

Rosero, D. S., E. van Heugten, J. Odle, R. Cabrera, C. Arellano, and R. D. Boyd. 2012. Sow and litter response to supplemental dietary fat in lactation diets during high ambient temperatures. *Journal of Animal Science* 90:550-559.

Ross, C. F., and D. M. Smith. 2006. Use of volatiles as indicators of lipid oxidation in muscle foods. *Comprehensive Reviews in Food Science and Food Safety* 5:18-25.

Rossi, R., G. Pastorelli, S. Cannata, and C. Corino. 2010. Recent advances in the use of fatty acids as supplements in pig diets: A review. *Animal Feed Science and Technology* 162:1-11.

Sauvant, D., J. M. Perex, and G. Tran. 2004. *Tables of Composition and Nutritional Value of Feed Materials*, INRA, Paris, France, ed. Wageningen, The Netherlands: Wageningen Academic.

Schafer, K. 1998. Accelerated solvent extraction of lipids for determining the fatty acid composition of biological material. *Analytica Chimica Acta* 358:69-77.

Seerley, R. W. 1984. The use of fat in sow diets. Pp. 333-352 in *Fats in Animal Nutrition*, J. Wiseman, ed. London: Butterworths.

Seppanen, C. M., and A. Sarri Csallany. 2002. Formation of 4-hydroxynonenal, a toxic aldehyde, in soybean oil at frying temperature. *Journal of the American Oil Chemists Society* 79:1033-1038.

Shahidi, F., and U. N. Wanasundara. 1996. Methods for evaluation of the oxidative stability of lipid-containing foods. *Food Science and Technology International* 2:73-81.

Shurson, G. C., M. G. Hogberg, N. DeFever, S. V. Radecki, and E. R. Miller. 1986. Effects of adding fat to the sow lactation diet on lactation and breeding performance. *Journal of Animal Science* 62:672-680.

Silva, H. O., R. V. Sousa, E. T. Fialho, J. A. F. Lima, and L. F. Silva. 2009. Digestible and metabolizable energy of oils and lards for growing pigs. *Journal of Animal Science* 87(E-Suppl. 2):63 (Abstr.).

Smits, R. J., B. G. Luxford, M. Mitchell, and M. B. Nottle. 2011. Sow litter size is increased in the subsequent parity when lactating sows are fed diets containing n-3 fatty acids from fish oil. *Journal of Animal Science* 89:2731-2738.

Stahly, T. S. 1984. Use of fats in diets for growing pigs. Pp. 313-331 in *Fats in Animal Nutrition*, J. Wiseman, ed. London: Butterworths.

Suomela, J. P., M. Ahotupa, and H. Kallio. 2005. Triacylglycerol oxidation in pig lipoproteins after a diet rich in oxidized sunflower seed oil. *Lipids* 40:437-444.

Takahashi, K., and Y. Akiba. 1999. Effect of oxidized fat on performance and some physiological responses in broiler chickens. *Japan Poultry Science* 36:304-310.

Tavarez, M. A., D. D. Boler, K. N. Bess, J. Zhao, F. Yan, A. C. Dilger, F. K. McKeith, and J. Killefer. 2011. Effect of antioxidant inclusion and oil quality on broiler performance, meat quality, and lipid oxidation. *Poultry Science* 90:922-930.

Thiel-Cooper, R. L., F. C. Parrish, Jr., J. C. Sparks, B. R. Wiegand, and R. C. Ewan. 2001. Conjugated linoleic acid changes swine performance and carcass composition. *Journal of Animal Science* 79:1821-1828.

Thies, F., L. D. Peterson, J. R. Powell, G. Nebe-von-Caron, T. L. Hurst, K. R. Matthews, E. A. Newsholme, and P. C. Calder. 1999. Manipulation of the type of fat consumed by growing pigs affects plasma and mononuclear cell fatty acid compositions and lymphocyte and phagocyte functions. *Journal of Animal Science* 77:137-147.

Thiex, N. 2009. Evaluation of analytical methods for the determination of moisture, crude protein, crude fat, and crude fiber in distillers dried grains with solubles. *Journal of AOAC International* 92:61-73.

Thiex, N. J., S. Anderson, and B. Gildemeister. 2003a. Crude fat, diethyl ether extraction, in feed, cereal grain, and forage (Randall/Soxtec/submersion method): Collaborative study. *Journal of AOAC International* 86:888-898.

Thiex, N. J., S. Anderson, and B. Gildemeister. 2003b. Crude fat, hexanes extraction, in feed, cereal grain, and forage (Randall/Soxtec/sumbersion method): Collaborative study. *Journal of AOAC International* 86:899-908.

Toschi, T. G., A. Bendini, A. Ricci, and G. Lercker. 2003. Pressurized solvent extraction of total lipids in poultry meat. *Food Chemistry* 83:551-555.

Turek, J. J., I. A. Schoenlein, B. A. Watkins, W. G. Van Alstine, L. K. Clark, and K. Knox. 1996. Dietary polyunsaturated fatty acids modulate responses of pigs to *Mycoplasma hyopneumoniae* infection. *Journal of Nutrition* 126:1541-1548.

Uchida, K. 2003. 4-Hydroxy-2-nonenal: A product and mediator of oxidative stress. *Progress in Lipid Research* 42:318-343.

Uchida, T., N. Gotoh, and S. Wada. 2002. Method for analysis of 4-hydroxy-2-(E)-nonenal with solid-phase microextraction. *Lipids* 37:621-626.

USDA (U.S. Department of Agriculture). 2011. Nutrient Data Laboratory. Available online at: http://www.nal.usda.gov/fnic/foodcomp/search/. Accessed on November 16, 2011.

van Kempen, T. A. T. G., and J. Odle. 1993. Medium-chain fatty acid oxidation in colostrum-deprived newborn piglets: Stimulatory effect of L-carnitine supplementation. *Journal of Nutrition* 123:1531-1537.

van Milgen, J., J. Noblet, and S. Dubois. 2001. Energetic efficiency of starch, protein and lipid utilization in growing pigs. *Journal of Nutrition* 131:1309-1318.

Wall, R., R. P. Ross, G. F. Fitzgerald, and C. Stanton. 2010. Fatty acids from fish: The anti-inflammatory potential of long-chain omega-3 fatty acids. *Nutrition Reviews* 68:280-289.

Wanasundara, J. P. D., and F. Shahidi. 2005. Antioxidants: Science, technology, and applications. In *Bayley's Industrial Oil and Fat Products*, 6th Ed., F. Shahidi, ed. Hoboken, NJ: John Wiley & Sons.

Warnants, N., M. J. Van Oeckel, and M. De Paepe. 2001. Fat in pork: Image, dietary modification and pork quality. *Pig News and Information* 22:107N-113N.

Weber, T. E., B. T. Richert, M. A. Belury, Y. Gu, K. Enright, and A. P. Schinckel. 2006. Evaluation of the effects of dietary fat, conjugated linoleic acid, and ractopamine on growth performance, pork quality, and fatty acid profiles in genetically lean gilts. *Journal of Animal Science* 84:720-732.

White, H. M., B. T. Richert, J. S. Radcliffe, A. P. Schinckel, J. R. Burgess, S. L. Koser, S. S. Donkin, and M. A. Latour. 2009. Feeding conjugated linoleic acid partially recovers carcass quality in pigs fed dried corn distillers grains with solubles. *Journal of Animal Science* 87:157-166.

Wiecek, J., A. Rekiel, and J. Skomial. 2010. Effect of feeding level and linseed oil on some metabolic and hormonal parameters and on fatty acid profile of meat and fat in growing pigs. *Archiv fur Tierzucht/Archives Animal Breeding* 53:37-49.

Wiegand, B. R., J. C. Sparks, F. C. Parrish, Jr., and D. R. Zimmerman. 2002. Duration of feeding conjugated linoleic acid influences growth performance, carcass traits, and meat quality of finishing barrows. *Journal of Animal Science* 80:637-643.

Wiseman, J., and F. Salvador. 1991. The influence of free fatty acid content and degree of saturation on the apparent metabolizable energy value of fat fed to broilers. *Poultry Science* 70:573-582.

Wiseman, J., D. J. A. Cole, and B. Hardy. 1990. The dietary energy values of soya-bean oil, tallow, and their blends for growing/finishing pigs. *Animal Production* 50:513-518.

Wood, J. D., M. Enser, A. V. Fisher, G. R. Nute, P. R. Sheard, R. I. Richardson, S. I. Hughes, F. M. Whittington. 2008. Fat deposition, fatty acid composition and meat quality: A review. *Meat Science* 78:343-358.

Xi, L., K. Brown, J. Woodworth, K. Shim, B. Johnson, and J. Odle. 2008. Maternal dietary L-carnitine supplementation influences fetal carnitine status and stimulates carnitine palmitoyltransferase and pyruvate dehydrogenase complex activities in swine. *Journal of Nutrition* 138:2356-2362.

Xing, J. J., E. van Heugten, D. F. Li, K. J. Touchette, J. A. Coalson, R. L. Odgaard, and J. Odle. 2004. Effects of emulsification, fat encapsulation, and pelleting on weanling pig performance and nutrient digestibility. *Journal of Animal Science* 82:2601-2609.

Xu, G., S. K. Baidoo, L. J. Johnston, D. Bibus, A. J. E. Cannon, and G. C. Shurson. 2010. Effects of feeding diets containing increasing content of corn distillers dried grains with solubles to grower-finisher pigs on growth performance, carcass composition, and pork fat quality. *Journal of Animal Science* 88:1398-1410.

Yun, M. R., D. S. Im, S. J. Lee, H. M. Park, S. S. Bae, W. S. Lee, and C. D. Kim. 2009. 4-Hydroxynonenal enhances CD36 expression on murine macrophages via p38 MAPK-mediated activation of 5-lipoxygenase. *Free Radical Biology and Medicine* 46:692-698.

Zanardi, E., C. G. Jagersma, S. Ghidini, and R. Chizzolini. 2002. Solid phase extraction and liquid chromatography-tandem mass spectrometry for the evaluation of 4-hydroxy-2-nonenal in pork products. *Journal of Agricultural and Food Chemistry* 50(19):5268-5272.

第4章 碳水化合物

引言

猪对日粮中的碳水化合物并没有特定的需要，但日粮中的能量却绝大部分来自于植物性饲料原料中的碳水化合物。一般根据碳水化合物的化学特性，如聚合度、键的类型及构成碳水化合物单糖的特性（Cummings and Stephen, 2007），对饲料中的碳水化合物进行基本的分类。饲料中的碳水化合物包括由单糖经糖苷键连接而成的二糖、寡糖（低聚糖）和多糖（图4-1）。根据连接单糖间的碳原子的位置，将连接单糖的糖苷键分为α-糖苷键和β-糖苷键。例如，一个单糖上的1位碳原子与另一个单糖上的4位碳原子以α-糖苷键形式连接，则将这个连接命名为α-(1, 4)糖苷键。

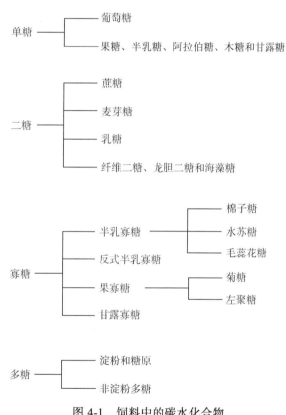

图4-1 饲料中的碳水化合物

在所有的碳水化合物中，只有单糖可以在猪的肠道吸收，并且吸收仅在小肠发生。因此，当碳水化合物通过小肠时，猪的消化酶必须将它的糖苷键分解，释放出单糖。然而，猪体内分泌的碳水化合物消化酶仅能水解有限的糖苷键，在小肠内许多碳水化合物未被酶解。这些未消化的碳水化合物可以经小肠或大肠内的肠道微生物发酵，产生短链脂肪酸并被吸收。所以，饲料中的碳水化合物不是在小肠段以单糖的形式被吸收，就是在小肠或大肠以短链脂肪酸的形式被吸收。这两类产物都可以给猪只供应能量。而未经酶解消化和肠道微生物发酵分解的部分碳水化合物，则随动物粪便排出体外，不能给猪只供应能量。

单糖

自然界中存在 20 种以上的单糖，而通常存在于动物饲料原料中的单糖不超过 10 种。一般依据单糖中碳原子的数量对单糖进行分类，含有 5 个碳原子的单糖被称为戊糖，含有 6 个碳原子的单糖被称为己糖。阿拉伯糖、核糖和木糖都属于戊糖，葡萄糖、果糖和半乳糖则属于己糖。目前已知的植物性饲料原料中含量最多的单糖是葡萄糖，但是猪日粮中也存在较高含量的果糖、半乳糖、阿拉伯糖、木糖和甘露糖，这主要取决于日粮中植物性原料的组成。葡萄糖和半乳糖在小肠内通过被动吸收或依赖其转运子载体以主动吸收的方式被吸收（Englyst and Hudson, 2000; Yen, 2011），而果糖、阿拉伯糖、木糖和甘露糖只能通过被动吸收的方式在小肠内被吸收（Englyst and Hudson, 2000; IOM, 2001）。植物性饲料原料中游离的单糖数量有限，日粮中几乎所有的单糖都能聚合形成二糖、寡糖（低聚糖）或多糖。

二糖

两个单糖通过糖苷键连接构成二糖。猪日粮中两种主要的二糖是蔗糖和乳糖（图 4-1）。蔗糖存在于许多植物性饲料原料中，而乳糖则只存在于乳制品中。因此，只有当日粮中含有乳制品，如脱脂奶粉、乳清粉、无蛋白乳清、液体乳清或乳糖时，日粮中才含有乳糖。在一些饲料原料中也存在少量的麦芽糖，并且麦芽糖也是淀粉消化的中间产物。葡萄糖和果糖通过 α-(1, 2) 糖苷键连接在一起形成蔗糖，两个葡萄糖单位通过 α-(1, 4) 糖苷键连接起来形成麦芽糖，而葡萄糖和半乳糖通过 β-(1, 4) 糖苷键连接起来形成乳糖。蔗糖、麦芽糖和乳糖中的糖苷键可以分别被蔗糖酶、麦芽糖酶和乳糖酶酶解消化。蔗糖酶的活力由小肠中的蔗糖酶-异麦芽糖酶复合物表达，该复合物在小肠中还表达其主要的麦芽糖酶活力（Treem, 1995; Van Beers et al., 1995）。麦芽糖酶的活力也可由麦芽糖酶-葡糖淀粉酶复合物表达，而乳糖酶却仅由乳糖酶基因表达（Van Beers et al., 1995）。小肠刷状缘中存在较高水平的蔗糖酶、麦芽糖酶和乳糖酶（Fan et al., 2001）。因此，蔗糖、麦芽糖和乳糖很容易被消化成游离的单糖并被吸收。这些二糖分解后产生的葡萄糖被吸收后，表现为血糖浓度迅速升高，因此，这些二糖也被称为血糖碳水化合物（glycemic carbohydrates）（Englyst and Englyst, 2005）。

除了蔗糖、麦芽糖和乳糖外，自然界中还存在其他的二糖，如纤维二糖、龙胆二糖和海藻糖。这些二糖都是两个葡萄糖单位通过 β-(1, 4) 糖苷键（纤维二糖）、β-(1, 6) 糖苷键（龙胆二糖）或 β-(1, 1) 糖苷键（海藻糖）连接而成的。猪不能分泌消化分解纤维二糖和龙胆二糖的酶，因此，这两种二糖仅能通过后肠发酵方式被利用。猪日粮中可能存在一些纤维二糖，但是通常并不含龙胆二糖。海藻糖作为存储型二糖，存在于昆虫和包括酵母在内的真菌中，所以如果猪日粮中添加酵母或酵母产品时，日粮中可能存在海藻糖。海藻糖可以被猪的小肠刷状缘分泌的海藻糖酶分解成单糖（Van Beers et al., 1995）。

寡糖（低聚糖）

寡糖是由一些单糖残基组成的具有特定结构的一类化合物。其中的单糖通过糖苷键连接在一起，不能被猪小肠中腺体分泌的酶消化。因此，这类寡糖被称为膳食纤维（dietary fiber），可在动物的小肠或大肠中通过微生物发酵产生短链脂肪酸被吸收。膳食纤维也包括非淀粉多糖，而寡糖以其在 80%（V/V）乙醇中的溶解度与多糖区分开来（Englyst and Englyst, 2005）。"不可消化寡糖"、"抗性寡糖"和"抗性短链碳水化合物"等术语均属同义词，指的是那些能够抵抗胰脏和小肠分泌液的消化，并且可溶于80%乙醇的碳水化合物（Englyst et al., 2007）。根据这一定义，寡糖包括半乳寡糖（包含反式半乳寡糖）、果寡糖和甘露寡糖等。

半乳寡糖

最大的一类半乳寡糖（也称为 α-半乳糖）是存在于豆科植物中的寡糖，包括棉子糖、水苏糖和毛蕊花糖（Cummings and Stephen, 2007; Martinez-Villaluenga et al., 2008）。棉子糖是由一个半乳糖单位与蔗糖通过 α-(1, 6)糖苷键连接组成的三糖。水苏糖是由两个半乳糖单位与蔗糖通过 α-(1, 6) 糖苷键连接组成的四糖，而毛蕊花糖则是由三个半乳糖单位通过 α-(1, 6) 糖苷键与蔗糖连接组成的五糖（Cummings and Stephen, 2007）。半乳寡糖主要存在于豆类植物的种子中，如豌豆和黄豆（Cummings and Stephen, 2007）。半乳寡糖中连接单糖的糖苷键可以被 α-半乳糖苷酶消化。同其他许多动物一样，猪的小肠中也不能分泌 α-半乳糖苷酶，因此在小肠中半乳寡糖不能通过酶解消化。然而，这些半乳寡糖可以很容易地被肠道微生物发酵分解，并且大多数的发酵是在小肠中进行的（Bengala Freire et al., 1991; Smiricky et al., 2002）。而一些未经小肠微生物发酵的半乳寡糖则进入大肠，可以发挥益生作用（prebiotic effect）（Meyer, 2004）。猪的日粮中添加 α-半乳糖苷酶和其他碳水化合物酶可以改善小肠中寡糖的消化率（Kim et al., 2003），但是并不一定能提高猪的生长性能（Jones et al., 2010）。一些植物，如大麦，表达 α-半乳糖苷酶的活力，其不仅可以参与棉子糖的代谢，而且还与叶片发育和抗逆性有关（Chrost et al., 2007）。

另一类半乳寡糖是反式半乳寡糖。这类寡糖不能在自然状态下合成，但是可以利用乳糖作为底物通过转糖基作用进行商业化生产（Houdijk et al., 1999; Meyer, 2004）。在 α-半乳糖苷酶的催化作用下，将乳糖转化为由 β-(1, 6) 键连接的半乳糖单位，末端再由 α-(1, 4) 键连接一个葡萄糖单位。该化合物的聚合度为 2~5（Meyer, 2004）。反式半乳寡糖被认为可以作为益生元（prebiotics），有助于改善猪的肠道健康，但这一作用还没有确凿的证据证实。

果寡糖

果寡糖或果聚糖主要是由果糖经过不同程度的聚合后形成的（BeMiller, 2007）。果寡糖被分为菊糖（inulins）或左聚糖（levans）。

菊糖是存储型碳水化合物，存在于多种水果和蔬菜中，包括洋葱、洋姜、小麦和菊苣（Englyst et al., 2007）。菊糖的链长度为 2~60 不等，聚合度平均为 12（Roberfroid, 2005）。利用菊苣商业化水解生产菊糖，产生菊糖型果聚糖，它是由果糖单位通过 β-(2, 1) 糖苷键连接起来的线性聚合物，并且在还原端连接一个蔗糖单位（BeMiller, 2007）。在一些菊糖型果聚糖中也可能存在由 β-(2, 6) 键连接的一个葡萄糖分子侧链（Meyer, 2004; Roberfroid, 2005）。

左聚糖是分泌果聚糖蔗糖酶的一些细菌和真菌合成的由 β-(2, 6) 糖苷键连接的果聚糖（Franck, 2006）。果聚糖蔗糖酶催化转糖基反应，将蔗糖转化为可能含有 β-(2, 1) 键侧链的左聚糖（BeMiller, 2007）。高聚合度（>10^7 Da）的果聚糖主要是左聚糖型的（Franck, 2006），但是它们未被商业化生产（Meyer, 2004）。除作为膳食纤维的来源外，果寡糖还可以作为益生元，促进双歧杆菌（Franck, 2006）和乳酸杆菌（Mul and Perry, 1994）的生长，同时抑制有害菌如梭状芽孢杆菌的生长（Franck, 2006），从而改善动物的肠道健康。

甘露寡糖

甘露寡糖是甘露糖的聚合物。用于猪日粮中的绝大部分甘露寡糖来源于酵母细胞壁（Zentek et al., 2002）。酵母细胞壁是由甘露聚糖、β-葡聚糖和壳多糖组成的一个网状结构（Cid et al., 1995）。甘露糖单位位于酵母细胞壁的外表面，通过 β-(1, 6) 和 β-(1, 3) 糖苷键与细胞壁内侧的 β-葡聚糖连接（Cid et al.,1995）。甘露寡糖不能被机体胃肠道分泌的消化酶消化（Zentek et al., 2002），当饲喂给动物后，甘露寡糖可作为益生元和免疫调节剂发挥作用。甘露寡糖为具有甘露聚糖特异性凝集素的细菌（即大肠杆菌）提供替代性的受体位点，从而提升动物胃肠道对病原微生物的抵抗力（Mul and Perry, 1994; Swanson et al., 2002）。

多糖

多糖可分为两类：淀粉和糖原；非淀粉多糖。在实际使用的猪日粮中，这两类碳水化合物都大量存在。

淀粉和糖原

淀粉

淀粉是大部分日粮中最重要的碳水化合物，因为淀粉是谷物中的主要存储型碳水化合物。淀粉全部由葡萄糖单位组成，它以颗粒的形式在直链和支链淀粉聚合物中天然存在，因而在碳水化合物中是独一无二的（BeMiller, 2007）。大部分谷物淀粉中包含 25%的直链淀粉和 75%的支链淀粉。直链淀粉（图 4-2）主要是葡萄糖残基以 α-(1, 4) 糖苷键连接形成的一条直链，也可能包含少量以 α-(1, 6) 糖苷键连接的侧链（Cummings and Stephen, 2007）。支链淀粉（图 4-3）是含 α-(1, 4) 糖苷键和 α-(1, 6) 糖苷键的高度分支的大分子聚合物（Cummings and Stephen, 2007）。全部或者几乎完全由支链淀粉构成的淀粉被称为蜡质淀粉（waxy starch）（BeMiller, 2007）。

当饲料与口腔分泌的唾液淀粉酶混合时，淀粉的消化就开始了（Englyst and Hudson, 2000）。由于饲料被吞食进入胃中后，唾液淀粉酶在胃内低 pH 条件下会很快失活，因此这段消化过程非常短暂

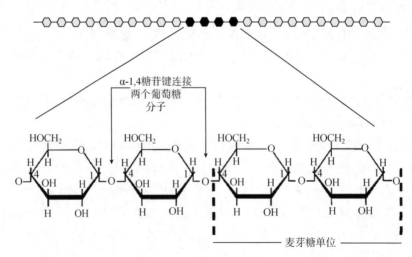

图 4-2 直链淀粉的结构

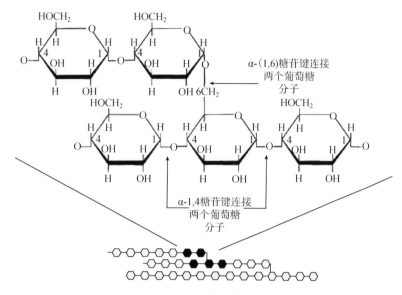

图 4-3 支链淀粉的结构

(Englyst and Hudson, 2000)。淀粉的消化主要在小肠中进行, 在胰腺和小肠分泌的 α-淀粉酶及异麦芽糖酶的作用下水解成麦芽糖、麦芽三糖和异麦芽糖（也称 α-糊精）的二级单位（Groff and Gropper, 2000）。麦芽糖酶将麦芽糖和麦芽三糖水解成葡萄糖单体, 而异麦芽糖酶（也称 α-糊精酶）将以 α-(1, 6) 糖苷键连接的异麦芽糖水解成葡萄糖分子（Groff and Gropper, 2000), 这些葡萄糖单体可以轻易地通过小肠的主动或被动运输途径被吸收。尽管这些碳水化合物酶理论上能够彻底消化淀粉, 但是淀粉实际在小肠中消化的速率和程度存在差异, 主要受三个方面因素影响：①淀粉颗粒的结晶结构或淀粉的来源；②直链淀粉和支链淀粉的比例；③淀粉加工处理的类型和程度（Cummings et al., 1997; Englyst and Hudson, 2000; Svihus et al., 2005)。可以根据淀粉的消化速率和葡萄糖在血液中出现的速率将淀粉分为快消化淀粉和慢消化淀粉（Englyst et al., 2007）。尽管如此, 淀粉的消化是一个高效的过程, 大部分谷物所含淀粉在小肠中的消化率都大于 95%（Bach Knudsen, 2001), 而紫花豌豆中的淀粉的回肠消化率约为 90%（Canibe and Bach Knudsen, 1997; Sun et al., 2006; Stein and Bohlke, 2007）。由于豌豆中的一部分淀粉被纤维性细胞壁成分包裹, 无法接触到消化酶, 因此其淀粉消化率低于谷物淀粉（Bach Knudsen, 2001)。同时, 豌豆中直链淀粉和支链淀粉的比例也比谷物高, 也可能降低其淀粉消化率（Bach Knudsen, 2001)。

在小肠内不能被消化的淀粉被称为抗性淀粉（resistant starch）（Brown, 2004）。实际上所有含淀粉的饲料原料中都存在抗性淀粉, 但抗性淀粉的含量与淀粉的来源、饲料的加工工艺, 以及饲喂前饲料的储存条件密切相关（Livesey, 1990; Brown, 2004; Goldring, 2004）。

抗性淀粉有 4 种类型。Ⅰ型抗性淀粉是指那些被难以消化的基质包裹而不能与淀粉酶接触的淀粉（BeMiller, 2007）。完整的谷物或部分碾磨的谷物所含的抗性淀粉属于这一类（Brown, 2004）。Ⅱ型抗性淀粉是指那些天然（未经熟化）具有抗消化性的淀粉颗粒, 它们抗消化的原因是这类淀粉颗粒的构象和结构（Brown, 2004）。这类淀粉通过加工可以被酶解。但是, 高直链淀粉比较特殊, 其淀粉颗粒即使经过加工处理也难于被消化酶水解（Brown, 2004）。Ⅲ型抗性淀粉是回生淀粉（retrograded starches）, 即经过糊化和冷却后形成结晶而难以被消化的淀粉（Brown, 2004）。Ⅳ型抗性淀粉是化学改性淀粉, 指经过特定的化学反应变性而产生的抗酶解淀粉（Brown, 2004）。抗性淀粉在大肠中极易发酵, 产生短链脂肪酸被机体利用。所以, 只有很少量的淀粉通过粪便排出体外。

糖原

动物体内的葡萄糖以糖原的形式储存在肌肉和肝脏中。糖原的结构与支链淀粉相似，包含葡萄糖残基以 α-(1, 4) 和 α-(1, 6) 糖苷键连接形成的直链。其消化的方式和所需的酶系也与支链淀粉相同，消化产生的葡萄糖通过小肠吸收。动物通常只在体内储存相对少量的糖原，大部分的能量以脂质（主要是甘油三酯）的形式储存。因此，只有猪日粮中含有肉粉或其他含糖原的动物源制品时，猪才会直接摄入糖原。但是大多数猪饲料中糖原的含量很少或不含糖原。

非淀粉多糖（NSP）

非淀粉多糖是被称为膳食纤维（dietary fiber）的一类碳水化合物，主要指那些在小肠内不能被消化酶消化或消化率很低，但是却能被肠道微生物部分或全部发酵的碳水化合物（De Vries, 2004）。在小肠内难于被消化的这部分碳水化合物也被称为"不能利用的碳水化合物（unavailable carbohydrates）"和"非血糖型碳水化合物（nonglycemic carbohydrates）"（Englyst et al., 2007）。与二糖、淀粉和糖原不同的是，组成非淀粉多糖的单糖不是由 α-(1, 4) 糖苷键或其他可被小肠消化酶消化的糖苷键连接形成的（Englyst et al., 2007）。所以猪日粮中的非淀粉多糖不会在小肠中消化成单糖被吸收，而只能被小肠或大肠微生物发酵以短链脂肪酸的形式被吸收。非淀粉多糖又可分为细胞壁成分和非细胞壁成分。

细胞壁成分

纤维素和半纤维素是细胞壁中最常见的非淀粉多糖，但是细胞壁中还包含阿拉伯木聚糖、木葡聚糖、阿拉伯半乳聚糖、半乳聚糖和 β-葡聚糖（Bach Knudsen, 2011）。纤维素是由葡萄糖单位以 β-(1, 4) 糖苷键连接形成的无分支型直链聚合物，所以这些链状分子紧密集合并形成微纤维，从而保证植物细胞和组织的结构完整性（Cummings and Stephen, 2007; Englyst et al., 2007）。正是由于所含糖苷键的特性，纤维素不能被猪小肠分泌的消化酶消化，但是可以被小肠或大肠中的微生物发酵分解。

半纤维素不同于纤维素，它是由不同类型的己糖和戊糖构成的支链型多糖（Cummings and Stephen, 2007）。包括谷物在内的一年生植物中最常见的半纤维素是木聚糖（BeMiller, 2007），构成木聚糖的木糖主链以直链或高度分支的形式存在（BeMiller, 2007）。侧链会出现在直链或分支的核心结构区，一般由阿拉伯糖、甘露糖、半乳糖和葡萄糖构成（Cummings and Stephen, 2007）。有些半纤维素也含有由葡萄糖和半乳糖产生的糖醛酸（分别为葡萄糖醛酸和半乳糖醛酸）（Southgate and Spiller, 2001）。这类含糖醛酸的半纤维素能够和金属离子如钙和锌结合形成盐类（Cummings and Stephen, 2007）。

木质素不是碳水化合物，但是它与植物细胞壁紧密联系在一起，所以在膳食纤维分析指标中也包含了木质素（Lunn and Buttriss, 2007）。木质素是由香豆醇、愈创木基醇、松柏醇和芥子醇等交联聚合而成的苯基丙烷类聚合物（Kritchevsky, 1988）。随着植物的成熟，木质素会穿透植物的多糖基质，在细胞壁基质中形成三维结构（Southgate, 2001）。木质素能够抵抗酶和细菌的降解，所以木质素含量高的植物很难被消化（Southgate, 2001; Wenk, 2001）。

非细胞壁成分

不是细胞壁组成成分，但被认为是非淀粉多糖的碳水化合物包括果胶、树胶和抗性淀粉。市场上常见的果胶通常是从柑橘果皮或苹果渣中提取的，当然也有其他来源的果胶（Fernandez, 2001）。果胶的一个重要特性是其主要由半乳糖醛酸以 α-(1, 4) 键连接形成直链多聚物（BeMiller, 2007）。果胶中也可能含有鼠李糖、半乳糖和阿拉伯糖的侧链（Cummings and Stephen, 2007）。

树胶是天然的植物多糖，但也可以通过发酵生产。天然树胶来源于植物或灌木伤裂处的分泌物（如阿拉伯树胶）或种子的胚乳（如瓜尔豆胶）（BeMiller, 2007）。黄原胶和普鲁兰多糖是通过发酵生产的树胶。

阿拉伯树胶（或金合欢树胶）是一种由多种成分组成的物质，半乳糖通过 β-(1, 3) 键连接形成分支型主链，树胶醛糖、鼠李糖、半乳糖和葡糖糖醛酸通过 1-6 位点连接成分支侧链（Osman et al., 1995; Williams and Phillips, 2001）。瓜尔豆胶是一种半乳甘露聚糖，由甘露糖以 β-(1, 4) 键连接形成线性主链，部分甘露糖单位会连接一个半乳糖形成侧链（BeMiller, 2007）。

碳水化合物的分析

饲料原料中的碳水化合物（图 4-4）可以通过不同的检测方法分析，并且每一种方法都可以检测碳水化合物的特定成分。单糖的含量可以通过酶法或高效液相色谱法（HPLC）分析（McCleary et al., 2006），二糖、寡糖和淀粉的含量通过酶-重量法分析，而非淀粉多糖的分析方法则有多种。最古老的分析方法是 Wende 分析法，利用该方法可将碳水化合物分为无氮浸出物和粗纤维。粗纤维的含量经酸消化后可利用重量分析法测定，其中包括大部分的木质素、不等含量的纤维素和少量的半纤维素（Grieshop et al., 2001; Mertens, 2003）。由于饲料原料中纤维素和半纤维素的回收率不稳定，粗纤维的含量分析值不能准确地反映饲料原料的营养价值，因此，该方法很少用于分析猪饲料的原料。

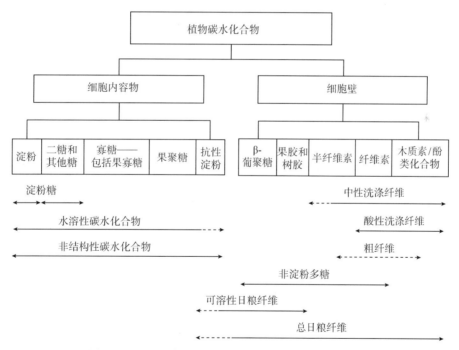

图 4-4　基于目前的分析方法对日粮碳水化合物的分类

洗涤纤维分析法属于化学-重量分析法，该方法将非淀粉多糖分为中性洗涤纤维（NDF）、酸性洗涤纤维（ADF）和木质素（Robertson and Horvath, 2001）。木质素和酸性洗涤纤维的含量之差为纤维素的含量，酸性洗涤纤维和中性洗涤纤维的含量之差为半纤维素的含量。尽管洗涤纤维分析法已被广泛应用，但是由于该方法不能回收可溶性的膳食纤维成分，如果胶、树胶和 β-葡聚糖，所以该方法不能总是准确地估计饲料原料中纤维性组分的含量（Grieshop et al., 2001）。由此可见，饲料原料中可溶性纤维的含量越高，洗涤纤维分析法测定的饲料原料中总纤维性成分的准确度就越低。

总膳食纤维（total dietary fiber, TDF）的分析可以克服洗涤纤维法中的一些局限性。该方法可以对饲料原料中所有的纤维组分进行定量分析，还将纤维分为可溶性膳食纤维和不可溶性膳食纤维两种（AOAC, 2007）。用该方法获得的结果比用洗涤纤维分析法更接近饲料原料中总膳食纤维的实际含量（Mertens, 2003）。但该方法存在的一个主要问题是其结果比洗涤纤维分析法结果重现性差，因此在营养学实验室并未得到广泛应用。

饲料原料中的非淀粉多糖也可以用酶-化学分析法定量分析，常用的方法主要有两种：Uppsala 分析法和 Englyst 分析法。Uppsala 分析法测定的非淀粉多糖为淀粉酶抗性多糖、糖醛酸和木质素含量的总和（AOAC, 2007）。利用 80%乙醇可将残渣分为可溶性和不可溶性成分，然后可以测定中性糖和糖醛的含量（Theander and Aman, 1979）。与 Uppsala 分析法不同的是，Englyst 分析法测定非淀粉多糖的最终结果中不含木质素和抗性淀粉含量（Englyst et al., 1996; Grieshop et al., 2001）。

参 考 文 献

AOAC (AOAC International). 2007. *Official Methods of Analysis of AOAC International*. 18th Ed., Rev. 2, W. Horwitz and G. W. Latimer, Jr., eds. Gaithersburg, MD: AOAC.

Bach Knudsen, K. E. 2001. The nutritional significance of "dietary fiber" analysis. *Animal Feed Science and Technology* 90:3-20.

Bach Knudsen, K. E. 2011. Triennial Growth Symposium: Effects of polymeric carbohydrates on growth and development in pigs. *Journal of Animal Science* 89:1965-1980.

BeMiller, J. 2007. *Carbohydrate Chemistry for Food Scientist,* 2nd Ed. St. Paul, MN: AACC International, Inc.

Bengala Freire, J., A. Aumaitre, and J. Peiniau. 1991. Effects of feeding raw and extruded peas on ileal digestibility, pancreatic enzymes and plasma glucose and insulin in early weaned pigs. *Journal of Animal Physiology and Animal Nutrition* 65:154-164.

Brown, I. L. 2004. Applications and uses of resistant starch. *Journal of AOAC International* 87:727-732.

Canibe, N., and K. E. Bach Knudsen. 1997. Digestibility of dried and toasted peas in pigs. 1. Ileal and total tract digestibilities of carbohydrates. *Animal Feed Science and Technology* 64:293-310.

Chrost, B., U. Kolukisaoglu, B. Schulz, and K. Krupinska. 2007. An alpha-galactosidase with an essential function during leaf development. *Planta* 225:311-320.

Cid, V. J., A. Durán, F. del Rey, M. P. Snyder, C. Nombela, and M. Sánchez. 1995. Molecular basis of cell integrity and morphogenesis in *Saccharomyces cerevisiae*. *Microbiological Reviews* 59:345-386.

Cummings, J. H., and A. M. Stephen. 2007. Carbohydrate terminology and classification. *European Journal of Clinical Nutrition* 61:S5-S18.

Cummings, J. H., M. B. Roberfroid, H. Andersson, C. Barth, A. Ferro-Luzzi, Y. Ghoos, M. Gibney, K. Hermonsen, W. P. T. James, O. Korver, D. Lairon, G. Pascal, and A. G. S. Voragen. 1997. A new look at dietary carbohydrate: Chemistry, physiology and health. *European Journal of Clinical Nutrition* 51:417-423.

De Vries, J. W. 2004. Dietary fiber: The influence of definition on analysis and regulation. *Journal of AOAC International* 87:682-706.

Englyst, K. N., and H. N. Englyst. 2005. Carbohydrate bioavailability. *British Journal of Nutrition* 94:1-11.

Englyst, K. N, and G. J. Hudson. 2000. Carbohydrates. Pp. 61-76 in *Human Nutrition and Dietetics*, 10th Ed., J. S. Garrow, W. P. T. James, and A. Ralph, eds. Edinburgh, UK: Churchill Livingston.

Englyst, K. N., S. M. Kingman, G. J. Hodsun, and J. H. Cummings. 1996. Measurement of resistant starch in vitro and in vivo. *British Journal of Nutrition* 75:749-755.

Englyst, K. N., S. Liu, and H. N. Englyst. 2007. Nutritional characterization and measurement of dietary carbohydrates. *European Journal of Clinical Nutrition* 61:S19-S39.

Fan, M. Z., B. Stoll, R. Jiang, and D. G. Burrin. 2001. Enterocyte digestive enzyme activity along the crypt-villus and longitudinal axes in the neonatal pig small intestine. *Journal of Animal Science* 79:371-381.

Fernandez, M. L. 2001. Pectin: Composition, chemistry, physicochemical properties, food applications, and physiological effects. Pp. 583-601 in *Handbook of Dietary Fiber*, S. S. Cho and M. L. Dreher, eds. New York: Marcel Dekker, Inc.

Franck, A. 2006. Inulin. Pp. 335-352 in *Food Polysaccharides and Their Applications,* 2nd Ed., A. M. Stephen, G. O. Phillips, and P. A. Williams, eds. Boca Raton, FL: CRC Press.

Goldring, J. M. 2004. Resistant starch: Safe intakes and legal status. *Journal of AOAC International* 87:733-739.

Grieshop, C. M., D. E. Reese, and G. C. Fahey, Jr. 2001. Non-starch polysaccharides and oligosaccharides in swine nutrition. Pp. 107-130 in *Swine Nutrition*, A. J. Lewis and L. L. Southern, eds. Boca Raton, FL: CRC Press.

Groff, J. L., and S. S. Gropper, eds. 2000. *Advanced Nutrition and Human Metabolism*, 3rd Ed. Belmont, CA: Wadsworth.

Houdijk, J. G. M., M. W. Bosch, S. Taminga, M. W. A. Verstegen, E. B. Berenpas, and H. Knoop. 1999. Apparent ileal and toral-tract nutrient digestion by pigs as affected by dietary nondigestible oligosaccharides. *Journal of Animal Science* 77:148-158.

IOM (Institute of Medicine). 2001. *Dietary Reference Intakes: Proposed Definition of Dietary Fiber*. Washington, DC: National Academy Press.

Jones, C. K., J. R. Bergstrom, M. D. Tokach, J. M. DeRouchey, R. D. Goodband, J. L. Nelssen, and S. S. Dritz. 2010. Efficacy of commercial enzymes in diets containing various concentrations and sources of dried distillers grains with solubles for nursery pigs. *Journal of Animal Science* 88:2084-2091.

Kim, S. W., D. A. Knabe, K. J. Hong, and R. A. Easter. 2003. Use of carbohydrases in corn-soybean meal-based nursery diets. *Journal of Animal Science* 81:2496-2504.

Kritchevsky, D. 1988. Dietary fiber. *Annual Review of Nutrition* 8:301-328.

Livesey, G. 1990. Energy values of unavailable carbohydrates and diets: An inquiry and analysis. *American Journal of Clinical Nutrition* 51:617-637.

Lunn, J., and J. L. Buttriss. 2007. Carbohydrates and dietary fibre. *Nutrition Bulletin* 32:21-64.

Martínez-Villaluenga, C., J. Frias, and C. Vidal-Valverde. 2008. Alpha-galactosides: Antinutritional factors or functional ingredients? *Critical Reviews in Food Science and Nutrition* 48:301-316.

McCleary, B., S. J. Charnock, P. C. Rossiter, M. F. O'Shea, A. M. Power, and R. M. Loyd. 2006. Measurement of carbohydrates in grain, feed and food. *Journal of the Science of Food and Agriculture* 86:1648-1661.

Mertens, D. R. 2003. Challenges in measuring insoluble dietary fiber. *Journal of Animal Science* 81:3233-3249.

Meyer, P. D. 2004. Nondigestible oligosaccharides as dietary fiber. *Journal of AOAC International* 87:718-726.

Mul, A. J., and F. G. Perry. 1994. The role of fructo-oligosaccharides in animal nutrition. Pp. 57-79 in *Recent Advances in Animal Nutrition*, P. C. Garnsworthy, J. H. Pemberton, and R. G. Cole, eds. Loughborough, UK: Nottingham University Press.

Osman, M. E., A. R. Menzles, B. A. Martin, P. A. Williams, G. O. Phillips, and T. C. Baldwin. 1995. Characterization of gum arabic fractions obtained by anion-exchange chromatography. *Phytochemistry* 38:409-417.

Roberfroid, M. B. 2005. Introducing inulin-type fructans. *British Journal of Nutrition* 93:S13-S25.

Robertson, J. B., and P. J. Horvath. 2001. Detergent analysis of foods. P. 63 in *CRC Handbook of Dietary Fiber in Human Nutrition*, 3rd Ed., G. A. Spiller, ed. Boca Raton, FL: CRC Press.

Smiricky, M. R., C. M. Grieshop, D. M. Albin, J. E. Wubben, V. M. Gabert, and G. C. Fahey, Jr. 2002. The influence of soy oligosaccharides on apparent and true ileal amino acid digestibilities and fecal consistency in growing pigs. *Journal of Animal Science* 80:2433-2441.

Southgate, D. A. T. 2001. Food components associated with dietary fiber. Pp. 19-21 in *CRC Handbook of Dietary Fiber in Human Nutrition*, 3rd Ed., G. A. Spiller, ed. Boca Raton, FL: CRC Press.

Southgate, D. A. T., and G. A. Spiller. 2001. Polysaccharide food additives that contribute to dietary fiber. Pp. 27-31 in *CRC Handbook of Dietary Fiber in Human Nutrition*, 3rd Ed., G. A. Spiller, ed. Boca Raton, FL: CRC Press.

Stein, H. H., and R. A. Bohlke. 2007. The effects of thermal treatment of field peas (*Pisum sativum L.*) on nutrient and energy digestibility by growing pigs. *Journal of Animal Science* 85:1424-1431.

Sun, T., H. N. Lærke, H. Jørgensen, and K. E. Bach Knudsen. 2006. The effect of extrusion cooking of different starch sources on the in vitro and in vivo digestibility in growing pigs. *Animal Feed Science and Technology* 131:66-85.

Svihus, B., A. K. Uhlen, and O.M. Harstad. 2005. Effect of starch granule structure, associated components and processing on nutritive value of cereal starch: A review. *Animal Feed Science and Technology* 122:303-320.

Swanson, K. S., C. M. Grieshop, E. A. Flickinger, L. L. Bauer, H. Healy, K. A. Dawson, N. R. Merchen, and G. C. Fahey, Jr. 2002. Supplemental fructooligosaccharides and mannanoligosaccharides influence immune function, ileal and total tract nutrient digestibilities, microbial populations and concentrations of protein catabolites in the large bowel of dogs. *Journal of Nutrition* 132:980-989.

Theander, O., and P. Aman. 1979. The chemistry, morphology, and analysis of dietary fiber components. Pp. 215-244 in *Dietary Fiber Chemistry and Nutrition*, G. E. Inglett and S. I. Falkehag, eds. New York: Academic Press.

Treem, W. R. 1995. Congenital sucrase-isomaltase deficiency. *Journal of Pediatric Gastroenterology and Nutrition* 21:1-14.

Van Beers, E. H., H. A. Büller, R. J. Grand, A. W. C. Einerhand, and J. Dekker. 1995. Intestinal brush border glycohydrolases: Structure, function, and development. *Critical Reviews in Biochemistry and Molecular Biology* 30:197-262.

Wenk, C. 2001. The role of dietary fibre in the digestive physiology of the pig. *Animal Feed Science and Technology* 90:21-33.

Williams, P. A., and G. O. Phillips. 2001. Gum arabic: Production, safety and physiological effects, physicochemical characterization, functional properties, and food applications. Pp. 375-396 in *Handbook of Dietary Fiber*, S. S. Cho and M. L. Dreher, eds. New York: Marcel Dekker, Inc.

Yen, J. T. 2011. Nutrients: Digestion and absorption. Pp. 834-836 in *Encyclopedia of Animal Science*, 2nd Ed., D. E. Ullrey, C. Kirk Baer, and W. G. Pond, eds. Boca Raton, FL: CRC Press.

Zentek, J., B. Marquart, and T. Pietrzak. 2002. Intestinal effects of mannanoligosaccharides, transgalactooligosaccharides, lactose and lactulose in dogs. *Journal of Nutrition* 132:1682S-1684S.

第 5 章 水

引言

水被广泛认为是一种重要的营养素，但在猪对水的需要量方面的研究却少得惊人。由于可用于猪生产的水资源有限（Deutsch et al., 2010），以及在不同地区的废弃物处理和利用方面的问题，所以将来很有必要开展猪对水的生理和代谢需要的相关研究。

水的功能

水维持着动物许多生命必需的生理功能（Roubicek, 1969）。水是动物通过细胞膨度维持机体形态的一种主要结构性物质，在机体温度调节过程中也起着重要的作用。水具有高比热的特点，这使得它成为散发各种代谢过程中产生的多余热量的理想物质。当 1 g 液态水变为气态水时，大约 580 cal 的热量被释放（Thulin and Brumm, 1991）。水在养分进入机体组织细胞的过程中以及代谢废物从组织细胞清除的过程中扮演着重要的角色。高的介电常数使得水能够溶解大量的物质，并能通过循环系统在全身转运这些物质。此外，水几乎在机体发生的每一个化学反应中均扮演着重要的角色。碳水化合物、脂肪和蛋白质的氧化都会生成水，这些物质代谢产生能量的过程需要经过一系列复杂的反应，最终除了能量以外，还产生二氧化碳和水。此外，水作为关节润滑液和脑脊髓液的主要成分，对机体关节的润滑作用和神经系统的缓冲保护作用至关重要。

猪体内水的含量随着年龄而改变。对于体重约 1.5 kg 的新生仔猪，水的比例占空腹体重（总体重减去胃肠道内容物的质量）高达 82%，随后含量逐渐下降，当长到约 110 kg 出栏体重时，水的比例下降至其空腹体重的 48%~53%（与出栏肥猪的瘦肉率有关）（Shields et al., 1983; de Lange et al., 2001）。含水量随着年龄变化，主要是因为猪的脂肪含量随着日龄的增长而增加，而脂肪组织的含水量比肌肉组织的要低得多（Georgievskii, 1982）。

水的周转

猪可以从三种途径摄取到水：①直接通过饮水摄取；②饲料原料中的水，典型风干饲料的含水量为 10%~12%；③来源于碳水化合物、脂肪和蛋白质分解的代谢水。1 kg 的脂肪、碳水化合物和蛋白质氧化分解可以分别产生 1190 g、560 g 和 450 g 的水（NRC, 1981）。根据 Yang 等（1984）的报道，每消耗 1 kg 风干饲料，将产生 0.38~0.48 kg 的代谢水。

水会通过 4 种途径排出：①肺脏（呼吸）；②皮肤（蒸发）；③肠道（排泄）；④肾脏（排尿）。正常的呼吸过程中，水分持续从呼吸道排出。吸入的空气经过呼吸道的加温和加湿后，呼出时其水分饱和度可以达到 90% 左右（Roubicek, 1969）。据估计，在中立温度区的环境中，20 kg 和 60 kg 体重的猪通过呼吸损失的水分别为 0.29 L/d 和 0.58 L/d（Holmes and Mount, 1967）。损失的程度受温度和相对湿度的影响；水的损失随着温度的升高而增加，随着相对湿度的增加而减少。

因为猪的汗腺并不发达，所以通过皮肤的出汗和蒸发损失的水，并不是猪机体水损失的主要途径。

据估计，在中立温度区内，通过皮肤损失的水分为 12~16 g/m² (Morrison et al., 1967)。随着环境温度从-5℃升高到30℃，水的损失由 7 g/m² 增加到 32 g/m² (Ingram, 1964)。但是，相对湿度提高对皮肤的水损失没有明显的影响 (Morrison et al., 1967)。

大量的水经由粪便损失。工厂化饲养母猪每天产生的粪便质量为其体重的 8%~9%，粪便中水的含量为 62%~79% (Brooks and Carpenter, 1993)。通过肠道损失的水量与日粮的特性有关。通常情况下，不可消化物质所占比例越高，损失的水越多 (Maynard et al., 1979)。水损失的量随着纤维摄入量的增加而增加 (Cooper and Tyler, 1959)，并随具有轻泻特性的饲料摄入而增加 (Sohn et al., 1992; Darroch et al., 2008)。腹泻时，通过粪便损失的水也会增加 (Thulin and Brumm, 1991)。

虽然经由尿液排泄的水量变化很大，但排尿是猪排泄水的主要途径。肾脏通过排泄水的多少来调节体液体积和组成，也取决于饮水量和其他排泄途径。一般认为，猪日粮中矿物质和蛋白质含量高时，水的排泄会增加。据 Wahlstrom 等（1970）报道，日粮中蛋白质浓度越高，水的损失越大，导致对水的需要量也会越高。同样，据 Sinclair（1939）报道，食盐摄入的增加会导致水的摄入增加，并伴随尿液排泄的增加。然而，Shaw 等（2006）在一个商业养殖企业进行的研究发现，日粮中蛋白质和矿物质浓度的较大变化并未导致饮水量的明显变化。由此他们认为，除了日粮中蛋白质和矿物质浓度，以及每天摄入蛋白质和矿物质的量以外的其他因素，如饮水设备的设计或猪的行为差异等，可能对耗水量有着更大的影响。因此，如果忽略其他重要的影响因素，仅仅通过日粮调控的策略来调节水的用量，其效果可能是有限的。

水的需要量

许多因素，包括日粮、生理状态和环境等均能影响猪对水的需要量 (NRC, 1981; Mroz et al., 1995)。由于猪机体的含水量在任何特定的年龄阶段都是相对恒定的，猪每天必须摄取足够的水用以补充水的流失，以达到机体水的平衡。因此，任何已知的可能增加水流失的因素将增加水的需要量。水的最小需要量包括需要平衡水的损失、生产乳汁，以及生长和妊娠过程中形成新组织细胞的需水量。

在确定水的需要量时，区分水的需要量（摄入量）和水的使用量是非常重要的 (Fraser et al., 1993)。因为浪费的水一般不予考虑，所以，猪真实的水需要量通常会被高估。采用氚标记水的方法测定水的周转率，在圈养和正常干喂的饲喂条件下，据估算，生长猪（30~40 kg 体重）和非泌乳成年猪（157 kg 体重）对水的需要量分别为约 120 ml/kg 体重和 80 ml/kg 体重 (Yang et al., 1981)。

但是，由于开展这种研究的难度很大，一般用水的消耗量来估计水的需要量。除了猪的代谢需要，还有其他许多因素能影响养猪生产中总的用水量。这些因素包括环境温度，因为环境温度会影响猪主动浪费水的量，如果采用滴水/喷雾系统降温，也会增加水的用量。猪场设备的选择和安装，以及饮水器的数量和水的流速等，是影响水的消耗量的管理及与硬件设备相关的因素。Brumm（2010）详细地综述了这类因素对水用量影响的信息。

哺乳仔猪

由于乳汁中含有约 80%的水，所以，人们通常认为哺乳仔猪通过摄入乳汁便能完全满足它们对水的需要而不需要饮水 (Pond and Houpt, 1978)。然而，事实上，哺乳仔猪会在出生后的 1~2 天开始饮水 (Aumaitre, 1964)。另外，因为乳汁是一种含有高浓度蛋白质和矿物质的食物，乳汁的摄入会增加尿液的排泄，从而可能导致事实上水的缺乏 (Lloyd et al., 1978)。

猪营养需要

Fraser等（1988）测定了51窝哺乳仔猪分娩后前4天的饮水量，结果发现在不同窝之间饮水量变化很大，每天的饮水量为0~200 ml/d，每天平均用水量为46 ml。该数据明显高于先前研究的结果，先前的研究表明平均饮水量接近10 ml/d。Fraser等（1993）推测，新近研究中饮水量较高的原因可能与越来越重视分娩舍温度控制有关，目前分娩舍温度有所提高，这可能会导致猪只的水分损失增加。他们的研究数据表明，与置于20℃分娩舍的哺乳仔猪相比，分娩舍温度为28℃的哺乳仔猪饮水量增加了几乎4倍。

Fraser等（1988）指出，为哺乳仔猪提供额外的饮水可能有助于减少断奶前死亡。这些数据显示，营养不良的仔猪，特别是当置于温暖环境中时，在分娩后头几天会更加容易脱水，其中也有一些发育比较成熟的仔猪会通过饮水来补偿机体水的损失。暴露的水表面（水碗或者水杯等）会比乳头式饮水器更加适合于新生的哺乳仔猪（Phillips and Fraser, 1990, 1991）。

出生1周后，哺乳仔猪饮水最主要的考虑因素是其在刺激仔猪采食教槽料中所起的作用。虽然在前3周，仔猪教槽料的采食量通常很低，但是，如果没有提供饮水，随后的采食量会更少（Friend and Cunningham, 1966）。仔猪的健康状况也是影响它饮水量的因素之一。腹泻的仔猪比健康的仔猪饮水量少约15%（Baranyiova and Holub, 1993）。

断奶仔猪

Gill等（1986）测定了3~6周龄断奶仔猪的饮水量。仔猪断奶后1周、2周和3周的日均饮水量分别为每头0.49 L、0.89 L和1.46 L。Brooks等（1984）采用如下方程式描述采食量与饮水量的关系：

$$\text{饮水量（L/d）} = 0.149 + (3.053 \times \text{每天干饲料的摄入量/kg}) \quad (式5\text{-}1)$$

McLeese等（1992）观测到了两种不同的饮水模式。在第一个阶段，持续到大约断奶后5天，饮水出现了与生理需求无关的波动，这似乎与生长、采食和腹泻的严重程度无关。在第二个阶段，饮水变得稳定，与生长和采食保持同步。作者推测，断奶后前几天，饮水量可能比较高，这样断奶仔猪在不进食情况下得到了饱腹感。Torrey等（2008）认为早期断奶的仔猪不能通过饮水得到饱腹感。他们也观察到，虽然饮水器的类型能够影响早期断奶仔猪的行为和浪费水的程度，但这并不影响总的采食量和生长性能。Brooks等（1984）进行的关于采食模式的研究中，作者观察到饮水量在一天内的变化，持续光照条件下，断奶仔猪上午8:30至下午17:00的饮水量明显高于下午17:00至上午8:30的饮水量。

Nienaber和Hahn（1984）研究了限制水流量对断奶仔猪生长性能的影响。他们的研究结果表明，当水流速在0.1~1.1 L/min变化时，对生长性能的影响非常小。然而，用水量却随着水流速的增加而急剧增加，这是因为水的浪费增加。同样，当乳头式饮水器向上翘起（45°）时，用水量也会比饮水器向下倾斜（45°）增加（Carlson and Peo, 1982）。栏养断奶仔猪，当乳头式饮水器向下倾斜时，与饮水器向上翘起相比较，日增重增加6.5%，饲料转化效率改善7%，用水量也会减少63%。而采用滴水式饮水器与非滴水式饮水器相比没有明显的优势（Ogunbameru et al., 1991）。

生长育肥猪

对于生长育肥猪，建议在料槽附近提供能自由使用的饮水设施，通常对于干饲系统是这样的。消化物或水从胃排空的速度（g/h）随着饮水量的增加而增加（Low et al., 1985）。这一机制调节胃肠道内容物的干物质含量，特别是采食后1 h。

饲料采食量、日粮的原料组成、环境温度和湿度、猪只的健康状况以及应激等因素均影响水的需要量。饮水量通常与采食量和体重呈正相关。对于体重为20~90 kg的生长育肥猪，对水的需要量最小约为 2 kg/kg 饲料。在自由采食和饮水的情况下，生长猪的饮水量约为 2.5 kg/kg 饲料；在限饲的情况下，猪的饮水量约为 3.7 kg/kg 饲料（Cumby，1986）。自由采食和限饲条件下，生长猪对水的消耗量不同可能是当猪采食饲料后仍然感觉饥饿时，会趋向于通过饮水得到饱腹感。

Braude 等（1957）给猪只自由采食和饮水，每头猪每天采食干饲料的量最大为 3 kg。10~22 周龄，料水比为 1:2.56；16~18 周龄，每头猪平均日饮水量和采食量分别为 7.0 kg 和 2.7 kg。

Olsson 和 Andersson（1985）用鼻子触发的饮水装置进行的研究表明，生长猪在采食过程中的饮水量具有很大的阶段性，在进食开始和结束时出现饮水的高峰。进食之间的饮水量峰值分别出现在上午进食后 2 h 和下午进食后 1 h。这些研究结果支持了 Yang 等（1984）的结论，即生长猪趋向于在限饲时增加总的饮水量，这可能是为了增加饱腹感。总之，这些研究结果表明，如果限制采食时，生长猪会在下午饮水以增加饱腹感。

Barber 等（1988）研究了水的流速和饮水器的数量对生长猪饮水量的影响。与低水流速（300 ml/min）相比较，高水流速（900 ml/min）时，生长猪的饮水量从 1.9 L/d 增加到 3.8 L/d。然而，水流速增加对猪的生长性能并没有明显影响。当将饮水器的数量从每栏（8 头猪）1 个增加到 2 个时，对猪的饮水量和生长性能没有影响。

Mount 等（1971）研究表明，不同的环境温度（7℃、9℃、12℃、20℃和 22℃）下生长猪的饮水量的差异比较小，尽管在同一温度下猪与猪之间的饮水量也有差异。其中，环境温度为 30℃和 33℃时，猪的饮水量增加了 25%~50%（取决于特定的比较）。Close 等（1971）研究了温度上升对猪行为反应的影响，发现环境温度在 30℃及以上时，猪会把尿液和粪便排泄到整个栏的全部区域，也会将水溅出水槽外面，这可能是猪要通过这种方式降低体表的温度。

水温也会影响饮水量，因为需要额外的能量对低于机体温度的水进行加热。澳大利亚的研究（Vajrabukka et al.，1981）表明，猪在 45~90 kg 阶段分别饲养在凉爽猪舍（温度保持 22℃恒定），或环境温度每 12h 在 35℃和 24℃之间变化的炎热猪舍中：凉爽的猪舍时，猪对温度为 11℃的冷水饮水量为 3.3 L/d，对温度为 30℃的温水饮水量约为 4.0 L/d；与此相对照，饲养在炎热猪舍中的猪对温度为 11℃的冷水饮水量为 10.5 L/d，对温度为 30℃的温水饮水量则仅为 6.6 L/d。

Hagsten 和 Perry（1976）的研究表明，当生长猪日粮中总食盐（NaCl）或食盐总当量低于 0.20% 时，与食盐浓度为 0.27% 和 0.48% 相比，其饮水量和日增重分别减少 20% 和 38%。

抗生素的使用也可能影响饮水量。一些学者认为抗生素的使用会增加饮水量，而其他学者则报道会减少饮水量。Brooks 和 Carpenter（1993）等认为，抗生素是增加还是减少饮水量，取决于抗生素通过控制腹泻减少水的损失量和机体通过肾脏清除抗生素及残留物对水的需求的增加量。

在湿料饲喂系统中，料水比在 1:1.5 到 1:3.0 时，对生长猪的生长性能和胴体品质没有影响（Barber et al.，1963；Holme and Robinson，1965）。然而，在湿料饲喂系统中，为了防止猪舍温度的突然改变或日粮组成的改变（如高盐、蛋白质浓度的变化），必须提供额外的淡水以确保猪有足够的饮水。

妊娠母猪

头胎妊娠母猪的饮水量会随着干物质采食量的增加而增加（Friend，1971）。后备母猪在发情期的采食量和饮水量都会减少（Friend，1973；Friend and Wolynetz，1981）。Bauer（1982）的研究表明，后备母猪配种前的饮水量为 11.5 L/d，妊娠后期的饮水量为 20 L/d。这与 Riley（1978）的研究结果很接近，在他们的研究中，妊娠和泌乳母猪的日饮水量分别为 13.5 L 和 25.1 L（Riley，1978），以及 10.0 L 和 17.7 L（Lightfoot and Armsby，1984）。母猪中膀胱炎、感染、碱性尿以及尿道炎症等泌尿系统疾病非常常见，这些疾病常常伴随饮水量的减少（Madec，1984）。限饲的妊娠母猪为了补偿饱腹感，可能会增加饮水量。妊娠母猪日粮的纤维水平的提高也可能导致所需水料比提高。

哺乳母猪

哺乳母猪需要大量的水，因为哺乳母猪不仅需要补偿每天分泌 8~16 kg 的乳汁所消耗的水，而且还需要通过尿液排泄大量的代谢终产物（例如，乳汁形成过程中氨基酸代谢生成尿素，因为乳汁的氨基酸成分与机体组织和饲料均不同）。哺乳母猪日饮水量为 12~40 L，日平均饮水量为 18 L（Lightfoot，1978）。相似地，Seynaeve 等（1996）研究表明，日饮水量最小可能会小于 11 L，最大可能大于 17 L，这与日粮中食盐的浓度有关。这些数据与其他报道的哺乳母猪日饮水量相似。Bauer（1982）报道的日饮水量为 20 L，Riley（1978）报道的日饮水量为 25.1 L，Lightfoot 和 Armsby（1984）报道的日饮水量为 17.7 L，Peng 等（2007）报道的日饮水量为 17.3 L。

Phillips 等（1990）研究表明，哺乳母猪产床采用乳头式饮水器饮水，水流速在高流速（2 L/min）和低流速（0.6 L/min）时，母猪的饮水量没有明显的差异。Peng 等（2007）报道乳头饮水器离地面的高度（300 mm 和 600 mm）对母猪的饮水模式没有影响。同样，Peng 等（2007）也报道，由于湿料/干料-水自动喂料模式可以让母猪自由选择进食的时间和采食量，以及可以选择在进食时是否让干料与水混合，所以，自动湿料/干料-水自动喂料模式可以提高哺乳母猪的采食量和窝增重，水的浪费也较料-水的人工饲喂模式要少。

Jeon 等（2006）研究表明，在猪舍温度持续高于 25℃ 条件下，哺乳母猪在热应激时，如果提供 10℃ 或者 15℃ 凉水而不是 22℃ 的水将可以缓解热应激。与给予 22℃ 水的母猪相比，给予 10℃ 或者 15℃ 凉水的母猪摄入更多的饲料（5.3 kg/d 比较 3.8 kg/d）和水（38.1 L/d 比较 31.2 L/d），也会有更低的直肠温度和呼吸频率，仔猪窝断奶体重和日增重也更大。

公猪

关于公猪的水需要量的研究非常少，但是一般建议自由饮水。Straub 等（1976）观察到体重为 70~110 kg 的后备公猪饮水量在环境温度为 25℃ 时最大可以达到 15 L/d，而环境温度为 15℃ 时最大约为 10 L/d。

水的品质

水中含有的无机物和有机物达到一定浓度时能对猪产生毒害作用（NRC，1974）。水可能包含各种微生物，其中包括细菌和病毒。细菌中以沙门氏菌属、钩端螺旋体属和大肠杆菌属最为常见（Fraser et al.，1993）。水中也可能含致病性原生动物以及肠道蠕虫的卵和囊虫（Fraser et al.，1993）。这些微生物的存在是否造成危害很大程度上取决于其种类和浓度。美国国家事务局（The Bureau of National Affairs）于 1973 年建议，家畜饮水中的细菌数不能超过 5000 个/100 ml。不过，这一建议只能作为一个指导，因为一些病原体低于该浓度时可能是有害的，而其他的无害微生物可能在远高于该浓度时也是无害的。地表水中微生物污染通常比深井水和承压水等地下水更为常见（MDH，2011；Skipton et al.，2008）。

水中总可溶性固形物（TDS），即水样中总的可溶性无机物含量指标。钙、镁和钠的碳酸盐、氯盐或硫酸盐形式是高 TDS 水中最常见的盐类（Thulin and Brumm，1991）。水中 TDS 含量超过 6000 ppm[①]

---

① 1 ppm=1×10^{-6}

时，尽管一般不会影响动物的健康和生长性能，但可能会造成暂时性的腹泻和日饮水量增加。Paterson 等（1979）研究表明，当母猪配种后 30 天至仔猪 28 天断奶期间内供给 TDS 含量为 5060 ppm 的饮水时，对母猪的繁殖性能没有显著影响。断奶仔猪饮水中添加高达 6000 ppm 的 TDS 时，对其日增重和饲料转化效率没有影响。然而，报道显示，当饮水中 TDS 升高时，猪只饮水量会增加，并伴随轻度腹泻和粪便松软的现象（Anderson and Stothers, 1978; Paterson et al., 1979）。

TDS 并不是一个评价动物饮水品质的确切指标。作为一个基本评价标准，水中 TDS 含量<1000 ppm 被认为是安全的，而 TDS 含量>7000 ppm 被认为对怀孕和哺乳母猪以及处于应激状态的猪存在健康风险，不适合猪饮用（NRC, 1974）。加拿大环境部长理事会（1987）推荐家畜饮用水的 TDS 含量不得超过 3000 ppm。因为有如此多的不同无机物可能导致 TDS 含量过高，所以有必要对可溶性盐类进行进一步的化学分析，以确定是否存在健康风险。TDS 值作为猪饮用水品质评价标准的指导见表 5-1。

表 5-1 基于水中总溶解固体对猪饮用水的评估

总可溶性固形物/ppm	水质评价	反应
<1000	安全	对猪无危害
1000~2999	满意	不适应的猪会出现轻度腹泻
3000~4999	满意	可能导致猪出现暂时性拒绝饮水
5000~6999	可接受	不适于种猪
>7000	不适合	不适于种猪和处于热应激的猪

资料来源：NRC（1974）

由于几乎所有水样的 pH 均在 6.5~8.5 的可接受范围内，所以水的 pH 与水质的直接相关性比较小（Fraser et al., 1993）。但是，水的 pH 变化会对水处理过程中的化学反应产生重要的影响。水的 pH 过高会影响氯化作用的效率，pH 过低可能会造成经饮水给药的某些抗菌物质发生沉淀，特别是对磺胺类药物风险较大（Russell, 1985），并可能导致胴体中磺胺类药物残留的隐患，这是因为沉淀在水管中的药物可能在停止加药后重新溶解进入饮水。

水的硬度是由多价金属阳离子造成的，主要是钙离子和镁离子。如果水中多价阳离子浓度<60 ppm，被认为是软水；水中多价阳离子浓度在 120 和 180 ppm 之间，为硬水；水中多价阳离子浓度>180 ppm，则为极硬水（Durfor and Becker, 1964）。即使是极硬水，也很少对猪造成危害（NRC, 1980），尽管极硬水会导致输水系统中水垢的积累。如果水垢积累到一定程度影响到水线的畅通，则会造成问题。在加拿大魁北克地区的一项调查中，极硬水能够为妊娠母猪提供高达每日需要量的 29% 的钙源（Filpot and Ouellet, 1988）。

在北美的许多地区，井水的水质问题主要是硫酸盐。在加拿大大草原地区进行的一项调查表明，25% 的井水中硫酸盐超量（>1000 ppm）（McLeese et al., 1991）。而在美国俄亥俄州的一项调查表明，水中硫酸盐的浓度范围为 6~1629 ppm，其含量与地理位置、水井深度和 TDS 含量相关（Veenhuizen, 1993）。猪肠道对硫酸盐的耐受性并不好，当水中硫酸盐浓度超过 7000 ppm 时，会导致猪只腹泻和生长性能降低（Anderson et al., 1994）。然而，水中含有低浓度（不高于 2650 ppm）的硫酸盐不会对猪的生长性能产生负面影响（Veenhuizen et al., 1992; Maenz et al., 1994; Patience et al., 2004）。经过一段时间的适应，猪只似乎会逐渐适应水中高水平的硫酸盐。这也可以解释为什么断奶仔猪对硫酸盐最为敏感，主要是因为仔猪在断奶前的饮水很少，所以随着断奶后饮水量的增加，导致对水中硫酸盐的不适应。另外，水的气味不是水质差的指标。虽然当水中含 1900 ppm 的硫酸盐时，会出现明显的"臭鸡蛋"气味，但对猪的生长性能没有影响（DeWit et al., 1987）。

亚硝酸盐通过将血红蛋白氧化成高铁血红蛋白的方式减低血液的携氧能力。土地上氮肥的过度施洒和动物排泄物对径流水的污染会增加水体中的氮含量。Winks 等（1950）研究表明，水体中硝酸盐转变为亚硝酸盐是产生毒害作用的必要条件，猪饮用含有 290~490 ppm 硝态氮的井水会出现死亡。Seerley 等（1965）分析认为，除非硝酸盐的初始浓度超过 300 ppm 硝态氮，否则水中不可能形成足够的亚硝酸盐并对饮用的猪造成损害。亚硝态氮的浓度超过 10 ppm 时需要引起关注（美国水质监控专家组，1987）。水中含有硝酸盐和亚硝酸盐还可能影响维生素 A 的利用（Wood et al., 1967）。其他的离子偶尔也会在水样中检测到。家畜饮用水质量安全标准指南见表 5-2，关于其他离子的具体信息见 NRC（2005）的报告。

表 5-2 家畜饮用水质量标准指南

指标	推荐的最大值/ppm	
	TFWQG[a]	NRC[b]
总可溶性固形物	3000	
常量离子		
钙	1000	—
硝态氮和亚硝态氮	100	100
亚硝态氮	10	10
硫酸盐	1000	—
重金属离子和微量元素离子		
铝	5.0	—
砷	0.5	0.2
铍	0.1	—
硼	5.0	—
镉	0.02	0.05
铬	1.0	1.0
钴	1.0	1.0
铜	5.0	0.5
氟化物	2.0	2.0
铅	0.1	0.1
汞	0.003	0.01
钼	0.5	—
镍	1.0	1.0
硒	0.05	—
铀	0.2	—
钒	0.1	0.1
锌	50.0	25.0

a 美国水质监控专家组（1987）；b NRC（1974）。

在水质差的地区，非常有必要评估它对动物生长性能的影响。生产者经常会过度担忧那些猪只生长性能没有受到损害而出现腹泻的情况。渗透压升高（如摄入的硫酸盐和其他一些矿物质增加）导致粪便含水量的增加（即"腹泻"），绝对与微生物感染和疾病导致的腹泻不同。但当水质较差导致生产性能降低时，有许多途径可用以有效地缓解这些问题（将在以下三段中叙述）。

氯化作用会杀灭和清除致病微生物。与细菌相比，原生动物和肠病毒对氯化作用的抵抗力强得多（Fraser et al., 1993）。水中亚硝酸盐、铁离子、硫化氢、氨以及有机物的含量会影响消毒的有效性和水中需要添加的氯的量。水中存在的有机物会将游离氯转变为氯胺，氯胺的消毒作用会减弱。次氯酸钠和漂白剂（5.25%的氯溶液）通常被用于氯化消毒。水中的 pH 越高，则达到相同的消毒效果所需要的氯越多。

为了应对水质的问题，日粮也需要作出一些改变。通常情况下，水中矿物质含量（TDS）较高的农场需要降低日粮中盐的含量。由于大部分日粮具有一定水平的安全阈量，日粮中的一部分盐即使不加也不会造成影响。但必须注意的是，要确保日粮中含有足够量的氯，因为品质较差的水中一般不会含有高浓度的氯。

硬水的水质可用通过软化器得到改善。最常用的软化器是一个离子交换装置，可以用钠离子替换水中的钙和镁。由于这种方法处理后钠离子浓度会升高，所以，水的硬度虽然降低了，但对水的总矿物质含量（TDS）没有影响。反渗透装置在一定程度上可以去除水中的硫酸盐和硝酸盐。此外，除了要考虑水处理系统的有效性外，水处理系统的建设和运行成本也是决定在大多数养殖场能否使用水处理系统的重要因素。

参 考 文 献

Anderson, D. M., and S. C. Stothers. 1978. Effects of saline water high in sulfates, chlorides and nitrates on the performance of young weanling pigs. *Journal of Animal Science* 47:900-907.

Anderson, J. S., D. M. Anderson, and J. M. Murphy. 1994. The effect of water quality on nutrient availability for grower/finisher pigs. *Canadian Journal of Animal Science* 74:141-148.

Aumaitre, A. 1964. Le besoin en eau du porcelet: Étude de la consommation d'eau avant le sevrage (Water requirements of suckling piglets). *Annales de Zootechnie* 13:183-198.

Baranyiova, E., and A. Holub. 1993. Effect of diarrhoea on water consumption of piglets weaned on the first day after birth. *Acta Veterinaria Brno* 62:27-32.

Barber, J., P. H. Brooks, and J. L. Carpenter. 1988. The effect of water delivery rate and drinker number on the water use of growing pigs. *Animal Production* 46:521 (Abstr.).

Barber, R. S., R. Braude, and K. G. Mitchell. 1963. Further studies on the water requirements of the growing pig. *Animal Production* 5:277-282.

Bauer, W. 1982. Der Tränkwasserverbrauch güster, hochtragender und laktierender Jungsauen (Consumption of drinking water by nonpregnant, pregnant and lactating gilts). *Archiv Fur Experimentelle Veterinarmedizin* 36:823-827.

Braude, R., P. M. Clarke, K. G. Mitchell, A. S. Cray, A. Franke, and P. H. Sedgwick. 1957. Unrestricted whey for fattening pigs. *Journal of Agricultural Science (Cambridge)* 49:347-356.

Brooks, P. H., and J. L. Carpenter. 1993. The water requirement of growing/finishing pigs: Theoretical and practical considerations. Pp. 179-200 in *Recent Developments in Pig Nutrition 2*, D. J. Cole, W. Haresign, and P. C. Garnsworthy, eds. Loughborough, UK: Nottingham University Press.

Brooks, P. H., S. J. Russell, and J. L. Carpenter. 1984. Water intake of weaned piglets from three to seven weeks old. *Veterinary Record* 115:513-515.

Brumm, M. 2010. Water recommendations and systems for swine. Pp. 58-64 in *National Swine Nutrition Guide*, D. J. Meisinger, ed. Ames, IA: U.S. Pork Center of Excellence.

Bureau of National Affairs. 1973. EPA drafts water quality criteria as required under federal order law. *Environment Reporter* 4:663-670.

Canadian Council of Ministers of the Environment. 1987. *Canadian Water Quality Guidelines*. Ottawa: Environment Canada, Water Quality Branch, Inland Water Directorates.

Carlson, R. L., and E. R. Peo. 1982. Nipple waterer position: Up or down? Pp. 8-9 in *Nebraska Swine Report*, Lincoln, NE: University of Nebraska.

Close, W. H., L. E. Mount, and I. B. Start. 1971. The influence of environmental temperature and plane of nutrition on heat losses from groups of growing pigs. *Animal Production* 13:285-302.

Cooper, P. H., and C. Tyler. 1959. Some effects of bran and cellulose on the water relationships in the digesta and faeces of pigs. Part 1. The effect of including bran and two forms of cellulose in otherwise normal rations. *Journal of Agricultural Science (Cambridge)* 52:332-347.

Cumby, T. R. 1986. Design requirements of liquid feeding systems for pigs: A review. *Journal of Agricultural Engineering and Resources* 34:153-172.

Darroch, C. S., C. R. Dove, C. V. Maxwell, Z. B. Johnson, and L. L. Southern. 2008 A regional evaluation of the effect of fiber type in gestation diets on sow reproductive performance. *Journal of Animal Science* 86:1573-1578.

de Lange, C. F. M., S. H. Birkett, and P. C. H. Morel. 2001. Protein, fat, and bone tissue growth in swine. Pp. 65-81 in *Swine Nutrition*, 2nd Ed., A. J. Lewis and L.L. Southern, eds. Boca Raton, FL: CRC Press.

Deutsch, L., M. Falkenmark, L. Gordon, J. Rockström, and C. Folke. 2010. Water-mediated ecological consequences of intensification and expansion of livestock production. Pp. 97-110 in *Livestock in a Changing Landscape, Volume 1: Drivers, Consequences, and Responses*, H. Steinfeld, H. A. Mooney, F. Schneider, and L. E. Neville, eds. Washington DC: Island Press.

DeWit, P., L. G. Young, R. Wenzell, R. Friendship, and D. Peer. 1987. Water quality and pig performance. *Canadian Journal of Animal Science* 67:1196 (Abstr.).

Durfor, C. M., and E. Becker. 1964. *USGS Water-Supply Paper 1812*. Washington, DC: U.S. Government Printing Office.

Filpot, P. M., and G. Ouellet. 1988. Mineral and nitrate content of swine drinking water in four Quebec regions. *Canadian Journal of Animal Science* 68:997-1000.

Fraser, D., P. A. Phillips, B. K. Thompson, and W. B. Peeters Weem. 1988. Use of water by piglets in the first days after birth. *Canadian Journal of Animal Science* 68:603-610.

Fraser, D., J. F. Patience, P. A. Phillips, and J. M. McLeese. 1993. Water for piglets and lactating sows: Quantity, quality and quandaries. Pp. 200-224 in *Recent Developments in Pig Nutrition 2*, D. J. Cole, W. Haresign, and P. C. Garnsworthy, eds. Loughborough, UK: Nottingham University Press.

Friend, D. W. 1971. Self-selection of feeds and water by swine during pregnancy and lactation. *Journal of Animal Science* 32:658-666.

Friend, D. W. 1973. Self-selection of feeds and water by unbred gilts. *Journal of Animal Science* 37:1137-1141.

Friend, D. W., and H. M. Cunningham. 1966. The effect of water consumption on the growth, feed intake, and carcass composition of suckling piglets. *Canadian Journal of Animal Science* 46:203-209.

Friend, D. W., and M. S. Wolynetz. 1981. Self-selection of salt by gilts during pregnancy and lactation. *Canadian Journal of Animal Science* 61:429-438.

Georgievskii, V. I. 1982. Water metabolism and the animal's water requirements. Pp. 79-89 in *Mineral Nutrition of Animals*, V. I. Georgievskii, B. N. Annenkov, and V. I. Samokhin, eds. London: Butterworths.

Gill, B. P., P. H. Brooks, and J. L. Carpenter. 1986. The water intake of weaned pigs from 3 to 6 weeks of age. *Animal Production* 42:470 (Abstr.).

Hagsten, I., and T. W. Perry. 1976. Evaluation of dietary salt levels for swine. 1. Effect on gain, water consumption and efficiency of feed conversion. *Journal of Animal Science* 42:1187-1190.

Holme, D. W., and K. L. Robinson. 1965. A study of water allowances for the bacon pig. *Animal Production* 7:377-384.

Holmes, C. W., and L. E. Mount. 1967. Heat loss from groups of growing pigs under various conditions of environmental temperature and air movement. *Animal Production* 9:435-452.

Ingram, D. L. 1964. The effect of environmental temperature on heat loss and thermal insulation in the young pig. *Research in Veterinary Science* 5:357-364.

Jeon, J. H., S. C. Yeon, Y. H. Choi, W. Min, S. Kim, P. J. Kim, and H. H. Chang. 2006. Effects of chilled drinking water on the performance of lactating sows and their litters during high ambient temperatures under farm conditions. *Livestock Science* 105:86-93.

Lightfoot, A. L. 1978. Water consumption of lactating sows. *Animal Production* 26:386 (Abstr.).

Lightfoot, A. L., and A. W. Armsby. 1984. Water consumption and slurry production of dry and lactating sows. *Animal Production* 38:541 (Abstr.).

Lloyd, L. E., B. E. McDonald, and E. W. Crampton. 1978. Water and its metabolism. Pp. 22-34 in *Fundamentals of Nutrition*, 2nd Ed. San Francisco: W. H. Freeman and Co.

Low, A. G., R. T. Pittman, and R. J. Elliott. 1985. Gastric emptying of barley-soya-bean diets in the pig: Effects of feeding level, supplementary maize oil, sucrose or cellulose, and water intake. *British Journal of Nutrition* 54:437-447.

Madec, F. 1984. Urinary disorders in intensive pig herds. *Pig News Info* 5:89-93.

Maenz, D. D., J. F. Patience, and M. S. Wolynetz. 1994. The influence of the mineral level in drinking water and thermal environment on the performance and intestinal fluid flux of newly-weaned pigs. *Journal of Animal Science* 72:300-308.

Maynard, L. A., J. K. Loosli, H. F. Hintz, and R. G. Warner. 1979. *Animal Nutrition*, 7th Ed. New York: McGraw-Hill.

McLeese, J. M., J. F. Patience, M. S. Wolynetz, and G. I. Christison. 1991. Evaluation of the quality of ground water supplies used on Saskatchewan swine farms. *Canadian Journal of Animal Science* 71:191-203.

McLeese, J. M., M. L. Tremblay, J. F. Patience, and G. I. Christison. 1992. Water intake patterns in the weanling pig: Effect of water quality, antibiotics and probiotics. *Animal Production* 54:135-142.

MDH (Minnesota Department of Health). 2011. Well Management: Bacterial Safety of Well Water. Available online at http://www.health.state.mn.us/divs/eh/wells/waterquality/bacteria.html. Accessed on November 15, 2011.

Morrison, S. R., T. E. Bond, and H. Heitman. 1967. Skin and lung moisture loss from swine. *Transactions of the American Society of Agricultural Engineers* 10:691-697.

Mount, L. E., C. W. Holmes, W. H. Close, S. R. Morrison, and I. B. Start. 1971. A note on the consumption of water by the growing pig at several environmental temperatures and levels of feeding. *Animal Production* 13:561-563.

Mroz, Z., A. W. Jongbloed, N. P. Lenis, and K. Vreman. 1995. Water in pig nutrition: Physiology, allowances and environmental implications. *Nutrition Research Reviews* 8:137-164.

Nienaber, J. A., and G. L. Hahn. 1984. Effects of water flow restriction and environmental factors on performance of nursery-age pigs. *Journal of Animal Science* 59:1423-1429.

NRC (National Research Council). 1974. *Nutrient and Toxic Substances in Water for Livestock and Poultry*. Washington, DC: National Academy Press.

NRC. 1980. *Mineral Tolerance of Domestic Animals*. Washington, DC: National Academy Press.

NRC. 1981. Water-environment interactions. Pp. 39-50 in *Effect of Environment on Nutrient Requirements of Domestic Animals*. Washington, DC: National Academy Press.

NRC. 2005. *Mineral Tolerance of Animals*, 2nd Rev. Ed. Washington, DC: National Academies Press.

Ogunbameru, B. O., E. T. Kornegay, and C. M. Wood. 1991. A comparison of drip and non-drip nipple waterers used by weanling pigs. *Canadian Journal of Animal Science* 71:581-583.

Olsson, O., and T. Andersson. 1985. Biometric considerations when designing value drinking systems for growing-finishing pigs. *Acta Agriculturae Scandinavica* 35:55-66.

Paterson, D. W., R. C. Wahlstrom, G. W. Libal, and O. E. Olson. 1979. Effects of sulfate in water on swine reproduction and young pig performance. *Journal of Animal Science* 49:664-667.

Patience, J. F., A. D. Beaulieu, and D. A. Gillis. 2004. The impact of ground water high in sulfates on the growth performance, nutrient utilization, and tissue mineral levels of pigs housed under commercial conditions. *Journal of Swine Health and Production* 12:228-236.

Peng, J. J., S. A. Somes, and D. W. Rozeboom. 2007. Effect of system of feeding and watering on performance of lactating sows. *Journal of Animal Science* 85:853-860.

Phillips, P. A., and D. Fraser. 1990. Water bowl size for newborn pigs. *Applied Engineering in Agriculture* 6:79-81.

Phillips, P. A., and D. Fraser. 1991. Discovery of selected water dispensers by newborn pigs. *Canadian Journal of Animal Science* 71:233-236.

Phillips, P. A., D. Fraser, and B. K. Thompson. 1990. The influence of water nipple flow rate and position and room temperature on sow water intake and spillage. *Applied Engineering in Agriculture* 6:75-78.

Pond, W. G., and K. A. Houpt. 1978. Lactation and the mammary gland. Pp. 181-191 in *The Biology of the Pig*. Ithaca, NY: Cornell University Press.

Riley, J. E. 1978. Drinking "straws": A method of watering housed sows during pregnancy and lactation. *Animal Production* 26:386 (Abstr.).

Roubicek, D. 1969. Water metabolism. Pp. 353-373 in *Animal Growth and Nutrition*, H. Hafez and I. Dyer, eds. Philadelphia: Lea and Febiger.

Russell, I. D. 1985. Some fundamentals of water medications. *Poultry Digest* 44:422-423.

Seerley, R. W., Jr., R. J. Emerick, L. B. Emery, and O. E. Olson. 1965. Effect of nitrate or nitrite administered continuously in drinking water for swine and sheep. *Journal of Animal Science* 24:1014-1019.

Seynaeve, M., R. De Wilde, G. Janssens, and B. De Smet. 1996. The influence of dietary salt level on water consumption, farrowing, and reproductive performance of lactating sows. *Journal of Animal Science* 74:1047-1055.

Shaw, M. I., A. D. Beaulieu, and J. F. Patience. 2006. Effect of diet composition on water consumption in growing pigs. *Journal of Animal Science* 84:3123-3132.

Shields, R. G., Jr., D. C. Mahan, and P. L. Graham. 1983. Changes in swine body composition from birth to 145 kg. *Journal of Animal Science* 57:43-54.

Sinclair, R. D. 1939. The salt requirements of growing pigs. *Scientia Agricola* 20:109-119.

Skipton, S. O., B. I. Dvorak, W. Woldt, and S. Wirth. 2008. *Drinking water: Bacteria*. University of Nebraska–Lincoln: Extension Publication G1826.

Sohn, K. S., T. M. Fakler, and C. V. Maxwell. 1992. Effect of psyllium fed during late gestation and lactation period on reproductive performance and stool consistency. *Journal of Animal Science* 70(Suppl. 1):19 (Abstr.).

Straub, G., J. H. Weniger, E. S. Tawfik, and D. Steinhauf. 1976. The effects of high environmental temperatures on fattening performance and growth of boars. *Livestock Production Science* 3:65-74.

Task Force on Water Quality Guidelines. 1987. Livestock watering. Pp. 4-23–4-37 in *Canadian Water Quality Guidelines*. Ottawa, Ontario: Inland Waters Directorate.

Thulin, A. J., and M. C. Brumm. 1991. Water: The forgotten nutrient. Pp. 315-324 in *Swine Nutrition*, E. R. Miller, D. E. Ullrey, and A. J. Lewis, eds. Stoneham, MA: Butterworth-Heinemann.

Torrey, S., E. L. M. T. Tamminga, and T. M. Widowski. 2008. Effect of drinker type on water intake and waste in newly weaned piglets. *Journal of Animal Science* 86:1438-1445.

Vajrabukka, C., C. J. Thwaites, and D. J. Farrell. 1981. Overcoming the effects of high temperature on pig growth. Pp. 99-114 in *Recent Advances in Animal Nutrition in Australia*, D. J. Farrell and P. Vohra, eds. Armidale, Australia: University of New England Publishing Unit.

Veenhuizen, M. F. 1993. Association between water sulfate and diarrhea in swine on Ohio farms. *Journal of the American Veterinary Medical Association* 202:1255-1260.

Veenhuizen, M. F., G. C. Shurson, and E. M. Kohler. 1992. Effect of concentration and source of sulfate on nursery pig performance and health. *Journal of the American Veterinary Medical Association* 201:1203-1208.

Wahlstrom, R. C., A. R. Taylor, and R. W. Seerley. 1970. Effects of lysine in the drinking water of growing swine. *Journal of Animal Science* 30:368-373.

Winks, W. R., A. K. Sutherland, and R. M. Salisbury. 1950. Nitrite poisoning of pigs. *Queensland Journal of Agricultural Science* 7:1-14.

Wood, R. D., C. H. Chaney, D. G. Waddill, and G. W. Garrison. 1967. Effect of adding nitrate or nitrite to drinking water on the utilization of carotene by growing swine. *Journal of Animal Science* 26:510-513.

Yang, T. S., B. Howard, and W. V. McFarlane. 1981. Effects of food on drinking behaviour of growing pigs. *Applied Animal Ethology* 7:259-270.

Yang, T. S., M. A. Price, and F. X. Aherne. 1984. The effect of level of feeding on water turnover in growing pigs. *Applied Animal Behaviour Science* 12:103-109.

第6章 矿 物 质

引言

猪的生长需要多种无机元素，包括钙（Ca）、氯（Cl）、铬（Cr）、铜（Cu）、碘（I）、铁（Fe）、镁（Mg）、锰（Mn）、磷（P）、钾（K）、硒（Se）、钠（Na）、硫（S）和锌（Zn）。胃肠道内维生素 B_{12} 的合成也需要钴（Co）的参与，但吸收后可能就不需要了。另外还有许多微量元素（如砷[As]、硼[B]、溴[Br]、钼[Mo]、镍[Ni]、硅[Si]、锡[Sn]和钒[V]）可能是猪需要的，研究显示这些元素在一种或多种动物体内具有某种生理功效（Underwood, 1977; Nielsen, 1984）。然而，即使这些元素都是动物生长所必需，但由于需要量如此之低，以致目前也未能确定日粮需要量。饲料和体组织中矿物元素的测定一般是先用酸消解，然后通过原子吸收分光光度法或电感耦合等离子体光谱法测定，这些检测方法一般都不难，但许多元素在测定过程中必须很小心，因为环境中到处存在污染因素，如果不注意就会在采集、处理和操作样本的过程中造成污染；阴离子的测定需要专业的实验室技术。

这些无机元素的生理功能尤为多种多样。它们或是某些组织的结构成分，或参与其他组织多种多样的理化调节，作为许多酶的组成要素（辅酶）或者辅基，可以提高动物对蛋白质和能量的利用效率。因此，尽管在数量和成本上，这些元素可能是日粮中很小的一部分，但它们在猪生产中，对猪的健康生长、生物学功能的发挥和经济效益都有重要影响。动物生长周期内不同生长阶段对每种元素最低需要推荐量参见第 16 章的表格。原料中矿物元素的生物学利用率影响猪机体对矿物元素的需要量。Ammerman 等（1995）编著的《动物对养分的生物利用率》（*Bioavailability of Nutrients for Animals*）一书中对矿物质的生物利用率进行了详述。

锑（Sb）、砷（As）、镉（Cd）、氟（F）、铅（Pb）、汞（Hg）等几种矿物质元素可能对猪有毒性（Carson, 1986）。对于必需矿物元素与其他矿物元素的毒性和日粮最大耐受量，《动物对矿物元素的耐受量》（*Mineral Tolerance of Animals*）（NRC, 2005）一书对此有详细说明。

常量元素

钙和磷

钙（Ca）和磷（P）在骨骼系统的生长发育和维持，以及许多其他生理功能的发挥中起主要作用（Hays, 1976; Peo, 1976, 1991; Kornegay, 1985; Crenshaw, 2001）。本版中 Ca/P 需要量的确定不是直接评估的试验结果，而是通过营养需要模型得到的。将源于模型的钙和磷需求量与文献中实验测得结果进行对比，以此来评估两者是否一致。首先评定机体各生长阶段的全消化道可消化磷的标准（STTD P），然后根据动物各生长阶段对应的 Ca/STTD P 比值估算钙的需要量。在确定猪群动态生长过程中的需要量时，评定法的改善及 STTD P 的运用会使结果更加精确，将粪便中磷的排泄水平降至最低。第 16 章中表 16-1A 至表 16-4B、表 16-6A 至 16-7B、表 16-9 分别列出了 5~135 kg 阶段生长猪、妊娠期和哺乳期母猪、公猪生长速度最高、饲料报酬最好时对钙磷的需要量。随后列出的文献资料对该模型的方法原理进行了简述；钙磷需要量模型的详细描述参见第 8 章。

Peo（1991）指出，确定不同类型猪适宜的钙磷需要量基于以下几点：① 日粮中每种元素以可利用形式的适宜供给量；② 日粮中适宜的、可利用的钙磷比例；③ 足量的维生素 D。Ca/P 比值过高会降低磷的吸收率，从而减缓生长速度，影响骨骼的钙化，尤其当日粮中磷处于临界水平时（Vipperman et al., 1974; Doige et al., 1975; van Kempen et al., 1976; Reinhart and Mahan, 1986; Hall et al., 1991; De Wilde and Jourquin, 1992; Eeckhout et al., 1995）。如果日粮中含有过量的磷，那么该比值就不那么关键（Prince et al., 1984; Hall et al., 1991）。对于谷物-豆粕型日粮，总钙/总磷的推荐比例在 1:1 至 1.25:1 之间，较窄的 Ca/P 比值可促进磷的有效利用。足量的维生素 D 有利于钙磷的正常代谢，而超量维生素 D 会动员骨骼中钙磷的过量释放（Hancock et al., 1986; Jongbloed, 1987）。最近研究（Lauridsen et al., 2010）表明，母猪对维生素 D 的需要量被低估了。因此，本版修订了维生素 D 的需要量，它也会影响骨骼的发育，而以前可能一直被认为是由于日粮中钙和（或）磷不足所造成的。

关于断奶仔猪钙磷的需要量（Rutledge et al., 1961; Combs and Wallace, 1962; Combs et al., 1962, 1966; Miller et al., 1962, 1964a, b, 1965a, b, c; Menehan et al., 1963; Zimmerman et al., 1963; Blair and Benzie, 1964; Mudd et al., 1969; Coalson et al., 1972, 1974; Mahan et al., 1980; Mahan, 1982）及生长育肥猪的钙磷需要量（Chapman et al., 1962; Libal et al., 1969; Cromwell et al., 1970, 1972; Stockland and Blaylock, 1973; Doige et al., 1975; Pond et al., 1975, 1978; Fammatre et al., 1977; Kornegay and Thomas, 1981; Thomas and Kornegay, 1981; Maxson and Mahan, 1983; Combs et al., 1991a, b; Ekpe et al., 2002; Ruan et al., 2007; Hu et al., 2010; Partanen et al., 2010; Saraiva et al., 2011）已进行了大量的研究。尽管有关评价生长猪钙磷营养的研究报道很多，但其中仅有少数能用来确定磷需要量的资料，这些研究数据包括 3 个或 3 个以上磷水平的日粮，并在平均日增重（ADG）与磷水平呈曲线相关的条件下所获得的需要量估测结果；根据这些数据，获得评定试验的配合日粮，再用配合日粮中各饲料营养成分（本版中所示）的全消化道表观消化率（ATTD）和 STTD 值确定 ATTD P 和 STTD P 含量，具体试验操作参照第 2 章中氨基酸测定的方法。表 6-1 显示了各种基于平均体重的研究数据，并给出 ADG、ADFI、日粮 ME（kcal/kg）、与增重相对应的 ATTD P 和 STTD P 值。图 6-1 显示了 ATTD P 和 STTD P 需求曲线，其中每千克增重分别需要 5.7g ATTD P 和 6.7g STTD P。

表 6-1 体重对生长育肥猪磷需要量试验估测值的影响

文献	体重/kg			生产性能		日粮	ATTD		STTD	
	均值	始重	末重	ADG	ADFI	ME	%	g/kg 增重	%	g/kg 增重
Coalson 等（1972）	11.4	2.9	19.8	410	683	3555	0.334	5.56	0.372	6.20
Mahan 等（1980）	13.5	7.0	20.0	350	680	3312	0.285	5.55	0.335	6.51
Ruan 等（2007）	30.4	21.4	39.3	668	1640	3274	0.292	7.18	0.356	8.75
Maxson 和 Mahan（1983）	37.5	18.3	56.7	620	1690	3345	0.223	6.07	0.263	7.18
Ekpe 等（2002）	42.4	23.7	61.1	895	1916	3216	0.238	5.09	0.277	5.94
Partanen 等（2010）	45.0	25.0	65.0	864	1814	2868	0.256	5.38	0.294	6.18
Hastad 等（2004）	45.9	33.8	57.9	861	1514	3319	0.249	4.37	0.289	5.09
Cromwell 等（1970）	55.2	18.1	92.2	783	2470	3324	0.185	5.82	0.221	6.98
Bayley 等（1975a）	57.5	25.0	90.0	823	2410	3324	0.185	5.41	0.223	6.52
Thomas 和 Kornegay（1981）	64.0	25.0	103.0	800	2510	3291	0.196	6.13	0.231	7.25
Thomas 和 Kornegay（1981）	66.0	25.0	107.0	810	2520	3291	0.196	6.08	0.231	7.19
Hastad 等（2004）	98.9	88.5	109.3	742	2143	3314	0.206	5.96	0.240	6.93

能使动物满足最快生长速率的钙、磷水平不一定能满足骨骼的最大矿化。要使骨骼强度和骨灰质含量达到最大化，所需要的钙、磷量至少要比满足最快生长和增重所需量高 0.1 个百分点（Cromwell et al., 1970, 1972; Mahan et al., 1980; Crenshaw et al., 1981; Kornegay and Thomas, 1981; Mahan, 1982; Maxson and Mahan, 1983; Koch et al., 1984; Combs et al., 1991a, b）。然而，给生长猪饲喂大量的钙、磷，

即使能使骨骼强度最大化，却不一定能促进其结构正常（Pointillart and Gueguen, 1978; Kornegay and Thomas, 1981; Kornegay et al., 1981a, b, 1983; Calabotta et al., 1982; Brennan and Aherne, 1984; Lepine et al., 1985; Eeckhout et al., 1995）。

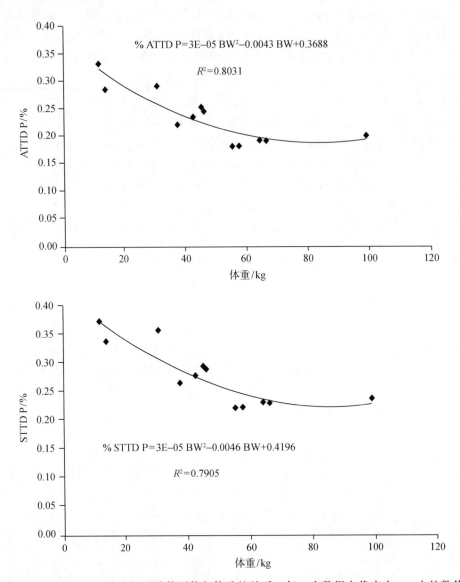

图 6-1　ATTD、STTD P 的试验估测值与体重的关系，每一个数据点代表表 6-1 中的数值

以百分含量来表示日粮钙、磷需要量时，青年母猪的需要量会稍微比阉公猪高些（Thomas and Kornegay, 1981; Calabotta et al., 1982）。生长期的公猪对钙、磷的需要量比青年母猪和阉公猪都高（Hickman et al., 1983; Kesel et al., 1983; Hansen et al., 1987）。猪日粮中添加生长激素会增加瘦肉率，降低日采食量，并由此使以百分含量计的日粮营养需要增加（Weeden et al., 1993a, b; Carter and Cromwell, 1998a, b）。也有充分的证据显示，与未添加的猪相比，添加生长激素会增加猪每日的钙、磷需要量，以满足最优生长性能、骨骼钙化和胴体瘦肉率的需要（Carter and Cromwell, 1998a, b）。

Kornegay 等（1973）、Harmon 等（1974b, 1975）、Nimmo 等（1981a, b）、Mahan 和 Fetter（1982）、Arthur 等（1983a, b）、Grandhi 和 Strain（1983）、Kornegay 和 Kite（1983）、Maxson 和 Mahan（1986）、Mahan 等（2009），以及 Everts 等（1998a, b）许多学者研究过种猪对钙、磷的需要。有研究（Nimmo et al., 1981a, b）表明，在生长早期通过给青年母猪饲喂充足的钙、磷，可以使骨骼钙化达到最佳水平，从而延长母猪的繁殖寿命，但也有其他一些研究（Arthur et al., 1983a, b; Kornegay et al., 1984）不这样认为。妊娠期间，为满足胎儿的生长，母猪对钙、磷的需要量也会相应增加，并在怀孕后期达到最高水平（Mahan et al., 2009）。在哺乳期，这种需要受母猪泌乳水平的影响。一般来说，钙、磷的需要量是基于一定的采食量的：妊娠期日采食量为 1.8~2.0 kg，哺乳期日采食量为 5~6 kg。如果妊娠期母猪的日采食量低于 1.8 kg，那么就要提高日粮钙、磷含量以满足需要。相对地，如果母猪通过增加采食量来获取更多的蛋白质和能量以维持良好的体况，这时就可以适当降低日粮中的钙、磷水平。哺乳母猪在高温环境下，采食量降低，如果要保持泌乳量不变，那么就要重新调整泌乳日粮以满足每日的钙、磷需要。由于骨骼生长的需要，充足的钙、磷营养对初产母猪的影响比经产母猪更为关键（Giesemann et al., 1998）。

饲料原料中磷的存在形式影响其利用率。在谷物及其副产物和油籽粕中，60%~75%的磷以植酸磷或植酸钙镁磷（肌醇 1, 2, 3, 4, 5, 6 六磷酸二氢盐与不同的阳离子、蛋白质和碳水化合物形成的复合物）形式存在（Nelson et al., 1968; Lolas et al., 1976; Angel et al., 2002），这些磷是猪难以利用的（Taylor, 1965; Peeler, 1972; Erdman, 1979; Jongbloed and Kemme, 1990; Pallauf and Rimbach, 1997）。各种谷物中的磷的有效利用率不同，其范围可从玉米的低于 15%（Barley and Thomson, 1969; Miracle et al., 1977; Calvert et al., 1978; Trotter and Allee, 1979a, b; Huang and Allee, 1981; Ross et al., 1983）到小麦的接近 50%（Miracle et al., 1977; Trotter and Allee, 1979a; Cromwell, 1992）。小麦和小麦副产物（Stober et al., 1980a; Hew et al., 1982）中磷的高利用率源于小麦中本身存在的植酸酶（McCance and Widdowson, 1944; Mollgaard, 1946; Pointillart et al., 1984）；高水分的玉米或高粱中的磷比干的籽实中的磷更容易利用（Trotter and Allee, 1979b; Boyd et al., Ross et al., 1983）；低植酸含量的玉米中的磷相对利用率较高（77%）（Cromwell et al., 1998b），所有低植酸型原料也被认为如此。

油籽粕中磷的利用率很低（Tonroy et al., 1973; Miracle et al., 1977; Trotter and Allee, 1979a; Stober et al., 1980b; Harrold, 1981; Ross et al., 1982; Cromwell, 1992）。与此相对照，动物来源蛋白原料中的磷主要是以无机形式存在，大部分动物蛋白原料（包括牛奶和血液副产物）中磷的利用率较高（Cromwell et al., 1976; Hew et al., 1982; Coffey and Cromwell, 1993）；肉骨粉中磷的生物利用率变异较大。研究表明，肉骨粉中磷的利用率稍低于其他大部分动物性原料（67%）（Cromwell, 1992），但也有研究显示利用率较高（90%）（Traylor et al., 2005）。一些研究（Bayley and Thompson, 1969; Bayley et al., 1975b）认为，蒸气制粒有利于植酸磷的利用，但另外一些研究者不这么认为（Trotter and Allee, 1979c; Corley et al., 1980; Ross et al., 1983）。

在含高植酸的谷物饼粕型日粮中，添加微生物植酸酶是改善植酸磷利用率的主要方法（Nasi, 1990; Simons et al., 1990; Jongbloed et al., 1992; Pallauf et al., 1992a, b; Cromwell et al., 1993b, 1995; Lei et al., 1993b）。该法降低了日粮无机磷的添加水平，从而可减少 30%~60%的磷排泄。影响微生物植酸酶作用的主要因素是：日粮中可利用磷和总磷（包括植酸磷）的比例、植酸酶添加量、钙磷比（或钙含量）及维生素 D 含量（Jongbloed et al., 1993; Düngelhoef et al., 1994; Lei et al., 1994; Kornegay, 1996; Adeda et al., 1998; Johansen and Poulsen, 2003; Selle and Ravindran, 2008; Selle et al., 2009; Kerr et al., 2010; Letourneau-Montminy et al., 2010）。微生物植酸酶也可以改善钙（Pallauf et al., 1992b; Lei et al., 1993b; Young et al., 1993; Mroz et al., 1994）、铁（Stahl et al., 1999）和锌（Pallauf et al., 1992a, 1994a, b; Lei et al., 1993a; Revy et al., 2004）的利用率，对日粮蛋白的消化也有促进作用（Ketaren et al., 1993; Mroz et al., 1994; Kemme et al., 1995; Biehl and Baker, 1996）。由于锌对铜的吸收有拮抗作用（Zacharias et al., 2003），添加植酸酶可促进植酸复合物中锌的释放，所以添加植酸酶的同时要提高铜的添加量。由于制粒过程中需要高温，高温将降低或破坏植酸酶的活性。有研究显示，温度超过 60℃，植酸酶就会失活（Jongbloed and Kemme, 1990; Nunes, 1993）；植酸酶的失活会降低钙、磷的消化率（Jongbloed and Kemme, 1990）。

各种无机磷原料中的磷利用率是不同的。磷酸铵、磷酸钙、磷酸钠中磷的利用率很高（Kornegay, 1972b; Hays, 1976; Clawson and Armstrong, 1981; Partridge, 1981; Tunmire et al., 1983; Cromwell, 1992）。蒸气干燥的骨粉中的磷比磷酸氢钙和磷酸二氢钙中的利用率低（Cromwell, 1992）；一般来说，脱氟磷酸盐中的磷比磷酸二氢钙或磷酸氢钠中的利用率低（Cromwell, 1992; Coffey et al., 1994b），但也可能因来源和加工过程不同而变化（Kornegay and Radcliffe, 1997）；磷酸钙中磷的利用率可能随特殊形式和水合度不同而变动（Eeckhout and De Paepe, 1997）。某些磷源物质中的磷是很难利用的，包括高氟磷酸石、软磷酸盐、胶质黏土及库拉索磷酸盐（Chapman et al., 1955; Plumlee et al., 1958; Harmon et al., 1974b; Hays, 1976）。

对天然饲料原料中钙的利用率的研究甚少。由于植酸的存在，谷物型日粮、苜蓿、各种牧草和干草中的钙的利用率相对较低（Soares, 1995）。由于大部分饲料原料含钙低，因而钙的利用率受限。方解石、石膏、贝壳粉、鱼骨粉、脱脂奶粉、文石及白垩粉等原料中钙有很高的利用率（Pond et al., 1981; Ross et al., 1984; Pointillart et al., 2000; Malde et al., 2010;），但白云石中可利用的钙只有50%~75%（Ross et al., 1984）。颗粒大小（直径0.5mm）对钙利用率的影响可能很小（Ross et al., 1984）。来自家禽的研究数据显示，磷酸氢钙、磷酸钙、脱氟磷酸盐、葡萄糖酸钙、硫酸钙和骨粉等原料中的钙具有高效利用性，与碳酸钙相比，利用率达到90%~100%（Baker, 1991; Soares, 1995），但没有在猪上获得类似的应用数据。

钙、磷缺乏症状类似于维生素D的缺乏症状。这包括生长缓慢和骨骼钙化异常，造成仔猪的软骨病和大猪的骨软化。母猪的钙或磷不足会造成后肢瘫痪，称为产后瘫痪。这种症状经常发生在母猪泌乳高峰期的后期或者泌乳高峰期过后。

过量的钙、磷可能会降低猪的生产性能（Reinhart and Mahan, 1986; Hall et al., 1991），当Ca/P比例加大时，这种影响更大。过量的钙不但会降低磷的利用率，还会因植酸盐的存在而增加锌的需要量（Luecke et al., 1956; Whiting and Bezeau, 1958; Morgan et al., 1969; Oberleas, 1983）。当阳离子（Zn, Ca）与植酸盐的物质的量比达到2:1或3:1时，则更容易形成不溶性复合物（Oberleas and Harland, 1996）。

通过析因法估测钙磷需要的基础

在本次修订中我们采用了一种模型来评估生长育肥猪和母猪日粮中STTD磷和总钙的需要量。此模型的基本原理之前已有详尽描述（Jongbloed et al., 1999, 2003; Jondreville and Dourmad, 2005; GfE, 2008）。决定磷需要量的因素包括：①机体磷沉积的最大速率；②妊娠母猪胎儿的磷沉积；③乳汁中排出的磷；④基础的内源性肠道磷损失；⑤最低的尿磷损失；⑥用STTD磷摄入量估测生长育肥猪磷沉积的边际效率；⑦获得最大生产性能时的磷需要量和最大机体磷沉积时的磷需要量的比例关系。由于缺乏相关的数据，钙的需要量是直接根据生长育肥猪、妊娠母猪和泌乳母猪对STTD磷和总钙需要量之间各自特定的比率而从STTD磷的需要量中简单推算出来的。这样的比率最好是可消化钙和可消化磷之间的比率，但由于缺乏相关数据，这里采用了总钙和STTD磷之间的比率。用来计算钙和磷的利用和需要量的参数和方程见第8章。对模型方法估测的钙磷的需要量的评估也在第8章。

生长育肥猪机体磷的总量和磷的沉积最大速率是从体蛋白总量估算出来的（如Hendriks and Moughan, 1993; Pettey, 2004; Hinson, 2005）。这与Jongbloed等（1999, 2003）、Jondreville和Dourmad（2005）及GfE（2008）表述的方法不同，他们用活体或空腹体重来估算机体磷的总量。根据查到的文献综述数据，机体总磷和机体总氮之间可建立一个明显相近的关联（图6-2; Cromwell et al., 1970; Coalson et al., 1972; Fammatre et al., 1977; Mahan et al., 1980; Crenshaw et al., 1981; Mahan and Fetter, 1982; Maxson and Mahan, 1983; Reinhart and Mahan, 1986; Coffey et al., 1994b; Eeckhout et al., 1995; O'Quinn et al., 1997; Ekpe et al., 2002; Hastad et al., 2004; Pettey et al., 2006; Ruan et al., 2007; Hinson et al., 2009），该关联不受猪的基因型和性别影响。用这种方法评估的磷沉积和需要量，与在以前章节所涉及的估测不同性别和瘦肉生长速度潜力对磷需要量的估测结果一致。

如 Jongbloed 等（1999，2003；比率 0.0096）概述的那样，母猪体内磷的沉积与其蛋白质总量的变化有关，一般以肌肉蛋白中磷和蛋白的比例为基础，根据这个关系也可以评估母猪哺乳期间在蛋白质负平衡时体内磷的动员量；还可以用来评估妊娠母猪骨骼组织中磷的沉积，结果表明，沉积量随胎次而降低，由第一胎的 2.0 g/d 下降到第四胎或更高胎次的 0.8 g/d。这些结果要比 Jongbloed 等（2003）根据有限的数据得出的结果稍高一些（从第一胎 1.5 g/d 到第四胎 0.2 g/d）。胚胎磷的沉积参照 Jongbloed 等（1999）及 Jondreville 和 Dourmad（2005）的表述。如前所述，已确定满足最大生产性能的日粮磷需要量比满足最大磷沉积的磷需要量要低（大约低 0.10%），因此估测到满足生长育肥猪最大生产性能的 STTD 磷的需要量，是满足最大磷沉积量时的 0.85。这个 0.85 估测值的出发点就是总磷需要量的 0.1% 的差异。计算机模型的结果和各种估测表明，这个 0.85 数值与目前可得到的经验值非常吻合。

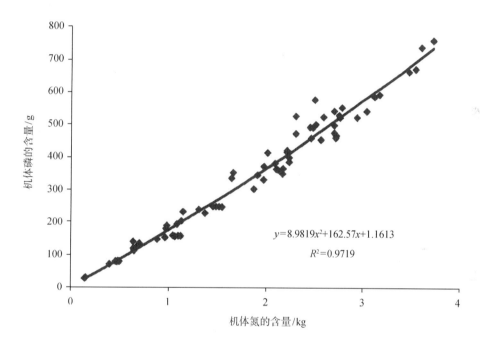

图 6-2　生长育肥猪中机体氮和磷含量的关系。每个数据各点代表处理组的均值

与 Jondreville 和 Dourmad（2005）的处理方式一致，乳汁中排出的磷可根据乳汁中排出的氮来预测。基于文献综述，乳汁中磷和氮的比率一般稳定在 0.196（Boyd et al., 1982; Coffey et al., 1982; Mahan and Fetter, 1982; Hill et al., 1983b; Kalinowski and Chavez, 1984; Miller et al., 1994; Park et al., 1994; Farmer et al., 1996; Seynaeve et al., 1996; Jurgens et al., 1997; Giesemann et al., 1998; Tilton et al., 1999; Lyberg et al., 2007; Peters and Mahan, 2008; Leonard et al., 2010; Peters et al., 2010），这个数值与 Jondreville 和 Dourmad（2005）得出的 0.194 非常接近。

为了降低日粮磷水平对磷全消化道消化率的影响，我们采用了 STTD 这个概念，这与采用标准化回肠可消化氨基酸（第 13 章）的方式一致。基于文献综述和对猪饲喂无磷日粮的试验观察，基础内源性粪磷损失估测为 190 mg/kg 干物质摄入量（第 13 章）。除了基础粪磷损失外，最小的尿磷损失也是磷维持需要量的一部分，Jongbloed 等（1999，2003）及 Jondreville 和 Dourmad（2005）指出，最小的尿磷损失与体重相关，在生长育肥猪和母猪中常采用 7 mg/kg 体重（Jondreville and Dourmad, 2005）。

生长猪平衡试验结果表明，当摄入磷稍微低于最大磷沉积的需要量时，用可消化磷摄入量估测机

猪营养需要

体磷沉积的最大边际效率为95%（Rodehutscord et al., 1998; Pettey et al., 2006; Nieto et al., 2008; Stein et al., 2008），过量摄入的磷最终会通过内源性粪磷和尿磷的损失排出。然而由于动物间的个体差异，这个数值群体研究的结果差异比个体的差异更小（如Pomar et al., 2003）。因此，把最大的效率值降到0.77，这与育肥猪和妊娠母猪对氨基酸利用率的调整在某种程度上保持数量上的一致（第2章），由于缺乏相关信息，我们假定这个效率值在生长育肥猪、妊娠母猪和泌乳母猪上是相似的。

钠和氯

钠和氯分别是机体细胞外主要的阳离子和阴离子，氯离子是胃液分泌的主要阴离子。Mahan等（1996，1999b）报道，饲喂含有干燥乳清粉或血浆蛋白粉（二者钠含量均相对较高）日粮的断奶仔猪，对钠（由氯化钠或磷酸钠形式提供）及氯（以盐酸形式提供）的添加均有反应。钠离子与氯离子，尤其是氯离子对断奶仔猪生长和饲料效率产生影响，消化试验表明增加氯离子可改善氮消化率，早期（特别是断奶后7~14天内）断奶仔猪需要更多的钠离子和氯离子。Monegue等（2011）发现，刚断奶仔猪，特别是阉公猪，喜采食高钠盐日粮，断奶后两周对这种高钠盐日粮的采食偏好降低。因此，估计5~7 kg、7~11 kg和11~25 kg体重的猪对钠和氯的需要量分别需要增至0.40%/0.50%、0.35%/0.45%和0.28%/0.32%。

过去一直认为，生长育肥猪日粮中钠的需要量不超过日粮的0.08%~0.10%（Meyer et al., 1950; Alcantara et al., 1980; Honeyfield and Froseth, 1985; Honeyfield et al., 1985; Kornegay et al., 1991）；生长猪日粮中对氯的需要量没有明确规定，但认为不超过0.08%（Honeyfield and Froseth, 1985; Honeyfield et al., 1985）。由此可见，生长育肥猪玉米-豆粕型日粮中添加0.20%~0.25%的氯化钠就可以满足钠和氯的需要量（Hagsten and Perry, 1976a, b; Hagsten et al., 1976）。不过，Yin等（2008）将氯化钠由0.1%逐步增至0.6%，发现0.4%水平时磷的表观消化率和真消化率最大。因此，和断奶仔猪一样，在生长猪阶段，较高的磷的消化率需要较高水平的钠或氯元素，估计育肥猪也不例外。

种猪对钠和氯的需要量尚未确定。有研究认为，日粮中添加0.3%氯化钠（含0.12%钠）对妊娠母猪是不足的（Friend and Wolynetz, 1981）；一项区域性研究结果表明，两胎或多胎次的母猪在妊娠和泌乳阶段氯化钠添加量从0.50%减至0.25%时，仔猪出生重和断奶重均降低（Cromwell et al., 1989a）；考虑到母猪乳中钠的含量为0.03%~0.04%（ARC, 1981），泌乳阶段日粮中钠的需要量应比妊娠阶段高约0.05%，本版建议在母猪妊娠和泌乳期日粮中分别添加0.4%和0.5%的氯化钠。

多数饲料原料中钠和氯的利用率是90%~100%（Miller, 1980），水（沿海地区水中钠的含量高达184 mg/L）和脱氟磷酸盐中钠的利用率均较高（Kornegay et al., 1991）。

钠或氯缺乏会降低猪的生长速度和效率，但只要能提供充足的淡水，猪可耐受日粮中含高水平的氯化钠（NRC, 2005），如果饮用水有限或者水中氯化钠含量很高，会导致中毒。显然，钠离子因干扰动物体内的水平衡而引起其不良生理反应；钠中毒症状包括神经症状、虚弱、行走不稳、癫痫、瘫痪和死亡（Bohstedt and Grummer, 1954; Carson, 1986）。

钠、钾和氯是影响动物电解质平衡及酸碱平衡的主要离子。在大多数情况下，日粮中矿物质平衡以钠+钾-氯毫当量表示（mEq）（Mongin, 1981），也叫做电解质平衡。Patience和Wolynetz（1990）建议，计算电解质平衡时，钙、镁、硫和磷离子也应包括在内。日粮中最佳的电解质平衡大约是250 mEq的阳离子（钠+钾-氯）/kg日粮（Austic and Calvert, 1981; Golz and Crenshaw, 1990; Haydon et al., 1993; Dersjant-Li et al., 2001）。但是最佳生长可以在0~600 mEq/kg日粮的范围内出现（Patience et al., 1987; Kornegay et al., 1994）。如果日粮中钠、钾、氯中任一种离子缺乏，那么钠+钾-氯的关系不能精确预测最佳生长所需的日粮离子水平（Mongin, 1981）。

镁

镁是很多酶系的辅助因子和骨骼的组成成分。用以奶粉为基础的半纯合日粮人工饲养的仔猪对镁的需要量为 300~500 mg/kg（即 0.03%~0.05%）日粮（Mayo et al., 1959; Bartley et al., 1961; Miller et al., 1965b, c, d）；猪乳中含有充足的镁可满足仔猪的需要量（Miller et al., 1965b, c）。在断奶-生长-育肥整个阶段，猪对镁需要量变化不大。尽管某些研究表明，饲料原料中的镁只有 50%~60% 可被利用（Miller, 1980; Nuoranne et al., 1980），但是玉米-豆粕型日粮中的镁含量（0.14%~0.18%）显然可满足猪的需要（Svajgr et al., 1969; Krider et al., 1975）。

种猪对镁的需要量尚未确定。Harmon 等（1976）在单一胎次的研究中，母猪妊娠阶段饲喂含 0.04% 和 0.09% 镁的半纯合日粮，随后在泌乳阶段饲喂含 0.015% 和 0.065% 镁的日粮，结果在繁殖或哺乳性能方面均无差异。但在一个平衡试验中，饲喂低水平镁的母猪在泌乳阶段出现镁的负平衡。

镁缺乏的猪依次出现的症状为：高度应激、肌肉痉挛、不愿站立、踝部异常、共济失调、四肢抽搐，继而死亡（Mayo et al., 1959; Miller et al., 1965b）；高锰日粮会加重镁的缺乏症（Miller et al., 2000）。

钾

钾（K）是猪体内第三丰富的矿物元素，仅次于钙和磷（Manners and McCrea, 1964），钾也是肌肉组织中最丰富的矿物元素（Stant et al., 1969），参与维持电解质平衡和神经肌肉的功能。钾通过提供一价阳离子来维持细胞内阴离子的平衡，作为钠钾泵的一部分发挥其生理功能。

体重 1~4 kg、5~10 kg、16 kg、20~35 kg 的猪日粮中钾的需要量分别为 0.27%~0.39%（Manners and McCrea, 1964）、0.26%~0.33%（Jensen et al., 1961; Combs et al., 1985）、0.23%~0.28%（Meyer et al., 1950）和小于 0.15%（Hughes and Ittner, 1942; Mraz et al., 1958）；种猪和育肥猪的需要量尚未确定。常规日粮中钾的含量较为充足，可满足各阶段猪的需要量。玉米和豆粕中钾的利用率为 90%~97%（Combs and Miller, 1985）。

日粮钾与日粮钠和氯是相互作用的。在含 0.1% 钾的纯合日粮中，将氯的含量由 0.03% 逐渐增至 0.60%，结果仔猪的生长速度反而降低；在含 1.1% 钾的日粮中提高氯的含量，则猪的生长速度提高（Golz and Crenshaw, 1990）。日粮中钾和氯的相互作用似乎对额外的阴、阳离子，尤其是铵离子和磷酸盐的排出和沉积有影响。钾对生长的影响是通过与肾铵离子代谢有关的机制来介导的（Golz and Crenshaw, 1991）。

钾缺乏表现为：食欲不振、被毛粗糙、消瘦、精神萎靡和运动失调（Jensen et al., 1961）。心电图显示，钾缺乏的猪心率降低，心跳间隔期增加（Cox et al., 1966）。病猪尸检无特征性病理症状。

钾中毒水平尚不清楚。如果提供充足饮用水，猪可耐受 10 倍需要量以上的钾（Farries, 1958）。然而，养猪生产中应用的一些液态副产物中钾水平较高，容易造成猪的采食量和生长速率下降；尽管饲料利用效率和胴体指标等可能不受影响，在实践中仍要小心使用，过高的钾可损伤肾脏（如肾斑和钙盐沉积）（Guimaraes et al., 2009）。对猪静脉注射氯化钾，可造成心电图异常（Coulter and Swenson, 1970）。

硫

硫（S）是一种必需元素。普遍认为含硫氨基酸提供的硫足以满足猪合成含硫化合物（如牛磺酸、谷胱甘肽、硫辛酸和硫酸软骨素）的需要（Miller, 1975; Baker, 1977），因为在低蛋白日粮中额外添加无机硫酸盐并无益处。但是，越来越多的研究者担心日粮硫过量的问题，因为使用各类谷物副产物后增加了日粮中总含硫量，造成硫酸盐还原菌可利用的底物量增加而生成过量的硫化氢（Kerr et al.，2008），从而对胃肠道功能和健康不利。Kerr 等（2011）在给 13 kg 猪饲喂 0.21%~1.21%含硫无机酸盐的两个试验中均发现，随着硫含量的增加，日增重呈线性降低；硫水平较高的日粮确实会改变炎症介质和肠道内细菌。Perez 等（2011b） 给 9 kg 猪饲喂 0.2%~0.6%含硫无机酸盐，也得到硫含量与日增重呈线性负相关的关系。究其原因，主要是由于硫可影响采食量从而造成猪的生长速度降低。

微量/痕量元素

铬

铬（Cr）与碳水化合物、脂类、蛋白质和核酸的代谢有关（Nielsen，1994）。具有生物活性的铬通过改变血清中葡萄糖或胰岛素水平来改变组织对胰岛素的敏感性。含铬的"葡萄糖耐受因子"可增强猪的胰岛素活性，并具有生物学活性（Steele et al.，1977）。Evock-Clover 等（1993）报道，生长猪日粮中添加铬（吡啶甲酸铬形式），可降低血清中的胰岛素和葡萄糖浓度。给禁食的妊娠母猪饲喂吡啶甲酸铬，与对照母猪相比，摄食后血清胰岛素水平、胰岛素与葡萄糖的比率均降低（Lindemann et al.，1995）。Garcia 等（1997）也证明，猪采食添加吡啶甲酸铬的日粮可提高其胰岛素效率。但 Page 等（1993）的研究却表明，在正常的饲喂状况下，铬对组织对胰岛素敏感性的影响不明显，且未观察到血清中葡萄糖浓度的变化。采用传统方法进行静脉注射葡萄糖耐量试验（IVGTT）和静脉注射胰岛素耐量试验（IVICT）所获得的结果大多一致。这些试验均表明，在日粮中添加吡啶甲酸铬（Amoikon et al.，1995；Matthews et al.，2001）、酵母铬（Guan et al.，2000）、丙酸铬（Matthews et al.，2001）或蛋氨酸铬（Fakler et al.，1999）等含铬产品均可改变葡萄糖和胰岛素的水平。铬对葡萄糖和胰岛素的影响是通过一种低分子质量且具有多种功能的铬调蛋白来调控的（Vincent，2001；Davis et al.，1996；Davis and Vincent，1997），有生物活性的铬还可影响生长激素的分泌（Wang et al.，2008，2009）。

铬对断奶仔猪作用的研究较生长育肥猪少。添加有机形式的铬一般不能改善仔猪的生长性能，对免疫系统的影响也不一致（van Heugten and Spears，1997；Lee et al.，2000a, b；Tang et al.，2001；van de Ligt et al.，2002a, b；Lien et al.，2005）。生长育肥猪日粮中添加有机形式的铬可提高胴体瘦肉率（如增加肌肉和/或减少脂肪含量）（Page et al.，1993；Boleman et al.，1995；Lindemann et al.，1995；Mooney and Cromwell，1995，1997；Min et al.，1997；Lien et al.，2001；Urbanczyk et al.，2001；Xi et al.，2001；Wang and Xu，2004；Jackson et al.，2009；Park et al.，2009）。但也有一些相反结果的报道（Harris et al.，1995；Mooney and Cromwell，1996；Lemme et al.，1999）。添加丙酸铬除了改善猪胴体的整体效果外，也有报道可改善猪肉品质 （Matthews et al.，2003，2005；Shelton et al.，2003；Jackson et al.，2009）。但这些研究中铬对日增重和饲料报酬的影响结果不一致，其中有两个研究表明有机铬可改善营养物质的消化率（Kornegay et al.，1997；Park et al.，2009），引起试验结果不一致的原因可能与日粮中铬的水平、铬的添加形式、铬的营养状况和日粮中氨基酸水平有关（Lindemann，2007）。玉米-豆粕型日粮中总铬含量为 750~3000 ppb，但是大部分不能被动物利用，特别是无机铬在胃肠道吸收率很低，一般为 0.4%~3%（Anderson，1987 的综述）。

Lindemann 等（1995）报道，在母猪日粮中添加 200 ppb 的吡啶甲酸铬可获得较高的产仔数，随后 Hagen 等（2000）、Lindemann 等（2000，2004）和 Real 等（2008）的试验也获得类似结果，但 Campbell （1998）没有观测到相似的结果；蛋氨酸铬也能提高母猪的产仔数（Perez-Mendoza et al.，2003）。铬对其他繁殖指标的影响不一致，包括断奶-发情天数、受孕率、分娩率和淘汰率等。由于肌肉是胰岛素作用的靶组织，也是构成身体的最大单一组织，Lindemann 等（2004）评估了每单位体重摄入的铬对繁殖性能的影响。试验采用生长猪评估 IVGTTs 和 IVICTs 对铬添加的反应来确定铬的水平，这个量大概是每天 7.5 μg/kg 体重，当这个值被用到繁殖动物（以体重和采食量为基础）时，在日粮中需添加 500~600 ppb 的铬才可与生长猪每单位体重获得的铬相当。然后用繁殖母猪评估不同水平吡啶甲酸铬来源的铬（0、200 ppb、600 ppb 和 1000 ppb）对其繁殖性能的影响，试验持续了至少两个胎次，结果表明，产仔数对铬添加量呈二次反应，在添加 600 ppb 铬时获得了最大的产仔数，这证实了评估限饲的繁殖动物对补充养分反应时，不能仅仅以每单位日粮中的含量或者每天的补充量为准。

三价铬和六价铬是最常见的两种形式，两者均稳定。六价铬比三价铬毒性大，三价铬被认为是必需微量元素（Anderson, 1987; Mertz, 1993）。猪日粮中最高的铬耐受水平为 3000 ppm（氧化铬来源）和 100 ppm（可溶性三价铬）（NRC, 2005）；六价铬有毒，不适宜用在猪日粮中。在屠宰前 75 天内的试验猪日粮中添加 5000 ppb 铬（以吡啶甲酸铬、丙酸铬、酵母铬或者蛋氨酸铬形式提供），结果对猪的生长性能、胴体和临床化学上等指标无任何不良影响。Tan 等（2008）在日粮中添加含 3200 ppb 铬（吡啶甲酸铬来源）80 天（几乎整个生长育肥期），仅观察到一些抗氧化物酶活力的改变。该结果表明，长期饲喂不同水平的吡啶甲酸铬，并未增加生长育肥猪体内氧化损伤的生物分子标记物的生成，说明添加 200 ppb 的铬（允许添加的最高量）没有问题。

猪对铬需要量的定量评估还不完善。在很多国家规定了家畜日粮中铬的添加形式和允许水平，饲料配方师必须知道这些影响猪日粮的限制规定。NRC（1997）发表了一篇有关铬的综述；关于家畜日粮中铬的最新综述，可见 Lindemann（2007）的研究。

钴

钴（Co）是维生素 B_{12} 的组成成分（Rickes et al., 1948）。日粮中的钴在肠道内被微生物利用合成维生素 B_{12}，如果日粮维生素 B_{12} 不足，那么肠道内的合成就显得更重要（Klosterman et al., 1950; Kline et al., 1954）。因为饲料中添加维生素 B_{12} 很普遍，关于钴需要量的研究比较少。

除用于维生素 B_{12} 合成外，目前尚无证据表明猪的生长绝对需要钴，但是钴能代替羧肽酶中的锌及部分代替碱性磷酸酶的锌。Hoekstra（1970）报道，补充钴能预防因锌缺乏所引起的损伤。Stangl 等（2000）报道，未添加维生素 B_{12} 的日粮中添加 1 ppm 的钴，结果血清或肝脏中维生素 B_{12} 水平未见改变，却改变了因日粮中缺乏维生素 B_{12} 而导致的肝脏过氧化氢酶和血清谷胱甘肽过氧化物酶的活性，这也说明尚需对钴代谢进行进一步研究。

日粮中添加 400 ppm 的钴对仔猪有毒（Huck and Clawson, 1976），可能引起食欲不振、四肢僵硬、弓背、运动失调、肌肉颤抖和贫血。Kornegay 等（1995）报道，在钴含量小于 2 ppm 的基础日粮中分别添加 0 ppm、150 ppm 和 300 ppm 水平的钴，4~5 周内肝脏和肾脏钴的浓度呈直线上升，但生长速率则直线下降。Van Vleet 等（1977）报道，硒、维生素 E 和半胱氨酸在某种程度上能有效防止因日粮钴过量而引起的毒副作用，但促生长水平的铜则加剧因钴中毒导致的生长抑制（Kornegay et al., 1995）。

铜

猪需要铜（Cu）来合成血红蛋白，合成与激活正常代谢所必需的一些氧化酶类（Miller et al., 1979）。初生仔猪日粮铜水平 5~6 ppm 即已足够（Okonkwo et al., 1979; Hill et al., 1983a），生长后期的需要量可能也不会超过 5~6 ppm。妊娠期和哺乳期，铜需要量方面的确切资料很少。Lillie 和 Frobish（1978）认为，母猪饲喂 60 ppm 的铜，可提高仔猪初生重和断奶重，这可能是日粮中高铜的药理作用所致。Kirchgessner 等（1980）发现饲喂 2 ppm 铜的妊娠母猪比饲喂 9.5 ppm 铜的妊娠母猪血浆铜蓝蛋白降低，产死胎数多。Kirchgessner 等（1981）通过平衡试验估计怀孕母猪对铜的需要量大约为 6 ppm。Yen 等（2005）在哺乳母猪日粮铜添加量的研究中发现，以蛋白质-铜化合物的形式额外补充 14 mg/d 的铜，可提高断奶后 7 天的配种率。

生物学效价较高的铜盐有硫酸铜、碳酸铜和氯化铜（Miller, 1980; Cromwell et al., 1998a）。硫化铜和氧化铜中的铜对猪效价很低（Cormwell et al., 1978, 1989b）。试验表明，铜的有机复合物似乎与硫酸铜有同等的生物学效价（Bunch et al., 1965; Zoubek et al., 1975; Stansbury et al., 1990; Coffey et al., 1994a; Apgar et al., 1995; Apgar and Kornegay, 1996）。但 Coffey 等（1994a）和 Zhou 等（1994a）的报道称，饲喂促生长水平的赖氨酸铜较硫酸铜猪的生长性能更佳。

铜缺乏可导致铁动员差，造血功能异常，角质化作用差，胶原蛋白、弹性蛋白和骨髓合成变差。铜缺乏症包括：小红细胞低色素性贫血、弓腿、自发性骨折、心血管异常及毛褪色（Hart et al., 1930; Elvehjem and Hart, 1932; Teague and Carpenter, 1951; Follis et al., 1955; Carnes et al., 1961; Hill et al., 1983a）。

长期饲喂超过 250 ppm 铜的日粮，可能引起铜中毒（NRC, 1980），中毒症状为血红蛋白水平降低和黄疸，这是过量铜在肝脏和其他重要器官中蓄积的结果。降低日粮锌和铁的水平或者提高日粮钙水平都能加重铜中毒（Suttle and Mills, 1966a, b; Hedges and Kornegay, 1973; Prince et al., 1984）。猪对日粮中铜的最大耐受量为 250 ppm（NRC, 2005）。

饲喂 100~250 ppm 的铜（以硫酸铜形式提供）可促进猪的生长（Barber et al., 1955a; Braude, 1967; Wallace, 1967; Cromwell et al., 1981; Kornegay et al., 1989; Cromwell, 1997）。铜对仔猪的促生长作用不依赖于其他抗菌剂，但与其他抗菌剂之间有加性效应（Stahly et al., 1980; Roof and Mahan, 1982; Edmonds et al., 1985; Cromwell, 1997）。日粮中添加脂肪可增强猪对高铜的反应（Dove and Haydon, 1992; Dove, 1993a, 1995）。母猪连续 6 个妊娠-哺乳周期饲喂高铜（常规日粮中含 9 ppm 铜的前提下，额外添加 250 ppm 铜），虽然肝和肾铜浓度大幅度增加，但对繁殖性能无任何明显副作用（Cormwell et al., 1993a）。事实上，饲喂高铜的母猪，其产仔数、仔猪初生重、断奶窝重、断奶重均有所提高，断奶后发情间隔缩短；这次历时两年多的试验中，除了发现上述益处外，并未发现高铜对母猪的副作用，这可能是由于限饲，母猪单位体重获得的营养水平远远低于自由采食的生长猪。Lillie 和 Frobish（1978）也发现，母猪饲喂高铜能提高哺乳仔猪的体增重。但另一些研究表明，在妊娠后期和整个哺乳期（Thacker, 1991）或者哺乳期间（Roos and Easter, 1986; Dove, 1993b）的日粮中加铜，对哺乳仔猪体重无影响。

通过添加超过常规水平的铜所获得的有益效应的机制尚不清楚。日粮铜的促生长作用一直被认为与它的抗菌作用有关（Fuller et al., 1960），但尚缺乏支持这一假说的证据。Bowland 等（1961）和 Cromwell 等（1989b）认为铜的可利用率与其促生长作用相关。Zhou 等（1994b）报道，给仔猪隔天静脉注射组氨酸铜 18 天，可提高仔猪的体增重和血清有丝分裂活性。因为本次研究绕过了胃肠道，说明铜可系统地发挥促生长的作用。最新研究表明，175~250 ppm 的铜可影响下丘脑食欲调节基因的 mRNA 表达水平（Zhu et al., 2011）。饲喂 250 ppm 的铜可刺激断奶仔猪脂肪酶和磷脂酶 A 的活性，从而提高日粮中脂肪的消化率（Luo and Dove, 1996）。虽然日粮中高水平的铜也增加了粪便中铜的排泄量，但 Payne 等（1988）研究发现，将饲喂 250 ppm 高铜的猪的粪便（粪铜含量为 1550 ppm）作为肥料，施于三种不同类型的土壤 8 年，并未降低谷物的产量，植株中铜的含量也处于正常范围。这表明，植物无法利用这些额外添加的铜。Cabral 等（1998）也证实，猪粪中的铜对植株组织无影响，这与铁、锰和锌都不同。不过对于那些在用粪便施肥的作物上放牧的牲畜而言，粪便的潜在毒性是一个有争议的话题（Prince et al., 1975; Suttle and Price, 1976），这一争议也许取决于粪便的施用量。

碘

猪体内大部分碘（I）作为一、二、三和四碘甲状腺原氨酸（甲状腺素）的组成成分存在于甲状腺中，这些激素对调节机体新陈代谢速度至关重要。Hart 和 Steenbock（1918）、Kalkus（1920）、Welch（1928）证明，美国西北部和五大湖地区饲养的猪存在甲状腺机能减退的症状，是那里的低碘土壤生产的缺碘饲料所致。

日粮碘需要量尚未很好确定，因为某些饲料原料（如油菜籽、亚麻籽、小扁豆、花生及大豆等）中存在的致甲状腺肿素会使碘的需要量升高（McCarrison, 1933; Underwood, 1977; Schone et al., 1997a, b, 2001）。玉米-豆粕型日粮中 0.14 ppm 的碘就足以防止生长猪甲状腺肿大（Cromwell et al., 1975）。日粮中添加 0.35 ppm 的碘可防止母猪缺碘（Andrews et al., 1948）。

从营养的角度来讲，碘酸钙、碘酸钾和正过碘酸钙都是碘的可利用形式，且在碘盐混合物中比碘化钠或碘化钾更稳定（Kuhajek and Andelfinger, 1970）。在谷物-豆粕日粮中添加 0.2%的含碘食盐（0.007%碘）即可满足生长猪对碘的需要（0.14 ppm）。

严重缺碘可导致猪生长停滞、昏睡和甲状腺肿大（Beeson et al., 1947; Braude and Cotchin, 1949; Sihombing et al., 1974）。饲喂缺碘、含致甲状腺肿素日粮的母猪，产被毛稀少的弱仔或死胎，出现黏液性水肿、甲状腺肿大和出血症状（Hart and Steenbock, 1918; Slatter, 1955; Devilat and Skoknic, 1971）。

800 ppm 的日粮碘降低生长猪的生长速度、血红蛋白水平及肝铁浓度（Newton and Clawson, 1974）。在哺乳期和妊娠期最后 30 天，日粮添加 1500~2500 ppm 的碘对母猪无害（Arrington et al., 1965）。

铁

铁（Fe）是红细胞中血红蛋白的组成成分。铁还以肌红蛋白形式存在于肌肉中，以运铁蛋白形式存在于血清中，以子宫运铁蛋白形式存在于胎盘中，以乳铁蛋白形式存在于乳中，以铁蛋白和血铁黄素的形式存在于肝脏中（Zimmerman, 1980; Ducsay et al., 1984）；在体内还作为多种代谢酶的成分发挥着重要作用（Hill and Spears, 2001）。

新生仔猪体内约含 50 mg 铁，且大部分以血红蛋白形式存在（Venn et al., 1947）。在妊娠后期给母猪补饲高水平的铁（Brady et al., 1978）或在妊娠期经非肠道途径给予母猪右旋糖酐铁（Rydberg et al., 1959; Pond et al., 1961; Ducsay et al., 1984）均未明显增加铁从胎盘向胎儿的转移。哺乳仔猪必须留存 7~16 mg 铁/d 或者 21 mg 铁/kg 体增重，以维持体内充足的血红蛋白和铁储备（Venn et al., 1947; Braude et al., 1962）。每升猪乳平均仅含铁 1 mg（Brady et al., 1978），因此，仅摄食母乳的仔猪很快就会出现贫血（Hart et al., 1930; Venn et al., 1947）。给妊娠和哺乳母猪饲喂高水平的各种形式的铁，包括硫酸铁和铁螯合物，都不能使乳中铁含量提高到可以防止仔猪铁缺乏的程度，但哺乳仔猪采食饲喂高铁母猪的粪便可防止缺铁（Chaney and Barnhart, 1963; Veum et al., 1965; Spruill et al., 1971; Brady et al., 1978; Sansom and Gleed, 1981; Gleed and Sansom, 1982）。

许多研究表明，仔猪出生后头 3 天以右旋糖酐铁、糊精铁或葡庚糖酐铁的形式一次性肌肉注射 100~200 mg 铁是有效的（Barber et al., 1955b; McDonald et al., 1955; Maner et al., 1959; Rydberg et al., 1959; Ullrey et al., 1959; Zimmerman et al., 1959; Kernkamp et al., 1962; Pollmann et al., 1983）。新生仔猪

的小肠黏膜能主动吸收铁（Furugouri and Kawabata, 1975, 1976, 1979），出生后最初几小时内口服有生物活性的无机或有机铁也能满足哺乳仔猪对铁的需要，但关键是要在肠道关闭对大分子物质吸收之前及早服用（Harmon et al., 1974a; Thoren-Tolling, 1975）。应避免注射或口服过量的铁（超过200 mg），因为未结合的血清铁能促进细菌生长，从而易导致感染和腹泻（Weinberg, 1978; Klasing et al., 1980; Knight et al., 1983; Kadis et al., 1984）。

饲喂乳或纯合液态日粮的仔猪铁需要量为50~150 mg/kg乳固形物（Matrone et al., 1960; Ullrey et al., 1960; Manners and McCrea, 1964; Harmon et al., 1967; Hitchcock et al., 1974）。Miller 等（1982）建议，常规或无菌环境下，仔猪铁的需要量为100 mg/kg乳固形物。饲喂以酪蛋白为基础的固态日粮与饲喂相应的液态日粮相比，猪对铁的需要量要高出约50%（以干物质计）（Hitchcock et al., 1974）。

一些研究者认为仔猪断奶后日粮铁需要量为80 ppm（Pickett et al., 1960），而另一些研究者则认为这个需要量高达200 ppm（Rincker et al., 2005; Lee et al., 2008）。在生长后期和成熟阶段，随着血容量增加速度的减慢，铁的需要量也随之减少。天然的饲料原料一般能提供足量的铁以满足仔猪断奶后的需要。饲料级脱氟磷酸盐和磷酸氢钙含铁0.6%~1.0%，也能提供大量的铁。

不同来源铁的利用率差异很大（Zimmerman, 1980）。硫酸亚铁、氯化铁、柠檬酸铁、柠檬酸胆碱铁、柠檬酸铵铁都能有效预防缺铁性贫血（Harmon et al., 1967; Ammerman and Miller, 1972; Ullrey et al., 1973; Miller et al., 1981），而溶解度低的铁化合物（如氧化铁）则无效（Ammerman and Miller, 1972）。碳酸亚铁中铁的生物学效价比硫酸亚铁低，且变幅大（Harmon et al., 1969; Ammerman et al., 1974）。据报道，蛋氨酸螯合铁和甘氨酸螯合铁中铁的生物学效价是硫酸铁的68%~180%（Lewis et al., 1995; Kegley et al., 2002; Feng et al., 2007, 2009）。脱氟磷酸盐中的铁对猪的效价约为硫酸亚铁的65%（Kornegay, 1972a）。豆粕中含有175~200 ppm的铁，根据对鸡进行的血红蛋白耗竭重新补充试验，估计豆粕中铁的生物利用率仅为38%（Biehl et al., 1997）。

血液中血红蛋白浓度是反映猪体内铁营养状况的可靠指标，且易于测定。全血中血红蛋白水平只要达到10 g/dl，可认为是足够的；8 g/dl是贫血的临界值；7 g/dl或者更少，则表示贫血（Zimmerman, 1980）。缺铁造成的贫血叫做小细胞低色素性贫血，主要症状包括生长缓慢、精神萎靡、被毛粗乱、皮肤褶皱及黏膜苍白。生长快的贫血猪会因缺氧而猝死，一个特征性症状是少量运动后呼吸困难或膈肌抽搐痉挛，"猪肺病"一词由此而来。剖检发现肝脏肥大，伴有脂肪肝、血液稀薄、心脏明显扩张、脾脏肿大且硬化。贫血猪更易患传染性疾病（Osborne and Davis, 1968）。虽然补充铁能改善猪的血红细胞总数、血红蛋白浓度、血浆和肝脏中铁的状况，但要避免肆意添加，因为这也可能会导致腹泻增多、生长变缓（Lee et al., 2008）。

3~10日龄仔猪口服硫酸亚铁来源的铁的中毒剂量大约为600 mg/kg体重（Campbell, 1961），口服后1~3 h可见临床中毒症状（Nilsson, 1960; Arpi and Tollerz, 1965）。Lannek 等（1962）和Patterson 等（1967, 1969）发现，给缺维生素E的猪注射铁（右旋糖酐铁来源铁100 mg）是有毒的。猪缺铁会增加十二指肠金属转运载体（DMT1和ZIP14）的基因表达，而以硫酸亚铁的形式补充500 ppm的铁则会降低这些转运载体的表达（Hansen et al., 2009）。日粮铁水平达5000 ppm时会产生佝偻病症状，但这可以通过提高日粮磷水平来预防（O'Donovan et al., 1963; Furugouri, 1972）。

锰

锰（Mn）作为许多与碳水化合物、脂类、蛋白质代谢有关的酶的组成成分发挥作用。锰是线粒体的超氧化物歧化酶（SOD）必不可少的组分，也是合成硫酸软骨素（骨有机质黏多糖的成分之一）所必需的成分（Leach and Muenster, 1962）。

日粮中锰的需要量很显然相当的低，且尚未很好确定（Johnson, 1944）。Leibholz 等（1962）报道，仅仅 0.4 ppm 的锰对仔猪来说已经足够。但缺锰母猪所产仔猪锰的需要量是 3~6 ppm（Kayongo-Male et al., 1975）。玉米-豆粕型日粮含有足量的锰来满足生长-育肥猪正常生长和骨骼形成（Svajgr et al., 1969）。

给母猪长期饲喂含锰仅 0.5 ppm 的日粮会导致骨骼生长异常、脂肪沉积增加、发情周期异常或消失、胎儿重吸收、初生仔猪弱小和产奶量下降（Plumlee et al., 1956）。锰易穿过胎盘，所以母猪体内锰的营养状况可影响新初生仔猪体内锰的状况（Newland and Davis, 1961; Gamble et al., 1971）。基于锰的沉积状况，Kirchgessner 等（1981）估计妊娠母猪锰需要量为 25 ppm。饲喂玉米-豆粕型低锰（10 ppm）基础日粮的母猪比饲喂添加了 84 ppm 锰的基础日粮的母猪的总初生窝重低（Rheaume and Chavaz, 1989）。饲喂加锰日粮母猪的初乳和常乳锰浓度较高，但锰的存留仅表现在数字上的升高。Christianson 等（1989, 1990）报道，饲喂 10 ppm 或 20 ppm 锰比饲喂 5 ppm 锰的母猪所产仔猪初生重大，而且饲喂 20 ppm 的锰也有助于母猪再发情。

虽然锰的中毒水平并未很好确定，但当猪饲喂 4000 ppm 锰时可观察到采食量下降，生长速度降低（Leibholz et al., 1962）。日粮锰水平为 2000 ppm 时，会导致血红蛋白水平降低（Matrone et al., 1959）；500 ppm 时，生长猪生长速度降低，肢体僵硬（Grummer et al., 1950）。

硒

硒（Se）是谷胱甘肽过氧化物酶的组成成分（Rotruck et al., 1973），该酶可消除脂质过氧化物的毒性作用，保护细胞及亚细胞膜免受过氧化物的损伤。因此，硒与维生素 E 的相互节省效应源于二者均有抗氧化作用。但高水平维生素 E 并不能完全替代对硒的需要（Ewan et al., 1969; Bengtsson et al., 1978a, b; Hakkarainen et al., 1978）。已有研究表明，硒在甲状腺代谢中发挥着作用，因为碘甲腺原氨酸 5'-脱碘酶已被确定是一种含硒蛋白质（Arthur, 1994）。

日粮硒的需要量从断奶仔猪的 0.3 ppm 到育肥猪和母猪的 0.15 ppm 范围内变化（Groce et al., 1971, 1973a, b; Ku et al., 1973; Mahan et al., 1973; Ullrey, 1974; Young et al., 1976; Glienke and Ewan, 1977; Wilkinson et al., 1977a, b; Mahan and Moxon, 1978a, b, 1984; Piatkowski et al., 1979; Meyer et al., 1981; Lei et al., 1998）。日粮硒的需要量受日粮磷水平的影响（Lowry et al., 1985b），但不受日粮钙水平影响（Lowry et al., 1985a）。几种形态的硒，包括富硒酵母、亚硒酸钠和硒酸钠，均能有效地满足日粮硒的需要（Mahan and Magee, 1991; Suomi and Alaviuhkola, 1992; Mahan and Kim, 1996; Mahan and Parrett, 1996）。母猪体内硒营养状况影响其繁殖性能及哺乳和断奶仔猪体内硒的状况（Van Vleet et al., 1973; Mahan et al., 1977; Piatkowski et al., 1979; Chavez, 1985; Ramisz et al., 1993）。随着日粮硒水平的升高（0.1~0.3 ppm 或 0.5 ppm），机体中硒的总沉积量，以及生长猪、育肥猪、繁殖母猪及其哺乳仔猪的血清和组织中硒水平也随之升高；而且在各水平下，富硒酵母来源硒的存留量往往要高于亚硒酸钠（Mahan, 1995; Mahan and Kim, 1996; Mahan and Parrett, 1996; Mahan and Peters, 2004）。但无论是富硒酵母来源的硒还是亚硒酸钠来源的硒，超过 0.1 ppm 的硒都不能提高繁殖母猪血清谷胱甘肽过氧化酶的活性，二者的作用程度相似（Mahan and Kim, 1996）。当死胎率较高时，添加酵母硒或亚硒酸钠都可以降低死胎率（Yoon and McMillan, 2006）。在生长育肥猪中，不管是以富硒酵母形式还是亚硒酸钠形式补充至日粮硒水平达 0.1 ppm 时，血清硒浓度和谷胱甘肽过氧化物酶活性达到平台期，但在低水平添加时，酵母硒所引起的应答反应要低于亚硒酸钠，这提示富硒酵母中硒的生物学效价低于亚硒酸钠（Mahan and Parrett, 1996; Mahan et al., 1999a）。富硒酵母产品中约 50%的硒是蛋氨酸硒，其余为几种硒-氨基酸或其类似物中的一种（Mahan，1995）。维生素 E 和硒对公猪繁殖力各方面的影响已有所研究（Marin-Guzman et al., 1997, 2000a, b; Jacyno et al., 2002; Kolodziej and Jacyno, 2005; Echeverria-Alonzo et al., 2009）。在这些研究中，许多方面的指标［组织（血清、肝脏和睾丸）的谷胱甘肽过氧化物酶活性、硒和 α-生育酚的浓度、睾丸精子储备量、滋养细胞数量、次级精母细胞、每次射精的精子数、精子活力、正常精子比率、精子头部畸形情况以及细胞质滴的存留］均有改善。总的来讲，添加硒比维生素 E 的效果更明显。

美国和加拿大一些地区土壤含硒量较低。当仅使用这些地区生长的饲料原料配合日粮时，如果不额外添加硒就可能出现缺硒（Grant et al., 1961; Trapp et al., 1970; Ewan, 1971; Groce et al., 1971; Sharp et al., 1972a, b; Ku et al., 1973; Mahan et al., 1973, 1974; Diehl et al., 1975; Doornenbal, 1975; Piper et al., 1975; Wilkinson et al., 1977b; Bengtsson et al., 1978b）。然而，即使额外添加硒，动物组织中硒含量还是会更多地受到这些地区的饲料原料自身含硒量的影响（Mahan et al., 2005）。环境应激能增加缺硒的发生率和程度（Michel et al.,1969; Mahan et al., 1975）。

1974 年，美国食品和药物管理局（FDA）批准所有猪日粮中可添加 0.1 ppm 的硒。1982 年，FDA 批准对 20 kg 以内的猪日粮可添加 0.3 ppm 硒，因为添加 0.1 ppm 并不总能防止断奶仔猪出现硒缺乏症（Mahan and Moxon, 1978b; Meyer et al., 1981）。目前的法规允许所有阶段的猪日粮最高可添加 0.3 ppm 的硒（FDA, 1987a, b）。如 Ullrey（1992）所综述的，考虑到硒对环境的污染，应努力将硒的添加水平降低到 0.1 ppm，但 0.3 ppm 的水平仍一直保持着。

缺硒引起的主要生化变化是谷胱甘肽过氧化物酶活性降低（Thompson et al., 1976; Young et al., 1976; Fontaine and Valli, 1977）。所以，血浆谷胱甘肽过氧化酶水平是考察猪体内硒营养状况的可靠指标（Chavez, 1979a, b; Wegger et al.,1980; Adkins and Ewan, 1984）。猝死是缺硒病的典型特征（Ewan et al.,1969; Groce et al., 1971, 1973a, b）。缺硒剖检症状与维生素 E 缺乏症相同，包括：大面积肝坏死（营养性肝功能障碍），结肠袢、肺、皮下组织和胃黏膜下层水肿，骨骼肌苍白与营养不良（白肌病），心肌出现斑点和营养不良（桑葚心病），繁殖障碍，泌乳量下降和免疫应答变差（Orstadius et al., 1959; Lindberg and Siren, 1963, 1965; Trapp et al., 1970; Sharp et al., 1972a, b; Ruth and Van Vleet, 1974; Ullrey, 1974; Fontaine et al., 1977a, b, c; Nielsen et al., 1979; Sheffy and Schultz, 1979; Peplowski et al., 1980; Spallholz, 1980; Larsen and Tollersrud, 1981; Simesen et al., 1982）。

生长猪日粮中以亚硒酸钠、硒酸钠、蛋氨酸硒或含硒玉米形式添加低于 5 ppm 的硒都不会产生毒性，但添加 5 ppm（Mahan and Moxon, 1984; Kim and Mahan, 2001a, b）以及更高水平的硒时（Wahlstrom et al., 1955; Trapp et al., 1970; Herigstad et al., 1973; Goehring et al., 1984a, b）会引起中毒，且亚硒酸钠比酵母硒产生的硒中毒症状更快、更严重（Kim and Mahan, 2001a, b）。中毒症状包括食欲减退、掉毛、肝脏脂肪样浸润、肝与肾的退变、水肿、有时蹄壳从蹄冠处与皮分离（Miller, 1938; Miller and Williams, 1940; Wahlstrom et al., 1955; Orstadius, 1960; Lindberg and lannek, 1965; Herigstad et al., 1973），以及对称的空泡状病灶区和神经元坏死（Stowe and Herdt, 1992）。日粮中的砷有助于缓解硒中毒（Wahlstrom et al., 1955）。

锌

锌（Zn）是许多金属酶（包括 DNA 和 RNA 合成酶和转移酶及许多消化酶）的组成成分，也可与激素、胰岛素结合。所以，它在蛋白质、碳水化合物和脂类代谢中发挥着重要作用。另外，锌与锌指结构的转录、细胞内与细胞间信号向细胞核的传导都有关联。高剂量的锌通过促进胃分泌胃饥饿素（ghrelin）而刺激采食（Yin et al., 2009），也有报道称锌能提高几种胰酶的活力（Hedemann, 2006），增加大肠黏液面积，改变小肠上皮形态（Li et al., 2001）。

许多与日粮有关的因素都会影响锌的需要量（Miller et al., 1979），包括植酸或植酸盐（Oberleas et al., 1962; Oberleas, 1983）、钙（Tucker and Salmon, 1955; Hoekstra et al., 1956; Lewis, 1956, 1957a, b; Luecke, 1956, 1957; Stevenson and Earle, 1956; Bellis and Philp, 1957; Newland et al., 1958; Whiting and Bezeau, 1958; Berry et al., 1961; Hansard and Itoh, 1968; Morgan et al., 1969; Norrdin et al., 1973; Oberleas, 1983）、铜（Hoefer et al., 1960, O'Hara et al., 1960; Ritchie et al., 1963; Kirchgessner and Grassman, 1970）、镉（Pond et al., 1966）、钴（Hoekstra, 1970）、乙二胺四乙酸（EDTA）（Owen et al., 1973）、组氨酸（Dahmer et al., 1972a），以及蛋白质水平与来源（Smith et al., 1962; Dahmer et al., 1972b）。

饲喂酪蛋白-葡萄糖型日粮的仔猪对锌的需要量低（15 ppm），因为这种饲粮不含植酸盐等会降低锌利用率的因素（Smith et al., 1962; Shanklin et al., 1968）。但是，对饲喂含有植酸盐的传统日粮的断奶仔猪，需要添加 80 ppm 锌才能满足其需要（Van Heugten et al., 2003）。生长猪饲喂含大豆分离蛋白的半纯合日粮或含推荐钙水平的玉米-豆粕型日粮（二者都含有大量的植酸盐）时，锌的需要量约为 50 ppm（Lewis et al., 1956, 1957a, b; Luecke et al., 1956; Stevenson and Earle, 1956; Smith, 1958; 1962; Miller et al., 1970）。公猪锌需要量高于母猪，母猪又高于阉公猪（Liptrap et al., 1970; Miller et al., 1970）。饲喂过量水平的钙会使锌的需要量增加（Lewis et al., 1956; Forbes, 1960; Hoefer et al., 1960; Pond and Jones, 1964; Pond et al., 1964; Oberleas, 1983）。种猪对锌的需要量尚未很好确定，但由于胎儿的生长、乳汁的合成、子宫复旧过程中的组织修复及公猪精子的生成，种猪对锌的需要量可能高于生长猪。连续 5 胎饲喂含 33 ppm 锌的玉米-豆粕型日粮的母猪可获得最佳的妊娠表现，但对泌乳不然（Hedges, 1976）。Kirchgessner 等（1981）通过一个平衡试验估算出怀孕母猪的锌需要量是 25 ppm。然而，Payne 等（2006）证实，在含锌（以硫酸锌形式提供）100 ppm 的基础日粮中额外补充 100 ppm 有机锌，增加了仔猪初生窝重和断奶重。

生长猪缺锌的典型症状是皮肤角化过度，即所谓副皮炎（Kernkamp and Ferrin, 1953; Tucker and Salmon, 1955）。缺锌会降低生长速度和效率，以及血清中的锌、碱性磷酸酶和白蛋白的水平（Hoekstra et al., 1956, 1967; Luecke et al., 1957; Theuer and Hoekstra, 1966; Miller et al., 1968, 1970; Prasad et al., 1969, 1971; Ku et al., 1970）。若妊娠期的最后 4 周日粮锌水平较低（13 ppm）会延长分娩时间（Kalinowski and Chavez, 1984）。在妊娠和泌乳期采食缺锌日粮的母猪产仔数少而且个小，血清和组织锌水平降低（Pond and Jones, 1964; Hoekstra et al., 1967; Hill et al., 1983a, c, d），这些母猪所产乳的锌浓度也降低（Pond and Jones, 1964）。缺锌阻碍公猪的睾丸发育，耗竭生精上皮细胞，改变睾丸滋养细胞形态，并影响仔猪胸腺发育（Miller et al., 1968; Liptrap et al., 1970; Cigankova et al., 2008）。

日粮中各种锌盐中锌的生物学效价各不相同，且受日粮原料类型的影响（Miller, 1991）。硫酸锌、碳酸锌、氯化锌和金属锌粉中锌的利用率非常高（100%）。生物学效价的估测值是以占认可标准的百分比来表示，而不是吸收或者存留的百分比。吸收或存留的锌通常不到摄入量的 50%。氧化锌的锌利用率较低（50%~80%），硫化锌的锌利用率则更低（Miller, 1991）。有机复合物中的锌与硫酸锌中的锌的生物学效价近乎相同（Hill et al., 1986; Hahn and Baker, 1993; Wedekind et al., 1994; Schell and Kornegay, 1996; Swinkels et al., 1996; Cheng et al., 1998）。谷物与植物蛋白质中的锌利用率较低（Miller, 1991），但可以通过向日粮中添加微生物植酸酶得以改善（Kornegay, 1996）。

研究发现，教槽料中添加 3000 ppm 的锌（以氧化锌的形式提供），持续 14 天，可降低仔猪断奶后的腹泻率，并且提高了体增重（Poulsen, 1989），这一报道使人们对锌的药理作用产生了很大兴趣。一些研究已经证实这一发现结果，即氧化锌具有降低腹泻/下痢的作用（Rutkowska-Pejsak et al., 1998; Heo et al., 2010）。另有研究表明，即使在没有发生腹泻的情况下，氧化锌也能提高体增重（Hahn and

Baker, 1993; McCully, 1995; Hill et al, 1996; Case and Carlson, 2002; Hollis et al., 2005; Han and Thacker, 2009)。在有些研究中，锌的添加水平在2000 ppm到6000 ppm之间变化，饲喂时间长达5周。Ward等（1996）发现仔猪教槽料中添加蛋氨酸锌来源的锌250 ppm与添加氧化锌来源的锌2000 ppm相比，仔猪的生长性能有同等程度的改善；另一些研究也表明其他形式的锌与氧化锌也有类似的作用（Mavromichalis et al., 2001; Case and Carlson, 2002）。但也有研究未发现药理水平锌的有益影响（Fryer et al., 1992; Tokach et al., 1992; Schell and Kornegay, 1996）。在断奶仔猪日粮中添加高锌（3000 ppm，氧化锌）和高铜（250 ppm，硫酸铜）的试验研究中发现，两者在单独使用时都具有促生长作用，但同时添加未见加性效应（Smith et al., 1997; Hill et al., 2000）。不过，也有其他研究表明可利用形式的高锌、高铜的促生长作用具有加性效应（Perez et al., 2011a）。Hill等（2001）认为高锌可协同抗生素提高猪的生长性能。

玉米-豆粕型日粮中添加2000~4000 ppm碳酸锌来源的锌，导致生长猪出现锌中毒，表现为嗜睡、关节炎、腋下出血、胃炎和死亡。但1000 ppm的日粮锌无毒（Brink et al., 1959）。也有报道称生长猪饲喂氧化锌来源的锌2000~4000 ppm不会出现中毒症状（Cox and Hale, 1962; Hsu et al., 1975; Hill et al., 1983c）。然而，饲粮添加1000 ppm乳酸锌来源的锌，猪在2个月内变得跛足（Grimmett et al., 1937）。日粮高钙可减缓锌的毒性（Hsu et al., 1975）。连续2胎给母猪饲喂含锌（以氧化锌形式提供）5000 ppm的日粮，窝产仔数和仔猪断奶重均降低，并引发母猪软骨病（Hill and Miller, 1983; Hill et al., 1983a）。饲喂高锌日粮的母猪所产仔猪组织中铜水平降低，若饲喂低铜日粮便很快出现贫血症（Hill et al., 1983c, d）。因此，锌的毒性取决于锌的来源、日粮锌的水平、饲喂持续时间以及日粮中其他矿物元素的水平。猪日粮锌的最高耐受水平设定为1000 ppm，氧化锌除外，它可以更高水平添加并持续数周（NRC, 2005）。

参 考 文 献

Adeola, O., J. I. Orban, D. Ragland, T. R. Cline, and A. L. Sutton. 1998. Phytase and cholecalciferol supplementation of low-calcium and low-phosphorus diets for pigs. *Canadian Journal of Animal Science* 78:307-313.

Adkins, R. S., and R. C. Ewan. 1984. Effect of selenium on performance, serum selenium concentration and glutathione peroxidase activity in pigs. *Journal of Animal Science* 58:346-350.

Alcantara, P. F., L. E. Hanson, and J. D. Smith. 1980. Sodium requirements, balance and tissue composition of growing pigs. *Journal of Animal Science* 50:1092-1101.

Ammerman, C. B., and S. M. Miller. 1972. Biological availability of minor mineral ions: A review. *Journal of Animal Science* 35:681-694.

Ammerman, C. B., J. F. Standish, C. E. Holt, R. H. Houser, S. M. Miller, and G. E. Combs. 1974. Ferrous carbonates as sources of iron for weanling pigs and rats. *Journal of Animal Science* 38:52-58.

Ammerman, C. B., D. H. Baker, and A. J. Lewis, eds. 1995. *Bioavailability of Nutrients for Animals. Amino Acid, Minerals and Vitamins*. New York: Academic Press.

Amoikon, E. K., J. M. Fernandez, L. L. Southern, D. L. Thompson, Jr., T. L. Ward, and B. M. Olcott. 1995. Effect of chromium tripicolinate on growth, glucose tolerance, insulin sensitivity, plasma metabolites, and growth hormone in pigs. *Journal of Animal Science* 73:1123-1130.

Anderson, R. A. 1987. Chromium in animal tissues and fluids. Pp. 225-244 in *Trace Elements in Human and Animal Nutrition*, 5th Ed., Volume 1, W. Mertz, ed. New York: Academic Press.

Andrews, F. N., C. L. Shrewsbury, C. Harper, C. M. Vestal, and L. P. Doyle. 1948. Iodine deficiency in newborn sheep and swine. *Journal of Animal Science* 7:298-310.

Angel, R., N. M. Tamim, T. J. Applegate, A. S. Dhandu, and L. E. Ellestad. 2002. Phytic acid chemistry: Influence on phytin-phosphorus availability and phytase efficacy. *Journal of Applied Poultry Research* 11:471-480.

Apgar, G. A., and E. T. Kornegay. 1996. Mineral balance of finishing pigs fed copper sulfate or a copper-lysine complex at growth-stimulating levels. *Journal of Animal Science* 74:1594-1600.

Apgar, G. A., E. T. Kornegay, M. D. Lindemann, and D. R. Notter. 1995. Evaluation of copper sulfate and a copper lysine complex as growth promotants for weanling swine. *Journal of Animal Science* 73:2640-2646.

ARC (Agricultural Research Council). 1981. *The Nutrient Requirements of Pigs*, Technical Review Ed. Slough, UK: Commonwealth Agricultural Bureaux.

Arpi, T., and G. Tollerz. 1965. Iron poisoning in piglets: Autopsy findings in spontaneous and experimental cases. *Acta Veterinaria Scandinavica* 6:360-373.

Arrington, L. R., R. N. Taylor, Jr., C. B. Ammerman, and R. L. Shirley. 1965. Effects of excess dietary iodine upon rabbits, hamsters, rats and swine. *Journal of Nutrition* 87:394-398.

Arthur, J. R. 1994. The biochemical functions of selenium: Relationships to thyroid metabolism and antioxidant systems. Pp. 11-20 in *Rowett Research Institute Annual Report for 1993*. Aberdeen, UK: Rowett Research Institute, Bucksburn.

Arthur, S. R., E. T. Kornegay, H. R. Thomas, H. P. Veit, D. R. Notter, and R. A. Barczewski. 1983a. Restricted energy intake and elevated calcium and phosphorus intake for gilts during growth. III. Characterization of feet and limbs and soundness scores of sows during three parities. *Journal of Animal Science* 56:876-886.

Arthur, S. R., E. T. Kornegay, H. R. Thomas, H. P. Veit, D. R. Notter, K. E. Webb, Jr., and J. L. Baker. 1983b. Restricted energy intake and elevated calcium and phosphorus intake for gilts during growth. IV. Characterization of metacarpal, metatarsal, femur, humerus and turbinate bones of sows during three parities. *Journal of Animal Science* 57:1200-1214.

Austic, R. E., and C. C. Calvert. 1981. Nutritional interrelationships of electrolytes and amino acids. *Federation Proceedings* 40:63-67.

Baker, D. H. 1977. *Sulfur in Nonruminant Nutrition*. West Des Moines, IA: National Feed Ingredients Association.

Baker, D. H. 1991. Bioavailability of minerals and vitamins. Pp. 341-359 in *Swine Nutrition*, E. R. Miller, D. E. Ullrey, and A. J. Lewis, eds. Boston: Butterworth-Heinemann.

Barber, R. S., R. Braude, and K. G. Mitchell. 1955a. Studies on anemia in pigs. 1. The provision of iron by intramuscular injection. *Veterinary Record* 67:348-349.

Barber, R. S., R. Braude, K. G. Mitchell, and J. Cassidy. 1955b. High copper mineral mixtures for fattening pigs. *Chemistry and Industry* 21:601-603.

Bartley, J. C., E. F. Reber, J. W. Yusken, and H. W. Norton. 1961. Magnesium balance study in pigs three to five weeks of age. *Journal of Animal Science* 20:137-141.

Bayley, H. S., and R. G. Thomson. 1969. Phosphorus requirements of growing pigs and effect of steam pelleting on phosphorus availability. *Journal of Animal Science* 28:484-491.

Bayley, H. S., D. Arthur, G. H. Bowman, J. Pos, and R. G. Thomson. 1975a. Influence of dietary phosphorus level on growth and bone development in boars and gilts. *Journal of Animal Science* 40:864-870.

Bayley, H. S., J. Pos, and R. G. Thomson. 1975b. Influence of steam pelleting and dietary calcium level on the utilization of phosphorus by the pig. *Journal of Animal Science* 40:857-863.

Beeson, W. M., F. N. Andrews, H. L. Witz, and T. W. Perry. 1947. The effect of thyroprotein and thiouracil on the growth and fattening of swine. *Journal of Animal Science* 6:482 (Abstr.).

Bellis, D. B., and J. M. Philp. 1957. Effect of zinc, calcium and phosphorus on the skin and growth of pigs. *Journal of the Science of Food and Agriculture* 8:119-127.

Bengtsson, G., J. Hakkarainen, L. Jonsson, N. Lannek, and P. Lindberg. 1978a. Requirement for selenium (as selenite) and vitamin E (as alpha-tocopherol) in weaned pigs. I. The effect of varying alpha-tocopherol levels in a selenium deficient diet on the development of the VESD syndrome. *Journal of Animal Science* 47:143-152.

Bengtsson, G., J. Hakkarainen, L. Jonsson, N. Lannek, and P. Lindberg. 1978b. Requirement for selenium (as selenite) and vitamin E (as alpha-tocopherol) in weaned pigs. II. The effect of varying selenium levels in a vitamin E deficient diet on the development of the VESD syndrome. *Journal of Animal Science* 46:153-160.

Berry, R. K., M. C. Bell, R. B. Crainger, and R. G. Buescher. 1961. Influence of dietary calcium and zinc on calcium-45, phosphorus-32, and zinc-65 metabolism in swine. *Journal of Animal Science* 20:433-439.

Biehl, R. R., and D. H. Baker. 1996. Efficacy of supplemental 1 α-hydroxycholecalciferol and microbial phytase for young pigs fed phosphorus- or amino acid-deficient corn-soybean meal diets. *Journal of Animal Science* 74:2960-2966.

Biehl, R. R., J. L. Emmert, and D. H. Baker. 1997. Iron bioavailability in soybean meal as affected by supplemental phytase and 1 α-hydroxycholecalciferol. *Poultry Science* 76:1424-1427.

Blair, R., and D. Benzie. 1964. The effect of level of dietary calcium and phosphorus on skeletal development in the young pig to 25-lb live weight. *British Journal of Nutrition* 18:91-101.

Bohstedt, G., and R. H. Grummer. 1954. Salt poisoning of pigs. *Journal of Animal Science* 13:933-939.

Boleman, S. L., S. J. Boleman, T. D. Bidner, L. L. Southern, T. L. Ward, J. E. Pontif, and M. M. Pike. 1995. Effect of chromium picolinate on growth, body composition, and tissue accretion in pigs. *Journal of Animal Science* 73:2033-2042.

Bowland, J. P., R. Braude, A. G. Chamberlain, R. F. Glascock, and K. G. Mitchell. 1961. The absorption, distribution and excretion of labelled copper in young pigs given different quantities, as sulphate or sulphide, orally or intravenously. *British Journal of Nutrition* 15:59-72.

Boyd, R. D., B. D. Moser, E. R. Peo, Jr., A. J. Lewis, and R. K. Johnson. 1982. Effect of tallow and choline chloride addition to the diet of sows milk composition, milk yield and preweaning pig performance. *Journal of Animal Science* 54:1-7.

Boyd, R. D., D. Hall, and J. F. Wu. 1983. Plasma alkaline phosphatase as a criterion for determining biological availability of phosphorus for swine. *Journal of Animal Science* 57:396-401.

Brady, P. S., P. K. Ku, D. E. Ullrey, and E. R. Miller. 1978. Evaluation of an amino acid-iron chelate hematinic for the baby pig. *Journal of Animal Science* 47:1135-1140.

Braude, R. 1967. Copper as a stimulant in pig feeding (cuprum pro pecunia). *World Review of Animal Production* 3:69-82.

Braude, R., and E. Cotchin. 1949. Thiourea and methylthiouracil as supplements in rations of fattening pigs. *British Journal of Nutrition* 3:171-186.

Braude, R., A. G. Chamberlain, M. Kotarbinska, and K. G. Mitchell. 1962. The metabolism of iron in piglets given labeled iron either orally or by injection. *British Journal of Nutrition* 16:427-449.

Brennan, J. J., and F. X. Aherne. 1984. Effect of calcium and phosphorus levels in the diet on the incidence of leg weakness in swine. Pp. 8-10 in *Agriculture and Forestry Bulletin Special Issue*.

Brink, M. F., D. E. Becker, S. W. Terrill, and A. H. Jensen. 1959. Zinc toxicity in the weanling pig. *Journal of Animal Science* 18:836-842.

Bunch, R. J., J. T. McCall, V. C. Speer, and V. W. Hays. 1965. Copper supplementation for weanling pigs. *Journal of Animal Science* 24:995-1000.

Cabral, F., E. Vasconcelos, and C. M. D. S. Cordovil. 1998. Effects of solid phase from pig slurry on iron, copper, zinc, and manganese content of soil and wheat plants. *Journal of Plant Nutrition* 21:1955-1966.

Calabotta, D. F., E. T. Kornegay, H. R. Thomas, J. W. Knight, D. R. Notter, and H. P. Veit. 1982. Restricted energy intake and elevated calcium and phosphorus intake for gilts during growth. I. Feedlot performance and foot and leg measurements and scores during growth. *Journal of Animal Science* 54:565-575.

Calvert, C. C., R. J. Besecker, M. P. Plumlee, T. R. Cline, and D. M. Forsyth. 1978. Apparent digestibility of phosphorus in barley and corn for growing swine. *Journal of Animal Science* 47:420-426.

Campbell, E. A. 1961. Iron poisoning in the young pig. *Australian Veterinary Journal* 37:78-83.

Campbell, R. G. 1998. Chromium and its role in pig meat production. Pp. 229-237 in *Proceedings of Alltech's Fourteenth Annual Symposium*, T. P. Lyons and K. A. Jacques, eds. Loughborough, UK: Nottingham University Press.

Carnes, W. H., C. S. Shields, C. E. Cartwright, and M. M. Winthrop. 1961. Vascular lesions in copper-deficient swine. *Federation Proceedings* 20:118 (Abstr.).

Carson, T. L. 1986. Toxic chemicals, plants, metals and mycotoxins. Pp. 688-701 in *Diseases of Swine*, 6th Ed, A. D. Leman, B. Straw, R. D. Glock, W. L. Mengeling, R. H. C. Penny, and E. Scholl, eds. Ames: Iowa State University Press.

Carter, S. D., and G. L. Cromwell. 1998a. Influence of porcine somatotropin on the phosphorus requirement of finishing pigs. I. Performance and bone characteristics. *Journal of Animal Science* 76:584-595.

Carter, S. D., and G. L. Cromwell. 1998b. Influence of porcine somatotropin on the phosphorus requirement of finishing pigs. II. Carcass characteristics, tissue accretion rates, and chemical composition of the ham. *Journal of Animal Science* 76:596-605.

Case, C. L., and M. S. Carlson. 2002. Effect of feeding organic and inorganic sources of additional zinc on growth performance and zinc balance in nursery pigs. *Journal of Animal Science* 80:1917-1924.

Chaney, C. H., and C. E. Barnhart. 1963. Effect of iron supplementation of sow rations on the prevention of baby pig anemia. *Journal of Nutrition* 81:187-192.

Chapman, H. L., Jr., J. Kastelic, C. C. Ashton, and D. V. Catron. 1955. A comparison of phosphorus from different sources for growing and finishing swine. *Journal of Animal Science* 14:1073-1085.

Chapman, H. L., Jr., J. Kastelic, G. C. Ashton, P. G. Homeyer, C. Y. Roberts, D. V. Catron, V. W. Hays, and V. C. Speer. 1962. Calcium and phosphorus requirements for growing-finishing swine. *Journal of Animal Science* 21:112-118.

Chavez, E. R. 1979a. Effects of dietary selenium depletion and repletion on plasma glutathione peroxidase activity and selenium concentration in blood and body tissue of growing pigs. *Canadian Journal of Animal Science* 59:761-771.

Chavez, E. R. 1979b. Effects of dietary selenium on glutathione peroxidase activity in piglets. *Canadian Journal of Animal Science* 59:67-75.

Chavez, E. R. 1985. Nutritional significance of selenium supplementation in a semipurified diet fed during gestation and lactation to first-litter gilts and their piglets. *Canadian Journal of Animal Science* 64:497-506.

Cheng, J., E. T. Kornegay, and T. Schell. 1998. Influence of dietary lysine on the utilization of zinc from zinc sulfate and a zinc-lysine complex by young pigs. 1998. *Journal of Animal Science* 76:1064-1074.

Christianson, S. L., E. R. Peo, Jr., and A. J. Lewis. 1989. Effects of dietary manganese levels on reproductive performance of sows. *Journal of Animal Science* 67(Suppl. 1):251 (Abstr.).

Christianson, S. L., E. R. Peo, Jr., A. J. Lewis, and M. A. Giesemann. 1990. Influence of dietary manganese levels on reproduction, serum cholesterol and milk manganese concentration of sows. *Journal of Animal Science* 68(Suppl. 1):368 (Abstr.).

Cigankova, V., P. Mesaros, V. Almasiova, and J. Bire. 2008. Morphological changes of testes in zinc deficient boars. *Acta Veterinaria (Beograd)* 58:89-97.

Clawson, A. J., and W. D. Armstrong. 1981. Ammonium polyphosphate as a source of phosphorus and nonprotein nitrogen for monogastrics. *Journal of Animal Science* 52:1-7.

Coalson, J. A., C. V. Maxwell, J. C. Hillier, R. D. Washam, and E. C. Nelson. 1972. Calcium and phosphorus requirements of young pigs reared under controlled environmental conditions. *Journal of Animal Science* 35:1194-1200.

Coalson, J. A., C. V. Maxwell, J. C. Hillier, and E. C. Nelson. 1974. Calcium requirement of the Cesarean derived colostrum-free pig from 3 through 9 weeks of age. *Journal of Animal Science* 38:772-777.

Coffey, M. T., R. W. Seerley, and J. W. Mabry. 1982. The effect of source of supplemental dietary energy on sow milk yield, milk composition and litter performance. *Journal of Animal Science* 55:1388-1394.

Coffey, R. D., and G. L. Cromwell. 1993. Evaluation of the biological availability of phosphorus in various feed ingredients for growing pigs. *Journal of Animal Science* 71(Suppl. 1):66 (Abstr.).

Coffey, R. D., G. L. Cromwell, and H. J. Monegue. 1994a. Efficacy of a copper-lysine complex as a growth promotant for weanling pigs. *Journal of Animal Science* 72:2880-2886.

Coffey, R. D., K. W. Mooney, G. L. Cromwell, and D. K. Aaron. 1994b. Biological availability of phosphorus in defluorinated phosphates with different phosphorus solubilities in neutral ammonium citrate for chicks and pigs. *Journal of Animal Science* 72:2653-2660.

Combs, C. E., and H. D. Wallace. 1962. Growth and digestibility studies with young pigs fed various levels and sources of calcium. *Journal of Animal Science* 21:734-737.

Combs, G. E., J. M. Vandepopuliere, H. D. Wallace, and M. Koger. 1962. Phosphorus requirement of young pigs. *Journal of Animal Science* 21:3-8.

Combs, G. E., T. H. Berry, H. D. Wallace, and R. C. Crum, Jr. 1966. Levels and sources of vitamin D for pigs fed diets containing varying quantities of calcium. *Journal of Animal Science* 25:827-830.

Combs, N. R., and E. R. Miller. 1985. Determination of potassium availability in K_2CO_3, $KHCO_3$, corn and soybean meal for the young pig. *Journal of Animal Science* 60:715-719.

Combs, N. R., E. R. Miller, and P. K. Ku. 1985. Development of an assay to determine the bioavailability of potassium in feedstuffs for the young pig. *Journal of Animal Science* 60:709-714.

Combs, N. R., E. T. Kornegay, M. D. Lindemann, and D. R. Notter. 1991a. Calcium and phosphorus requirement of swine from weaning to market: I. Development of response curves for performance. *Journal of Animal Science* 69:673-681.

Combs, N. R., E. T. Kornegay, M. D. Lindemann, D. R. Notter, J. W. Wilson, and J. P. Mason. 1991b. Calcium and phosphorus requirement of swine from weaning to market: II. Development of response curves for bone criteria and comparison of bending and shear bone testing. *Journal of Animal Science* 69:682-693.

Corley, J. R., D. H. Baker, and R. A. Easter. 1980. Biological availability of phosphorus in rice bran and wheat bran as affected by pelleting. *Journal of Animal Science* 50:286-292.

Coulter, D. B., and M. J. Swenson. 1970. Effects of potassium intoxication on porcine electrocardiograms. *American Journal of Veterinary Research* 31:2001-2011.

Cox, D. H., and O. M. Hale. 1962. Liver iron depletion without copper loss in swine for excess zinc. *Journal of Nutrition* 77:225-228.

Cox, J. L., D. E. Becker, and A. H. Jensen. 1966. Electrocardiographic evaluation of potassium deficiency in young swine. *Journal of Animal Science* 25:203-206.

Crenshaw, T. D. 2001. Calcium, phosphorus, vitamin D, and vitamin K in swine nutrition. Pp. 187-212 in *Swine Nutrition*, 2nd Ed., A. J. Lewis and L. L. Southern, eds. Boca Raton, FL: CRC Press.

Crenshaw, T. D., E. R. Peo, Jr., A. J. Lewis, B. D. Moser, and D. Olson. 1981. Influence of age, sex and calcium and phosphorus levels on the mechanical properties of various bones in swine. *Journal of Animal Science* 52:1319-1329.

Cromwell, G. L. 1985. Phosphorus requirements of swine. Pp. 48-65 in *Proceedings of the 8th Annual International Minerals Conference*. Mundelein, IL: International Minerals and Chemical Corp.

Cromwell, G. L. 1992. The biological availability of phosphorus in feedstuffs for pigs. *Pig News and Information* 13:75N.

Cromwell, G. L. 1997. Copper as a nutrient for animals. Pp. 177-202 in *Handbook of Copper Compounds and Applications*, H. W. Richardson, ed. New York: Marcel Dekker.

Cromwell, G. L., V. W. Hays, C. H. Chaney, and J. R. Overfield. 1970. Effects of dietary phosphorus and calcium level on performance, bone mineralization and carcass characteristics of swine. *Journal of Animal Science* 30:519-525.

Cromwell, G. L., V. W. Hays, C. W. Scherer, and J. R. Overfield. 1972. Effects of dietary calcium and phosphorus on performance and carcass, metacarpal and turbinate characteristics of swine. *Journal of Animal Science* 34:746-751.

Cromwell, G. L., D. T. H. Sihombing, and V. W. Hays. 1975. Effects of iodine level on performance and thyroid traits of growing pigs. *Journal of Animal Science* 41:813-818.

Cromwell, G. L., V. W. Hays, J. R. Overfield, and J. L. Krug. 1976. Meat and bone meal as a source of phosphorus for growing swine. *Journal of Animal Science* 42:1350 (Abstr.).

Cromwell, G. L., V. W. Hays, and T. L. Clark. 1978. Effects of copper sulfate, copper sulfide and sodium sulfide on performance and copper stores of pigs. *Journal of Animal Science* 46:692-698.

Cromwell, G. L., T. S. Stahly, and W. D. Williams. 1981. Efficacy of copper as a growth promotant and its interrelation with sulfur and antibiotics for swine. *Proceedings of the Distillers Feed Conference, Distillers Feed Research Council* 36:15-29.

Cromwell, G. L., D. D. Hall, G. E. Combs, O. M. Hale, D. L. Handlin, J. P. Hitchcock, D. A. Knabe, E. T. Kornegay, M. D. Lindemann, C. V. Maxwell, and T. J. Prince. 1989a. Effects of dietary salt level during gestation and lactation on reproductive performance of sows: A cooperative study. *Journal of Animal Science* 67:374-385.

Cromwell, G. L., H. J. Monegue, and T. S. Stahly. 1989b. Effects of source and level of copper on performance and liver copper stores in weanling pigs. *Journal of Animal Science* 67:2996-3002.

Cromwell, G. L., H. J. Monegue, and T. S. Stahly. 1993a. Long-term effects of feeding a high copper diet to sows during gestation and lactation. *Journal of Animal Science* 71:2996-3002.

Cromwell, G. L., T. S. Stahly, R. D. Coffey, H. J. Monegue, and J. H. Randolph. 1993b. Efficacy of phytase in improving the bioavailability of phosphorus in soybean meal and corn-soybean meal diets for pigs. *Journal of Animal Science* 71:1831-1840.

Cromwell, G. L., R. D. Coffey, G. R. Parker, H. J. Monegue, and J. H. Randolph. 1995. Efficacy of a recombinant-derived phytase in improving the bioavailability of phosphorus in corn-soybean meal diets for pigs. *Journal of Animal Science* 73:2000-2008.

Cromwell, G. L., M. D. Lindemann, H. J. Monegue, D. D. Hall, and D. E. Orr, Jr. 1998a. Tribasic copper chloride and copper sulfate as copper sources for weanling pigs. *Journal of Animal Science* 76:118-123.

Cromwell, G. L., J. L. Pierce, T. E. Sauber, D. W. Rice, D. S. Etrl, and V. Raboy. 1998b. Bioavailability of phosphorus in low-phytic acid corn for growing pigs. *Journal of Animal Science* 76(Suppl. 2):54 (Abstr.).

Dahmer, E. J., B. W. Coleman, R. H. Grummer, and W. G. Hoekstra. 1972a. Alleviation of parakeratosis in zinc-deficient swine by high levels of dietary histidine. *Journal of Animal Science* 35:1181-1189.

Dahmer, E. J., R. H. Grummer, and W. G. Hoekstra. 1972b. Prevention of zinc deficiency in swine by feeding blood meal. *Journal of Animal Science* 35:1176-1180.

Davis, C. M., and J. B. Vincent. 1997. Chromium oligopeptide activates insulin receptor tyrosine kinase activity. *Biochemistry* 36:4382-4385.

Davis, C. M., K. H. Sumrall, and J. B. Vincent. 1996. The biologically active form of chromium may activate a membrane phosphotyrosine phosphatase (PTP). *Biochemistry* 35:12963-12969.

Dersjant-Li, Y., H. Schulze, J. W. Schrama, J. A. Verreth, and M. W. A. Verstegen. 2001. Feed intake, growth, digestibility of dry matter and nitrogen in young pigs as affected by dietary cation–anion difference and supplementation of xylanase. *Journal of Animal Physiology and Animal Nutrition* 85:101-109.

Devilat, J., and A. Skoknic. 1971. Feeding high levels of rapeseed meal to pregnant gilts. *Canadian Journal of Animal Science* 51:715-719.

De Wilde, R. O., and J. Jourquin. 1992. Estimation of digestible phosphorus requirements in growing-finishing pigs by carcass analysis. *Journal of Animal Physiology and Animal Nutrition* 68:218.

Diehl, J. S., D. C. Mahan, and A. L. Moxon. 1975. Effects of single intramuscular injections of selenium at various levels to young swine. *Journal of Animal Science* 40:844-850.

Doige, C. E., B. D. Owen, and J. H. L. Mills. 1975. Influence of calcium and phosphorus on growth and skeletal development of growing swine. *Canadian Journal of Animal Science* 55:147-164.

Doornenbal, H. 1975. Tissue selenium content of the growing pig. *Canadian Journal of Animal Science* 55:325-330.

Dove, C. R. 1993a. The effect of adding Cu and various fat sources to the diets of weanling swine on growth performance and serum fatty acid profiles. *Journal of Animal Science* 71:2187-2192.

Dove, C. R. 1993b. The effect of Cu supplementation during lactation on sow and pig performance and the subsequent nursery performance of pigs. *Journal of Animal Science* 71(Suppl. 1):173 (Abstr.).

Dove, C. R. 1995. The effect of Cu level on nutrient utilization of weanling pigs. *Journal of Animal Science* 73:166-171.

Dove, C. R., and K. D. Haydon. 1992. The effect of Cu and fat addition to the diets of weanling swine on growth performance and serum fatty acids. *Journal of Animal Science* 70:805-810.

Ducsay, C. A., W. C. Buhi, F. W. Bazer, R. M. Roberts, and C. E. Combs. 1984. Role of uteroferrin in placental iron transport: Effect of maternal iron treatment on fetal iron and uteroferrin content and neonatal hemoglobin. *Journal of Animal Science* 59:1303-1308.

Düngelhoef, M., M. Rodehutschord, H. Spiekers, and E. Pfeffer. 1994. Effects of supplemental microbial phytase on availability of phosphorus contained in maize, wheat and triticale to pigs. *Animal Feed Science and Technology* 49:1-10.

Echeverria-Alonzo S., R. Santos-Ricalde, F. Centurion-Castro, R. Ake-Lopez, M. Alfaro-Gamboa, and J. Rodriguez-Buenfil. 2009. Effects of dietary selenium and vitamin E on semen quality and sperm morphology of young boars during warm and fresh season. *Journal of Animal and Veterinary Advances* 8:2311-2317.

Edmonds, M. S., O. A. Izquierdo, and D. H. Baker. 1985. Feed additive studies with newly weaned pigs: Efficacy of supplemental copper, antibiotics and organic acids. *Journal of Animal Science* 60:462-469.

Eeckhout, W., and M. De Paepe. 1997. The digestibility of three calcium phosphates for pigs as measured by difference and by slope-ratio assay. *Journal of Animal Physiology and Animal Nutrition* 77:53-60.

Eeckhout, W., M. De Paepe, N. Warnants, and H. Bekaert. 1995. An estimation of the minimal P requirements for growing-finishing pigs, as influenced by the Ca level of the diet. *Animal Feed Science and Technology* 52:29-40.

Ekpe, E. D., R. T. Zijlstra, and J. F. Patience. 2002. Digestible phosphorus requirement of grower pigs. *Canadian Journal of Animal Science* 82:541-549.

Elvehjem, C. A., and E. B. Hart. 1932. The necessity of copper as a supplement to iron for hemoglobin formation in the pig. *Journal of Biological Chemistry* 95:363-370.

Erdman, J. W., Jr. 1979. Oilseed phytates: Nutritional implications. *Journal of the American Oil Chemists' Society* 56:736-741.

Everts, H., A. Jongbloed, and R. A. Dekker. 1998a. Calcium, magnesium and phosphorus balance of sows during lactation for three parities. *Livestock Production Science* 55:109-115.

Everts, H., A. Jongbloed, and R. A. Dekker. 1998b. Calcium, phosphorus and magnesium retention and excretion in pregnant sows during three parities. *Livestock Production Science* 55:113-121.

Evock-Clover, C. M., M. M. Polansky, R. A. Anderson, and N. C. Steele. 1993. Dietary chromium supplementation with or without somatotropin treatment alters serum hormones and metabolites in growing pigs without affecting growth performance. *Journal of Nutrition* 123:1504-1512.

Ewan, R. C. 1971. Effect of vitamin E and selenium on tissue composition of young pigs. *Journal of Animal Science* 32:883-887.

Ewan, R. C., M. E. Wastell, E. J. Bicknell, and V. C. Speer. 1969. Performance and deficiency symptoms of young pigs fed diets low in vitamin E and selenium. *Journal of Animal Science* 29:912-915.

Fakler, T. M., T. L. Ward, E. B. Kegley, M. T. Socha, A. B. Johnson, and C. V. Maxwell. 1999. Metabolic effects of dietary chromium-L-methionine in growing pigs. P. 110 in *Proceedings of the 10th International Symposium on Trace Elements in Man and Animal*, May 2-7, Evian, France.

Fammatre, C. A., D. C. Mahan, A. W. Fetter, A. P. Grifo, Jr., and J. K. Judy. 1977. Effects of dietary protein, calcium and phosphorus levels for growing and finishing swine. *Journal of Animal Science* 44:65-71.

Farmer, C., S. Robert, and J. J. Matte. 1996. Lactation performance of sows fed a bulky diet during gestation and receiving growth hormone-releasing factor during lactation. *Journal of Animal Science* 74:1298-1306.

Farries, F. E. 1958. Kali-Briefe (Tierzucht) no. 3. 1981. P. 290 in *The Nutrient Requirements of Pigs*. Slough, UK: Commonwealth Agricultural Bureaux.

FDA (Food and Drug Administration). 1987a. Food additives permitted in feed and drinking water of animals: Selenium. *Federal Register* 52 (April 6):10887.

FDA. 1987b. Food additives permitted in feed and drinking water of animals: Selenium; Correction. *Federal Register* 52 (June 4):21001.

Feng, J., W. Q. Ma, Z. R. Xu, Y. Z. Wang, and J. X. Liu. 2007. Effects of iron glycine chelate on growth, haematological and immunological characteristics in weanling pigs. *Animal Feed Science and Technology* 134:261-272.

Feng, J., W. Q. Ma, Z. R. Xu, J. X. He, Y. Z. Wang, and J. X. Liu. 2009. The effect of iron glycine chelate on tissue mineral levels, fecal mineral concentration, and liver antioxidant enzyme activity in weanling pigs. *Animal Feed Science and Technology* 150:106-113.

Follis, R. H., Jr., J. A. Bush, G. E. Cartwright, and M. M. Wintrobe. 1955. Studies on copper metabolism. XVIII. Skeletal changes associated with copper deficiency in swine. *Bulletin of the Johns Hopkins Hospital* 97:405.

Fontaine, M., and V. E. O. Valli. 1977. Studies on vitamin E and selenium deficiency in young pigs. II. The hydrogen peroxide hemolysis test and the measure of red cell lipid peroxides as indices of vitamin E and selenium status. *Canadian Journal of Comparative Medicine* 41:52-56.

Fontaine, M., V. E. O. Valli, and L. G. Young. 1977a. Studies on vitamin E and selenium deficiency in young pigs. IV. Effect on coagulation system. *Canadian Journal of Comparative Medicine* 41:64-76.

Fontaine, M., V. E. O. Valli, and L. G. Young. 1977b. Studies on vitamin E and selenium deficiency in young pigs. III. Effect on kinetics of erythrocyte production and destruction. *Canadian Journal of Comparative Medicine* 41:57-63.

Fontaine, M., V. E. O. Valli, L. G. Young, and J. H. Lumsden. 1977c. Studies on vitamin E and selenium deficiency in young pigs. I. Hematological and biochemical changes. *Canadian Journal of Comparative Medicine* 41:41-51.

Forbes, R. M. 1960. Nutritional interactions in zinc and calcium. *Federation Proceedings* 19:643-647.

Friend, D. W., and M. S. Wolynetz. 1981. Self-selection of salt by gilts during pregnancy and lactation. *Canadian Journal of Animal Science* 61:429-438.

Fryer, A. J., P. K. Ku, E. R. Miller, and D. E. Ullrey. 1992. Effect of elevated dietary zinc on growth performance of weanling swine. *Journal of Animal Science* 70(Suppl. 1):62 (Abstr.).

Fuller, R., L. G. M. Newland, C. A. E. Briggs, R. Braude, and K. G. Mitchell. 1960. The normal intestinal flora of the pig. IV. The effect of dietary supplements of penicillin, chlortetracyline or copper sulphate on the faecal flora. *Journal of Applied Bacteriology* 23:195-205.

Furugouri, K. 1972. Effect of elevated dietary levels of iron on iron store in liver, some blood constituents and phosphorus deficiency in young swine. *Journal of Animal Science* 34:573-577.

Furugouri, K., and A. Kawabata. 1975. Iron absorption in nursing piglets. *Journal of Animal Science* 41:1348-1354.

Furugouri, K., and A. Kawabata. 1976. Iron absorption by neonatal pig intestine in vivo. *Journal of Animal Science* 42:1460-1464.

Furugouri, K., and A. Kawabata. 1979. Iron absorptive function of neonatal pig intestine. *Journal of Animal Science* 49:715-723.

Gamble, C. T., S. L. Hansard, B. R. Moss, D. J. Davis, and E. R. Lidvall. 1971. Manganese utilization and placental transfer in the gravid gilt. *Journal of Animal Science* 32:84-87.

Garcia, M. R., M. D. Newcomb, and W. E. Trout. 1997. Effects of dietary chromium picolinate supplementation on glucose tolerance and ovarian and uterine function in gilts. *Journal of Animal Science* 75:82 (Abstr.).

GfE (Society of Nutrition Physiology). 2008. *Energy and Nutrient Requirements of Livestock, No. 11: Recommendations for the Supply of Energy and Nutrients to Pigs.* Committee for Requirement Standards of the GfE. Frankfurt am Main, Germany: DLG-Verlag.

Giesemann, M. A., A. J. Lewis, P. S. Miller, and M. P. Akhter. 1998. Effects of the reproductive cycle and age on calcium and phosphorus metabolism and bone integrity of sows. *Journal of Animal Science* 76:796-807.

Gleed, P. T., and B. F. Sansom. 1982. Ingestion of iron in sow's faeces by piglets reared in farrowing crates with slotted floors. *British Journal of Nutrition* 47:113-117.

Glienke, L. R., and R. C. Ewan. 1977. Selenium deficiency in the young pig. *Journal of Animal Science* 45:1334-1340.

Goehring, T. B., I. S. Palmer, O. E. Olson, C. W. Libal, and R. C. Wahlstrom. 1984a. Effects of seleniferous grains and inorganic selenium on tissue and blood composition and growth performance of rats and swine. *Journal of Animal Science* 59:725-732.

Goehring, T. B., I. S. Palmer, O. E. Olson, C. W. Libal, and R. C. Wahlstrom. 1984b. Toxic effects of selenium on growing swine fed corn-soybean meal diets. *Journal of Animal Science* 59:733-737.

Golz, D. I., and T. D. Crenshaw. 1990. Interrelationships of dietary sodium, potassium and chloride on growth in young swine. *Journal of Animal Science* 68:2736-2747.

Golz, D. I., and T. D. Crenshaw. 1991. The effect of dietary potassium and chloride on cation-anion balance in swine. *Journal of Animal Science* 69:2504-2515.

Grandhi, R. R., and J. H. Strain. 1983. Dietary calcium-phosphorus levels for growth and reproduction in gilts and sows. *Canadian Journal of Animal Science* 63:443-454.

Grant, C. A., B. Thafvelin, and R. Christell. 1961. Retention of selenium by pig tissues. *Acta Pharmacologica et Toxicologica* 18:285-297.

Grimmett, R. E. R., I. G. McIntosh, E. M. Wall, and C. S. M. Hopkirk. 1937. Chromium zinc poisoning of pigs. Results of experimental feeding of pure zinc lactate. *New Zealand Journal of Agriculture* 54:216-223.

Groce, A. W., E. R. Miller, K. K. Keahey, D. E. Ullrey, and D. J. Ellis. 1971. Selenium supplementation of practical diets for growing-finishing swine. *Journal of Animal Science* 32:905-911.

Groce, A. W., E. R. Miller, J. P. Hitchcock, D. E. Ullrey, and W. T. Magee. 1973a. Selenium balance in the pig as affected by selenium source and vitamin E. *Journal of Animal Science* 37:942-947.

Groce, A. W., E. R. Miller, D. E. Ullrey, P. K. Ku, K. K. Keahey, and D. J. Ellis. 1973b. Selenium requirements in corn-soy diets for growing-finishing swine. *Journal of Animal Science* 37:948-956.

Grummer, R. H., O. G. Bentley, P. H. Phillips, and G. Bohstedt. 1950. The role of manganese in growth, reproduction and lactation of swine. *Journal of Animal Science* 9:170-175.

Guan, X., J. J. Matte, P. K. Ku, J. L. Snow, J. L. Burton, and N. L. Trottier. 2000. High chromium yeast supplementation improves glucose tolerance in pigs by decreasing hepatic extraction of insulin. *Journal of Nutrition* 130:1274-1279.

Guimaraes, J., C. L. Zhu, and C. F. M. de Lange. 2009. High dietary potassium levels appear to limit co-products usage in grower-finisher pig diets. *Canadian Journal of Animal Science* 89:60 (Abstr.).

Hagen, C. D., M. D. Lindemann, and K. W. Purser. 2000. Effect of dietary chromium tripicolinate on productivity of sows under commercial conditions. *Swine Health and Production* 8:59-63.

Hagsten, I., and T. W. Perry. 1976a. Evaluation of dietary salt levels for swine. I. Effect on gain, water consumption and efficiency of feed conversion. *Journal of Animal Science* 42:1187-1190.

Hagsten, I., and T. W. Perry. 1976b. Evaluation of dietary salt levels for swine. II. Effect on blood and excretory patterns. *Journal of Animal Science* 42:1191-1195.

Hagsten, I., T. R. Cline, T. W. Perry, and M. P. Plumlee. 1976. Salt supplementation of corn-soy diets for swine. *Journal of Animal Science* 42:12-15.

Hahn, J. D., and D. H. Baker. 1993. Growth and plasma zinc responses of young pigs fed pharmacologic levels of zinc. *Journal of Animal Science* 71:3020-3024.

Hakkarainen, J., P. Lindberg, G. Bengtsson, L. Jonsson, and N. Lannek. 1978. Requirement for selenium (as selenite) and vitamin E (as alpha-tocopherol) in weaned pigs. III. The effect on the development of the VESD syndrome of varying selenium levels with a low-tocopherol diet. *Journal of Animal Science* 46:1001-1008.

Hall, D. D., G. L. Cromwell, and T. S. Stahly. 1991. Effects of dietary calcium, phosphorus, calcium:phosphorus ratio and vitamin K on performance, bone strength and blood clotting status of pigs. *Journal of Animal Science* 69:646-655.

Han, Y. K., and P. A. Thacker. 2009. Performance, nutrient digestibility and nutrient balance in weaned pigs fed diets supplemented with antibiotics or zinc oxide. *Journal of Animal and Veterinary Advances* 8:868-875.

Hancock, J. E., E. R. Peo, Jr., A. J. Lewis, J. D. Crenshaw, and B. D. Moser. 1986. Vitamin D toxicity in young pigs. *Journal of Animal Science* 63(Suppl. 1):268 (Abstr.).

Hansard, S. L., and H. Itoh. 1968. Influence of limited dietary calcium upon zinc absorption, placental transfer and utilization by swine. *Journal of Nutrition* 95:23-30.

Hansen, B. C., A. J. Lewis, and E. R. Peo, Jr. 1987. Bone traits of growing boars, barrows and gilts fed different levels of dietary protein and available phosphorus. *Journal of Animal Science* 65(Suppl. 1):126 (Abstr.).

Hansen, S. L., N. Trakooljul, H. C. Liu, A. J. Moeser, and J. W. Spears. 2009. Iron transporters are differentially regulated by dietary iron, and modifications are associated with changes in manganese metabolism in young pigs. *Journal of Nutrition* 139:1474-1479.

Harmon, B. G., D. E. Becker, and A. H. Jensen. 1967. Efficacy of ferric ammonium citrate in preventing anemia in young swine. *Journal of Animal Science* 26:1051-1053.

Harmon, B. G., D. E. Hoge, A. H. Jensen, and D. H. Baker. 1969. Efficacy of ferrous carbonate as a hematinic for young swine. *Journal of Animal Science* 29:706-710.

Harmon, B. C., S. G. Cornelius, J. Totsch, D. H. Baker, and A. H. Jensen. 1974a. Oral iron dextran and iron from steel slats as hematinics for swine. *Journal of Animal Science* 39:699-702.

Harmon, B. G., C. T. Liu, S. G. Cornelius, J. E. Pettigrew, D. H. Baker, and A. H. Jensen. 1974b. Efficacy of different phosphorus supplements for sows during gestation and lactation. *Journal of Animal Science* 39:1117-1123.

Harmon, B. G., C. T. Liu, A. H. Jensen, and D. H. Baker. 1975. Phosphorus requirements of sows during gestation and lactation. *Journal of Animal Science* 40:660-664.

Harmon, B. G., C. T. Liu, A. H. Jensen, and D. H. Baker. 1976. Dietary magnesium levels for sows during gestation and lactation. *Journal of Animal Science* 42:860-865.

Harris, J. E., S. D. Crow, and M. D. Newcomb. 1995. Effect of chromium picolinate on growth performance and carcass characteristics on pigs fed adequate and low-protein diets. *Journal of Animal Science* 73(Suppl. 1):194 (Abstr.).

Harrold, R. L. 1981. Digestible energy and available phosphorus content of sunflower seed products. *Journal of Animal Science* 53(Suppl. 1):516 (Abstr.).

Hart, E. B., and H. Steenbock. 1918. Hairless pigs: The cause and remedy. *Wisconsin Agricultural Experiment Station Bulletin* 297:1-11.

Hart, E. B., C. A. Elvehjem, H. Steenbock, A. R. Kemmerer, G. Bohstedt, and J. M. Fargo. 1930. A study of the anemia of young pigs and its prevention. *Journal of Nutrition* 2:277-294.

Hastad, C. W., S. S. Dritz, M. D. Tokach, R. D. Goodband, J. L. Nelssen, J. M. DeRouchey, R. D. Boyd, and M. E. Johnston. 2004. Phosphorus requirements of growing-finishing pigs reared in a commercial environment. *Journal of Animal Science* 82:2945-2952.

Haydon, K. K., J. W. West, and M. N. McCarter. 1993. Effect of dietary electrolyte balance on performance and blood parameters of growing-finishing swine fed in high ambient temperatures. *Journal of Animal Science* 68:2400-2406.

Hays, V. W. 1976. *Phosphorus in Swine Nutrition*. West Des Moines, IA: National Feed Ingredients Association.

Hedemann, M. S., B. B. Jensen, and H. D. Poulsen. 2006. Influence of dietary zinc and copper on digestive enzyme activity and intestinal morphology in weaned pigs. *Journal of Animal Science* 84:3310-3320.

Hedges, J. D., and E. T. Kornegay. 1973. Interrelationship of dietary copper and iron as measured by blood parameters, tissue stores and feedlot performance of swine. *Journal of Animal Science* 37:1147-1154.

Hedges, J. D., E. T. Kornegay, and H. R. Thomas. 1976. Comparison of dietary zinc levels for reproducing sows and the effect of dietary zinc and calcium on the subsequent performance of their progeny. *Journal of Animal Science* 43:453-463.

Hendriks, W. H., and P. J. Moughan. 1993. Whole-body mineral composition of entire male and female pigs depositing protein at maximum rates. *Livestock Production Science* 33:161-170.

Heo, J. M., J. C. Kim, C. F. Hansen, B. P. Mullan, D. J. Hampson, H. Maribo, N. Kjeldsen, J. R. Pluske. 2010. Effects of dietary protein level and zinc oxide supplementation on the incidence of post-weaning diarrhoea in weaner pigs challenged with an enterotoxigenic strain of *Escherichia coli*. *Livestock Science* 133:210-213.

Herigstad, R. R., C. K. Whitehair, and O. E. Olson. 1973. Inorganic and organic selenium toxicosis in young swine: Comparison of pathologic changes with those in swine with vitamin E-selenium deficiency. *American Journal of Veterinary Research* 34:1227-1238.

Hew, V. F., G. L. Cromwell, and T. S. Stahly. 1982. The bioavailability of phosphorus in some tropical feedstuffs for pigs. *Journal of Animal Science* 55(Suppl. 1):277 (Abstr.).

Hickman, D. S., D. C. Mahan, and J. H. Cline. 1983. Dietary calcium and phosphorus for developing boars. *Journal of Animal Science* 56:431-437.

Hill, D. A., E. R. Peo, Jr., A. J. Lewis, and J. D. Crenshaw. 1986. Zinc amino acid complexes for swine. *Journal of Animal Science* 63:121-130.

Hill, G. M., and E. R. Miller. 1983. Effect of dietary zinc levels on the growth and development of the gilt. *Journal of Animal Science* 57:106-113.

Hill, G. M., and J. W. Spears. 2001. Trace and ultratrace elements in swine nutrition. Pp. 229-261 in *Swine Nutrition*, 2nd Ed., A. J. Lewis and L. L. Southern, eds. Boca Raton, FL: CRC Press.

Hill, G. M., P. K. Ku, E. R. Miller, D. E. Ullrey, T. A. Losty, and B. L. O'Dell. 1983a. A copper deficiency in neonatal pigs induced by a high zinc maternal diet. *Journal of Nutrition* 113:867-872.

Hill, G. M., E. R. Miller, and P. K. Ku. 1983b. Effect of dietary zinc levels on mineral concentration in milk. *Journal of Animal Science* 57:123-129.

Hill, G. M., E. R. Miller, and H. D. Stowe. 1983c. Effect of dietary zinc levels on health and productivity of gilts and sows through two parities. *Journal of Animal Science* 57:114-122.

Hill, G. M., E. R. Miller, P. A. Whetter, and D. E. Ullrey. 1983d. Concentrations of minerals in tissues of pigs from dams fed different levels of dietary zinc. *Journal of Animal Science* 57:130-138.

Hill, G. M., G. L. Cromwell, T. D. Crenshaw, R. C. Ewan, D. A. Knabe, A. J. Lewis, D. C. Mahan, G. C. Shurson, L. L. Southern, and T. L. Veum. 1996. Impact of pharmacological intakes of zinc and (or) copper on performance of weanling pigs. *Journal of Animal Science* 74(Suppl. 1):181 (Abstr.).

Hill, G. M., G. L. Cromwell, T. D. Crenshaw, C. R. Dove, R. C. Ewan, D. A. Knabe, A. J. Lewis, G. W. Libal, D. C. Mahan, G. C. Shurson, L. L. Southern, and T. L. Veum. 2000. Growth promotion effects and plasma changes from feeding high dietary concentrations of zinc and copper to weanling pigs (regional study). *Journal of Animal Science* 78:1010-1016.

Hill, G. M., D. C. Mahan, S. D. Carter, G. L. Cromwell, R. C. Ewan, R. L. Harrold, A. J. Lewis, P. S. Miller, G. C. Shurson, and T. L. Veum. 2001. Effect of pharmacological concentrations of zinc oxide with or without the inclusion of an antibacterial agent on nursery pig performance. *Journal of Animal Science* 79:934-941.

Hinson, R. B. 2005. The effect of low nutrient excretion diets on growth performance, carcass characteristics, and nutrient mass balance in swine reared under both research and commercial settings, Ph.D. Dissertation, Purdue University, West Lafayette, IN.

Hinson, R. B., A. P. Schinckel, J. S. Radcliffe, G. L. Allee, A. L. Sutton, and B. T. Richert. 2009. Effect of feeding reduced crude protein and phosphorus diets on weaning-finishing pig growth performance, carcass characteristics and bone characteristics. *Journal of Animal Science* 87:1502-1517.

Hitchcock, J. P., P. K. Ku, and E. R. Miller. 1974. Factors influencing iron utilization by the baby pig. Pp. 598-600 in *Trace Element Metabolism in Animals, Volume A*, W. G. Hoekstra, J. W. Suttie, H. E. Ganther, and W. Mertz, eds. Baltimore, MD: University Park Press.

Hoefer, J. A., E. R. Miller, D. E. Ullrey, H. D. Ritchie, and R. W. Luecke. 1960. Interrelationships between calcium, zinc, iron and copper in swine feeding. *Journal of Animal Science* 19:249-259.

Hoekstra, W. G. 1970. The complexity of dietary factors affecting zinc nutrition and metabolism in chicks and swine. Pp. 347-353 in *Trace Element Metabolism in Animals*, C. F. Mills, ed. Edinburgh, UK: E. & S. Livingstone.

Hoekstra, W. G., P. K. Lewis, P. H. Phillips, and R. H. Grummer. 1956. The relationship of parakeratosis, supplemental calcium and zinc to the zinc content of certain body components of swine. *Journal of Animal Science* 15:752-764.

Hoekstra, W. G., E. C. Faltin, C. W. Lin, H. F. Roberts, and R. H. Grummer. 1967. Zinc deficiency in reproducing gilts fed a diet high in calcium and its effect on tissue zinc and blood serum alkaline phosphatase. *Journal of Animal Science* 26:1348-1357.

Hollis, G. R., S. D. Carter, T. R. Cline, T. D. Crenshaw, G. L. Cromwell, G. M. Hill, S. W. Kim, A. J. Lewis, D. C. Mahan, P. S. Miller, H. H. Stein, and T. L. Veum. 2005. Effects of replacing pharmacological levels of dietary zinc oxide with lower dietary levels of various organic zinc sources for weanling pigs. *Journal of Animal Science* 83:2123-2129.

Honeyfield, D. C., and J. A. Froseth. 1985. Effects of dietary sodium and chloride on growth, efficiency of feed utilization, plasma electrolytes and plasma basic amino acids in young pigs. *Journal of Nutrition* 115:1366-1371.

Honeyfield, D. C., J. A. Froseth, and R. J. Barke. 1985. Dietary sodium and chloride levels for growing-finishing pigs. *Journal of Animal Science* 60:691-698.

Hsu, F. S., L. Krook, W. G. Pond, and J. R. Duncan. 1975. Interactions of dietary calcium with toxic levels of lead and zinc in pigs. *Journal of Nutrition* 105:112-118.

Hu, Q., L. Yang, J. Fang, S. X. Wang, X. G. Shu, Z. Y. Deng, G. Liu, M. Z. Fan, and Z. Ruan. 2010. Estimating an optimal ratio of true digestible Ca:P in corn-rough-soybean meal-based diets for 20-50 kg growing pigs. *Journal of Food, Agriculture and Environment* 8:556-562.

Huang, K. C., and G. L. Allee. 1981. Bioavailability of phosphorus in selected feedstuffs for young chicks and pigs. *Journal of Animal Science* 53(Suppl. 1):248 (Abstr.).

Huck, D. W., and A. J. Clawson. 1976. Cobalt toxicity in pigs. *Journal of Animal Science* 43:253 (Abstr.).

Hughes, E. H., and N. R. Ittner. 1942. The potassium requirement of growing pigs. *Journal of Agricultural Research* 64:189-192.

Jackson, A. R., S. Powell, S. L. Johnston, J. O. Matthews, T. D. Bidner, F. R. Valdez, and L. L. Southern. 2009. The effect of chromium as chromium propionate on growth performance, carcass traits, meat quality, and the fatty acid profile of fat from pigs fed no supplemented dietary fat, choice white grease, or tallow. *Journal of Animal Science* 87:4032-4041.

Jacyno, E., M. Kawecka, M. Kamyczek, A. Kolodziej, J. Owsianny, and B. Delikator. 2002. Influence of inorganic Se + vitamin E and organic Se + vitamin E on reproductive performance of young boars. *Agricultural and Food Science Finland* 11:175-184.

Jensen, A. H., S. W. Terrill, and D. E. Becker. 1961. Response of the young pig to levels of dietary potassium. *Journal of Animal Science* 20:464-467.

Johansen, K., and H. D. Poulsen. 2003. Substitution of inorganic phosphorus in pig diets by microbial phytase supplementation—A review. *Pig News Information* 24:77N-82N.

Johnson, S. R. 1944. Studies with swine on low manganese rations of natural foodstuffs. *Journal of Animal Science* 3:136-142.

Jondreville, C., and J. Y. Dourmad. 2005. Phosphorus in pig nutrition. *Revue INRA Productions Animales* 18:183-192.

Jongbloed, A. W. 1987. *Phosphorus in the Feeding of Pigs: Effect of Diet on the Absorption and Retention of Phosphorus by Growing Pigs*, XVI. Lelystad, The Netherlands: Instituut voor Veevoedingsanderzoek.

Jongbloed, A. W., and P. A. Kemme. 1990. Effect of pelleting mixed feeds on phytase activity and the apparent absorbability of phosphorus and calcium in pigs. *Animal Feed Science and Technology* 28:233-242.

Jongbloed, A. W., Z. Mroz, and P. A. Kemme. 1992. The effect of supplementary *Aspergillus niger* phytase in diets for pigs on concentration and apparent digestibility of dry matter, total phosphorus, and phytic acid in different sections of the alimentary tract. *Journal of Animal Science* 70:1159-1168.

Jongbloed, A. W., Z. Mroz, P. A. Kemme, C. Geerse, and Y. Van Der Honing. 1993. The effect of dietary calcium levels on microbial phytase efficacy in growing pigs. *Journal of Animal Science* 71(Suppl. 1):166 (Abstr.).

Jongbloed, A. W., H. Everts, P. A. Kemme, and Z. Mroz, 1999. Quantification of absorbability and requirements of macroelements. Pp. 275-298 in *Quantitative Biology of the Pig*, I. Kyriazakis, ed. Wallingford, UK: CABI.

Jongbloed, A. W., J. Th. M. Diepen, and P. A. van Kemme. 2003. *Phosphorus Requirements of Pigs:* Revision 2003 (in Dutch). CVB Documentation Report No. 30. Lelystad, The Netherlands: CVB.

Jurgens, M. H., R. A. Rikabi, and D. R. Zimmerman. 1997. The effect of dietary active dry yeast supplement on performance of sows during gestation-lactation and their pigs. *Journal of Animal Science* 75:593-597.

Kadis, S., F. A. Udeze, J. Polanco, and D. W. Dreesen. 1984. Relationship of iron administration to susceptibility of newborn pigs to enterotoxic colibacillosis. *American Journal of Veterinary Research* 45:255-259.

Kalinowski, J., and E. R. Chavez. 1984. Effect of low dietary zinc during late gestation and early lactation on the sow and neonatal piglets. *Canadian Journal of Animal Science* 64:749-758.

Kalkus, J. W. 1920. A study of goiter and associated conditions in domestic animals. *Washington Agricultural Experiment Station Bulletin* 156:1-48.

Kayongo-Male, H., D. E. Ullrey, and E. R. Miller. 1975. The Mn requirement of the baby pig from sows fed a low Mn diet. *East African Agricultural and Forestry Journal* 41(2):157-164.

Kegley, E. B., J. W. Spears, W. L. Flowers, and W. D. Schoenherr. 2002. Iron methionine as a source of iron for the neonatal pig. *Nutrition Research* 22:1209-1217.

Kemme, P. A., A. W. Jongbloed, Z. Mroz, and M. Makinen. 1995. Apparent ileal digestibility of protein and amino acids from a maize-soybean meal diet with or without extrinsic phytate and phytase in pigs. P. 6 in *International Symposium on Nutrient Management of Food Animals to Enhance the Environment Abstracts*, June 4-8.

Kernkamp, H. C. H., and E. F. Ferrin. 1953. Parakeratosis in swine. *Journal of the American Veterinary Medical Association* 123:217-220.

Kernkamp, H. C. H., A. J. Clawson, and R. H. Ferneyhough. 1962. Preventing iron deficiency anemia in baby pigs. *Journal of Animal Science* 21:527-532.

Kerr, B. J., C. J. Ziemer, T. E. Weber, S. L. Trabue, B. L. Bearson, G. C. Shurson, and M. H. Whitney. 2008. Comparative sulfur analysis using thermal combustion on inductively coupled plasma methodology and mineral composition of common livestock feeds. *Journal of Animal Science* 86:2377-2384.

Kerr, B. J., T. E. Weber, P. S. Miller, and L. L. Southern. 2010. Effect of phytase on apparent total tract digestibility of phosphorus in corn-soybean meal diets fed to finishing pigs. *Journal of Animal Science* 88:238-247.

Kerr, B. J., T. E. Weber, C. J. Ziemer, C. Spence, M. A. Cotta, and T. R. Whitehead. 2011. Effect of dietary inorganic sulfur level on growth performance, fecal composition, and measures of inflammation and sulfate-reducing bacteria in the intestine of growing pigs. *Journal of Animal Science* 89:426-437.

Kesel, G. A., J. W. Knight, E. T. Kornegay, J. P. Veit, and D. R. Notter. 1983. Restricted energy and elevated calcium and phosphorus intake for boars during growth. 1. Feedlot performance and bone characteristics. *Journal of Animal Science* 57:82-98.

Ketaren, P. P., E. S. Batterham, E. B. Dettmann, and D. J. Farrell. 1993. Phosphorus studies in pigs. 3. Effect of phytase supplementation on the digestibility and availability of phosphorus in soy-bean meal for grower pigs. *British Journal of Nutrition* 70:289-311.

Kim, Y. Y., and D. C. Mahan. 2001a. Comparative effects of high dietary levels of organic and inorganic selenium on selenium toxicity of growing-finishing pigs. *Journal of Animal Science* 79:942-948.

Kim, Y. Y., and D. C. Mahan. 2001b. Effect of dietary selenium source, level, and pig hair color on various selenium indices. *Journal of Animal Science* 79:949-955.

Kirchgessner, M., and E. Grassman. 1970. The dynamics of copper absorption. Pp. 277-287 in *Trace Element Metabolism in Animals*, C. F. Mills, ed. Edinburgh, UK: E. & S. Livingstone.

Kirchgessner, M., H. Mader, and E. Grassman. 1980. Zur Fruchtbarkeitsleistung von Saven bei unterschiedlicher Cu-Versorgung. *Zuchtungskunde* 52:46-53.

Kirchgessner, M., D. A. Roth-Maier, and R. Sporl. 1981. Untersuchungen zum Trachtigkeitsanabolismus der spurenelemente Kupfer, Zin, Nickel und Mangan bei Zuchtsaver. *Archiv fur Tierernahrung-Archives of Animal Nutrition* 31:21-34.

Klasing, K. C., C. D. Knight, and D. M. Forsyth. 1980. Effects of iron on the anti-coli capacity of sow's milk in vitro and in ligated intestinal segments. *Journal of Nutrition* 110:1914-1921.

Kline, E. A., J. Kastelic, C. C. Ashton, P. G. Homeyer, L. Quinn, and D. V. Catron. 1954. The effect on the growth performance of young pigs of adding cobalt, vitamin B$_{12}$ and antibiotics to semipurified rations. *Journal of Nutrition* 53:543-555.

Klosterman, E. W., W. E. Dinusson, E. L. Lasley, and M. L. Buchanan. 1950. Effect of trace minerals on growth and fattening of swine. *Science* 112:168-169.

Knight, C. D., K. C. Klasing, and D. M. Forsyth. 1983. *E. coli* growth in serum of iron dextran-supplemented pigs. *Journal of Animal Science* 57:387-395.

Koch, M. E., D. C. Mahan, and J. R. Corley. 1984. An evaluation of various biological characteristics in assessing low phosphorus intake in weanling swine. *Journal of Animal Science* 59:1546-1556.

Kolodziej, A., and E. Jacyno. 2005. Effect of selenium and vitamin E supplementation on reproductive performance of young boars. *Archiv fur Tierzucht-Archives of Animal Breeding* 48:68-75.

Kornegay, E. T. 1972a. Availability of iron contained in defluorinated phosphate. *Journal of Animal Science* 34:569-572.

Kornegay, E. T. 1972b. Supplementation of lysine, ammonium polyphosphate and urea in diets for growing-finishing pigs. *Journal of Animal Science* 34:55-63.

Kornegay, E. T. 1985. Calcium and phosphorus in animal nutrition. Pp. 1-106 in *Calcium and Phosphorus in Animal Nutrition*. West Des Moines, IA: National Feed Ingredients Association.

Kornegay, E. T. 1996. Nutritional, environmental and economical considerations for using phytase in pig and poultry diets. Pp. 279-304 in *International Symposium on Nutrient Management of Food Animals to Enhance the Environment*, E. T. Kornegay, ed. Boca Raton, FL: CRC Press, Inc.

Kornegay, E. T., and B. Kite. 1983. Phosphorus in swine. VI. Utilization of nitrogen, calcium and phosphorus and reproductive performance of gravid gilts fed two dietary phosphorus levels for five parities. *Journal of Animal Science* 57:1463-1473.

Kornegay, E. T., and J. S. Radcliffe. 1997. Relative bioavailability of phosphorus sources with different solubilities in neutral ammonium citrate (NAC) for young pigs. *Journal of Animal Science* 75(Suppl. 1):188 (Abstr.).

Kornegay, E. T., and H. R. Thomas. 1981. Phosphorus in swine. II. Influence of dietary calcium and phosphorus levels and growth rate on serum minerals, soundness scores and bone development in barrows, gilts and boars. *Journal of Animal Science* 52:1049-1059.

Kornegay, E. T., H. R. Thomas, and T. N. Meacham. 1973. Evaluation of dietary calcium and phosphorus for reproducing sows housed in total confinement on concrete or in dirt lots. *Journal of Animal Science* 37:493-500.

Kornegay, E. T., H. R. Thomas, and J. L. Baker. 1981a. Phosphorus in swine. IV. Influence of dietary calcium and phosphorus and protein levels on feedlot performance, serum minerals, bone development and soundness scores in boars. *Journal of Animal Science* 52:1070-1084.

Kornegay, E. T., H. R. Thomas, J. H. Carter, L. B. Allen, C. C. Brooks, and K. H. Hinklemann. 1981b. Phosphorus in swine. V. Interrelationships of various feedlot performance, serum minerals, structural soundness and bone parameters in barrows, boars and gilts. *Journal of Animal Science* 52:1085-1090.

Kornegay, E. T., H. P. Veit, J. W. Knight, D. R. Notter, H. S. Bartlett, and D. F. Calabotta. 1983. Restricted energy intake and elevated calcium and phosphorus intake for boars during growth. II. Foot and leg measurements and toe and soundness scores. *Journal of Animal Science* 57:1182-1199.

Kornegay, E. T., B. G. Diggs, O. M. Hale, D. L. Handlin, J. P. Hitchcock, and R. A. Barczwski. 1984. Reproductive performance of sows fed elevated calcium and phosphorus levels during growth and development. A cooperative study. Report S-145 of the Committee on Nutritional Systems for Swine to Increase Reproductive Efficiency. *Journal of Animal Science* 59(Suppl. 1):253 (Abstr.).

Kornegay, E. T., P. H. G. van Heugten, M. D. Lindemann, and D. J. Blodgett. 1989. Effects of biotin and high copper levels on performance and immune response of weanling pigs. *Journal of Animal Science* 67:1471-1477.

Kornegay, E. T., M. D. Lindemann, and H. S. Bartlett. 1991. The influence of sodium supplementation of two phosphorus sources on performance and bone mineralization of growing-finishing swine evaluated at two geographical locations. *Canadian Journal of Animal Science* 71:537-547.

Kornegay, E. T., J. L. Evans, and V. Ravidndran. 1994. Effects of diet acidity and protein level or source of calcium on the performance, gastrointestinal content measurements, bone measurements and carcass composition of gilt and barrow weanling pigs. *Journal of Animal Science* 72:2670-2680.

Kornegay, E. T., W. Zhou, J. W. G. M. Swinkels, and C. R. Risley. 1995. Characterization of cobalt-copper antagonism in the study of copper-stimulated growth in weanling pigs. *Journal of Animal and Feed Sciences* 4:21-33.

Kornegay, E. T., Z. Wang, C. M. Wood, and M. D. Lindemann. 1997. Supplemental chromium picolinate influences nitrogen balance, dry matter digestibility, and carcass traits in growing-finishing pigs. *Journal of Animal Science* 75:1319-1323.

Krider, J. L., J. L. Albright, M. P. Plumlee, J. H. Conrad, C. L. Sinclair, L. Underwood, R. G. Jones, and R. B. Harrington. 1975. Magnesium supplementation, space and docking effects on swine performance and behavior. *Journal of Animal Science* 40:1027-1033.

Ku, P. K., D. E. Ullrey, and E. R. Miller. 1970. Zinc deficiency and tissue nucleic acid and protein concentration. Pp. 158-164 in *Trace Element Metabolism in Animals*, C. F. Mills, ed. Edinburgh, UK: E. & S. Livingstone.

Ku, P. K., W. T. Ely, A. W. Groce, and D. E. Ullrey. 1973. Natural dietary selenium, A-tocopherol and effect on tissue selenium. *Journal of Animal Science* 37:501-505.

Kuhajek, E. J., and G. F. Andelfinger. 1970. A new source of iodine for salt blocks. *Journal of Animal Science* 31:51-58.

Lannek, N., P. Lindberg, and G. Tollerz. 1962. Lowered resistance to iron in vitamin E-deficient piglets and mice. *Nature* 195:1006-1007.

Larsen, H. J., and S. Tollersrud. 1981. Effect of dietary vitamin E and selenium on photohemagglutinin response of the pig lymphocytes. *Research in Veterinary Science* 31:301-305.

Lauridsen, C., U. Halekoh, T. Larsen, and S. K. Jensen. 2010. Reproductive performance and bone status markers of gilts and lactating sows supplemented with two different forms of vitamin D. *Journal of Animal Science* 88:202-213.

Leach, R. M., Jr., and A. M. Muenster. 1962. Studies on the role of manganese in bone formation. 1. Effect upon the mucopolysaccharide content of chick bone. *Journal of Nutrition* 78:51-56.

Lee, D. N., T. F. Shen, H. T. Yen, C. F. Weng, and B. J. Chen. 2000a. Effects of chromium supplementation and lipopolysaccharide injection on the immune responses of weanling pigs. *Asian-Australasian Journal of Animal Sciences* 13(10):1414-1421.

Lee, D. N., C. F. Weng, H. T. Yen, T. F. Shen, and B. J. Chen. 2000b. Effects of chromium supplementation and lipopolysaccharide injection on physiological responses of weanling pigs. *Asian-Australasian Journal of Animal Sciences* 13(4):528-534.

Lee, S. H., P. Shinde, J. Choi, M. Park, S. Ohh, I. K. Kwon, S. I. Pak, and B. J. Chae. 2008. Effects of dietary iron levels on growth performance, hematological status, liver mineral concentration, fecal microflora, and diarrhea incidence in weanling pigs. *Biological Trace Element Research* 126:S57-S68.

Lei, X. G., P. K. Ku, E. R. Miller, D. E. Ullrey, and M. T. Yokoyama. 1993a. Supplemental microbial phytase improves bioavailability of dietary zinc to weanling pigs. *Journal of Nutrition* 123:1117-1123.

Lei, X. G., P. K. Ku, E. R. Miller, and M. T. Yokoyama. 1993b. Supplementing corn–soybean meal diets with microbial phytase linearly improves phytate phosphorus utilization by weanling pigs. *Journal of Animal Science* 71:3359-3367.

Lei, X. G., P. K. Ku, E. R. Miller, M. T. Yokoyama, and D. E. Ullrey. 1994. Calcium level affects the efficacy of supplemental microbial phytase in corn–soybean meal diets of weanling pigs. *Journal of Animal Science* 72:139-143.

Lei, X. G., H. M. Dann, D. A. Ross, W. H. Cheng, G. F. Combs, Jr., and K. R. Roneker. 1998. Dietary selenium supplementation is required to support full expression of three selenium-dependent glutathione peroxidases in various tissues of weanling pigs. *Journal of Nutrition* 128:130-135.

Leibholz, J. M., V. C. Speer, and V. W. Hays. 1962. Effect of dietary manganese on baby pig performance and tissue manganese levels. *Journal of Animal Science* 21:772-776.

Lemme, A., C. Wenk, M. Lindemann, and G. Bee. 1999. Chromium yeast affects growth performance but not whole carcass composition of growing-finishing pigs. *Annales de Zootechnie* 48(6):457-468.

Leonard, S. G., T. Sweeney, B. Bahar, B. P. Lynch, and J. V. O'Doherty. 2010. Effect of maternal fish oil and seaweed extract supplementation on colostrum and milk composition, humoral immune response, and performance of suckled piglets. *Journal of Animal Science* 88:2988-2997.

Lepine, A. J., E. T. Kornegay, D. R. Notter, H. P. Veit, and J. W. Knight. 1985. Foot and leg measurements, toe lesions, soundness scores and feedlot performance of crossbred boars as influenced by nutrition and age. *Canadian Journal of Animal Science* 65:459-472.

Letourneau-Montminy, M. P., A. Narch, M. Magnin, D. Sauvant, J. F. Bernier, C. Pomar, and C. Jondreville. 2010. Effect of reduced dietary calcium concentration and phytase supplementation on calcium and phosphorus utilization in weanling pigs with modified mineral status. *Journal of Animal Science* 88:1706-1717.

Lewis, A. J., P. S. Miller, and C. K. Wolverton. 1995. Bioavailability of iron in iron methionine for weanling pigs. *Journal of Animal Science* 73(Suppl. 1):172 (Abstr.).

Lewis, P. K., Jr., W. C. Hoekstra, R. H. Grummer, and P. H. Phillips. 1956. The effects of certain nutritional factors including calcium, phosphorus and zinc on parakeratosis. *Journal of Animal Science* 15:741-751.

Lewis, P. K., Jr., R. H. Grummer, and W. G. Hoekstra. 1957a. The effect of method of feeding upon the susceptibility of the pig to parakeratosis. *Journal of Animal Science* 16:927-936.

Lewis, P. K., Jr., W. G. Hoekstra, and R. H. Grummer. 1957b. Restricted calcium feeding versus zinc supplementation for the control of parakeratosis in swine. *Journal of Animal Science* 16:578-588.

Li, B. T., A. G. Van Kessel, W. R. Caine, S. X. Huang, and R. N. Kirkwood. 2001. Small intestinal morphology and bacterial populations in ileal digesta and feces of newly weaned pigs receiving a high dietary level of zinc oxide. *Canadian Journal of Animal Science* 81:511-516.

Libal, G. W., E. R. Peo, Jr., R. P. Andrews, and P. E. Vipperman, Jr. 1969. Levels of calcium and phosphorus for growing-finishing swine. *Journal of Animal Science* 28:331-335.

Lien, T. F., C. P. Wu, B. J. Wang, M. S. Shiao, T. Y. Shiao, B. H. Lin, J. J. Lu, and C. Y. Hu. 2001. Effect of supplemental levels of chromium picolinate on the growth performance, serum traits, carcass characteristics and lipid metabolism of growing-finishing pigs. *Animal Science* 72:289-296.

Lien, T. F., K. H. Yang, and K. J. Lin. 2005. Effects of chromium propionate supplementation on growth performance, serum traits and immune response in weaned pigs. *Asian-Australasian Journal of Animal Sciences* 18(3):403-408.

Lillie, R. J., and L. T. Frobish. 1978. Effect of copper and iron supplements on performance and hematology of confined sows and their progeny through four reproductive cycles. *Journal of Animal Science* 46:678-685.

Lindberg, P., and N. Lannek. 1965. Retention of selenium in kidneys, liver and striated muscle after prolonged feeding of therapeutic amounts of sodium selenite to pigs. *Acta Veterinaria Scandinavica* 6:217-223.

Lindberg, P., and M. Siren. 1963. Selenium concentration in kidneys of normal pigs and pigs affected with nutritional muscular dystrophy and liver dystrophy (hepatosis dietetica). *Life Sciences* 2:326-330.

Lindberg, P., and M. Siren. 1965. Fluorometric selenium determinations in the liver of normal pigs and in pigs affected with nutritional muscular dystrophy and liver dystrophy. *Acta Veterinaria Scandinavica* 6:59-64.

Lindemann, M. D. 2007. Use of chromium as an animal feed supplement. Pp. 85-118 in *The Nutritional Biochemistry of Chromium (III)*, J. B. Vincent, ed. Amsterdam: Elsevier Press.

Lindemann, M. D., C. M. Wood, A. F. Harper, E. T. Kornegay, and R. A. Anderson. 1995. Dietary chromium picolinate additions improve gain:feed and carcass characteristics in growing-finishing pigs and increase litter size in reproducing sows. *Journal of Animal Science* 73:457-465.

Lindemann, M. D., R. E. Hall, and K. W. Purser. 2000. Use of chromium tripicolinate to improve pigs born alive confirmed in multiparous sows. Pp. 133-137 in *Proceedings of the 31st Annual Meeting of the American Association of Swine Practitioners*, March 11-14, Indianapolis, IN.

Lindemann, M. D., S. D. Carter, L. I. Chiba, C. R. Dove, F. M. Lemieux, and L. L. Southern. 2004. A regional evaluation of chromium tripicolinate supplementation of diets fed to reproducing sows. *Journal of Animal Science* 82: 2972-2977.

Lindemann, M. D., G. L. Cromwell, H. J. Monegue, and K. W. Purser. 2008. Effect of chromium source on tissue concentration of chromium in pigs. *Journal of Animal Science* 86:2971-2978.

Liptrap, D. O., E. R. Miller, D. E. Ullrey, D. L. Whitenack, B. L. Schoepke, and R. W. Luecke. 1970. Sex influence on the zinc requirement of developing swine. *Journal of Animal Science* 30:736-741.

Lolas, G. M., N. Palamidis, and P. Markakis. 1976. The phytic acid-total phosphorus relationship in barley, oats, soybeans and wheat. *Cereal Chemistry* 53:867-871.

Lowry, K. R., D. C. Mahan, and J. R. Corley. 1985a. Effect of dietary calcium on selenium retention in postweaning swine. *Journal of Animal Science* 60:1429-1437.

Lowry, K. R., D. C. Mahan, and J. R. Corley. 1985b. Effect of dietary phosphorus on selenium retention in postweaning swine. *Journal of Animal Science* 60:1438-1446.

Luecke, R. W., J. A. Hoefer, W. G. Brammell, and F. Thorp, Jr. 1956. Mineral interrelationships in parakeratosis of swine. *Journal of Animal Science* 15:247-251.

Luecke, R. W., J. A. Hoefer, W. S. Brammell, and D. A. Schmidt. 1957. Calcium and zinc in parakeratosis of swine. *Journal of Animal Science* 16:3-11.

Luo, X. G., and C. R. Dove. 1996. Effect of dietary copper and fat on nutrient utilization, digestive enzyme activities, and tissue mineral levels in weanling pigs. *Journal of Animal Science* 74:1888-1896.

Lyberg, K., H. K. Andersson, A. Simonsson, and J. E. Lindberg. 2007. Influence of different phosphorus levels and phytase supplementation in gestation diets on sow performance. *Journal of Animal Physiology and Animal Nutrition (Berlin)* 91:304-311.

Mahan, D. C. 1982. Dietary calcium and phosphorus levels for weanling swine. *Journal of Animal Science* 54:559-564.

Mahan, D. C. 1995. Selenium metabolism in animals: What role does selenium yeast have? Pp. 257-267 in *Biotechnology in the Feed Industry, Proceedings of Alltech's 11th Annual Symposium*, T. P. Lyons and K. A. Jacques, eds. Nottingham, UK: Nottingham University Press.

Mahan, D. C., and A. W. Fetter. 1982. Dietary calcium and phosphorus levels for reproducing sows. *Journal of Animal Science* 54:285-291.

Mahan, D. C., and Y. Y. Kim. 1996. Effect of inorganic selenium at two dietary levels on reproductive performance and tissue selenium concentrations in first parity gilts and their progeny. *Journal of Animal Science* 74:2711-2718.

Mahan, D. C., and P. L. Magee. 1991. Efficacy of dietary sodium selenite and calcium selenite provided in the diet at approved, marginally toxic, and toxic levels to growing swine. *Journal of Animal Science* 69:4722-4725.

Mahan, D. C., and A. L. Moxon. 1978a. Effect of adding inorganic or organic selenium sources to the diets of young swine. *Journal of Animal Science* 47:456-466.

Mahan, D. C., and A. L. Moxon. 1978b. Effect of increasing the level of inorganic selenium supplementation in the postweaning diets of swine. *Journal of Animal Science* 46:384-390.

Mahan, D. C., and A. L. Moxon. 1984. Effect of inorganic selenium supplementation on selenosis in postweaning swine. *Journal of Animal Science* 58:1216-1221.

Mahan, D. C., and N. A. Parrett. 1996. Evaluating the efficacy of selenium-enriched yeast and sodium selenite on tissue selenium retention and serum glutathione peroxidase activity in grower and finisher diets. *Journal of Animal Science* 74:2967-2974.

Mahan, D. C., and J. C. Peters. 2004. Long-term effects of dietary organic and inorganic selenium sources and levels on reproducing sows and their progeny. *Journal of Animal Science* 82:1343-1358.

Mahan, D. C., J. E. Jones, J. H. Cline, R. F. Cross, H. S. Teague, and A. P. Grifo, Jr. 1973. Efficacy of selenium and vitamin E injections in the prevention of white muscle disease in young swine. *Journal of Animal Science* 36:1104-1108.

Mahan, D. C., L. H. Penhale, J. H. Cline, A. L. Moxon, A. W. Fetter, and J. T. Yarrington. 1974. Efficacy of supplemental selenium in reproductive diets on sow and progeny performance. *Journal of Animal Science* 39:536-543.

Mahan, D. C., A. L. Moxon, and J. H. Cline. 1975. Efficacy of supplemental selenium in reproductive diets on sow and progeny serum and tissue selenium values. *Journal of Animal Science* 40:624-631.

Mahan, D. C., A. L. Moxon, and M. Hubbard. 1977. Efficacy of inorganic selenium supplementation to sow diets on resulting carry-over to their progeny. *Journal of Animal Science* 45:738-746.

Mahan, D. C., K. E. Ekstrom, and A. W. Fetter. 1980. Effect of dietary protein, calcium and phosphorus for swine from 7 to 20 kilograms body weight. *Journal of Animal Science* 50:309-314.

Mahan, D. C., E. A. Newton, and K. R. Cera. 1996. Effect of supplemental sodium chloride, sodium phosphate, or hydrochloric acid in starter pig diets containing dried whey. *Journal of Animal Science* 74:1217-1222.

Mahan, D. C., T. R. Cline, and B. Richert. 1999a. Effects of dietary levels of selenium-enriched yeast and sodium selenite as selenium sources fed to growing-finishing pigs on performance, tissue selenium, serum glutathione peroxidase activity, carcass characteristics, and loin quality. *Journal of Animal Science* 77:2172-2179.

Mahan, D. C., T. D. Wiseman, E. M. Weaver, and L. E. Russell. 1999b. Effect of supplemental sodium chloride and hydrochloric acid added to initial starter diets containing spray-dried blood plasma and lactose on resulting performance and nitrogen digestibility of 3-week-old weaned pigs. *Journal of Animal Science* 77:3016-3021.

Mahan, D. C., J. H. Brendemuhl, S. D. Carter, L. I. Chiba, T. D. Crenshaw, G. L. Cromwell, C. R. Dove, A. F. Harper, G. M. Hill, G. R. Hollis, S. W. Kim, M. D. Lindemann, C. V. Maxwell, P. S. Miller, J. L. Nelssen, B. T. Richert, L. L. Southern, T. S. Stahly, H. H. Stein, E. van Heugten, and J. T. Yen. 2005. Comparison of dietary selenium fed to grower-finisher pigs from various regions of the United States on resulting tissue Se and loin mineral concentrations. *Journal of Animal Science* 83:852-857.

Mahan, D. C., M. R. Watts, and N. St-Pierre. 2009. Macro- and micromineral composition of fetal pigs and their accretion rates during fetal development. *Journal of Animal Science* 87:2823-2832.

Malde, M. K., I. E. Graff, H. Siljander-Rasi, E. Venalainen, K. Julshamn, J. I. Pedersen, and J. Valaja. 2010. Fish bones—A highly available calcium source for growing pigs. *Journal of Animal Physiology and Animal Nutrition* 94:e66-e76.

Maner, J. H., W. G. Pond, and R. S. Lowrey. 1959. Effect of method and level of iron administration on growth, hemoglobin and hematocrit of suckling pigs. *Journal of Animal Science* 18:1373-1377.

Manners, M. J., and M. R. McCrea. 1964. Estimates of the mineral requirements of 2-day weaned piglets derived from data on mineral retention by sow-reared piglets. *Annales de Zootechnie* 13:29-38.

Marin-Guzman, J., D. C. Mahan, Y. K. Chung, J. L. Pate, and W. F. Pope. 1997. Effects of dietary selenium and vitamin E on boar performance and tissue responses, semen quality, and subsequent fertilization rates in mature gilts. *Journal of Animal Science* 75:2994-3003.

Marin-Guzman, J., D. C. Mahan, and J. L. Pate. 2000a. Effect of dietary selenium and vitamin E on spermatogenic development in boars. *Journal of Animal Science* 78:1537-1543.

Marin-Guzman, J., D. C. Mahan, and R. Whitmoyer. 2000b. Effect of dietary selenium and vitamin E on the ultrastructure and ATP concentration of boar spermatozoa, and the efficacy of added sodium selenite in extended semen on sperm motility. *Journal of Animal Science* 78:1544-1550.

Matrone, G., R. H. Hartman, and A. J. Clawson. 1959. Studies of a manganese-iron antagonism in the nutrition of rabbits and baby pigs. *Journal of Nutrition* 67:309-317.

Matrone, G., E. L. Thomason, and C. R. Bunn. 1960. Requirement and utilization of iron by the baby pig. *Journal of Nutrition* 72:459-465.

Matthews, J. O., L. L. Southern, J. M. Fernandez, J. E. Pontif, T. D. Bidner, R. L. Odgaard. 2001. Effect of chromium picolinate and chromium propionate on glucose and insulin kinetics of growing barrows and on growth and carcass traits of growing-finishing barrows. *Journal of Animal Science* 79:2172-2178.

Matthews, J. O., A. D. Higbie, L. L. Southern, D. F. Coombs, T. D. Bidner, and R. L. Odgaard. 2003. Effect of chromium propionate and metabolizable energy on growth, carcass traits, and pork quality of growing-finishing pigs. *Journal of Animal Science* 81:191-196.

Matthews, J. O., A. C. Guzik, F. M. LeMieux, L. L. Southern, and T. D. Bidner. 2005. Effects of chromium propionate on growth, carcass traits, and pork quality of growing-finishing pigs. *Journal of Animal Science* 83:858-862.

Mavromichalis, I., D. M. Webel, E. N. Parr, and D. H. Baker. 2001. Growth-promoting efficacy of pharmacological doses of tetrabasic zinc chloride in diets for nursery pigs. *Canadian Journal of Animal Science* 81:387-391.

Maxson, P. F., and D. C. Mahan. 1983. Dietary calcium and phosphorus levels for growing swine from 18 to 57 kilograms body weight. *Journal of Animal Science* 56:1124-1134.

Maxson, P. F., and D. C. Mahan. 1986. Dietary calcium and phosphorus for lactating swine at high and average production levels. *Journal of Animal Science* 63:1163-1172.

Mayo, R. H., M. P. Plumlee, and W. M. Beeson. 1959. Magnesium requirement of the pig. *Journal of Animal Science* 18:264-273.

McCance, R. A., and E. M. Widdowson. 1944. Activity of the phytase in different cereals and its resistance to dry heat. *Nature* 153:650.

McCarrison, R. 1933. The goitrogenic action of soybean and groundnut. *Indian Journal of Medical Research* 7:189.

McCully, G. A., G. M. Hill, J. E. Link, R. L. Weavers, M. S. Carlson, and D. W. Rozeboom. 1995. Evaluation of zinc sources for the newly weaned pig. *Journal of Animal Science* 74(Suppl. 1):72 (Abstr.).

McDonald, F. F., D. Dunlop, and C. M. Bates. 1955. An effective treatment for anemia of piglets. *British Veterinary Journal* 111:403-407.

Menehan, L. A., P. A. Knapp, W. G. Pond, and J. R. Jones. 1963. Response of early-weaned pigs to variations in dietary calcium level with and without lactose. *Journal of Animal Science* 22:501-505.

Mertz, W. 1993. Chromium in human nutrition: A review. *Journal of Nutrition* 123:626-633.

Meyer, J. H., R. H. Grummer, P. H. Phillips, and G. Bohstedt. 1950. Sodium, chlorine, and potassium requirements of growing pigs. *Journal of Animal Science* 9:300-306.

Meyer, W. R., D. C. Mahan, and A. L. Moxon. 1981. Value of dietary selenium and vitamin E for weanling swine as measured by performance and tissue selenium and glutathione peroxidase activities. *Journal of Animal Science* 52:302-311.

Michel, R. L., C. K. Whitehair, and K. K. Keahey. 1969. Dietary hepatic necrosis associated with selenium-vitamin E deficiency in swine. *Journal of the American Veterinary Medical Association* 155:50-59.

Miller, E. R. 1975. Utilization of inorganic sulfate by growing-finishing swine. *Michigan Agricultural Experiment Station Research Report* 289:100-104.

Miller, E. R. 1980. Bioavailability of minerals. P. 144 in *Proceedings of the Minnesota Nutrition Conference*. St. Paul: University of Minnesota Press.

Miller, E. R. 1991. Iron, copper, zinc, manganese, and iodine in swine nutrition. Pp. 267-284 in *Swine Nutrition*, E. R. Miller, D. E. Ullrey, and A. J. Lewis, eds. Stoneham, MA: Butterworth-Heinemann.

Miller, E. R., D. E. Ullrey, C. L. Zutaut, B. V. Baltzer, D. A. Schmidt, J. A. Hoefer, and R. W. Luecke. 1962. Calcium requirement of the baby pig. *Journal of Nutrition* 77:7-16.

Miller, E. R., D. E. Ullrey, C. L. Zutaut, B. V. Baltzer, D. A. Schmidt, J. A. Hoefer, and R. W. Luecke. 1964a. Phosphorus requirement of the baby pig. *Journal of Nutrition* 82:34-39.

Miller, E. R., D. E. Ullrey, C. L. Zutaut, J. A. Hoefer, and R. W. Luecke. 1964b. Mineral balance studies with the baby pig: Effects of dietary phosphorus level upon calcium and phosphorus balance. *Journal of Nutrition* 82:111-114.

Miller, E. R., D. E. Ullrey, C. L. Zutaut, B. V. Baltzer, D. A. Schmidt, J. A. Hoefer, and R. W. Luecke. 1965a. Magnesium requirement of the baby pig. *Journal of Nutrition* 85:13-20.

Miller, E. R., D. E. Ullrey, C. L. Zutaut, J. A. Hoefer, and R. W. Luecke. 1965b. Comparisons of casein and soy proteins upon mineral balance and vitamin D_2 requirement of the baby pig. *Journal of Nutrition* 85:347-353.

Miller, E. R., D. E. Ullrey, C. L. Zutaut, J. A. Hoefer, and R. W. Luecke. 1965c. Mineral balance studies with the baby pig: Effects of dietary magnesium level upon calcium, phosphorus, and magnesium balance. *Journal of Nutrition* 86:209-212.

Miller, E. R., D. E. Ullrey, C. L. Zutaut, J. A. Hoefer, and R. W. Luecke. 1965d. Mineral balance studies with the baby pig: Effects of dietary vitamin D_2 level upon calcium, phosphorus, and magnesium balance. *Journal of Nutrition* 85:255-258.

Miller, E. R., R. W. Luecke, D. E. Ullrey, B. V. Baltzer, B. L. Bradley, and J. A. Hoefer. 1968. Biochemical, skeletal and allometric changes due to zinc deficiency in the baby pig. *Journal of Nutrition* 95:278-286.

Miller, E. R., D. O. Liptrap, and D. E. Ullrey. 1970. Sex influence on zinc requirement of swine. Pp. 377-379 in *Trace Element Metabolism in Animals*, C. F. Mills, ed. Edinburgh, UK: E. & S. Livingstone.

Miller, E. R., H. D. Stowe, P. K. Ku, and G. M. Hill. 1979. Copper and zinc in swine nutrition. P. 109 in *National Feed Ingredients Association Literature Review on Copper and Zinc in Animal Nutrition*. West Des Moines, IA: National Feed Ingredients Association.

Miller, E. R., M. J. Parsons, D. E. Ullrey, and P. K. Ku. 1981. Bioavailability of iron from ferric choline citrate and a ferric copper cobalt choline citrate complex for young pigs. *Journal of Animal Science* 52:783-787.

Miller, E. R., G. L. Waxler, P. K. Ku, D. E. Ullrey, and C. K. Whitehair. 1982. Iron requirements of baby pigs reared in germ-free or conventional environments on a condensed milk diet. *Journal of Animal Science* 54:106-115.

Miller, K. B., J. S. Caton, D. M. Schafer, D. J. Smith, and J. W. Finley. 2000. High dietary manganese lowers heart magnesium in pigs fed a low-magnesium diet. *Journal of Nutrition* 130:2032-2035.

Miller, M. B., T. G. Hartsock, B. Erez, L. Douglass, and B. Alston-Mills, 1994. Effect of dietary calcium concentrations during gestation and lactation in the sow on milk composition and litter growth. *Journal of Animal Science* 72:1315-1319.

Miller, W. T. 1938. Toxicity of selenium fed to swine in the form of sodium selenite. *Journal of Agricultural Research* 56:831-842.

Miller, W. T., and K. T. Williams. 1940. Minimum lethal dose of selenium, as sodium selenite, for horses, mules, cattle and swine. *Journal of Agricultural Research* 60:163-173.

Min, J. K., W. Y. Kim, B. J. Chae, I. B. Chung, I. S. Shin, Y. J. Choi, and I. K. Han. 1997. Effects of chromium picolinate (CrP) on growth performance, carcass characteristics and serum traits in growing-finishing pigs. *Asian-Australasian Journal of Animal Sciences* 10(1):8-14.

Miracle, G. L., G. L. Cromwell, T. S. Stahly, and D. D. Kratzer. 1977. Availability of phosphorus in corn, wheat, and soybean meal for pigs. *Journal of Animal Science* 45(Suppl. 1):101 (Abstr.).

Mollgaard, H. 1946. On phytic acid, its importance in metabolism and its enzymic cleavage in bread supplemented with calcium. *Biochemical Journal* 40:589-603.

Monegue, J. S., M. D. Lindemann, H. J. Monegue, and G. L. Cromwell. 2011. Growth performance and diet preference of nursery pigs fed varying levels of salt. *Journal of Animal Science* 89(Suppl. 2):66-67 (Abstr.).

Mongin, P. 1981. Recent advances in dietary cation-anion balance: Applications in poultry. *Proceedings of the Nutrition Society* 40:285-294.

Mooney, K. W., and G. L. Cromwell. 1995. Effects of dietary chromium picolinate supplementation on growth, carcass characteristics, and accretion rates of carcass tissues in growing-finishing swine. *Journal of Animal Science* 73:3351-3357.

Mooney, K. W., and G. L. Cromwell. 1996. Effects of chromium picolinate on performance and tissue accretion in pigs with different lean gain potential. *Journal of Animal Science* 74(Suppl. 1):65 (Abstr.)

Mooney, K. W., and G. L. Cromwell. 1997. Efficacy of chromium picolinate and chromium chloride as potential carcass modifiers in swine. *Journal of Animal Science* 75:2661-2671.

Morgan, D. P., E. P. Young, I. P. Earle, R. J. Davey, and J. W. Stevenson. 1969. Effects of dietary calcium and zinc on calcium, phosphorus and zinc retention in swine. *Journal of Animal Science* 29:900-905.

Mraz, F. R., A. M. Johnson, and H. Patrick. 1958. Metabolism of cesium and potassium in swine as indicated by cesium-134 and potassium-42. *Journal of Nutrition* 64:541-548.

Mroz, Z., A. W. Jongbloed, and P. A. Kemme. 1994. Apparent digestibility and retention of nutrients bound to phytate complexes as influenced by microbial phytase and feeding regimen in pigs. *Journal of Animal Science* 72:126-132.

Mudd, A. J., W. C. Smith, and D. G. Armstrong. 1969. The influence of dietary concentration of calcium and phosphorus on their retention in the body of growing pigs. *Journal of Agricultural Science (Cambridge)* 73:189-195.

Nasi, M. 1990. Microbial phytase supplementation for improving availability of plant phosphorus in the diet of the growing pig. *Journal of Agricultural Science Finland* 62:435-443.

Nelson, T. S., L. W. Ferrara, and N. L. Storer. 1968. Phytate phosphorus content of feed ingredients derived from plants. *Poultry Science* 47:1372-1374.

Newland, H. W., and G. K. Davis. 1961. Placental transfer of manganese in swine. *Journal of Animal Science* 20:15-17.

Newland, H. W., D. E. Ullrey, J. A. Hoefer, and R. W. Luecke. 1958. The relationship of dietary calcium to zinc metabolism in pigs. *Journal of Animal Science* 17:886-892.

Newton, G. L., and A. J. Clawson. 1974. Iodine toxicity: Physiological effects of elevated dietary iodine on pigs. *Journal of Animal Science* 39:879-884.

Nielsen, F. H. 1984. Ultratrace elements in nutrition. *Annual Review of Nutrition* 4:21-41.

Nielsen, F. H. 1994. Chromium. Pp. 264-268 in *Modern Nutrition in Health and Disease*, 8th Ed., M. E. Shils, J. A. Olson, and M. Shike, eds. Philadelphia: Lea & Febiger.

Nielsen, H. E., V. Danielsen, M. G. Simesen, G. Grissel-Nielsen, W. Hjarde, T. Leth, and A. Basse. 1979. Selenium and vitamin E deficiency in pigs. *Acta Veterinaria Scandinavica* 20:276-288.

Nieto, R., I. Seiquer, and J. F. Aguilera. 2008. The effect of dietary protein content on calcium and phosphorus retention in the growing Iberian pig. *Livestock Science* 116:275-288.

Nilsson, P. O. 1960. Acute iron poisoning with myocardial degeneration in piglets. *Nordisk Veterinaer Medicin* 12:113-119.

Nimmo, R. D., E. R. Peo, Jr., J. D. Crenshaw, B. D. Moser, and A. J. Lewis. 1981a. Effects of level of dietary calcium-phosphorus during growth and gestation on calcium-phosphorus balance and reproductive performance of first-litter sows. *Journal of Animal Science* 52:1343-1349.

Nimmo, R. D., E. R. Peo, Jr., B. D. Moser, and A. J. Lewis. 1981b. Effect of level of dietary calcium-phosphorus during growth and gestation on performance, blood, and bone parameters of swine. *Journal of Animal Science* 52:1330-1342.

Norrdin, R. W., L. Krook, W. G. Pond, and E. F. Walker. 1973. Experimental zinc deficiency in weanling pigs on high- and low-calcium diets. *Cornell Veterinarian* 63:264-290.

NRC (National Research Council). 1980. *Mineral Tolerance of Domestic Animals*. Washington, DC: National Academy Press.

NRC. 1997. *The Role of Chromium in Animal Nutrition*. Washington, DC: National Academy Press.

NRC. 2005. *Mineral Tolerance of Animals*, 2nd Rev. Ed. Washington, DC: National Academies Press.

Nunes, C. S. 1993. Evaluation of phytase resistance in swine diets to different pelleting temperatures. Pp. 269-271 in *Proceedings of the First Symposium on Enzymes in Animal Nutrition*, Kartause Ittingen, Switzerland.

Nuoranne, P. J., R. P. Raunio, P. Saukko, and H. Karppanen. 1980. Metabolic effects of a low-magnesium diet in pigs. *British Journal of Nutrition* 44:53-60.

Oberleas, D. 1983. The role of phytate in zinc bioavailability and homeostasis. Pp. 145-158 in *Nutritional Bioavailability of Zinc*, American Chemical Society Symposium Series No. 210, G. E. Inglett, ed. Washington, DC: American Chemical Society.

Oberleas, D., and B. F. Harland. 1996. Impact of phytate on nutrient availability. Pp. 77-84 in *Phytase in Animal Nutrition and Waste Management*, M. B. Coelho, and E. T. Kornegay, eds. Mount Olive, NJ: BASF Corporation.

Oberleas, D., M. E. Muhrer, and B. L. O'Dell. 1962. Effects of phytic acid on zinc availability and parakeratosis in swine. *Journal of Animal Science* 21:57-61.

O'Donovan, P. B., R. A. Pickett, M. P. Plumlee, and M. W. Beeson. 1963. Iron toxicity in the young pig. *Journal of Animal Science* 22:1075-1080.

O'Hara, P. J., A. P. Newman, and R. Jackson. 1960. Parakeratosis and copper poisoning in pigs fed a copper supplement. *Australian Veterinary Journal* 36:225-229.

O'Quinn, P. R., D. A. Knabe, and E. J. Gregg. 1997. Digestible phosphorus needs in terminal-cross growing-finishing pigs. *Journal of Animal Science* 75:1308-1318.

Okonkwo, A. C., P. K. Ku, E. R. Miller, K. K. Keahey, and D. E. Ullrey. 1979. Copper requirement of baby pigs fed purified diets. *Journal of Nutrition* 109:939-948.

Orstadius, K. 1960. Toxicity at a single subcutaneous dose of sodium selenite in pigs. *Nature* 188:1117.

Orstadius, K., B. Wretlind, P. Lindberg, C. Nordstrom, and N. Lannek. 1959. Plasma transaminase and transferase activities in pigs affected with muscular and liver dystrophy. *Zentralblatt für Veterinärmedizin* 6:971-980.

Osborne, J. C., and J. W. Davis. 1968. Increased susceptibility to bacterial endotoxin of pigs with iron deficiency anemia. *Journal of the American Veterinary Medical Association* 152:1630-1632.

Owen, A. A., E. R. Peo, Jr., P. J. Cunningham, and B. D. Moser. 1973. Effect of EDTA on utilization of dietary zinc by G-F swine. *Journal of Animal Science* 37:470-478.

Page, T. G., L. L. Southern, T. L. Ward, and D. L. Thompson, Jr. 1993. Effect of chromium picolinate on growth and serum and carcass traits of growing-finishing pigs. *Journal of Animal Science* 71:656-662.

Pallauf, J., and G. Rimbach. 1997. Nutritional significance of phytic acid and phytase. *Archives of Animal Nutrition* 50:301-319.

Pallauf, V. J., D. Hohler, and G. Rimbach. 1992a. Effect of microbial phytase supplementation to a maize-soya-diet on the apparent absorption on Mg, Fe, Cu, Mn and Zn and parameters of Zn-status in piglets. *Journal of Animal Physiology and Animal Nutrition* 68:1-9.

Pallauf, V. J., D. Hohler, G. Rimbach, and H. Neusser. 1992b. Effect of microbial phytase supplementation to a maize-soy-diet on the apparent absorption of phosphorus and calcium in piglets. *Journal of Animal Physiology and Animal Nutrition* 67:30-40.

Pallauf, J., G. Rimbach, S. Pippig, B. Schindler, and E. Most. 1994a. Dietary effect of phytogenic phytase and an addition of microbial phytase to a diet based on field beans, wheat, peas and barley on the utilization of phosphorus, calcium, magnesium, zinc and protein in piglets. *Zeitschrift für Ernahrungswissenschaft* 33:128-135.

Pallauf, J., G. Rimbach, S. Pippig, B. Schindler, and E. Most. 1994b. Effect of phytase supplementation to a phytate-rich diet based on wheat, barley and soya on the bioavailability of dietary phosphorus, calcium, magnesium, zinc and protein in piglets. *Agribiological Research—Zeitschrift für Agrarbiologie Agrikulturchemie Okologie* 47:39-48.

Park, J. K., J. Y. Lee, B. J. Chae, and S. J. Ohh. 2009. Effects of different sources of dietary chromium on growth, blood profiles and carcass traits in growing-finishing pigs. *Asian-Australasian Journal of Animal Sciences* 22(11):1547-1554.

Park, Y. W., M. Kandeh, K. B. Chin, W. G. Pond, and L. D. Young. 1994. Concentrations of inorganic elements in milk of sows selected for high and low serum cholesterol. *Journal of Animal Science* 72:1399-1402.

Partanen, K., H. Siljander-Rasi, M. Karhapää, K. Ylivainio, and T. Tupasela. 2010. Responses of growing pigs to different levels of dietary phosphorus–performance, bone characteristics, and solubility of faecal phosphorus. *Livestock Science* 134(1-3):109-112.

Partridge, I. G. 1981. A comparison of defluorinated rock phosphate and dicalcium phosphate, in diets containing either skim milk powder or soya bean meal as the main protein supplement, for early-weaned pigs. *Animal Production* 32:67-73.

Patience, J. F., and M. S. Wolynetz. 1990. Influence of dietary undetermined anion on acid-base status and performance in pigs. *Journal of Nutrition* 120:579-587.

Patience, J. F., R. E. Austic, and R. D. Boyd. 1987. Effect of dietary electrolyte balance on growth and acid-base status in swine. *Journal of Animal Science* 64:457-466.

Patterson, D. S. P., W. M. Allen, D. C. Thurley, and J. T. Done. 1967. The role of tissue peroxidation in iron-induced myodegeneration of piglets. *Biochemical Journal* 104:2P-3P.

Patterson, D. S. P., W. M. Allen, S. Berrett, D. Sweasy, D. C. Thurley, and J. T. Done. 1969. A biochemical study of the pathogenesis of iron-induced myodegeneration of piglets. *Zentralblatt für Veterinärmedizin* 16:199-214.

Payne, G. G., D. C. Martens, E. T. Kornegay, and M. D. Lindemann. 1988. Availability and form of copper in three soils following eight annual applications of copper-enriched swine manure. *Journal of Environmental Quality* 14:740-746.

Payne, R. L., T. D. Bidner, T. M. Fakler, and L. L. Southern. 2006. Growth and intestinal morphology of pigs from sows fed two zinc sources during gestation and lactation. *Journal of Animal Science* 84:2141-2149.

Peeler, H. T. 1972. Biological availability of nutrients in feeds: Availability of major mineral ions. *Journal of Animal Science* 35:695-712.

Peo, E. R., Jr. 1976. *Calcium in Swine Nutrition*. West Des Moines, IA: National Feed Ingredients Association.

Peo, E. R., Jr. 1991. Calcium, phosphorus, and vitamin D in swine nutrition. Pp. 165-182 in *Swine Nutrition*, E. R. Miller, D. E. Ullrey, and A. J. Lewis, eds. Stoneham, MA: Butterworth-Heinemann.

Peplowski, M. A., D. C. Mahan, F. A. Murray, A. L. Moxon, A. H. Cantor, and K. E. Ekstrom. 1980. Effect of dietary and injectable vitamin E and selenium in weanling swine antigenically changed with sheep red blood cell. *Journal of Animal Science* 51:344-351.

Pérez, V. G., A. M. Waguespack, T. D. Bidner, L. L. Southern, T. M. Fakler, T. L. Ward, M. Steidinger, and J. E. Pettigrew. 2011a. Additivity of effects from dietary copper and zinc on growth performance and fecal microbiota of pigs after weaning. *Journal of Animal Science* 89:414-425.

Pérez, V. G., H. Yang, T. R. Radke, and D. P. Holzgraefe. 2011b. Sulfur addition in corn-soybean meal diets reduced nursery pig performance. *Journal of Animal Science* 89(Suppl. 1):334 (Abstr.).

Perez-Mendoza, V. G., J. A. Cuaron, C. J. Rapp, and T. M. Fakler. 2003. Lactating and rebreeding sow performance in response to chromium-L-methionine. *Journal of Animal Science* 81(Suppl. 2):71.

Peters, J. C., and D. C. Mahan. 2008. Effects of dietary organic and inorganic trace mineral levels on sow reproductive performances and daily mineral intakes over six parities. *Journal of Animal Science* 86:2247-2260.

Peters, J. C., D. C. Mahan, T. G. Wiseman, and N. D. Fastinger. 2010. Effect of dietary organic and inorganic micromineral source and level on sow body liver, colostrum, mature milk and progeny mineral compositions over six parities. *Journal of Animal Science* 88:626-637.

Pettey, L. A., 2004. The factorial estimation of dietary phosphorus requirements for growing and finishing pigs. Ph.D. Dissertation. University of Kentucky, Lexington.

Pettey, L. A., G. L. Cromwell, and M. D. Lindemann. 2006. Estimation of endogenous phosphorus loss in growing and finishing pigs fed semi-purified diets. *Journal of Animal Science* 84:618-626.

Piatkowski, T. L., D. C. Mahan, A. H. Cantor, A. L. Moxon, J. H. Cline, and A. P. Grifo, Jr. 1979. Selenium and vitamin E in semipurified diets for gravid and nongravid gilts. *Journal of Animal Science* 48:1357-1365.

Pickett, R. A., M. P. Plumlee, W. H. Smith, and W. M. Beeson. 1960. Oral iron requirement of the early-weaned pig. *Journal of Animal Science* 19:1284 (Abstr.)

Piper, R. C., J. A. Froseth, C. R. McDowell, G. H. Kroening, and I. A. Dyer. 1975. Selenium-vitamin E deficiency in swine fed peas (*Pisum sativum*). *American Journal of Veterinary Research* 36:273-281.

Plumlee, M. P., D. M. Thrasher, W. M. Beeson, F. N. Andrews, and H. E. Parker. 1956. The effects of a manganese deficiency upon the growth, development and reproduction of swine. *Journal of Animal Science* 15:352-368.

Plumlee, M. P., C. E. Jordan, M. H. Kennington, and W. M. Beeson. 1958. Availability of the phosphorus from various phosphate materials for swine. *Journal of Animal Science* 17:73-88.

Pointillart, A., and L. Gueguen. 1978. Osteochondrose et faiblesse des pattes chez le porc. *Annales de Biologie Animale Biochimie Biophysique* 18:201-210.

Pointillart, A., N. Fontaine, and M. Thomasset. 1984. Phytate phosphorus utilization and intestinal phosphatases in pigs fed low phosphorus: Wheat or corn diets. *Nutrition Reports International* 29:473-483.

Pointillart, A., V. Coxam, B. Seve, C. Colin, C. H. Lacroix, and L. Gueguen. 2000. Availability of calcium from skim milk, calcium sulfate and calcium carbonate for bone mineralization in pigs. *Reproduction Nutrition Development* 40:49-61.

Pollmann, D. S., J. E. Smith, J. S. Stevenson, D. A. Schoneweis, and R. H. Hines. 1983. Comparison of gleptoferron with iron dextran for anemia prevention in young pigs. *Journal of Animal Science* 56:640-644.

Pomar, C., I. Kyriazakis, G. C. Emmans, and P. W. Knap. 2003. Modeling stochasticity: Dealing with populations rather than individual pigs. *Journal of Animal Science* 81:E178-E186.

Pond, W. G., and J. R. Jones. 1964. Effect of level of zinc in high-calcium diets on pigs from weaning through one reproductive cycle and on subsequent growth of their offspring. *Journal of Animal Science* 23:1057-1060.

Pond, W. G., R. S. Lowrey, J. H. Maner, and J. K. Loosli. 1961. Parenteral iron administration to sows during gestation or lactation. *Journal of Animal Science* 20:747-750.

Pond, W. G., J. R. Jones, and G. H. Kroening. 1964. Effect of level of dietary zinc and source and level of corn on performance and incidence of parakeratosis in weanling pigs. *Journal of Animal Science* 23:16-20.

Pond, W. G., P. Chapman, and E. Walker. 1966. Influence of dietary zinc, corn oil and cadmium on certain blood components, weight gain and parakeratosis in young pigs. *Journal of Animal Science* 25:122-127.

Pond, W. G., E. F. Walker, Jr., and D. Kirkland. 1975. Weight gain, feed utilization and bone and liver mineral composition of pigs fed high or normal Ca-P diets from weaning to slaughter weight. *Journal of Animal Science* 41:1053-1056.

Pond, W. G., E. F. Walker, Jr., and D. Kirkland. 1978. Effect of dietary Ca and P levels from 40 to 100 kg body weight on weight gain and bone and soft tissue mineral concentrations. *Journal of Animal Science* 46:686-691.

Pond, W. G., J. T. Yen, D. A. Hill, and W. E. Wheeler. 1981. Dietary Ca source and level: Effects on weanling pigs. *Journal of Animal Science* 53(Suppl. 1):91 (Abstr.).

Poulsen, H. D. 1989. *Zinc Oxide for Pigs During Weaning*. Meddelelse No. 746. Denmark: Statens Husdyrbrugsforsoeq.

Prasad, A. S., D. Oberleas, P. Wolf, J. P. Horwitz, E. R. Miller, and R. W. Luecke. 1969. Changes in trace elements and enzyme activities in tissues of zinc-deficient pigs. *American Journal of Clinical Nutrition* 22:628-637.

Prasad, A. S., D. Oberleas, E. R. Miller, and R. W. Luecke. 1971. Biochemical effects of zinc deficiency: Changes in activities of zinc-dependent enzymes and ribonucleic acid and deoxyribonucleic acid content of tissues. *Journal of Laboratory and Clinical Medicine* 77:144-152.

Prince, T. J., V. W. Hays, and G. L. Cromwell. 1975. Environmental effects of high copper pig manure on pasture for sheep. *Journal of Animal Science* 41:326 (Abstr.).

Prince, T. J., V. W. Hays, and G. L. Cromwell. 1984. Interactive effects of dietary calcium, phosphorus and copper on performance and liver copper stores of pigs. *Journal of Animal Science* 58:356-361.

Ramisz, A., A. Balicka-Laurans, and G. Ramisz. 1993. The influence of selenium on production, reproduction and health in pigs. *Advances in Agricultural Science* 2:67.

Real, D. E., J. L. Nelssen, M. D. Tokach, R. D. Goodband, S. S. Dritz, J. C. Woodworth, and K. Q. Owen. 2008. Additive effects of L-carnitine and chromium picolinate on sow reproductive performance. *Livestock Science* 116(1-3):63-69.

Reinhart, G. A., and D. C. Mahan. 1986. Effect of various calcium:phosphorus ratios at low and high dietary phosphorus for starter, grower and finisher swine. *Journal of Animal Science* 63:457-466.

Revy, P. S., C. Jondreville, J. Y. Dourmad, and Y. Nys. 2004. Effect of zinc supplemented as either an organic or an inorganic source and of microbial phytase on zinc and other minerals utilisation by weanling pigs. *Animal Feed Science and Technology* 116:93-112.

Rheaume, J. A., and E. R. Chavez. 1989. Trace mineral metabolism in nongravid, gestating and lactating gilts fed two dietary levels of manganese. *Journal of Trace Elements in Experimental Medicine* 3:231-242.

Rickes, E. L., N. G. Brink, F. R. Koniuszy, T. R. Wood, and K. Folkers. 1948. Vitamin B_{12}, cobalt complex. *Science* 108:134.

Rincker, M. J., G. M. Hill, J. E. Link, A. M. Meyer, and J. E. Rowntree. 2005. Effects of dietary zinc and iron supplementation on mineral excretion, body composition, and mineral status of nursery pigs. *Journal of Animal Science* 83:2762-2774.

Ritchie, H. D., R. W. Luecke, B. V. Baltzer, E. R. Miller, D. E. Ullrey, and J. A. Hoefer. 1963. Copper and zinc interrelationships in the pig. *Journal of Nutrition* 79:117-123.

Rodehutscord, M., R. Haverkamp, and E. Pfeffer. 1998. Inevitable losses of phosphorus in pigs, estimated from balance data using diets deficient in phosphorus. *Archives of Animal Nutrition* 51:27-38.

Roof, M. D., and D. C. Mahan. 1982. Effect of carbadox and various dietary copper levels for weanling swine. *Journal of Animal Science* 55:1109-1117.

Roos, M. A., and R. A. Easter. 1986. Effect on sow and piglet performance of feeding a diet containing 250 ppm copper during lactation. *Journal of Animal Science* 63(Suppl. 1):115 (Abstr.).

Ross, R. D., G. L. Cromwell, and T. S. Stahly. 1982. Biological availability of the phosphorus in regular and dehulled soybean meal for growing pigs. *Journal of Animal Science* 55(Suppl. 1):93 (Abstr.).

Ross, R. D., G. L. Cromwell, and T. S. Stahly. 1983. Biological availability of the phosphorus in high-moisture and pelleted corn. *Journal of Animal Science* 57(Suppl. 1):96 (Abstr.).

Ross, R. D., G. L. Cromwell, and T. S. Stahly. 1984. Effects of source and particle size on the biological availability of calcium in calcium supplements for growing pigs. *Journal of Animal Science* 59:125-134.

Rotruck, J. T., A. L. Pope, H. E. Canther, A. B. Swanson, D. C. Hafeman, and W. G. Hoekstra. 1973. Selenium: Biochemical role as a component of glutathione peroxidase. *Science* 179:588-590.

Ruan, Z., Y. G. Zhang, Y. L. Yin, T. J. Li, R. L. Huang, S. W. Kim, G. Y. Wu, and Z. Y. Deng. 2007. Dietary requirement of true digestible phosphorus and total calcium for growing pigs. *Asian-Australasian Journal of Animal Sciences* 20:1236-1242.

Ruth, C. R., and J. F. Van Vleet. 1974. Experimentally induced selenium-vitamin E deficiency in growing swine: Selective destruction of type I skeletal muscle fibers. *American Journal of Veterinary Research* 35:237-244.

Rutkowska-Pejsak, B., A. Mokrzycka, and J. Szkoda. 1998. Influence of zinc oxide in feed on the health status of weaned pigs. *Medycyna Weterynaryjna* 54:194-200.

Rutledge, E. A., L. E. Hanson, and R. J. Meade. 1961. A study of the calcium requirements of pigs weaned at three weeks of age. *Journal of Animal Science* 20:243-245.

Rydberg, M. E., H. L. Self, T. Kowalczyk, and R. H. Grummer. 1959. The effect of prepartum intramuscular iron treatment of dams on litter hemoglobin levels. *Journal of Animal Science* 18:415-419.

Sansom, B. F., and P. T. Gleed. 1981. The ingestion of sow's faeces by suckling piglets. *British Journal of Nutrition* 46:451-456.

Saraiva, A., J. L. Donzele, R. F. M. de Oliveira, M. L. T. de Abreu, F. C. D. Silva, R. A. Vianna, and A. L. Lima. 2011. Available phosphorus levels in diets for 30 to 60 kg female pigs selected for meat deposition by maintaining calcium and available phosphorus ratio. *Revista Brasileira de Zootecnia—Brazilian Journal of Animal Science* 40:587-592.

Schell, T. C., and E. T. Kornegay. 1996. Zinc concentration in tissues and performance of weanling pigs fed pharmacological levels of zinc from ZnO, Zn-methionine, Zn-lysine, and ZnSO$_4$. *Journal of Animal Science* 74:1584-1593.

Schone, F., B. Groppel, A. Hennig, G. Jahreis, and R. Lange. 1997a. Rapeseed meals, methimazole, thiocyanate and iodine affect growth and thyroid. Investigations into glucosinolate tolerance in the pig. *Journal of the Science of Food and Agriculture* 74:69-80.

Schone, F., M. Leiterer, G. Jahreis, and B. Rudolph. 1997b. Effect of rapeseed feedstuffs with different glucosinolate content and iodine administration on gestating and lactating sow. *Journal of Veterinary Medicine Series A* 44:325-339.

Schone, F., M. Leiterer, H. Hartung, G. Jahreis, and F. Tischendorf. 2001. Rapeseed glucosinolates and iodine in sows affect the milk iodine concentration and the iodine status of piglets. *British Journal of Nutrition* 85:659-670.

Selle, P. H., and V. Ravindran. 2008. Phytate degrading enzymes in pig nutrition. *Livestock Science* 113:99-122.

Selle, P. H., A. J. Cowieson, and V. Ravindran. 2009. Consequences of Ca interactions with phytate and phytase for poultry and pigs. *Livestock Science* 124:126-141.

Seynaeve, M., R. De Wilde, G. Janssens, and B. De Smet. 1996. The influence of dietary salt level on water consumption, farrowing, and reproductive performance of lactating sows. *Journal of Animal Science* 74:1047-1055.

Shanklin, S. H., E. R. Miller, D. E. Ullrey, J. A. Hoefer, and R. W. Luecke. 1968. Zinc requirement of baby pigs on casein diets. *Journal of Nutrition* 96:101-108.

Sharp, B. A., L. C. Young, and A. A. van Dreummel. 1972a. Dietary induction of mulberry heart disease and hepatosis dietetica in pigs. 1. Nutritional aspects. *Canadian Journal of Comparative Medicine* 36:371-376.

Sharp, B. A., L. C. Young, and A. A. van Dreummel. 1972b. Effect of supplemental vitamin E and selenium in high-moisture corn diets on the incidence of mulberry heart disease and hepatosis dietetica in pigs. *Canadian Journal of Comparative Medicine* 36:393-397.

Sheffy, B. E., and R. D. Schultz. 1979. Influence of vitamin E and selenium on immune response mechanisms. *Federation Proceedings* 38:2139-2143.

Shelton, J. L., R. L. Payne, S. L. Johnston, T. D. Bidner, L. L. Southern, R. L. Odgaard, and T. G. Page. 2003. Effect of chromium propionate on growth, carcass traits, pork quality, and plasma metabolites in growing-finishing pigs. *Journal of Animal Science* 81:2515-2524.

Sihombing, D. T. H., G. L. Cromwell, and V. W. Hays. 1974. Effects of protein source, goitrogens and iodine level on performance and thyroid status of pigs. *Journal of Animal Science* 39:1106-1112.

Simesen, M. C., P. T. Jensen, A. Basse, C. Cissel-Nielsen, T. Leth, V. Danielsen, and H. E. Nielsen. 1982. Clinicopathologic findings in young pigs fed different levels of selenium, vitamin E and antioxidants. *Acta Veterinaria Scandinavica* 23:295-308.

Simons, P. C. M., H. A. J. Versteegh, A. W. Jongbloed, P. A. Kemme, P. Slump, K. D. Bos, M. G. E. Wolters, R. F. Beudeker, and G. J. Verschoor. 1990. Improvement of phosphorus availability by microbial phytase in broilers and pigs. *British Journal of Nutrition* 64:525-540.

Slatter, E. E. 1955. Mild iodine deficiency and losses of newborn pigs. *Journal of the American Veterinary Medical Association* 127:149-152.

Smith, J. W., M. D. Tokach, R. D. Goodband, J. L. Nelssen, and B. T. Richert. 1997. Effects of the interrelationship between zinc oxide and copper sulfate on growth performance of early-weaned pigs. *Journal of Animal Science* 75:1861-1866.

Smith, K. 1966. *Inactivation of Gossypol with Mineral Salts*. Memphis, TN: National Cottonseed Production Association.

Smith, W. H., M. P. Plumlee, and W. M. Beeson. 1958. Zinc requirement for growing swine. *Science* 128:1280-1281.

Smith, W. H., M. P. Plumlee, and W. M. Beeson. 1962. Effect of source of protein on zinc requirement of the growing pig. *Journal of Animal Science* 21:399-405.

Soares, J. H. 1995. Calcium bioavailability. Pp. 95-118 in *Bioavailability of Nutrients for Animals*, C. B. Ammerman, D. H. Baker, and A. J. Lewis, eds. New York: Academic Press.

Spallholz, J. E. 1980. Selenium: What role in immunity and immune cytoxicity. Pp. 103-117 in *Proceedings of the Second International Symposium on Selenium in Biology and Medicine*, J. E. Spallholz, J. L. Martin, and H. E. Ganther, eds. Westport, CT: AVI.

Spruill, D. C., V. W. Hays, and G. L. Cromwell. 1971. Effects of dietary protein and iron on reproduction and iron-related blood constituents in swine. *Journal of Animal Science* 33:376-384.

Stahl, C. H., Y. M. Han, K. R. Roneker, W. A. House, and X. G. Lei. 1999. Phytase improves iron bioavailability for hemoglobin synthesis in young pigs. *Journal of Animal Science* 77:2135-2142.

Stahly, T. S., G. L. Cromwell, and H. J. Monegue. 1980. Effects of the dietary inclusion of Cu and (or) antibiotics on the performance of weanling pigs. *Journal of Animal Science* 51:1347-1351.

Stangl, G. I., D. A. Roth-Maier, and M. Kirchgessner. 2000. Vitamin B-12 deficiency and hyperhomocysteinemia are partly ameliorated by cobalt and nickel supplementation in pigs. *Journal of Nutrition* 130:3038-3044.

Stansbury, W. F., L. F. Tribble, and D. E. Orr, Jr. 1990. Effect of chelated copper sources on performance of nursery and growing pigs. *Journal of Animal Science* 68:1318-1322.

Stant, E. C., T. C. Martin, and W. V. Kassler. 1969. Potassium content of the porcine body and carcass at 23, 46, 68 and 91 kilograms live weight. *Journal of Animal Science* 29:547-556.

Steele, N. C., T. G. Althen, and L. T. Frobish. 1977. Biological activity of glucose tolerance factor in swine. *Journal of Animal Science* 45:1341-1345.

Stein, H. H., C. T. Kadzere, S. W. Kim, and P. S. Miller. 2008. Influence of dietary phosphorus concentration on the digestibility of phosphorus in monocalcium phosphate by growing pigs. *Journal of Animal Science* 86:1861-1867.

Stevenson, J. W., and I. P. Earle. 1956. Studies on parakeratosis in swine. *Journal of Animal Science* 15:1036-1045.

Stober, C. R., G. L. Cromwell, and T. S. Stahly. 1980a. Biological availability of the phosphorus in cottonseed meal for growing pigs. *Journal of Animal Science* 51(Suppl. 1):49 (Abstr.).

Stober, C. R., G. L. Cromwell, and T. S. Stahly. 1980b. Biological availability of the phosphorus in oats, wheat middlings, and wheat bran for pigs. *Journal of Animal Science* 51(Suppl. 1):80 (Abstr.).

Stockland, W. L., and L. C. Blaylock. 1973. Influence of dietary calcium and phosphorus levels on the performance and bone characteristics of growing-finishing swine. *Journal of Animal Science* 37:906-912.

Stowe, H. D., and T. H. Herdt. 1992. Clinical assessment of selenium status of livestock. *Journal of Animal Science* 70:3928-3933.

Suomi, K., and T. Alaviuhkola. 1992. Responses to organic and inorganic selenium in the performance and blood selenium content of growing pigs. *Agricultural Sciences Finland* 1:211.

Suttle, N. F., and C. F. Mills. 1966a. Studies of toxicity of copper to pigs. I. Effects of oral supplements of zinc and iron salts on the development of copper toxicosis. *British Journal of Nutrition* 20:135-148.

Suttle, N. F., and C. F. Mills. 1966b. Studies of toxicity of copper to pigs. II. Effect of protein source and other dietary components on the response to high and moderate intakes of copper. *British Journal of Nutrition* 20:149-161.

Suttle, N. F., and J. Price. 1976. The potential toxicity of copper-rich animal excreta to sheep. *Animal Production* 23:233-241.

Svajgr, A. J., E. R. Peo, Jr., and P. E. Vipperman, Jr. 1969. Effects of dietary levels of manganese and magnesium on performance of growing-finishing swine raised in confinement and on pasture. *Journal of Animal Science* 29:439-443.

Swinkels, J. W. G. M., E. T. Kornegay, W. Zhou, M. D. Lindemann, K. E. Webb, Jr., and M. W. A. Verstegen. 1996. Effectiveness of a zinc amino acid chelate and zinc sulfate in restoring serum and soft tissue zinc concentrations when fed to zinc-depleted pigs. *Journal of Animal Science* 74:2420-2430.

Tan, G. Y., S. S. Zheng, M. H. Zhang, J. H. Feng, P. Xie, and J. M. Bi. 2008. Study of oxidative damage in growing-finishing pigs with continuous excess dietary chromium picolinate intake. *Biological Trace Element Research* 126:129-140.

Tang, L., D. F. Li, F. L. Wang, J. J. Xing, and L. M. Gong. 2001. Effects of different sources of organic chromium on immune function in weaned pigs. *Asian-Australasian Journal of Animal Sciences* 14:1164-1169.

Taylor, T. G. 1965. The availability of the calcium and phosphorus of plant materials for animals. *Proceedings of the Nutrition Society* 24:105-112.

Teague, H. S., and L. E. Carpenter. 1951. The demonstration of copper deficiency in young growing pigs. *Journal of Nutrition* 43:389-399.

Thacker, P. A. 1991. Effect of high levels of copper or dichlorvos during late gestation and lactation on sow productivity. *Canadian Journal of Animal Science* 71:227-248.

Theuer, R. C., and W. C. Hoekstra. 1966. Oxidation of ^{14}C-labeled carbohydrate, fat and amino acid substrates by zinc-deficient rats. *Journal of Nutrition* 89:448-454.

Thomas, H. R., and E. T. Kornegay. 1981. Phosphorus in swine. I. Influence of dietary calcium and phosphorus levels and growth rate on feedlot performance of barrows, gilts and boars. *Journal of Animal Science* 52:1041-1048.

Thompson, R. H., C. H. McMurray, and W. J. Blanchflower. 1976. The levels of selenium and glutathione peroxidase activity in blood of sheep, cows and pigs. *Research in Veterinary Science* 20:229-231.

Thoren-Tolling, K. 1975. Studies on the absorption of iron after oral administration in piglets. *Acta Veterinaria Scandinavica* 54 (Suppl.):1-121.

Tilton, S. L., P. S. Miller, A. J. Lewis, D. E. Reese, and P. M. Ermer. 1999. Addition of fat to the diets of lactating sows: I. Effects on milk production and composition and carcass composition of the litter at weaning. *Journal of Animal Science* 77:2491-2500.

Tokach, L. M., M. D. Tokach, R. D. Goodband, J. L. Nelssen, S. C. Henry, and T. A. Marsteller. 1992. Influence of zinc oxide in starter diets on pig performance. P. 411 in the *Proceedings of the American Association of Swine Practitioners*.

Tonroy, B., M. P. Plumlee, J. H. Conrad, and T. R. Cline. 1973. Apparent digestibility of the phosphorus in sorghum grain and soybean meal for growing swine. *Journal of Animal Science* 36:669-673.

Trapp, A. L., K. K. Keahey, D. L. Whitenack, and C. K. Whitehair. 1970. Vitamin E-selenium deficiency in swine. Differential diagnosis and nature of field problem. *Journal of the American Veterinary Medical Association* 157:289-300.

Traylor, S. L., G. L. Cromwell, and M. D. Lindemann. 2005. Bioavailability of phosphorus in meat and bone meal for swine. *Journal of Animal Science* 83:1054-1061.

Trotter, M., and G. L. Allee. 1979a. Availability of phosphorus in corn, soybean meal and wheat. *Journal of Animal Science* 49(Suppl. 1):255 (Abstr.).

Trotter, M., and G. L. Allee. 1979b. Availability of phosphorus in dry and high-moisture grain for pigs and chicks. *Journal of Animal Science* 49(Suppl. 1):98 (Abstr.).

Trotter, M., and G. L. Allee. 1979c. Effects of steam pelleting and extruding sorghum grain-soybean meal diets on phosphorus availability for swine. *Journal of Animal Science* 49(Suppl. 1):255 (Abstr.).

Tucker, H. F., and W. D. Salmon. 1955. Parakeratosis or zinc deficiency disease in the pig. *Proceedings of the Society for Experimental Biology and Medicine* 88:613-616.

Tunmire, D. L., D. E. Orr, Jr., and L. F. Tribble. 1983. Ammonium polyphosphate vs. dicalcium phosphate as a phosphorus supplement for growing-finishing swine. *Journal of Animal Science* 57:632-637.

Ullrey, D. E. 1974. The selenium deficiency problem in animal agriculture. Pp. 275-293 in *Trace Element Metabolism in Animals*, Volume 2, W. C. Hoekstra, J. W. Suttie, H. E. Ganther, and W. Mertz, eds. Baltimore, MD: University Park Press.

Ullrey, D. E. 1992. Basis for regulation of selenium supplements in animal diets. *Journal of Animal Science* 70:3922-3927.

Ullrey, D. E., E. R. Miller, D. R. West, D. A. Schmidt, R. W. Seerley, J. A. Hoefer, and R. W. Luecke. 1959. Oral and parenteral administration of iron in the prevention and treatment of baby pig anemia. *Journal of Animal Science* 18:256-263.

Ullrey, D. E., E. R. Miller, O. A. Thompson, I. M. Ackermann, D. A. Schmidt, J. A. Hoefer, and R. W. Luecke. 1960. The requirement of the baby pig for orally administered iron. *Journal of Nutrition* 70:187-192.

Ullrey, D. E., E. R. Miller, J. P. Hitchcock, P. K. Ku, R. L. Covert, J. Hegenauer, and P. Saltman. 1973. Oral ferric citrate vs. ferrous sulfate for prevention of baby pig anemia. *Michigan Agricultural Experiment Station Research Report* 232:34-38.

Underwood, E. J. 1971. *Trace Elements in Human and Animal Nutrition*, 3rd Ed. New York: Academic Press.

Underwood, E. J. 1977. *Trace Elements in Human and Animal Nutrition*, 4th Ed. New York: Academic Press.

Urbanczyk, J., E. Hanczakowska, and M. Swiatkiewicz. 2001. Effect of energy concentration on the efficiency of betaine and chromium picolinate as dietary supplements for fattening pigs. *Journal of Animal and Feed Sciences* 10(3):471-484.

van de Ligt, J. L. G., M. D. Lindemann, R. J. Harmon, H. J. Monegue, and G. L. Cromwell. 2002a. Effect of chromium tripicolinate supplementation on porcine immune response during the periparturient and neonatal period. *Journal of Animal Science* 80(2):456-466.

van de Ligt, J. L. G., M. D. Lindemann, R. J. Harmon, H. J. Monegue, and G. L. Cromwell. 2002b. Effect of chromium tripicolinate supplementation on porcine immune response during the postweaning period. *Journal of Animal Science* 80(2):449-455.

van Heugten, E., and J. W. Spears. 1997. Immune response and growth of stressed weanling pigs fed diets supplemented with organic or inorganic forms of chromium. *Journal of Animal Science* 75(2):409-416.

van Heugten, E., J. W. Spears, E. B. Kegley, J. D. Ward, and M. A. Qureshi. 2003. Effects of organic forms of zinc on growth performance, tissue zinc distribution, and immune response of weanling pigs. *Journal of Animal Science* 81:2063-2071.

van Kempen, G. J. M., P. van der Kerk, and A. H. M. Crimbergen. 1976. The influence of the phosphorus and calcium content of feeds on growth, feed conversion and slaughter quality and on the chemical, mechanical and histological parameters on the bone tissue of pigs. *The Netherlands Journal of Agricultural Science* 24:120-139.

Van Vleet, J. F., K. B. Meyer, and H. J. Olander. 1973. Control of selenium-vitamin E deficiency in growing swine by parenteral administration of selenium-vitamin E preparations to baby pigs or to pregnant sows and their baby pigs. *Journal of the American Veterinary Medical Association* 163:452-456.

Van Vleet, J. F., A. H. Rebar, and V. J. Ferns. 1977. Acute cobalt and isoproterenol cardiotoxicity in swine: Protection by selenium-vitamin E supplementation and potentiation by stress-susceptible phenotype. *American Journal of Veterinary Research* 38:991-1002.

Venn, J. A. J., R. A. McCance, and E. M. Widdowson. 1947. Iron metabolism in piglet anemia. *Journal of Comparative Pathology and Therapeutics* 57:314-325.

Veum, T. L., J. T. Gallo, W. G. Pond, L. D. Van Vleck, and J. K. Loosli. 1965. Effect of ferrous fumarate in the lactation diet on sow milk iron, pig hemoglobin and weight gain. *Journal of Animal Science* 24:1169-1173.

Vincent, J. B. 2001. The bioinorganic chemistry of chromium (III). *Polyhedron* 20:1-26.

Vipperman, P. E., Jr., E. R. Peo, Jr., and P. J. Cunningham. 1974. Effect of dietary calcium and phosphorus level upon calcium, phosphorus and nitrogen balance in swine. *Journal of Animal Science* 38:758-765.

Wahlstrom, R. C., L. D. Kamstra, and O. E. Olson. 1955. The effect of arsanilic acid and 3-nitro-4-hydroxyphenylarsonic acid on selenium poisoning in the pig. *Journal of Animal Science* 14:105-110.

Wallace, H. D. 1967. *High Level Cu in Swine Feeding*. New York: International Cu Research Association, Inc.

Wang, M. Q., and Z. R. Xu. 2004. Effect of chromium nanoparticle on growth performance, carcass characteristics, pork quality and tissue chromium in finishing pigs. *Asian-Australasian Journal of Animal Sciences* 17(8):1118-1122.

Wang, M. Q., Y. D. He, Z. R. Xu, and W. F. Li. 2008. Effects of chromium picolinate supplementation on growth hormone secretion and pituitary mRNA expression in finishing pigs. *Asian-Australasian Journal of Animal Sciences* 21(7):1033-1037.

Wang, M. Q., Z. R. Xu, W. F. Li, and Z. G. Jiang. 2009. Effect of chromium nanocomposite supplementation on growth hormone pulsatile secretion and mRNA expression in finishing pigs. *Journal of Animal Physiology and Animal Nutrition* 93(4):520-525.

Ward, T. L., G. L. Asche, G. F. Louis, and D. S. Pollmann. 1996. Zinc-methionine improves growth performance of starter pigs. *Journal of Animal Science* 74(Suppl. 1):182 (Abstr.).

Wedekind, K. J., A. J. Lewis, M. A. Giesemann, and P. S. Miller. 1994. Bioavailability of zinc from inorganic and organic sources for pigs fed corn-soybean meal diets. *Journal of Animal Science* 72:2681-2689.

Weeden, T. L., J. L. Nelssen, R. D. Goodband, J. A. Hansen, G. E. Fitzner, K. G. Fiesen, and J. L. Laurin. 1993a. Effects of porcine somatotropin and dietary phosphorus on growth performance and bone properties of gilts. *Journal of Animal Science* 71:2674-2682.

Weeden, T. L., J. L. Nelssen, R. D. Goodband, J. A. Hansen, K. G. Fiesen, and B. T. Richert. 1993b. The interrelationship of porcine somatotropin administration and dietary phosphorus on growth performance and bone properties in developing gilts. *Journal of Animal Science* 71:2683-2692.

Wegger, I., K. Rasmussen, and P. F. Jorgensen. 1980. Glutathione peroxidase activity in liver and kidney as indicator of selenium status in swine. *Livestock Production Science* 7:175-180.

Weinberg, E. D. 1978. Iron and infection. *Microbiological Research* 42:45-66.

Welch, H. 1928. Goiter in farm animals. *Montana Agricultural Experiment Station Bulletin* 214:1-27.

Whiting, F., and L. M. Bezeau. 1958. The calcium, phosphorus and zinc balance in pigs as influenced by the weight of pig and the level of calcium, zinc and vitamin D in the ration. *Canadian Journal of Animal Science* 38:109-117.

Wilkinson, J. E., M. C. Bell, J. A. Bacon, and F. B. Masincupp. 1977a. Effects of supplemental selenium on swine. I. Gestation and lactation. *Journal of Animal Science* 44:224-228.

Wilkinson, J. E., M. C. Bell, J. A. Bacon, and C. C. Melton. 1977b. Effects of supplemental selenium on swine. II. Growing-finishing. *Journal of Animal Science* 44:229-233.

Xi, G., Z. R. Xu, S. H. Wu, and S. J. Chen. 2001. Effect of chromium picolinate on growth performance, carcass characteristics, serum metabolites and metabolism of lipid in pigs. *Asian-Australasian Journal of Animal Sciences* 14(2):258-262.

Yen, J. T., J. J. Ford, and J. Klindt. 2005. Effect of supplemental copper proteinate on reproductive performance of first- and second-parity sows. *Canadian Journal of Animal Science* 85:205-210.

Yin, J., X. Li, D. Li, T. Yue, Q. Fang, J. Ni, X. Zhou, and G. Wu. 2009. Dietary supplementation with zinc oxide stimulates ghrelin secretion from the stomach of young pigs. *Journal of Nutritional Biochemistry* 20:783-790.

Yin, Y., C. Huang, X. Wu, T. Li, R. Huang, P. Kang, Q. Hu, W. Chu, and X. Kong. 2008. Nutrient digestibility response to graded dietary levels of sodium chloride in weanling pigs. *Journal of the Science of Food and Agriculture* 88:940-944.

Yoon, I., and E. McMillan. 2006. Comparative effects of organic and inorganic selenium on selenium transfer from sows to nursing pigs. *Journal of Animal Science* 84:1729-1733.

Young, L. G., J. H. Lumsden, A. Lun, J. Claxton, and D. E. Edmeades. 1976. Influence of dietary levels of vitamin E and selenium on tissue and blood parameters in pigs. *Canadian Journal of Comparative Medicine* 40:92-97.

Young, L. G., M. Leunissen, and J. L. Atkinson. 1993. Addition of microbial phytase to diets of young pigs. *Journal of Animal Science* 71:2147-2151.

Zacharias, B., H. Ott, and W. Drochner. 2003. The influence of dietary microbial phytase and copper on copper status in growing pigs. *Animal Feed Science and Technology* 106:139-148.

Zhou, W., E. T. Kornegay, and M. D. Lindemann. 1994a. The role of feed intake and copper source on copper-stimulated growth in weanling pigs. *Journal of Animal Science* 72:2385-2394.

Zhou, W., E. T. Kornegay, M. D. Lindemann, J. W. G. M. Swinkels, M. K. Welten, and E. A. Wong. 1994b. Stimulation of growth by intravenous injection of copper in weanling pigs. *Journal of Animal Science* 72:2395-2043.

Zhu, D., B. Yu, C. Ju, S. Mei, and D. Chen. 2011. Effect of high copper on the expression of hypothalamic appetite regulators in weanling pigs. *Journal of Animal and Feed Sciences* 20:60-70.

Zimmerman, D. R. 1980. Iron in swine nutrition. In *National Feed Ingredients Association Literature Review on Iron in Animal and Poultry Nutrition*. Des Moines, Iowa: National Feed Ingredients Association.

Zimmerman, D. R., V. C. Speer, V. W. Hays, and D. V. Catron. 1959. Injectable iron dextran and several oral iron treatments for the prevention of iron deficiency anemia of baby pigs. *Journal of Animal Science* 18:1409-1415.

Zimmerman, D. R., V. C. Speer, V. W. Hays, and D. V. Catron. 1963. Effect of calcium and phosphorus levels on baby pig performance. *Journal of Animal Science* 22:658-661.

Zoubek, G. L., E. R. Peo, Jr., B. D. Moser, T. Stahly, and P. J. Cunningham. 1975. Effects of source on copper uptake by swine. *Journal of Animal Science* 40:880-884.

第7章 维 生 素

引言

"维生素"这一术语描述的是区别于氨基酸、碳水化合物和脂类的一类有机化合物，只需少量就可维持正常的生长和繁殖。有些维生素可能不需要在日粮中添加，因为它们可由其他饲料、代谢产物或者肠道微生物合成。维生素一般分为脂溶性和水溶性两类。脂溶性维生素包括维生素 A、维生素 D、维生素 E 和维生素 K。水溶性维生素包括 B 族维生素（生物素、胆碱、叶酸、烟酸、泛酸、核黄素、硫胺素、维生素 B_6 和维生素 B_{12}）和维生素 C（抗坏血酸）。

维生素主要在营养代谢中作为辅酶。饲料原料中的维生素主要以前体复合物或辅酶的形式结合或复合存在。因此，消化过程是释放或转变维生素前体或复合物使之成为可利用与可吸收的形式。第 16 章中的表格列举了猪生命周期的各个阶段对单个维生素的需要量。为弥补实际日粮中维生素的不足，已研发出可常规添加于猪日粮中的维生素预混料。不同阶段猪预混料中维生素添加量（考虑到在最终日粮的添加比例）可能会大大高于其需要量，这是因为维生素含量会因预混料储存的时间与方式的不同而有所损失。不同维生素对各种因素如水分/湿度、光、热、pH 和氧化剂的敏感性不同。此外，在猪采食饲料前的饲料加工工艺，如膨化或制粒，会进一步加剧维生素的损失。Shurson 等（2011）检测复合维生素储存 120 天后各种维生素的损失，发现各种维生素间的损失差异显著；同时也注意到，对含维生素和微量元素的预混料而言，与无机微量元素相比，与特定氨基酸络合的微量元素提高了预混料中维生素的稳定性。

日粮中添加过量的维生素 A 和维生素 D 可以引起猪中毒（Crenshaw, 2000; Darroch, 2000）。但有关 B 族维生素、维生素 E 和维生素 K 中毒症状的报道甚少（NRC, 1987; Crenshaw, 2000; Dove and Cook, 2000; Mahan, 2000）。

有一些研究表明，猪日粮中一种或几种常用 B 族维生素（核黄素、烟酸、泛酸和维生素 B_{12}）的添加量不足以最大限度地发挥其生产性能（Lindemann et al., 1999; Stahly et al., 2007），但另一些研究结果不支持上述观点（Mahan et al., 2007）。确实，给猪日粮添加估测需要量的 2～10 倍 B 族维生素有改善猪的生长和饲料转化率的趋势，然而，还不知道哪个水平［高出 NRC（1988）和 NRC（1998）的标准推荐量］更合适。Lindemann 等（1995）发现，给断奶仔猪日粮中添加 5 倍 NRC（1988）推荐量的常用维生素（包括脂溶性维生素），有提高增重和采食量的趋势，但添加过高水平的维生素会使饲料转化率趋于降低。尽管目前猪的基因型与过去的不同，日粮中能量浓度比过去更高（会影响饲料采食量及需要的养分浓度），并且先前均是添加复合维生素进行研究，根本无法单独修订各个 B 族维生素的需要量。但这些研究确实引发了对添加超过 NRC 推荐量维生素效果的兴趣，并且表明有必要更多研究单个维生素的推荐添加量。

在商业生产条件下的研究中也得出了一些与维生素需要量相关的有趣结果。Coelho 和 Cousins（1997）开展了一项涉及养猪行业 23 个单位有关从断奶到育肥猪维生素添加量的调查研究。该调查涵盖了猪各生长阶段（包括由天然的饲料原料来满足而不需要添加维生素的阶段），除了饲料原料提供的可利用维生素量，最低四分位数的维生素添加量高于 NRC（1998）估测的需要量至少 2～15 倍。最高四分位数的维生素水平经常是最低四分位数的 2～10 倍。生产性能试验中，处理包括 NRC 估测的维生素水平组和业界添加水平的最低四分位数组、中分位数组、最高四分位数组、最高的5%组，同时结合一个应激因子，以模拟一些正常猪生产中所遇到的应激因素。应激以饲养密度/提供的地面空间、大肠杆菌感染、沙门氏菌感染、霉菌毒素感染、日粮的营养浓度来分为低、中、高三个水平。正如预料，随着应激的加大，生长速度和饲料转化率降低，死亡率增加。在低应激条件下，提高维生素水平对生长性能无明显影响。但是，在高应激条件下，随着维生素添加水平的提高，则对各项生产指标（生长速度、饲料转化率和死亡率）具有显著影响。这类研究显然把各种维生素和各种应激源混在一起了，因而无法用于确定单个维生素的需要量。然而，这确实反映出商业条件下与研究条件下的维生素需要量存在差异，如果要把研究条件下估测的需要量用于商业条件下，研究条件必须能反映商业条件。

Boyd等（2008）报道，额外添加维生素对繁殖母猪性能有潜在的益处。种猪群由各胎次的母猪组成，存在给非常大体重的母猪（妊娠期限制饲喂以限制能量摄入，防止过度生长）饲喂较少的维生素和矿物质（按每天每单位体重计算）的情况。调查者发现，限制能量摄入后，所有养分的摄入量受限，最大的母猪每单位体重的养分供应量最少。对超过50 000窝母猪进行一年的生产试验，结果表明，以单位体重计算，当6胎的母猪摄入维生素和微量元素的量与3胎的母猪相同时，提高维生素和微量元素预混料浓度使4~10胎母猪的窝产仔数增加[平均0.60头断奶仔猪/窝或1.44头/（母猪/年）]，部分抵消了随着胎次增加生产力的自然下降。但是，这种试验还是不能估测单个维生素的需要量，它仅能说明有必要在需要量的研究中考虑到特定的情况需要，并且当将研究条件下估测的需要量延伸到商业生产条件下，有必要用不同的方式表示种用动物的需要量。

Audet等（2004）研究了种公猪对维生素的潜在需要量。试验中在常规维生素添加量基础上，超量添加维生素C（1000 mg/kg日粮）、水溶性维生素（商业调查的行业平均用量的10倍）或脂溶性维生素（行业平均用量的3~5倍），研究在正常和密集采精条件下补饲维生素对青年公猪维生素水平、性欲及精液质量的潜在影响。在密集采精阶段，发现添加水溶性维生素提高了公猪精液产量，且在恢复阶段这些公猪的运动精子百分率也较高。与对照公猪相比，采食添加脂溶性维生素日粮的公猪也出现上述反应，但幅度较低。各组公猪的精子形态和性欲不受日粮处理的影响。因此在密集采精阶段，日粮中超量添加水溶性和脂溶性维生素提高了精液产量，但从本试验中不能确定是所有维生素还是仅某单一维生素的需要量需要增加。超量添加维生素C对种公猪无影响。随后，Audet等（2009）同时给公猪日粮添加与上述实验量相同的水溶性和脂溶性维生素，尽管公猪在一个密集采精期间精液量增加了，但未发现强化补饲维生素对精子产量或精子质量有影响。

脂溶性维生素

维生素A

维生素A为视觉、繁殖、生长与维持不同的上皮细胞以及黏液分泌所必需（Wald，1968；Goodman，1979，1980）。有证据表明，维生素A参与基因转录、胚胎发育、骨骼代谢、血细胞生成和各种免疫过程（Combs，1999）。

维生素A的命名法（Anonymous，1990）规定，"维生素A"这一术语适用于所有β-芷香酮衍生物，而不是维生素A原类胡萝卜素，前者有全反视黄醇（如维生素A醇或视黄醇）的生物学活性。维生素A存在于动物组织、蛋及全奶中。植物组织中只含有维生素A的前体物，需经肠道或肝脏的作用才能形成视黄醇。不论自然存在的维生素A还是合成的视黄醇类似物，通常都被称为视黄醇类。以大鼠数据为基础，1 IU的维生素A等于0.3 μg的晶体维生素A醇，或等于0.344 μg的维生素A醋酸酯，或等于0.55 μg

的维生素 A 棕榈酸酯。目前通常用视黄醇当量（RE）来描述食物或饲料中的维生素 A 活性。一个 RE 被定义为 1μg 的全反视黄醇。

猪将类胡萝卜素前体物转化为维生素 A 的效率次于家禽和大鼠。这种转化主要发生于肠黏膜中（Fidge et al., 1969）。如果把全反 β-胡萝卜素的生物学活性定义为 100%，玉米-豆粕型日粮中的类胡萝卜素（Wellenreiter et al., 1969）活性分别为：β-玉米胡萝卜素 25%，玉米黄质 57%（Petzold et al., 1959; Duel et al., 1945; Greenberg et al., 1950）。因此，Ullrey（1972）计算，全反式 β-胡萝卜素当量仅相当于化学测定胡萝卜素值的 52%。然后，Ullrey 计算出猪日粮中 β-胡萝卜素当量仅为 16%，因为它转化为可利用维生素 A 的效率仅为大鼠的 30%（Braude et al., 1941）。将此值乘以 1667 IU（1667 IU 代表大鼠将 1 mg 全反式 β-胡萝卜素转化为可利用维生素 A 的理论效价），玉米-豆粕型日粮中 1 mg 化学测定的胡萝卜素转化为维生素 A 的计算效价为 267 IU，或 80 mg 的维生素 A 醇。

Chew 等（1982）、Brief 和 Chew（1985）提出，β-胡萝卜素在繁殖中的作用与维生素 A 无关。他们研究认为，注射 β-胡萝卜素可提高母猪血浆中维生素 A 或 β-胡萝卜素含量，促进胚胎成活，这可能是因为母猪分泌了更多的子宫特异蛋白质。但日粮中添加 β-胡萝卜素却没有这种效应，这可能是由于猪对完整的 β-胡萝卜素的吸收能力差引起的（Poor et al., 1987）。猪能在肝脏中储存维生素 A，摄入不足时可动用这部分维生素 A。维生素 A 的需要量取决于评定指标，体增重不如脑脊髓液压、肝储或血浆水平敏感。出生后前 8 周内的猪，对维生素 A 的需要量为 75～605μg 维生素 A 醋酸酯/kg 日粮，这取决于所采用的反应指标（Sheffy et al., 1954; Frape et al., 1959）。对生长育肥猪而言，以体增重为指标时，维生素 A 的需要量为 35～130 μg/kg 日粮；而用肝储和脑脊髓液压为指标时，维生素 A 需要量为 344～930 μg/kg 日粮（Guilbert et al., 1937; Braude et al., 1941; Hentges et al., 1952; Myers et al., 1959; Hjarde et al., 1961; Nelson et al., 1962; Ullrey et al., 1965）。若饲料或饮水中存在硝酸盐或亚硝酸盐，会提高维生素 A 的需要量（Seerley et al., 1965; Wood et al., 1967; Hutagalung et al., 1968）。

由于母猪能储存维生素 A，所以确定其需要量很困难。Braude 等（1941）报道，采食不添加维生素 A 日粮的成熟母猪能正常完成 3 次妊娠，在第 4 次妊娠时才出现缺乏症。给青年母猪饲喂含足够维生素 A 的日粮直到 9 月龄，随后饲喂不含维生素 A 的日粮，母猪能完成两轮繁殖而不出现维生素 A 缺乏症（Hjarde et al., 1961; Selke et al., 1967）。Heaney 等（1963）给耗竭了维生素 A 的青年母猪按每日每千克体重饲喂 16μg、5μg 或 2.5 μg 的视黄醇棕榈酸酯，结果窝产仔数、初生重和成活率未受影响。Parrish 等（1951）认为，妊娠和泌乳期间，每日摄入 2100 IU 的维生素 A 足以维持血清和肝脏中的正常水平。最近 Lindemann 等（2008）在多个实验站对不同遗传背景的母猪进行研究，结果发现，在断奶和配种时给 1 和 2 胎次的青年母猪肌内注射高剂量（250 000IU 或 500 000IU）维生素 A，随后的窝产仔猪和断奶仔猪的数量直线增加。对 3~6 胎次的母猪来说，采用此处理则对窝产仔数无影响。该试验基础日粮中添加 11 000IU/kg 维生素 A。因此，为获得最高性能所需要的维生素 A 数量随日龄而变化，且仅通过日粮添加也许不能满足猪对维生素 A 的需要。

猪缺乏维生素 A 会导致体增重下降、运动失调、后躯麻痹、失明、脑脊髓液压升高、血浆维生素 A 水平下降和肝脏维生素 A 储存量减少（Guilbert et al., 1937; Braude et al., 1941; Hentges et al., 1952; Frape et al., 1959; Hjarde et al., 1961; Nelson et al., 1962, 1964）。

过量维生素 A 中毒的症状大致包括：被毛粗糙、鳞状皮肤、对触摸反应过度和敏感、蹄周围裂纹处出血、血尿、血粪、腿失控并伴随不能站立以及周期性震颤（Anderson et al., 1966）。给小猪饲喂每千克含视黄醇棕榈酸酯 605 000μg、484 000μg、363 000μg 或 242 000μg 的日粮，分别在喂后 16 天、17.5 天、32 天和 43 天出现维生素 A 中毒症状。但给猪连续 8 周饲喂每千克添加 121 000 μg 视黄醇棕榈酸酯的日粮，未表现中毒症状（Anderson et al., 1966）。Wolke 等（1968）观察到，饲喂过量维生素 A 的猪，在 5 周内出现软骨病变和膜内骨化症状。NRC（1987）确定的生长猪与种猪维生素 A 的安全上限水平分别是 20 000 IU/kg 日粮和 40 000 IU/kg 日粮。

在日粮或预混料中，维生素 A 酯比视黄醇更稳定。视黄醇侧链上的羟基及 4 个双键易氧化损失，因此，维生素 A 醇的酯化不能完全保护其免遭氧化。目前，商品维生素 A 一般都是含有诸如乙氧基喹啉（EMQ）或二丁基羟基甲苯（BHT）抗氧化剂的包被酯类（1 IU 维生素 A=0.344 μg 视黄醇醋酸酯或 1 IU 维生素 A=0.549 μg 视黄醇棕榈酸酯）。

预混料和饲料原料中的水分对维生素 A 的稳定性有负作用（Baker，1995）。水会引起维生素 A 微粒胶囊软化，并使氧气易于渗透到胶囊中。因此，高湿度和游离氯化胆碱（吸湿性强）的存在都会加速维生素 A 的破坏。暴露于潮湿空气中的预混料中的微量元素也会加剧维生素 A 的损失。为最大限度保持维生素 A 的活性，预混料应尽可能防潮，且 pH 应高于 5。低 pH 会使全反式维生素 A 异构化为低活性的顺式，并使维生素 A 酯去酯化形成更不稳定的视黄醇（De Ritter，1976）。

维生素 D

维生素 D 主要以麦角钙化醇（维生素 D_2）和胆钙化醇（维生素 D_3）两种形式存在。在紫外线的作用下，存在于植物中的麦角固醇形成麦角钙化醇。动物皮肤中的 7-脱氢胆固醇经光化学作用转化为胆钙化醇。1 IU 维生素 D 定义为 0.025 μg 胆钙化醇的生物学活性。麦角钙化醇和胆钙化醇在肝脏中羟基化形成 25-羟钙化醇。25-(OH)-D_3 在肾脏中进一步羟化形成 1,25-$(OH)_2$-D_3 或 24,25-$(OH)_2$-D_3。控制二羟基化代谢物的合成与反应的几种机制符合确定激素的标准，因此，二羟基 D_3 代谢物本身也被看成是激素（Schnoes and DeLuca，1980; Kormann and Weiser，1984）。

维生素 D 及其激素性代谢物作用于小肠黏膜细胞，引起钙结合蛋白的形成。这些蛋白质促进钙、镁和磷的吸收。维生素 D 代谢物、甲状旁腺素和降钙素的共同作用维持体内钙和磷的平衡。Braidman 和 Anderson（1985）综述了维生素 D 的内分泌功能。

Bethke 等（1946）指出，维生素 D_2 和维生素 D_3 在满足猪对维生素 D 的需要上等效。但 Horst 等（1982）证明，这两种形式的维生素 D 在猪体内的代谢存在差异。维生素 D_2 和维生素 D_3 在吸收和利用上的差异尚需进一步的量化研究。

饲喂酪蛋白-葡萄糖日粮的仔猪对维生素 D_2 的需要量为 100 IU/kg 日粮（Miller et al.，1964，1965）。如果饲喂分离大豆蛋白的话，则需要量增加（Miller et al.，1965; Hendricks et al.，1967）。维生素 D 缺乏降低钙、磷和镁的存留（Miller et al.，1965）。Bethke 等（1946）建议，生长猪维生素 D 的最低需要量为 200 IU/kg 日粮。但另一些研究表明，添加维生素 D 并不提高猪的增重（Wahlstrom and Stolte，1958; Combs et al.，1966）。

Weisman 等（1976）、Boass 等（1977）、Noff 和 Edelstein（1978）、Halloran 和 DeLuca（1979）及 Pike 等（1979）的研究表明，维生素 D 与大鼠的繁殖和泌乳有关。在分娩前对母猪进行肠外注射的胆钙化醇处理可以有效地给仔猪补充胆钙化醇（通过母乳）及其二羟代谢物（通过胎盘转运）（Goff et al.，1984）。Lauridsen 等（2010）使用维生素 D_3 或一种新研发的维生素 D 产品（25-羟基胆钙化醇）按 4 个水平（200 IU/kg、800 IU/kg、1400 IU/kg 和 2000 IU/kg 维生素 D）对母猪进行补饲。该试验只进行了一胎，补饲不同浓度的维生素 D 对母猪的繁殖性能几乎无影响。但是补饲高剂量维生素 D 组（1400 IU 和 2000 IU 维生素 D/kg）母猪产死胎的数量（分别是 1.17 个/窝和 1.13 个/窝）少于低剂量组（200 IU/kg 和 800 IU/kg，死胎数量分别是 1.98 个/窝和 1.99 个/窝），不过出生和断奶时活仔的数量不受浓度水平的影响。在同时开展的试验中，对后备母猪在妊娠头 28 天饲喂不同水平的 D_3 或 25-羟基胆钙化醇，发

现饲喂维生素 D_3 组的最终骨骼强度和灰分含量均高于同一饲喂水平的 25-羟基胆钙化醇组，且这种差异在添加量为 800 IU 时最为明显。作者推荐繁殖母猪日粮的维生素 D 添加量约为 1400 IU。

维生素 D 缺乏引起钙和磷的吸收和代谢紊乱，导致骨骼钙化不全。幼龄生长猪维生素 D 缺乏会导致佝偻病，而成年猪则引起骨骼矿物质含量减少（骨软化）。严重缺乏维生素 D 时，表现钙和镁缺乏症状，包括痉挛。饲喂缺乏维生素 D 日粮的猪表现缺乏症需要 4～6 个月（Johnson and Palmer, 1939; Quarterman et al., 1964）。维生素 D 参与许多重要的生理功能，其缺乏的一个主要影响是钙代谢和骨骼发育紊乱。维生素 D 也是软组织生长发育和健康所必需，目前在哺乳动物的 33 个器官里发现有 $1,25\text{-}(OH)_2D_3$ 的受体（Zempleni et al., 2007），已知它对免疫、内分泌、神经系统和繁殖均发挥作用。Viganò 等（2003）研究表明，维生素 D 对着床和胚胎形成至关重要。美国医学研究所（IOM, 1999）建议将 $25\text{-}(OH)D_3$ 作为测定人类维生素 D 状况的指标。当血浆 $25\text{-}(OH)D_3$ 浓度低于 25 nmol/L 时，表现维生素 D 缺乏。浓度达到 50 nmol/L 时，则表现边缘性缺乏症（Mosekilde, 2005）。如果这些数值最终被证明可以应用在猪上的话，采食维生素 D 含量低于 1400 IU/kg 的母猪，特别是哺乳期头 2 周的母猪会出现缺乏症。

断奶仔猪连续 4 周每日口服 6250 μg 维生素 D_3 可导致维生素 D 中毒（Quarterman et al., 1964），主要表现为：采食量、生长速度及肝脏、桡骨和尺骨的重量降低。尸检时，可见猪动脉、心脏、肾脏和肺钙化。给 20～25 kg 猪每日饲喂 11 825 μg 的维生素 D_3，4 天后猪死亡（Long, 1984）。对包括猪在内的许多种动物来说，维生素 D_3 的毒性高于维生素 D_2（NRC, 1987）。血浆中维生素 D 及其代谢产物测定方法的改进有助于进一步揭示维生素 D_2 和维生素 D_3 毒性方面差异的可能机制（Horst et al., 1981; NRC, 1987）。现在认为，长期饲喂条件下（超过 60 天）生长猪的维生素 D_3 最大安全水平是 2200 IU/kg 日粮；短期条件下（少于 60 天），它最高可耐受 33 000 IU/kg 日粮（NRC, 1987）。

维生素 E

天然存在的维生素 E 有 8 种形式：α、β、γ 和 δ 生育酚（Evans et al., 1936; Emerson et al., 1937; Stern et al., 1947），以及 α、β、γ 和 δ 生育三烯酚（Green et al., 1960; Pennock et al., 1964; Whittle et al., 1966），其中，D-α-生育酚生物学活性最高（Brubacher and Wiss, 1972; Ames, 1979; Bieri and McKenna, 1981）。1 IU 维生素 E 的活性相当于 1 mg DL-α-生育酚醋酸酯的活性。维生素 E 的 D-异构体比 L-异构体的活性高。在主要以大鼠为对象进行生物学测定的基础上，以 DL-α-生育酚醋酸酯为标准（1mg=1IU），计算出 1 mg DL-α-生育酚相当于 1.1 IU、1mg D-α-生育酚醋酸酯相当于 1.36 IU、1 mg D-α-生育酚相当于 1.49 IU 的维生素 E。Chung 等（1992）报道，对幼龄猪而言，1 mg 的 D-α-生育酚相当于 2.44 IU 的维生素 E。Anderson（1995a）认为，猪对 D-α-生育酚醋酸酯的利用效率比大鼠高。Wilburn 等（2008）也表明，对于幼龄猪而言，天然维生素 E（RRR-α-生育酚醋酸酯）优于合成的维生素 E（全消旋-α-生育酚醋酸酯），说明生物等效值低估了天然维生素 E 对猪的作用。在母猪（Mahan et al., 2000）和育肥猪（Yang et al., 2009）上的试验结果表明，如果以等量 IU 添加不同来源的维生素 E，D-α-生育酚醋酸酯比 DL-α-生育酚醋酸酯生物效价高。Lauridsen 等（2002）用氘标记的维生素 E 饲喂母猪，这些母猪可以区分 RRR-和全消旋 α-生育酚，并且发现血浆中 RRR 来源的 α-生育酚浓度大约比全消旋的高 2 倍。目前美国药典 USP 将 RRR-与全消旋 α-生育酚的效价比值定义为 1.36:1.00，对猪而言，这两者的效价比值可能以 2:1 更合适。天然来源的维生素 E 较合成来源的生物效价高，在猪上比大鼠上差异更大。同时也需考虑到随着日粮中铜、铁、锌或锰含量的增加，天然生育酚的氧化速度加快（Dove and Ewan, 1991）。

多年来，饲料中维生素 E 的主要来源是绿色植物及种子中的生育酚。氧化作用可迅速破坏天然维生素 E，而热、湿、酸败脂肪和微量元素可加速氧化的作用。因此，很难预测饲料中活性维生素 E 的数量。在 32℃的环境下储存 12 周的苜蓿，其维生素 E 的损失量达 50%～70%；脱水过程可造成苜蓿 5%~30%维生素 E 的损失（Livingston et al., 1968）。水分含量高的谷物籽实经储存或用有机酸处理，其维生素 E 含量大大减少（Madsen et al., 973; Lynch et al., 1975; Young et al., 1975, 1978）。日粮中高剂量的维生素 A 也会减少维生素 E 的吸收（Hoppe et al., 1992）。但 Anderson 等（1995b）观察到，日粮中维生素 A 水平高达需要量的 15 倍时，生长猪体内的维生素 E 状况未受影响。

在 20 世纪 70 年代，对猪维生素 E 的需要量开展了许多研究。ARC（1981）和 Ullrey（1981）对这些研究进行了综述。许多日粮因素影响维生素 E 的需要量，包括日粮中硒、不饱和脂肪酸、含硫氨基酸、视黄醇、铜、铁以及合成抗氧化剂的含量。Michel 等（1969）在每千克含 5~8 mg 维生素 E 和 0.04~0.06 mg 硒的玉米-豆粕型日粮中添加 22 mg/kg 的维生素 E，可预防猪的死亡。含有 5 mg/kg 维生素 E 和 0.04 mg/kg 硒的玉米-豆粕型日粮不能满足生长育肥猪的需要，会出现缺乏症和死亡。但是，在有足够硒的情况下，每千克日粮中添加 10~15 mg 的维生素 E 可防止死亡和缺乏症，并维持正常的生产性能（Groce et al., 1971, 1973; Sharp et al., 1972a, b; Ullrey, 1974; Wilkinson et al., 1977b; Hitchcock et al., 1978; Mahan and Moxon, 1978; Meyer et al., 1981）。防止发生缺乏症的维生素 E 需要量因日粮中硒（ARC, 1981; Ullrey, 1981）、抗氧化剂（Tollerz, 1973; Simesen et al., 1982）和脂类（Nielsen et al., 1973; Tiege et al., 1977, 1978）含量的不同而变异很大。

日粮中添加高水平维生素 E 可提高免疫反应（Ellis and Vorhies, 1976; Tiege, 1977; Nockels, 1979; Peplowski et al., 1980; Wuryastuti et al., 1993）。但 Bonnette 等（1990）给仔猪饲喂高水平的维生素 E 时未发现体液或细胞介导的免疫反应有提高。Pinelli-Saavedra 等（2008）在妊娠和泌乳母猪日粮（已添加了 36 IU/kg 维生素 E）中添加 500 mg/kg 的 α-生育酚醋酸酯和 10 g/d 的维生素 C（抗坏血酸），结果泌乳期第 21 天的总免疫球蛋白和免疫球蛋白 G 的浓度显著增加（维生素 E 和维生素 C 如不同时添加，则无效果）。Mavromatis 等（1999）对采食添加 20 mg/kg α-生育酚和 0.45 mg/kg 硒日粮的母猪额外添加 30mg/kg 的 α-生育酚和/或在妊娠 30 天、60 天和 90 天肌肉注射 30 mg 硒，结果发现维生素 E 与硒之间具有协同作用。额外添加维生素 E 使母猪在分娩时以及出生后 24 h 和 28 日龄的仔猪血清中 IgG 水平提高，而同时添加维生素 E 和注射硒的母猪血清 IgG 的浓度增加更多。

维生素 E 在细胞膜水平上发挥抗氧化剂的功能，还是细胞膜中的重要组成成分。有些维生素 E 缺乏症对维生素 E、硒或抗氧化剂有反应。维生素 E 缺乏导致大量病理变化，包括骨骼肌和心肌变性、退化性血栓性血管受损、胃角化不全、胃溃疡、贫血、肝坏死、脂肪组织黄染和猝死（Obel, 1953; Davis and Gorham, 1954; Hove and Seibold, 1955; Dodd and Newling, 1960; Grant, 1961; Lannek et al., 1961; Nafstad, 1965, 1973; Nafstad and Nafstad, 1968; Reid et al., 1968; Ewan et al., 1969; Michel et al., 1969; Nafstad and Tollersrud, 1970; Trapp et al., 1970; Baustad and Nafstad, 1972; Sharp et al., 1972a, b; Sweeney and Brown, 1972; Wastell et al., 1972; Piper et al., 1975; Bengtsson et al., 1978a, b; Hakkarainen et al., 1978; Tiege and Nafstad, 1978; Simesen et al., 1982）。此外，维生素 E 可能与母猪乳房炎-子宫炎-无乳综合征（MMA）有关（Ringarp, 1960; Ullrey et al., 1971; Whitehair et al., 1984）有关。

有一些关于繁殖对维生素 E 需要量的文献报道（Hanson and Hathaway, 1948; Adamstone et al., 1949; Cline et al., 1974; Malm et al., 1976; Young et al., 1977, 1978; Wilkinson et al., 1977a; Nielsen et al., 979; Piatkowski et al., 1979; Mahan, 1991, 1994）。维生素 E 通过胎盘由母猪转运至胎儿的量很小，因此，初生仔猪必须依赖初乳和常乳以满足它对维生素 E 的每日需要（Pinelli-Saavedraa and Scaifeb, 2005）。母猪初乳和常乳中维生素 E 的含量取决于母猪日粮中维生素 E 的水平（Mahan, 1991）。多数研究表明，日粮中添加 5~7 mg/kg 维生素 E 和 0.1 mg/kg 无机硒可防止出现维生素 E 缺乏症，并可维持正常的繁殖

性能。但是，为了维持组织中维生素E水平，似乎有必要添加0.1 mg/kg无机硒和22 mg/kg维生素E（Piatkowski et al., 1979）。另外，20世纪90年代的一些研究（Mahan, 1991, 1994; Wuryastuti et al., 1993）表明，为获得最大窝产仔数和免疫能力，妊娠和泌乳母猪日粮有必要含有44~60 mg/kg维生素E。

目前已经进行了一些添加维生素E和硒对公猪繁殖能力影响的研究（Marin-Guzman et al., 1997; 2000a,b; Jacyno et al., 2002; Kolodziej and Jacyno, 2005; Echeverria-Alonzo et al., 2009），发现二者对组织（血清、肝脏和睾丸）中谷胱甘肽过氧化物酶活性、硒和α-生育酚浓度、睾丸中精子存量、睾丸支持细胞数量、次级精母细胞数量、每次射精的精液中精子总量、精子活力、正常精子与头部畸形精子的比例以及细胞质滴的保留等具有积极作用。但是，由于选用未添加的对照日粮、试验处理数量有限，或是试验处理含维生素E和硒的混杂作用，造成无法确定使公猪发挥最佳繁殖能力的添加量。总体而言，添加硒比添加维生素E效果更明显。

普遍认为维生素E是所有维生素中毒性最小的一种。目前尚未证实维生素E对猪的毒性。给生长猪饲喂高达550 mg/kg维生素E的日粮，未出现毒性反应（Bonnette et al., 1990）。维生素E过多症在大鼠、家禽和人上已有研究，从这些有限的数据中可知最大耐受量在1000~2000IU/kg日粮（NRC, 1987）。

维生素K

虽然维生素K是4种脂溶性维生素中最晚发现的一种，但其代谢作用却比维生素A、D、E要明确得多（Suttie, 1980; Kormann and Weiser, 1984）。维生素K为合成凝血酶原、因子VII、因子IX和因子X所必需，而这些蛋白质为正常凝血所必需，它们在肝脏中合成，以无生物学活性的前体形式存在。通过维生素K的作用，将这些前体转化为有生物学活性的化合物（Suttie and Jackson, 1977; Suttie, 1980），由特定谷氨酸残基的酶促γ-羧化作用使之激活。生成的羧化谷氨酸残基是血凝所必需的钙离子的强螯合剂。维生素K缺乏或抗凝血化合物的存在会减少羧化谷氨酸残基的数量，导致血凝因子蛋白失活，使血凝时间延长。除凝血作用外，有证据表明维生素K依赖蛋白和肽可能参与了钙的代谢（Suttie, 1980; Kormann and Weiser, 1984）。

维生素K有K_1、K_2、K_3三种形式，叶绿醌（K_1）存在于植物中，甲基萘醌类（K_2）由微生物发酵形成，甲萘醌（K_3）是合成的。甲萘醌（2-甲基-1,4-萘醌）是维生素K的合成形式，与维生素K_1和K_2有相同的环状结构。维生素K的三种形态都有生物学活性。

猪日粮中添加的维生素K一般是水溶性的维生素K_3，主要有亚硫酸氢钠甲萘醌（MSB）、二甲基嘧啶醇亚硫酸甲萘醌（MPB）和亚硫酸氢钠甲萘醌复合物（MSBC）。维生素K的活性取决于这些产品中甲萘醌的含量。MSB、MSBC、MPB中甲萘醌的含量分别为50%、33%和45%。亚硫酸氢钠烟酰胺甲萘醌是应用在家禽上的一种维生素K的合成形式，具有维生素K和烟酸活性，其生物学活性与MPB相似，甲萘醌含量为46%（Oduho et al., 1993）。

维生素K缺乏使凝血酶原含量升高，凝血时间延长，并且可能会导致内出血和死亡（Schendel and Johnson, 1962; Brooks et al., 1973; Seerley et al., 1976; Hall et al., 1986, 1991）。Schendel和Johnson（1962）报道，将1~2日龄仔猪饲养于铁丝底笼子中，仔细清扫（使食粪的机会减少到了最小程度），并饲喂液态纯合日粮，其中含有减少肠道合成维生素K的磺胺噻唑（ST）和土霉素两种药物，结果发现，它对甲萘醌磷酸钠需要量为5 μg/kg体重。Seerley等（1976）报道，每千克日粮含1.1 mg的MPB可消除抗凝新戊酰对断奶仔猪的影响。Hall等（1986）认为，每千克日粮需要2 mg MPB形式的甲萘醌才能抵消新戊酰对生长猪的影响。

细菌合成的维生素K通过动物食粪随即被吸收，可减少或完全免除日粮添加维生素K的必要性。

高剂量的抗生素会降低肠道菌群合成维生素 K。尚无研究测定添加维生素 K 是否对种猪有益。

Muhrer 等（1970）、Osweiler（1970）和 Fritschen 等（1971）报道了猪在农场条件下的出血现象。他们怀疑是由霉菌毒素污染饲料所致，日粮中添加 2.0 mg/kg 的甲萘醌可以预防这种症状。一些研究表明，日粮中存在抗凝血香豆素可增加维生素 K 的需要量。过量的钙也可增加猪对维生素 K 的需要量（Hall et al., 1991）。即使短期采食维生素 K 缺乏的日粮，肝脏储存的维生素 K 也将很快被耗竭（Kindberg and Suttie, 1989）。由于猪日粮中普遍含有霉菌毒素（BIOMIN, 2010）以及一些副产物广泛用于日粮［副产物中霉菌毒素含量被浓缩（Schaafsma et al., 2009）］，因此进一步深入了解维生素 K 的可能作用对猪更有益。

水分、氯化胆碱、微量元素和碱性条件会破坏预混合饲料及日粮中添加的水溶性甲萘醌的稳定性。Coelho（1991）研究表明，含有胆碱的维生素微量元素预混料储存 3 个月，MSBC 和 MPB 的活性损失高达 80%；而不含胆碱的相同预混料储存同样的时间，则活性损失要少得多。目前，已有一些经包被处理的甲萘醌补充物，这种产品可改善维生素 K_3 在日粮或预混料中的稳定性。

动物对大剂量的维生素 K_3 复合物有很好的耐受性。Seerley 等（1976）给猪饲喂含有 110 mg/kg MPB 的日粮，Oduho 等（1993）给雏鸡饲喂含有 300 mg/kg MPB 的日粮，均未发现中毒症状。雏鸡采食含 3000 mg/kg MPB 的日粮 14 天，体增重和血红蛋白含量未减少。可见，动物对 1000 倍于营养需要量的甲萘醌有很好的耐受力（NRC, 1987; Oduho et al., 1993）。

水溶性维生素

生物素

生物素作为参与二氧化碳固定功能的几种酶的辅助因子而在代谢中发挥重要作用。作为丙酮酸羧化酶和丙酰辅酶 A 羧化酶的一部分，生物素在糖异生和柠檬酸循环中起重要作用。乙酰辅酶 A 羧化酶也是一种生物素依赖酶，它启动脂肪酸的生物合成。Whitehead 等（1980）、Misir 和 Blair（1986）认为，血浆生物素浓度和血浆丙酮酸羧化酶活性可作为评价猪生物素营养状况的方法。D-异构体是生物素的活性形式。

大多数常用饲料原料的生物素含量充足，但不同原料间的生物学效价差异很大。对雏鸡来说，黄玉米和豆粕中生物素的生物学效价高，但大麦、高粱、燕麦和小麦中的生物学效价则较低（Frigg, 1976; Anderson et al., 1978; Kopinski et al., 1989）。饲料原料中的很多生物素是以 ε-N-生物素酰-L-赖氨酸（生物细胞素）的形式存在的，生物细胞素是蛋白质的组成成分，其生物学效价（相对于结晶 D-生物素）差异很大，并取决于其所存在蛋白质的消化率。据推测，相当一部分猪所需要的生物素来自于肠道细菌的合成。

一般来说，在多种饲粮和条件下给 2~28 日龄断奶的仔猪或生长育肥猪补充生物素，其生长性能并未改善。给 2~28 日龄仔猪饲喂含 10 μg/kg 干物质的生物素的过滤脱脂奶粉（相当于猪乳中生物素含量的 15%）日粮，增重和饲料转化效率与添加 50 μg 生物素/kg 饲粮的同窝仔猪一样（Newport, 1981）。同样，给 21~28 日龄断奶仔猪或生长育肥猪饲喂 110~880 μg/kg 日粮的生物素，其增重速度和饲料转化效率也未见改善（Peo et al., 1970; Hanke and Meade, 1971; Meade, 1971; Washam et al., 1975; Simmins and Brooks, 1980; Easter et al., 1983; Bryant et al., 1985b; Hamilton and Veum, 1986）。但是，Adams 等（1967）对生长猪和 Peo 等（1970）对 28 日龄断奶仔猪的报道例外。Partridge 和 McDonald（1990）观察到在小麦-大麦-豆粕型生长猪日粮中添加生物素，其饲料转化效率得到改善。

有报道认为，采食添加生物素日粮的母猪，蹄部的硬度和致密度、承压强度、皮肤和被毛状况明显改善，蹄裂和脚垫损伤减少（Grandhi and Strain, 1980; Webb et al., 1984; Bryant et al., 1985a, b; Simmins and Brooks, 1985; Misir and Blair, 1986）。但 Hamilton 和 Veum（1984）、Tribble 等（1984）的研究未观察到上述指标的改善。

Lewis 等（1991）报道，在妊娠和泌乳母猪玉米-豆粕型日粮中添加 0.33 mg/kg 日粮的生物素，结果断奶仔猪数增加，母猪蹄部健康未见改善。Watkins 等（1991）对妊娠和泌乳母猪进行大规模的生物素评估试验，结果表明，日粮中添加 0.44 mg/kg 日粮的生物素，母猪繁殖性能、仔猪发育和母猪蹄部健康等指标没有变化。其他几位研究者应用各种谷物日粮进行的研究得出了不一致的结果（Brooks et al., 1977; Penny et al., 1981; Easter et al., 1983; Simmins and Brooks, 1983; Hamilton and Veum, 1984; Tribble et al., 1984; Bryant et al., 1985c; Kornegay, 1986; Misir and Blair, 1984）。由于不同试验间的结果缺乏一致性，以及生物素添加水平范围大（0.1~0.55 mg/kg 日粮），因此难以给出母猪特定的生物素需要量。

生物素缺乏的症状包括过度脱毛、皮肤溃烂和皮炎、眼液渗出、口腔黏膜炎症、蹄横裂，以及脚垫裂缝和出血（Cunha et al., 1946, 1948; Lindley and Cunha et al., 1946; Lehrer et al., 1952）。用含有磺胺药物的合成日粮可导致猪出现生物素缺乏症，这可能是由于磺胺药减少了肠道生物素的合成（Lindley and Cunha, 1946; Cunha et al., 1948; Lehrer et al., 1952）。在合成日粮中加入大量脱水鸡蛋清，会加重猪生物素缺乏症状（Cunha et al., 1946; Hamilton et al., 1983）。生鸡蛋清中含有抗生物素蛋白，可与肠道中的生物素形成复合物，使生物素不能被猪利用。

胆碱

虽然胆碱的需要量大大超过"微量有机营养素"这一维生素的定义，但仍将其归属于水溶性维生素之列。在猪日粮中，胆碱通常以氯化胆碱的形式添加，氯化胆碱中含有 74.6% 的胆碱活性（Emmert et al., 1996）。胆碱是一些机体生化反应所必需的：①磷脂（即卵磷脂）合成；②乙酰胆碱形成；③高半胱氨酸向蛋氨酸的转甲基作用，这一过程是通过胆碱的氧化产物——甜菜碱实现的。当胆碱严重缺乏时，磷脂和乙酰胆碱的合成优先于胆碱的供甲基功能。然而，谷物-油籽粕型日粮含有足够的胆碱，所以，等摩尔甜菜碱或胆碱在满足胆碱供甲基功能上有相同的功效（Lowry et al., 1987）。

猪通过三步反应将磷脂酰乙醇胺甲基化合成胆碱，该过程涉及 S-腺苷甲硫氨酸转甲基。因此，含有过量蛋氨酸的猪日粮无需再添加胆碱（Neumann et al., 1949; Nesheim and Johnson, 1950; Kroening and Pond, 1967）。

据估测，豆粕中胆碱的生物利用率大约相当于氯化胆碱中胆碱的 65%~83%（Molitoris and Baker, 1976; Emmert and Baker, 1997）。化学分析和用雏鸡进行的生物学利用率测定的结果表明，每千克去皮豆粕中含有 2218 mg 总胆碱和 1855 mg 可利用胆碱；花生粕中胆碱的生物学利用率（71%）略低于豆粕（83%），而双低菜籽粕（Canola 品种）中胆碱的生物学利用率仅为 24%（Emmert and Baker, 1997）。由于大豆产品中富含可利用胆碱，因此，在玉米-豆粕型或玉米-大豆分离蛋白型日粮中额外添加的胆碱不会对仔猪、生长猪和育肥猪起作用（Russett et al., 1979a; North Central Region-42 Committee on Swine Nutrition, 1980）。在饲料原料和未加工的油脂中，部分胆碱以磷脂结合型胆碱形式存在。该型胆碱具有很好的利用率（Emmert et al., 1996），但油脂精炼过程中需要脱胶处理，而此过程基本会造成磷脂结合型胆碱的完全流失（Anderson et al., 1979）。

在妊娠青年母猪和经产母猪的谷物-豆粕型日粮中添加 434~880 mg/kg 日粮的胆碱，通常可增加窝产活仔数和断奶仔猪数（Kornegay and Meacham, 1973; Stockland and Blaylock, 1974; North Central Region-42 Committee on Swine Nutrition, 1976; Grandhi and Strain, 1980）。在一个长期的繁殖试验中，Stockland 和 Blaylock（1974）也发现在玉米–豆粕型日粮中添加胆碱能提高母猪受胎率。Luce 等（1985）通过 4 个试验发现，妊娠青年母猪日粮中添加胆碱后，仔猪初生重提高，但仔猪后肢外张的发病率未降低。泌乳期间，在含 8%~10%脂肪或油的日粮中添加胆碱，母猪的泌乳性能未见改善（Seerley et al., 1981; Boyd et al., 1982）。

胆碱缺乏会导致猪的增重缓慢、被毛粗糙、红细胞计数减少、红细胞压积和血红蛋白浓度降低、血浆碱性磷酸酶活性增加、步履蹒跚不稳；肝脏和肾脏有脂肪浸润；严重缺乏时，肾小球因脂肪的大量浸润而堵塞（Wintrobe et al., 1942; Johnson and James, 1948; Neumann et al., 1949; Russett et al., 1979a）。

在由 30%无维生素酪蛋白、37%葡萄糖、26.6%猪油、2%磺胺苯二甲硫组成的含 0.8%蛋氨酸的新生仔猪日粮中，添加 260 mg/kg 胆碱，可预防胆碱缺乏（Johnson and James, 1948）。2 日龄仔猪的日粮中胆碱添加水平达到 1000 mg/kg 饲料固形物时，增重和饲料转化效率最佳，并能有效预防肝脏和肾脏的脂肪浸润（Neumann et al., 1949）。在含 1000 mg/kg 胆碱的同一日粮中另外添加 0.8%的 DL-蛋氨酸，则未能改善同等条件下饲养猪的生产性能（Nesheim and Johnson, 1950）。Kroening 和 Pond（1967）给 5 kg 体重的仔猪分别饲喂含 0、0.11%、0.22%水平 DL-蛋氨酸的低蛋白（12%）日粮，在此基础上，添加 1646 mg/kg 胆碱有改善 0、0.11%组仔猪的增重和饲料转化效率的趋势，但对 0.22%组仔猪无影响。Russett 等（1979a, b）报道，采食含 0.31%蛋氨酸和 0.33%胱氨酸半纯合日粮的 6~14 kg 仔猪对胆碱的最低需要量为 330mg/kg。

尚未有关于猪胆碱中毒症状的报道（NRC, 1987），但已观察到，日粮中添加 2000mg/kg 胆碱时，仔猪、生长和育肥猪日增重降低（Southern et al., 1986）。

叶酸

叶酸是包括了具有叶酸活性的一组化合物。叶酸的化学结构由蝶啶环、对氨基苯甲酸（PABA）和谷氨酸组成。动物细胞不能合成 PABA，也不能使谷氨酸与蝶酸结合。叶酸缺乏会引起一碳化合物代谢紊乱，包括甲基、丝氨酸、嘌呤和嘧啶的合成。叶酸参与由丝氨酸向甘氨酸、高半胱氨酸向蛋氨酸的转化。

饲料中的叶酸主要以多谷氨酸共价结合物的形式存在，该结合物是一条由 7 个谷氨酸残基以 γ 键连接的多肽链。一组肠道轭合酶（conjugase）（即叶酰多聚谷氨酸水解酶）脱掉除最后一个残基外的 6 个谷氨酸残基。只有单谷氨酰形式的叶酸可被吸收进入肠细胞。大部分经刷状缘吸收的叶酸被还原为四氢叶酸（FH_4），接着被甲基化为 5N-甲基四氢叶酸。与硫胺素一样，叶酸的蝶啶环上有一组游离氨基酸，这种结构使其对热不稳定，特别是日粮中含有像葡萄糖或乳糖一类的还原糖时。

除 Matte 等（1984a，b; 1992）、Lindemann 和 Kornegay（1989）的研究外，其他研究结果都表明，猪常用饲料原料提供的叶酸，再加上肠道微生物合成的叶酸足以满足各类猪对叶酸的需要。

在玉米-豆粕型妊娠母猪日粮中添加 200 mg/kg 的叶酸不会增加窝产活仔数和断奶仔猪数

（Easter et al., 1983）。Matte 等（1984a）给断奶至妊娠 60 天期间的母猪肌肉注射叶酸，每次 15 mg，共注射 10 次，结果发现窝产仔数显著增加。随后的研究中，Matte 等（1992）观察到妊娠母猪日粮添加 5 mg/kg 或 15 mg/kg 的叶酸，窝增重提高，但给泌乳母猪日粮添加叶酸，仔猪性能未见改善。Lindemann 和 Kornegay（1989）也报道，给玉米-豆粕型母猪日粮添加 1 mg/kg 的叶酸，窝产仔数增加，但没有增加断奶仔猪数。在 Tremblay 等（1986）所进行的一个试验中发现，向原本含 0.62 mg/kg 叶酸的母猪日粮中添加 4.3 mg/kg 叶酸，可使其血清中叶酸含量保持与在断奶至配种后 56 天期间间隔注射叶酸（每头一次 15 mg，共注射 10 次）的母猪一样高。一个以 393 头经产母猪为对象的大型研究中，Harper 等（1994）在母猪配种前、妊娠和泌乳期间的标准玉米-豆粕型日粮中添加 1 mg/kg、2 mg/kg 和 4 mg/kg 的叶酸，结果显示母猪繁殖性能未有明显改善。根据最近的这些研究结果，将妊娠和泌乳母猪的叶酸需要量增加到 1.3 mg/kg 日粮。

叶酸缺乏会导致猪体增重缓慢、被毛褪色、巨红细胞性或正常红细胞性贫血、血液中白细胞和血小板数量减少、红细胞压积降低以及骨髓增生。给猪饲喂含 1%~2% 磺胺类药物或叶酸拮抗物的合成日粮可导致叶酸缺乏症状（Cunha et al., 1948; Heinle et al., 1948; Cartwright et al., 1949, 1950; Johnson et al., 1950）。磺胺类药物可能降低肠道细菌对叶酸的合成。对于饲喂一种含 2%磺胺苯二甲硫的合成日粮的 4 日龄仔猪（Johnson et al., 1948）或饲喂一种合成日粮的 8 周龄仔猪（Cunha et al., 1947），补充叶酸对其生产性能没有影响。Newcomb 和 Allee（1986）报道，在 17~27 日龄断奶仔猪的玉米-豆粕-乳清粉型日粮中添加 1.1 mg/kg 叶酸无有益的影响。但 Lindemann 和 Kornegay（1986）观察到，在相似日龄仔猪的玉米-豆粕型日粮中添加 0.5 mg/kg 的叶酸，日增重得到改善。给仔猪、生长猪、育肥猪的玉米-豆粕型日粮中添加 200 μg/kg 或 360μg/kg 的叶酸，其增重和饲料转化率与不添加叶酸的猪相同（Easter et al., 1983; Gannon and Liebholz, 1989）。

烟酸

烟酸或尼克酸是辅酶烟酰胺腺嘌呤二核苷酸（NAD）和烟酰胺腺嘌呤二核苷酸磷酸（NADP）的组成成分。这两种辅酶是碳水化合物、蛋白质和脂类代谢所必需的。

日粮中过量色氨酸向烟酸的代谢性转化使烟酸需要量的确定复杂化（Luecke et al., 1948; Powick et al., 1948）。Firth 和 Johnson（1956）估测，日粮中每 50 mg 超过需要量的色氨酸可产生 1 mg 的烟酸。某些饲料原料中烟酸有限的生物学效价使烟酸状况的确定进一步复杂化。黄玉米、燕麦、小麦和高粱中的烟酸是以结合状态存在的，这种形式的烟酸大部分不能被幼猪利用（Kodicek et al., 1956; Luce et al., 1966, 1967; Harmon et al., 1969, 1970）。但豆粕中的烟酸可被雏鸡很好地利用，猪可能也同样能有效地利用（Yen et al., 1977）。

游离烟酸或游离烟酰胺是具有烟酸活性的商业化产品。相对于烟酸，烟酰胺在雏鸡上的生物学效价为 124%（Oduho and Baker, 1993），在大鼠上则为 109%（Carter and Carpenter, 1982）。

Firth 和 Johnson（1956）估测，日粮中无过剩色氨酸时，1~8 kg 仔猪对有效烟酸的需要量为 20 mg/kg。当日粮色氨酸水平接近其需要量时，10~50 kg 的生长猪对有效烟酸的需要量为 10~15 mg/kg（Braude et al., 1946; Kodicek et al., 1959; Harmon et al., 1969）。生长育肥猪日粮通常需要强化烟酸，但有试验在 45 kg 猪的玉米-豆粕型日粮中添加烟酸，其生产性能并未改善（Yen et al., 1978; Copelin et al., 1980）。在这些试验中，所用日粮色氨酸的计算值均超过了猪对色氨酸的需要量。Real 等（2002）在一个商业试验中，给猪分别饲喂添加 0、13 mg/kg、28 mg/kg、55 mg/kg、110 mg/kg 和 550 mg/kg 烟酸的日粮，结果发现，随烟酸添加量提高，猪的增重与采食量之比（二次曲线性，$P < 0.01$）、主观肉色评分以及最终 pH（直线性，$P < 0.01$）均明显提高，胴体损耗和滴水损失百分率减少（直线性，$P < 0.04$）。这些结果表明 13 mg/kg 是改善增重与采食量之比所需要的烟酸添加量，而要考虑到改善胴体和猪肉质量则需要更高水平的添加量。

目前对怀孕和哺乳母猪烟酸需要量的研究较少。Ivers 等（1993）用不添加烟酸的含 12.80% 粗蛋白的玉米-豆粕-燕麦型日粮连续饲喂 67 头母猪超过 5 个胎次（共产仔达 240 窝），发现此种日粮提供的烟酸足以满足妊娠和哺乳期的需要。Mosnier 等（2009）发现，在泌乳早期烟酸和维生素 B_6 的水平会暂时低于最佳需求量。分娩后 1 周血浆中色氨酸和烟酸的浓度下降，而犬尿氨酸（色氨酸转化为烟酸的中间产物）浓度增加。泌乳的第 2 和第 3 周，血浆色氨酸和犬尿氨酸浓度恢复到产前时的水平，而在整个泌乳期内烟酸的浓度提高。维生素 B_6（参与烟酸转化与利用的维生素）水平在分娩后 1 周也逐步增加，随后一直保持较高水平。目前仍需进一步研究确定在产仔的第 1 周母猪是否需要烟酸，以及在色氨酸刚能满足需要的情况下，烟酸是否会影响蛋白质的利用效率。

对雏鸡的研究表明，日粮铁缺乏会损害色氨酸作为烟酸前体的效率（Oduho et al., 1994）。这种关系在猪上是否存在还不清楚。铁是色氨酸合成烟酸单核苷酸途径中两种酶的辅助因子。

烟酸缺乏的症状包括增重减缓、厌食、呕吐、皮肤干燥、皮炎、被毛粗糙、脱毛、腹泻、黏膜溃疡、溃疡性胃炎、盲肠及结肠炎症和坏死，以及正常红细胞性贫血症状（Hughes, 1943; Braude et al., 1946; Wintrobe et al., 1946; Luecke et al., 1947; Powick et al., 1947a, b; Cartwright et al., 1948; Burroughs et al., 1950; Kodicek et al., 1956）。烟酸缺乏时，血液红细胞 NAD 活性与尿液中 N-甲基-尼克酰胺和 N'-甲基-2-吡啶酮-5-羧基酰胺的排泄量减少（Luce et al., 1966, 1967）。

泛酸

属于 B 族维生素的泛酸是由泛解酸和 β-丙氨酸经酰胺键连接而组成的。泛酸作为辅酶 A 的组成成分，对碳水化合物和脂肪代谢中二碳单位的分解与合成起重要作用。大麦、小麦和高粱中泛酸的生物学效价低，但玉米和豆粕中的泛酸的生物学效价较高（Southern and Baker, 1981）。饲料原料中的泛酸大多以辅酶 A、酰基辅酶 A 合成酶和酰基载体蛋白的形式存在。只有 D-泛酸异构体才有生物学活性。合成的泛酸一般是以泛酸钙的形式添加到各种猪的日粮中，泛酸钙比泛酸更稳定。D-泛酸钙的泛酸活性为 92%，泛酸钙的消旋混合物只有 46%。DL-泛酸钙与氯化钙的复合物也可利用，但其泛酸活性只有 32%。

采食纯合日粮的 2~10 kg 仔猪对泛酸的需要量为 15.0 mg/kg（Stothers et al., 1955）；5~50 kg 猪对泛酸的需要量估计为 4.0~9.0 mg/kg（Luecke et al., 1953; Barnhart et al., 1957; Sewell et al., 1962; Palm et al., 1968）。20~90 kg 猪对泛酸的需要量估计为 6.0~10.5 mg/kg（Catron et al., 1952; Pond et al., 1960; Davey and Stevenson, 1963; Palm et al., 1968; Roth-Maier and Kirchgessner, 1977）。近期，Groesbeck 等（2007）研究表明，玉米-豆粕型日粮中的泛酸含量可能足以满足 25~120 kg 猪的需要。

Ullrey 等（1955）、Davey 和 Stevenson（1963）及 Teague 等（1970）进行的三个试验都表明，当日粮泛酸水平低于 5.9 mg/kg 时，猪的繁殖性能差。然而，Bowland 和 Owen（1952）则表示，该泛酸水平下，猪的繁殖性能正常。Ullrey 等（1955）以及 Davey 和 Stevenson（1963）估测，母猪达到最佳繁殖性能对泛酸的需要量为 12.0~12.5 mg/kg。

泛酸缺乏的症状包括生长缓慢、厌食、腹泻、皮肤干燥、被毛粗糙、脱毛、免疫反应降低以及后肢运动异常，即所谓的"鹅步"（Hughes and Ittner, 1942; Wintrobe et al., 1943b; Luecke et al., 1948, 1950, 1952;

Wiese et al., 1951; Stothers et al., 1955; Harmon et al., 1963）。对泛酸缺乏的猪进行尸检，发现肠黏膜水肿与坏死、黏膜下层结缔组织浸入增多、神经髓脂质损失及背根神经节细胞退化（Wintrobe et al., 1943b; Follis and Wintrobe, 1945）。

核黄素

作为黄素单核苷酸（FMN）和黄素腺嘌呤二核苷酸（FAD）两个辅酶的组成成分，核黄素在蛋白质、脂肪和碳水化合物的代谢中起着重要作用。饲料原料中核黄素的活性大多以 FAD 形式存在。

采食纯合日粮的 2~20 kg 的猪对核黄素的需要量估计为 2~3 mg/kg（Forbes and Haines, 1952; Miller et al., 1954）。采食纯合日粮的生长猪对核黄素需要量估计为 1.1~2.9 mg/kg（Hughes, 1940a; Krider et al., 1949; Mitchell et al., 1950; Terrill et al., 1955），而采食常规日粮时则估计需要 1.8~3.1 mg/kg（Krider et al., 1949; Miller and Ellis, 1951）。Seymour 等（1968）报道，对于 5~17 kg 的猪，核黄素水平与环境温度之间无一致性的互作，此结果与早期 Mitchell 等（1950）的报道相矛盾。玉米-豆粕型日粮中有效核黄素含量不足。Chung 和 Baker（1990）从雏鸡的试验估测，玉米-豆粕型日粮中核黄素的生物学效价相当于晶体核黄素的 59%。

核黄素缺乏导致青年母猪不发情（Esch et al., 1981）和繁殖失败（Miller et al., 1953; Frank et al., 1984）。根据产仔性能和红细胞谷胱甘肽还原酶（FAD 依赖酶）的活性，Frank 等（1984）估测，妊娠母猪对有效核黄素的需要量大约为 6.5 mg/d。但是，Pettigrew 等（1996）观察到，从配种到妊娠 21 天期间，60 mg/d 的核黄素比 10 mg/d 的核黄素提高了母猪产仔率。Frank 等（1988）建议，依据红细胞谷胱甘肽还原酶的活性和产仔性能，哺乳母猪对核黄素的需要量大约为 16 mg/d。

幼龄生长猪核黄素缺乏的症状包括生长缓慢、白内障、步态僵硬、脂溢性皮炎、呕吐和脱毛（Wintrobe et al., 1944; Miller and Ellis, 1951; Lehrer and Wiese, 1952; Miller et al., 1954）。严重缺乏时，血液嗜中性粒细胞增多、免疫反应降低、肝和肾组织褪色、脂肪肝、卵泡萎缩、卵细胞退化、坐骨和臂神经髓磷脂退化（Wintrobe et al., 1944; Krider et al., 1949; Mitchell et al., 1950; Forbes and Haines, 1952; Lehrer and Wiese, 1952; Miller et al., 1954; Terrill et al., 1955; Harmon et al., 1963）。

硫胺素

硫胺素是碳水化合物和蛋白质代谢所必需的。辅酶-硫胺素焦磷酸盐是 α-酮酸氧化脱羧作用所必需的。硫胺素对热很不稳定，因此，过热或高压蒸煮会降低饲粮中硫胺素的含量，尤其当还原糖存在时。

Miller 等（1955）估计，2~10 kg 猪对硫胺素的需要量为 1.5 mg/kg。3 周龄断奶至 40 kg 猪对硫胺素的需要量大约为 1.0 mg/kg 日粮（Van Etten et al., 1940; Ellis and Madsen, 1944）。日粮脂肪水平增加至 28%会延长硫胺素缺乏猪的存活时间（Ellis and Madsen, 1944）。这个发现表明，用较高水平的脂肪替代碳水化合物作为日粮能量来源可使硫胺素需要量降低。30~90 kg 猪的日粮硫胺素水平增加至 1.1 mg/kg 时，猪体重增加，但硫胺素为 0.85 mg/kg 时，采食量才最大（Peng and Heitman, 1974）。Peng 和 Heitman（1973）将硫胺素焦磷酸盐加到体外溶剂中，通过测定红细胞转酮酶的活性的增加来评价生长育肥猪硫胺素营养状况。结果表明，用这一指标估测的硫胺素需要量是获得最大增重的需要量的 4 倍。而且当环境温度由 20℃提高到 35℃时，用这个指标测定的硫胺素需要量也随之增加（Peng and Heitman, 1974）。这种变化可能与采食量的减少有关。目前，尚无关于妊娠和哺乳母猪硫胺素需要量的文献报道。

用二氧化硫处理饲料原料会使硫胺素失活。在早期关于硫胺素需要量的研究中，曾用这种处理方法来制备硫胺素缺乏的日粮（Van Etten et al., 1940; Ellis and Madsen, 1944）。许多种类淡水鱼体内含有被称为硫胺素酶I的抗硫胺素因子（Tanphaichitr and Wood, 1984）。将适量未加工处理的淡水鱼制备物饲喂给其他动物可造成硫胺素缺乏（Green et al., 1941; Krampitz and Woolley, 1944）。

硫胺素缺乏的猪表现为食欲减退、增重减少、体温和心率降低，偶尔表现呕吐。另外一些缺乏症状包括心脏肥大松弛、心肌退化及心脏衰竭而猝死。硫胺素缺乏还会导致动物血浆丙酮酸盐浓度升高（Hughes, 1940b; Van Etten et al., 1940; Follis et al., 1943; Wintrobe et al., 1943a; Ellis and Madsen, 1944; Heinemann et al., 1946; Miller et al., 1955）。大多数猪日粮中常用的谷物籽实的硫胺素含量丰富。因此，饲喂给各类猪的谷物-油籽粕型日粮含有足够的硫胺素，一般不需要额外添加。

维生素 B_6（吡哆醇）

饲料原料中的维生素 B_6 以吡哆醇、吡哆醛、吡哆胺和磷酸吡哆醛的形式存在。磷酸吡哆醛是许多氨基酸酶系统，包括转氨酶、脱羧酶、脱氢酶、合成酶和消旋酶的重要辅助因子。维生素 B_6 在中枢神经系统的功能中起关键作用，它参与为神经递质和神经抑制剂的合成所需的氨基酸衍生物的脱羧作用。

对于雏鸡，玉米和豆粕中维生素 B_6 的生物学效价分别为40%和60%（Yen et al., 1976）。虽然尚无试验数据，但推测这两种饲料原料中维生素 B_6 对猪的生物学效价与雏鸡相似。Miller 等（1957）、Kösters 和 Kirchgessner（1976a, b）建议，2~10 kg 猪对日粮维生素 B_6 的需要量为1.0~2.0 mg/kg。以前对10~20 kg 猪的维生素 B_6 需要量的估测值为1.2~1.8 mg/kg（Sewall et al., 1964; Kösters and Kirchgessner, 1976a, b），而最近研究表明，采食常规日粮（Woodworth et al., 2000）和半纯合日粮（Zhang et al., 2009）的仔猪在断奶早期对 B_6 需要量接近7 mg/kg，高于先前的估测值。

Ritchie 等（1960）从妊娠第二个月到哺乳35天期间给青年和经产母猪饲喂含总吡哆醇1.0 mg/kg 或10.0 mg/kg 的日粮，结果表明不同处理组母猪繁殖或哺乳性能无差异。Easter 等（1983）报道，在妊娠青年母猪玉米-豆粕型日粮中添加1.0 ppm 的吡哆醇，窝产仔数和断奶仔猪数增加。在另一项研究中，每天采食0.45 mg 或2.1 mg 维生素 B_6 的性成熟青年母猪，其红细胞谷草转氨酶活性系数比采食过量维生素 B_6（83 mg/d）的青年母猪高。维生素 B_6 缺乏的青年母猪，其肌肉谷草转氨酶总活性降低，这表明青年母猪对维生素 B_6 的每日需要量可能大于2.1 mg（Russell et al., 1985a, b）。Knights 等（1998）评估日粮添加1.0 ppm 和15 ppm 维生素 B_6 对猪的影响，结果表明，增加维生素 B_6 添加量似乎有缩短母猪从断奶到发情间隔以及促进氮代谢的趋势。由于试验用维生素 B_6 添加量范围较宽，因而很难确定维生素 B_6 的需要量。

维生素 B_6 缺乏降低食欲和生长速度，严重缺乏导致眼周围渗出液体、抽搐、共济失调、昏迷和死亡。维生素 B_6 缺乏猪的血液血红蛋白、红细胞及淋巴细胞数量降低，血清铁和 γ 球蛋白水平上升。维生素 B_6 缺乏的典型特征是：感觉神经元的外周髓磷脂和轴突退化、小红细胞低色素性贫血及肝脏被脂肪浸润（Hughes and Squibb, 1942; Wintrobe et al., 1942, 1943c; Follis and Wintrobe, 1945; Lehrer et al., 1951; Miller et al., 1957; Harmon et al., 1963）。采用色氨酸向烟酸的转化受损的色氨酸负荷试验可测定维生素 B_6 的营养状况，这种损害导致尿中黄尿酸和犬尿酸浓度升高（Cartwright et al., 1944）。一般来说，由于饲料原料中有生物学活性的维生素 B_6 水平能满足猪的需要，谷物-豆粕日粮不需要添加维生素 B_6。

维生素 B_{12}

维生素 B_{12} 或称氰钴胺素，其分子中含有微量元素钴，这在所有维生素中是独一无二的。维生素 B_{12} 作为辅酶，参与了由甲酸盐、甘氨酸或丝氨酸衍生而来的活性甲基的从头合成，以及这些甲基转移到同型胱氨酸合成蛋氨酸的反应。维生素 B_{12} 在尿嘧啶甲基化形成胸腺嘧啶的过程中也很重要，胸腺嘧啶再转化为胸腺嘧啶脱氧核苷，后者用于 DNA 的合成。猪需要维生素 B_{12}，但对日粮添加维生素 B_{12} 后的反应不一致。外界环境和肠道中的微生物能合成维生素 B_{12}，加之猪有食粪的习性，使得维生素 B_{12} 供给量足以满足其需要（Bauriedel et al., 1954; Hendricks et al., 1964）。植物性饲料缺乏维生素 B_{12}，但动物性饲料及发酵副产物则含此维生素，并以甲基化形式（甲基钴胺素）或 $5'$-脱氧腺苷形式（腺苷钴胺素）存在，这两种化合物一般结合在蛋白质上。商用维生素 B_{12} 添加剂通过微生物发酵生产，并通常在谷物-豆粕型日粮中使用。

结合维生素 B_{12} 的受体位于回肠。在吸收前，钴胺素与通常称为"内因子"的糖蛋白结合。内因子来源于胃黏膜壁细胞。维生素 B_{12} 在体内能有效地储存。因此，过量摄入并储存于组织（主要是肝脏）中的维生素 B_{12}，能使采食维生素 B_{12} 缺乏日粮的猪推迟多月才出现缺乏症（Combs, 1999）。

采食合成乳日粮并饲养于铁丝网底式笼子的 1.5~20kg 猪对维生素 B_{12} 需要量的估测值为 15~20 μg/kg 日粮干物质（Anderson and Hogan, 1950b; Nesheim et al., 1950; Frederick and Brisson, 1961），但在 Neumann 等（1950）的研究中，需要量则高达 50 μg/kg 日粮干物质。体重 10~45 kg 猪对维生素 B_{12} 的需要量为 8.8~11.0 μg/kg 日粮干物质（Richardson et al., 1951; Catron et al., 1952）。这些试验中的猪均饲养于铁丝网底式笼子中。如果用获得最低血浆高半胱氨酸浓度的方法作为测定维生素 B_{12} 营养需要量的话，日粮中添加 30~35 μg/kg 可能较为合适（House and Fletcher, 2003）。

Anderson 和 Hogan（1950a）、Frederick 和 Brisson（1961）、Teague 和 Grifo（1966）在日粮中添加 11~1100 μg/kg 维生素 B_{12}，母猪的繁殖性能提高。Teague 和 Grifo（1966）比较了全植物性日粮中添加（110~1100 μg/kg）或不添加维生素 B_{12} 对母猪繁殖性能的影响，结果表明，直到第 3 和第 4 胎，产仔数、断奶仔猪数、初生重和断奶重均未降低。Simard 等（2007）在妊娠母猪日粮中添加 5 个水平（0、20 μg/kg、100 μg/kg、200 μg/kg 或 400 μg/kg）的维生素 B_{12}，研究其对血浆维生素 B_{12} 和高半胱氨酸（依赖维生素 B_{12} 的再甲基化途径中一种有害的中间代谢产物）的影响。根据折线回归模型可看出，血浆维生素 B_{12} 浓度达到最大时的日粮中维生素 B_{12} 浓度是 164 μg/kg，血浆高半胱氨酸浓度最低时日粮中维生素 B_{12} 浓度约为 93 μg/kg。维生素 B_{12} 似乎对窝产仔数有积极作用，但是研究者认为维生素 B_{12} 在这些浓度下的作用还需要通过增加动物数量并经过多胎次的生产性能试验验证。由于试验中添加水平变化幅度大，试验次数少，故难以确定繁殖和哺乳母猪的维生素 B_{12} 需要量，但估计为 15 μg/kg 日粮。

维生素 B_{12} 缺乏的猪表现体增重降低、食欲丧失、皮肤和被毛粗糙、烦躁、过敏及后腿共济失调。缺乏维生素 B_{12} 猪血液样品表现正常红细胞性贫血、嗜中性白细胞数增加、淋巴细胞数减少（Anderson and Hogan, 1950b; Neumann and Johnson, 1950; Neumann et al., 1950; Cartwright et al., 1951; Richardson et al., 1951; Catron et al., 1952）。维生素 B_{12} 和叶酸缺乏导致巨红细胞性贫血和骨髓增生，这两个症状与人的恶性贫血相似（Johnson et al., 1950; Cartwright et al., 1952）。由于叶酸盐代谢需要维生素 B_{12}，所以叶酸缺乏症通常伴有维生素 B_{12} 缺乏。叶酸或维生素 B_{12} 中任何一个缺乏，都会妨碍胸腺嘧啶合成过程中甲基的正常转移。

维生素 C

维生素 C（抗坏血酸）是一种水溶性抗氧化剂，它参与芳香族氨基酸的氧化、去甲肾上腺素和肉毒碱的合成，以及细胞铁蛋白铁还原以转移到体液中。抗坏血酸也为脯氨酸和赖氨酸羟化所必需，脯氨酸和赖氨酸是胶原的组成成分，而胶原则为软骨和骨骼生长所必需。维生素 C 促进骨基质和牙齿牙质的形成。维生素 C 缺乏时，全身有出血斑。饲料来源的维生素 C 为灵长类动物和豚鼠所必需，但包括猪在内的家畜均能利用 D-葡萄糖和其他几种相关化合物合成维生素 C（Braude et al., 1950; Dvorak, 1974; Brown and King, 1977）。Strittmatter 等（1978）、Cleveland 等（1983）和 Nakano 等（1983）研究了维生素 C 在预防和缓解猪骨软骨病中的作用。这些研究者推测，骨软骨病可能与由赖氨酸羟化减少而引起的胶原交联不足有关。但在日粮中添加维生素 C 对预防此疾病无效。

在某些情况下，猪不能快速合成足够的维生素 C 以满足其需要。Riker 等（1967）报道，在 29℃ 的环境温度下饲养的猪比在 18℃ 下饲养的猪血浆抗坏血酸浓度低。但给饲养在 19℃ 或 27℃ 下的猪补充维生素 C 均未改善增重的速度和效率（Kornegay et al., 1986）。Brown 等（1970）发现，能量摄入量和血清抗坏血酸水平间显著相关。随后，Brown 等（1975）报道，添加维生素 C 显著改善了 3 周龄仔猪的增重速度。动物在低能量摄入时对添加维生素 C 的反应比中等或高能量摄入时大。1 日龄或 40 日龄禁食仔猪比吮乳仔猪肝脏维生素 C 浓度和总量低（Dvorak, 1974）。还有报道认为，在没有施加刻意的应激的条件下，添加维生素 C 改善了猪的增重。Jewell 等（1981）报道，在一个试验中添加维生素 C 改善了 1 日龄断奶仔猪的增重，但在另一个试验中却无此效应。Brown 等（1975）、Yen 和 Pond（1981）以及 Mahan 等（1994）报道，日粮中添加维生素 C 改善了 3~4 周龄断奶仔猪的增重。Mahan 等（1966）观察到，注射或日粮中添加维生素 C，使初始体重 24 kg 的猪体增重增加。Cromwell 等（1970）进行了三个试验，其中两个试验的结果表明，由 15~27 kg 养至 90 kg 的生长猪对日粮添加维生素 C 有反应。其他研究者发现，给吮乳仔猪、3~4 周龄断奶猪和生长育肥猪补充维生素 C，未能改善其生产性能（Hutagalung et al., 1969; Leibbrandt, 1977; Strittmatter et al., 1978; Mahan and Saif, 1983; Nakano et al., 1983; Yen and Pond, 1984; Yen et al., 1985; Kornegay et al., 1986）。Mahan 等（1994）在生长育肥猪玉米-豆粕型日粮中添加维生素 C，未发现有益作用。Chiang 等（1985）综述了添加维生素 C 对断奶仔猪和生长育肥猪的影响。Bhar 等（2003）报道，添加维生素 C（每天 50 mg/头）对伤后猪的伤口治愈、抗体反应及生长反应有积极作用。

Sandholm 等（1979）报道，在预产前 5 天开始每天给妊娠母猪饲喂 1.0 g 维生素 C，结果其初生仔猪脐带出血迅速停止。补喂抗坏血酸母猪所产仔猪 3 周龄体重显著高于对照母猪所产仔猪。将水溶性维生素 K 制剂加入水中让母猪饮用，却未能防止初生仔猪的脐带出血问题。Lynch 和 O'Grady（1981）、Chavez 等（1983）、Yen 和 Pond（1983）从妊娠后期开始每天在母猪日粮中添加 1.0~10.0 g 抗坏血酸，结果仔猪成活率和生长速度并未得到改善。在后来的几个试验中，新生仔猪脐带出血并不是一个问题。

当动物处于应激条件下采食量受限的时候，或许短暂需要添加维生素 C。然而，由于有关维生素 C 添加发挥有益作用的条件尚未很好确定，而且具有短暂需要的特点，因此，未能给出猪的维生素 C 需要量的估测值。

参 考 文 献

Adams, C. R., C. E. Richardson, and T. J. Cunha. 1967. Supplemental biotin and vitamin B₆ for swine. *Journal of Animal Science* 26:903 (Abstr.).

Adamstone, P. B., J. D. Krider, M. F. James, and C. A. Blomquist. 1949. Response of swine to vitamin E-deficient rations. *Annals of the New York Academy of Sciences* 52:260-268.

Agricultural Research Council. 1981. *The Nutrient Requirements of Pigs*. Slough, UK: Commonwealth Agricultural Bureaux.

Ames, S. R. 1979. Biopotencies in rats of several forms of alpha-tocopherol. *Journal of Nutrition* 109:2198-2204.

Anderson, G. C., and A. C. Hogan. 1950a. Adequacy of synthetic diets for reproduction of swine. *Proceedings of the Society for Experimental Biology and Medicine* 75:288-290.

Anderson, G. C., and A. C. Hogan. 1950b. Requirements of the pig for vitamin B$_{12}$. *Journal of Nutrition* 40:243-250.

Anderson, L. E., R. O. Myer, J. H. Brendemuhl, and L. R. McCowell. 1995a. Bioavailability of various vitamin E compounds for finishing swine. *Journal of Animal Science* 73:490-495.

Anderson, L. E., R. O. Myer, J. H. Brendemuhl, and L. R. McCowell. 1995b. The effect of excessive dietary vitamin A on performance and vitamin E status in swine fed diets varying in dietary vitamin E. *Journal of Animal Science* 73:1093-1098.

Anderson, M. D., V. C. Speer, J. T. McCall, and V. W. Hays. 1966. Hypervitaminosis A in the young pig. *Journal of Animal Science* 25:1123-1127.

Anderson, P. A., D. H. Baker, and S. P. Mistry. 1978. Bioassay determination of the biotin content of corn, barley, sorghum and wheat. *Journal of Animal Science* 47:654-659.

Anderson, P. A., D. H. Baker, P. A. Sherry, and J. E. Corbin. 1979. Choline-methionine interrelationship in feline nutrition. *Journal of Animal Science* 49:522-527.

Anonymous. 1990. Nomenclature policy: Generic descriptors and trivial names for vitamins and related compounds. *Journal of Nutrition* 120:12-20.

Audet, I., J.-P. Laforest, G. P. Martineau, and J. J. Matte. 2004. Effect of vitamin supplements on some aspects of performance, vitamin status, and semen quality in boars. *Journal of Animal Science* 82:626-633.

Audet, I., N. Bérubé, J. L. Bailey, J.-P. Laforest, and J. J. Matte. 2009. Effects of dietary vitamin supplementation and semen collection frequency on reproductive performance and semen quality in boars. *Journal of Animal Science* 87:1960-1970.

Baker, D. H. 1995. Vitamin bioavailability. Pp. 399-431 in *Bioavailability of Nutrients for Animals: Amino Acids, Minerals, and Vitamins*, C. B. Ammerman, D. H. Baker, and A. J. Lewis, eds. San Diego, CA: Academic Press.

Barnhart, C. E., D. V. Catron, G. C. Ashton, and L. Y. Quinn. 1957. Effects of dietary pantothenic acid levels on the weanling pig. *Journal of Animal Science* 16:396-403.

Bauriedel, W. R., A. B. Hoerlein, J. C. Picken, Jr., and L. A. Underkofler. 1954. Selection of diet for studies of vitamin B_{12} depletion using unsuckled baby pigs. *Journal of Agricultural and Food Chemistry* 2:468-471.

Baustad, B., and I. Nafstad. 1972. Hematologic response to vitamin E in piglets. *British Journal of Nutrition* 28:183-190.

Bengtsson, G., J. Hakkarainen, L. Jonsson, J. Lannek, and P. Lindberg. 1978a. Requirement for selenium (as selenite) and vitamin E (as α-tocopherol) in weaned pigs. I. The effect of varying α-tocopherol levels in a selenium deficient diet on the development of the VESD syndrome. *Journal of Animal Science* 46:143-152.

Bengtsson, G., J. Hakkarainen, L. Jonsson, J. Lannek, and P. Lindberg. 1978b. Requirement for selenium (as selenite) and vitamin E (as α-tocopherol) in weaned pigs. II. The effect of varying selenium levels in a vitamin E deficient diet on the development of the VESD syndrome. *Journal of Animal Science* 46:153-160.

Bethke, R. M., W. Burroughs, O. H. M. Wilder, B. H. Edgington, and W. L. Robison. 1946. *The Comparative Efficiency of Vitamin D from Irradiated Yeast and Cod Liver Oil for Growing Pigs, with Observations on Their Vitamin D Requirements*. Ohio Agricultural Experiment Station Bulletin 667:1-29. Wooster: Ohio Agricultural Experiment Station.

Bhar, R., S. K. Maiti, T. K. Goswami, R. C. Patra, A. K. Garg, and A. K. Chhabra. 2003. Effect of dietary vitamin C and zinc supplementation on wound healing, immune response and growth performance in swine. *Indian Journal of Animal Science* 73:674-677.

Bieri, J. G., and M. C. McKenna. 1981. Expressing dietary values for fat-soluble vitamins: Changes in concepts and terminology. *American Journal of Clinical Nutrition* 34:289-295.

BIOMIN. 2010. BIOMIN Newsletter, Vol 8, No. 83, Special Edition, I. Rodrigues and K. Griessler, eds. Herzogenburg, Austria: BIOMIN Holding GmbH.

Boass, A., S. U. Toverud, T. A. McCain, J. W. Pike, and M. R. Haussler. 1977. Elevated serum levels of 1,γ-25-dihydroxycholecalciferol in lactating rats. *Nature* 267:630-632.

Bonnette, E. D., E. T. Kornegay, M. D. Lindemann, and C. Hammerberg. 1990. Humoral and cell-mediated immune response and performance of weaned pigs fed four supplemental vitamin E levels and housed at two nursery temperatures. *Journal of Animal Science* 68:1337-1345.

Bowland, J. P., and B. D. Owen. 1952. Supplemental pantothenic acid in small grain rations for swine. *Journal of Animal Science* 11:757 (Abstr.).

Boyd, R. D., B. D. Moser, E. R. Peo, Jr., A. J. Lewis, and R. K. Johnson. 1982. Effect of tallow and choline chloride addition to the diet of sows on milk composition, milk yield and preweaning pig performance. *Journal of Animal Science* 54:1-7.

Boyd, R. D., N. Williams, and G. L. Allee. 2008. Segregated parity structure in sow farms to capture nutrition, management and health opportunities. Pp. 45-50 in *Proceedings of the Midwest Swine Nutrition Conference*, September 4, 2008, Indianapolis, IN.

Braidman, I. P., and D. C. Anderson. 1985. Extra-endocrine functions of vitamin D. *Clinical Endocrinology* 23:445-460.

Braude, R., A. S. Foot, K. M. Henry, S. K. Kon, S. Y. Thompson, and T. H. Mead. 1941. Vitamin A studies with rats and pigs. *Biochemical Journal* 35:693-707.

Braude, R., S. K. Kon, and E. G. White. 1946. Observations on the nicotinic acid requirements of pigs. *Biochemical Journal* 40:843-855.

Braude, R., S. K. Kon, and J. W. G. Porter. 1950. Studies in the vitamin C metabolism of the pig. *British Journal of Nutrition* 4:186-197.

Brief, S., and B. P. Chew. 1985. Effects of vitamin A and β-carotene on reproductive performance in gilts. *Journal of Animal Science* 60:998-1004.

Brooks, C. C., R. M. Nakamura, and A. Y. Miyahara. 1973. Effect of menadione and other factors on sugar-induced heart lesions and hemorrhagic syndrome in the pig. *Journal of Animal Science* 37:1344-1350.

Brooks, P. H., D. A. Smith, and V. C. R. Irwin. 1977. Biotin supplementation of diets: The incidence of foot lesions and the reproductive performance of sows. *Veterinary Record* 101:46-50.

Brown, R. G., and G. J. King. 1977. Ascorbic acid synthesis in pigs. *Canadian Journal of Animal Science* 57:831 (Abstr.).

Brown, R. G., V. D. Sharma, and L. G. Young. 1970. Ascorbic acid metabolism in swine. Interrelationships between the level of energy intake and serum ascorbate levels. *Canadian Journal of Animal Science* 50:605-609.

Brown, R. G., J. G. Buchanan-Smith, and V. D. Sharma. 1975. Ascorbic acid metabolism in swine. The effects of frequency of feeding and level of supplementary ascorbic acid on swine fed various energy levels. *Canadian Journal of Animal Science* 55:353-358.

Brubacher, G., and O. Wiss. 1972. Vitamin E active compounds, synergists and antagonists. Pp. 255-258 in *The Vitamins*, Volume V, W. H. Sebrell and R. S. Harris, eds. New York: Academic Press.

Bryant, K. L., E. T. Kornegay, J. W. Knight, H. P. Veit, and D. R. Notter. 1985a. Supplemental biotin for swine. III. Influence of supplementation to corn- and wheat-based diets on the incidence and severity of toe lesions, hair and skin characteristics and structural soundness of sows housed in confinement during four parities. *Journal of Animal Science* 60:154-162.

Bryant, K. L., E. T. Kornegay, J. W. Knight, K. E. Webb, Jr., and D. R. Notter. 1985b. Supplemental biotin for swine. I. Influence on feedlot performance, plasma biotin and toe lesions in developing gilts. *Journal of Animal Science* 60:136-144.

Bryant, K. L., E. T. Kornegay, J. W. Knight, K. E. Webb, Jr., and D. R. Notter. 1985c. Supplemental biotin for swine. II. Influence of supplementation to corn- and wheat-based diets on reproductive performance and various biochemical criteria of sows during four parities. *Journal of Animal Science* 60:145-153.

Burroughs, W., B. H. Edgington, W. L. Robison, and R. M. Bethke. 1950. Niacin deficiency and enteritis in growing pigs. *Journal of Nutrition* 41:51-62.

Carter, E. G. A., and Carpenter, K. J. 1982. The available niacin values of food for rats and their relation to analytical values. *Journal of Nutrition* 112:2091-2103.

Cartwright, G. E., M. M. Wintrobe, P. Jones, M. Lauritsen, and S. Humphreys. 1944. Tryptophane derivatives in urine of pyridoxine deficient swine. *Bulletin of the Johns Hopkins Hospital* 75:35.

Cartwright, G. E., B. Tatting, and M. M. Wintrobe. 1948. Niacin deficiency anemia in swine. *Archives of Biochemistry* 19:109-118.

Cartwright, G. E., B. Tatting, H. Ashenbrucker, and M. M. Wintrobe. 1949. Experimental production of nutritional macrocytic anemia in swine. *Blood* 4:301-323.

Cartwright, G. E., J. G. Palmer, B. Tatting, H. Ashenbrucker, and M. M. Wintrobe. 1950. Experimental production of nutritional macrocytic anemia in swine. III. Further studies on pteroylglutamic acid deficiency. *Journal of Laboratory and Clinical Medicine* 36:675-693.

Cartwright, G. E., B. Tatting, J. Robinson, N. M. Fellows, F. D. Gunn, and M. M. Wintrobe. 1951. Hematologic manifestations of vitamin B_{12} deficiency in swine. *Blood* 6:867-891.

Cartwright, G. E., B. Tatting, D. Kurth, and M. M. Wintrobe. 1952. Experimental production of nutritional macrocytic anemia in swine. V. Hematologic manifestations of a combined deficiency of vitamin B_{12} and pteroylglutamic acid. *Blood* 7:992-1004.

Catron, D. V., D. Richardson, L. A. Underkofler, H. M. Maddock, and W. C. Friedland. 1952. Vitamin B_{12} requirement of weanling pigs. II. Performance on low level of vitamin B_{12} and requirements for optimum growth. *Journal of Nutrition* 47:461-468.

Chavez, E. R. 1983. Supplemental value of ascorbic acid during late gestation on piglet survival and early growth. *Canadian Journal of Animal Science* 63:683-687.

Chew, B. P., H. Rasmussen, M. H. Pubols, and R. L. Preston. 1982. Effects of vitamin A and β-carotene on plasma progesterone and uterine protein secretions in gilts. *Theriogenology* 18:643-654.

Chiang, S. H., J. E. Pettigrew, R. L. Moser, S. G. Cornelius, K. P. Miller, and T. R. Heeg. 1985. Supplemental vitamin C in swine diets. *Nutrition Reports International* 31:573-581.

Chung, T. K., and D. H. Baker. 1990. Riboflavin requirement of chicks fed purified amino acid and conventional corn-soybean meal diets. *Poultry Science* 69:1357-1363.

Chung, Y. K., D. C. Mahan, and A. J. Lepine. 1992. Efficacy of D-α-tocopherol and DL-α-tocopheryl acetate for weanling pigs. *Journal of Animal Science* 70:2485-2492.

Cleveland, E. R., G. L. Newton, B. G. Mullinix, O. M. Hale, and T. M. Frye. 1983. Foot-leg and performance traits of boars fed two levels of ascorbic acid. *Journal of Animal Science* 57(Suppl. 1):387 (Abstr.).

Cline, J. H., D. C. Mahan, and A. L. Moxon. 1974. Progeny effects of supplemental vitamin E in sow diets. *Journal of Animal Science* 39:974 (Abstr.).

Coelho, M. B. 1991. Vitamin stability. *Feed Management* 42(10):24.

Coelho, M. B., and B. Cousins. 1997. Vitamin supplementation supports higher performance. *Feedstuffs*, January 27:10-12, 20-21.

Combs, G. E., T. H. Berry, H. D. Wallace, and R. C. Crum, Jr. 1966. Levels and sources of vitamin D for pigs fed diets containing varying quantities of calcium. *Journal of Animal Science* 25:827-830.

Combs, G. F., Jr. 1999. *The Vitamins, Fundamental Aspects in Nutrition and Health*, 2nd Ed. San Diego, CA: Academic Press.

Copelin, J. L., H. Monegue, and G. E. Combs. 1980. Niacin levels in growing-finishing swine diets. *Journal of Animal Science* 51 (Suppl. 1):190 (Abstr.).

Crenshaw, T. D. 2000. Calcium, phosphorus, vitamin D, and vitamin K in swine production. Pp. 187-212 in *Swine Nutrition*, 2nd Ed., A. J. Lewis and L. L. Southern, eds. Boca Raton, FL: CRC Press.

Cromwell, G. L., V. W. Hays, and J. R. Overfield. 1970. Effect of dietary ascorbic acid on performance and plasma cholesterol levels of growing swine. *Journal of Animal Science* 31:63-66.

Cunha, T. J., D. C. Lindley, and M. E. Ensminger. 1946. Biotin deficiency syndrome in pigs fed desiccated egg white. *Journal of Animal Science* 5:219-225.

Cunha, T. J., L. K. Bustad, W. E. Ham, D. R. Cordy, E. C. McCullock, I. F. Woods, G. H. Corner, and M. A. McGregor. 1947. Folic acid, para-aminobenzoic acid and anti-pernicious anemia liver extract in swine nutrition. *Journal of Nutrition* 34:173-187.

Cunha, T. J., R. W. Colby, L. K. Bustad, and J. F. Bone. 1948. The need for and interrelationship of folic acid, anti-pernicious anemia liver extract, and biotin in the pig. *Journal of Nutrition* 36:215-229.

Darroch, C. S. 2000. Vitamin A in swine nutrition. Pp. 263-280 in *Swine Nutrition*, 2nd Ed., A. J. Lewis and L. L. Southern, eds. Boca Raton, FL: CRC Press.

Davey, R. J., and J. W. Stevenson. 1963. Pantothenic acid requirement of swine for reproduction. *Journal of Animal Science* 22:9-13.

Davis, C. L., and J. R. Gorham. 1954. The pathology of experimental and natural cases of yellow fat disease in swine. *American Journal of Veterinary Research* 15:55-59.

De Ritter, E. 1976. Stability characteristics of vitamins in processed foods. *Food Technology* 30:48-54.

Dodd, D. C., and P. E. Newling. 1960. Muscle degeneration and liver necrosis in the pig. Report of a natural outbreak. *New Zealand Veterinary Journal* 8:95-98.

Dove, C. R., and D. C. Cook. 2000. Water-soluble vitamins in swine nutrition. Pp. 315-356 in *Swine Nutrition*, 2nd Ed., A. J. Lewis and L.L. Southern, eds. Boca Raton, FL: CRC Press.

Dove, C. R., and R. C. Ewan. 1991. Effect of trace minerals on the stability of vitamin E in swine grower diets. *Journal of Animal Science* 69:1994-2000.

Duel, H. J., Jr., E. R. Meserve, C. H. Johnston, A. Polgar, and L. Zechmeister. 1945. Reinvestigation of the relative provitamin A potency of cryptoxanthin and β-carotene. *Archives of Biochemistry* 7:447-450.

Dvorak, M. 1974. Effects of corticotrophin, starvation and glucose on ascorbic acid levels in the blood plasma and liver of piglets. *Nutrition & Metabolism* 16:215-222.

Easter, R. A., P. A. Anderson, E. J. Michel, and J. R. Corley. 1983. Response of gestating gilts and starter, grower and finisher swine to biotin, pyridoxine, folacin and thiamine additions to a corn-soybean meal diet. *Nutrition Reports International* 28:945-953.

Echeverria-Alonzo S., R. Santos-Ricalde, F. Centurion-Castro, R. Ake-Lopez, M. Alfaro-Gamboa, and J. Rodriguez-Buenfil. 2009. Effects of dietary selenium and vitamin E on semen quality and sperm morphology of young boars during warm and fresh season. *Journal of Animal and Veterinary Advances* 8:2311-2317.

Ellis, N. R., and L. L. Madsen. 1944. The thiamine requirements of pigs as related to the fat content of the diet. *Journal of Nutrition* 27:253-292.

Ellis, R. P., and M. M. Vorhies. 1976. Effect of supplemental dietary vitamin E on the serologic response of swine to an *Esherichia coli* bacterin. *Journal of the American Veterinary Medical Association* 168:231-232.

Emerson, O. H., G. A. Emerson, A. Mohammad, and H. M. Evans. 1937. The chemistry of vitamin E: Tocopherols from various sources. *Journal of Biological Chemistry* 122:99-107.

Emmert, J. L., and D. H. Baker. 1997. A chick bioassay approach for determining the bioavailable choline concentration in normal and overheated soybean meal, canola meal, and peanut meal. *Journal of Nutrition* 127:745-752.

Emmert, J. L., T. A. Garrow, and D. H. Baker. 1996. Development of an experimental diet for determining bioavailable choline concentration and its application in studies with soybean lecithin. *Journal of Animal Science* 74:2738-2744.

Esch, M. W., R. A. Easter, and J. M. Bahr. 1981. Effect of riboflavin deficiency on estrous cyclicity in pigs. *Biology of Reproduction* 25:659-665.

Evans, H. M., O. H. Emerson, and G. A. Emerson. 1936. The isolation from wheat germ oil of an alcohol, α-tocopherol, having the properties of vitamin E. *Journal of Biological Chemistry* 113:319-332.

Ewan, R. C., M. E. Wastell, E. J. Bicknell, and V. C. Speer. 1969. Performance and deficiency symptoms of young pigs fed diets low in vitamin E and selenium. *Journal of Animal Science* 29:912-915.

Fidge, N. H., F. R. Smith, and D. S. Goodman. 1969. Vitamin A and carotenoids. The enzymatic conversion of β-carotene into retinal in hog intestinal mucosa. *Biochemical Journal* 114:689-694.

Firth, J., and B. C. Johnson. 1956. Quantitative relationships of tryptophan and nicotinic acid in the baby pig. *Journal of Nutrition* 59:223-234.

Follis, R. H., Jr., and M. M. Wintrobe. 1945. A comparison of the effects of pyridoxine and pantothenic acid deficiencies on the nervous tissues of swine. *Journal of Experimental Medicine* 81:539-551.

Follis, R. H., M. H. Miller, M. M. Wintrobe, and H. J. Stein. 1943. Development of myocardial necrosis and absence of nerve degeneration in thiamine deficiency in pigs. *American Journal of Pathology* 19:341-357.

Forbes, R. M., and W. T. Haines. 1952. The riboflavin requirement of the baby pig. *Journal of Nutrition* 47:411-424.

Frank, G. R., J. M. Bahr, and R. A. Easter. 1984. Riboflavin requirement of gestating swine. *Journal of Animal Science* 59:1567-1572.

Frank, G. R., J. M. Bahr, and R. A. Easter. 1988. Riboflavin requirement of lactating swine. *Journal of Animal Science* 66:47-52.

Frape, D. L., V. C. Speer, V. W. Hays, and D. V. Catron. 1959. The vitamin A requirement of the young pig. *Journal of Nutrition* 68:173-187.

Frederick, G. L., and G. J. Brisson. 1961. Some observations on the relationship between vitamin B_{12} and reproduction in swine. *Canadian Journal of Animal Science* 41:212-219.

Frigg, M. 1976. Bio-availability of biotin in cereals. *Poultry Science* 55:2310-2318.

Fritschen, R. D., O. D. Grace, and E. R. Peo, Jr. 1971. Bleeding pig disease. *Nebraska Swine Report* EC71 219:22-23.

Gannon, N. J., and J. Liebholz. 1989. The effects of folic acid supplementation on the performance of growing pigs. P. 136 in *Manipulating Pig Production* II, J. L. Barnett and D. P. Hennessy, eds. Werribee, Victoria, Australia: Australasian Pig Science Association.

Goff, J. P., R. L. Horst, and E. T. Littledike. 1984. Effect of sow vitamin D status at parturition on the vitamin D status of neonatal pigs. *Journal of Nutrition* 114:163-169.

Goodman, D. S. 1979. Vitamin A and retinoids: Recent advances. *Federation Proceedings* 38:2501-2503.

Goodman, D. S. 1980. Vitamin A metabolism. *Federation Proceedings* 39:2716-2722.

Grandhi, R. R., and J. H. Strain. 1980. Effect of biotin supplementation on reproductive performance and foot lesions in swine. *Canadian Journal of Animal Science* 60:961-969.

Grant, C. A. 1961. Morphological and etiological studies of dietetic microangiopathy in pigs ("mulberry heart"). *Acta Veterinaria Scandinavica* 2(Suppl. 3):1 (Abstr.).

Green, J., P. Mamalis, S. Marcinkiewicz, and D. McHale. 1960. Structure of β-tocopherol. *Chemistry and Industry* (*London*), 16(January):73.

Green, R. G., W. E. Carlson, and C. A. Evans. 1941. A deficiency disease of foxes produced by feeding fish. *Journal of Nutrition* 21:243-256.

Greenberg, S. M., A. Chatterjee, C. E. Calbert, H. J. Duel, Jr., and L. Zechmeister. 1950. A comparison of the provitamin A activity of β-carotene and cryptoxanthin in the chick. *Archives of Biochemistry* 25:61-65.

Groce, A. W., E. R. Miller, K. K. Keahey, D. E. Ullrey, and D. J. Ellis. 1971. Selenium supplementation of practical diets for growing-finishing swine. *Journal of Animal Science* 32:905-911.

Groce, A. W., E. R. Miller, D. E. Ullrey, P. K. Ku, K. K. Keahey, and D. J. Ellis. 1973. Selenium requirements in corn-soy diets for growing-finishing swine. *Journal of Animal Science* 37:948-956.

Groesbeck, C. N., R. D. Goodband, M. D. Tokach, S. S. Dritz, J. L. Nelssen, and J. M. DeRouchey. 2007. Effects of pantothenic acid on growth performance and carcass characteristics of growing-finishing pigs fed diets with or without ractopamine hydrochloride. *Journal of Animal Science* 85:2492-2497.

Guilbert, H. R., R. F. Miller, and E. H. Hughes. 1937. The minimum vitamin A and carotene requirements of cattle, sheep, and swine. *Journal of Nutrition* 13:543-564.

Hakkarainen, J., P. Lindberg, G. Bengtsson, L. Jonsson, and N. Lannek. 1978. Requirement for selenium (as selenite) and vitamin E (as α-tocopherol) in weaned pigs. III. The effect on the development of the VESD syndrome of varying selenium levels in a low-tocopherol diet. *Journal of Animal Science* 46:1001-1008.

Hall, D. D., G. L. Cromwell, and T. S. Stahly. 1986. The vitamin K requirement of the growing pig. *Journal of Animal Science* 63(Suppl. 1):268 (Abstr.).

Hall, D. D., G. L. Cromwell, and T. S. Stahly. 1991. Effects of dietary calcium, phosphorus, calcium:phosphorus ratio and vitamin K on performance, bone strength and blood clotting status of pigs. *Journal of Animal Science* 69:646-655.

Halloran, B. P., and H. F. DeLuca. 1979. Vitamin D deficiency and reproduction in rats. *Science* 204:73-74.

Hamilton, C. R., and T. L. Veum. 1984. Response of sows and litters to added dietary biotin in environmentally regulated facilities. *Journal of Animal Science* 59:151-157.

Hamilton, C. R., and T. L. Veum. 1986. Effect of biotin and (or) lysine additions to corn-soybean meal diets on the performance and nutrient balance of growing pigs. *Journal of Animal Science* 62:155-162.

Hamilton, C. R., T. L. Veum, D. E. Jewell, and J. A. Siwecki. 1983. The biotin status of weanling pigs fed semipurified diets as evaluated by plasma and hepatic parameters. *International Journal for Vitamin and Nutrition Research* 53:44-50.

Hanke, H. E., and R. J. Meade. 1971. *Biotin and Pyridoxine Additions to Diets for Pigs Weaned at an Early Age.* 1970-71 Minnesota Swine Research Report H-120. St. Paul, MN: University of Minnesota Press.

Hanson, L. E., and I. L. Hathaway. 1948. The fertility of boars fed a vitamin E deficient ration. *Journal of Animal Science* 7:528 (Abstr.).

Harmon, B. G., E. R. Miller, J. A. Hoefer, D. E. Ullrey, and R. W. Luecke. 1963. Relationship of specific nutrient deficiencies to antibody production in swine. II. Pantothenic acid, pyridoxine or riboflavin. *Journal of Nutrition* 79:263-268.

Harmon, B. G., D. E. Becker, A. H. Jensen, and D. H. Baker. 1969. Nicotinic acid-tryptophan relationship in the nutrition of the weanling pig. *Journal of Animal Science* 28:848-852.

Harmon, D. G., D. E. Becker, A. H. Jensen, and D. H. Baker. 1970. Nicotinic acid-tryptophan nutrition and immunologic implications in young swine. *Journal of Animal Science* 31:339-342.

Harper, A. F., M. D. Lindemann, L. I. Chiba, G. E. Combs, D. L. Handlin, E. T. Kornegay, and L. L. Southern. 1994. An assessment of dietary folic acid levels during gestation and lactation on reproductive and lactational performance of sows: A cooperative study. *Journal of Animal Science* 72:2338-2344.

Heaney, D. P., J. A. Hoefer, D. E. Ullrey, and E. R. Miller. 1963. Effects of marginal vitamin A intake during gestation in swine. *Journal of Animal Science* 22:925-928.

Heinemann, W. W., M. E. Ensminger, T. J. Cunha, and E. C. McCulloch. 1946. The relation of the amount of thiamine in the ration of the hog to the thiamine and riboflavin content of the tissue. *Journal of Nutrition* 31:107-125.

Heinle, R. W., A. D. Welch, and J. A. Pritchard. 1948. Essentiality of both the anti-pernicious anemia factor of liver and pteroylglutamic acid for hematopoiesis in swine. *Journal of Laboratory and Clinical Medicine* 33:1647 (Abstr.).

Hendricks, D. G., E. R. Miller, D. E. Ullrey, R. D. Struthers, B. V. Baltzer, J. A. Hoefer, and R. W. Luecke. 1967. β-Carotene vs. retinyl acetate for the baby pig and the effect upon ergocalciferol requirement. *Journal of Nutrition* 93:37-43.

Hendricks, H. K., H. S. Teague, D. R. Redman, and A. P. Grifo, Jr. 1964. Absorption of vitamin B_{12} from the colon of the pig. *Journal of Animal Science* 23:1036-1038.

Hentges, J. F., Jr., R. H. Grummer, P. H. Phillips, and G. Bohstedt. 1952. The minimum requirement of young pigs for a purified source of carotene. *Journal of Animal Science* 11:266-272.

Hitchcock, J. P., E. R. Miller, K. K. Keahey, and D. E. Ullrey. 1978. Effects of arsanilic acid and vitamin E upon utilization of natural or supplemental selenium by swine. *Journal of Animal Science* 46:425-435.

Hjarde, W., A. Neimann-Sorensen, B. Palludan, and P. H. Sorensen. 1961. Investigations concerning vitamin A requirement, utilization and deficiency symptoms in pigs. *Acta Agriculturae Scandinavica* 11:13-53.

Hoppe, P. P., F. J. Schoner, and M. Frigg. 1992. Effects of dietary retinol on hepatic retinol storage and on plasma and tissue α-tocopherol in pigs. *International Journal for Vitamin and Nutrition Research* 62:121-129.

Horst, R. L., E. T. Littledike, J. L. Riley, and J. L. Napoli. 1981. Quantitation of vitamin D and its metabolites and their plasma concentrations in five species of animals. *Analytic Biochemistry* 116:189-203.

Horst, R. L., J. L. Napoli, and E. T. Littledike. 1982. Discrimination in the metabolism of orally dosed ergocalciferol and cholecalciferol by the pig, rat, and chick. *Biochemical Journal* 204:185-189.

House, J. D., and C. M. T. Fletcher. 2003. Response of early weaned piglets to graded levels of dietary cobalamin. *Canadian Journal of Animal Science* 83:247-255.

Hove, E. L., and H. R. Seibold. 1955. Liver necrosis and altered fat composition in vitamin E-deficient swine. *Journal of Nutrition* 56:173-186.

Hughes, E. H. 1940a. The minimum requirement of riboflavin for the growing pig. *Journal of Nutrition* 20:233-238.

Hughes, E. H. 1940b. The minimum requirement of thiamine for the growing pig. *Journal of Nutrition* 20:239-241.

Hughes, E. H. 1943. The minimum requirement of nicotinic acid for the growing pig. *Journal of Animal Science* 2:23.

Hughes, E. H., and N. R. Ittner. 1942. The minimum requirement of pantothenic acid for the growing pig. *Journal of Animal Science* 1:116-119.

Hughes, E. H., and R. L. Squibb. 1942. Vitamin B_6 (pyridoxine) in the nutrition of the pig. *Journal of Animal Science* 1:320-325.

Hutagalung, R. I., C. H. Chaney, R. D. Wood, and D. G. Waddill. 1968. Effects of nitrates and nitrites in feed on utilization of carotene in swine. *Journal of Animal Science* 27:79-82.

Hutagalung, R. I., G. L. Cromwell, V. W. Hays, and C. H. Chaney. 1969. Effect of dietary fat, protein, cholesterol and ascorbic acid on performance, serum and tissue cholesterol levels and serum lipid levels of swine. *Journal of Animal Science* 29:700-705.

IOM (Institute of Medicine). 1999. *Dietary Reference Intakes for Calcium, Phosphorus, Magnesium, Vitamin D, and Fluoride.* Washington, DC: National Academy Press.

Ivers, D. J., S. L. Rodhouse, M. R. Ellersieck, and T. L. Veum. 1993. Effect of supplemental niacin on sow reproduction and sow and litter performance. *Journal of Animal Science* 71:651-655.

Jacyno, E., M. Kawecka, M. Kamyczek, A. Kolodziej, J. Owsianny, and B. Delikator. 2002. Influence of inorganic Se + vitamin E and organic Se + vitamin E on reproductive performance of young boars. *Agricultural and Food Science Finland* 11:175-184.

Jewell, D. E., J. A. Siwecki, and T. L. Veum. 1981. The effect of dietary vitamin C on performance and tissue vitamin C levels in neonatal pigs. *Journal of Animal Science* 53(Suppl. 1):98 (Abstr.).

Johnson, B. C., and M. F. James. 1948. Choline deficiency in the baby pig. *Journal of Nutrition* 36:339-344.

Johnson, B. C., M. F. James, and J. L. Krider. 1948. Raising newborn pigs to weaning age on a synthetic diet with attempts to produce a pteroylglutamic acid deficiency. *Journal of Animal Science* 7:486-493.

Johnson, B. C., A. L. Neuman, R. O. Nesheim, M. F. James, J. L. Krider, A. S. Dana, and J. B. Thiersch. 1950. The interrelationship of vitamin B_{12} and folic acid in the baby pig. *Journal of Laboratory and Clinical Medicine* 36:537-546.

Johnson, D. W., and L. S. Palmer. 1939. Individual and breed variations in pigs on rations devoid of vitamin D. *Journal of Agricultural Research* 58:929-940.

Kindberg, C. G., and J. W. Suttie. 1989. Effect of various intakes of phylloquinone on signs of vitamin K deficiency and serum liver phylloquinone concentrations in the rat. *Journal of Nutrition* 119:175-180.

Knights, T. E. N., R. R. Grandhi, and S. K. Baidoo. 1998. Interactive effects of selection for lower backfat and dietary pyridoxine levels on reproduction, and nutrient metabolism during the gestation period in Yorkshire and Hampshire sows. *Canadian Journal of Animal Science* 78:167-173.

Kodicek, E., R. Braude, S. K. Kon, and K. G. Mitchell. 1956. The effect of alkaline hydrolysis of maize on the availability of its nicotinic acid to the pig. *British Journal of Nutrition* 10:51-66.

Kodicek, E., R. Braude, S. K. Kon, and K. G. Mitchell. 1959. The availability to pigs of nicotinic acid in tortilla baked from maize treated with limewater. *British Journal of Nutrition* 13:363-384.

Kolodziej, A., and E. Jacyno. 2005. Effect of selenium and vitamin E supplementation on reproductive performance of young boars. *Archiv fur Tierzucht—Archives of Animal Breeding* 48:68-75.

Kopinski, J. S., J. Leibholz, and L. Bryden. 1989. Biotin studies in pigs. 4. Biotin availability in feedstuffs for pigs and chickens. *British Journal of Nutrition* 62:773-780.

Kormann, A. W., and H. Weiser. 1984. Protective functions of fat-soluble vitamins. Pp. 201-222 in *Proceedings of the 37th Nottingham Feed Manufacturer's Conference*, Nottingham, UK. London: Butterworth.

Kornegay, E. T. 1986. Biotin in swine production: A review. *Livestock Production Science* 14:65-89.

Kornegay, E. T., and T. N. Meacham. 1973. Evaluation of supplemental choline for reproducing sows housed in total confinement on concrete or in dirt lots. *Journal of Animal Science* 37:506-509.

Kornegay, E. T., J. B. Meldrum, G. Schurig, M. D. Lindemann, and F. C. Gwazdauskas. 1986. Lack of influence of nursery temperature on the response of weanling pigs to supplemental vitamins C and E. *Journal of Animal Science* 63(2):484-491.

Kösters, W. W., and M. Kirchgessner. 1976a. Gewichtenwicklung und Futterverwertung Friiheutwijhnter Ferkel bei unterschiedlicher Vitamin B_6-Versorgung. (Growth rate and feed efficiency of early-weaned piglets with varying vitamin B_6 supply.) *Zeitschrift fur Tierphysiologie Tierernahrung und Futtermittelkunde—Journal of Animal Physiology and Animal Nutrition* 37:235-246.

Kösters, W. W., and M. Kirchgessner. 1976b. Zur Veranderugn des Futterverzeha Friiheutwohnter Ferkel bei Untersehiedlicher Vitamin B_6Versorgung. (Change in feed intake of early-weaned piglets in response to different vitamin B_6 supply.) *Zeitschrift fur Tierphysiologie Tierernahrung und Futtermittelkunde—Journal of Animal Physiology and Animal Nutrition* 37:247-254.

Krampitz, L. O., and D. W. Woolley. 1944. The manner of inactivation of thiamine by fish tissue. *Journal of Biological Chemistry* 152:9-17.

Krider, J. L., S. W. Terrill, and R. F. VanPoucke. 1949. Response of weanling pigs to various levels of riboflavin. *Journal of Animal Science* 8:121-125.

Kroening, G. H., and W. G. Pond. 1967. Methionine, choline and threonine interrelationships for growth and lipotropic action in the baby pig and rat. *Journal of Animal Science* 26:352-357.

Lannek, N., P. Lindberg, G. Nilsson, G. Nordstrom, and K. Orstadius. 1961. Production of vitamin E deficiency and muscular dystrophy in pigs. *Research in Veterinary Science* 2:67-72.

Lauridsen, C., H. Engel, A. M. Craig, and M. G. Traber. 2002. Relative bioactivity of dietary *RRR*- and all-rac-alpha-tocopheryl acetates in swine assessed with deuterium-labeled vitamin E. *Journal of Animal Science* 80:702-707.

Lauridsen, C., U. Halekoh, T. Larsen, and S. K. Jensen. 2010. Reproductive performance and bone status markers of gilts and lactating sows supplemented with two different forms of vitamin D. *Journal of Animal Science* 88:202-213.

Lehrer, W. P., Jr., and A. C. Wiese. 1952. Riboflavin deficiency in baby pigs. *Journal of Animal Science* 11:244-250.

Lehrer, W. P., Jr., A. C. Wiese, P. R. Moore, and M. E. Ensminger. 1951. Pyridoxine deficiency in baby pigs. *Journal of Animal Science* 10:65-72.

Lehrer, W. P., Jr., A. C. Wiese, and P. R. Moore. 1952. Biotin deficiency in suckling pigs. *Journal of Nutrition* 47:203-212.

Leibbrandt, V. D. 1977. Influence of ascorbic acid on suckling pig performance. *Journal of Animal Science* 45(Suppl. 1):98 (Abstr.).

Lewis, A. J., G. L. Cromwell, and J. E. Pettigrew. 1991. Effects of supplemental biotin during gestation and lactation on reproductive performance of sows: A cooperative study. *Journal of Animal Science* 69:207-214.

Lindemann, M. D., and E. T. Kornegay. 1986. Folic acid additions to weanling pig diets. *Journal of Animal Science* 63(Suppl. 1):35 (Abstr.).

Lindemann, M. D., and E. T. Kornegay. 1989. Folic acid supplementation to diets of gestating-lactating swine over multiple parities. *Journal of Animal Science* 67:459-464.

Lindemann, M. D., G. L. Cromwell, and H. J. Monegue. 1995. Effects of inadequate and high levels of vitamin fortification on performance of weanling pigs. *Journal of Animal Science* 73(Suppl. 1):16 (Abstr.).

Lindemann, M. D., G. L. Cromwell, J. L. G. van de Ligt, and H. J. Monegue. 1999. Higher levels of selected B-vitamins improve performance and lean deposition in growing/finishing swine. *Journal of Animal Science* 77(Suppl. 1):58.

Lindemann, M. D., J. H. Brendemuhl, L. I. Chiba, C. S. Darroch, C. R. Dove, M. J. Estienne, and A. F. Harper. 2008. A regional evaluation of injections of high levels of vitamin A on reproductive performance of sows. *Journal of Animal Science* 86:333-338.

Lindley, D. C., and T. J. Cunha. 1946. Nutritional significance of inositol and biotin for the pig. *Journal of Nutrition* 32:47-59.

Livingston, A. L., J. W. Nelson, and G. O. Kohler. 1968. Stability of alpha-tocopherol during alfalfa dehydration and storage. *Journal of Agricultural and Food Chemistry* 16:492-495.

Long, G. G. 1984. Acute toxicosis in swine associated with excessive intake of vitamin D. *Journal of the American Veterinary Medical Association* 184:164-170.

Lowry, K. R., O. A. Izquierdo, and D. H. Baker. 1987. Efficacy of betaine relative to choline as a dietary methyl donor. *Poultry Science* 66(Suppl. 1):120 (Abstr.).

Luce, W. G., E. R. Peo, Jr., and D. B. Hudman. 1966. Availability of niacin in wheat for swine. *Journal of Nutrition* 88:39-44.

Luce, W. G., E. R. Peo, Jr., and D. B. Hudman. 1967. Availability of niacin in corn and milo for swine. *Journal of Animal Science* 26:76-84.

Luce, W. G., D. S. Buchanan, C. V. Maxwell, H. E. Jordan, and R. O. Bates. 1985. Effect of supplemental choline and dichlorvos on reproductive performance of gilts. *Nutrition Reports International* 32:245-251.

Luecke, R. W., W. N. McMillen, F. Thorpe, Jr., and C. Tull. 1947. The relationship of nicotinic acid, tryptophane and protein in the nutrition of the pig. *Journal of Nutrition* 33:251-261.

Luecke, R. W., W. N. McMillen, F. Thorpe, Jr., and C. Tull. 1948. Further studies on the relationship of nicotinic acid, tryptophane and protein in the nutrition of the pig. *Journal of Nutrition* 36:417-424.

Luecke, R. W., W. N. McMillen, and F. Thorpe, Jr. 1950. Further studies of pantothenic acid deficiency in weanling pigs. *Journal of Animal Science* 9:78-82.

Luecke, R. W., J. A. Hoefer, and F. Thorpe, Jr. 1952. The relationship of protein to pantothenic acid and vitamin B_{12} in the growing pig. *Journal of Animal Science* 11:238-243.

Luecke, R. W., J. A. Hoefer, and F. Thorpe, Jr. 1953. The supplementary effects of calcium pantothenate and aureomycin in a low-protein ration for weanling pigs. *Journal of Animal Science* 12:605-610.

Lynch, P. B., and J. F. O'Grady. 1981. Effect of vitamin C (ascorbic acid) supplementation on sows in late pregnancy on piglet mortality. *Irish Journal of Agricultural Research* 20:217-219.

Lynch, P. B., G. E. Hall, L. D. Hill, E. E. Hatfield, and A. H. Jensen. 1975. Chemically preserved high-moisture corns in diets for growing-finishing swine. *Journal of Animal Science* 40:1063-1069.

Madsen, A., H. P. Mortensen, W. Hjarde, E. Leebeck, and T. Leth. 1973. Vitamin E in barley treated with propionic acid with special reference to the feeding of bacon pigs. *Acta Agriculturae Scandinavica* Suppl. 19:169-173.

Mahan, D. C. 1991. Assessment of the influence of dietary vitamin E on sows and offspring in three parities: Reproductive performance, tissue tocopherol, and effects on progeny. *Journal of Animal Science* 69:2904-2917.

Mahan, D. C. 1994. Effects of dietary vitamin E on sow reproductive performance over a five-parity period. *Journal of Animal Science* 72:2870-2879.

Mahan, D. C. 2000. Selenium and vitamin E in swine nutrition. Pp. 281-314 in *Swine Nutrition*, 2nd Ed., A. J. Lewis and L. L. Southern, eds. Boca Raton, FL: CRC Press.

Mahan, D. C., and A. L. Moxon. 1978. Effect of increasing the level of inorganic selenium supplementation in the postweaning diets of swine. *Journal of Animal Science* 46:384-390.

Mahan, D. C., and L. J. Saif. 1983. Efficacy of vitamin C supplementation for weanling pigs. *Journal of Animal Science* 56:631-639.

Mahan, D. C., R. A. Pickett, T. W. Perry, T. M. Curtin, W. R. Featherston, and W. M. Beeson. 1966. Influence of various nutritional factors and physical form of feed on esophagogastric ulcers in swine. *Journal of Animal Science* 25:1019-1023.

Mahan, D. C., A. J. Lepine, and K. Dabrowski. 1994. Efficacy of magnesium-L-ascorbyl-2-phosphate as a vitamin C source for weanling and growing-finishing swine. *Journal of Animal Science* 72:2354-2361.

Mahan, D. C., Y. Y. Kim, and R. L. Stuart. 2000. Effect of vitamin E sources (*RRR*- or all-rac-alpha-tocopheryl acetate) and levels on sow reproductive performance, serum, tissue, and milk alpha-tocopherol contents over a five-parity period, and the effects on the progeny. *Journal of Animal Science* 78:110-119.

Mahan, D. C., S. D. Carter, T. R. Cline, G. M. Hill, S. W. Kim, P. S. Miller, J. L. Nelssen, H. H. Stein, T. L. Veum, and the North Central Coordinating Committee on Swine Nutrition (NCCC-42). 2007. Evaluating the effects of supplemental B vitamins in practical swine diets during the starter and grower-finisher periods—A regional study. *Journal of Animal Science* 85:2190-2197.

Malm, A., W. G. Pond, E. F. Walker, Jr., M. Homan, A. Aydin, and D. Kirtland. 1976. Effect of polyunsaturated fatty acids and vitamin E level of the sow gestation diet on reproductive performance and on level of alpha-tocopherol in colostrum, milk and dam and progeny blood serum. *Journal of Animal Science* 42:393-399.

Marin-Guzman, J., D. C. Mahan, Y. K. Chung, J. L. Pate, and W. F. Pope. 1997. Effects of dietary selenium and vitamin E on boar performance and tissue responses, semen quality, and subsequent fertilization rates in mature gilts. *Journal of Animal Science* 75:2994-3003.

Marin-Guzman, J., D. C. Mahan, and J. L. Pate. 2000a. Effect of dietary selenium and vitamin E on spermatogenic development in boars. *Journal of Animal Science* 78:1537-1543.

Marin-Guzman, J., D. C. Mahan, and R. Whitmoyer. 2000b. Effect of dietary selenium and vitamin E on the ultrastructure and ATP concentration of boar spermatozoa, and the efficacy of added sodium selenite in extended semen on sperm motility. *Journal of Animal Science* 78:1544-1550.

Matte, J. J., C. L. Girard, and G. J. Brisson. 1984a. Folic acid and reproductive performance of sows. *Journal of Animal Science* 59:1020-1025.

Matte, J. J., C. L. Girard, and G. J. Brisson. 1984b. Serum folates during the reproductive cycle of sows. *Journal of Animal Science* 59:158-163.

Matte, J. J., C. L. Girard, and G. J. Brisson. 1992. The role of folic acid in the nutrition of gestating and lactating primaparous sows. *Livestock Production Science* 32:131-148.

Mavromatis, J., G. Koptopoulos, S. C. Kyriakis, A. Papasteriadis, and K. Saoulidis. 1999. Effects of alpha-tocopherol and selenium on pregnant sows and their piglets' immunity and performance. *Journal of Veterinary Medicine Series A-Physiology Pathology Clinical Medicine* 46:545-553.

Meade, R. J. 1971. *Biotin and Pyridoxine Supplementation of Diets for Growing Pigs*. 1970-71 Minnesota Swine Research Report H-218. St. Paul: University of Minnesota Press.

Meyer, W. R., D. C. Mahan, and A. L. Moxon. 1981. Value of dietary selenium and vitamin E for weanling swine as measured by performance and tissue selenium and glutathione peroxidase activities. *Journal of Animal Science* 52:302-311.

Michel, R. L., C. K. Whitehair, and K. K. Keahey. 1969. Dietary hepatic necrosis associated with selenium-vitamin E deficiency in swine. *Journal of the American Veterinary Medical Association* 155:50-59.

Miller, C. O., and N. R. Ellis. 1951. The riboflavin requirement of growing swine. *Journal of Animal Science* 10:807-812.

Miller, C. O., N. R. Ellis, J. W. Stevenson, and R. Davey. 1953. The riboflavin requirement of swine for reproduction. *Journal of Nutrition* 51:163-170.

Miller, E. R., R. L. Johnston, J. A. Hoefer, and R. W. Luecke. 1954. The riboflavin requirement of the baby pig. *Journal of Nutrition* 52:405-413.

Miller, E. R., D. A. Schmidt, J. A. Hoefer, and R. W. Luecke. 1955. The thiamine requirement of the baby pig. *Journal of Nutrition* 56:423-430.

Miller, E. R., D. A. Schmidt, J. A. Hoefer, and R. W. Luecke. 1957. The pyridoxine requirement of the baby pig. *Journal of Nutrition* 62:407-419.

Miller, E. R., D. E. Ullrey, C. L. Zutaut, B. V. Baltzer, D. A. Schmidt, B. H. Vincent, J. A. Hoefer, and R. W. Luecke. 1964. Vitamin D_2 requirement of the baby pig. *Journal of Nutrition* 83:140-148.

Miller, E. R., D. E. Ullrey, C. L. Zutaut, J. A. Hoefer, and R. W. Luecke. 1965. Comparisons of casein and soy proteins upon mineral balance and vitamin D_2 requirement of the baby pig. *Journal of Nutrition* 85:347-353.

Misir, R., and R. Blair. 1984. Effect of biotin supplementation of a barley-wheat diet on restoration of healthy feet, legs, and skin of biotin deficient sows. *Journal of Research in Veterinary Science* 40:212-218.

Misir, R., and R. Blair. 1986. Reproductive performance of gilts and sows as affected by induced biotin deficiency and subsequent dietary biotin supplementation. *Journal of Animal Physiology and Animal Nutrition* 55:196-208.

Mitchell, H. H., B. C. Johnson, T. S. Hamilton, and W. T. Haines. 1950. The riboflavin requirement of the growing pig at two environmental temperatures. *Journal of Nutrition* 41:317-337.

Molitoris, B. A., and D. H. Baker. 1976. Assessment of the quantity of biologically available choline in soybean meal. *Journal of Animal Science* 42:481-489.

Mosekilde, L. 2005. Vitamin D and the elderly. *Clinical Endocrinology (Oxford)* 62:265-281.

Mosnier, E., J. J. Matte, M. Etienne, P. Ramaekers, B. Sève, and N. Le Floc. 2009. Tryptophan metabolism and related B vitamins in the multiparous sow fed ad libitum after farrowing. *Archives of Animal Nutrition* 63:467-478.

Muhrer, M. E., R. G. Cooper, C. N. Cornell, and R. D. Thomas. 1970. Diet related hemorrhagic syndrome in swine. *Journal of Animal Science* 31:1025 (Abstr.).

Myers, G. S., Jr., H. D. Eaton, and J. E. Rousseau, Jr. 1959. Relative value of carotene from alfalfa and vitamin A from a dry carrier fed to lambs and pigs. *Journal of Animal Science* 18:288-297.

Nafstad, I. 1965. Studies of hematology and bone marrow morphology in vitamin E-deficient pigs. *Pathologia Veterinaria* 2:277-287.

Nafstad, I. 1973. Some aspects of vitamin E deficiency in pigs. *Acta Agriculturae Scandinavica* 19:31-34.

Nafstad, I., and H. J. Nafstad. 1968. An electron microscopic study of blood and bone marrow in vitamin E-deficient pigs. *Pathologia Veterinaria* 5:520-537.

Nafstad, I., and S. Tollersrud. 1970. The vitamin E-deficiency syndrome in pigs. I. Pathological changes. *Acta Veterinaria Scandinavica* 11:452-480.

Nakano, T., F. X. Aherne, and J. R. Thompson. 1983. Effect of dietary supplementation of vitamin C on pig performance and the incidence of osteochondrosis in elbow and stifle joints in young growing swine. *Canadian Journal of Animal Science* 63:421-428.

Nelson, E. C., B. A. Dehority, H. S. Teague, V. L. Sanger, and W. D. Pounden. 1962. Effect of vitamin A on some biochemical and physiological changes in swine. *Journal of Nutrition* 76:325-332.

Nelson, E. C., B. A. Dehority, H. S. Teague, A. P. Grifo, Jr., and V. L. Sanger. 1964. Effect of vitamin A and vitamin A acid on cerebrospinal fluid pressure and blood and liver vitamin A concentration in the pig. *Journal of Nutrition* 82:263-268.

Nesheim, R. O., and B. C. Johnson. 1950. Effect of a high level of methionine on the dietary choline requirement of the baby pig. *Journal of Nutrition* 41:149-152.

Nesheim, R. O., J. L. Krider, and B. C. Johnson. 1950. The quantitative crystalline B_{12} requirement of the baby pig. *Archives of Biochemistry* 27:240-242.

Neumann, A. L., and B. C. Johnson. 1950. Crystalline vitamin B_{12} in the nutrition of the baby pig. *Journal of Nutrition* 40:403-414.

Neumann, A. L., J. L. Krider, M. R. James, and B. C. Johnson. 1949. The choline requirement of the baby pig. *Journal of Nutrition* 38:195-214.

Neumann, A. L., J. B. Thiersch, J. L. Krider, M. F. James, and B. C. Johnson. 1950. Requirement of the baby pig for vitamin B_{12} fed as a concentrate. *Journal of Animal Science* 9:83-89.

Newcomb, M. D., and G. L. Allee. 1986. Water-soluble vitamins for weanling pigs. *Journal of Animal Science* 63(Suppl. 1):108 (Abstr.).

Newport, M. J. 1981. A note on the effect of low levels of biotin in milk substitutes for neonatal pigs. *Animal Production* 33:333-335.

Nielsen, H. E., N. J. Hojgaard-Olsen, W. Hjarde, and E. Leerbeck. 1973. Vitamin E content in colostrum and sow's milk and sow milk yield at two levels of dietary fats. *Acta Agriculturae Scandinavica* 19(Suppl.):35-38.

Nielsen, H. E., V. Danielsen, M. G. Simesen, C. Gissel-Nielsen, W. Hjarde, T. Leth, and A. Basse. 1979. Selenium and vitamin E deficiency in pigs. I. Influence on growth and reproduction. *Acta Veterinaria Scandinavica* 20:276-288.

Nockels, C. F. 1979. Protective effects of supplemental vitamin E against infection. *Federation Proceedings* 38:2134-2138.

Noff, D., and S. Edelstein. 1978. Vitamin D and its hydroxylated metabolites in the rat. *Hormone Research* 9:292-300.

North Central Region-42 Committee on Swine Nutrition. 1976. Effects of supplemental choline on reproductive performance of sows: A cooperative regional study. *Journal of Animal Science* 42:1211-1216.

North Central Region-42 Committee on Swine Nutrition. 1980. Effect of supplemental choline on performance of starting, growing and finishing pigs: A cooperative regional study. *Journal of Animal Science* 50:99-102.

NRC (National Research Council). 1987. *Vitamin Tolerance of Animals*. Washington, DC: National Academy Press.

NRC. 1988. *Nutrient Requirements of Swine*, 9th Rev. Ed. Washington, DC: National Academy Press.

NRC. 1998. *Nutrient Requirements of Swine*, 10th Rev. Ed. Washington, DC: National Academy Press.

Obel, A. L. 1953. Studies on the morphology and etiology of so-called toxic liver dystrophy (hepatosis dietetica) in swine. *Acta Pathologica et Microbiologica Scandinavica* 94(Suppl.):1-87.

Oduho, G., and D. H. Baker. 1993. Quantitative efficacy of niacin sources for chicks: Nicotinic acid, nicotinamide, NAD and tryptophan. *Journal of Nutrition* 123:2201-2206.

Oduho, G. W., T. K. Chung, and D. H. Baker. 1993. Menadione nicotinamide bisulfite is a bioactive source of vitamin K and niacin activity for chicks. *Journal of Nutrition* 123:737-743.

Oduho, G. W., Y. Han, and D. H. Baker. 1994. Iron deficiency reduces the efficacy of tryptophan as a niacin precursor for chicks. *Journal of Nutrition* 124:444-450.

Osweiler, G. D. 1970. Porcine hemorrhagic disease. No. AS3531 in *Proceedings of Pork Producers Day*. Ames: Iowa State University Press.

Palm, B. W., R. J. Meade, and A. L. Melliere. 1968. Pantothenic acid requirement of young swine. *Journal of Animal Science* 27:1596-1601.

Parrish, D. B., C. E. Aubel, J. S. Hughes, and J. D. Wheat. 1951. Relative value of vitamin A and carotene for supplying the vitamin A requirements of swine during gestation and beginning lactation. *Journal of Animal Science* 10:551-559.

Partridge, I. G., and M. S. McDonald. 1990. A note on the response of growing pigs to supplemental biotin. *Animal Production* 50:195-197.

Peng, C. L., and H. Heitman, Jr. 1973. Erythrocyte transketolase activity and the percentage stimulation by thiamin pyrophosphate as criteria of thiamin status in the pig. *British Journal of Nutrition* 30:391-399.

Peng, C. L., and H. Heitman, Jr. 1974. The effect of ambient temperature on the thiamin requirement of growing-finishing pigs. *British Journal of Nutrition* 32:1-9.

Pennock, J. F., F. W. Hemming, and J. D. Kerr. 1964. A reassessment of tocopherol chemistry. *Biochemical and Biophysical Research Communications* 17:542-548.

Penny, R. H. C., R. H. A. Cameron, S. Johnson, P. J. Kenyan, H. A. Smith, A. W. P. Bell, J. P. L. Cole, and J. Taylor. 1981. The influence of biotin supplementation on sow reproductive efficiency. *Veterinary Record* 109:80-81.

Peo, E. R., Jr., G. F. Wehrebein, B. Moser, P. J. Cunningham, and P. E. Vipperman, Jr. 1970. Biotin supplementation of baby pig diets. *Journal of Animal Science* 31:209 (Abstr.).

Peplowski, M. A., D. C. Mahan, F. A. Murray, A. L. Moxon, A. H. Cantor, and K. E. Ekstrom. 1980. Effect of dietary and injectable vitamin E and selenium in weanling swine antigenically challenged with sheep red blood cells. *Journal of Animal Science* 51:344-351.

Pettigrew, J. E., S. M. El-Kandelgy, L. J. Johnston, and G. C. Shurson. 1996. Riboflavin nutrition of sows. *Journal of Animal Science* 74:2226-2230.

Petzold, E. N., F. W. Quackenbush, and M. McQuistan. 1959. Zeacarotenes, new provitamins from corn. *Archives of Biochemistry and Biophysics* 82:117-124.

Piatkowski, T. L., D. C. Mahan, A. H. Cantor, A. L. Moxon, J. H. Cline, and A. P. Grifo, Jr. 1979. Selenium and vitamin E in semipurified diets for gravid and nongravid gilts. *Journal of Animal Science* 48:1357-1365.

Pike, W. J., J. B. Parker, M. R. Haussler, A. Boass, and S. U. Toverud. 1979. Dynamic changes in circulating 1,25-dihydroxyvitamin D during reproduction in rats. *Science* 204:1427-1429.

Pinelli-Saavedraa, A., and J. R. Scaifeb. 2005. Pre- and postnatal transfer of vitamins E and C to piglets in sows supplemented with vitamin E and vitamin C. *Livestock Production Science* 97:231-240.

Pinelli-Saavedra, A., A. M. Calderón de la Barca, J. Hernández, R. Valenzuela, and J. R. Scaife. 2008. Effect of supplementing sows' feed with α-tocopherol acetate and vitamin C on transfer of α-tocopherol to piglet tissues, colostrum, and milk: Aspects of immune status of piglets. *Research in Veterinary Science* 85:92-100.

Piper, R. C., J. A. Froseth, L. R. McDowell, G. H. Kroening, and I. A. Dyer. 1975. Selenium-vitamin E deficiency in swine fed peas (*Pisum sativum*). *American Journal of Veterinary Research* 36:273-281.

Pond, W. G., E. Kwong, and J. K. Loosli. 1960. Effect of level of dietary fat, pantothenic acid, and protein on performance of growing-fattening swine. *Journal of Animal Science* 19:1115-1122.

Poor, C. L., S. D. Miller, G. C. Fahey, R. A. Easter, and J. W. Erdman. 1987. Animal models for carotenoid utilization studies: Evaluation of the chick and the pig. *Nutrition Reports International* 36:229-234.

Powick, W. C., N. R. Ellis, and C. N. Dale. 1947a. Relationship of corn diets to nicotinic acid deficiency in growing pigs. *Journal of Animal Science* 6:395-400.

Powick, W. C., N. R. Ellis, L. L. Madsen, and C. N. Dale. 1947b. Nicotinic acid deficiency and nicotinic acid requirement of young pigs on a purified diet. *Journal of Animal Science* 6:310-324.

Powick, W. C., N. R. Ellis, and C. N. Dale. 1948. Relationship of tryptophane to nicotinic acid in the feeding of growing pigs. *Journal of Animal Science* 7:228-232.

Quarterman, J., A. C. Dalgarno, A. Adams, B. F. Fell, and R. Boyne. 1964. The distribution of vitamin D between the blood and the liver in the pig, and observations on the pathology of vitamin D toxicity. *British Journal of Nutrition* 18:65-77.

Real, D. E., J. L. Nelssen, J. A. Unruh, M. D. Tokach, R. D. Goodband, S. S. Dritz, J. M. DeRouchey, and E. Alonso. 2002. Effects of increasing dietary niacin on growth performance and meat quality in finishing pigs reared in two different environments. *Journal of Animal Science* 80:3203-3210.

Reid, I. M., R. H. Barnes, W. G. Pond, and L. Krook. 1968. Methionine-responsive liver damage in young pigs fed a diet low in protein and vitamin E. *Journal of Nutrition* 95:499-508.

Richardson, D., D. V. Catron, L. A. Underkofler, H. M. Maddock, and W. C. Friedland. 1951. Vitamin B_{12} requirement of male weanling pigs. *Journal of Nutrition* 44:371-381.

Riker, J. T., III, T. W. Perry, R. A. Pickett, and C. J. Heidenreich. 1967. Influence of controlled temperatures on growth rate and plasma ascorbic acid values in swine. *Journal of Nutrition* 92:99-103.

Ringarp, N. 1960. Clinical and experimental investigations into a postparturient syndrome with agalactia in sows. *Acta Agriculturae Scandinavica* 7(Suppl.):1-166.

Ritchie, H. D., E. R. Miller, D. E. Ullrey, J. A. Hoefer, and R. W. Luecke. 1960. Supplementation of the swine gestation diet with pyridoxine. *Journal of Nutrition* 70:491-496.

Roth-Maier, D. A., and M. Kirchgessner. 1977. Utersuchungen zum optimalen Pantothensaurebedarf von Mastschweinen. (Studies on the optimal pantothenic acid requirement of market pigs.) *Zeitschrift fur Tierphysiologie Tierernahrung und Futtermittelkunde—Journal of Animal Physiology and Animal Nutrition* 38:121-131.

Russell, L. E., P. J. Bechtel, and R. A. Easter. 1985a. Effect of deficient and excess dietary vitamin B_6 on amino transaminase and glycogen phosphorylase activity and pyridoxal phosphate content in two muscles from postpubertal gilts. *Journal of Nutrition* 115:1124-1135.

Russell, L. E., R. A. Easter, and P. J. Bechtel. 1985b. Evaluation of the erythrocyte aspartate aminotransferase activity coefficient as an indicator of the vitamin B_6 status of postpubertal gilts. *Journal of Nutrition* 115:1117-1123.

Russett, J. C., J. L. Krider, T. R. Cline, and L. B. Underwood. 1979a. Choline requirement of young swine. *Journal of Animal Science* 48:1366-1373.

Russett, J. C., J. L. Krider, T. R. Cline, H. L. Thacker, and L. B. Underwood. 1979b. Choline-methionine interactions in young swine. *Journal of Animal Science* 49:708-714.

Sandholm, M., T. Honkanen-Buzalski, and V. Rasi. 1979. Prevention of navel bleeding in piglets by preparturient administration of ascorbic acid. *Veterinary Record* 104:337-338.

Schaafsma, A. W., V. Limay-Rios, D. E. Paul, and J. D. Miller. 2009. Mycotoxins in fuel ethanol co-products derived from maize: A mass balance for deoxynivalenol. *Journal of the Science of Food and Agriculture* 89:1574-1580.

Schendel, H. E., and B. C. Johnson. 1962. Vitamin K deficiency in the baby pig. *Journal of Nutrition* 76:124-130.

Schnoes, H. K., and H. F. Deluca. 1980. Recent progress in vitamin D metabolism and the chemistry of vitamin D metabolites. *Federation Proceedings* 39:2723-2729.

Seerley, R. W., R. J. Emerick, L. B. Embry, and O. E. Olsen. 1965. Effect of nitrate or nitrite administered continuously in drinking water for swine and sheep. *Journal of Animal Science* 24:1014-1019.

Seerley, R. W., O. W. Charles, H. C. McCampbell, and S. P. Bertch. 1976. Efficacy of menadione dimethylpyrimidinol bisulfite as a source of vitamin K in swine diets. *Journal of Animal Science* 42:599-607.

Seerley, R. W., R. A. Snyder, and H. C. McCampbell. 1981. The influence of sow dietary lipids and choline on piglet survival, milk and carcass composition. *Journal of Animal Science* 52:542-550.

Selke, M. R., C. E. Barnhart, and C. H. Chaney. 1967. Vitamin A requirement of the gestating and lactating sow. *Journal of Animal Science* 26:759-763.

Sewell, R. F., D. G. Price, and M. C. Thomas. 1962. Pantothenic acid requirement of the pig as influenced by dietary fat. *Federation Proceedings* 21:468.

Sewell, R. F., D. Nugara, R. L. Hill, and W. A. Knapp. 1964. Vitamin B-6 requirement of early-weaned pigs. *Journal of Animal Science* 23:694-699.

Seymour, E. W., V. C. Speer, and V. W. Hays. 1968. Effect of environmental temperature on the riboflavin requirement of young pigs. *Journal of Animal Science* 27:389-393.

Sharp, B. A., L. G. Young, and A. A. van Dreumel. 1972a. Dietary induction of mulberry heart disease and hepatosis dietetica in pigs. I. Nutritional aspects. *Canadian Journal of Comparative Medicine* 36:371-376.

Sharp, B. A., L. G. Young, and A. A. van Dreumel. 1972b. Effect of supplemental vitamin E and selenium in high-moisture corn diets on the incidence of mulberry heart disease and hepatosis dietetica in pigs. *Canadian Journal of Comparative Medicine* 36:393-397.

Sheffy, B. E., N. Drouliscos, J. K. Loosli, and J. P. Willman. 1954. Vitamin A requirements of baby pigs. *Journal of Animal Science* 13:999 (Abstr.).

Shurson, G. C., T. M. Salzer, D. D. Koehler, and M. H. Whitney. 2011. Effect of metal specific amino acid complexes and inorganic trace minerals on vitamin stability in premixes. *Animal Feed Science and Technology* 163(2-4):200-206.

Simard, F., F. Guay, C. L. Girard, A. Giguere, J. P. Laforest, J. J. Matte. 2007. Effects of concentrations of cyanocobalamin in the gestation diet on some criteria of vitamin B-12 metabolism in first-parity sows. *Journal of Animal Science* 85:3294-3302.

Simesen, M. G., P. T. Jensen, A. Basse, G. Gissel-Nielsen, T. Leth, V. Danielson, and H. E. Nielsen. 1982. Clinico-pathologic findings in young pigs fed different levels of selenium, vitamin E and antioxidant. *Acta Veterinaria Scandinavica* 23:295-308.

Simmins, P. H., and P. H. Brooks. 1980. The effect of dietary biotin level on the physical characteristics of pig hoof tissue. *Animal Production* 30:469 (Abstr.).

Simmins, P. H., and P. H. Brooks. 1983. Supplementary biotin for sows: Effect on reproductive characteristics. *Veterinary Record* 112:425-429.

Simmins, P. H., and P. H. Brooks. 1985. Effect of different levels of dietary biotin intake on the hoof horn hardness of the gilt. *Animal Production* 40:544-545 (Abstr.).

Southern, L. L., and D. H. Baker. 1981. Bioavailable pantothenic acid in cereal grains and soybean meal. *Journal of Animal Science* 53:403-408.

Southern, L. L., D. R. Brown, D. D. Werner, and M. C. Fox. 1986. Excess supplemental choline for swine. *Journal of Animal Science* 62:992-996.

Stahly, T. S., N. H. Williams, T. R. Lutz, R. C. Ewan, and S. G. Swenson. 2007. Dietary B vitamin needs of strains of pigs with high and moderate lean growth. *Journal of Animal Science* 85:188-195.

Stern, M. H., C. D. Robeson, L. Weisler, and J. G. Blaxter. 1947. δ-Tocopherol. I. Isolation from soybean oil and properties. *Journal of the American Chemical Society* 69:869-874.

Stockland, W. L., and L. G. Blaylock. 1974. Choline requirement of pregnant sows and gilts under restricted feeding conditions. *Journal of Animal Science* 39:1113-1116.

Stothers, S. C., D. A. Schmidt, R. L. Johnston, J. A. Hoefer, and R. W. Luecke. 1955. The pantothenic acid requirement of the baby pig. *Journal of Nutrition* 57:47-54.

Strittmatter, J. E., D. J. Ellis, M. G. Hogberg, A. L. Trapp, M. J. Parsons, and E. R. Miller. 1978. Effects of vitamin C on swine growth and osteochondrosis. *Journal of Animal Science* 47(Suppl. 1):16 (Abstr.).

Suttie, J. W. 1980. The metabolic role of vitamin K. *Federation Proceedings* 39:2730-2735.

Suttie, J. W., and C. M. Jackson. 1977. Prothrombin structure, activation and biosynthesis. *Physiological Reviews* 57:1-70.

Sweeney, P. R., and R. G. Brown. 1972. Ultrastructure changes in muscular dystrophy. 1. Cardiac tissue of piglets deprived of vitamin E and selenium. *American Journal of Pathology* 68:479-485.

Tanphaichitr, V., and B. Wood. 1984. *Thiamin. Present Knowledge in Nutrition*. Washington, DC: The Nutrition Foundation.

Teague, H. S., and A. P. Grifo, Jr. 1966. *Vitamin B_{12} in Sow Rations*. Swine Research. Research Summary 13. Ohio Agricultural Research Development Center. Wooster: Ohio State University Press.

Teague, H. S., W. M. Palmer, and A. P. Grifo, Jr. 1970. *Pantothenic Acid Deficiency in the Reproducing Sow*. Ohio Agricultural Research Development Center Animal Science Mimeograph 200. Wooster: Ohio State University Press.

Terrill, S. W., C. B. Ammerman, D. E. Walker, R. M. Edwards, H. W. Norton, and D. E. Becker. 1955. Riboflavin studies with pigs. *Journal of Animal Science* 14:593-603.

Tiege, J., Jr. 1977. The generalized Shwartzman reaction induced by a single injection of endotoxin in pigs fed a vitamin E-deficient commercial diet. *Acta Veterinaria Scandinavica* 18:140-142.

Tiege, J., Jr., and H. J. Nafstad. 1978. Ultrastructure of colonic epithelial cells in vitamin E- and selenium-deficient pigs. *Acta Veterinaria Scandinavica* 19:549-560.

Tiege, J., Jr., K. Nordstoga, and J. Aurjo. 1977. Influence of diet on experimental swine dysentery. I. Effects of vitamin E- and selenium-deficient diet supplemented with 6.8% cod liver oil. *Acta Veterinaria Scandinavica* 18:384-396.

Tiege, J., Jr., F. Saxegaard, and A. Froslie. 1978. Influence of diet on experimental swine dysentery. 2. Effects of a vitamin E and selenium-deficient diet supplemented with 3% cod liver oil, vitamin E or selenium. *Acta Veterinaria Scandinavica* 19:133-146.

Tollerz, G. 1973. Vitamin E, selenium and some related compounds and tolerance toward iron in piglets. *Acta Agriculturae Scandinavica* 19(Suppl.):184-187.

Trapp, A. L., K. K. Keahey, D. L. Whitenack, and C. K. Whitehair. 1970. Vitamin E-selenium deficiency in swine: Differential diagnosis and nature of field problem. *Journal of the American Medical Association* 157:289-300.

Tremblay, G. F., J. J. Matte, L. Lemieux, and G. J. Brisson. 1986. Serum folates in gestating swine after folic acid addition to diet. *Journal of Animal Science* 63:1173-1178.

Tribble, L. R., J. D. Hancock, and D. E. Orr, Jr. 1984. Value of supplemental biotin on reproductive performance of sows in confinement. *Journal of Animal Science* 59(Suppl. 1):245 (Abstr.).

Ullrey, D. E. 1972. Biological availability of fat-soluble vitamins: Vitamin A and carotene. *Journal of Animal Science* 35:648-657.

Ullrey, D. E. 1974. The selenium deficiency problem in animal agriculture. Pp. 275-293 in *Trace Element Metabolism in Animals—2*, W. G. Hoekstra, J. W. Suttie, H. E. Ganther, and W. Mertz, eds. Baltimore, MD: University Park Press.

Ullrey, D. E. 1981. Vitamin E for swine. *Journal of Animal Science* 53:1039-1056.

Ullrey, D. E., D. E. Becker, S. W. Terrill, and R. A. Notzold. 1955. Dietary levels of pantothenic acid and reproductive performance of female swine. *Journal of Nutrition* 57:401-414.

Ullrey, D. E., E. R. Miller, R. D. Struthers, R. E. Peterson, J. A. Hoefer, and H. M. Hall. 1965. Vitamin A activity of fermentation β-carotene for swine. *Journal of Nutrition* 85:375-385.

Ullrey, D. E., E. R. Miller, D. J. Ellis, D. E. Orr, J. P. Hitchcock, K. K. Keahey, and A. L. Trapp. 1971. Vitamin E (selenium and choline), reproduction and MMA. Pp. 48-51 in *Report of Swine Research 148*, Michigan State University Agricultural Experiment Station. East Lansing: Michigan State University Press.

Van Etten, C., N. R. Ellis, and L. L. Madsen. 1940. Studies on the thiamine requirement of young swine. *Journal of Nutrition* 20:607-624.

Viganò, P., S. Mangioni, F. Pompei, and I. Chiodo. 2003. Maternal-conceptus cross talk—a review. *Placenta* 24:556-561.

Wahlstrom, R. C., and D. E. Stolte. 1958. The effect of supplemental vitamin D in rations for pigs fed in the absence of direct sunlight. *Journal of Animal Science* 17:699-705.

Wald, G. 1968. Molecular basis of visual excitement. *Science* 162:230-239.

Washam, R. D., J. E. Sowers, and L. W. DeGoey. 1975. Effect of zinc-proteinate or biotin in swine starter rations. *Journal of Animal Science* 40:179 (Abstr.).

Wastell, M. E., D. C. Ewan, M. W. Vorhies, and V. C. Speer. 1972. Vitamin E and selenium for growing and finishing pigs. *Journal of Animal Science* 34:969-973.

Watkins, K. L., L. L. Southern, and J. E. Miller. 1991. Effect of dietary biotin supplementation on sow reproductive performance and soundness and pig growth and mortality. *Journal of Animal Science* 69:201-206.

Webb, N. G., R. H. C. Penny, and A. M. Johnston. 1984. The effect of a dietary supplement of biotin on pig hoof horn strength and hardness. *Veterinary Record* 114:185-189.

Weisman, Y., R. Sapir, A. Harell, and S. Edelstein. 1976. Maternal perinatal interrelationships of vitamin D metabolism in rats. *Biochimica et Biophysica Acta* 428:388-395.

Wellenreiter, R. H., D. E. Ullrey, E. R. Miller, and W. T. Magee. 1969. Vitamin A activity of corn carotenes for swine. *Journal of Nutrition* 99:129-136.

Whitehair, C. K., E. R. Miller, M. Loudenslager, and M. G. Hogberg. 1984. MMA in sows—A vitamin E-selenium deficiency. *Journal of Animal Science* 59(Suppl. 1):106 (Abstr.).

Whitehead, C. C., D. W. Bannister, and J. P. F. D'Mello. 1980. Blood pyruvate carboxylase activity as a criterion of biotin status in young pigs. *Research in Veterinary Science* 29:126-128.

Whittle, K. J., P. J. Dunphy, and J. F. Pennock. 1966. The isolation and properties of δ-tocotrienol from Heuca latex. *Biochemical Journal* 100:138-145.

Wiese, A. C., W. P. Lehrer, Jr., P. R. Moore, O. F. Pahnish, and W. V. Hartwell. 1951. Pantothenic acid deficiency in baby pigs. *Journal of Animal Science* 10:80-87.

Wilburn, E. E., D. C. Mahan, D. A. Hill, T. E. Shipp, and H. Yang. 2008. An evaluation of natural (*RRR*-α-tocopheryl acetate) and synthetic (all-racα-tocopheryl acetate) vitamin E fortification in the diet or drinking water of weanling pigs. *Journal of Animal Science* 86:584-591.

Wilkinson, J. E., M. C. Bell, J. A. Bacon, and F. B. Masincupp. 1977a. Effects of supplemental selenium on swine. I. Gestation and lactation. *Journal of Animal Science* 44:224-228.

Wilkinson, J. E., M. C. Bell, J. A. Bacon, and C. C. Melton. 1977b. Effects of supplemental selenium on swine. II. Growing-finishing. *Journal of Animal Science* 44:229-233.

Wintrobe, M. M., M. H. Miller, R. H. Follis, Jr., H. J. Stein, C. Mushatt, and S. Humphreys. 1942. Sensory neuron degeneration in pigs. IV. Protection afforded by calcium pantothenate and pyridoxine. *Journal of Nutrition* 24:345-366.

Wintrobe, M. M., R. Alcayaga, S. Humphreys, and R. H. Follis, Jr. 1943a. Electrocardiographic changes associated with thiamine deficiency in pigs. *Bulletin of the Johns Hopkins Hospital* 73:169.

Wintrobe, M. M., R. H. Follis, Jr., R. Alcayaga, M. Paulson, and S. Humphreys. 1943b. Pantothenic acid deficiency in swine with particular reference to the effects on growth and on the alimentary tract. *Bulletin of the Johns Hopkins Hospital* 73:313.

Wintrobe, M. M., R. H. Follis, Jr., M. H. Miller, H. J. Stein, R. Alcayaga, S. Humphreys, A. Suksta, and G. E. Cartwright. 1943c. Pyridoxine deficiency in swine with particular reference to anemia, epileptiform convulsions and fatty liver. *Bulletin of the Johns Hopkins Hospital* 72:1-25.

Wintrobe, M. M., W. Buschke, R. H. Follis, Jr., and S. Humphreys. 1944. Riboflavin deficiency in swine with special reference to the occurrence of cataracts. *Bulletin of the Johns Hopkins Hospital* 75:102-110.

Wintrobe, M. M., H. J. Stein, R. H. Follis, Jr., and S. Humphreys. 1946. Nicotinic acid and the level of protein intake in the nutrition of the pig. *Journal of Nutrition* 30:395-412.

Wolke, R. E., S. W. Nielsen, and J. E. Rousseau. 1968. Bone lesions of hypervitaminosis A in the pig. *American Journal of Veterinary Research* 29:1009-1024.

Wood, R. D., C. H. Chaney, D. G. Waddill, and G. W. Garrison. 1967. Effect of adding nitrate or nitrite to drinking water on the utilization of carotene by growing swine. *Journal of Animal Science* 26:510-513.

Woodworth, J. C., R. D. Goodband, J. L. Nelssen, M. D. Tokach, and R. E. Musser. 2000. Added dietary pyridoxine, but not thiamin, improves weanling pig growth performance. *Journal of Animal Science* 78:88-93.

Wuryastuti, H., H. D. Stowe, R. W. Bull, and E. R. Miller. 1993. Effects of vitamin E and selenium on immune responses of peripheral blood, colostrum, and milk leukocytes of sows. *Journal of Animal Science* 71:2464-2472.

Yang, H., D. C. Mahan, D. A. Hill, T. E. Shipp, T. R. Radke, and M. J. Cecava. 2009. Effect of vitamin E source, natural versus synthetic, and quantity on serum and tissue α-tocopherol concentrations in finishing swine. *Journal of Animal Science* 87:4057-4063.

Yen, J. T., and W. G. Pond. 1981. Effect of dietary vitamin C addition on performance, plasma vitamin C and hematic iron status in weanling pigs. *Journal of Animal Science* 53:1292-1296.

Yen, J. T., and W. G. Pond. 1983. Response of swine to periparturient vitamin C supplementation. *Journal of Animal Science* 56:621-624.

Yen, J. T., and W. G. Pond. 1984. Responses of weanling pigs to dietary supplementation with vitamin C or carbadox. *Journal of Animal Science* 58:132-137.

Yen, J. T., A. H. Jensen, and D. H. Baker. 1976. Assessment of the concentration of biologically available vitamin B_6, in corn and soybean meal. *Journal of Animal Science* 42:866-870.

Yen, J. T., A. H. Jensen, and D. H. Baker. 1977. Assessment of the availability of niacin in corn, soybeans and soybean meal. *Journal of Animal Science* 45:269-278.

Yen, J. T., R. Lauxen, and T. L. Veum. 1978. Effect of supplemental niacin on finishing pigs fed soybean meal supplemented diets. *Journal of Animal Science* 47(Suppl. 1):325 (Abstr.).

Yen, J. T., P. K. Ku, W. G. Pond, and E. R. Miller. 1985. Response to dietary supplementation of vitamins C and E in weanling pigs fed low vitamin E-selenium diets. *Nutrition Reports International* 31:877-885.

Young, L. G., A. Lun, J. Pos, R. P. Forshaw, and D. Edmeades. 1975. Vitamin E stability in corn and mixed feed. *Journal of Animal Science* 40:495-499.

Young, L. G., R. B. Miller, D. E. Edmeades, A. Lun, G. C. Smith, and G. J. King. 1977. Selenium and vitamin E supplementation of high-moisture corn diets for swine reproduction. *Journal of Animal Science* 45:1051-1060.

Young, L. G., R. B. Miller, D. E. Edmeades, A. Lun, G. C. Smith, and G. J. King. 1978. Influence of method of corn storage and vitamin E and selenium supplementation on pig survival and reproduction. *Journal of Animal Science* 47:639-647.

Zempleni, J., R. B. Rucker, J. W. Suttie, and D. B. McCormick, eds. 2007. *Handbook of Vitamins*, 4th Ed. Boca Raton, FL: CRC Press.

Zhang, Z., E. Kebreab, M. Jing, J. C. Rodriguez-Lecompte, R. Kuehn, M. Flintoft, and J. D. House. 2009. Impairments in pyridoxine-dependent sulphur amino acid metabolism are highly sensitive to the degree of vitamin B_6 deficiency and repletion in the pig. *Animal* 3:826-837.

第8章 猪营养需要量估测模型

引言

研究表明，不同猪群之间的营养需要量不同，并受动物生理状态、生产潜力、环境条件等的影响（NRC，1998）。为了估测 20~140 kg 活重（BW）的生长育肥猪、妊娠母猪及泌乳母猪的回肠标准可消化（SID）氨基酸、氮（N）、全消化道标准可消化（STTD）磷（P）及总钙（Ca）的需要量，新版已对上一版 NRC（1998）中的这三个数学模型进行了更新和调整。模型的构建要在易用性、透明性和简易性与预测的精确性以及与生产实际相关性之间取得平衡。回肠表观可消化（AID）氨基酸和全消化道表观可消化（ATTD）磷需要量的估测分别源于回肠标准可消化（SID）氨基酸和全消化道标准可消化（STTD）磷需要量。新版还建立了针对玉米-豆粕型基础日粮中，总日粮氨基酸和磷需要量的估测。体重低于 20 kg 猪的营养需要、猪对维生素以及除了磷和钙以外的矿物质的需要量，也通过实证评估并用于预测模型中，以保证模型的完整性。以一个简单的饲料配方程序作为补充，这些模型可以直接将计算日粮养分浓度与模型生成的需要量估测值进行比较。

生长育肥猪、妊娠母猪及泌乳母猪三个模型在从整体水平描述动物营养代谢原理和能量利用率等方面具有规律性、动态性和确定性。这些模型的规律性在于可以精确地用数学方法表示影响动物营养需要量的基本生物学原理。这些基本原理已经在第 1 章（能量）、第 2 章（蛋白质和氨基酸）和第 6 章（矿物质）中进行了概述。但是，这些模型中也必然包含实证经验因素，使模型生成的营养需要量的估测值与实验观测值相一致。动物的阶段累积生产性能（生长、妊娠和泌乳）是通过对一个动态的时间段，以 1 天为间隔进行叠加计算得来，阶段的长短是由使用者决定的。通过运行动态模拟，使用者可以研究单天或连续几天的营养需要量。连续几天的营养需要量可以通过单天营养需要量的平均值简单地计算得到。这些模型还可以在没有明确描述动物个体之间差异的条件下，对动物的营养需要量进行估测。在这些模型中，动物个体间的差异则通过调整养分吸收后的利用效率的估测值来考虑，这些估测值在单个动物中已经建立（Pomar et al., 2003），正如第 2 章（蛋白质和氨基酸）中所概述。

为了估测不同阶段猪只的营养需要，模型的使用者需要设定能量摄入量和动物的生产性能水平。对于生长育肥猪和泌乳母猪，增加一些程序来对能量摄入水平进行简单的预测。基于这些输入模型中的数据，模型可以生成动物每日整体蛋白质沉积（Pd）、整体脂肪沉积（Ld）和体重变化的估测值。对于妊娠母猪，则同时考虑了孕体和繁殖组织的蛋白质、脂肪沉积和总体增重；而对于泌乳母猪，则通过评估窝产仔数和仔猪平均日生长速率来评价乳中养分和能量的输出量，从而生成所观察动物生产性能的营养需要。因为动物对能量摄入量的应答是估测得来的，所以，这些模型不能被直接用于估测能量需要量。动物对未达最佳标准水平（不论是绝对的还是相对的）的营养摄入量的应答没有在这些模型中体现出来。因此，模型没有估测在经过一段时间限饲后，可能会受到潜在补偿生长影响的动物的营养需要量。

模型生成的营养需要量来自相对来说没有疾病和应激挑战的动物表现出的生物学性能，而且并没有考虑经济效益。除了热环境对估测能量摄入量和维持能量需要量的影响以外，模型没有考虑疾病挑战或环境条件对营养需要量的潜在影响。产生最大经济效益或最佳养分利用效率的日粮养分摄入量，可能与模型生成的营养需要量估测值不同。

在这些模型中，能量计量单位为"有效的"代谢能（ME）。在本文和所有的公式中用代谢能（ME）

来表示有效代谢能，"有效的"消化能（DE）能够基于典型的玉米-豆粕型日粮净能（NE）按照固定转化系数计算得到；这些典型的日粮代表那些用于生成部分能量效率的估测值时所使用的日粮。这个概念在第 1 章（能量）中已经详细阐述。

在这三个模型中，有一个选项用来输入观测到的体组成（如背膘厚）和体重（如生长育肥猪的生长性能、妊娠期的总体重变化，或者是泌乳期间母猪的体重变化）的改变值，进而将模型估测值和观测值进行对比或匹配。当观测值和模型估测值接近时，使用者能够对模型生成的营养需要的估测值更有信心。更多关于怎样将观测到的体组成和体重的改变值与模型预测值匹配的细节在模型使用指南中（与模型同时提供/分发）列出。

在本章中，还给出了生成这些营养需要量的数学方法。这些公式中的一部分已经在第 1 章、第 2 章和第 6 章中出现过，但在这里给出以保证信息的全面。更多的关于全部模型输入输出的详细描述、主屏打印出来的资料以及简单的教程在使用者指南中列出（附录 A）。

生长育肥猪模型

主要概念

生长通常用猪机体的蛋白质和脂肪的每日沉积速率来表示，蛋白质和脂肪日沉积速率的变化则可引起体蛋白质（BP）和机体脂肪（BL）的总量变化。本模型中采用蛋白质沉积量（Pd）来描述猪的类型（基因型和性别）及生产性能水平，并且蛋白质沉积被认为是比瘦肉组织生长更为客观和常用的衡量标准。空腹体重（EBW）和体重（BW）可以由体蛋白质量和机体脂肪量预测得到。能量摄入量需要在机体维持功能需要、蛋白质沉积和脂肪沉积之间进行分配。由于维持能量需要量由饲喂含蛋白质日粮的动物确定，因此，蛋白质能量被认为是能量摄入量的一部分，用于蛋白质维持的蛋白质并没有从维持能量需要量中减掉。维持能量需要量可以由体重和环境温度来预测，并且可以由模型使用者对其进行校正从而获得特定情况下的需要量。猪的生产性能或生长潜力可根据蛋白质沉积曲线来描述，蛋白质沉积曲线可以由模型使用者根据能量摄入量进行定义，也可以由观测到的生产性能进行估测。未用作机体维持功能和蛋白质沉积的摄入能量都将用作脂肪沉积。回肠标准可消化（SID）氨基酸和氮需要量由蛋白质沉积、体重和采食量进行估算。全消化道标准可消化（STTD）磷需要量由采食量、蛋白质沉积和体重得到；而总钙需要量则可通过 STTD 磷推算得到。回肠表观可消化（AID）和总氨基酸的需要量，以及全消化道表观可消化（ATTD）磷和总磷的需要量，由含 3%预混料和 0.1%赖氨酸盐酸盐且配制满足 SID 氨基酸和 STTD 磷需要量的玉米-豆粕型基础日粮的 SID 氨基酸和 STTD 磷值计算而来。

饲喂莱克多巴胺（RAC）和对未去势公猪进行针对促性腺激素释放激素（GnRH）的免疫对营养物质需要量的影响，通过它们对代谢能摄入量、维持代谢能需要量、蛋白质沉积以及随后的脂肪沉积的影响而估测得到。分别跟踪 RAC 诱导的蛋白质沉积，用来表达它对蛋白质沉积的氨基酸组成和体组成的影响。

这个动力学模型包括能够描述能量摄入量、蛋白质沉积和体增重随体重增加而变化的数学方程式。两个可供选择的方程式可以用来表达其中的任一关系。其中的多项式方程使用容易，且能通过运用诸如 Microsoft Excel 之类的电子表格比较容易地使其参数化。另一个则是渐近线或 S 型函数方程式，在表达生物学关系时更具有代表性，但这需要更为先进的统计软件包以完成参数化。针对小母猪、阉公猪、未去势公猪的典型能量摄入量和蛋白质沉积曲线作为默认值包含在模型中。

体组成

近来的一篇综述文章（de Lange et al., 2003）用数学方法较为准确地阐述了机体的化学成分和物理成分。4 种化学体成分——体脂肪量（BL）、体蛋白量（BP）、机体总水分量（Wat）和机体总灰分量（Ash）的总和组成了空腹体重（EBW）（式 8-1）。其中，机体总水分量和总灰分量与体蛋白质量直接相关，且都用 kg 表示（式 8-2 和式 8-3）。在机体总水分量与体蛋白质量的关系中，也考虑到了猪蛋白质沉积的可操作上限（Pd_{Max}；蛋白质沉积曲线的最高值；g/d）。肠道内容物重量可以由体重（初始体重，kg；式 8-4）或者空腹体重（后来的体重，kg；式 8-5）预测得来。肠道内容物重量和空腹体重组成了体重（BW）。由于机体总水分量和体蛋白量之间的异速增长关系，不管是体增重还是瘦肉组织增重的化学成分，都随着猪的生长阶段和类型的不同而不同（Emmans and Kyriazakis, 1995）。

空腹体重（EBW, kg）=体蛋白量+体脂肪量+机体总水分量+机体总灰分量 （式 8-1）

机体总水分（kg）=（4.322+0.0044 × Pd_{Max}）× $P^{0.855}$ （式 8-2）

机体总灰分（kg）= 0.189 × 体蛋白量 （式 8-3）

肠道内容物（kg）= 0.277 × $BW^{0.612}$ （式 8-4）

肠道内容物（kg）= 0.3043 × $EBW^{0.5977}$ （式 8-5）

一个迭代步骤（Newton-Raphson 法；Arfken, 1995）可根据体重来估测初始体重时的化学体成分，且是在已估算出体脂肪量（BL）/体蛋白量（BP）比值（式 8-6）的基础上实现的。

BL/BP（初始体重）=（0.305 − 0.000 875 × Pd_{Max}）× $BW^{0.45}$ （式 8-6）

在估测胴体瘦肉含量的时候使用了一种标准的背膘厚度的测定方法。世界上很多地区都会通过探头对背膘厚度进行例行检测，且这种方法在北美使用也越来越普遍（Fortin et al., 2004; Schinckel et al., 2010b）。背膘厚的典型测定方法是使用光学探头在已屠宰的热胴体倒数第三、第四根肋骨距背中线 7cm 处测得。化学体成分和探头背膘厚之间的关系（式 8-7）则在对大量数据做进一步分析的基础上得到（Wagner et al., 1999; Schinckel et al., 2001, 2010b），并且用 Quiniou（1995）的数据进行测试（初始分析是由新西兰梅西大学的 P. Morel 完成的）。考虑到在测定背膘厚时可能出现的错误，以及这些错误对预测胴体瘦肉含量的影响，这个参数必须小心解读（Johnson et al., 2004; Schinckel et al., 2006）。探头背膘厚与胴体瘦肉含量之间的关系随着胴体瘦肉含量的估算定义和方法的不同而不同，并且受猪的基因型和性别影响。美国全国猪肉生产商理事会（NPPC; National Pork Board, 2000）认为，此模型中默认的方程式（式 8-8）为胴体无脂瘦肉组织含量提供了一个合理的预测，但是在特定条件下可能需要对它进行校正。基于这个方程式，胴体无脂瘦肉增重可以预测为蛋白质沉积×2.55（NRC, 1998）。然而，这种关系只在一个较宽的体重范围内（如 25~125 kg 体重）适用，对 Pd_{Max} 值较高的猪，会低估无脂瘦肉组织增重。模型使用者应该根据实际情况对式 8-8 中的参数以及无脂瘦肉增重/Pd 的比值进行适当校正。

探头背膘厚（mm）= −5+12.3 × BL/BP+0.13 × BP （式 8-7）

NPPC 胴体无脂瘦肉含量（%）=62.073+0.0308 × 胴体重−1.0101 × 探头背膘厚+0.007 74 ×（探头背膘厚）^{2}

（式 8-8）

能量摄入量与采食量

针对不同的体重，生长育肥猪模型包括了三种选择以完成对代谢能摄入量的估测。第一，作为体重（kg）的一个函数，可以生成一个简单的代谢能摄入量估测值，这个公式考虑了：性别；机体的采食量能力；环境温度（可选）；猪群的密度（可选）。第二，在特定的体重范围内通过观测的采食量生成代谢能摄入量曲线，后者可以与代谢能摄入量参考曲线结合使用。第三，模型使用者可以输入两种类型方程式的参数，从而将代谢能摄入量和体重联系起来。

在使用者自己设定的日粮代谢能含量的基础上，将代谢能摄入量与采食量联系起来。需要估测饲料浪费量，由模型使用者以饲料采食量/（饲料采食量+饲料浪费量）表示，将估测饲料采食量与估测饲料消耗量，或观测饲料消耗量与饲料采食量和代谢能摄入量联系起来。通常情况下，饲料浪费量占饲料供应量的5%，变异范围可在3%到超过10%之间。对饲料浪费量的输入值进行调整可以说明它对营养需要量的影响以及降低饲料浪费量的重要性。

代谢能摄入量参考曲线（式8-9）作为一个基准，可用于将特定体重下的观测代谢能摄入量推导成其他体重下的代谢能摄入量。这个代谢能摄入量参考曲线相当于 NRC（1987，1998 年版 NRC 同样采用）建议值的 83.6%。它是基于 Bridges 函数建立的（Schinckel et al., 2009b），相当于小母猪（式8-10）和阉公猪（式8-11）摄入量的平均值，且已通过校正来代表实际生产条件下猪的典型日粮采食量水平。然而需要特别注意的是，这个参考曲线并没有包括饲料浪费量。未阉公猪的能量摄入量被认为比小母猪低3%（式8-12）。

$$\text{参考代谢能摄入量（kcal/d）} = 10\,563 \times \{1 - \exp[-\exp(-4.04) \times BW]\} \quad \text{（式8-9）}$$

针对这三种性别类型，需要分别使用不同的默认代谢能摄入量曲线（图 8-1）：

$$\text{默认代谢能摄入量，小母猪（kcal/d）} = 10\,967 \times \{1 - \exp[-\exp(-3.803) \times BW^{0.9072}]\} \quad \text{（式8-10）}$$

$$\text{默认代谢能摄入量，阉公猪（kcal/d）} = 10\,447 \times \{1 - \exp[-\exp(-4.283) \times BW^{1.0843}]\} \quad \text{（式8-11）}$$

$$\text{默认代谢能摄入量，未阉公猪（kcal/d）} = 10\,638 \times \{1 - \exp[-\exp(-3.803) \times BW^{0.9072}]\} \quad \text{（式8-12）}$$

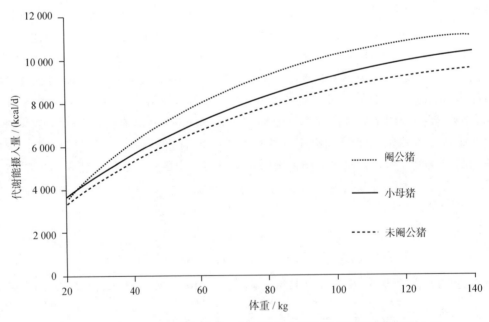

图 8-1　体重 20~140 kg 的阉公猪、小母猪及未阉公猪的典型代谢能摄入量

为了表示有效环境温度（T）对代谢能摄入量的影响（Bruce and Clark, 1979；Quiniou et al., 2000；Noblet et al., 2001），模型估算出了低临界温度（LCT）（式8-13）。假定在 LCT 和 LCT+3℃之间，有效

环境温度不会影响代谢能摄入量。当有效环境温度高于LCT+3℃时，代谢能摄入量随着有效环境温度的升高而降低（从Quiniou et al., 2000校正得来，式8-14）。当有效环境温度低于LCT时，代谢能摄入量随着有效环境温度的降低而线性增加。模型定义了体重为25 kg和90 kg的猪在有效环境温度低于LCT时，代谢能摄入量和有效环境温度之间的线性关系，且将体重对有效环境温度和预测代谢能摄入量之间关系的影响进行了线性调整。对体重为25 kg的猪而言，环境温度每低于LCT 1℃，估测代谢能摄入量将增加1.5%。对体重为90 kg的猪而言，环境温度每低于LCT 1℃，估测代谢能摄入量将增加3%。

$$\text{低临界温度（LCT, ℃）} = 17.9 - 0.0375 \times BW \quad \text{（式8-13）}$$

$$\text{代谢能摄入量百分数} = 1 - 0.012\,914 \times [T-(LCT+3)] - 0.001\,179 \times [T-(LCT+3)]^2 \quad \text{（式8-14）}$$

为了预测猪群密度对估测代谢能摄入量的影响，可以根据体重计算出代谢能摄入量最大时的最小地板空间要求（式8-15），当地板面积每降低1%时，预计代谢能摄入量将降低0.252%（Gonyou et al., 2006）。

$$\text{代谢能摄入量最大时所需的最小地板面积（m}^2\text{/头）} = 0.0336 \times BW^{0.667} \quad \text{（式8-15）}$$

特别需要注意的是，幼龄的生长猪摄入饲料的物理能力有限。当饲料的物理性摄入能力成为限制时，日粮能量或者养分浓度的降低并不会像式8-9和式8-12所显示的那样引起日采食量的增加，而是会直接导致养分日摄入量的降低。这个观点可由作为体重函数的最大日采食量的限制来阐明（Black, 2009；式8-16）。这个方程式同样表明，当有效环境温度低于LCT时，物理性采食能力是增加的。

$$\text{最大日采食量（g/d）} = 111 \times BW^{0.803} + 111 \times BW^{0.803} \times (LCT-T) \times 0.025 \quad \text{（式8-16）}$$

需要强调的是，这种预测代谢能摄入量的方法很大程度上是根据实证经验得来的，并不能反映环境和动物这两个被认为对能量摄入量有影响因素的作用。环境和动物因素包括：地板类型，空气质量和空气流通情况，猪的品种，日粮的养分，抗营养因子的水平（如Torrallardona and Roura, 2009）。这里介绍的方法的应用，仅仅是用来阐明某些环境因素与估测养分需要量之间的潜在相互关系，并且让使用者能够定量地考查这些因素对估测的养分需要量的影响。

当模型使用者设定了实际的饲料消耗水平（包括饲料浪费量）和相应的平均体重时，考虑日粮代谢能含量和饲料浪费量，就能够计算出观测的代谢能摄入水平。在该体重下的观测代谢能摄入量，作为代谢能摄入量的一个部分，可根据代谢能摄入量参考曲线计算得出。这一比例又可以用于估算其他体重下的代谢能摄入量。

作为体重（kg）的一个函数，两种类型的数学方程式（Bridges式8-17；多项式8-18）可用来定义代谢能摄入量曲线，a、b、c、d是这两个方程式的参数。

$$\text{观测代谢能摄入量+浪费量（kcal/d）} = a\{1-\exp[-\exp(b) \times BW^c]\} \quad \text{（式8-17）}$$

$$\text{观测代谢能摄入量+浪费量（kcal/d）} = a + b \times BW + c \times BW^2 + d \times BW^3 \quad \text{（式8-18）}$$

摄入代谢能的分配

在这个模型中，首先要考虑的是满足猪的维持能量需要。通过猪体重（kg）可估测标准维持代谢能需要（式8-19）。如果考虑有效环境温度的影响，当温度低于LCT时，标准维持代谢能需要呈线性增加（式8-20）。

$$\text{标准维持代谢能需要（kcal/d）} = 197 \times BW^{0.60} \quad \text{（式8-19）}$$

$$\text{生热代谢能需要（kcal/d）} = 0.074\,25 \times (LCT-T) \times \text{（标准的维持代谢能需要）} \quad \text{（式8-20）}$$

模型使用者可以通过提高一定比例的标准维持代谢能需要量，对维持能量需要进行校正，从而考虑猪只的活动量或特定品种影响导致的变异，然后计算总的维持代谢能需要量（式8-21）。

维持代谢能需要量（kcal/d）= 标准维持代谢能需要量+生热代谢能需要量
+活动量增加或品种差异的校正的代谢能需要 （式8-21）

　　动物摄入的代谢能超过维持需要以外的部分，用于体蛋白和体脂肪的沉积。特定体重的蛋白质沉积比例通过使用者设定的蛋白质沉积曲线或能量摄入量确定。确定蛋白质沉积曲线有三种可供选择的方法：①输入一个25~125 kg体重之间蛋白质沉积量的平均值；②确定一个表示体蛋白或蛋白质沉积与体重关系的数学方程的参数；③输入蛋白质沉积曲线的最高值（Pd_{Max}）开始下降时的Pd_{Max}和体重的值。

　　对于方法①，将平均蛋白质沉积量与特定性别的标准蛋白质沉积曲线相结合，生成特定体重的蛋白质沉积量（式8-22~式8-24）。这些标准曲线是在上一版NRC（1998）基础上修订而成，并反映了性别对生长模式的典型影响（Hendriks and Moughan, 1993；Wager et al., 1999；BSAS, 2003；van Milgen et al., 2008；Schinckel et al., 2009a, b）。基于这些曲线类型和三种动物性别的典型平均蛋白质沉积值（体重范围25~125 kg的小母猪、阉公猪和未阉公猪，蛋白质沉积分别为137 g/d、133 g/d、151 g/d）的整体蛋白沉积曲线见图8-2。

蛋白质沉积（小母猪）（g/d）=（137）×（$0.7066+0.013\,289 \times BW$
$-0.000\,131\,20 \times BW^2 + 2.8627 \times 10^{-7} \times BW^3$） （式8-22）

蛋白质沉积（阉公猪）（g/d）=（133）×（$0.7078+0.013\,764 \times BW$
$-0.000\,142\,11 \times BW^2 + 3.2698 \times 10^{-7} \times BW^3$） （式8-23）

蛋白质沉积（未阉公猪）（g/d）=（151）×（$0.6558+0.012\,740 \times BW$
$-0.000\,103\,90 \times BW^2 + 1.640\,01 \times 10^{-7} \times BW^3$）（式8-24）

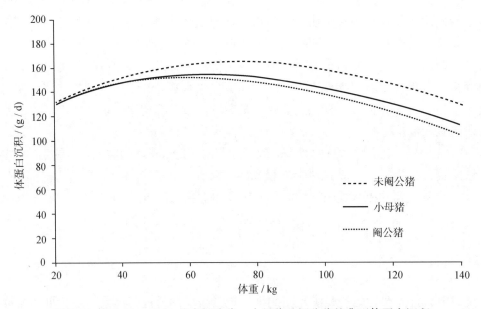

图8-2　体重20~140 kg的未阉公猪、小母猪及阉公猪的典型体蛋白沉积

　　对于方法②，采用广义米氏动力学函数（generalized Michaelis-Menton kinetics function，式8-25），需要使用者依据体重变化建立一个体重增长的曲线模型，从而计算每日的蛋白质沉积量。多项式方程（式8-26）给出了蛋白质沉积量和体增重之间的直接关系。

体蛋白（BP）(kg) = $BP_{初始}$ + {[($BP_{末期}$ − $BP_{初始}$) × (BW/a)^b]/[1 + (BW/a)^b]}　　（式8-25）

蛋白质沉积（Pd）(g/d) = a + b × BW + c × BW^2 + d × BW^3　　（式8-26）

对于方法③，假定猪的最大蛋白质沉积量是恒定的，不受体重的影响，直到最大蛋白质沉积量（Pd_{Max}）开始下降时的体重为止。因此，在这个方法中，只要观测蛋白质沉积量随体重的增长而增加，蛋白质沉积量就由能量摄入量决定。当体重大于 Pd_{Max} 开始下降的体重时，则采用 Gompertz 公式来表示随着体蛋白的增加蛋白质沉积量下降的趋势（式8-27~式8-29）。

成熟时体蛋白（kg）=（Pd_{Max} 开始下降时体重的 BP）× 2.7182　　（式8-27）

速率常数 = [Pd_{Max}/（成熟时体蛋白 × 1000）] × 2.7182　　（式8-28）

Pd_{Max} 开始下降的体重以后的最大 Pd（g/d）=（当前体重下的 BP）× 1000
× 速率常数 × ln（成熟时 BP/当前体重下的 BP）　　（式8-29）

在这个模型中，可以模拟计算由能量摄入量决定的每一天的潜在蛋白质沉积量（式8-30，根据 Black 等，1986 和 NRC，1998 修订）。方程中能量摄入量和蛋白质沉积量之间呈线性关系，而这个关系的斜率随着猪体重的增长而变小（图8-3）。这个数学方程也表明，当能量摄入量反推到维持能量需要摄入量时，生长猪会通过动用体脂肪来沉积体蛋白。这一推论与试验发现相一致（Black et al., 1986）。公式还显示，针对于瘦肉组织生长潜力高的猪，方程的斜率也高。当考虑环境温度的时候，方程的斜率随环境温度升高而下降。模型的使用者可以通过采用一个校正因子对斜率进行调整，以确保针对特定猪群的实际测定体增重与模型估测的体增重相匹配。如果根据能量摄入量设定的蛋白质沉积量低于使用者设定的蛋白质沉积量，则假定真实的蛋白质沉积量等于根据能量摄入量生成的蛋白质沉积量。通过能量摄入量决定蛋白质沉积量的假设可适用于前面提到的三个用来确定蛋白质沉积曲线的方法。

依据能量摄入量生成的蛋白质沉积量（kg/d）= {30 + [21 + 20 × exp$^{(-0.021×BW)}$]
× （代谢能摄入量 − 1.3 × 维持代谢能需要量）× （Pd_{Max} 或平均蛋白质沉积/125）
× [1 + 0.015 × （20 − T）]} × 校正系数　　（式8-30）

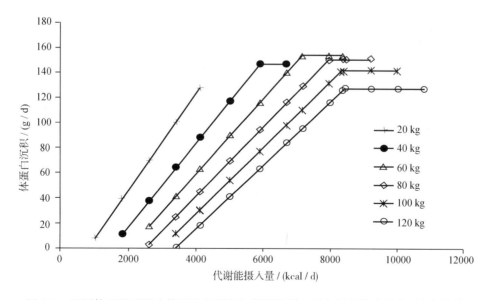

图8-3　不同体重的母猪中体蛋白沉积量与代谢能摄入量之间的关系及典型生长性能

一旦确定了蛋白质沉积数值,可以利用超过维持需要的能量用于蛋白质沉积和脂肪沉积的利用效率来计算脂肪沉积量(式8-31),式中的10.6和12.5分别表示蛋白质沉积和脂肪沉积消耗的代谢能(第1章,能量)。

体脂肪沉积(g/d)=(ME摄入量–维持ME需要量–Pd×10.6)/12.5　　(式8-31)

表8-1列出了三种不同性别类型的猪的典型生长性能。这些性能水平的设定是根据默认的ME摄入量曲线(式8-10~式8-12,图8-1)和Pd曲线(式8-22~式8-24,图8-2)。为了使模型模拟的猪生长性能和背膘厚与末期体重时观测的生长性能和背膘厚一致,可能需要改变饲料采食量曲线和Pd曲线。而且,模型使用者还可以改变维持能量需要量(式8-21)和蛋白质沉积与能量摄入量线性关系的斜率(式8-30)。

表8-1　模型估测的小母猪、阉公猪、未阉公猪(20~130 kg体重)的典型生长性能 a

项目	小母猪	阉公猪	未阉公猪
预测末期体重/kg	130.6	130.5	130.2
代谢能摄入量/(kcal/d)	6825	7345	6583
饲料采食量+浪费量/(g/d)	2177	2343	2100
体增重/(g/d)	819	857	841
体蛋白沉积/(g/d)	132	130	143
体脂肪沉积/(g/d)	234	277	207
增重:饲料(采食量+浪费量)	0.376	0.366	0.401
末期体重时背膘厚/mm	17.5	20.9	14.3

a 这些评定结果是基于默认的代谢能摄入量曲线(式8-10~式8-12,图8-1)和蛋白质沉积曲线(式8-22~式8-24,图8-2);日粮代谢能浓度为3300 kcal/kg,饲料浪费量为5%。

饲喂莱克多巴胺(RAC)和公猪进行针对促性腺激素释放激素(GnRH)的免疫注射对养分分配的影响

为表示饲喂RAC对养分分配的影响,采用Schinckel等(2006)提出的模型计算法则。简而言之,模型考虑了饲喂RAC的水平和持续时间对能量摄入量和蛋白质沉积的影响,以及RAC引起的蛋白质沉积(Pd)对Pd的氨基酸组成和体组成的影响。

如果日粮中含20 mg/kg的RAC,使用RAC后最初20 kg的猪体重增重期间,ME摄入量将出现一定比例减少(MEIR),减少量假定为不添加RAC的对照猪代谢能摄入量的0.036。然后,当RAC引起的猪体增重(BWG_{RAC})接近40 kg时,代谢能摄入量减少(MEIR)逐步增加到不添加RAC的对照猪代谢能摄入量的0.078(式8-32)。

$$MEIR = -0.191263 + (0.019013 \times BWG_{RAC}) - (0.000443 \times BWG_{RAC}^2) + (0.000003539 \times BWG_{RAC}^3)$$
(式8-32)

如果日粮中RAC的添加水平低于20 mg/kg,代谢能摄入量(Mcal/d)可根据式8-33计算。

ME摄入量(kcal/d) = {1 – [MEIR × (日粮RAC水平/20)$^{0.7}$]} × ME摄入量(未添加RAC的对照猪)
(式8-33)

对饲养期为28天的RAC引起平均估测蛋白质沉积增加量进行计算,计算表示为与未添加RAC对照组猪蛋白质沉积量的比例，效果随着 RAC 水平的增加而下降（式 8-34；对 Schinckel 等，2006 进行了微小的修订）。公式估测当 RAC 添加水平分别为 5 mg/kg 和 10 mg/kg 时，效果分别是添加 20 mg/kg 的 RAC 时的 65% 和 80%。

$$\text{RAC 引起的蛋白质沉积平均相对增加量} = 0.33 \times (\text{日粮 RAC 水平}/20)^{0.33} \quad (\text{式 8-34})$$

基于 RAC 引起的体增重（BWG_{RAC}）和饲喂 RAC 的天数，对饲喂一段时间 RAC 后 RAC 引起的平均蛋白质沉积进行校正，如式 8-35 和式 8-36 所示，两个式中猪的体重相等。

$$\text{RAC 引起的相对蛋白质沉积量} = 1.73 + (0.0076 \times BWG_{RAC}) - (0.00205 \times BWG_{RAC}^2)$$
$$+ (0.000017 \times BWG_{RAC}^3) + \{[(0.1 \times \text{日粮 RAC 水平}) - 1] \times (BWG_{RAC} \times 0.001875)\} \quad (\text{式 8-35})$$

$$\text{RAC 引起的相对蛋白质沉积量} = [1.714 + (0.0147 \times \text{饲喂天数}_{RAC}) - (0.00361 \times \text{饲喂天数}_{RAC}^2)$$
$$+ (0.000055 \times \text{饲喂天数}_{RAC}^3) \quad (\text{式 8-36})$$

为解释日粮 RAC 水平递增（日粮 RAC 水平随着时间逐渐提高）引起的效应，蛋白质沉积量基于当天（即第 n 天）日粮 RAC 的水平和当天以前 21 天到当天以前 7 天（$n-21$ 天和 $n-7$ 天）之间的平均 RAC 水平进行校正（式 8-37）。

$$\text{RAC 引起的蛋白质沉积相对增加（递增模式）}$$
$$= 6.73 \, (\text{当天日粮 RAC 水平与平均 RAC 水平的差异})^{0.50}/100 \quad (\text{式 8-37})$$

在这个模型中，添加 RAC 诱发的蛋白质沉积被作为一个单独的蛋白质库，这是对 Schinckel 等（2006）模型的修正。这一调整可以显示 RAC 引起的蛋白质沉积的独特氨基酸组成、RAC 对所有氨基酸和氮需要量，以及机体化学和物理性组成的影响。

$$\text{RAC 引起的无脂瘦肉组织增重（g/d）} = \text{RAC 引起 Pd}/0.2 \quad (\text{式 8-38})$$

据推测，饲喂 RAC 不会改变猪对能量和氨基酸的利用效率，包括维持能量的需要，RAC 的效应本身不受猪的品种和环境条件的影响。

通过研究 RAC 对探头背膘厚的影响，目前已知的是 RAC 对机体不同脂肪库中体脂分布的影响（式 8-39）。在这个方程中，饲喂 RAC 天数不能超过 10，这表明饲喂 RAC 在 10 天能完全实现它对背膘厚的全部影响。当日粮中 RAC 水平为 20 mg/kg 时，预测探针背膘厚将会增加 5%。

$$\text{针对 RAC 校正的探针背膘厚（mm）} = \text{探针背膘厚} \times (1 + 0.05 \times \text{饲喂 RAC 天数}/10)$$
$$\times (\text{日粮 RAC 水平}/20)^{0.7} \quad (\text{式 8-39})$$

在本文献的准备阶段，还尚无有意义的实证研究以证实对未去势公猪进行针对促性腺激素释放激素（GnRH）的免疫注射对动物养分需要的影响。然而，基于用 Improvest™（异普威）针对 GnRH 进行第二次免疫注射 4~5 周后，对能量摄入量、体增重和估测化学体成分的变化的逆向建模，生成了对营养需要量的估测值（第 1 章，能量）。据估计，经过一个过渡期后，免疫注射导致能量摄入量增加 21%，维持能量需要减少 12%，蛋白质沉积减少 8%。另外，基于每天采食量的变化，假定未去势公猪在第二次免疫注射后转化为针对 GnRH 免疫去势的公猪，有 10 天的过渡期。对于营养需要量的估测，一般认为针对未去势公猪进行 GnRH 免疫去势对于主要机体功能的能量和氨基酸的利用效率不会造成影响，而且免疫注射的效应不受猪的品种和环境条件的影响。在这些计算公式中，没有考虑针对 GnRH 的免疫注射对肠道充盈度的影响；但是当屠宰时根据活体重计算无脂瘦肉增重时，还是需要考虑针对 GnRH 的免疫注射对肠道充盈度和屠宰率的影响。

氨基酸需要量

正如第 2 章（蛋白质和氨基酸）中概述，必需氨基酸和氮的营养需要量的估测模型从 Moughan

（1999）修改所得。该模型中确定必需氨基酸和氮的营养需要量时考虑的决定因素为：①与采食量相关的基础内源性消化道损失；②以代谢体重（$BW^{0.75}$，kg）的函数形式表示的表皮损失；③蛋白质沉积量；④摄入 SID 氨基酸用于前面三种功能的利用率。无效氨基酸（inefficiency of amino acid utilization）指的是最低+必然（不可避免的）氨基酸分解代谢量（minimum plus inevitable amino acid catabolism），以及动物个体间蛋白质沉积（Pd）上的差异。由于动物个体间采食量和蛋白质沉积上的差异，致使群饲条件下猪的氨基酸利用率低于单个饲喂条件下猪的氨基酸利用率（Pomar et al.，2003）。

这里给出了赖氨酸的计算公式。根据维持机体主要功能的氨基酸之间最佳比例和估测的氨基酸利用率，估测了其他必需氨基酸（表 2-12）和总氮的需要量。

回肠末端收集的基础内源赖氨酸损失量估测必然（不可避免的）为 0.417 g/kg 干物质饲料采食量；这些损失量与采食量（假设 88%的饲料干物质）和全胃肠道损失（假设大肠损失占到回肠末端回收的胃肠道损失量的 10%）相关（式 8-40）。表皮赖氨酸损失估测为 4.5 mg/ kg $BW^{0.75}$（式 8-41）。

基础内源性胃肠道赖氨酸损失（g/d）=采食量×（0.417×1000）×0.88×1.1 （式 8-40）
表皮赖氨酸损失（g/d）=0.0045×$BW^{0.75}$ （式 8-41）

为了评估这两种机体功能对 SID 赖氨酸的需要量，对最低+必然氨基酸分解代谢量进行了估测（式 8-42），这与 Moughan（1999）提到的方法有所不同。假定最低+必然氨基酸分解代谢量占到 SID 赖氨酸摄入量的 25%，相当于 SID 赖氨酸用于支持基础胃肠道赖氨酸损失和表皮赖氨酸损失的利用效率为 0.75。必然+最低的氨基酸分解代谢量的值来自单个饲养猪只和严格控制的连续屠宰试验，猪只体重为 30~70 kg（Bikker et al.，1994；Moehn et al.，2000）。氨基酸利用率似乎与体重无关，随着猪只生长潜力的改善而提高。最大蛋白质沉积量（Pd）每增加 1 g，小母猪和阉公猪的最低+必然赖氨酸分解代谢值与典型平均值相比会降低 0.002（Moehn et al.，2004）。

胃肠道加表皮损失的 SID 赖氨酸需要量 （g/d）=（式 8-40+式 8-41）/（0.75+0.002×（Pd 的最大值-147.7））
  （式 8-42）

假设蛋白质沉积中的赖氨酸含量为 7.10%，而 RAC 介导的蛋白质沉积中赖氨酸含量为 8.22%（第 2 章；式 8-43）。

蛋白质沉积中赖氨酸存留量（g/d）=非 RAC 介导的蛋白质沉积×7.10/100
  + RAC 介导的蛋白质沉积×8.22/100 （式 8-43）

考虑到动物个体间的差异，超过维持需要的 SID 赖氨酸进食量的边际利用效率被（从 0.75）调低，使之与实证赖氨酸需要量研究得到的估测 SID 赖氨酸需要量相匹配，如在第 2 章（蛋白质和氨基酸）中所提到的。这些分析表明赖氨酸的边际利用效率随着体重增加而下降。这个效率系数在猪体重 20 kg 时为 0.682（相当于蛋白质沉积的赖氨酸需要量增加 9.9%），在 120 kg 体重时为 0.568（相当于蛋白质沉积的赖氨酸需要量增加 32.05%），对于其他体重则可根据这个效率系数与体重的线性关系推导。基于前面提到蛋白质沉积中赖氨酸的含量，对于没有饲喂 RAC 和最大蛋白质沉积为 147.7 g/d 的猪，在 20 kg 和 120 kg 体重时每沉积 100 g 蛋白质所需要的 SID 赖氨酸分别为 10.4 g 和 12.5 g。蛋白质沉积（Pd）的 SID 赖氨酸需要量和每日总 SID 赖氨酸需要量根据式 8-44 和式 8-45 计算。图 8-4 所示为特定性别猪的 SID 赖氨酸需要量曲线。

Pd 所需的 SID 赖氨酸需要量（g/d）={Pd 中赖氨酸沉积量/[0.75+0.002×（最大 Pd - 147.7）]}
  × （1+0.0547+0.002 215×BW） （式 8-44）

总的 SID 赖氨酸需要量（g/d）= 肠道和表皮损失需要量+Pd 需要量 （式 8-45）

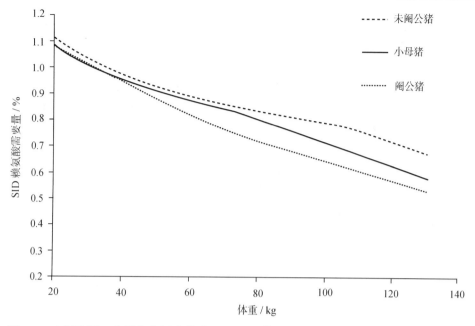

图 8-4 未阉公猪、小母猪和阉公猪（20~130 kg 体重）的模拟 SID 赖氨酸需要量（%）

根据赖氨酸与其他氨基酸需要量之间的比率，上述计算方法可用于所有其他必需氨基酸和总氮（第 2 章，表 2-5~表 2-12）。对于每种氨基酸的最低+必然分解代谢的绝对需要量（如式 8-43 和式 8-44 中的 0.75），则通过不同氨基酸的实证估测值与模型生成的 SID 氨基酸需要量相匹配（第 2 章，蛋白质和氨基酸）。对于几种没有实证经验值可参考的氨基酸（如亮氨酸、苯丙氨酸和苯丙氨酸+酪氨酸），则对这些氨基酸的最低+必然分解代谢量绝对比例进行调整，使之与 NRC（1998）中 65 kg 体重、典型生产性能的生长猪的需要量相匹配。对组氨酸而言，最低+必然分解代谢量的比率设定为 1，使得 SID 组氨酸需要量超过了前一版 NRC（1998）的需要量。对于精氨酸，其最低+必然分解代谢量的比率设定为 1.47，意味着有部分内源性精氨酸的合成。

唯一额外的规则是可发酵 SID 苏氨酸损失的计算（式 8-46），式 8-46 是每日可消化纤维含量的函数（第 2 章，蛋白质和氨基酸；Zhu et al., 2005）。

可发酵 SID 苏氨酸损失（g/d）=（采食量/1000）× 日粮发酵纤维含量 ×（4.2/1000）

（式 8-46）

钙和磷的需要量

决定全消化道标准可消化（STTD）磷和总钙需要的析因估计值是在 Jongbloed 等（1999）、Jondreville 和 Dourmad（2005）的建议基础上调整得来的，在第 6 章（矿物质）已进行过阐述。影响 STTD 磷需要量的因素包括：①体内磷最大沉积率，是体蛋白变化的函数；②基础内源性胃肠道磷损失，是饲料干物质摄入量的函数；③尿磷的最低损失量，是体重的函数；④摄入 STTD 磷用于磷沉积的边际效率；⑤最佳生长性能时的磷需要量占机体最大磷沉积时磷需要量的比例。钙需要量直接由 STTD 磷需要量推导而得到。

为了考虑猪的基因型和性别因素对磷需要量的影响，整个机体的磷含量被直接与体蛋白（BP）联系起来（式 8-47；体蛋白以 kg 表示；第 6 章，矿物质，图 6-1）。假设饲喂 RAC 或针对公猪进行 GnRH

免疫都不会影响整个机体磷含量与 BP 之间的关系。

$$机体磷含量（g）= 1.1613+26.012 \times BP+0.2299 \times BP^2 \qquad (式8-47)$$

基础内源性胃肠道磷损失估计为 190 mg 磷/kg 摄入饲料干物质，而每天最小尿磷损失量大约为 7 mg/kg 体重（第 6 章，矿物质）。将摄入 STTD 磷用于整个机体磷沉积的边际利用效率估计为 0.77；边际效率反映了伴随摄入 STTD 磷的增加和磷摄入量接近最大磷沉积需要量时内源粪尿磷损失的增加，同时也反映机体代谢效率低下以及动物个体之间的差异（第 6 章，矿物质）。在这个模型中，假定猪在最大生长性能时磷需要量是机体磷最大沉积时磷需要量的 0.85（式 8-48）。

$$STTD 磷需要量（g/d）=0.85 \times [整个机体磷的最大沉积量/0.77+0.19 \times 饲料干物质摄入量+0.007 \times BW]$$

$$(式8-48)$$

根据 STTD 磷需要量的 2.15 倍这个固定比率来计算钙的需要量。

在建立这些需要量时，假设日粮常量矿物元素之间，特别是钙和磷之间没有比例不平衡的情况。有证据表明，钙摄入过量将降低磷的利用效率和提高磷的需要量。在第 6 章（矿物质）中对此有更深入的讨论。植酸酶对 STTD 磷和钙需要量的影响未作考虑。因此，假定植酸酶仅仅影响磷的消化率，但不会影响前面提到的 STTD 磷和钙的需要量。

妊娠母猪模型

主要概念

妊娠母猪营养需要量模型以 Dourmad 等（1999，2008）提出的为基础。每日能量摄入量必须由模型使用者设定，而且可能依据不同妊娠阶段而变化。孕体（胎儿、胎盘和子宫液）的重量、蛋白质以及能量增量作为预期窝产仔数、仔猪平均初生重和出生时间的函数明确表示。空子宫和乳腺组织的增重以及能量沉积被认为是母体的组成部分。在这个模型中，分为 6 种不同的蛋白库，即胎儿、胎盘和羊水、子宫、乳腺组织、时间依赖性的母体蛋白沉积、能量摄入依赖性的母体蛋白沉积，这与 Dourmad 等（1999，2008）的研究报道有一定差异，在第 2 章（蛋白质和氨基酸）有详细描述。在这个模型中，假设能量依赖的母体蛋白沉积量随着能量摄入量的增加而呈线性增加，同时假设这种线性关系会随着胎次的变化而变化，而在妊娠期的各个阶段则完全相同。妊娠母猪摄入的能量中，未用于机体维持需要、孕体增重和母体蛋白沉积（包括子宫和乳腺）的部分，则用于母体脂肪的沉积。当能量摄入不能满足维持需要、孕体增重和母体蛋白沉积所需时，母体脂肪将被动员并用作能量来源。母体体重的变化根据每天母体蛋白（不包括孕体，但包括子宫和乳腺）和母体脂肪的变化进行推测。P2 点背膘厚测量值被用于估计母猪的肥度。SID 氨基酸需要量根据 6 个不同蛋白库的蛋白质沉积、体重和采食量进行推测。STTD 磷需要量根据采食量、体重、母体和孕体增重，以及胎次依赖性的骨骼矿化（或重新矿化）的磷需要量推导而来。总钙需要量则根据 STTD 磷需要量估测得来。

体组成

体组成用 Dourmand 等（1999，2008）的方法以数学模型表述。总体重（BW, kg）是母体重和孕体重之和。体重与空腹体重（empty body weight, EBW）的差值等于肠道内容物重量，其重量约占母体体重的 4%（式 8-49）。EBW 和 P2 点背膘厚用于估计妊娠开始时的母体体脂肪（BL）和母体体蛋白（BP）（式 8-50 和式 8-51）。在动态模拟模型中，通过跟踪母体 BL 和母体 BP 含量来预测 EBW（式 8-52）、

P2 点背膘厚（式 8-53）和总体重的日变化量。

$$母体 EBW（kg）= 0.96 \times 母体体重 \qquad （式8\text{-}49）$$
$$母体 BL 含量（kg）= -26.4+0.221 \times 母体 EBW+1.331 \times P2 点背膘厚 \qquad （式8\text{-}50）$$
$$母体 BP 含量（kg）= 2.28+0.178 \times 母体 EBW-0.333 \times P2 点背膘厚 \qquad （式8\text{-}51）$$
$$母体 EBW（kg）= 119.457+4.5249 \times 母体 BP-6.0226 \times 母体 BL \qquad （式8\text{-}52）$$
$$P2 点背膘厚（mm）=16.76-0.7117 \times 母体 BP+0.5732 \times 母体 BL \qquad （式8\text{-}53）$$

孕体生长和蛋白库

孕体的重量和能量含量用自然对数值来评估，并用时间（t，妊娠天数）和预期窝产仔数（ls，窝总产仔数）的函数表示（式 8-54 和式 8-55，Dourmad et al., 1999, 2008）。胎儿的蛋白质含量也用相似的方法估算（式 8-56），而胎盘和羊水中蛋白质含量则基于第 2 章（蛋白质和氨基酸）总结的数据，采用米氏动力学函数（Michaelis-Menton kinetics function），用时间和预期窝产仔数的函数表示（式 8-57）。孕体的日增重、蛋白质或能量沉积量是通过连续两天（n 和 $n+1$）值的差异来计算。

$$孕体重（g）=\exp[8.621-21.02 \times \exp(-0.053 \times t) +0.114 \times ls] \qquad （式8\text{-}54）$$
$$孕体能量含量（kcal）=\{\exp[11.72-8.62 \times \exp(-0.0138 \times t) +0.0932 \times ls]\}/4.184 \qquad （式8\text{-}55）$$
$$胎儿蛋白质含量（g）=\exp[8.729-12.5435 \times \exp(-0.0145 \times t)+0.0867 \times ls] \qquad （式8\text{-}56）$$
$$胎盘及羊水蛋白质含量（g）=[(38.54) \times (t/54.969)^{7.5036}]/[1+(t/54.969)^{7.5036}] \qquad （式8\text{-}57）$$

根据仔猪实际初生窝重与预期初生窝重的比值，对以上 4 个计算模型用平均仔猪初生重校正后用于预测平均仔猪初生重，预期仔猪初生窝重是根据预测妊娠期的长短和窝产仔数来估测的（比值，式 8-85；假设妊娠期为 114 天）。

$$比值=（ls \times 平均仔猪初生重，g）/1.12 \times \exp\{[9.095-17.69\exp(-0.0305 \times 114) +0.0878 \times ls]\} \qquad （式8\text{-}58）$$

在这些计算中，假设能量摄入不会影响孕体的生长，这与 Dourmad 等（1999）观察的结果一致，只有当严重限饲的时候孕体增重才会减少。

基于第 2 章（蛋白质和氨基酸）总结的数据，用自然对数值来估测子宫和乳腺组织中的蛋白质含量，用妊娠天数的函数来表示（式 8-59 和式 8-60）。

$$子宫蛋白质含量（g）=\exp[6.6361-2.4132 \times \exp(-0.0101 \times t)] \qquad （式8\text{-}59）$$
$$乳腺组织蛋白质含量（g）= \exp\{8.4827-7.1786 \times \exp[-0.0153 \times (t-29.18)]\} \qquad （式8\text{-}60）$$

时间依赖性母体蛋白沉积代表在氮（N）平衡研究中观测到的、不能归入其他任何蛋白库的剩余的蛋白质沉积。因为这个蛋白库中的蛋白质沉积仅仅发生在妊娠初期，妊娠 56 天后的蛋白质沉积值强制设定为 0，蛋白质增量用米氏动力学函数（式 8-61）计算。

$$时间依赖性母体蛋白沉积（g）=\{[1522.48 \times (56-t)]/36\}^{2.2}/\{1+(56-t)/36\}^{2.2}\} \qquad （式8\text{-}61）$$

依赖每日能量摄入量的母体蛋白沉积量与妊娠第一天高于维持需要量的代谢能摄入量呈线性关系（式 8-62），其斜率（a）随着胎次的增加而降低，但不可能低于 0（式 8-63）。该斜率是参照 Dourmad 等（2008）的结果校正得来，并随胎次不同而变化，以使不同胎次母猪母体组成变化的实际观测值与估计值能匹配（见本章模型的评估部分）。模型使用者可调整这个线性关系的斜率，以使母猪体重变化以及背膘厚度变化的观测值与预测值相匹配。各种蛋白库的蛋白质沉积模型见图 2-1 和图 2-2，并在图 8-5 中总结描述。

依赖能量摄入的母体蛋白沉积(g/d)=a×(ME 摄入量−母猪妊娠第一天维持代谢能需要量，kcal/d)×校正值

（式 8-62）

式 8-62 中的系数 a =（2.75−0.5×胎次）× 校正值；a > 0　　　　（式 8-63）

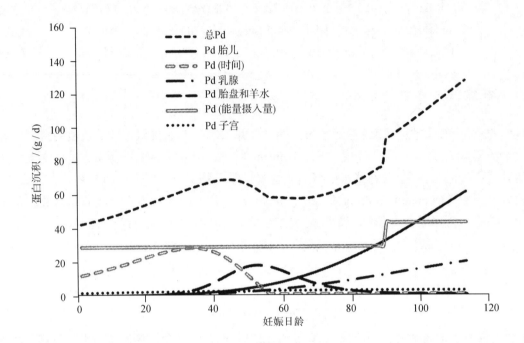

图 8-5　二胎母猪妊娠期胎儿、乳腺组织、胎盘和羊水的典型蛋白沉积（PD）模式，母体蛋白是妊娠时间的函数，妊娠期母体蛋白是能量摄入量的函数，基于预期 13.5 头窝仔数和 1.4 kg 平均初生重。

摄入代谢能的分配

在模型中优先考虑的能量需要是满足机体维持、孕体生长和母体蛋白沉积（包括子宫和乳腺组织的蛋白质沉积）。标准维持能量需要量作为总体重（BW）（kg，式 8-64）的函数计算得到。模型考虑到了妊娠母猪的活动水平和温热环境对母体维持能量需要量的影响。另外，模型使用者可进行适当调整以满足特殊情况下的维持能量需要。

$$\text{标准维持代谢能需要量（kcal/d）}=100×（总 BW）^{0.75} \quad （式 8\text{-}64）$$

如果母猪每天的站立时间超出 4h，则站立时间每增加 1min，维持代谢能需要量增加 0.0717 kcal/(d·kg)(总 BW)$^{0.75}$（Dourmad et al., 2008）。在模型中，假设单独饲喂和群饲母猪的低临界温度（LCT）分别为 20℃和 16℃。对于有麦草垫料的群饲母猪，则其 LCT 比上述水平再降低 4℃（Bruce and Clark, 1979）。当温度低于 LCT 时，每降低 1℃，单独饲喂和群饲母猪的维持代谢能需要量分别增加 4.30 kcal/d/kg（总 BW）$^{0.75}$ 和 2.39 kcal/d/kg（总 BW）$^{0.75}$。

摄入的能量中未用于维持机体功能、孕体生长和母体蛋白沉积（Pd）的部分将用于母体脂肪沉积（Ld）（式 8-65；能量用 kcal 表示；第 1 章，能量）。如果摄入的能量不足以满足机体维持、孕体生长和母体蛋白沉积的需要，母体体脂肪（BL）将被动用并作为 ME 的来源，能量利用效率为 0.80。

母体脂肪沉积量（g/d）=（ME 摄入量−维持 ME 需要量−孕体能量沉积/0.5−母体 Pd×10.6)/(12.5)

（式 8-65）

氨基酸需要量

在妊娠母猪模型中决定氨基酸需要量的主要因素包括：①基础内源性消化道（GIT）的损失，这与饲料的摄入量有关；②上皮细胞脱落损失，是代谢体重（kg $BW^{0.75}$）的函数；③6个不同的蛋白质库中蛋白质的沉积量；④摄入的 SID 氨基酸用于上述功能的利用效率。基础内源性消化道的损失、表皮细胞脱落损失及 SID 氨基酸的利用效率是根据生长育肥猪模型调整得来的。

除了假设每千克摄入饲料的内源性消化道赖氨酸损失为 0.5053 g，以及式 8-42 中没有对猪只的性能潜力做任何调整以外（第 2 章，蛋白质和氨基酸），计算包括内源性肠道赖氨酸损失和表皮赖氨酸损失的 SID 赖氨酸的需要量的方法与生长育肥猪模型是一致的（式 8-40~式 8-42）。用于赖氨酸沉积的 SID 赖氨酸的需要量反映了六大蛋白质库的赖氨酸沉积量，以及最低+必然分解代谢量，并且进行调整以考虑不同动物个体之间的差异（式 8-66；第 2 章，蛋白质和氨基酸），此计算模型是从式 8-44 调整得来的。总 SID 赖氨酸需要量是各种形式赖氨酸需要量的总和，包括内源性肠道赖氨酸损失、表皮赖氨酸损失，以及用于赖氨酸沉积的 SID 赖氨酸需要量。妊娠期赖氨酸需要量的变化见图 8-6。

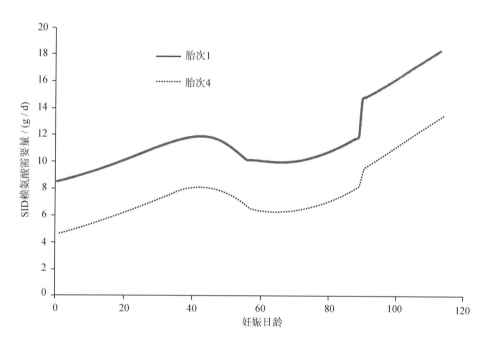

图 8-6　初产妊娠母猪（配种体重 140 kg；预期妊娠总增重 65 kg；窝产仔数为 12.5 头；仔猪初生重为 1.4 kg）和经产妊娠母猪 1~4 胎（配种体重 205 kg；预期妊娠总增重 45 kg；窝产仔数为 13.5 头；仔猪初生重为 1.4 kg）的模拟 SID 赖氨酸需要量（g/d）。

用于赖氨酸沉积的 SID 赖氨酸需要量（g/d）=[（总赖氨酸沉积）/0.75]×1.589　　　（式 8-66）

以上计算式可用于计算其他所有必需氨基酸及总氮的需要量，其依据是它们与赖氨酸的比例（第 2 章，表 2-5 和表 2-11）。对于赖氨酸以外的其他氨基酸，尚无能满足第 2 章（蛋白质和氨基酸）中给出的标准的需要量方面的研究。模型强制设定氨基酸最低+必然分解代谢的绝对量（如式 8-66 中的值 0.75；表 2-12），以使模型生成的需要量与 NRC（1998）中给出的妊娠母猪（第 3 胎母猪，初始体重为 175 kg）需要量相匹配。相对于生长育肥猪模型应用的氨基酸最低+必然分解代谢量的预测值而言，妊娠母猪模型中对于色氨酸和缬氨酸设定的参数被认为过高（分别为 0.752 和 0.934）；而对异亮氨酸设定的参数则被认为太低。因此，对于色氨酸、缬氨酸和异亮氨酸的最低+必然分解代谢量的估测值进行了额外的调整。这种调整反映了色氨酸、缬氨酸和异亮氨酸在孕体、乳腺组织和子宫代谢库内的沉积，与母体体蛋白库存在本质区别，而且在上版 NRC（1998）中没有这些氨基酸的组成比例。对于氮，模型使用的参数为 0.85，与生长育肥猪模型中使用的参数相同（表 2-12）。

钙和磷需要量

常用的用于估测妊娠母猪 STTD 磷需要量的方法与生长育肥猪的相似（第 6 章，矿物质），STTD 磷反映了：①母体和孕体的磷沉积；②基础内源性肠道磷损失（90mg/kg 摄入饲料干物质）；③最低尿磷损失（7 mg/kg 体重）；④STTD 磷用于磷沉积的边际效率（0.77）。

孕体（胎儿和胎盘）中磷的总量是按照 Jongbloed 等（1999）的研究结果表示的，所用方法与 Jondreville 和 Dourmad（2005）相一致。胎儿中磷的总量以妊娠天数和窝产仔数的函数计算（式 8-67）。胎盘和羊水中的磷总量根据蛋白质中磷的比例为 0.0096，通过蛋白质含量来计算（Jongbloed et al.，1999）。磷在胎儿、胎盘及羊水中的含量，根据仔猪初生重进行调整，与孕体中其他养分的计算方法相同（式 8-68）。

$$\text{胎儿磷含量}(g) = \exp\{4.591 - 6.389 \times \exp[-0.02398 \times (t-45)] + (0.0897 \times 1s)\} \quad \text{（式 8-67）}$$
$$\text{胎盘磷含量}(g) = 0.096 \times \text{胎盘及羊水中蛋白质含量} \quad \text{（式 8-68）}$$

母体（包括空子宫和乳腺组织）磷沉积根据母体蛋白沉积和胎次依赖性骨组织中磷日沉积量（第 1、2、3 和 4 及以上胎次母猪磷日沉积量分别为：2.0 g/d、1.6 g/d、1.2 g/d 和 0.8 g/d）计算，此方法从 Jongbloed 等（1999；式 8-69）的研究调整得来。利用固定的钙磷比例 2.3，根据 STTD 磷的需要量来计算钙的需要量。

$$\text{母体磷沉积量}(g/d) = 0.0096 \times \text{母体蛋白沉积量} + \text{胎次依赖性骨组织中磷日沉积量} \quad \text{（式 8-69）}$$

泌乳母猪模型

主要概念

泌乳母猪模型由 Dourmad 等（2008）提出的模型校正得来。每日能量摄入量可通过胎次和泌乳（分娩后）天数估测，或者由模型使用者定义。每日的奶能量和奶蛋白质的产出量通过窝仔数、整个泌乳期仔猪平均增重速率和标准产奶量曲线进行预测。未用于维持机体功能和产奶的摄入能量，则用于母体脂肪沉积（Ld）和蛋白质沉积（Pd）。当摄入能量不足以支持维持和产奶需要时，母体体脂肪（BL）和体蛋白（BP）都被动员并作为能量来源供能。母体体重变化通过每日母体体脂肪和体蛋白的变化来预测。P2 点背膘测量值用来估测母猪的肥度。SID 氨基酸需要量通过窝增重速率、母体体蛋白变化、体重和采食量进行预测。STTD 磷需要量通过饲料采食量、体重、窝增重速率和母体体重变化推导得来，而总钙需要量则通过 STTD 磷需要量来预测。

体组成

泌乳母猪体组成的表示方法与妊娠母猪的相同。

产奶

每日平均产奶能量和氮产出量依据 Dourmad 等（1999，2008）的窝仔日平均增重和窝仔数进行预测（式 8-70 和式 8-71）。在特定的时间，用标准泌乳曲线将这些平均值转化为奶中能量和氮产出量（式 8-72）。每日产奶量通过奶中氮产出量来计算，假设奶中含氮量为 8.0 g/kg（第 2 章）。

$$\text{平均奶中能量产出量（kcal/d）} = 4.92 \times \text{平均窝增重（g/d）} - 90 \times ls \qquad (\text{式 8-70})$$

$$\text{平均奶中氮产出量（g/d）} = 0.0257 \times \text{平均窝增重（g/d）} + 0.42 \times ls \qquad (\text{式 8-71})$$

$$\text{奶中能量或氮产出量（某一特定泌乳日，} t\text{）} = \text{平均产出量} \times (2.763 - 0.014 \times \text{泌乳时间})$$
$$\times \exp(-0.025 \times t) \times \exp[-\exp(0.5 - 0.1 \times t)] \qquad (\text{式 8-72})$$

摄入代谢能的分配

每日代谢能摄入量可由模型使用者定义，或者通过泌乳时间预测得到[式 8-73；从 Schinckel 等（2010a）给出的数据降低 7.5%，使 20 天泌乳期内日平均代谢能摄入值为 20.5 kcal/d]。对头胎母猪而言，预测的代谢能摄入量降低 10%（图 8-7）（Schinckel et al.，2010a）。另外，假设在高临界温度（UCT，22℃）基础上每提高 1℃，每日代谢能摄入量减少（22~25℃内，每增加 1℃，每日代谢能摄入量减少 1.6%；高于 25℃，每增加 1℃，每日代谢能摄入量减少 3.67%；第 1 章，能量）。

$$\text{经产母猪预测代谢能摄入量（kcal/d）} = 4921 + \{[(28\,000 - 4921) \times (\text{天数}/4.898)^{1.612}] / [1 + (\text{天数}/4.898)^{1.612}]\} \qquad (\text{式 8-73})$$

在此模型中，摄入代谢能优先用于满足维持能量需要（式 8-74）和产奶能量需要（式 8-75）。假设在这些模型中产奶对能量摄入量不敏感。

$$\text{标准维持代谢能需要量（kcal/d）} = 100 \times (BW, kg)^{0.75} \qquad (\text{式 8-74})$$

$$\text{产奶代谢能需要量（kcal/d）} = (\text{奶中能量产出量，kcal/d}) / 0.70 \qquad (\text{式 8-75})$$

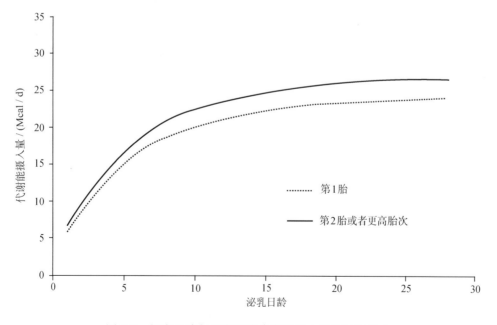

图 8-7 初产母猪与经产母猪代谢能摄入量的经典模式

如果能量摄入量超过维持和产奶的需要，则超出部分会用于母体增重（包括体脂肪和体蛋白），每增重 1 g 体脂肪和体蛋白分别需要 10.6 kcal 和 12.5 kcal 代谢能。在大多数情况下，摄入的代谢能不足以满足维持和产奶的需要。在这种情况下，动用机体能量储备用于奶中能量产出的能量利用效率假设为 0.87。在动员机体能量的前提下，来自体蛋白和体脂肪的能量与机体能量含量变化的比值默认为 0.12，相当于母体体重变化中 10% 为体蛋白变化（第 2 章，蛋白质和氨基酸）。这个比值是通过对发表的泌乳期母猪体重和背膘变化数据的综述，以及依据式 8-49~式 8-51 估测的体组成变化值推导得来的；对于机体能量处于正或者负平衡情况下的母猪，这个比值被认为是相同的。模型使用者可以对此默认比值进行校正，使泌乳期内体重和背膘厚度变化的观测值与预测值相匹配。

氨基酸需要量

必需氨基酸和氮需要量来源于维持机体主要功能的氨基酸最佳比例和预估的氨基酸利用效率（表 2-5、表 2-11 和表 2-12）。在泌乳母猪模型中，考虑两种效率，分别反映日粮摄入 SID 氨基酸或动用体蛋白用于奶中产出氨基酸的利用效率。

用于表示泌乳母猪内源胃肠道氨基酸损失和表皮氨基酸损失的氨基酸需要量的方法与妊娠母猪相同，除采食每千克日粮胃肠道赖氨酸损失量假定为 0.2827 g 外（第 2 章，蛋白质和氨基酸）。母体能量负平衡诱导的体蛋白动用被认为是为产奶提供必需氨基酸和氮。总 SID 赖氨酸需要量代表满足内源胃肠道赖氨酸损失、体表赖氨酸损失和用于产奶的 SID 赖氨酸需要量的总和。

用于产奶的日粮 SID 赖氨酸需要量通过每日奶中氮产出量和母体蛋白动员量来估测（式 8-76）。假定动员体蛋白产生用于产乳的氨基酸的效率对于所有必需氨基酸和氮均一致（0.868），与动员机体能量储备用于产奶的能量利用效率类似。用于产奶的 SID 赖氨酸需要量的估测值极易受到超过维持需要量的赖氨酸用于产奶的利用效率的影响。

正如在第 2 章所述，确定了这个参数（0.67；表示为了考虑不同品种动物的差异性，对参考值 0.75 的校正）（图 2-4）。图 8-8 给出了典型的 SID 赖氨酸需要量。

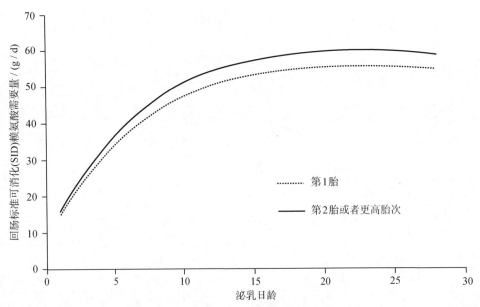

图 8-8　第 1 胎和第 2 胎或多胎次母猪回肠标准可消化（SID）赖氨酸需要量（g/d）的预测值。假设第 1 胎母猪开始泌乳时体重为 175 kg，哺乳 11 头仔猪共 28 天，平均日增重为 230 g/d。假设第 2 胎或者多胎次母猪开始泌乳时体重为 210 kg，哺乳 11.5 头仔猪共 28 天，平均日增重为 230 g/d

用于产奶的 SID 赖氨酸需要量（g/d）=［（每日奶氮产出量×6.38×0.0701
　　　　－母体蛋白动员量×0.0674/0.868）/0.75］×1.1197　　　　　（式 8-76）

上述计算式可以应用于所有其他必需氨基酸和总氮需要量的计算，根据它与赖氨酸的比值进行计算（第 2 章，表 2-5 和表 2-11）。对苏氨酸和色氨酸的最低+必然分解代谢绝对量（如式 8-76 中 0.75，表 2-12）进行校正，从而使模型生成的 SID 氨基酸需要量估测值与实证经验估测值相匹配（第 2 章，蛋白质和氨基酸）。对于其他氨基酸，强制使模型生成的最低+必然分解代谢绝对量与前一版 NRC（1998）中给出的泌乳母猪（母猪初始体重 175kg，10 头仔猪平均增重 250g/d，21 天泌乳期母猪体重损失 10 kg）的氨基酸需要量一致。对于蛋氨酸和蛋氨酸+半胱氨酸，最低+必然分解代谢量与用生长育肥猪模型预测的最低+必然分解代谢量相比较，被认为过高（分别为 0.778 和 0.823）。氮的最低+必然分解代谢量值为 0.85，与生长育肥猪模型中所使用的值相同。

钙和磷需要量

用于估测 STTD 磷需要量的常规方法与生长肥育和妊娠母猪相同（第 6 章，矿物质），反映了：①奶中磷产出量；②基础内源消化道磷损失量（190 mg/kg 摄入干物质）；③最低尿磷损失量（7 mg/kg 体重）；④摄入 STTD 磷摄用于奶中产出磷的边际效率（0.77）；⑤体蛋白动员引起的机体磷动员的影响。奶中磷产出量基于一个固定的比例 0.1955，通过奶中氮产出量来计算（第 6 章，矿物质）。假设母猪每损失 1 g 体蛋白，就会从体储备中动用 9.6 mg 的磷（Jongbloed et al., 1999）。根据 STTD 磷的需要量，以固定的比例 2.0 来计算钙的需要量（第 6 章，矿物质）。

断奶仔猪

生长模型不能用于预测体重低于 20 kg 仔猪的营养需要量，因为对这一阶段的动物缺乏足够的生物学关系的信息。其实可以使用一个相对简单的数学方法来估测其氨基酸需要量。

对于体重低于 20 kg 的仔猪，每日饲料采食量可以通过 NRC（1987）中的校正式估测得到（式 8-77）。当日粮能量浓度低时，仔猪的饲料采食能力会限制其养分的摄入量（式 8-15）。

$$ME 摄入量（kcal/d）= -783.5+315.9\times BW-5.7685\times BW^2 \qquad （式 8-77）$$

对于平均体重在 5~20kg 的仔猪，其 SID 赖氨酸需要量（日粮的百分比）的实证估测值与仔猪平均体重相关。回归方程代表了基于实证经验数据生成的需要量估测值的最佳拟合曲线（第 2 章，蛋白质和氨基酸；式 8-78）：6 kg 体重仔猪 SID 赖氨酸需要量为 1.50%，9 kg 体重仔猪 SID 赖氨酸需要量为 1.35%，18 kg 体重仔猪 SID 赖氨酸需要量为 1.23%。

$$SID 赖氨酸需要量（占日粮的百分比，\%）=1.871-0.22\times \ln BW \qquad （式 8-78）$$

为了计算其他氨基酸的需要量，每日 SID 赖氨酸需要量被划分为机体维持需要（利用式 8-40 和式 8-41）和生长需要（总 SID 赖氨酸需要量与机体维持 SID 赖氨酸需要量之差）。正如早先在生长育肥猪模型中描述的，基于不同机体功能所需要的氨基酸和氮的平衡（表 2-5、表 2-8 和表 2-12），计算对于其他氨基酸和氮的需要量。通过这种方法估测的日粮最佳氨基酸需要量与通过实验法估测的氨基酸需要量相当一致。

这种估测氨基酸需要量的方法没有考虑猪只的生长潜力或者健康状况的差异，这些因素会影响体重低于 20 kg 仔猪的养分需要量。同样，模型也没有考虑性别、温度和饲养密度的因素。

模型使用者还必须了解生长模型并不总是能够使氨基酸需要量从仔猪阶段末期（19.9 kg 体重）平

稳过渡到生长猪阶段早期（20.0 kg 体重），因为预测体重低于 20 kg 和高于 20 kg 仔猪养分需要量时使用的方法不同。

STTD 磷需要量（占日粮的百分比，%）与体重相关（式 8-79）。

STTD 磷需要量（占日粮的百分比，%）= 0.6418−0.1083×ln（BW） （式 8-79）

总钙需要量与 STTD 磷需要量的比例也因体重而异。

总钙需要量/STTD 磷需要量=1.548+0.9176×ln（BW） （式 8-80）

矿物质与维生素需要

没有用传统的模型方法估测除钙、磷以外的矿物质和维生素的需要量。实际上，估测是基于实证实验得出的。日粮按照养分浓度基础划分为 7 个体重阶段（5~7 kg、7~11 kg、11~25 kg、25~50 kg、50~75 kg、75~100 kg、100~135 kg 体重）、妊娠母猪和泌乳母猪阶段。仔猪（5~25 kg）和生长育肥猪（25~135 kg）阶段都运用指数方程以使曲线与体重范围的中间点吻合，式如下：

营养需要量 = a+b×ln（BW） （式 8-81）

这些参数的实际值见表 8-2。图 8-9 是由方程式生成的一种维生素（核黄素，维生素 B_2）的需要量与不同体重阶段（5~135 kg）估测值相比较的一个实例。注意，方程式给出的需要量曲线与估测值在猪不同体重范围中间点处相交。矿物质和维生素估测公式中所用的系数见表 8-2。每日需要量由预估的日粮营养浓度和基于典型营养浓度而决定的日粮采食量相乘得到（式 8-9；表 16-1）。如果实际采食量与典型采食量有偏差，就需要调整以日粮营养浓度基础表示的营养需要量使之与动物的每日需要量相符。

表 8-2 生长猪模型中预测不同体重阶段每日矿物质、维生素和亚油酸需要量所用的系数[a]

营养物质	断奶仔猪			生长育肥猪		
	系数		R^2	系数		R^2
	a	b		a	b	
矿物质						
钠/（g/d）	−1.3128	1.3339	0.9994	−2.5588	1.1335	0.9979
氯/（g/d）	−1.0885	1.3955	0.9789	2.0706	0.9068	0.9979
镁/（g/d）	−0.32	0.2349	0.9966	1.0353	0.4534	0.9979
钾/（g/d）	−1.7815	1.4257	0.9981	0.4591	1.0774	0.9827
铜/（mg/d）	−3.0925	2.6471	0.9974	0.8705	1.9286	0.9423
碘/（mg/d）	−0.112	0.0822	0.9966	0.3624	0.1587	0.9979
铁/（mg/d）	−79.992	58.718	0.9966	34.357	15.904	0.7342
锰/（mg/d）	−1.4927	1.4727	0.9810	5.1766	2.2669	0.9979
硒/（mg/d）	−0.1546	0.1324	0.9974	0.0924	0.1048	0.9043
锌/（mg/d）	−45.852	41.198	0.9932	70.251	43.634	0.9810
维生素						
维生素 A/（IU/d）	−991.67	897.61	0.9924	3364.8	1473.5	0.9979
维生素 D_3/（IU/d）	−141.84	111.66	1.000	388.24	170.02	0.9979
维生素 E/（IU/d）	−4.2638	5.015	0.9489	28.471	12.468	0.9979
维生素 K（甲萘醌）/（mg/d）	−0.4	0.2936	0.9966	1.2941	0.5667	0.9979
生物素/（mg/d）	−0.0225	0.0229	0.9166	0.1294	0.0567	0.9979

续表

营养物质	断奶仔猪			生长育肥猪		
	系数		R^2	系数		R^2
	a	b		a	b	
胆碱/(g/d)	−0.1709	0.1844	1.0000	−0.7765	0.34	0.9979
叶酸/(mg/d)	−0.24	0.1762	0.9966	0.7765	0.34	0.9979
烟酸（有效）/(mg/d)	−23.997	17.616	0.9966	77.649	34.004	0.9979
泛酸/(mg/d)	−5.124	4.5637	0.9943	12.202	6.6304	0.9933
核黄素（B_2）/(mg/d)	−1.5868	1.4702	0.9945	2.2184	1.615	0.9618
硫氨酸（B_1）/(mg/d)	−0.5079	0.4792	0.9403	2.5883	1.1335	0.9979
维生素 B_6/(mg/d)	1.2285	0.6063	0.2230	2.5883	1.1335	0.9979
维生素 B_{12}/(μg/d)	−8.2708	7.5456	0.9994	16.64	−0.852	0.0474
亚油酸(g/d)	−0.7999	0.5872	0.9966	−2.5883	1.1335	0.9979

a 估测需要量 = a + b × ln(BW)，BW 是以 kg 表示的体重。生成这些公式所用的体重表示不同猪只体重范围（断奶仔猪 5~7 kg、7~11 kg、11~25 kg；生长育肥猪 25~50 kg、50~75 kg、75~100 kg 和 100~135 kg）的中点。这些公式将给出表 16-5B 中列出的体重范围猪的矿物质和维生素需要量的粗略值。

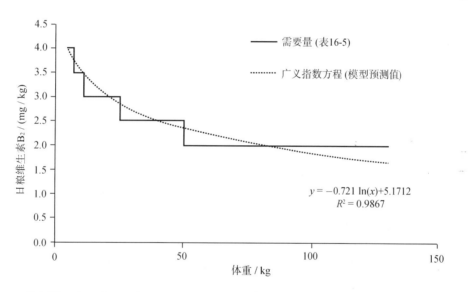

图 8-9 使用模型中的广义指数方程估测 5~135 kg 体重猪每日维生素 B_2 需要量（mg/kg 日粮）

在妊娠和哺乳母猪中没有运用指数方程来估测矿物质和维生素的需要量。母猪矿物质和维生素每日需要量用估测的日粮营养浓度与饲料日采食量相乘得到。

氮、磷和碳沉积效率的估测

在三个模型中，用总量平衡方法来分别计算生长育肥猪、妊娠母猪和泌乳母猪+哺乳仔猪对摄入日粮氮、磷和碳的沉积效率。无效沉积表示这些元素在粪、尿及呼出气体（对于碳来说）中的排出。这些元素的排出会造成环境恶化，应该在营养管理策略中考虑。

为了计算氮、磷和碳平衡，模型使用者必须设定饲料消耗量（饲料采食量和浪费量）、日粮原料组成和（阶段）饲喂程序。在饲喂程序中，应提供不同生产阶段所饲喂的不同日粮的信息。日粮中氮（粗蛋白×0.16）、磷和碳的水平依据日粮原料组成计算得出，而原料中碳的水平则依据其营养成分计算得到（式8-82），假定粗蛋白、粗脂肪、淀粉、糖和其他有机物分别含有53%、76%、44%、42%和45%的碳（Kleiber，1961）。氮、磷和碳的累计摄入量可依据饲料日采食量（包括浪费饲料量）和日粮养分浓度计算得出。

碳含量（g/kg）= 粗蛋白含量（g/kg）×0.53 + 粗脂肪含量（g/kg）×0.76 + 淀粉含量（g/kg）×0.44 + 糖含量（g/kg）×0.42 + 其他有机物含量（g/kg）×0.45 （式8-82）

计算每日氮（粗蛋白质×0.16）、磷和碳（蛋白质沉积量×0.53+脂肪沉积量×0.76）的沉积，并对整个生产阶段求和以推导出养分的沉积效率。每天蛋白质沉积（Pd）和脂肪沉积（Ld）的数值依据式8-31（生长育肥猪）和式8-65（妊娠母猪）中能量分配原则计算，正如在本章前面泌乳母猪"摄入代谢能的分配"部分所描述。对于妊娠母猪，孕体中蛋白质和脂肪的增加也是计算得出，脂肪沉积量是通过总能量沉积和蛋白质能量沉积的差值计算得出（式8-55~式8-57）。每日磷沉积的计算使用式8-47（生长育肥猪）、式8-67和式8-68（妊娠母猪，同时考虑母体磷的沉积），以及第六章"钙和磷"一节中关于泌乳母猪的描述。对于生长育肥猪，假定磷的存留量是最大的（式8-47）。根据文献综述，假定哺乳仔猪每100 g体增重沉积15.3 g蛋白质、16.5 g脂肪和0.0039 g磷（Zijlstra et al.，1996；Mathews，2004；Ebert et al.，2005；Birkenfeld et al.，2006；Canario et al.，2007；Bergsma et al.，2009；Losel et al.，2009；Pastorelli et al.，2009；Charneca et al.，2010）。

模型计算了整个生产周期的氮、磷和碳的平衡。对于生长育肥猪来说，各生长阶段的养分平衡也可以计算得到。在这些计算中，假定日粮养分摄入量不会限制动物的生长性能，即每种日粮中的必需养分水平都超出动物的营养需要量。饲喂不能满足动物营养需要的日粮会使氮、磷和碳平衡的计算无效。

模型评估

有4个途径来评估这些模型：
（1）专家通过输入值变化对模型预测反应的主观性评估（行为分析）；
（2）模型预测对选定模型参数变化的敏感性测试；
（3）估测的氨基酸和磷需要量与1998年版NRC中模型的直接比较；
（4）文献报道的试验数据的模拟，将模拟值与测定反应和需要量进行比较。

主要的建模概念和许多模型参数，特别是与摄入能量分配、体化学组成相关的参数，都是从现有模型得到的，因此之前已经评估过（Agricultural Research Council，1981；NRC，1998；de Lange et al.，2003；Jongbloed et al.，2003；Schinckel et al.，2006；Dourmad et al.，2008；GfE，2008；van Milgen et al.，2008；Bergsma et al.，2009）。这些模型都是经过同行评审，是公认合理的（能量摄入量和使用者设定的猪生长性能水平的变化，在模拟体重变化和养分需要量上产生相应的合理改变）。例如，饲喂莱克多巴胺（RAC）或针对促性腺激素释放激素（GnRH）进行免疫接种，对猪只生长性能和赖氨酸预测需要量的影响与专家的观点一致，饲喂RAC的结果与动物试验结果是一致的（例如，Apple et al.，2004，2007；Webster et al.，2007）。

基于敏感性分析，确定了关键模型参数，如每100 g蛋白质沉积的回肠标准可消化（SID）赖氨酸需要量、窝生长速率和乳汁中氮产出量的关系、内源胃肠道赖氨酸损失（蛋白质沉积、乳蛋白、胎儿和其他与繁殖有关的蛋白质沉积的）、氨基酸组成、氨基酸吸收后的利用率、乳汁和机体中氮和磷沉积的关系。这些关键参数的估测值是根据对大量文献资料的综述而获得的，如本章的前面部分第1章（能量)、第 2 章(蛋白质和氨基酸)和第 6 章(矿物质)所述。

在下面的章节中,将把模型模拟的结果与 1998 年版 NRC,或者单一研究中观察到的动物性能水平和营养需要量进行比较。这些对比与使用模型生成的结果一致,可看成是高度聚合水平的评估。这些比较反映了能量利用的累计效应、动物机体化学与物理组成的关系、与氨基酸和磷需要量有关的生化过程中养分的利用效率。

在一些情况下,通过试验观察值来生成估测模型参数并比较模拟的营养需要量。当确定一种特定养分的营养需要量的时候,只有少数严格控制的、发表的研究可以参考时,尤其是这样。因此,不能用在模型建立过程中未使用的数据对模型进行验证性测试。但是,这类分析确认了模型与试验观察结果和它的用途是一致的。

生长育肥猪模型

在图 8-10A、B、C、D 和 E 中,模型估测的回肠标准可消化(SID)需要量与精心选择的试验研究中观测到的赖氨酸、苏氨酸、蛋氨酸、蛋氨酸+半胱氨酸以及色氨酸的 SID 需要量是相关的,这在第 2 章中已做了描述。对每一种氨基酸来说,相关度是高度线性的,斜率和截距在 1 到 0 之间均没有差别,表明绝对需要量的预测是精确的。对其他一些必需氨基酸而言,研究数量还不足以做出这种分析。图 8-11 表明,模型预测的每千克体重 SID 赖氨酸需要量与需要量观测值接近。这就确保了新模型可以较好地表示随着体重的增加,SID 赖氨酸需要量和体组成变化的关系。

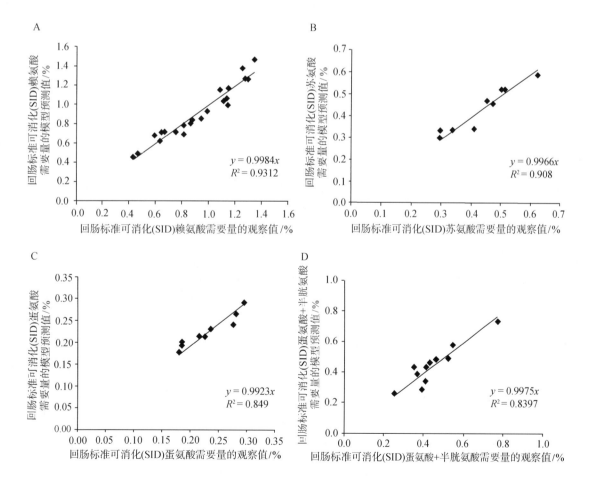

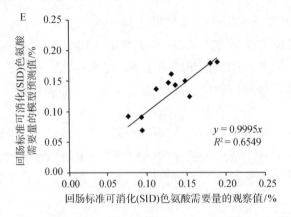

图 8-10　生长育肥猪 SID 赖氨酸（A）、苏氨酸（B）、蛋氨酸（C）、蛋氨酸+半胱氨酸（D）、色氨酸（E）需要量（日粮百分比，%）的模型预测值与观测值之间的关系。数据见表 2-2 和图 2-3A~E

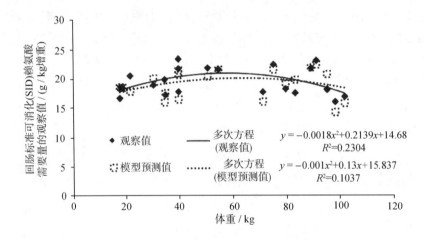

图 8-11　观测或模型预测 SID 赖氨酸需要量（g/kg 体增重）与平均体重之间的关系。数据见表 2-2 和图 2-3A

在表 8-3 中，模型生成的 SID 氨基酸、全消化道标准可消化（STTD）磷和总钙需要量的估测值与上版 NRC（1998）中所指定的生产性能水平的数值（表 10-1，1998 年 NRC）进行直接比较。为了评估 STTD 磷的需要量，依据 NRC（1998）中原料各种营养指标和有效磷需要量设计玉米-豆粕日粮配方。根据本书中给出的原料养分含量值，用得到的日粮原料组成计算 STTD 磷需要量。以这种比较为基础（与 1998 年版 NRC 相比），新模型得出的赖氨酸需要量估测值对 20~50 kg 体重的猪低 3%，对 100~130 kg 体重的猪则高 8%。这种差异与新模型中维持赖氨酸需要量估测值增加，以及随着体重增加每 100 g 蛋白质沉积（Pd）对赖氨酸的需要量增加是一致的。在 1998 年版 NRC 中，假定每 100 g Pd 的赖氨酸需要量与体重无关。通过这些调整，已经被注意到的 1998 年版 NRC 中对于 80~120 kg 体重猪的赖氨酸需要量估测值的表观低估问题得到了解决。

在新模型中，蛋氨酸和精氨酸相对于赖氨酸的需要量提高，而异亮氨酸和色氨酸的需要量则降低。这种需要量的变化与最近的研究结果是一致的（第 2 章，蛋白质和氨基酸）。尽管缺乏最新的、有意义的对组氨酸需要量的估测，在新模型中它的需要量也提高了。降低模型生成的组氨酸需要量估测值则需要组氨酸表观吸收利用率大于 100%，而这是不实际的。对于其他氨基酸，新模型中的需要量与赖氨酸的相对比值变化并不大。

表 8-3 针对 1998 年版 NRC（表 10-1）设定的生产性能水平，依据新生长育肥猪模型和 1998 年版 NRC 估测的生长育肥猪回肠标准可消化（SID）氨基酸、总钙和全消化道标准可消化（STTD）磷的需要量[a]

体重/kg	20~50		50~80		80~120	
日粮 ME 浓度/（kcal/kg）	3 265		3 265		3 265	
估测 ME 摄入量/（kcal/d）	6 050		8 410		10 030	
来源	NRC（1998）	新模型	NRC（1998）	新模型	NRC（1998）	新模型
估测饲料采食量/（g/d）	1855	1821	2575	2579	3075	3097
SID 赖氨酸/（日粮的%）	0.83	0.80	0.66	0.67	0.52	0.56
SID 赖氨酸/（g/d）	15.3	14.6	17.1	17.4	15.8	17.2
SID 氨基酸（相对于赖氨酸的比例%）						
精氨酸	39.8	45.9	36.4	46.0	30.8	46.1
组氨酸	31.3	34.4	31.8	34.4	30.8	34.4
异亮氨酸	54.2	50.8	56.1	51.3	55.8	52.0
亮氨酸	100.0	101.1	101.5	101.5	98.1	102.0
赖氨酸	**100.0**	**100.0**	**100.0**	**100.0**	**100.0**	**100.0**
蛋氨酸	26.5	28.9	27.3	28.8	26.9	28.8
蛋氨酸+半胱氨酸	56.6	57.0	59.1	57.8	59.6	58.9
苯丙氨酸	59.0	60.2	60.6	60.7	59.6	61.3
苯丙氨酸+酪氨酸	94.0	94.7	95.5	95.5	94.2	96.6
苏氨酸	62.7	62.5	65.2	64.5	65.4	67.2
色氨酸	18.1	17.4	18.2	17.7	19.2	18.2
缬氨酸	67.5	65.8	68.2	66.6	67.3	67.7
N × 6.25	—	1 367.5	—	1 391	—	1 424
总钙（日粮的%）	0.60	0.52	0.50	0.45	0.45	0.39
有效磷（日粮的%）	0.23	—	0.19	—	0.15	—
STTD 磷（日粮的%）	0.30	0.24	0.26	0.21	0.21	0.18

a 未考虑饲料浪费量，假定为 0%。

新模型中，STTD 磷的需要量降低了，这很大程度上是基于欧洲对磷需要量的综述研究（Jongbloed et al., 1999；BSAS, 2003；Jondreville and Dourmad, 2005, 2006；GfE, 2008）。与 1998 年版 NRC 中的模型不同，在新模型中，随着 Pd 变化而导致猪只生长速度变化，因而对日粮磷的需要量也变化。因此，未阉公猪磷需要量估测值要高于小母猪和阉公猪，这与试验观测数据相一致（第 6 章，矿物质）。若猪蛋白质沉积率较高，磷的需要量估测值就会接近 NRC（1998）的建议值，而高于 Jongbloed 等（1999）、Jondreville 和 Dourmad（2005，2006）、BSAS（2003）和 GfE（2008）推荐需要量。这些原则同样适用于钙的需要量，根据 STTD 磷的需要量直接估测得出。相对于磷的钙需要量，新模型比 1998 年 NRC 略有提高。

为了模拟个别营养物质需要量研究的性能数据，需要在模型中输入观测到的饲料和能量摄入量水平，以及与所测定营养需要量相对应的动物体重范围。假设饲料浪费量占记录的饲料采食量与浪费量之和的 5%。为了与观测及模拟的体增重和饲料转化效率相匹配，对平均蛋白质沉积（Pd）量进行调整。不同性别蛋白质沉积曲线的默认形状没有改变。如果有背膘厚度测定数据，这些信息也要输入；为了使背膘厚度的观测值与模拟值相匹配，维持能量的需要量也要进行调整。当模型校准后（例如，相对不同平均瘦肉组织和维持能量需要量，动物生长速度和背膘厚度的观测值和估测值相匹配），就可以模拟和比较营养需要量以确定需要量。例如，表 8-4 将估测赖氨酸需要量与 Coma 等（1995）和 Dourmad 等（1996）用试验方法测定的需要量进行比较。正如第 2 章中所描述，这些研究并未用于建立模型，这些比较可看成是对模型的独立测试。表 8-4 中总结的结果表明赖氨酸需要量的观测值和模型生成的估测值是基本一致的。模型似乎系统性地高估了单独饲喂猪只的赖氨酸需要量，这可能是为了反映动物间差异对营养需要量的影响（如 Pomar et al., 2003）而降低模型中赖氨酸吸收利用率造成的。这些结果同时说明新模型合理地考虑了饲喂水平和体重（Coma et al., 1995），以及性别和体重（Dourmad et al., 1996）的互作效应对赖氨酸需要量的影响。基于这些结论和其他分析（如图 8-10A），在预测群饲生长育肥猪赖氨酸需要量时未发现明显的、系统性的偏差。

表 8-4　生长育肥猪赖氨酸试验测定需要量与模型预测需要量的比较

性别	体重范围/kg	饲料采食量+浪费量/(kg/d)	观测体增重/(g/d)	估测平均瘦肉增重/(g/d)	赖氨酸需要量（占日粮的百分比，%）		
					测定值	预测值	差异 [a]/%
					总赖氨酸		
Coma 等，1995[b]							
阉公猪	27.1~35.4	1.864	—	325	0.97	0.95	-2
阉公猪	27.1~35.4	1.282	—	325	1.01	1.05	4
阉公猪	92.6~104	3.543	—	325	0.61	0.61	0
阉公猪	92.6~104	2.643	—	325	0.85	0.76	-10[b]
					SID 赖氨酸		
Dourmad 等，1996[c]							
阉公猪	50~80	2.251	779	329	0.68	0.78	15
小母猪	50~80	2.244	850	377	0.71	0.81	14
阉公猪	80~110	2.822	896	329	0.56	0.65	17
小母猪	80~110	2.841	950	377	0.68	0.71	4

a 100×（预测值−测定值）/（测定值）。
b 猪只限制饲喂玉米豆粕为基础的日粮，添加递增水平的赖氨酸；估测的日粮 ME 水平对较低和较高体重范围分别为 3261 kcal/kg 和 3271 kcal/kg；假定 5%的饲料浪费量；每个处理组的平均生长性能数据并未给出；在所有模拟中使用了先前针对这一阶段的猪只测定得出的瘦肉增重的平均恒定值。较高体重的猪只在采食量下降时测定的每日赖氨酸需要量提高（对于低和高采食量，分别为 22.5 g/d vs. 21.6 g/d）；这一异常现象部分解释了赖氨酸需要量测定值和预测值之间的差异。
c 单栏饲喂的猪被饲喂小麦为基础的日粮，添加递增水平的 L-赖氨酸盐酸盐；估测的日粮净能（NE）水平为 2342 kcal/kg；假定 5%的饲料浪费量；不同性别的猪只在两个体重范围的平均瘦肉增重性能恒定的数值，用此模型估测并使测定值与预测值相符。对赖氨酸需要量的系统性高估可能是因为测定值来自于单栏饲喂的猪，而不是群饲的猪只。

在使用模型估测赖氨酸或其他营养物质的需要量时，存在出现偏差的可能，尤其是当动物体重范围变化较大或猪只生产性能潜力变异极大时。赖氨酸需要量的经验性估测是建立在生长性能研究的基础上，这些研究持续一定周期，并且动物达到了一定的体增重。通常认为生长猪在试验初期对日粮中较高的赖氨酸浓度有反应，原因很简单，因为动物对赖氨酸需要量随体重的增加而降低（如图 2-3A）。因此，试验得出的需要量（以占日粮百分比表示）在接近初始体重时是适用的。但是，通常研究报道的采食量和生长性能都是整个试验期的。为此，模型计算每日日粮赖氨酸需要量的平均数，会低估接近初始体重时猪的需要量。同样，由于动物间性能潜力的差异，群饲动物的估测营养需要量要比单独饲喂的动物高（如 Poman et al., 2003）。在阐述赖氨酸需要量和校正赖氨酸利用效率时，在一定程度上考虑到了这些潜在的偏差，正如在本章前面和第 2 章（蛋白质和氨基酸）中所阐述。然而，当动物体重范围较大或猪只生产性能潜力变异大时，估测赖氨酸或其他营养素需要量时则会出现偏差。为了减少出现偏差的来源，如果营养需要量的研究涵盖生长猪 20 kg 以上、育肥猪 30 kg 以上的体增重范围，或报道的猪只潜在生产性能高度变异时，则必须审慎阐述，因此不在本评估的考虑范围之内。当运用模型估测营养需要量时，必须要考虑这些潜在的偏差。

妊娠母猪模型

正如1998年版NRC中第2章（蛋白质和氨基酸）、第6章（矿物质）中所述，在妊娠母猪营养需要量方面所做的严格控制的研究很少。因此，在对妊娠母猪氨基酸、磷和钙需要量的主要影响因素进行定量，以及对Dourmad等（2008）所详细描述的妊娠母猪模型进行完善时必须格外小心。Dourmad等（2008）模型所做的主要完善是在估测氨基酸需要量时考虑到了不同蛋白库中的氨基酸组成、在表示孕体产物生长时不仅考虑窝仔数还考虑到仔猪初生重、考虑了胎次对能量摄入量与母体蛋白沉积关系的影响，以及考虑了孕体产物磷沉积与母体的关系。

表8-5所示结果说明新的妊娠母猪模型略低估了不同胎次母猪妊娠期的体重和背膘变化。在妊娠母猪模型中，预测的生产性能对估测的维持能量需要非常敏感。例如，表8-5中第4胎母猪的结果-预测和观测到的生产性能差异最大，当把维持能量需要量仅降低13%时（默认值是100 kcal/kg代谢体重），预测母猪体重变化将达到39.7 kg，背膘变化将达到2.7 mm，从而接近观测值。但是，在商业条件下饲养的妊娠母猪维持能量需要量是变化的，很可能要高于87 kcal/kg代谢体重。因此，维持能量需要的默认值在模型中保持不变。模型使用者可审慎地使用维持能量需要量的校正值，使妊娠期母猪体重和背膘变化的观测值与预测值相符合。基于这些以及其他分析，可得出以下结论：模型可以有效地表现母猪机体和孕体产物对能量摄入的反应，以及在蛋白质增重和脂肪增重之间沉积能量的分配。

表8-5 妊娠期母猪体重和背膘变化观测值与模型预测值的对比 [a]

胎次	1[b]	2[c]	3[d]	4[e]
观测性能				
配种时体重/kg	135.4	158.3	196.4	184.8
妊娠期增重/kg	67.4	56.3	46.4	42.4
配种时背膘厚/mm	16.3	17.2	16.9	17.9
妊娠期背膘厚度增加/mm	4.5	2.5	2.6	1.7
窝仔数	10.7	10.8	11.4	11.1
饲料采食量+浪费量/（kg/d）	2.334	2.285	2.327	1.983
日粮ME浓度/（kcal/kg）	3100	3145	3240	3257
模型预测性能				
妊娠期体重增重/kg	61.8	51.8	44.9	33.1
妊娠期背膘厚度增加/mm	2.3	2.2	1.7	-0.6

a 模拟观测的每窝平均值。假设所有胎次的平均仔猪初生重为1.4 kg。假定饲料浪费量为5%。模型中对于两个模型校正指标（维持能量需要量；母体氮沉积和能量摄入量之间的关系）使用默认值。体重观测值和预测值，以及分娩时的背膘厚的吻合度可以通过调整这两个模型校正指标来改善。例如，对于第4胎母猪，如果维持能量需要量降低13%，妊娠增重提高到39.7 kg，妊娠期背膘厚增加2.7 mm。

b 对于第1胎的母猪，观测的性能值显示的是Mahan（1998）、Cooper等（2001）、van der Peet-Schwering等（2003）、Gill（2006）和Dourmad等（2008）（n=5）所观测的数值的平均值。

c 对于第2胎的母猪，观测的性能值显示的是Mahan（1998）、Cooper等（2001）、van der Peet-Schwering等（2003）和Veum等（2009）（n=4）所观测的数值的平均值。

d 对于第3胎的母猪，观测的性能值显示的是Mahan（1998）、Young等（2004；3个平均值）、van der Peet-Schwering等（2003）和Veum等（2009）（n=6）所观测的数值的平均值。

e 对于第4胎的母猪，观测的性能值显示的是Mahan（1998）、Musser等（2004）和Veum等（2009）（n=3）所观测的数值的平均值。

妊娠母猪模型需要被强制与三个精心选择的赖氨酸需要量研究保持一致，这要通过控制回肠标准可消化（SID）赖氨酸用于蛋白质沉积中的赖氨酸沉积的效率（如本章前文所述），并生成了比使用

Dourmad 等（2008）模型生成的赖氨酸需要量估测值略高的值。

在表 8-6 中，模型生成的回肠标准可消化（SID）氨基酸、全消化道标准可消化（STTD）磷和总钙需要量的估测值与 1998 年版 NRC 表 10-8 中给出的生产性能水平直接进行比较。基于这些比较，新模型产生了 114 天妊娠期平均赖氨酸需要量的估测值：与 1998 年 NRC 中的需要量相比，新模型第 1 胎母猪的值稍高，第 2 胎母猪的值稍低，第 3、第 4 胎母猪的值显著降低。这些差别很大程度上是由于不同胎次母体蛋白沉积量的变化，该值在新模型中比 1998 年版 NRC 中的大。相对于赖氨酸，新模型中对色氨酸、缬氨酸的需要量增加，异亮氨酸需要量则降低。这些需要量的变化与妊娠母猪各种不同蛋白库中氨基酸组成是一致的，特别是胎儿的蛋白质（第 2 章，蛋白质和氨基酸）。很可能建议的这三种氨基酸需要量的变化低估了所需要变化的真实值。然而一般认为，在对这三种和其他氨基酸需要量进行额外调整前应当借助试验估测其需要量。新模型中全消化道标准可消化（STTD）磷和钙的需要量降低了，很大程度上基于欧洲对于磷需要量的综述(Jongbloed et al., 1999; BSAS, 2003; Jondreville and Dourmad, 2005, 2006; GfE, 2008)。总之，新模型生成的 STTD 磷需要量估测值比欧洲估测值要稍高，这与摄入 STTD 磷用于磷沉积的边际效率相对较低是一致的。相对于磷、钙需要量比 1998 年版 NRC 略有提高。

表 8-6 针对 1998 年版 NRC（表 10-8）设定的生产性能水平，依据新妊娠母猪模型和 1998 年版 NRC 估测的妊娠母猪回肠标准可消化（SID）氨基酸、总钙和全消化道标准可消化（STTD）磷的需要量[a]

配种时体重/kg	125		150		175		200	
胎次	1		2		3		4	
妊娠期增重/kg	55		45		40		35	
窝仔数	11		12		12		12	
日粮 ME 浓度/(kcal/kg)	3265		3265		3265		3265	
来源	NRC(1998)	新模型	NRC(1998)	新模型	NRC(1998)	新模型	NRC(1998)	新模型
估测饲料采食量/(kg/d)	1.96	1.892	1.84	1.847	1.88	1.927	1.92	1.987
SID 赖氨酸（日粮的%）	0.50	0.56	0.49	0.47	0.46	0.40	0.44	0.35
SID 赖氨酸/(g/d)	9.7	10.6	9.0	8.6	8.7	7.7	8.4	6.9
SID 氨基酸（相对于赖氨酸需要量的比例）								
精氨酸	8.2	52.5	1.1	52.1	0.0	51.8	0.0	51.4
组氨酸	32.0	33.8	32.2	33.1	32.2	32.6	32.1	32.2
异亮氨酸	57.7	55.6	57.8	55.8	58.6	56.3	59.5	56.8
亮氨酸	96.9	91.4	96.7	93.2	95.4	94.5	94.0	95.8
赖氨酸	**100.0**	**100.0**	**100.0**	**100.0**	**100.0**	**100.0**	**100.0**	**100.0**
蛋氨酸	27.8	28.0	27.8	27.8	27.6	27.7	27.4	27.5
蛋氨酸+半胱氨酸	66.0	64.6	67.8	67.2	70.1	69.3	71.4	71.6
苯丙氨酸	58.8	54.8	57.8	56.1	57.5	57.1	57.1	58.1
苯丙氨酸+酪氨酸	97.9	95.6	98.9	97.1	98.9	98.8	100.0	100.1
苏氨酸	75.3	71.1	77.8	74.9	79.3	78.6	82.1	82.3
色氨酸	19.6	18.1	20.0	19.3	19.5	20.1	20.2	21.0
缬氨酸	68.0	70.9	67.8	73.0	67.8	74.8	67.9	76.7
N×6.25	—	1589.3	—	1655.2	—		—	1770.3
总钙（日粮的%）	0.75	0.69	0.75	0.65	0.75	0.57	0.75	0.50
有效磷（日粮的%）	0.35		0.35		0.35		0.35	
STTD 磷（日粮的%）	0.40	0.30	0.40	0.28	0.40	0.25	0.40	0.22

a 未考虑饲料浪费量，假定为 0%。

新的妊娠母猪模型与1998年版NRC相比的一个主要变化是对妊娠期的不同阶段设定了不同的营养需要量（表16-6A和表16-6B）。妊娠后期每天能量、氨基酸、磷和钙需要量显著增加与妊娠期不同组织的发育模式（第2章，蛋白质和氨基酸）、欧洲推荐标准（Dourmad et al., 2008；GfE, 2008）、现代母猪妊娠期氮沉积的变化（Srichana, 2006）、最近运用氨基酸氧化指示剂方法获得的赖氨酸需要量估测值（Moehn et al., 2011）是一致的。主要是因为妊娠后期营养需要量的快速变化，估测营养需要量的平均值对选择的时间阶段非常敏感。如果整个妊娠期只使用一个日粮，则建议配制的这种日粮要满足妊娠期90~114天的营养需要；对于多个胎次，营养需要量要比1998年版NRC中的需要量要高（表16-6A和表16-6B）。

泌乳母猪模型

图8-12显示了模型估测的泌乳母猪回肠标准可消化（SID）赖氨酸需要量与第2章（蛋白质与氨基酸）中所概述的精选的试验研究得出的观测需要量之间的关系。这种关系是高度线性的，其斜率与截距在1与0之间均无变化，表明模型对赖氨酸绝对需要量的预测很准确。对于其他必需氨基酸，研究的数量不足，还不能做出这样的分析。

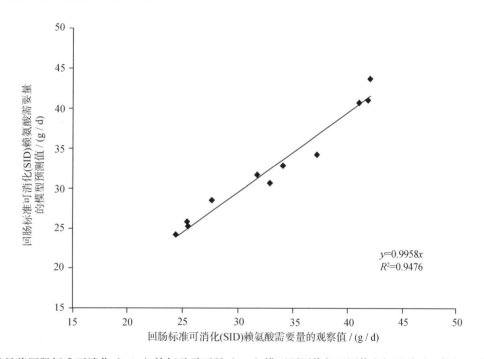

图8-12 泌乳母猪回肠标准可消化（SID）赖氨酸需要量（g/d）模型预测值与观测值之间的关系。数据见表2-3和图2-5

在表8-7中，模型生成的SID氨基酸、全消化道标准可消化（STTD）磷和总钙的需要量估测值与1998年版NRC中表10-10指定的生产性能水平下直接进行比较。这些结果显示，新的泌乳母猪模型中动物对能量摄入量的性能反应与1998年版NRC非常相似。然而，新模型对21天泌乳期平均赖氨酸需要量的估测值比1998年版NRC推荐的需要量低11%~15%，而且这种差异随着泌乳期母猪体重损失的增加而加大。后者是因为新模型中，对由于乳中赖氨酸产出造成的能量负平衡导致的母猪体蛋白损失的贡献的更多考虑（第2章，蛋白质和氨基酸）。新模型与1998年版NRC的差异，在某种程度上是由于在新模型的营养需要量研究中将每日养分摄入量的5%作为饲料浪费的校正，这种浪费直接影响赖氨酸需要量的估测。饲料浪费在1998年版NRC中没有考虑。当使用新模型时，建议把5%的饲料浪费设为默认值，这会使表 8-7 中以日粮浓度表示的赖氨酸需要量提高 5%。

表 8-7　针对 1998 年版 NRC（表 10-10）设定的生产性能水平，依据新泌乳母猪模型和 1998 年 NRC 估测的泌乳母猪回肠标准可消化（SID）氨基酸、总钙和全消化道标准可消化（STTD）磷的需要量[a]

母猪分娩后体重/kg		175		175
预计泌乳体重变化/kg		0		−10
仔猪日增重/g		250		250
日粮 ME 浓度/（kcal/kg）		3265		3265
来源	NRC（1998）	新模型	NRC（1998）	新模型
估测饲料采食量/（kg/d）	6.4	6.462	5.66	5.477
SID 赖氨酸（日粮的%）	0.85	0.75	0.9	0.79
SID 赖氨酸/（g/d）	54.3	48.2	51.2	43.5
SID 氨基酸（相对于赖氨酸需要量的比例）				
精氨酸	57.3	57.8	54.7	54.5
组氨酸	40.0	40.1	39.6	39.7
异亮氨酸	55.4	55.7	55.7	55.7
亮氨酸	113.3	111.9	113.5	113.7
赖氨酸	**100.0**	**100.0**	**100.0**	**100.0**
蛋氨酸	26.0	26.8	25.8	26.6
蛋氨酸+半胱氨酸	47.9	52.8	47.9	53.3
苯丙氨酸	54.7	54.3	54.5	54.6
苯丙氨酸+酪氨酸	112.5	111.5	112.9	113.1
苏氨酸	61.3	64.3	61.5	64.4
色氨酸	17.9	19.0	18.4	19.5
缬氨酸	84.3	85.3	85.2	85.3
N × 6.25	—	1349.6	—	1339.8
总钙（日粮的%）	0.75	0.63	0.75	0.72
有效磷（日粮的%）	0.35	—	0.35	—
STTD 磷（日粮的%）	0.41	0.32	0.41	0.36

a 未考虑饲料浪费量，假定为 0%。

　　对 1998 年版 NRC 中使用的赖氨酸需要量研究的更新解释同样导致泌乳母猪赖氨酸需要量估测值的降低。例如，Boomgaardt 等（1972）的研究中没有观察到添加赖氨酸的效果。因而假定试验中最低的日粮赖氨酸水平为其需要量是不正确的，所以这个研究在新模型中没有被采用。另外，对 Johnston 等（1993）报道的数据的重新解释导致赖氨酸需要量估测值的大幅降低。后一个研究对 1998 年版 NRC 中采用的每单位窝增重估测赖氨酸需要量有相当大的影响。而且，新模型基于 Johnston 等（1993）的数据得到的赖氨酸需要量估测值，显著地改善了回肠标准可消化（SID）赖氨酸摄入量与日粮赖氨酸在乳中产出量的线性关系的契合度（第 2 章，蛋白质和氨基酸，图 2-4）。苏氨酸、色氨酸、蛋氨酸和蛋氨酸+半胱氨酸相对于赖氨酸的需要量在新模型中是提高的。对于苏氨酸和色氨酸来说，这些变化与氨基酸需要量研究结果是一致的（第 2 章，蛋白质和氨基酸）。对于蛋氨酸和蛋氨酸+半胱氨酸需要量来说，氨基酸吸收利用效率比符合 1998 年版 NRC 需要量的值有所降低，使之与生长育肥猪和妊娠母猪的利用率更接近。乳中含有一定量的来自半胱氨酸的牛磺酸（Wu and Knabe，1994），这降低了蛋氨酸+

半胱氨酸用于乳中蛋氨酸+半胱氨酸产出的效率。新模型生成的最佳日粮 SID 蛋氨酸及蛋氨酸+半胱氨酸相对于赖氨酸的比例的估测值与其他推荐值更接近（例如，BSAS，2003；Dourmad et al.，2008；GfE，2008）。似乎新模型建议的蛋氨酸及蛋氨酸+半胱氨酸需要量的变化要比实际需要的变化有所低估。然而，在对这些氨基酸和其他氨基酸的需要量做出进一步调整之前，需要获得其需要量的实证估测值。新模型中 STTD 磷和钙的需要量与 1998 年版 NRC 相比降低，主要基于欧洲对于磷需要量的综述（Jongbloed et al.，1999，2003；BSAS，2003；Jondreville and Dourmad，2005，2006；GfE，2008）。总之，STTD 磷需要量估测值比欧洲的估测值要稍高，这与摄入 STTD 磷用于磷沉积的边际效率相对较低是一致的。相对于磷的钙需要量比 1998 年版 NRC 略有提高。

泌乳母猪模型被用于模拟三个未用于建立模型的赖氨酸需要量的研究（表 8-8）。在这三个研究中，母猪饲喂玉米豆粕基础日粮，基于总日粮赖氨酸浓度基础上模拟。对每个赖氨酸需要量的研究，采食量（校正 5%浪费）、日粮代谢能水平、母猪产后体重、泌乳期长度、窝仔数和平均日增重都输入模型中。如有必要，对维持能量需要量进行调整以使观察值与模型预测母猪体重变化相符。因为没有可用的信息来估测母猪体重变化的组成，应用模型默认值来估测体蛋白和体脂肪变化在机体能量平衡中的相对贡献。在其中两个研究中（Stahly et al.，1990；Monegue et al.，1993），随着日粮赖氨酸水平提高到最高水平，生产性能随之不断提高。在这种情况下，最高的饲喂水平被认定为测定的需要量，尽管达到最高生产性能的需要量可能会更高。用这种方法用来评估模型是合适的，因为模型能估测出达到试验中生产性能所需的赖氨酸的水平。这两个研究中，模型估测的以日粮中浓度来表达的赖氨酸需要量都略高。在 Srichana（2006）的研究中，泌乳母猪饲喂 5 种不同赖氨酸水平（0.95%~1.35%）的日粮。结果表明，母猪泌乳性能在赖氨酸水平最高时达到最佳，而下一胎次的繁殖性能并未受到日粮赖氨酸水平的影响。在此研究中，窝增重和母体增重都随着日粮赖氨酸水平提高而显著线性提高，尽管额外提高赖氨酸摄入量的边际效应较小。正如式 8-71 和式 8-76 所阐述，基于对乳和母体增重中的估测赖氨酸含量，摄入 SID 赖氨酸的边际利用效率的估测值在所有日粮赖氨酸水平下恒定不变且低于 15%，这要比第 2 章（蛋白质和氨基酸）中报道的其他需要量研究中的观测到的结果要低很多。基于这些考虑，表 8-8 只给出两个最低日粮赖氨酸水平的生产性能结果。模拟结果显示，修正的模型高估了支持饲喂 0.99%总赖氨酸的母猪的泌乳性能的赖氨酸需要量，低估了饲喂 1.04%总赖氨酸的母猪的生产性能，而这两个处理组间母猪的泌乳性能相差很小。依据这三个研究可看出，泌乳母猪模型可为泌乳母猪实证得出的赖氨酸需要量提供合理预测。

表 8-8 泌乳母猪试验测定与模型预测赖氨酸需要量的对比

来源	饲料采食量+浪费量/（kg/d）	断奶仔猪数	仔猪增重/（g/d）	总赖氨酸需要量（日粮的%）		差异 [a]
				测定值	预测值	
Monegue 等（1993）[b]	6.070	11.1	210	0.90	0.94	4%
Stahly 等（1990）[c]	5.404	10.76	194	0.86	0.89	3%
Srichana（2006）[d]	5.400	9.1	251	0.99	1.01	2%
Srichana（2006）[e]	5.700	9.3	248	1.04	0.95	−9%

a 100 ×（预测值−测定值）/（测定值）。
b 泌乳期长度 28 天；分娩后体重 198 kg；断奶时体重 201.6 kg；估测日粮 ME 浓度为 3265 kcal/kg。
c 泌乳期长度 27 天；分娩后体重 186 kg；断奶时体重 181.5 kg；估测日粮 ME 浓度为 3368 kcal/kg。
d 处理组 1；泌乳期长度 19.5 天；分娩后体重 190 kg；断奶时体重 194.1 kg；估测日粮 ME 浓度为 3460 kcal/kg。
e 处理组 2；泌乳期长度 19.2 天；分娩后体重 190.8 kg；断奶时体重 194.8 kg；估测日粮 ME 浓度为 3460 kcal/kg。

参 考 文 献

Agricultural Research Council. 1981. *The Nutrient Requirements of Pigs, Technical Review,* 2nd Ed. Slough, UK: Commonwealth Agricultural Bureaux.

Apple, J. K., C. V. Maxwell, D. C. Brown, K. G. Friesen, R. E. Musser, Z. B. Johnson, and T. A. Armstrong. 2004. Effects of dietary lysine and energy density on performance and carcass characteristics of finishing pigs fed ractopamine. *Journal of Animal Science* 82:3277-3287.

Apple, J. K., P. J. Rincker, F. K. McKeith, S. N. Carr, T. A. Armstrong, and P. D. Matzat. 2007. Review: Meta-analysis of the ractopamine response in finishing swine. *The Professional Animal Scientist* 23:179-196.

Arfken, G. 1985. *Mathematical Methods for Physicists,* 3rd Ed. Orlando, FL: Academic Press.

Bergsma, R., E. Kanis, M. W. A. Verstegen, C. M. C. van der Peet Schwering, and E. F. Knol. 2009. Lactation efficiency as a result of body composition dynamics and feed intake in sows. *Livestock Science* 125:208-222.

Bikker, P., M. W. Verstegen, R. G. Campbell, and B. Kemp. 1994. Digestible lysine requirement of gilts with high genetic potential for lean gain, in relation to the level of energy intake. *Journal of Animal Science* 72:1744-1753.

Birkenfeld, C., J. Doberenz, H. Kluge, and K. Eder. 2006. Effect of L-carnitine supplementation of sows on L-carnitine status, body composition and concentrations of lipids in liver and plasma of their piglets at birth and during the suckling period. *Animal Feed Science and Technology* 129:23-38.

Black, J. L. 2009. Models to predict feed intake. Pp. 323-351 in *Voluntary Feed Intake in Pigs,* D. Torrallardona and E. Roura, eds. Wageningen, The Netherlands: Wageningen Academic.

Black, J. L., R. G. Campbell, I. H. Williams, K. J. James, and G. T. Davies. 1986. Simulation of energy and protein utilization in the pig. *Research and Development in Agriculture* 3:121-145.

Boomgaardt, J., D. H. Baker, A. H. Jensen, and B. G. Harmon. 1972. Effect of dietary lysine levels on 21-day lactation performance of first-litter sows. *Journal of Animal Science* 34:408-410.

Bruce, J. M., and J. J. Clark. 1979. Models of heat production and critical temperature for growing pigs. *Animal Production* 28:353-369.

BSAS (British Society of Animal Science). 2003. *Nutrient Requirement Standards of Pigs,* C. T. Whittemore, M. J. Hazzledine, and W. H. Close, authors. Penicuik, UK: British Society of Animal Science.

Canario, L., M. C. Père, T. Tribout, F. Thomas, C. David, J. Gogué, P. Herpin, J. P. Bidanel, and J. Le Dividich. 2007. Estimation of genetic trends from 1977 to 1998 of body composition and physiological state of Large White pigs at birth. *Animal* 1:1409-1413.

Charneca, R., J. L. T. Nunes, and J. Le Dividich. 2010. Body composition and blood parameters of newborn piglets from Alentejano and conventional (Large White × Landrace) genotype. *Spanish Journal of Agricultural Research* 8:317-325.

Coma, J., D. R. Zimmerman, and D. Carrion. 1995. Interactive effects of feed intake and stage of growth on the lysine requirement of pigs. *Journal of Animal Science* 73:3369-3375.

Cooper, D. R., J. F. Patience, R. T. Zijlstra, and M. Rademacher. 2001. Effect of energy and lysine intake in gestation on sow performance. *Journal of Animal Science* 7:2367-2377.

de Lange, C. F. M., P. C. H. Morel, and S. H. Birkett. 2003. Modeling chemical and physical body composition of the growing pig. *Journal of Animal Science* 81:E159-E165.

Dourmad, J. Y., D. Guillou, B. Sève, and Y. Henry. 1996. Response to dietary lysine supply during the finisher period in pigs. *Livestock Production Science* 45:179-186.

Dourmad, J. Y., J. Noblet, M. C. Père, and M. Etienne. 1999. Mating, pregnancy and prenatal growth. Pp. 129-152 in *Quantitative Biology of the Pig,* I. Kyriazakis, ed. Wallingford, UK: CABI.

Dourmad, J. Y., M. Étienne, A. Valancogne, S. Dubois, J. van Milgen, and J. Noblet. 2008. InraPorc: A model and decision support tool for the nutrition of sows. *Animal Feed Science and Technology* 143:372-386.

Ebert, A. R., A. S. Berman, R. J. Harrell, A. M. Kessler, S. G. Cornelius, and J. Odle. 2005. Vegetable proteins enhance growth of milk fed piglets, despite lower apparent ileal digestibility. *Journal of Nutrition* 135:2137-2143.

Emmans, G. C., and I. Kyriazakis. 1995. A general method for predicting the weight of water in the empty bodies of pigs. *Animal Science* 61:103-108.

Fortin, A., A. K. W. Tong, and W. M. Robertson. 2004. Evaluation of three ultrasound instruments, CVT-2, UltraFom 300 and AutoFom for predicting salable meat yield and weight of lean in the primals of pork carcasses. *Meat Science* 68:537-549.

GfE (Society of Nutrition Physiology). 2008. Energy and nutrient requirements of livestock, Nr. 11: Recommendations for the supply of energy and nutrients to pigs. Committee for Requirement Standards of the GfE. Frankfurt am Main, Germany: DLG-Verlag.

Gill, B. P. 2006. Body composition of breeding gilts in response to dietary protein and energy balance from thirty kilograms of body weight to completion of first parity. *Journal of Animal Science* 84:1926-1934.

Gonyou, H. W., M. C. Brumm, E. Bush, J. Deen, S. A. Edwards, T. Fangman, J. J. McGlone, M. Meunier-Salaun, R. B. Morrison, H. Spoolder, P. L. Sundberg, and A. K. Johnson. 2006. Application of broken-line analyses to assess floor space requirements of nursery and grower-finisher pigs expressed on an allometric basis. *Journal of Animal Science* 84:229-235.

Hendriks, W. H., and P. J. Moughan. 1993. Whole-body mineral composition of entire male and female pigs depositing protein at maximum rates. *Livestock Production Science* 33:161-170.

Johnson, R. K., E. P. Berg, R. Goodwin, J. W. Mabry, R. K. Miller, O. W. Robison, H. Sellers, and M. D. Tokach. 2004. Evaluation of procedures to predict fat-free lean in swine carcasses. *Journal of Animal Science* 82:2428-2441.

Johnston, L. J., J. E. Pettigrew, and J. W. Rust. 1993. Response of maternal-line sows to dietary protein concentration during lactation. *Journal of Animal Science* 71:2151-2156.

Jondreville, C., and J. Y. Dourmad. 2005. Le phosphore dans la nutrition des porcs. *INRA Productions Animales* 18:183-192.

Jondreville, C., and J. Y. Dourmad. 2006. Phosphorus in pig nutrition. *Proceedings of the AAAP Animal Science Congress.* Busan, Korea: Bexco.

Jongbloed, A. W., H. Everts, P. A. Kemme, and Z. Mroz. 1999. Quantification of absorbability and requirements of macroelements. Pp. 275-298 in *Quantitative Biology of the Pig,* I. Kyriazakis, ed. Wallingford, UK: CABI.

Jongbloed, A. W., J. Th. M. Van Diepen, and P. A. Kemme. 2003. Fosfornormen voor varkens: herziening 2003 (Phosphorus allowances for pigs: revision 2003). Lelystad, The Netherlands: Centraal Veevoederbureau.

Kleiber, M. 1961. *The Fire of Life: An Introduction to Animal Energetics.* New York: John Wiley & Sons, Inc.

Losel, D., C. Kalbe, and C. Rehfeldt. 2009. L-Carnitine supplementation during suckling intensifies the early postnatal skeletal myofiber formation in piglets of low birth weight. *Journal of Animal Science* 87:2216-2226.

Mahan, D. C. 1998. Relationship of gestation protein and feed intake level over a five-parity period using a high-producing sow genotype. *Journal of Animal Science* 76:533-541.

Mathews, S. A. 2004. Investigating the effects of long chain polyunsaturated fatty acids on lipid metabolism and body composition in the neonatal pig. Ph.D. Dissertation, North Carolina State University, Raleigh, NC. Available online at http://www.lib.ncsu.edu/resolver/1840.16/3419.

Moehn, S., A. M. Gillis, P. J. Moughan, and C. F. M. de Lange. 2000. Influence of dietary lysine and energy intakes on body protein deposition and lysine utilization in the growing pig. *Journal of Animal Science* 78:1510-1519.

Moehn, S., R. O. Ball, M. F. Fuller, A. M. Gillis, and C. F. M. de Lange. 2004. Growth potential, but not body weight or moderate limitation of lysine intake, affects inevitable lysine catabolism in growing pigs. *Journal of Nutrition* 134:2287-2292.

Moehn, S., D. Franco, C. Levesque, R. Samual, and R. O. Ball. 2011. New energy and amino acid requirements for gestating sows. Advances in Pork Production. Pp. 1-10 in *Proceedings of the 22 Annual Banff Pork Seminar,* University of Alberta, Edmonton, Alberta, Canada.

Monegue, H. J., G. L. Cromwell, R. D. Coffey, S. D. Carter, and M. Cervantes. 1993. Elevated dietary lysine levels for sows nursing large litters. *Journal of Animal Science* 71(Suppl. 1):67 (Abstr.).

Moughan, P. J. 1999. Protein metabolism in the growing pig. Pp. 299-331 in *Quantitative Biology of the Pig*, I. Kyriazakis, ed. Wallingford, UK: CABI.

Musser, R. E., D. L. Davis, S. S. Dritz, M. D. Tokach, J. L. Nelssen, J. E. Minton, and R. D. Goodband. 2004. Conceptus and maternal responses to increased feed intake during early gestation in pigs. *Journal of Animal Science* 82:3154-3161.

National Pork Board. 2000. *Pork Composition and Quality Assessment Procedures*. Des Moines, IA: National Pork Board.

Noblet, J., L. Le Dividich, and J. van Milgen. 2001. Thermal environment and swine nutrition. Pp. 519-544 in *Swine Nutrition*, A. J. Lewis and L. L. Southern, eds. Boca Raton, FL: CRC Press.

NRC (National Research Council). 1987. *Predicting Feed Intake of Food-Producing Animals*. Washington, DC: National Academy Press.

NRC. 1998. *Nutrient Requirements of Swine*, 10th Rev. Ed. Washington, DC: National Academy Press.

Pastorelli, G., M. Neil, and I. Wigren. 2009. Body composition and muscle glycogen contents of piglets of sows fed diets differing in fatty acids profile and contents. *Livestock Science* 123:329-334.

Pauly, C., P. Spring, J. V. O'Doherty, S. Ampuero Kragten, and G. Bee. 2009. Growth performance, carcass characteristics and meat quality of group-penned surgically castrated, immunocastrated (Improvac®) and entire male pigs and individually penned entire male pigs. *Animal* 3:1057-1066.

Pomar, C., I. Kyriazakis, G. C. Emmans, and P. W. Knap. 2003. Modeling stochasticity: Dealing with populations rather than individual pigs. *Journal of Animal Science* 81:E178-E186.

Quiniou, N. 1995. Utilisation de l'énergie chez le porc selon son potential de croissance: Contribution a la modelisation des besoins nutritionnels et de la composition corporelle. Ph.D. Dissertation. Rennes, France: INRA (French National Institute for Agricultural Research).

Quiniou, N., S. Dubois, and J. Noblet. 2000. Voluntary feed intake and feeding behaviour of group-housed growing pigs are affected by ambient temperature and body weight. *Livestock Production Science* 63:245-253.

Schinckel, A. P., J. R. Wagner, J. C. Forrest, and M. E. Einstein. 2001. Evaluation of alternative measures of pork carcass composition. *Journal of Animal Science* 79:1093-1119.

Schinckel, A. P., N. Li, B. T. Richert, P. V. Preckel, K. Foster, and M. E. Einstein. 2006. Development of a model to describe the compositional growth and dietary lysine requirements of pigs fed increasing dietary concentrations of ractopamine. *The Professional Animal Scientist* 22:438-449.

Schinckel, A. P., M. E. Einstein, S. Jungst, C. Booher, and S. Newman. 2009a. Evaluation of different mixed model nonlinear functions to describe the body weight growth of pigs of different sire and dam lines. *The Professional Animal Scientist* 25:307-324.

Schinckel, A. P., M. E. Einstein, S. Jungst, C. Booher, and S. Newman. 2009b. Evaluation of different mixed model nonlinear functions to describe the feed intakes of pigs of different sire and dam lines. *The Professional Animal Scientist* 25:345-359.

Schinckel, A. P., C. R. Schwab, V. M. Duttlinger, and M. E. Einstein. 2010a. Analyses of feed and energy intakes during lactation for three breeds of sows. *The Professional Animal Scientist* 26:35-50.

Schinckel, A. P., J. R. Wagner, J. C. Forrest, and M. E. Einstein. 2010b. Evaluation of the prediction of alternative measures of pork carcass composition by three optical probes. *Journal of Animal Science* 88:767-794.

Srichana, P. 2006. Amino acid nutrition in gestating and lactating sows. Ph.D. Dissertation, University of Missouri-Columbia.

Stahly, T. S., G. L. Cromwell, and H. J. Monegue. 1990. Lactational responses of sows nursing large litters to dietary lysine levels. *Journal of Animal Science* 68(Suppl. 1):369 (Abstr.).

Torrallardona, D., and E. Roura, eds. 2009. *Voluntary Feed Intake in Pigs*. Wageningen, The Netherlands: Wageningen Academic.

van der Peet-Schwering, C. M., B. Kemp, G. P. Binnendijk, L. A. den Hartog, H. A. Spoolder, and M. W. A. Verstegen. 2003. Performance of sows fed high levels of nonstarch polysaccharides during gestation and lactation over three parities. *Journal of Animal Science* 81:2247-2258.

van Milgen, J., J. Noblet, A. Valancogne, S. Dubois, and J. Y. Dourmad. 2008. InraPorc: A model and decision support tool for the nutrition of growing pigs. *Animal Feed Science and Technology* 143:387-405.

Veum, T. L., J. D. Crenshaw, T. D. Crenshaw, G. L. Cromwell, R. A. Easter, R. C. Ewan, J. L. Nelssen, E. R. Miller, J. E. Pettigrew, and M. R. Ellersieck. 2009. The addition of ground wheat straw as a fiber source in the gestation diet of sows and the effect on sow and litter performance for three successive parities. *Journal of Animal Science* 87:1003-1012.

Wagner, J. R., A. P. Schinckel, W. Chen, J. C. Forrest, and B. L. Coe. 1999. Analysis of body composition changes of swine during growth and development. *Journal of Animal Science* 77:1442-1466.

Webster, M. J., R. D. Goodband, M. D. Tokach, J. L. Nelssen, S. S. Dritz, J. A. Unruh, K. R. Brown, D. E. Real, J. M. Derouchey, J. C. Woodworth, C. N. Groesbeck, and T. A. Marsteller. 2007. Interactive effects between ractopamine hydrochloride and dietary lysine on finishing pig growth performance, carcass characteristics, pork quality, and tissue accretion. *The Professional Animal Scientist* 23:597-611.

Wu, G., and D. A. Knabe. 1994. Free and protein-bound amino acids in sow's colostrum and milk. *Journal of Nutrition* 124:415-424.

Young, M. G., M. D. Tokach, F. X. Aherne, R. G. Main, S. S. Dritz, R. D. Goodband, and J. L. Nelssen. 2004. Comparison of three methods of feeding sows in gestation and subsequent effects on lactation performance. *Journal of Animal Science* 82:3058-3070.

Zhu, C. L., M. Rademacher, and C. F. M. de Lange. 2005. Increasing dietary pectin level reduces utilization of digestible threonine intake, but not lysine intake, for body protein deposition in growing pigs. *Journal of Animal Science* 83:1044-1053.

Zijlstra, R. T., K. Y. Whang, R. A. Easter, and J. Odle. 1996. Effect of feeding a milk replacer to early-weaned pigs on growth, body composition, and small intestinal morphology, compared with suckled littermates. *Journal of Animal Science* 74:2948-2959.

第9章 玉米和大豆加工副产物

引言

自从20世纪50年代早期开始使用玉米-豆粕型日粮以来，在美国，猪饲料就以玉米-豆粕型为主（Cromwell，2000）。玉米和豆粕的氨基酸组成互补性很好，即玉米中的含硫氨基酸水平相对较高，而含硫氨基酸是豆粕对畜禽的第一限制性氨基酸；豆粕中的赖氨酸和色氨酸水平相对较高，而赖氨酸和色氨酸则是玉米对畜禽的第一限制性氨基酸。尽管玉米-豆粕型日粮已广泛应用，但猪并不是一定需要这两种原料。事实上，猪需要的是能量和某些特定的营养素，有时通过一些非玉米或者非豆粕原料提供这些营养素可能会更经济。例如，湿法碾磨和干法碾磨玉米酒精工业生产大量玉米副产物，另外从大豆中也生产出很多除了豆粕以外的其他副产物。这些大部分是食品工业的副产物，都能很好地应用到猪的日粮配制中。

本章主要对猪饲料中使用的玉米和大豆工业副产物在能量和养分的组成及消化率方面的区别进行了描述，但本章无法全面概述所有副产物的使用。有关各种副产物的使用量和实际使用的具体建议已有大量已发表的综述可以参考，本章中引用了其中部分文章。

玉米副产物

玉米酒精糟（DDG）、玉米酒精糟及可溶物（DDGS）、低脂DDGS和脱脂DDGS

如果用玉米制作酒精或酒精饮料，玉米是通过发酵和蒸馏生成二氧化碳和酒精或酒精饮料。玉米未发酵的部分（即蛋白质、脂肪、纤维和灰分）就是其副产物。这些副产物通常被分为酒精糟和可溶物部分。酒精糟可干燥制成DDG单独出售，可溶物部分也可以加入酒精糟中，然后干燥制成DDGS（Shurson and Alghamdi，2008；Belyea et al.，2010；Liu，2011；Stein，2012）。DDG和DDGS中的粗脂肪含量为9%~14%，但一些酒精加工厂会将可溶物先离心提取脂肪，然后加入酒精糟，这样就制成了低脂DDGS，这种产品的粗脂肪含量为5%~8%。到目前为止还没有文献报道低脂DDGS的营养价值，但是估计低脂DDGS的消化能值和代谢能值要比常规DDGS低。

还可以通过溶剂提取工艺提取DDGS中的脂肪，其产物就是脱脂DDGS，它的粗脂肪含量为2%~6%（Jacela et al.，2011）。脱脂DDGS的能值远低于常规DDGS，但其氨基酸的含量和消化率与常规DDGS的相当（Jacela et al.，2011）。

常规DDGS的粗蛋白含量为25%~30%。由于其蛋白质主要来源于玉米，所以赖氨酸（0.5%~1.0%）和色氨酸（0.10%~0.34%）的含量很低（Spiehs et al.，2002；Stein and Shurson，2009；Liu，2011）。DDGS中赖氨酸的含量比其他氨基酸含量变异更大（Shurson and Alghamdi，2008），因为过度热处理会破坏DDGS中的赖氨酸，或者将赖氨酸转变成其他不能用于蛋白质合成的化合物（Fastinger and Mahan，2006；Pahm et al.，2008a，b；Stein and Shurson，2009；见本书第2章）。过度热处理对DDG中赖氨酸的破坏程度要低于对DDGS中的，因为将可溶物加入酒精糟中会提高发生美拉德反应的概率，因而导致赖氨酸的破坏（Pahm et al.，2008b）。DDGS中的美拉德反应也降低了赖氨酸的回肠表观和标准消化率。因此，DDGS中的赖氨酸消化率比其他氨基酸变异性更大（Fastinger and Mahan，2006；Stein and Shurson，2009）。

DDGS 的中性洗涤纤维（NDF）含量为 30%~35%（Spiehs et al., 2002），但因其脂肪和蛋白质的浓度相对较高，所以它的消化能和代谢能与玉米相当（Pedersen et al., 2007; Stein et al., 2009）。DDGS 的磷含量为 0.37%~0.88%（Shurson and Alghamdi, 2008）。在 DDGS 的生产过程中一部分植酸盐被水解，这可能是由于在发酵过程中加入的酵母菌产生了少量的植酸酶（Liu, 2011）。DDGS 中的磷只有 43%与植酸结合，而玉米粉中的磷却有 73%与植酸结合（Liu and Han, 2011）。所以，DDGS 的磷消化率为 50%~70%，而玉米的磷消化率低于 40%（Pedersen et al., 2007; Almeida and Stein, 2010, 2012）。不过添加微生物植酸酶可提高玉米的磷消化率，但不能进一步提高 DDGS 的磷消化率（Almeida and Stein, 2010, 2012）。

DDGS 中大部分矿物质的含量是玉米的 3 倍，而硫、钠和钙的含量比玉米中的 3 倍还多，这是因为在 DDGS 的生产过程中加入了外源的这些矿物元素（Liu and Han, 2011）。与玉米相比，DDGS 的硫含量更高，这可能会导致配制的日粮中硫的含量也会高于玉米-豆粕型日粮，但适口性和生长性能似乎均不受影响（Kim et al., 2012）。

多篇文献综述了 DDGS 在生长猪和种猪日粮中的使用（Shurson et al., 2004; Patience et al., 2007; Shurson and Alghamdi, 2008; Stein and Shurson, 2009; Stein, 2012）。在泌乳母猪日粮中可添加至 30%的 DDGS 而不影响母猪和仔猪的生产性能（Hill et al., 2008; Song et al., 2010），怀孕母猪的日粮中可添加高达 44%的 DDGS（Thong et al., 1978）。虽然有报道指出添加 20%的 DDGS 会对断奶仔猪生长性能产生不良影响（Kim et al., 2012），但是大量报道仍认为在断奶仔猪日粮中可添加 20%~30%的 DDGS 而不影响其生长性能（Whitney and Shurson, 2004; Almeida and Stein, 2010; Jones et al., 2012a）。

大量的试验表明在生长育肥猪日粮中添加 30%的 DDGS 而不降低猪的生长性能（Widyaratne and Zijlstra, 2007; Widmer et al., 2008; Xu et al., 2010a; Yoon et al., 2010; McDonnell et al., 2011），但也有报道指出在生长育肥猪日粮中添加 30%的 DDGS 会降低其生长性能（Whitney et al., 2006; Linneen et al., 2008; Leick et al., 2010; Kim et al., 2012）。最新的一个试验表明，添加 45%的 DDGS 会对生长育肥猪的平均日增重产生轻微的不良影响，但不影响采食量和饲料转化率（Cromwell et al., 2011）。

许多试验研究报道了 DDGS 对胴体组成和猪肉品质的影响。其中约 50%的研究表明，添加 DDGS 降低屠宰率，但是在另外 50%的研究中并没有观察到这一现象（Stein and Shurson, 2009）。DDGS 对瘦肉率和背膘厚的影响极小，但研究一致表明 DDGS 会导致育肥猪脂肪组织中不饱和脂肪酸的含量增加（Benz et al., 2010; Leick et al., 2010; Xu et al., 2010a,b; Cromwell et al., 2011）。不饱和脂肪酸含量的增加会使猪腹部变软，因而降低熏肉的切片质量（Whitney et al., 2006; Leick et al., 2010; Cromwell et al., 2011）。在屠宰前 3~4 周的日粮中停止使用 DDGS，可部分改善猪腹部脂肪的硬度（Xu et al., 2010b）。

虽然不是全部，但有部分研究结果表明，断奶仔猪和生长育肥猪日粮中添加 DDGS 会导致采食量下降（Stein and Shurson, 2009; Stein, 2012）。采食量下降的原因可能是，与添加 DDGS 的日粮相比，猪更喜欢不添加 DDGS 的日粮（Seabolt et al., 2010; Kim et al., 2012）。

在猪日粮中添加 DDGS 对猪其他方面的影响包括增加排粪量，这是由于 DDGS 的干物质消化率低于玉米和豆粕的（Shurson et al., 2004; McDonnell et al., 2011）。同时，在猪日粮中添加 DDGS 会使粪中氮的浓度升高（McDonnell et al., 2011），提高的幅度取决于配方技术。相反，与玉米相比，DDGS 的磷消化率更高，粪中磷的浓度会下降（Hill et al., 2008; Almeida and Stein, 2010）。

高蛋白 DDG、高蛋白 DDGS 和玉米胚芽

在一些酒精厂，玉米在发酵和蒸馏之前先进行去皮和脱胚处理。这个过程的目的是为了降低不发

酵物质（即纤维和脂肪）的浓度，提高进入发酵过程的淀粉浓度以提高酒精的产量（Rausch and Belyea, 2006; Rosentrater et al., 2012）。通过这种工艺制得的酒精糟，其粗蛋白（40%~48%）和粗灰分含量均高于常规的酒精糟，但是脂肪的浓度会下降至6%以下（Widmer et al., 2007; Kim et al., 2009; Jacela et al., 2010）。如果使用这种工艺，可溶物通常不加入酒精糟中，这样得到的酒精糟就叫做高蛋白 DDG（HP-DDG），但如果可溶物加入酒精糟中就叫做高蛋白 DDGS（HP-DDGS）（Stein, 2012）。HP-DDG 的消化能和代谢能比玉米和常规的 DDGS 高，氨基酸消化率则与常规 DDGS 相当（Widmer et al., 2007; Kim et al., 2009; Jacela et al., 2010）。HP-DDG 的磷含量比常规的 DDGS 低，但磷消化率与常规 DDGS 相当（Widmer et al., 2007; Almeida and Stein, 2012）。和 DDGS 相似，微生物植酸酶只能略微提高 HP-DDG 的磷消化率（Almeida and Stein, 2012）。

在必需氨基酸平衡的生长猪日粮中，至少可以使用40%的HP-DDG（Widmer et al., 2008），并可全部取代育肥猪日粮中的豆粕（Widmer et al., 2008; Kim et al., 2009）。目前还没有相关数据显示 HP-DDG 在断奶仔猪、怀孕母猪或泌乳母猪日粮中的使用量。

玉米胚芽是在最初的脱胚过程中制得的，也可作为猪饲料的原料。玉米胚芽的粗脂肪含量为16%~20%，粗蛋白含量大约为15%，而粗纤维的含量相对比较高（Widmer et al., 2007）。玉米胚芽的消化能和代谢能与玉米相当（Widmer et al., 2007）。玉米胚芽的磷含量在1.1%以上，但大部分与磷酸相结合，因此，玉米胚芽的磷消化率很低（Widmer et al., 2007; Almeida and Stein, 2012）。在使用玉米胚芽的日粮中添加微生物植酸酶能在一定程度上提高磷的消化率，使它接近于 HP-DDGS 和 DDGS 的水平（Almeida and Stein, 2012）。在生长育肥猪日粮中可添加30%的玉米胚芽而不会影响猪的生长性能（Lee, 2011）。但由于玉米胚芽中含有相对较高浓度的不饱和油脂，如果猪只饲喂含玉米胚芽的日粮，其中高浓度的不饱和脂肪酸将沉积于背膘和腹脂中，使猪只腹部变软（Lee, 2011）。目前在断奶仔猪、怀孕母猪或泌乳母猪日粮中使用玉米胚芽的影响尚无相关数据发表。

玉米蛋白粉、玉米蛋白饲料、玉米胚芽粕和玉米糁

玉米蛋白粉是玉米通过湿法工艺除去大部分的淀粉、胚芽和部分纤维后得到的副产物（Stock et al., 2000）。全部的蛋白质留在了玉米蛋白粉中，其粗蛋白含量约60%，并且中性洗涤纤维含量很低（de Godoye et al., 2009; Almeida et al., 2011）。对生长育肥猪来说，玉米蛋白粉的大部分氨基酸消化率高于玉米（Knabe et al., 1989; Almeida et al., 2011），代谢能和消化能值也高于玉米（Young et al., 1977）。

相对于猪的营养需要来说，玉米蛋白粉的氨基酸组成并不理想，因此在猪日粮中的玉米蛋白粉的用量相对较少。如果在使用玉米蛋白粉的日粮中添加晶体赖氨酸和色氨酸，就能配制出必需氨基酸平衡的日粮。在断奶仔猪日粮中可添加高达15%的玉米蛋白粉而不影响猪的生长性能（Mahan, 1993）。

玉米蛋白饲料也是湿法工艺的副产物，是玉米提取淀粉、胚芽和谷朊用于制作玉米淀粉和玉米糖浆后的部分。它主要由玉米皮、玉米胚芽和玉米浆组成（Honeyman and Zimmerman, 1991; Stock et al., 2000）。因此，玉米蛋白饲料中纤维含量高，其中性洗涤纤维含量高于30%，粗蛋白含量为20%~25%。玉米蛋白饲料大部分氨基酸的消化率接近玉米（Almeida et al., 2011），对于生长育肥猪来说，其消化能和代谢能低于玉米（Yen et al., 1974; Young et al., 1977），但如果用于饲喂怀孕母猪，其代谢能和消化能与玉米相当（Honeyman and Zimmerman, 1991）。玉米蛋白饲料通常不用于断奶仔猪和生长猪日粮中，但可在怀孕母猪日粮中大量使用而不影响其繁殖和仔猪生产性能（Honeyman and Zimmerman, 1990）。

玉米胚芽是玉米湿磨工艺中，在分离出淀粉之前的初始步骤中将胚芽从玉米谷粒中分离获得的

（Stock et al., 2000），或者在玉米生产玉米粉、玉米糁或其他玉米产品前通过干磨法得到的。玉米胚芽通过提取，生产出来的油脂可作为人类的食用油。脱脂后的玉米胚芽就叫做玉米胚芽粕，其脂肪含量一般低于3%（Stock et al., 2000; Weber et al., 2010）。因此玉米胚芽粕与玉米胚芽的组成差别很大。玉米胚芽粕的中性洗涤纤维的含量高于50%，粗蛋白含量大约为20%（Weber et al., 2010）。对生长育肥猪来说，玉米胚芽粕的大部分氨基酸消化率略低于玉米（Almeida et al., 2011）。在生长猪日粮中使用38%的玉米胚芽粕可能不会影响其生长性能，但可能降低其饲料转化率（Weber et al., 2010）。

玉米糁饲料是玉米通过干磨法工艺提取玉米淀粉、玉米糁和珍珠米后的副产物，它由玉米糠、破碎玉米粒、提油后的胚芽渣，以及胚芽、种皮和胚乳碎片组成（Larson et al., 1993; Stock et al., 2000）。玉米糁饲料的粗蛋白含量为6%~10%，乙醚浸出物的含量大于4%，淀粉和中性洗涤纤维的含量是变异的，但是大部分来源的玉米糁饲料的淀粉含量高于50%，中性洗涤纤维含量低于30%（Larson et al., 1993）。对猪来说，玉米糁饲料的能值与玉米相当（Stanley and Ewan, 1982），但是其大部分氨基酸的消化率低于玉米（Almeida et al., 2011）。玉米糁饲料的适口性很好，很容易被猪采食，因此可以应用于各阶段猪的日粮中。但是，目前还未有相关的梯度试验研究不同阶段猪的日粮中玉米糁饲料的最适宜添加量。

大豆制品

全脂大豆

典型的美国大豆含有15%~20%的乙醚浸出物和35%~37%的粗蛋白（Grieshop et al., 2003; Karr-Lilienthal et al., 2005）。由于大豆中含有胰蛋白酶抑制因子，在饲喂给猪前需要热处理，这一过程通常在使用前进行挤压膨胀完成（Baker, 2000）。在生大豆中，胰蛋白酶抑制因子的含量大约为35个胰蛋白酶抑制因子单位，但经热处理后这一指标可降低到小于4个单位（Lallès, 2000; Goebel and Stein, 2011a）。全脂大豆可以以完整的全脂大豆或者去皮全脂大豆的形式饲喂。完整的全脂大豆含有8%~12%的中性洗涤纤维，而去皮全脂大豆含有5%的中性洗涤纤维。完整的大豆中总碳水化合物含量为35%~40%，其中大约15%为非结构性碳水化合物（主要为蔗糖和寡糖），剩下的为结构性多糖，如酸性多糖、阿拉伯半乳聚糖和纤维素物质（Karr-Lilienthal et al., 2005）。大豆中淀粉含量低于1.0%。

近几年，育种选育出了高蛋白大豆。这些大豆的粗蛋白含量为44%~48%，而普通的大豆只含有35%~37%的粗蛋白（Cervantes-Pahm and Stein, 2008; Baker et al., 2011）。但粗蛋白含量的提高却是以降低乙醚浸出物和某些碳水化合物为代价的，大豆中粗蛋白和乙醚浸出物含量呈负相关（Yaklich, 2001）。高蛋白大豆的蔗糖和中性洗涤纤维的含量也通常低于普通大豆（Hartwig et al., 1997; Cervantes-Pahm and Stein, 2008）。

普通大豆含有大约15%的非结构性碳水化合物，如蔗糖、糖醛酸、寡糖和游离糖（Grieshop et al., 2003; Karr-Lilienthal et al., 2005）。普通大豆蔗糖的含量为4%~8%，寡糖（棉子糖、水苏糖、毛蕊花糖）的含量为4%~7%（Grieshop et al., 2003; Cervantes-Pahm and Stein, 2008; Goebel and Stein, 2011a）。由于寡糖对幼龄动物有不良营养作用，因此选育了寡糖含量低于2%的大豆品种（van Kempen et al., 2006; Baker and Stein, 2009; Baker et al., 2010）。人们相信，与普通豆粕相比，仔猪更容易耐受用低寡糖品种大豆制成的豆粕，但目前还没有相关数据证实这一假设。

豆粕

溶剂浸出豆粕

大部分的大豆经过溶剂浸出油脂后以脱脂豆粕的形式添加到猪的日粮中。大豆在浸出油脂之前先进行清洗和压片，提取的油脂应用于工业或者食品行业，而大部分豆粕则应用于畜禽饲料。脱脂豆粕通过脱溶剂处理去除残留的乙烷后，用蒸汽熟化使胰蛋白酶抑制因子灭活（Witte, 1995）。测定脲酶就可以知道豆粕中胰蛋白酶抑制因子的水平，如果在标准胰蛋白酶测定中 pH 升高低于 0.2，则说明胰蛋白酶抑制因子已经被灭活（Witte, 1995）。豆粕生产的最后一个步骤是将其粉碎成正常的粒度。通过溶剂浸出得到的豆粕的乙醚浸出物含量通常低于 3%（Wang and Johnson, 2001; Karr-Lilienthal et al., 2005）。

用于制作豆粕的豆子可以是完整的豆子，也可以是在压片前去皮的豆子（Ericson, 1995）。这两种工艺得到的豆粕分别为带皮和去皮豆粕。去皮豆粕的粗蛋白含量为 46%~48%（Grieshop et al., 2003; Baker and Stein, 2009），中性洗涤纤维的含量为 6%~8%；而带皮豆粕粗蛋白含量为 42%~44%，中性洗涤纤维含量为 12%~14%（Cervantes-Pahm and Stein, 2008）。

机械压榨豆粕

作为溶剂浸出工艺的一种替代方法，大豆还可以利用连续的螺旋压力通过机械提取或机械榨取将大豆脱脂。美国采用这一工艺生产的豆粕少于 1%（Ericson, 1995）。因为通常在榨油前先通过挤压来进行热处理，所以这种工艺生产的豆粕叫压榨豆粕（Wang and Johnson, 2001; Woodworth et al., 2001; Baker and Stein, 2009）。由于机械榨油工艺的效率低于溶剂浸出工艺，因此压榨豆粕的乙醚浸出物含量为 5%~10%（Wang and Johnson, 2001; Karr-Lilienthal et al., 2006）。用于压榨的大豆通常不去皮，因此相对于溶剂浸出的去皮豆粕，压榨豆粕的中性洗涤纤维较高而粗蛋白含量较低（Karr-Lilienthal et al., 2006; Baker and Stein, 2009）。

酶处理豆粕和发酵豆粕

普通豆粕中由于抗原的存在，阻碍了豆粕在仔猪日粮中的大量使用（Li et al., 1990）。但大豆中的抗原可通过酶处理或者发酵的方法去除。这两种工艺中都可以去除蔗糖和大部分的寡糖，所以酶处理或发酵工艺生产的豆粕抗原和寡糖的含量都很低（Cervantes-Pahm and Stein, 2010; Goebel and Stein, 2011b）。去除蔗糖和寡糖后的酶处理豆粕和发酵豆粕的总养分组成与普通大豆不同（Cervantes-Pahm and Stein, 2010）。酶处理豆粕和发酵豆粕粗蛋白的含量为 52%~57%，中性洗涤纤维（NDF）的含量也高于普通豆粕（Cervantes-Pahm and Stein, 2010; Goebel and Stein, 2011b）。

发酵豆粕和酶处理豆粕去除了抗原和寡糖，因此人们相信这两种来源的豆粕可以在断奶仔猪日粮中应用，不会像普通豆粕一样引起消化不良。最近的数据也证实这两种来源的豆粕可替代断奶仔猪日粮中的动物源性蛋白（Yang et al., 2007; Jones et al., 2010b; Kim et al., 2010）。

大豆浓缩蛋白和大豆分离蛋白

大豆浓缩蛋白是通过去除去皮脱脂豆粕中的水溶性或醇溶性的非蛋白成分（含有可溶性碳水化合

物）后获得的（Lusas and Rhee, 1995; Endres, 2001）。根据定义，大豆浓缩蛋白至少含有65%的粗蛋白（干物质基础）（Lusas and Rhee, 1995; Endres, 2001）。大豆浓缩蛋白可以通过酸浸，用含水酒精提取，或通过湿热法使蛋白质变性后用水提取生产（Endres, 2001）。大豆浓缩蛋白可在断奶仔猪的日粮中替代动物源性蛋白而对生产性能无任何不良影响（Lenehan et al., 2007; Yang et al., 2007）。同样，大豆浓缩蛋白也可作为代乳粉中的蛋白源（Endres, 2001）。

大豆分离蛋白是通过去除去皮脱脂大豆中大部分的非蛋白成分获得的（Endres, 2001）。把蛋白质溶于中性或弱碱性溶液，再通过酸化使提取物沉淀得到分离蛋白（Berk, 1992）。大豆分离蛋白干物质中的蛋白质含量高于90%。大豆分离蛋白价格相对比较高，很少用于商业生产的猪饲料中，但可添加到半合成猪日粮中用于科学研究。大豆分离蛋白的氨基酸消化率很高，与酪蛋白的氨基酸消化率相似（Cervantes-Pahm and Stein, 2010）。

大豆皮

大部分的大豆在提取油脂前先去皮，脱脂后的豆粕可作为去皮豆粕销售，在这个过程中生产出的大豆皮单独销售，可用于猪饲料的配制中。大豆皮中性洗涤纤维含量高于50%，粗蛋白含量为12%~15%（Kornegay, 1981; Jacela et al., 2007; Barbosa et al., 2008）。由于中性洗涤纤维含量较高，大豆皮的代谢能相对较低（Jacela et al., 2007），因此建议在生长育肥猪日粮中的用量不要超过15%（Kornegay, 1981）。同样，在日粮中添加大豆皮可能使某些氨基酸的消化率降低（Dilger et al., 2004）。

粗甘油

近年来生物柴油工业的生产规模不断扩大，粗甘油是生物柴油生产的副产物。生产1L的生物柴油就可产生80g的粗甘油（Thompson and He, 2006; Sharma et al., 2008）。粗甘油的化学成分组成变异比较大，主要成分为甘油、水分和灰分，以及少量的脂肪酸与甲醇。粗甘油通常含有78%~85%的甘油、8%~15%的水、2%~10%的盐（NaCl或KCl）、0.5%的游离脂肪酸和小于0.5%的甲醇（Hansen et al., 2009; Kerr et al., 2009）。粗甘油可作为猪日粮中的能量来源（Bartelt and Schneider, 2002; Lammers et al., 2008b; Zijlstra et al., 2009），粗甘油的能值直接取决于其甘油、脂肪酸和甲醇的含量（Kerr et al., 2009）。粗甘油可用于各个阶段猪的日粮中，而不影响猪的生长性能、胴体组成和肉质（Groesbeck et al., 2008; Lammers et al., 2008a; Della Casa et al., 2009; Hansen et al., 2009; Zijlstra et al., 2009）。需要根据粗甘油中盐的种类和含量对配方进行调整，以防止日粮中的钠（Na）、钾（K）和氯（Cl）超量。同时也应特别注意甲醇的含量，因为甲醇是一个潜在的毒性化合物。在美国，只要粗甘油的甘油含量高于80%、水分低于15%、甲醇低于0.15%、盐低于8%、硫低于0.1%，以及重金属低于5ppm（AAFCO, 2010），那么其在非反刍动物全价料中的使用量最高可达10%。在德国，法规规定粗甘油中的甲醇允许量为0.5%（Normenkommission fur Einzelfuttermittel im Zentralausschuss der Deutschen Landwirtschaf, 2006）。

参 考 文 献

AAFCO (Association of American Feed Control Officials, Inc.). 2010. *Official Publication*. West Lafayette, IN: AAFCO.

Almeida, F. N., and H. H. Stein. 2010. Performance and phosphorus balance of pigs fed diets formulated on the basis of values for standardized total tract digestibility of phosphorus. *Journal of Animal Science* 88:2968-2977.

Almeida, F. N., and H. H. Stein. 2012. Effects of graded levels of microbial phytase on the standardized total tract digestibility of phosphorus in corn and corn coproducts fed to pigs. *Journal of Animal Science* 90:1262-1269.

Almeida, F. N., G. I. Petersen, and H. H. Stein. 2011. Digestibility of amino acids in corn, corn co-products, and bakery meal fed to growing pigs. *Journal of Animal Science* 89:4109-4115.

Baker, D. H. 2000. Nutritional constraints to the use of soy products by animals. Pp. 1-12 in *Soy in Animal Nutrition*, J. K. Drackley, ed. Savoy, IL: Federation of Animal Science Societies.

Baker, K. M., and H. H. Stein. 2009. Amino acid digestibility and concentration of digestible and metabolizable energy in soybean meal produced from high protein or low oligosaccharide varieties of soybeans and fed to growing pigs. *Journal of Animal Science* 87:2282-2290.

Baker, K. M., B. G. Kim, and H. H. Stein. 2010. Amino acid digestibility in conventional, high protein, or low oligosaccharide varieties of full-fat soybeans and in soybean meal by weanling pigs. *Animal Feed Science and Technology* 162:66-73.

Barbosa, F. F., M. D. Tokach, J. M. DeRouchey, R. D. Goodband, J. L. Nelssen, and S. S. Dritz. 2008. Variation in chemical composition of soybean hulls. Pp. 158-165 in *Proceedings of Kansas State University Swine Day*. Manhattan: Kansas State University.

Bartelt, J., and D. Schneider. 2002. Investigation on the energy value of glycerol in the feeding of poultry and pigs. Pp. 15-36 in *Union for the Promotion of Oilseeds—Schriften Heft 17*. Berlin: Union Zur Förderung Von Oel-Und Proteinplafalzen E.V.

Belyea, R. L., K. D. Rauch, T. E. Clevenger, V. Singh, D. B. Johnston, and M. E. Tumbleson. 2010. Sources of variation in composition of DDGS. *Animal Feed Science and Technology* 159:122-130.

Benz, J. M., S. K. Linneen, M. D. Tokach, S. S. Dritz, J. L. Nelssen, J. M. DeRouchy, R. D. Goodband, R. C. Sulabo, and K. J. Prusa. 2010. Effects of dried distillers grains with solubles on carcass fat quality of finishing pigs. *Journal of Animal Science* 88:3666-3682.

Berk, Z. 1992. Technology of production of edible flours and protein products from soybean. Available online at http://www.fao.org/docrep/t0532e/t0532e00.htm. Accessed July 7, 2009.

Cervantes-Pahm, S. K., and H. H. Stein. 2008. Effect of dietary soybean oil and soybean protein concentrate on the concentration of digestible amino acids in soybean products fed to growing pigs. *Journal of Animal Science* 86:1841-1849.

Cervantes-Pahm, S. K., and H. H. Stein. 2010. Ileal digestibility of amino acids in conventional, fermented, and enzyme treated soybean meal and in soy protein isolate, fishmeal, and casein fed to weanling pigs. *Journal of Animal Science* 88:2674-2683.

Cromwell, G. L. 2000. Utilization of soy products in swine diets. Pp. 258-282 in *Soy in Animal Nutrition*, J. K. Drackley, ed. Savoy, IL: Federation of Animal Science Societies.

Cromwell, G. L., M. J. Azain, O. Adeola, S. K. Baidoo, S. D. Carter, T. D. Crenshaw, S. W. Kim, D. C. Mahan, P. S. Miller, and M. C. Shannon. 2011. Corn distillers grains with solubles in diets for growing-finishing pigs: A cooperative study. *Journal of Animal Science* 89:2801-2811.

de Godoye, M. R. C., L. L. Bauer, C. M. Parsons, and G. C. Fahey, Jr. 2009. Select corn coproducts from the ethanol industry and their potential as ingredients in pet foods. *Journal of Animal Science* 87:189-199.

Della Casa, G., D. Bochicchio, V. Faeti, G. Marchetto, E. Poletti, A. Rossi, A. Garavaldi, A. Panciroli, and N. Brogna. 2009. Use of pure glycerol in fattening heavy pigs. *Meat Science* 81:238-244.

Dilger, R. N., J. S. Sands, D. Ragland, and O. Adeola. 2004. Digestibility of nitrogen and amino acids in soybean meal with added soyhulls. *Journal of Animal Science* 82:715-724.

Endres, J. G. 2001. *Soy Protein Products: Characteristics, Nutritional Aspects, and Utilization*. Urbana, IL: American Oil Chemists' Society.

Ericson, D. E. 1995. Overview of modern soybean processing and links between processes. Pp. 56-64 in *Practical Handbook of Soybean Processing and Utilization*, D. E. Ericson, ed. Urbana, IL: American Oil Chemists' Society.

Fastinger, N. D., and D. C. Mahan. 2006. Determination of the ileal amino acid and energy digestibilities of corn distillers dried grains with solubles using grower-finisher pigs. *Journal of Animal Science* 84:1722-1728.

Goebel, K. P., and H. H. Stein. 2011a. Ileal digestibility of amino acids in conventional and low-Kunitz soybean products fed to weanling pigs. *Asian-Australian Journal of Animal Science* 24:88-95.

Goebel, K. P., and H. H. Stein. 2011b. Phosphorus and energy digestibility of conventional and enzyme treated soybean meal fed to weanling pigs. *Journal of Animal Science* 89:764-772.

Grieshop, C. M., C. T. Kadzere, G. M. Clapper, E. A. Flickinger, L. L. Bauer, R. L. Frazier, and G. C. Fahey, Jr. 2003. Chemical and nutritional characteristics of United States soybeans and soybean meals. *Journal of Agricultural and Food Chemistry* 51:7684-7691.

Groesbeck, C. N., L. J. McKinney, J. M. DeRouchey, M. D. Tokach, R. D. Goodband, S. S. Dritz, J. L. Nelssen, A. W. Duttlinger, A. C. Fahrenholz, and K. C. Behnke. 2008. Effect of crude glycerol on pellet mill production and nursery pig growth performance. *Journal of Animal Science* 86:2228-2236.

Hansen, C. F., A. Hernandez, B. P. Mullan, K. Moore, M. Trezona-Murray, R. H. King, and J. R. Pluske. 2009. A chemical analysis of samples of crude glycerol from the production of biodiesel in Australia, and the effects of feeding crude glycerol to growing-finishing pigs on performance, plasma metabolites and meat quality at slaughter. *Animal Production Science* 49:154-161.

Hartwig, E. E., T. M. Kuo, and M. M. Kenty. 1997. Seed protein and its relationship to soluble sugars in soybean. *Crop Science* 37:770-778.

Hill, G. M., J. E. Link, M. J. Rincker, D. L. Kirkpatrick, M. L. Gibson, and K. Karges. 2008. Utilization of distillers dried grains with solubles and phytase in sow lactation diets to meet the phosphorus requirement of the sow and reduce fecal phosphorus concentration. *Journal of Animal Science* 86:112-118.

Honeyman, M. S., and D. R. Zimmerman. 1990. Long-term effects of corn gluten feed on the reproductive performance and weight of gestating sows. *Journal of Animal Science* 68:1329-1336.

Honeyman, M. S., and D. R. Zimmerman. 1991. Metabolizable energy of corn (maize) gluten feed and apparent digestibility of the fibrous components for gestating sows. *Animal Feed Science and Technology* 35:131-137.

Jacela, J. Y., J. M. DeRouchey, M. D. Tokach, J. L. Nelssen, R. D. Goodband, S. S. Dritz, and R. C. Sulabo. 2007. Amino acid digestibility and energy content of two different soy hull sources for swine. Pp. 142-149 in *Proceedings of the Kansas State University Swine Day*. Manhattan: Kansas State University.

Jacela, J. Y., H. L. Frobose, J. M. DeRouchey, M. D. Tokach, S. S. Dritz, R. D. Goodband, and J. L. Nelssen. 2010. Amino acid digestibility and energy concentration of high-protein corn dried distillers grains and high-protein sorghum dried distillers grains with solubles for swine. *Journal of Animal Science* 88:3617-3623.

Jacela, J. Y., J. M. DeRouchey, S. S. Dritz, M. D. Tokach, R. D. Goodband, J. M. Nelssen, R. C. Sulabo, R. C. Thaler, L. Brandts, D. E. Little, and K. J. Prusa. 2011. Amino acid digestibility and energy content of deoiled (solvent extracted) corn distillers dried grains with solubles for swine and its effects on growth performance and carcass characteristics. *Journal of Animal Science* 89:1817-1829.

Jones, C. K., J. R. Bergstrom, M. D. Tokach, J. M. DeRouchey, R. D. Goodband, J. L. Nelssen, and S. S. Dritz. 2010a. Efficacy of commercial enzymes in diets containing various concentrations and sources of dried distillers grains with solubles for nursery pigs. *Journal of Animal Science* 88:2084-2091.

Jones, C. K., J. M. DeRouchey, J. L. Nelssen, M. D. Tokach, S. S. Dritz, and R. D. Goodband. 2010b. Effects of fermented soybean meal and specialty animal protein sources on nursery pig performance. *Journal of Animal Science* 88:1725-1732.

Karr-Lilienthal, L. K., C. T. Kadzere, C. M. Grieshop, and G. C. Fahey, Jr. 2005. Chemical and nutritional properties of soybean carbohydrates as related to nonruminants: A review. *Livestock Production Science* 97:1-12.

Karr-Lilienthal, L. K., L. L. Bauer, P. L. Utterback, K. E. Zinn, R. L. Frazier, C. M. Parsons, and G. C. Fahey, Jr. 2006. Chemical composition and nutritional quality of soybean meals prepared by extruder/expeller processing for use in poultry diets. *Journal of Agricultural and Food Chemistry* 54:8108-8114.

Kerr, B. J., T. E. Weber, W. A. Dozier, III, and M. T. Kidd. 2009. Digestible and metabolizable energy content of crude glycerin originating from different sources in nursery pigs. *Journal of Animal Science* 87:4042-4049.

Kim, B. G., G. I. Petersen, R. B. Hinson, G. L. Allee, and H. H. Stein. 2009. Amino acid digestibility and energy concentration in a novel source of high-protein distillers dried grains and their effects on growth performance of pigs. *Journal of Animal Science* 87:4013-4021.

Kim, B. G., Y. Zhang, and H. H. Stein. 2012. Sulfur concentration in diets containing corn, soybean meal, and distillers dried grains with solubles does not affect feed preference or growth performance of weanling or growing-finishing pigs, *Journal of Animal Science* 90:272-281.

Kim, S. W., E. van Heugten, F. Ji, C. H. Lee, and R. D. Mateos. 2010. Fermented soybean meal as a vegetable protein source for nursery pigs: I. Effects on growth performance of nursery pigs. *Journal of Animal Science* 88:214-224.

Knabe, D. A., D. C. LaRue, E. J. Gregg, G. M. Martinez, and T. D. Tanksley, Jr. 1989. Apparent digestibility of nitrogen and amino acids in protein feedstuffs by growing pigs. *Journal of Animal Science* 67:441-458.

Kornegay, E. T. 1981. Soybean hull digestibility by sows and feeding value for growing-finishing pigs. *Journal of Animal Science* 53:138-145.

Lallès, J. P. 2000. Soy products as protein sources for preruminants and young pigs. Pp. 106-126 in *Soy in Animal Nutrition*, J. K. Drackley, ed. Savoy, IL: Federation of Animal Science Societies.

Lammers, P. J., B. J. Kerr, T. E. Weber, K. Bregendahl, S. M. Lonergan, K. J. Prusa, D. U. Ahn, W. C. Stoffregen, W. A. Dozier, III, and M. S. Honeyman. 2008a. Growth performance, carcass characteristics, meat quality, and tissue histology of growing pigs fed crude glycerin-supplemented diets. *Journal of Animal Science* 86:2962-2970.

Lammers, P. J., B. J. Kerr, T. E. Weber, W. A. Dozier, III, M. T. Kidd, K. Bregendahl, and M. S. Honeyman. 2008b. Digestible and metabolizable energy of crude glycerol in pigs. *Journal of Animal Science* 86:602-608.

Larson, E. M., R. A. Stock, T. J. Klopfenstein, M. H. Sindt, and D. H. Shain. 1993. Energy value of hominy feed for finishing ruminants. *Journal of Animal Science* 71:1092-1099.

Lee, J. W. 2011. Evaluation of corn germ and distillers dried grains with solubles in diets fed to pigs. MS Thesis, University of Illinois, Urbana-Champaign.

Lenehan, N. A., J. M. DeRouchey, R. D. Goodband, M. D. Tokach, S. S. Dritz, J. L. Nelssen, C. N. Groesbeck, and K. R. Lawrence. 2007. Evaluation of soy protein concentrates in nursery pig diets. *Journal of Animal Science* 85:3013-3021.

Leick, C. M., C. L. Puls, M. Ellis, J. Killefer, T. R. Carr, S. M. Scramlin, M. B. England, A. M. Gaines, B. F. Wolter, S. N. Carr, and F. K. McKeith. 2010. Effect of distillers dried grains with solubles and ractopamine (Paylean) on quality and shelf-life of fresh pork and bacon. *Journal of Animal Science* 88:2751-2766.

Li, D. F., J. L. Nelssen, P. G. Reddy, F. Blecha, J. D. Hancock, G. L. Allee, R. D. Goodband, and R. D. Klemm. 1990. Transient hypersensitivity to soybean meal in the early-weaned pig. *Journal of Animal Science* 68:790-1799.

Linneen, S. K., J. M. DeRouchey, S. S. Dritz, R. D. Goodband, M. D. Tokach, and J. L. Nelssen. 2008. Effects of distillers grains with solubles on growing and finishing pig performance in a commercial environment. *Journal of Animal Science* 86:1579-1587.

Liu, K. 2011. Chemical composition of distillers grains, A review. *Journal of Agricultural and Food Chemistry* 59:1508-1526.

Liu, K., and J. Han. 2011. Changes in mineral concentrations and phosphorus profile during dry-grind processing of corn into ethanol. *Bioresource Technology* 102:3110-3118.

Lusas, E. W., and K. C. Rhee. 1995. Soy protein processing and utilization. Pp. 117-160 in *Practical Handbook of Soybean Processing and Utilization*, D. E. Ericson, ed. Urbana, IL: American Oil Chemists' Society.

Mahan, D. C. 1993. Evaluation of two sources of dried whey and the effects of replacing the corn and dried whey component with corn gluten meal and lactose in the diets of weanling pigs. *Journal of Animal Science* 71:2860-2866.

McDonnell, P., C. J. O'Shea, J. J. Callan, and J. V. O'Doherty. 2011. The response of growth performance, nitrogen, and phosphorus excretion of growing-finishing pigs to diets containing incremental levels of maize dried distiller's grains with solubles. *Animal Feed Science and Technology* 169:104-112.

Normenkommission fur Einzelfuttermitteln im Zentralausschuss der Deutschen Landwirtschaft. 2006. *Positivliste fur Einzelfuttermitteln*, 5. Auflage, #12.07.03:35.

Pahm, A. A., C. Pedersen, D. Hoehler, and H. H. Stein. 2008a. Factors affecting the variability in ileal amino acid digestibility in corn distillers dried grains with solubles fed to growing pigs. *Journal of Animal Science* 86:2180-2189.

Pahm, A. A., C. Pedersen, and H. H. Stein. 2008b. Application of the reactive lysine procedure to estimate lysine digestibility in distillers dried grains with solubles fed to growing pigs. *Journal of Agricultural and Food Chemistry* 56:9441-9446.

Patience, J. F., P. Leterme, A. D. Beaulieu, and R. T. Zijlstra. 2007. Utilization in swine diets of distillers dried grains with solubles derived from corn or wheat used in ethanol production. Pp. 89-102 in *Biofuels: Implications for the Feed Industry*, J. Doppenberg and P. van der Arr, eds. Wageningen, The Netherlands: Wageningen Academic Press.

Pedersen, C., M. G. Boersma, and H. H. Stein. 2007. Digestibility of energy and phosphorus in 10 samples of distillers dried grains with solubles fed to growing pigs. *Journal of Animal Science* 85:1168-1176.

Sharma, Y. C., B. Singh, and S. N. Upadhyay. 2008. Advancements in development and characterization of biodiesel: A review. *Fuel* 87:2355-2373.

Rausch, K. D., and R. L. Belyea. 2006. The future of coproducts from corn processing. *Applied Biochemistry and Biotechnology* 128:47-85.

Rosentrater, K. A., K. Ileleji, and D. B. Johnson. 2012. Manufacturing of fuel ethanol and distillers dried grains—current and evolving processes. Pp. 73-102 in *Distiller's Grains: Production, Properties and Utilization*, K. Liu and K. A. Rosentrater, eds. Urbana, IL: AOCS Publishing.

Seabolt, B. S., E. van Heugten, S. W. Kim, K. D. Ange-van Heugten, and E. Roura. 2010. Feed preferences and performance of nursery pigs fed diets containing various inclusion amounts and qualities of distillers coproducts and flavor. *Journal of Animal Science* 88:3725-3738.

Shurson, J., and A. S. Alghamdi. 2008. Quality and new technologies to create corn co-prodcuts from ethanol production. Pp. 231-259 in *Using Distillers Grains in the U.S. and International Livestock and Poultry Industries*, B. A. Babcock, D. J. Hayes, and J. D. Lawrence, eds. Ames: MATRIC, Iowa State University.

Shurson, G., M. Spiehs, and M. Whitney. 2004. The use of maize distiller's dried grains with solubles in pig diets. *Pig News and Information*. 25:75N-83N.

Song, M., S. K. Baidoo, G. C. Shurson, M. H. Whitney, L. J. Johnston, and D. D. Gallaher. 2010. Dietary effects of distillers dried grains with solubles on performance and milk composition of lactating sows. *Journal of Animal Science* 88:3313-3319.

Spiehs, M. J., M. H. Whitney, and G. C. Shurson. 2002. Nutrient database for distiller's dried grains with solubles produced from new ethanol plants in Minnesota and South Dakota. *Journal of Animal Science* 80:2639-2645.

Stanley, D. L., and R. C. Ewan. 1982. Utilization of hominy feed and alfalfa meal by young pigs. *Journal of Animal Science* 54:1175-1180.

Stein, H. H. 2012. Feeding distillers dried grains with solubles (DDGS) and other ethanol coproducts to swine. Pp. 297-315 in *Distiller's Grains:*

Production, Properties and Utilization, K. Liu and K. A. Rosentrater, eds. Urbana, IL: AOCS Publishing.

Stein, H. H., and G. C. Shurson. 2009. Board invited review: The use and application of distillers dried grains with solubles (DDGS) in swine diets. *Journal of Animal Science* 87:1292-1303.

Stein, H. H., S. P. Connot, and C. Pedersen. 2009. Energy and nutrient digestibility in four sources of distillers dried grains with solubles produced from corn grown within a narrow geographical area and fed to growing pigs. *Asian-Australian Journal of Animal Science* 22:1016-1025.

Stock, R. A., J. M. Lewis, T. J. Klopfenstein, and C. T. Milton. 2000. Review of new information on the use of wet and dry milling feed by-products in feedlot diets. *Journal of Animal Science* 77:1v-12v.

Thompson, J. C., and B. B. He. 2006. Characterization of crude glycerol from biodiesel production from multiple feedstocks. *Applied Engineering in Agriculture* 22:261-265.

Thong, L. A., A. H. Jensen, B. G. Harmon, and S. C. Cornelius. 1978. Distillers dried grains with solubles as a supplemental protein source in diets for gestating swine. *Journal of Animal Science* 46:674-677.

van Kempen, T. A. T. G., E. van Heugten, A. J. Moser, N. S. Muley, V. J. H. Sewalt. 2006. Selecting soybean meal characteristics preferred for swine nutrition. *Journal of Animal Science* 84:1387-1395.

Wang, T., and L. A. Johnson. 2001. Survey of soybean oil and meal qualities produced by different processes. *Journal of the American Oil Chemists Society* 78:311-318.

Weber, T. E., S. L. Trabue, C. J. Ziemer, and B. J. Kerr. 2010. Evaluation of elevated dietary corn fiber from corn germ meal in growing female pigs. *Journal of Animal Science* 88:192-201.

Whitney, M. H., and G. C. Shurson. 2004. Growth performance of nursery pigs fed diets containing increasing levels of corn distiller's dried grains with solubles originating from a modern Midwestern ethanol plant. *Journal of Animal Science* 82:122-128.

Whitney, M. H., G. C. Shurson, L. J. Johnston, D. M. Wulf, and B. C. Shanks. 2006. Growth performance and carcass characteristics of grower-finisher pigs fed high-quality corn distillers dried grain with solubles originating from a modern Midwestern ethanol plant. *Journal of Animal Science* 84:3356-3363.

Widmer, M. R., L. M. McGinnis, and H. H. Stein. 2007. Energy, amino acid, and phosphorus digestibility of high protein distillers dried grain and corn germ fed to growing pigs. *Journal of Animal Science* 85:2994-3003.

Widmer, M. R., L. M. McGinnis, D. M. Wulf, and H. H. Stein. 2008. Effects of feeding distillers dried grains with solubles, high protein distillers dried grains, and corn germ to growing-finishing pigs on pig performance, carcass quality, and the palatability of pork. *Journal of Animal Science* 86:1819-1831.

Widyaratne, G. P., and R. T. Zijlstra. 2007. Nutritional value of wheat and corn distiller's dried grain with solubles: Digestibility and digestible contents of energy, amino acids and phosphorus, nutrient excretion and growth performance of grower-finisher pigs. *Canadian Journal of Animal Science* 87:103-114.

Witte, N. H. 1995. Soybean meal processing and utilization. Pp. 93-116 in *Practical Handbook of Soybean Processing and Utilization*, D. E. Ericson, ed. Urbana, IL: American Oil Chemists' Society.

Woodworth, J. C., M. D. Tokach, R. D. Goodband, J. L. Nelssen, P. R. O'Quinn, D. A. Knabe, and N. W. Said. 2001. Apparent ileal digestibility of amino acids and the digestible and metabolizable energy content of dry extruded-expelled soybean meal and its effects on growth performance of pigs. *Journal of Animal Science* 79:1280-1287.

Xu, G., S. K. Baidoo, L. J. Johnston, D. Bibus, J. E. Cannon, and G. C. Shurson. 2010a. Effects of feeding diets containing increasing content of corn distillers dried grains with solubles to grower-finisher pigs on growth performance, carcass composition, and pork fat quality. *Journal of Animal Science* 88:1398-1410.

Xu, G., S. K. Baidoo, L. J. Johnston, D. Bibus, J. E. Cannon, and G. C. Shurson. 2010b. The effects of feeding diets containing corn distillers dried grains with solubles, and withdrawal period of distillers dried grains with solubles, on growth performance and pork quality in grower-finisher pigs. *Journal of Animal Science* 88:1388-1397.

Yaklich, R. W. 2001. β-Conglycinin and glycinin in high-protein soybean seeds. *Journal of Agricultural and Food Chemistry* 49:729-735.

Yang, Y. X., Y. G. Kim, J. D. Lohakare, J. H. Yun, J. K. Lee, M. S. Kwon, J. I. Park, J. Y. Choi, and B. J. Chae. 2007. Comparative efficacy of different soy protein sources on growth performance, nutrient digestibility and intestinal morphology in weaned pigs. *Asian-Australian Journal of Animal Science* 20:775-783.

Yen, J. T., J. D. Brooks, and A. H. Jensen. 1974. Metabolizable energy value of corn gluten feed. *Journal of Animal Science* 39:335-337.

Yoon, S. Y., Y. X. Yang, P. L. Shinde, J. Y. Choi, J. S. Kim, Y. W. Kim, K. Yun, J. K. Jo, J. H. Lee, S. J. Ohh, I. K. Kwon, and B. J. Chae. 2010. Effects of mannanase and distillers dried grain with solubles on growth performance, nutrient digestibility, and carcass characteristics of grower-finisher pigs. *Journal of Animal Science* 88:181-191.

Young, L. G., G. C. Ashton, and G. C. Smith. 1977. Estimating the energy value of some feeds for pigs using regression equations. *Journal of Animal Science* 44:765-771.

Zijlstra, R. T., K. Menjivar, E. Lawrence, and E. Beltranena. 2009. The effect of feeding crude glycerol on growth performance and nutrient digestibility in weaned pigs. *Canadian Journal of Animal Science* 89:85-89.

第10章 非营养性饲料添加剂

引言

非营养性饲料添加剂是猪本身并不需要，但又存在于猪日粮中的一类添加剂；其中，抗菌剂是应用最为广泛的一种非营养性添加剂。美国食品和药物管理局（FDA）将抗菌剂和驱虫药定义为"药物"，并明确规定了用量、配伍及屠宰前的休药期。

除抗菌剂和驱虫药外，猪日粮中可能还含有其他添加剂，这些添加剂已被证实或未被证实具有提高猪生长性能的作用。Stein（2007）对这些添加剂（酸化剂、直接饲喂微生物添加剂、不可消化寡糖和植物提取物）进行了综述，本章将对这一综述进行更新。

抗菌剂

猪日粮中添加低剂量（亚治疗剂量）抗菌剂的效益已被广泛证实（Cromwell, 2001）。目前，在美国已有11种抗生素和5种化学治疗药物被允许在猪日粮中应用（Cromwell, 2011）。所有化学治疗药物需要在动物屠宰前停药，而抗生素则没有休药期（Cromwell, 2011）。尽管已发表的研究报告中抗菌剂的作用效果存在较大差异，但就总体平均而言，日粮中添加抗菌剂可使断奶仔猪和生长猪的生长速度分别提高16.4%和10.6%，饲料利用率分别改善6.9%和4.5%（Cromwell, 2001, 2011）。育肥猪使用抗菌剂的效果不如幼龄猪，整个生长肥育阶段使用抗菌剂可使平均日增重提高4.2%，饲料利用率改善2.2%（Cromwell, 2001）。饲料中添加抗菌剂通常会降低死亡率（从4.3%降至2.0%），尤其在疾病压力大时，降低死亡率的效果会更明显（Cromwell, 2011）。母猪日粮中添加抗菌剂可提高母猪产仔率和窝产活仔数；泌乳母猪日粮中添加抗菌剂可提高仔猪断奶重和存活率（Cromwell, 2011）。

尽管抗菌剂的作用机理尚不清楚，但大量研究报告表明抗菌剂具有减少疾病发生率的作用（Ding et al., 2006; Hays, 2011），其主要是通过提高猪的免疫力和控制肠道病原菌发挥作用。同时，抗菌剂也可能提高日粮中能量和养分的消化率，从而为机体组织合成提供更多养分（Roth and Kirchgessner, 1993; Gaines et al., 2005; Agudelo et al., 2007; Stewart et al., 2010）。日粮中能量和养分消化率的提高可能是肠道中微生物菌群变化的结果（Stewart et al., 2010），猪日粮中添加抗菌剂还可以观察到肠壁变薄的现象。

驱虫药

体内寄生虫可能会降低猪的生长性能，并可能给养猪生产造成巨大经济损失（Myers, 1988; Urban et al., 1989; Jacela et al., 2009）；在极端情况下，感染还可能导致胴体废弃。因此，寄生虫控制是畜群保健方案中一个重要组成部分。寄生虫一般采用驱虫药去防治，驱虫药被称为"dewormers"。目前有8种驱虫药在美国被允许商业使用，其中6种规定有休药期，休药期为屠宰前24h至21天不等（Jacela et al., 2009）。

最常见的体内寄生虫有蛔虫、蛲虫、肾蠕虫、鞭虫和肺蠕虫。这些寄生虫可通过饲料中添加驱虫药（或通过饮水添加）防治，也有一些驱虫药被制成针剂。

目前批准在猪上使用的 8 种驱虫药属于 6 个不同组别的药物：敌敌畏、芬苯达唑、伊维菌素、左旋咪唑、哌嗪和噻嘧啶酒石酸（Jacela et al., 2009）。所有产品对全部或部分体内寄生虫有效，其中伊维菌素对外部寄生虫（如虱和疥癣）也有效。敌敌畏除了具有驱虫药的活性外，怀孕母猪饲喂含敌敌畏日粮还能够提高窝产活仔数（Siers et al., 1976），提高初乳中乳脂浓度，增加断奶窝重（Siers et al., 1976; Young et al., 1979）。驱虫药除了减少寄生虫的感染，还能够改善猪活体增重和饲料转化效率（Zimmerman et al., 1982; Southern et al., 1989; Urban et al., 1989）。驱虫药的促生长作用可能是减少宿主寄生虫感染的间接效果。

酸化剂

日粮酸化剂包括有机酸、无机酸和酸盐。添加有机酸如富马酸（Falkowski and Aherne, 1984; Giesting and Easter, 1985; Radecki et al., 1988; Giesting et al., 1991）、甲酸、柠檬酸和丙酸（Falkowski and Aherne, 1984; Henry et al., 1985; Manzanilla et al., 2004）能够提高猪的生长性能。添加丁酸可能通过调节断奶仔猪的免疫刺激反应（Weber and Kerr, 2008），进而改善饲料转化效率（Manzanilla et al., 2006），但它们对生长性能的影响通常比较小（de Lange et al., 2010）。

一些无机酸如磷酸或盐酸也可提高猪的生长性能（Mahan et al., 1996）；其他无机酸如硫酸反而会降低猪的生长性能。通常，为了取得正效果，有机酸添加量为 1%~2%，而无机酸仅需添加小于 0.5% 就可以了。

研究发现，断奶仔猪日粮中添加甲酸钠（Kirchgessner and Roth, 1987）、甲酸钙（Kirchgessner and Roth, 1990; Pallauf and Hüter, 1993）和二甲酸钾（Overland et al., 2000; Canibe et al., 2001）等酸盐具有正面效应，这些酸盐产品的添加量一般需要大于 1%。

商业酸化剂产品可能同时包含有机酸和无机酸，日粮中的添加水平通常较低。由于商业酸化剂产品中某种特定的酸的添加量属于商业机密，所以组合产品的效果较难预测，但已有文献报道这类复合产品具有正面效果（Walsh et al., 2007a, b）。生长育肥猪日粮中添加酸化剂还可降低尿液的 pH，这可能会减少养猪生产过程中氨的排放量（van Kempen, 2001）。

直接饲喂的微生物添加剂

直接饲喂的微生物添加剂又称为益生菌（probiotics），主要分为三类：
1. 芽孢杆菌（形成芽孢的革兰氏阳性细菌）；
2. 产乳酸细菌（乳酸杆菌、双歧杆菌、肠球菌）；
3. 酵母。

益生菌是指按适当剂量直接饲喂给动物时，能对宿主动物的健康具有有益作用的微生物（Kenny et al, 2011）。常用作益生菌使用的微生物包括乳酸菌（*Lactobacillus* spp.）、粪肠球菌（*Enterococci faecium*）、芽孢杆菌（*Bacillus subtillis*）、两歧双歧杆菌（*Bifidobacterium bifidum*）、嗜热双歧杆菌（*Bifidobacterium themophilus*）及其他（Jonson and Conway, 1992）。

益生菌培养物具备以下条件才能对猪的生长性能有促进作用：
- 可以在动物的胃肠道内定殖；

- 具有较快的生长速度；
- 能分泌对致病菌具有抑制作用的代谢产物；
- 在商业生产条件下能够生长；
- 具有良好的稳定性并在饲料中保持活性。

微生物添加剂的作用机理是基于它们在动物肠道内定殖成为肠道主导菌群，阻止病原菌的定殖（竞争性排除作用）。

许多直接饲喂的微生物添加剂产品都含有产乳酸细菌。乳酸菌常用于防止动物断奶期间肠道乳酸菌的减少（Doyle, 2001），多个研究报道了在断奶仔猪日粮中添加乳酸菌具有积极作用（Apgar et al., 1993; Zani et al., 1998; Kyriakis et al., 1999）。在生长育肥猪日粮中添加芽孢杆菌还能提高猪的生长性能（Davis et al., 2008）。在泌乳母猪日粮中添加粪肠球菌可以减少哺乳仔猪的腹泻率和死亡率（Taras et al., 2006）。仔猪从出生到断奶期间服用粪肠球菌可以减少腹泻的发生，并提高仔猪体增重（Zeyner and Boldt, 2006）。

酵母培养物可以以活酵母和干酵母的形式添加到猪日粮中，目前还没有资料表明哪种形式的酵母更好。酵母与酵母类产品可能含氨基酸、酶、核苷酸、维生素、多糖、矿物质和其他代谢产物。多个研究者报道（Mathew et al., 1998; van Heugten et al., 2003; van der Peet-Schwering et al., 2007; Shen et al., 2009），在断奶仔猪或生长猪日粮中添加酵母对生长性能具有促进作用；但也有文献指出，日粮中添加酵母对猪生长性能无影响（Kornegay et al., 1995; Sauerwein et al., 2007）。同样，在母猪日粮中添加益生菌可提高母猪生产性能（Kim et al., 2008, 2010），但也有与之不一致的研究报道（Veum et al., 1995; Jurgens et al., 1997）。猪日粮中添加酵母的正面影响可能是因为酵母可以抑制肠道中大肠型杆菌的数量（White et al., 2002）。但断奶仔猪或生长猪日粮中添加酵母或酵母培养物对肠道微生物区系的影响并不一致（Mathew et al., 1998; van Heugten et al., 2003; van der Peet-Schwering et al., 2007; Shen et al., 2009）。

寡糖

寡糖又称益生元（prebiotics 或 nutraceuticals），包括易发酵、不可消化的寡糖，如果寡糖、β-葡聚糖、半乳寡糖和反式半乳寡糖。寡糖通过促进大肠双歧杆菌的增殖，提高乳酸含量，降低肠道 pH，从而提高猪的生产性能（Houdijk et al., 2002）。只有有益菌（如双歧杆菌和乳酸杆菌）能够发酵利用寡糖，而致病菌（如沙门氏菌和大肠杆菌）则不能（Flickinger et al., 2003）。寡糖还能改善肠道分泌、促进消化道黏膜的生长，研究评估了多种不同纤维组分促进猪只生长和抑制病原微生物定殖的特定能力。半乳寡糖被认为能促进大肠中有益菌的生长，进而改善动物肠道健康（Smiricky-Tjardes et al., 2003），如双歧杆菌可以通过促进乙酸生产来抑制病原菌（如大肠杆菌）的生长，并进一步降低 pH 和减少腹泻率（Mosenthin et al., 1999）。因此，日粮中添加寡糖能够促进肠道内有益菌生长，从而提高营养物质利用率，或减少肠道内病原菌的定殖。

其他纤维组分，如甘露寡糖，能够改善动物的健康状况和生长性能。多个研究表明，在日粮中添加甘露寡糖可以改善猪的生长性能（LeMieux et al., 2003; Rozeboom et al., 2005），其作用机理可能是甘露寡糖能与病原微生物表面的特异性凝集素结合，阻止病原菌与动物肠上皮细胞表面的特异性凝集素受体结合，从而将病原菌从肠道中"冲洗"出去（LeMieux et al., 2003; Rozeboom et al., 2005）。此外，甘露寡糖还可通过直接诱发抗体反应，增强动物机体免疫能力（Davis et al., 2004）。

植物提取物

很久以来就发现中草药提取物和香料制剂具有抗菌作用。植物提取物的生物活性物质通常又称为"精油"（Zaika et al., 1983），尽管并非总是如此（Deans and Ritchie, 1987）。植物提取物的活性受诸多因素的影响，如植物品种和生长环境等（Deans and Ritchie, 1987; Piccaglia et al., 1993）。精油通过改变细菌表面脂类的溶解度（Dabbah et al., 1970）来发挥抗菌作用，当然，其他作用机理，如分解细胞外膜也已被证实。

猪日粮中常用的植物提取物有大蒜提取物、牛至提取物、百里香酚、香芹酚和肉桂提取物。尽管这些成分在体外具有很强的抗菌活性，但目前没有足够的证据表明它们对生长性能的促进作用。Namkung 等（2004）报道，断奶仔猪日粮中同时添加牛至油、百里香酚和肉桂提取物降低了仔猪的生长性能；而且，在使用其他植物提取物组合的研究中也没有发现积极效果（Manzanilla et al., 2004, 2006; Insley et al., 2005）。

混合的植物提取物被认为是猪饲料用抗生素的替代品，但是目前尚缺乏来自严密控制试验的数据以支持这一概念。

外源酶

碳水化合物酶

在含有大麦、小麦或燕麦的日粮中添加碳水化合物分解酶，可提高纤维素的消化率，但对猪的生长性能并不总是有正面效果（Inborr et al., 1993; Nonn et al., 1999; Thacker and Campbell, 1999; Carneiro et al., 2008; O'Shea et al., 2010）。大麦中主要的非淀粉多糖是 β-葡聚糖，小麦中主要的非淀粉多糖是阿拉伯木聚糖。因此，可以预期添加 β-葡聚糖酶可提高大麦及其副产物的利用率，而添加木聚糖酶可提高小麦及其副产物的饲喂价值。小麦型日粮中添加"鸡尾酒"复合酶（纤维素酶、半乳聚糖酶、甘露聚糖酶和果胶酶）饲喂 6 kg 仔猪可提高其生长性能（Omogbenigun et al., 2004）。同样，在断奶仔猪日粮中添加木聚糖酶可以减少断奶后结肠炎的发生率（Newbold and Hillman, 2011）。

在以玉米为基础的日粮中添加外源酶对养分的消化率或对猪生长性能的影响报道有限。玉米-豆粕型日粮中添加 β-葡聚糖酶对 6 kg 仔猪干物质（DM）、能量或粗蛋白（CP）的消化率没有影响（Li et al., 1996）；玉米-豆粕型日粮中添加 β-甘露聚糖酶对 93 kg 阉公猪 DM、能量或氮（N）的消化率也无影响（Pettey et al., 2002）。相反，Ji 等（2008）的研究发现，在玉米-豆粕型日粮中添加含 β-葡聚糖和蛋白酶的复合酶提高了 DM、能量、CP、总日粮纤维和磷的消化率。同样，日粮中添加 β-甘露聚糖酶可以提高 6~14 kg 猪饲料利用率，提高 23~110 kg 猪的增重和饲料利用率（Pettey et al., 2002）。小麦副产物日粮中添加木聚糖酶也可以提高生长育肥猪对能量、DM 和一些必需氨基酸的消化率（Nortey et al., 2007，2008）；而且小麦副产物日粮中添加木聚糖酶的生长猪较未添加酶的猪饲料转化率明显改善（Nortey et al., 2007）。这些报道证实在小麦及其副产物日粮中添加木聚糖酶能够提高日粮中养分和能量的消化率。在玉米-豆粕日粮中添加碳水化合物复合酶（α-1,6-半乳糖苷酶和 β-1,4-甘露聚糖酶）也可以提高断奶仔猪的饲料利用率（Kim et al., 2003）。

在含有 30%玉米酒精糟（DDGS）的日粮中添加酶可以提高保育猪生长性能（Spencer et al., 2007），但结果并非总是如此（Jones et al., 2010a）。在玉米-豆粕-DDGS 日粮中添加外源酶对育肥猪的生长性能没有影响（Jacela et al., 2010b），但 Yoon 等（2010）报道，在含高达 15% DDGS 的日粮中添加甘露聚糖酶能够提高生长育肥猪的增重及养分消化率。

外源酶对气体排放的影响尚不清楚，且结果也不一致（Garry et al., 2007a, b; O'Shea et al., 2010）。也就是说，目前尚不可能明确地预测酶对臭味和氨排放的影响。

磷酸酶

猪日粮中使用磷酸酶（也称为"植酸酶"）的效果已由大量研究所报道（Adeola et al., 2004, 2006; Almeida and Stein, 2010），磷酸酶以肌醇六磷酸分子中的 3-或 6-碳为起点，将磷水解释放出（Konietzny and Greiner, 2002）。植酸酶活性（FTU）定义为在 pH 5.5、37℃下，每分钟从 0.0051 mol/L 的植酸钠中释放 1 μmol 的无机磷所需要的植酸酶量（Engelen et al., 1994）。目前，植酸酶活性的标准测定方法是美国官方分析化学家协会（AOAC）的规定方法，即 AOAC 2000.12 方法（AOAC International, 2007）；尽管存在标准测定方法，但是各实验室内部以及不同试验室之间的测定方法都存在差异（Gizzi et al., 2008）。由于植酸酶的生化特性存在着差异，因此将最初的实验室测定方法进行修改就变得非常普遍（Kim and Lei, 2005; Selle and Ravindran, 2008）。所以，植酸酶活性的表达形式随植酸酶的来源和分析方法不同而不同（Jones et al., 2010b; Kerr et al., 2010）。

猪日粮添加植酸酶通常会提高磷的全消化道消化率和减少磷排放量（Selle and Ravindran, 2008; Almeida and Stein, 2010），但是改善程度受日粮组成（Düngelhoef et al., 1994; Johansen and Poulsen, 2003; Almeida and Stein, 2010）、植酸酶的添加量与来源（Selle and Ravindran, 2008; Jones et al., 2010b; Kerr et al., 2010），以及钙磷比（Adeola et al., 1998; Selle et al., 2009; Letourneau-Montminy et al., 2010）的影响。

植酸酶对日粮中其他组分的影响在多个试验中也有研究。在一些研究中，植酸酶对能量、氨基酸和矿物元素的消化率具有正效应，但在另外一些研究中并未发现这样的效果，这表明植酸酶的效果变异较大，可能受其他日粮因素的影响。

调味剂

调味剂、甜味剂、香味剂或它们的复合物是为了提高日粮的适口性、提高采食量和掩盖异味而添加到猪日粮中的饲料添加剂（Jacela et al., 2010a）。强有力的证据表明猪对甜味具有较强的偏好性（Kennedy and Baldwin, 1972; Danilova et al., 1999; Glaser et al., 2000）。过去，通常在仔猪日粮中使用蔗糖来改善适口性并提供能量，当然也可以使用人工合成的具有高强度甜味的糖精、新橙皮苷二氢查尔酮和索马甜等甜味剂。在成百上千的调味剂和复合调味剂中，断奶仔猪只对奶酪味、水果味、肉味和甜味有很强的偏好（McLaughlin et al., 1983）。

泌乳母猪日粮和教槽料中添加同一种特定的调味剂，通过仔猪对母猪乳汁的记忆，能够提高仔猪对教槽料的采食量（Campbell, 1976; King, 1979; Langendijk et al., 2007）。在仔猪哺乳阶段，可以在教槽料中添加调味剂以诱使仔猪接受固体饲料，提高采食量和断奶重；但是，其结果变异较大（Gatel and Guion, 1990）或几乎无影响（King, 1979; Millet et al., 2008; Sulabo et al., 2010）。调味剂常被添加到保育仔猪日粮中，用于提高刚断奶仔猪的采食量。但是，添加调味剂不会影响刚断奶仔猪的生长性能（Munro et al., 2000; Sterk et al., 2008; Seabolt et al., 2010; Sulabo et al., 2010）。一些研究发现，仔猪饲喂添加调味剂的简单日粮（玉米-豆粕为主）并不能获得与饲喂未添加调味剂的复杂日粮（包含鱼粉和乳制品等）仔猪相似的生长性能（Costa et al., 2003; Sulabo et al., 2010）；生长育肥猪日粮中添加调味剂对生长性能也没有任何益处（Koch et al., 1976, 1977; Johnston et al., 1989）。总之，这些结果表明多数情况下日粮中添加调味剂和甜味剂对猪的采食量和生长性能影响不大。

霉菌毒素吸附剂

产毒素霉菌及其所产生的各种霉菌毒素是对饲料原料及成品饲料不利的重要污染源。霉菌毒素是由曲霉菌属（*Aspergillus* spp.）、镰刀菌属（*Fusarium* spp.）及青霉菌属（*Penicillium* spp.）等丝状真菌产生的次级代谢产物，这些霉菌毒素被动物摄食后会表现出中毒症状（霉菌毒素中毒症）。猪日粮中最重要的霉菌毒素是黄曲霉毒素 B_1、玉米赤霉烯酮、呕吐毒素（DON）、T-2 毒素、烟曲霉毒素 B_1 和赭曲霉毒素 A。关于霉菌毒素对动物的生物化学作用机理和临床表现已有综述（Newberne and Butler, 1969; Fink-Gremmels and Malekinejad, 2007; Glenn, 2007; Pestka, 2007; Voss et al., 2007），虽然每种霉菌毒素都有其各自独特的危害，但霉菌毒素通常都会使动物拒食、饲料转化效率低下、体增重下降、免疫抑制、繁殖力低下，以及造成动物产品中毒素残留等问题，并由此造成经济损失。有关霉菌毒素的其他信息将在第 11 章中阐述。

在作物收获前、收获期间、储藏以及饲料原料加工期间可以通过一些物理和化学措施来预防、消除或降低饲料原料中霉菌毒素的污染（Samarajeewa et al., 1990; Jouany, 2007）。另外还可以用生物学方法来降低霉菌毒素的活性。通过在动物饲料中添加霉菌毒素吸附剂这类非营养添加剂，以抑制或减少饲料中霉菌毒素的吸收，或促进饲料中霉菌毒素的排泄。霉菌毒素吸附剂的作用机理主要是通过与霉菌毒素相结合而使毒素失活，但还有一些霉菌毒素抑制剂具有解毒和将毒素降解为低毒性的代谢产物的作用。

使用霉菌毒素吸附剂应对霉菌毒素中毒症的综述已有报道（Ramos et al., 1996; Ramos and Hernandez, 1997; Huwig et al., 2001; Avantaggiato et al., 2005; Diaz and Smith, 2005）。无机吸附剂产品包括硅铝酸盐黏土、活性炭和酚类吸附剂（PVPP）。黏土包括天然沸石粉（斜发沸石）、加工处理过的沸石粉、膨润土及水合铝硅酸钠钙（HSCAS）等硅铝酸盐。沸石作为霉菌毒素吸附剂在猪上的研究很少，而在肉鸡上开展的相关试验表明日粮中添加沸石粉可降低黄曲霉毒素的负作用（Miazzo et al., 2000; Oğuz and Kurtoglu, 2000; Oğuz et al., 2000a, b; Piva et al., 2005）。膨润土具有良好的离子交换能力，按照所携带金属离子的不同可分为钙、镁、钾和钠基膨润土，它们对猪都能有效地预防黄曲霉毒素中毒症（Schell et al., 1993a, b; Miazzo et al., 2005）。水合硅铝酸钠钙则是研究最多的霉菌毒素吸附剂，Phillips 等 1988 年首次报道了水合硅铝酸钠钙在肉鸡试验中对黄曲霉毒素 B_1 具有较高的亲和性及吸附能力。水合硅铝酸钠钙黏土颗粒表面具有黄曲霉毒素的多个结合位点，通过化学吸附作用，黄曲霉毒素被吸附在这些带负电荷的黏土表面（Grant and Phillips, 1998）。水合硅铝酸钠钙防治黄曲霉毒素中毒症效果的研究已有综述（Ramos and Hernández, 1997; Phillips, 1999; Bingham et al., 2003）。一般来说，水合硅铝酸钠钙对于减轻猪黄曲霉毒素中毒的危害有很好的效果（Colvin et al., 1989; Beaver et al., 1990; Lindemann et al., 1993; Schell et al., 1993b; Harvey et al., 1994），然而在污染其他霉菌毒素的猪日粮中使用硅铝酸盐黏土吸附剂并不能缓解霉菌毒素中毒症状（Patterson and Young, 1993; Williams et al., 1994; Doll et al., 2005）。

活性炭是一种由碳元素组成的无定形物，是在隔绝空气的条件下对碳进行加热并使之与氧气反应，从而使得每个碳原子之间产生无数的空隙结构（Diaz and Smith, 2005）。活性炭的吸附能力很强，通常被用作医疗处置手段（Huwig et al., 2001）。然而在被黄曲霉毒素 B_1 或其他霉菌毒素污染的猪和肉鸡日粮中添加活性炭却没能改善其生长性能、相对器官重和免疫功能（Dalvi and Ademoyero, 1984; Edrington et al., 1997; Cabassi et al., 2005; Piva et al., 2005）。

酚类吸附剂（PVPP）是一种化学惰性物质，是由乙烯基聚吡咯烷酮组成的交联聚合物。该物质不溶于水并具有较高的吸附性能，它能吸附一些极性分子如黄曲霉毒素（Çelik et al., 2000）。有人对PVPP在家禽上对霉菌毒素的吸附效果进行了评估（Kececi et al., 1998; Kiran et al., 1998; Çelik et al., 2000），而在猪上开展的相关研究很少。Friend 等（1984）报道了PVPP吸附剂不能缓解猪的DON中毒症状。

从酵母细胞壁中提取的葡甘露聚糖多聚物也被作为一种有机吸附剂使用。尽管葡甘露聚糖的具体吸附方式目前还没有完全阐明，但体外研究表明 β-D-葡聚糖可能是酵母细胞壁中吸附霉菌毒素的主要作用成分（Yiannikouris et al., 2006）。然而在自然污染镰刀菌毒素的断奶仔猪日粮（Swamy et al., 2002, 2003）、妊娠母猪日粮及哺乳母猪日粮（Díaz-Llano and Smith, 2006, 2007; Díaz-Llano et al., 2010）中添加 0.2%葡甘露聚糖多聚物，不能减轻霉菌毒素造成的不利影响。最近也有学者报道，优杆菌属细菌（*Eubacterium* BBSH 797）和（嗜）霉菌毒素毛孢子菌（*Trichosporon mycotoxinivorans*）通过使霉菌毒素在肠道内还未被吸收之前就被酶解，从而使赭曲霉毒素和玉米赤霉烯酮毒素失活（Schatzmayr et al., 2006）。然而，关于优杆菌属细菌（*Eubacterium* BBSH）在猪体内效果的研究还未见报道。

尽管目前有关霉菌毒素吸附剂的研究很多，但还没有被FDA批准的用于预防或处理霉菌毒素中毒症的吸附剂产品。硅铝酸盐黏土产品虽已获得美国的"公认安全使用物质"（GRAS）认证，但目前仅被批准在动物饲料中当作抗结块剂和颗粒黏合剂来使用（AAFCO，2010）。

抗氧化剂

氧化会使饲料及饲料原料产生臭味、酸败并破坏脂溶性维生素，所以常在饲料及饲料原料中添加抗氧化剂防止脂肪和维生素的氧化（Jacela et al., 2010a）。维生素 E、维生素 C 和硒是非常有效的抗氧化剂，有助于降低动物组织脂质过氧化的易感性（Mahan et al., 1994, 1996; Lauridsen et al., 1999）。如果饲料或饲料原料中没有充足的营养性抗氧化剂，则可添加非营养性抗氧化剂；有时也使用复合的抗氧化剂产品（Jacela et al., 2010a）。常用的抗氧化剂产品有乙氧基喹啉、二丁基羟基甲苯（BHT）、丁基羟基茴香醚（BHA）和没食子酸丙酯（Jacela et al., 2010a）。

如果含有不饱和脂肪酸的饲料或饲料原料（如鱼粉、玉米酒精糟和玉米副产物）储存于高温条件下，则建议额外添加抗氧化剂产品。因为这些饲料或饲料原料容易迅速氧化，添加抗氧化剂则能够推迟氧化的发生。

颗粒黏合剂

与饲喂粉料比较，饲喂颗粒饲料可以提高断奶仔猪和生长育肥猪的生长性能（Hansen et al., 1992; Traylor et al., 1996; Potter et al., 2010）。而颗粒饲料的饲喂效果取决于颗粒饲料的物理性状（Stark et al., 1994）。颗粒黏合剂作为饲料添加剂用于提高颗粒料的耐久性和降低饲料在生产、保存和运输过程中细粉的产生。黏合剂需要通过水来激活，用于提高饲料微粒间的内聚力和黏合力（Thomas and van der Poel, 1996）。虽然颗粒黏合剂可提高颗粒料的质量，但日粮的组成和饲料加工工艺也在决定特定颗粒饲料的质量中扮演着重要的角色（Thomas and van der Poel, 1996）。

动物饲料生产中常用的颗粒黏合剂包括无机黏土（如膨润土、海泡石和蒙脱石）、木质素磺酸盐、胶原蛋白衍生物（如明胶）和纤维素胶。黏土常用作制粒助剂，作为填充料起到降低颗粒料的孔隙率和润滑剂的作用（Thomas et al., 1998）。黏土可提高颗粒料的稳定性，尤其是当饲料中脂肪含量较高时（Salmon, 1985; Angulo et al., 1995）。然而，日粮中需添加较高剂量的黏合剂（日粮的2%~3%），才能取得较好的制粒效果。水溶性木质素磺酸盐是造纸行业的副产物，能够提高颗粒料的稳定性，并降低制粒过程中的能耗（Van Zuilichem et al., 1979a, b, 1980）。木质素磺酸盐的推荐添加量为0.5%~3%（Thomas et al., 1998）。黏土和木质素磺酸盐在成品饲料中的最大推荐使用量分别为2%和4%（AAFCO, 2010）。

流散剂

流散剂和防结块剂作为饲料添加剂用以防止结块，改善粒状或者粉状饲料原料及粉料在处理、储存和加工过程中的流散性。流散剂通常是由惰性非水溶性物质制成，这些物质由于含有很大表面积而具有较强的吸水性（Ganesan et al., 2008b）。用作颗粒黏合剂的无机黏土也是最常用的流散剂，在日粮中的最高添加量为2%（AAFCO, 2010）。尽管流散剂在颗粒固体和粉状固体上对流动性的影响已有研究（Chen and Chou, 1993; Onwulata et al., 1996; Jaya and Das, 2004），但是在常用原料中使用流散剂的发表数据很有限。然而，最近的研究结果表明，使用流散剂并未改善玉米酒精糟（DDGS）的流散性（Ganesan et al., 2008a; Johnston et al., 2009）。

莱克多巴胺

莱克多巴胺和盐酸莱克多巴胺属于β-肾上腺素受体兴奋剂类化合物。在美国被批准使用的唯一莱克多巴胺产品是美国礼来公司销售的"Paylean"。Mills（2002）对莱克多巴胺的作用机制进行了综述，它对猪只体组成的影响也有充分的研究报道（Watkins et al., 1990; Dunshea et al., 1993; See et al., 2004）。日粮中添加莱克多巴胺可以降低机体脂肪沉积，提高胴体瘦肉率（Mitchell et al., 1991; Moody et al., 2000）；但也有部分试验结果表明莱克多巴胺对降低脂肪沉积的作用效果不稳定或无任何效果（Dunshea et al., 1993）。莱克多巴胺对脂肪沉积作用效果不稳定，是因为长期使用莱克多巴胺下调了脂肪细胞中β-肾上腺素受体的缘故（Spurlock et al., 1994）。

使用莱克多巴胺对猪营养需要量的影响已有综述（NRC, 1994）。莱克多巴胺能提高猪的生长性能、胴体瘦肉率，以及胃肠道、肝和肾的重量，但对整个动物的产热并无影响（Yen et al., 1991）。其潜在机制可能是，使用β-兴奋剂增加了能量消耗，营养物质不沉积于脂肪组织，而定向沉积于瘦肉组织，这就解释了饲喂含有莱克多巴胺猪胴体组成改变的现象（Reeds and Mersmann, 1991）。由于提高了瘦肉率，饲喂莱克多巴胺的猪对日粮必需氨基酸需要量也高于未饲喂莱克多巴胺的猪，因此，含莱克多巴胺日粮中必需氨基酸与代谢能（ME）的比值也应提高（Schinckel et al., 2003; Apple et al., 2004）。

在美国，莱克多巴胺被批准用于生长育肥猪（>68 kg），在最后23~41 kg增重阶段的饲料中使用。批准使用量为5~10 ppm（1000 kg全价料中添加5~10 g）。此外，对于饲料标签的规定要求含有莱克多巴胺的日粮其粗蛋白（CP）水平不得低于16%。

猪生长模型（第8章）模拟了猪育肥后期阶段对使用莱克多巴胺的反应，并预测猪只对能量和养分的需要量。利用小母猪（90~120 kg）莱克多巴胺三阶段递增添加程序，预测了饲喂莱克多巴胺的120 kg小母猪（瘦肉日增重350 g/d）的回肠标准可消化（SID）赖氨酸需要量是19 g/d。饲喂不含莱克多巴胺的120 kg小母猪每天需要的SID赖氨酸仅为15 g/d,因此添加莱克多巴胺提高了26%的赖氨酸需要量。同样,与未饲喂莱克多巴胺的猪只相比,饲喂莱克多巴胺的猪每天对磷的需要量增加近29%,而预测的ME摄入量则降低了3%。

肉毒碱和共轭亚油酸

日粮中添加肉毒碱和共轭亚油酸对猪的影响在第3章中讨论。

臭味和氨气控制剂

日粮中添加臭味和氨气控制剂对猪的影响在第14章中讨论。

参 考 文 献

AAFCO (Association of American Feed Control Officials). 2010. *Official Publication 2010*. Oxford, IN: AAFCO.

Adeola, O., J. I. Orban, D. Ragland, T. R. Cline, and A. L. Sutton. 1998. Phytase and cholecalciferol supplementation of low-calcium and low-phosphorus diets for pigs. *Canadian Journal of Animal Science* 78:307-313.

Adeola, O., J. S. Sands, P. H. Simmins, and H. Schulze. 2004. The efficacy of an *Escherichia coli*-derived phytase preparation. *Journal of Animal Science* 82:2657-2666.

Adeola, O., O. A. Olukosi, J. A. Jendza, R. N. Dilger, and M. R. Bedford. 2006. Response of growing pigs to *Peniophora lycii*- and *Escherichia coli*-derived phytases or varying ratios of calcium to total phosphorus. *Animal Science* 82:637-644.

Agudelo, J. H., M. D. Lindemann, G. L. Cromwell, M. C. Newman, and R. D. Nimmo. 2007. Virginiamycin improves phosphorus digestibility and utilization by growing-finishing pigs fed a phosphorus-deficient, corn-soybean meal diet. *Journal of Animal Science* 85:2173-2182.

Almeida, F. N., and H. H. Stein. 2010. Performance and phosphorus balance of pigs fed diets formulated on the basis of values for standardized total tract digestibility of phosphorus. *Journal of Animal Science* 88:2968-2977.

Angulo, E., J. Brufau, and E. Esteve-Garcia. 1995. Effect of sepiolite on pellet durability in feeds differing in fat and fibre content. *Animal Feed Science and Technology* 53:233-241.

AOAC International. 2007. *Official Methods of Analysis of AOAC International*, 18th Ed., Rev. 2, W. Hortwitz and G. W. Latimer, Jr., eds. Gaithersburg, MD: AOAC International.

Apgar, G. A., E. T. Kornegay, M. D. Lindemann, and C. M. Wood. 1993. The effect of feeding various levels of *Bifidobacterium globosum A* on the performance, gastrointestinal measurements, and immunity measurements of growing-finishing pigs. *Journal of Animal Science* 71:2173-2179.

Apple, J. K., C. V. Maxwell, D. C. Brown, K. G. Friesen, R. E. Musser, Z. B. Johnson, and T. A. Armstrong. 2004. Effects of dietary lysine and energy density on performance and carcass characteristics of finishing pigs fed ractopamine. *Journal of Animal Science* 82:3277-3287.

Avantaggiato, G., M. Solfrizzo, and A. Visconti. 2005. Recent advances on the use of adsorbent materials for detoxification of *Fusarium* mycotoxins. *Food Additives and Contaminants* 22:379-388.

Beaver, R. W., D. M. Wilson, M. A. James, K. D. Haydon, B. M. Colvin, L. T. Sangster, A. H. Pikul, and J. D. Groopman. 1990. Distribution of aflatoxins in tissues of growing pigs fed an aflatoxin-contaminated diet amended with a high affinity aluminosilicate sorbent. *Veterinary and Human Toxicology* 32:16-18.

Bingham, A. K., T. D. Phillips, and J. E. Bauer. 2003. Potential for dietary protection against the effects of aflatoxins in animals. *Journal of the American Veterinary Medical Association* 222:591-596.

Cabassi, E., F. Miduri, and A. M. Cantoni. 2005. Intoxication with fumonisin B1 (FB1) in piglets and supplementation with granulated activated carbon: Cellular-mediated immunoresponse. *Veterinary Research Communications* 29:225-227.

Campbell, R. G. 1976. A note on the use of a feed flavor to stimulate the feed intake of weaner pigs. *Animal Production* 23:417-419.

Canibe, N., S. H. Steien, M. Overland, and B. B. Jensen. 2001. Effect of K-diformate in starter diets on acidity, microflora, and the amount of organic acids in the digestive tract of piglets, and on gastric alterations. *Journal of Animal Science* 79:2123-2133.

Carneiro, M. S. C., M. M. Lordelo, L. F. Cunha, and J. P. B. Freire. 2008. Effects of dietary fibre source and enzyme supplementation on faecal apparent digestibility, short chain fatty acid production and activity of bacterial enzymes in the gut of piglets. *Animal Feed Science and Technology* 146:124-136.

Çelik, I., H. Oğuz, Ö. Demet, H. H. Dönmez, M. Boydak, and E. Sur. 2000. Efficacy of polyvinylpolypyrrolidone in reducing the immunotoxicity of aflatoxin in growing broilers. *British Poultry Science* 41:430-439.

Chen, Y. L., and J. Y. Chou. 1993. Selection of anticaking agents through crystallization. *Powder Technology* 77:1-6.

Colvin, B. M., L. T. Sangster, K. D. Haydon, R. W. Bequer, and D. M. Wilson. 1989. Effect of high affinity aluminosilicate sorbent on prevention of aflatoxicosis in growing pigs. *Veterinary and Human Toxicology* 31:46-48.

Costa, L. L., J. A. de Freitas Lima, E. T. Fialho, A. I. Oliveira, L. D. S. Murgas, and E. P. Filgueiras. 2003. Flavours in the diets for piglets from 6 to 18 kg. *Revista Brasileira de Zootecnia* 32:1-8.

Cromwell, G. L. 2001. Antimicrobial and promicrobial agents. Pp. 421-426 in *Swine Nutrition*, 2nd Ed., A. J. Lewis and L. L. Southern, eds. Boca Raton, FL: CRC Press.

Cromwell, G. L. 2011. Feed supplements: Antibiotics. Pp. 391-393 in *Encyclopedia of Animal Science*, 2nd Ed., D. E. Ullrey, C. Kirk Baer, and W. G. Pond, eds. Boca Raton, FL: CRC Press.

Dabbah, R. V., M. Edwards, and W. A. Moats. 1970. Antimicrobial action of some citrus fruit oils on selected food-borne bacteria. *Applied Microbiology* 19:27-31.

Dalvi, R. R., and A. A. Ademoyero. 1984. Toxic effects of aflatoxin B1 in chickens given feed contaminated with *Aspergillus flavus* and reduction of the toxicity by activated charcoal and some chemical agents. *Avian Diseases* 28:61-69.

Danilova, V., T. Roberts, and G. Hellekant. 1999. Responses of single taste fibers and whole chorda tympani and glossopharyngeal nerve in the domestic pig, *Sus scrofa*. *Chemical Senses* 24:301-316.

Davis, M. E., C. V. Maxwell, G. F. Erf, D. C. Brown, and T. J. Wistuba. 2004. Dietary supplementation with phosphorylated mannans improves growth response and modulates immune function of weanling pigs. *Journal of Animal Science* 82:1882-1891.

Davis, M. E., T. Parrott, D. C. Brown, B. Z. de Rodas, Z. B. Johnson, C. V. Maxwell, and T. Rehberger. 2008. Effect of a *Bacillus*-based direct-fed microbial feed supplement on growth performance and pen cleaning characteristics of growing-finishing pigs. *Journal of Animal Science* 86:1459-1467.

Deans, S. G., and G. Ritchie. 1987. Antibacterial properties of plant essential oils. *International Journal of Food Science* 5:165-180.

de Lange, C. F. M., J. Pluske, J. Gong, and C. M. Nyachoti. 2010. Strategic use of feed ingredients and feed additives to stimulate gut health and development in young pigs. *Livestock Science* 134:124-134.

Diaz, D. E., and T. K. Smith. 2005. Mycotoxin sequestering agents: Practical tools for the neutralisation of mycotoxins. Pp. 323-339 in *The Mycotoxin Blue Book*, D. Diaz, ed. Nottingham, UK: Nottingham University Press.

Díaz-Llano, G., and T. K. Smith. 2006. Effects of feeding grains naturally contaminated with *Fusarium* mycotoxins with and without a polymeric glucomannan mycotoxin adsorbent on reproductive performance and serum chemistry of pregnant gilts. *Journal of Animal Science* 84:2361-2366.

Díaz-Llano, G., and T. K. Smith. 2007. The effects of feeding grains naturally contaminated with *Fusarium* mycotoxins with and without a polymeric glucomannan adsorbent on lactation, serum chemistry, and reproductive performance after weaning of first-parity lactating sows. *Journal of Animal Science* 85:1412-1423.

Díaz-Llano, G., T. K. Smith, H. J. Boermans, C. Caballero-Cortes, and R. Friendship. 2010. Effects of feeding diets naturally contaminated with *Fusarium* mycotoxins on protein metabolism in late gestation and lactation of first-parity sows. *Journal of Animal Science* 88:998-1008.

Ding, M. X., Z. H. Yuan, Y. L. Wang, H. L. Zhu, and S. X. Fan. 2006. Olaquindox and cyadox stimulate growth and decrease intestinal mucosal immunity of piglets orally inoculated with *Escherichia coli*. *Journal of Animal Physiology and Animal Nutrition* 90:238-243.

Doll, S., S. Gericke, S. Dänicke, J. Taila, K. H. Ueberschar, H. Valenta, U. Schnurrbusch, F. J. Schweigert, and G. Flachowsky. 2005. The efficacy of a modified aluminosilicate as a detoxifying agent in *Fusarium* toxin contaminated maize containing diets for piglets. *Journal of Animal Physiology and Animal Nutrition* 89:342-358.

Doyle, M. E. 2001. *Alternatives to antibiotic use for growth promotion in animal husbandry*. FRI Briefings. University of Wisconsin. Madison, WI: Food Research Institute.

Düngelhoef, M., M. Rodehutschord, H. Spiekers, and E. Pfeffer. 1994. Effects of supplemental microbial phytase on availability of phosphorus contained in maize, wheat and triticale to pigs. *Animal Feed Science and Technology* 49:1-10.

Dunshea, F. R., R. H. King, and R. G. Campbell. 1993. Interrelationships between dietary protein and ractopamine on protein and lipid deposition in finishing gilts. *Journal of Animal Science* 71:2931-2941.

Edrington, T. S., L. F. Kubena, R. B. Harvey, and G. E. Rottinghaus. 1997. Influence of a superactivated charcoal on the toxic effects of aflatoxin or T-2 toxin in growing broilers. *Poultry Science* 76:1205-1211.

Engelen, A. J., F. C. van der Heeft, P. H. G. Randsdorp, and E. L. C. Smit. 1994. Simple and rapid determination of phytase activity. *Journal of AOAC International* 77:760-764.

Falkowski, J. F., and F. X. Aherne. 1984. Fumaric and citric acid as feed additives in starter pig nutrition. *Journal of Animal Science* 58:935-938.

Fink-Gremmels, J., and H. Malekinejad. 2007. Clinical effects and biochemical mechanisms associated with exposure to the mycoestrogen zearalenone. *Animal Feed Science and Technology* 137:326-341.

Flickinger, E. A., J. V. Loo, and G. J. Fahey. 2003. Nutritional responses to the presence of inulin and oligofructose in the diets of domesticated animals: A review. *Critical Reviews in Food Science and Nutrition* 43:19-60.

Friend, D. W., H. L. Trenholm, J. C. Young, B. Thompson, and K. E. Hartin. 1984. Effects of adding potential vomitoxin (deoxynivalenol) detoxicants or a *F. graminearum* inoculated corn supplement to wheat diets to pigs. *Canadian Journal of Animal Science* 64:733-741.

Gaines, A. M., G. L. Allee, B. W. Ratliff, P. Srichana, R. D. Nimmo, B. R. Gramm. 2005. Determination of energy value of Stafac® (virginiamycin) in finishing pigs. P. 35 in *Proceedings of the Allen D. Leman Swine Conference*, St. Paul, MN, September 17-20, 2005.

Ganesan, V., K. Muthukumarappan, and K. A. Rosentrater. 2008a. Effect of flow agent addition on the physical properties of DDGS with varying moisture content and soluble levels. *Transactions of the ASABE* 51:591-601.

Ganesan, V., K. A. Rosentrater, and K. Muthukumarappan. 2008b. Flowability and handling characteristics of bulk solids and powders—a review with implications for DDGS. *Biosystems Engineering* 101:425-435.

Garry, B. P., M. Fogarty, T. P. Curran, M. J. O'Connell, and J. V. O'Doherty. 2007a. The effect of cereal type and enzyme addition on pig performance, intestinal microflora, and ammonia and odour emissions. *Animal* 1:751-757.

Garry, B. P., M. Fogarty, T. P. Curran, and J. V. O'Doherty. 2007b. Effect of cereal type and exogenous enzyme supplementation in pig diets on odour and ammonia emissions. *Livestock Science* 109:212-215.

Gatel, F., and P. Guion. 1990. Effects of monosodium L-glutamate on diet palatability and piglet performance during the suckling and weaning periods. *Animal Production* 50:365-372.

Giesting, D. W., and R. A. Easter. 1985. Response of starter pigs to supplementation of corn soybean meal diets with organic acids. *Journal of Animal Science* 60:1288-1294.

Giesting, D. W., M. A. Ross, and R. A. Easter. 1991. Evaluation of the effect of fumaric acid and sodium bicarbonate addition on performance of starter pigs fed diets of different types. *Journal of Animal Science* 69:2489-2496.

Gizzi, G., P. Thyregod, C. von Holst, G. Bertin, K. Vogel, M. Faurschou-Isaksen, R. Betz, R. Murphy, and G. G. Andersen. 2008. Determination of phytase activity in feed: Interlaboratory study. *Journal of AOAC International* 91:259-267.

Glaser, D., M. Wanner, J. M. Tinti, and C. Nofre. 2000. Gustatory responses of pigs to various natural and artificial compounds known to be sweet in man. *Food Chemistry* 68:375-385.

Glenn, A. E. 2007. Mycotoxigenic *Fusarium* species in animal feed. *Animal Feed Science and Technology* 137:213-240.

Grant, P. G., and T. D. Phillips. 1998. Isothermal adsorption of aflatoxin B1 on HSCAS clay. *Journal of Agricultural and Food Chemistry* 46:599-605.

Hansen, J. A., J. L. Nelssen, M. D. Tokach, R. D. Goodband, L. J. Kats, and K. G. Friesen. 1992. Effects of a grind and mix high nutrient density diet on starter pig performance. *Journal of Animal Science* 70(Suppl. 1):59(Abstr.).

Harvey, R. B., L. F. Kubena, M. H. Elissalde, D. E. Corrier, and T. D. Phillips. 1994. Comparison of two hydrated sodium calcium aluminosilicate compounds to experimentally protect growing barrows from aflatoxicosis. *Journal of Veterinary Diagnostic Investigation* 6:88-92.

Hays, V. W. 2011. Antibiotics: Sub-therapeutic levels. Pp. 40-42 in *Encyclopedia of Animal Science*, 2nd Ed., D. E. Ullrey, C. Kirk Baer, and W. G. Pond, eds. Boca Raton, FL: CRC Press.

Henry, R. W., D. W. Pickard, and P. E. Hughes. 1985. Citric acid and fumaric acid as food additives for early weaned piglets. *Animal Production* 40:505-509.

Houdijk, J. G. M., R. Hartemink, M. W. A. Verstegen, and M. W. Bosch. 2002. Effects of dietary non-digestible oligosaccharides on microbial characteristics of ileal chyme and faeces in weaner pigs. *Archives of Animal Nutrition* 56:297-307.

Huwig, A., S. Freimund, O. Kappeli, and H. Dutler. 2001. Mycotoxin detoxication of animal feed by different adsorbents. *Toxicology Letters* 122:179-188.

Inborr, J., M. Schmitz, and F. Ahrens. 1993. Effect of adding fibre and starch degrading enzymes to a barley/wheat based diet on performance and nutrient digestibility in different segments of the small intestine of early weaned pigs. *Animal Feed Science and Technology* 44:113-127.

Insley, S. E., H. M. Miller, and C. Kamel. 2005. Effects of dietary quillaja saponin and curcumin on the performance and immune status of weaned piglets. *Journal of Animal Science* 83:82-88.

Jacela, J. Y., J. M. DeRouchey, M. D. Tokach, R. D. Goodband, J. L. Nelssen, D. G. Renter, and S. S. Dritz. 2009. Feed additives for swine: Fact sheets—carcass modifiers, carbohydrate-degrading enzymes and proteases, and anthelmintics. *Journal of Swine Health and Production* 17:325-329.

Jacela, J. Y., J. M. DeRouchey, M. D. Tokach, R. D. Goodband, J. L. Nelssen, D. G. Renter, and S. S. Dritz. 2010a. Feed additives for swine: Fact sheets—flavors and mold inhibitors, mycotoxin binders, and antioxidants. *Journal of Swine Health and Production* 18:27-29.

Jacela, J. Y., S. S. Dritz, J. M. DeRouchey, M. D. Tokach, R. D. Goodband, and J. L. Nelssen. 2010b. Effects of supplemental enzymes in diets containing distillers dried grains with solubles on finishing pig growth performance. *Professional Animal Scientist* 26:412-424.

Jaya, S., and H. Das. 2004. Effect of maltodextrin, glycerol monostearate and tricalcium phosphate on vacuum dried mango powder properties. *Journal of Food Engineering* 63:125-134.

Ji, F., D. P. Casper, P. K. Brown, D. A. Spangler, K. D. Haydon, and J. E. Pettigrew. 2008. Effects of dietary supplementation of an enzyme blend on the ileal and fecal digestibility of nutrients in growing pigs. *Journal of Animal Science* 86:1533-1543.

Johansen, K., and H. D. Poulsen. 2003. Substitution of inorganic phosphorus in pig diets by microbial phytase supplementation—A review. *Pig News and Information* 24:77N-82N.

Johnston, L. J., J. Goihl, G. C. Shurson. 2009. Selected additives did not improve flowability of DDGS in commercial systems. *Applied Engineering in Agriculture* 25:75-82.

Johnston, M. E., J. L. Nelssen, and G. R. Stoner. 1989. Effects of a flavoring agent on finishing swine performance. Pp. 164-165 in *KSU Swine Day 1989*. Manhattan: Kansas State University.

Jones, C. K., J. R. Bergstrom, M. D. Tokach, J. M. DeRouchey, R. D. Goodband, J. L. Nelssen, and S. S. Dritz. 2010a. Efficacy of commercial enzymes in diets containing various concentrations and sources of dried distillers grains with solubles for nursery pigs. *Journal of Animal Science* 88:2084-2091.

Jones, C. K., M. D. Tokach. S. S. Dritz, B. W. Ratliff, N. L. Horn, R. D. Goodband, J. M. DeRouchey, R. C. Sulabo, and J. L. Nelssen. 2010b. Efficacy of different commercial phytase enzymes and development of an available phosphorus release curve for *Escherichia coli*-derived phytases in nursery pigs. *Journal of Animal Science* 88:3631-3644.

Jonsson, E., and P. Conway. 1992. Probiotics for pigs. Pp. 259-315 in *Probiotics, the Scientific Basis*, R. Fuller, ed. London: Chapman and Hall.

Jouany, J. P. 2007. Methods for preventing, decontaminating and minimizing the toxicity of mycotoxins in feeds. *Animal Feed Science and Technology* 137:342-362.

Jurgens, M. H., R. A. Rikabi, and D. R. Zimmerman. 1997. The effect of dietary active dry yeast supplement on performance of sows during gestation-lactation and their pigs. *Journal of Animal Science* 75:593-597.

Kececi, T., H. Oğuz, V. Kurtoglu, and O. Demet. 1998. Effects of polyvinylpolypyrrolidone, synthetic zeolite and bentonite on serum biochemical and haematological characters of broiler chickens during aflatoxicosis. *British Poultry Science* 39:452-458.

Kennedy, J. M., and B. A. Baldwin. 1972. Taste preferences in pigs for nutritive and non-nutritive sweet solutions. *Animal Behaviour* 20:706-718.

Kenny, M., H. Smidt, E. Mengheri, and B. Miller. 2011. Probiotics—Do they have a role in the pig industry? *Animal* 5:462-470.

Kerr, B. J., T. E. Weber, P. S. Miller, and L. L. Southern. 2010. Effect of phytase on apparent total tract digestibility of phosphorus in corn-soybean meal diets fed to finishing pigs. *Journal of Animal Science* 88:238-247.

Kim, S. W., D. A. Knabe, K. J. Hong, and R. A. Easter. 2003. Use of carbohydrases in corn-soybean meal-based nursery diets. *Journal of Animal Science* 81:2496-2504.

Kim, S. W., M. Brandherm, M. Freeland, B. Newton, D. Cook, and I. Yoon. 2008. Effects of yeast culture supplementation to gestation and lactation diets on growth of nursery piglets. *Asian-Australian Journal of Animal Science* 21:1011-1014.

Kim, S. W., M. Brandherm, B. Newton, D. R. Cook, I. Yoon, and G. Fitzner. 2010. Effect of supplementing *Saccharomyces cerevisiae* fermentation product in sow diets on reproductive performance in a commercial environment. *Canadian Journal of Animal Science* 90:229-232.

Kim, T. W., and X. G. Lei. 2005. An improved method for a rapid determination of phytase activity in animal feed. *Journal of Animal Science* 83:1062-1067.

King, R. H. 1979. The effect of adding a feed flavour to the diets of young pigs before and after weaning. *Australian Journal of Experimental Agriculture and Animal Husbandry* 19:695-697.

Kiran, M. M., O. Demet, M. Ortatali, and H. Oğuz. 1998. The preventive effect of polyvinylpolypyrrolidone on aflatoxicosis in broilers. *Avian Pathology* 27:250-255.

Kirchgessner, M., and F. X. Roth. 1987. Einsatz von Formiaten in der ferkelfutterung. 2. Mitteillung: Natriumformiat. *Landwirtschaftliche Forschung* 40:287-294.

Kirchgessner, M., and F. X. Roth., 1990. Nutritive effect of calcium formate in combination with free acids in the feeding of piglets. *Agribiological Research—Zeitschrift Fur Agrarbiologie Agrikulturchemie Okol* 43:53-64.

Koch, B. A., G. L. Allee, and R. H. Hines. 1976. Flavor enhancers in growing pig rations. Pp. 35-36 in *KSU Swine Day 1976*. Manhattan: Kansas State University.

Koch, B. A., G. L. Allee, and R. H. Hines. 1977. Flavor enhancers and/or vitamin C in growing-finishing rations. Pp. 27-28 in *KSU Swine Day 1977*. Manhattan: Kansas State University.

Konietzny, U., and R. Greiner. 2002. Molecular and catalytic properties of phytate-degrading enzymes (phytases). *International Journal of Food Science and Technology* 37:791-812.

Kornegay, E. T., D. Rhein-Welker, M. D. Lindemann, and C. M. Wood. 1995. Performance and nutrient digestibility in weanling pigs as influenced by yeast culture additions to starter diets containing dried whey or one of two fiber sources. *Journal of Animal Science* 73:1381-1389.

Kyriakis, S. C., V. K. Tsiloyiannis, J. Vlemmas, K. Sarris, A. C. Tsinas, C. Alexopoulos, and L. Jansegers. 1999. The effect of probiotic LSP 122 on the control of post weaning diarrhea syndrome of piglets. *Research in Veterinary Science* 67:223-228.

Langendijk, P., J. E. Bolhuis, and B. F. A. Laurenssen. 2007. Effects of pre- and post-natal exposure to garlic and aniseed flavor on pre- and post-weaning feed intake in pigs. *Livestock Science* 108:284-287.

Lauridsen, C., J. H. Nielsen, P. Henckel, and M. T. Sorensen. 1999. Antioxidative and oxidative status in muscles of pigs fed rapeseed oil, vitamin E, and copper. *Journal of Animal Science* 77:105-115.

LeMieux, F. M., L. L. Southern, and T. D. Bidner. 2003. Effect of mannan oligosaccharides on growth performance of weanling pigs. *Journal of Animal Science* 81:2482-2487.

Letourneau-Montminy, M. P., A. Narch, M. Magnin, D. Sauvant, J. F. Bernier, C. Pomar, and C. Jondreville. 2010. Effect of reduced dietary calcium concentration and phytase supplementation on calcium and phosphorus utilization in weanling pigs with modified mineral status. *Journal of Animal Science* 88:1706-1717.

Li, S., W. C. Sauer, R. Mosenthin, and B. Kerr. 1996. Effect of β-glucanase supplementation of cereal-based diets for starter pigs on the apparent digestibilities of dry matter, crude protein and energy. *Animal Feed Science and Technology* 59:223-231.

Lindemann, M. D., D. J. Blodgett, E. T. Kornegay, and G. G. Schurig. 1993. Potential ameliorators of aflatoxicosis in weanling/growing swine. *Journal of Animal Science* 71:171-178.

Mahan, D. C., A. J. Lepine, and K. Dabrowski. 1994. Efficacy of magnesium-L-ascorbyl-2-phosphate as a vitamin C source for weanling and growing-finishing swine. *Journal of Animal Science* 72:2354-2361.

Mahan, D. C., E. A. Newton, and K. R. Cera. 1996. Effect of supplemental sodium phosphate or hydrochloric acid in starter diets containing dried whey. *Journal of Animal Science* 74:1217-1222.

Manzanilla, E. G., J. F. Perez, M. Martin, C. Kamel, F. Baucells, and J. Gasa. 2004. Effect of plant extracts and formic acid on the intestinal equilibrium of early weaned pigs. *Journal of Animal Science* 82:3210-3218.

Manzanilla, E. G., M. Nofrarias, M. Anquita, M. Castillo, J. F. Perez, S. M. Martin-Orue, C. Kamel, and J. Gasa. 2006. Effects of butyrate, avilamycin, and a plant extract combination on the intestinal equilibrium of early weaned pigs. *Journal of Animal Science* 84:2743-2751.

Mathew, A. G., S. E. Chattin, C. M. Robbins, and D. A. Golden. 1998. Effects of a direct-fed yeast culture on enteric microbial populations, fermentation acids, and performance of weanling pigs. *Journal of Animal Science* 76:2138-2145.

McLaughlin, C. L., C. A. Baile, L. L. Buckholtz, and S. K. Freeman. 1983. Preferred flavors and performance of weaning pigs. *Journal of Animal Science* 56:1287-1293.

Miazzo, R., C. A. R. Rosa, E. C. De Queiroz Carvalho, C. Magnoli, S. M. Chiacchiera, G. Palacio, M. Saenz, A. Kikot, E. Basaldella, and A. Dalcero. 2000. Efficacy of synthetic zeolite to reduce the toxicity of aflatoxin in broiler chicks. *Poultry Science* 79:1-6.

Miazzo, R., M. F. Peralta, C. Magnoli, M. Salvano, S. Ferrero, S. M. Chiacchiera, E. C. Q. Carvalho, C. A. R. Rosa, and A. Dalcero. 2005. Efficacy of sodium bentonite as a detoxifier of broiler feed contaminated with aflatoxin and fumonisin. *Poultry Science* 84:1-8.

Millet, S., M. Aluwé, D. L. De Brabander, and M. J. van Oeckel. 2008. Effect of seven hours intermittent suckling and flavor recognition on piglet performance. *Archives of Animal Nutrition* 62:1-9.

Mills, S. E. 2002. Biological basis for the ractopamine response. *Journal of Animal Science* 80(E. Suppl. 2):E28-E32.

Mitchell, A. D., M. B. Solomon, and N. C. Steele. 1991. Influence of level of dietary protein or energy on effects of ractopamine in finishing swine. *Journal of Animal Science* 69:4487-4495.

Moody, D. E., D. L. Hancock, and D. B. Anderson. 2000. Phenethanolamine repartitioning agents. Pp. 65-96 in *Farm Animal Metabolism and Nutrition*, J. P. F. D'Mello, ed. Wallingford, Oxon, UK: CABI.

Mosenthin, R., E. Hambrecht, and W. C. Sauer. 1999. Utilization of different fibers in piglet feeds. Pp. 227-256 in *Recent Advances in Animal Nutrition*, P. C. Garnsworthy and J. Wiseman, eds. Nottingham, UK: Nottingham University Press.

Munro, P. J., A. Lirette, D. M. Anderson, and H. Y. Ju. 2000. Effects of a new sweetener, Stevia, on performance of newly weaned pigs. *Canadian Journal of Animal Science* 80:529-531.

Myers, G. H. 1988. Strategies to control internal parasites in cattle and swine. *Journal of Animal Science* 66:1555-1564.

Namkung, H., M. Li, J. Gong, H. Yu, M. Corttrill, and C. F. M. de Lange. 2004. Impact of feeding blends of organic acids and herbal extracts on growth performance, gut microbiota and digestive function in newly weaned pigs. *Canadian Journal of Animal Science* 84:697-704.

Newberne, P. M., and W. H. Butler. 1969. Acute and chronic effects of aflatoxin on the liver of domestic and laboratory animals: A review. *Cancer Research* 29:236-250.

Newbold, C. J., and K. Hillman. 2011. Feed supplements: Enzymes, probiotics, and yeasts. Pp. 398-400 in *Encyclopedia of Animal Science*, 2nd Ed., D. E. Ullrey, C. Kirk Baer, and W. G. Pond, eds. Boca Raton, FL: CRC Press.

Nonn, H., H. Kluge, H. Jeroch, and J. Broz. 1999. Effects of carbohydrate-hydrolysing enzymes in weaned piglets fed diets based on peas and wheat. *Agribiological Research—Zeitschrift Fur Agrarbiologie Agrikulturchemie Okol* 52:137-144.

Nortey, T. N., J. F. Patience, P. H. Simmins, N. L. Trottier, and R. T. Zijlstra. 2007. Effects of individual or combined xylanase and phytase supplementation on energy, amino acids, and phosphorus digestibility and growth performance of grower pigs fed wheat-based diets containing wheat millrun. *Journal of Animal Science* 85:1432-1443.

Nortey, T. N., J. F. Patience, J. S. Sands, N. L. Trottier, and R. T. Zijlstra. 2008. Effects of xylanase supplementation on the apparent digestibility and digestible content of energy, amino acids, phosphorus, and calcium in wheat and wheat by-products from dry milling fed to grower pigs. *Journal of Animal Science* 86:3450-3464.

NRC (National Research Council). 1994. *Metabolic Modifiers: Effects on the Nutrient Requirements of Food-Producing Animals*. Washington, DC: National Academy Press.

Oğuz, H., and V. Kurtoglu. 2000. Effect of clinoptilolite on performance of broiler chickens during experimental aflatoxicoses. *British Poultry Science* 41:512-517.

Oğuz, H., T. Keçeci, Y. O. Birdane, F. Önder, and V. Kurtoglu. 2000a. Effect of clinoptilolite on serum biochemical and haematological characters of broiler chickens during aflatoxicosis. *Research in Veterinary Science* 69:89-93.

Oğuz, H., V. Kurtoglu, and B. Coşkun. 2000b. Preventive efficacy of clinoptilolite in broilers during chronic aflatoxin (50 and 100 ppb) exposure. *Research in Veterinary Science* 69:197-201.

Omogbenigun, F. O., C. M. Nyachoti, and B. A. Slominski. 2004. Dietary supplementation with multienzyme preparations improved nutrient utilization and growth performance in weaned pigs. *Journal of Animal Science* 82:1053-1061.

Onwulata, C. I., R. P. Konstance, and V. H. Holsinger. 1996. Flow properties of encapsulated milk fat powders as affected by flow agent. *Journal of Food Science* 61:1211-1215.

O'Shea, C. J., T. Sweeney, M. B. Lynch, D. A. Gahan, J. J. Callan, and J. V. O'Doherty. 2010. Effect of β-glucans contained in barley- and oat-based diets and exogenous enzyme supplementation on gastrointestinal fermentation of finisher pigs and subsequent manure odor and ammonia emissions. *Journal of Animal Science* 88:1411-1420.

Overland, M., T. Granli, N. P. Kjos, O. Fjetland, S. H. Steien, and M. Stokstad. 2000. Effect of dietary formates on growth performance, carcass traits, sensory quality, intestinal microflora, and stomach alterations in growing-finishing pigs. *Journal of Animal Science* 78:1875-1884.

Pallauf, J., and J. Hüter. 1993. Studies on the influence of calcium formate on growth, digestibility of crude nutrients, nitrogen balance and calcium retention in weaned piglets. *Animal Feed Science and Technology* 43:65-76.

Patterson, R., and L. G. Young. 1993. Efficacy of hydrated sodium calcium aluminosilicates, screening and dilution in reducing the effects of mold contaminated corn in pigs. *Canadian Journal of Animal Science* 73:615-624.

Pestka, J. J. 2007. Deoxynivalenol: Toxicity, mechanisms and animal health risks. *Animal Feed Science and Technology* 137:283-298.

Pettey, L. A., S. D. Carter, B. W. Senne, and J. A. Shriver. 2002. Effects of beta-mannanase addition to corn-soybean meal diets on growth performance, carcass traits, and nutrient digestibility of weanling and growing-finishing pigs. *Journal of Animal Science* 80:1012-1019.

Phillips, T. D. 1999. Dietary clay in the chemoprevention of aflatoxin-induced disease. *Toxicological Sciences* 52:118-126.

Phillips, T. D., L. F. Kubena, R. B. Harvey, D. R. Taylor, and N. D. Heildebaugh. 1988. Hydrated sodium calcium aluminosilicate: A high affinity sorbent for aflatoxin. *Poultry Science* 67:243-247.

Piccaglia, R., M. Marotti, E. Giovanelli, S. G. Deans, and E. Eaglesham. 1993. Antibacterial and antioxidant properties of Mediterranean aromatic plants. *Industrial Crops and Products* 2:47-50.

Piva, A., G. Casadei, G. Pagliuca, E. Cabassi, F. Galvano, M. Solfrizzo, R. T. Riley, and D. E. Diaz. 2005. Activated carbon does not prevent the toxicity of culture material containing fumonisin B_1 when fed to weanling piglets. *Journal of Animal Science* 83:1939-1947.

Potter, M. L., S. S. Dritz, M. D. Tokach, J. M. DeRouchey, R. D. Goodband, and J. L. Nelssen. 2010. Effects of meal or pellet diet form on finishing pig performance and carcass characteristics. Pp. 245-251 in *KSU Swine Day 2010*. Manhattan: Kansas State University.

Radecki, S. V., M. R. Juhl, and E. R. Miller. 1988. Fumaric and citric acids as feed additives in starter diets: Effect on performance and nutrient balance. *Journal of Animal Science* 66:2598-2605.

Ramos, A. J., and E. Hernández. 1997. Prevention of aflatoxicosis in farm animals by means of hydrated sodium calcium aluminosilicates addition to feedstuffs: A review. *Animal Feed Science and Technology* 65:197-206.

Ramos, A. J., J. Fink-Gremmels, and E. Hernández. 1996. Prevention of toxic effects of mycotoxins by means of nonnutritive adsorbent compounds. *Journal of Food Protection* 59:631-641.

Reeds, P. J., and H. J. Mersmann. 1991. Protein and energy requirements of animals treated with beta-adrenergic agonists: A discussion. *Journal of Animal Science* 69:1532-1550.

Roth, F. X., and M. Kirchgessner. 1993. Influence of avilamycin and tylosin on retention and excretion of nitrogen in finishing pigs. *Journal of Animal Physiology and Animal Nutrition* 69:245-250.

Rozeboom, D. W., D. T. Shaw, R. J. Tempelman, J. C. Miquel, J. E. Pettigrew, and A. Connelly. 2005. Effects of mannan oligosaccharide and an antimicrobial product in nursery diets on performance of pigs reared on three different farms. *Journal of Animal Science* 83:2637-2644.

Salmon, R. E. 1985. Effects of pelleting, added sodium bentonite and fat in a wheat-based diet on performance and carcass characteristics of small white turkeys. *Animal Feed Science and Technology* 12:223-232.

Samarajeewa, V., A. E. Sen, M. D. Cohen, and C. I. Wey. 1990. Detoxification of aflatoxins in foods and feeds by physical and chemical methods. *Journal of Food Protection* 53:489-501.

Sauerwein, H., S. Schmitz, and S. Hiss. 2007. Effects of a dietary application of a yeast cell wall extract on innate and acquired immunity, on oxidative status and growth performance in weanling piglets and on the ileal epithelium in fattened pigs. *Journal of Animal Physiology and Animal Nutrition* 91:369-380.

Schatzmayr, G., F. Zehner, M. Täubel, D. Schatzmayr, A. Klimitsch, A. P. Loibner, and E. M. Binder. 2006. Microbiologicals for deactivating mycotoxins. *Molecular Nutrition and Food Research* 50:543-551.

Schell, T. C., M. D. Lindemann, E. T. Kornegay, and D. J. Blodgett. 1993a. Effects of feeding aflatoxin-contaminated diets with and without clay to weanling and growing pigs on performance, liver function, and mineral metabolism. *Journal of Animal Science* 71:1209-1218.

Schell, T. C., M. D. Lindemann, E. T. Kornegay, D. J. Blodgett, and J. A. Doerr. 1993b. Effectiveness of different types of clay for reducing the detrimental effects of aflatoxin-contaminated diets on performance and serum profiles of weanling pigs. *Journal of Animal Science* 71:1226-1231.

Schinckel, A. P., N. Li, B. T. Richert, P. V. Preckel, and M. E. Einstein. 2003. Development of a model to describe the compositional growth and dietary lysine requirements of pigs fed ractopamine. *Journal of Animal Science* 81:1106-1119.

Seabolt, B. S., E. van Heugten, S. W. Kim, K. D. Ange-van Heugten, and E. Roura. 2010. Feed preferences and performance of nursery pigs fed diets containing various inclusion amounts and qualities of distillers coproducts and flavor. *Journal of Animal Science* 88:3725-3738.

See, M. T., T. A. Armstrong, and W. C. Weldon. 2004. Effect of ractopamine feeding program on growth performance and carcass composition in finishing pigs. *Journal of Animal Science* 82:2474-2480.

Selle, P. H., and V. Ravindran. 2008. Phytate degrading enzymes in pig nutrition. *Livestock Science* 113:99-122.

Selle, P. H., A. J. Cowieson, and V. Ravindran. 2009. Consequences of calcium interactions with phytate and phytase for poultry and pigs. *Livestock Science* 124:126-141.

Shen, Y. B., X. S. Piao, S. W. Kim, L. Wang, P. Liu, I. Yoon, and Y. G. Zhen. 2009. Effects of yeast culture supplementation on growth performance, intestinal health, and immune response of nursery pigs. *Journal of Animal Science* 87:2614-2624.

Siers, D. G., D. E. DeKay, H. J. Mersmann, L. J. Brown, and H. C. Stanton. 1976. Late gestation feeding of dichlorvos: A physiological characterization of the neonate and a growth-survival response. *Journal of Animal Science* 42:381-392.

Smiricky-Tjardes, M. R., C. M. Grieshop, E. A. Flickinger, L. L. Bauer, and G. C. Fahey, Jr. 2003. Dietary galactooligosaccharides affect ileal and total tract nutrient digestibility, ileal and fecal bacteria concentrations, and ileal fermentative characteristics of growing pigs. *Journal of Animal Science* 81:2535-2545.

Southern, L. L., T. B. Stewart, E. Bodak-Koszalka, D. L. Leon, P. G. Hoyt, and M. E. Bessette. 1989. Effect of fenbendazole and pyrantel tartrate on the induction of protective immunity in pigs naturally or experimentally infected with *Ascaris suum*. *Journal of Animal Science* 67:628-634.

Spencer, J. D., G. I. Petersen, A. M. Gaines, and N. R. Augsburger. 2007. Evaluation of different strategies for supplementing distillers dried grains with solubles (DDGS) to nursery pig diets. *Journal of Animal Science* 85(Suppl. 2):96-97 (Abstr.).

Spurlock, M. E., J. C. Cusumano, S. Q. Ji, D. B. Anderson, C. K. Smith, II, D. L. Hancock, and S. E. Mills. 1994. The effect of ractopamine on beta-adrenergic density and affinity in porcine adipose and skeletal muscle tissue. *Journal of Animal Science* 72:75-80.

Stark, C. R., K. C. Behnke, J. D. Hancock, S. L. Traylor, and R. H. Hines. 1994. Effect of diet form and fines in pelleted diets on growth performance of nursery pigs. *Journal of Animal Science* 72(Suppl. 1):214 (Abstr.).

Stein, H. 2007. *Feeding the Pigs' Immune System and Alternatives to Antibiotics*. London Swine Conference, April 3-4, 2007, London, Ontario.

Sterk, A., P. Schlegel, A. J. Mul, M. Ubbink-Blanksma, and E. M. A. M. Bruininx. 2008. Effects of sweeteners on individual feed intake characteristics and performance in group-housed weanling pigs. *Journal of Animal Science* 86:2990-2997.

Stewart, L. L., B. G. Kim, B. R. Gramm, R. D. Nimmo, and H. H. Stein. 2010. Effect of virginiamycin on the apparent ileal digestibility of amino acids by growing pigs. *Journal of Animal Science* 88:1718-1724.

Sulabo, R. C., M. D. Tokach, J. M. DeRouchey, S. S. Dritz, R. D. Goodband, and J. L. Nelssen. 2010. Influence of feed flavors and nursery diet complexity on preweaning and nursery pig performance. *Journal of Animal Science* 88:3918-3926.

Swamy, H. V. L. N., T. K. Smith, E. J. MacDonald, H. J. Boermans, and E. J. Squires. 2002. Effects of feeding a blend of grains naturally contaminated with *Fusarium* mycotoxins on swine performance, brain regional neurochemistry, and serum chemistry and the efficacy of a polymeric glucomannan mycotoxin adsorbent. *Journal of Animal Science* 80:3257-3267.

Swamy, H. V. L. N., T. K. Smith, E. J. MacDonald, N. A. Karrow, B. Woodward, and H. J. Boermans. 2003. Effects of feeding a blend of grains naturally contaminated with *Fusarium* mycotoxins on growth and immunological measurements of starter pigs, and the efficacy of a polymeric glucomannan mycotoxin adsorbent. *Journal of Animal Science* 81:2792-2803.

Taras, D., W. Vahjen, M. Macha, and O. Simon. 2006. Performance, diarrhea incidence, and occurrence of *Escherichia coli* virulence genes during long-term administration of a probiotic *Enterococcus faecium* strain to sows and piglets. *Journal of Animal Science* 84:608-617.

Thacker, P. A., and G. L. Campbell. 1999. Performance of growing/finishing pigs fed untreated or micronized hulless barley-based diets with or without β-glucanase. *Journal of Animal and Feed Sciences* 8:157-170.

Thomas, M., and A. F. B. van der Poel. 1996. Physical quality of pelleted animal feeds. 1. Criteria for pellet quality. *Animal Feed Science and Technology* 61:89-112.

Thomas, M., T. van Vliet, and A. F. B. van der Poel. 1998. Physical quality of pelleted animal feed. 3. Contribution of feedstuff components. *Animal Feed Science and Technology* 70:59-78.

Traylor, S. L., K. C. Behnke, J. D. Hancock, P. Sorrell, and R. H. Hines. 1996. Effect of pellet size on growth performance in nursery and finishing pigs. *Journal of Animal Science* 74 (Suppl. 1):67(Abstr.).

Urban, J. F., Jr., R. D. Romanowski, and N. C. Steele. 1989. Influence of helminth parasite exposure and strategic application of anthelmintics on the development of immunity and growth in swine. *Journal of Animal Science* 67:1668-1677.

van der Peet-Schwering, C. M. C., A. J. M. Jansman, H. Smidt, and I. Yoon. 2007. Effects of yeast culture on performance, gut integrity, and blood cell composition of weanling pigs. *Journal of Animal Science* 85:3099-3109.

van Heugten, E., D. W. Funderburke, and K. L. Dorton. 2003. Growth performance, nutrient digestibility, and fecal microflora in weanling pigs fed live yeast. *Journal of Animal Science* 81:1004-1012.

van Kempen, T. A. 2001. Dietary adipic acid reduces ammonia emission from swine excreta. *Journal of Animal Science* 79:2412-2417.

Van Zuilichem, D. J., L. Moscicki, and W. Stolp. 1979a. Nieuwe bindmiddelen verbeteren brokkwaliteit. *De Molenaar* 82:1572.

Van Zuilichem, D. J., L. Moscicki, and W. Stolp. 1979b. Nieuwe bindmiddelen verbeteren brokkwaliteit. *De Molenaar* 82:1618-1622.

Van Zuilichem, D. J., L. Moscicki, and W. Stolp. 1980. Nieuwe bindmiddelen verbeteren brokkwaliteit. *De Molenaar* 83:12-18.

Veum, T. L., J. Reyes, and M. Ellersieck. 1995. Effect of supplemental yeast culture in sow gestation and lactation diets on apparent nutrient digestibilities and reproductive performance through one reproductive cycle. *Journal of Animal Science* 73:1741-1745.

Voss, K. A., G. W. Smith, and W. M. Haschek. 2007. Fumonisins: Toxicokinetics, mechanism of action and toxicity. *Animal Feed Science and Technology* 137:299-325.

Walsh, M. C., D. M. Sholly, R. B. Hinson, K. L. Saddoris, A. L. Sutton, J. S. Radcliffe, R. Odgaard, J. Murphy, and B. T. Richert. 2007a. Effects of water and diet acidification with and without antibiotics on weanling pig growth and microbial shedding. *Journal of Animal Science* 85:1799-1808.

Walsh, M. C., D. M. Sholly, R. B. Hinson, S. A. Trapp, A. L. Sutton, J. S. Radcliffe, J. W. Smith, II, and B. T. Richert. 2007b. Effects of Acid LAC and Kem-Gest acid blends on growth performance and microbial shedding in weanling pigs. *Journal of Animal Science* 85:459-467.

Watkins, L. E., D. J. Jones, D. H. Mowrey, D. B. Anderson, and E. L. Veenhuizen. 1990. The effects of various levels of ractopamine hydrochloride on the performance and carcass characteristics of finishing swine. *Journal of Animal Science* 68:3588-3595.

Weber, T. E., and B. J. Kerr. 2008. Effect of sodium butyrate on growth performance and response to lipopolysaccharide in weanling pigs. *Journal of Animal Science* 86:442-450.

White, L. A, M. C. Newman, G. L. Cromwell, and M. D. Lindemann. 2002. Brewers dried yeast as a source of mannan oligosaccharides for weanling pigs. *Journal of Animal Science* 80:2619-2628.

Williams, K. C., B. J. Blaney, and R. T. Peters. 1994. Pigs fed *Fusarium*-infected maize containing zearalenone and nivalenol with sweeteners and bentonite. *Livestock Production Science* 39:275-281.

Yen, J. T., J. A. Nienaber, J. Klindt, and J. D. Crouse. 1991. Effect of ractopamine on growth, carcass traits, and fasting heat production of U.S. contemporary crossbred and Chinese Meishan pure- and crossbred pigs. *Journal of Animal Science* 69:4810-4822.

Yiannikouris, A., G. André, A. Buléon, L. Poughon, J. François, C. G. Dussap, G. Jeminet, I. Canet, G. Bertin, and J. P. Jounary. 2006. Chemical and conformational study of the interactions involved in mycotoxins complexation with β-D-glucans. *Biomacromolecules* 7:1147-1155.

Yoon, S. Y., Y. X. Yang, P. L. Shinde, J. Y. Choi, J. S. Kim, Y. W. Kim, K. Yun, J. K. Jo, J. H. Lee, S. J. Ohh, I. K. Kwon, and B. J. Chae. 2010. Effects of mannanase and distillers dried grain with solubles on growth performance, nutrient digestibility, and carcass characteristics of grower-finisher pigs. *Journal of Animal Science* 88:181-191.

Young, R., Jr., D. K. Hass, and L. J. Brown. 1979. Effect of late gestation feeding of dichlorvos in non-parasitized and parasitized sows. *Journal of Animal Science* 48:45-51.

Zaika, L. L., J. C. Kissinger, and A. E. Wasserman. 1983. Inhibition of lactic acid bacteria by herbs. *Journal of Food Science* 48:1455-1459.

Zani, J. L., F. W. Dacruz, A. F. Dossantos, and C. Gilturnes. 1998. Effect of probiotic CenBiot on the control of diarrhea and feed efficiency in pigs. *Journal of Applied Microbiology* 84:68-71.

Zeyner, A., and E. Boldt. 2006. Effects of a probiotic *Enterococcus faecium* strain supplemented from birth to weaning on diarrhoea patterns and performance of piglets. *Journal of Animal Physiology and Animal Nutrition* 90:25-31.

Zimmerman, D. R., D. P. Conway, D. H. Bliss, D. O. Farrington, and H. J. Barnes. 1982. Effects of carbadox and pyrantel tartrate on performance and indices of *Mycoplasma hyopneumoniae* and *Ascaris suum* infection in pigs. *Journal of Animal Science* 55:733-740.

第 11 章 饲料污染物

引言

除了基于非营养目的专门添加到日粮中的非营养性饲料添加剂外（第 10 章），很多日粮可能还含有对猪和其他动物无害或有害的物质。这些物质即使无害，仍为污染物，可分为化学、生物及物理三类。在自然条件下，饲料污染很常见，虽然我们必须努力使它降至最低，但通常关注不够。然而，由于人为向饲料/食品中添加某些物质（如三聚氰胺，Sharma and Paradakar, 2010）所引发的动物健康事故，以及在收获季节发生的某些饲料/食品极端严重的天然污染事件频发（如霉菌毒素，Pollock, 2010），污染物正越来越受到严密监控。

本章首次出现在 NRC《动物营养需要》系列丛书中，增列本章不是因为仅限于猪料的已知或怀疑的问题，而是因为即使偶尔发生，种类众多的污染物会损害动物健康和福利，这一观点在各个地方的各种动物上得到证实。由于饲料原料的国际贸易属性和国内的动物生产，饲料/食品供应体系的安全就成了世界范围内的重要问题。在许多国家，提供安全饲料一直是饲料生产企业的首要任务并接受各种机构的协同管理，这些机构除了行业组织美国饲料管理协会（AAFCO）和美国饲料工业协会（AFIA），还包括政府组织美国食品和药物管理局（FDA）、美国农业部（USDA）动植物健康检查署。本章旨在通过总结目前存在的特别重要或大众关注的一些问题、解决这些问题的尝试以及在世界范围内涉及各种家畜的研究，维护和强化以确保饲料/食品供应公共安全为目标的努力。

化学性污染物

在通常情况下，化学污染受到关注的程度高于生物和物理污染。化学污染物主要分为三小类：农药及农药残留、霉菌毒素、重金属/放射性核素。农药和农药残留为数众多，其中被 FDA 动物饲料安全体系（FDA, 2011）列入具有潜在危险的化学污染物草拟名单的包括：艾氏剂（aldrin）、克灭杀（bezene hexachloride）、氯丹（chlordane）、毒死蜱（chlorpyrifos）、甲基毒死蜱（chlorpyrifos-methyl）、二嗪磷（diazinon）、狄氏剂（dieldrin）、滴滴涕（dichlorodiphenyltrichloroethane）+滴滴滴（tetrachlorodiphenylethane）+二氯二苯二氯乙烷（dichlorodiphenyldichloroethyene）（DDT+TDE+DDE）、开乐散（dicofol）、硫丹（endosulfan）、异狄氏剂（endrin）、乙硫磷（ethion）、α-六氯环己烷（六六六，α-hexachlorocyclohexane）（HCH）、β-六氯环己烷（β-HCH）、γ-六氯环己烷（γ-HCH）（林旦，lindane）、七氯（heptachlor）、七氯（heptachlor）+环氧七氯（heptachlor epoxide）、六氯苯（hexachlorobenzne）、马拉硫磷（四零四九，malathion）、甲氧滴滴涕（methoxychlor）、全氯五环葵烷（mirex）、硝酚硫磷酯（parathion）、毒杀芬［toxaphene（camphechlor）］和脱叶磷（tribuphos）。上述农药有的正在农业中使用，有的虽然先后禁用，但仍残留于自然界之中。列入 FDA（2011）文件的霉菌毒素有：黄曲霉毒素（B_1、B_2、G_1 和 G_2）、伏马毒素/烟曲霉毒素（B_1、B_2 和 B_3）、脱氧雪腐镰刀菌烯醇（DON 或呕吐毒素）、赭曲霉毒素和玉米赤霉烯酮。被列入的重金属/放射性核素有：砷、镉、铬、铅、^{241}镅（^{241}Am）、^{134}铯（^{134}Cs）、^{131}碘（^{131}I）、^{238}钚（^{238}Pu）、^{103}钌（^{103}Ru）、^{106}钌（^{106}Ru）和 ^{90}锶（^{90}Sr）。对于猪料，放射性核素不是受到密切关注的污染物，但 D'Mello（2000）指出自 1986 年切

尔诺贝利核电站事故后，释放出来的 ^{134}铯和 ^{137}铯广泛污染了牧场和储藏牧草，进而污染了牛奶和绵羊胴体，绵羊运输和屠宰也受到限制。除了上述主要三小类化学污染物外，FDA（2011）还列出了其他化学污染物，如乙氧喹、二噁英、汞、高氯酸盐、多氯联苯（PCB）、聚乙二醇和硒。

农药

在谷物生产中使用的化学药品为数众多，如除草剂、杀真菌剂及收获前后的杀虫剂。Van Barneveld（1999）综述了收获后用于家畜饲料的澳大利亚谷物中的多种农药的影响。蛋鸡研究表明，按单独无害剂量组合这些农药降低了蛋鸡生产性能和生产效率。猪采食经草甘膦和/或乙烯磷处理的大麦得到的结果是不一致的，一些研究表明没有负效应，而另一些研究则表明接受了上述某些处理的母猪的仔猪成活率降低。结果虽有差异，但却表明在某些情况下，饲料中的除草剂/杀虫剂残留可能引起有害反应。此外，在审查管理认证过程中不存在但在常规的作物生产中可能存在的多种农药及其残留组合可能导致动物的不良反应，这些组合在认证过程中是识别不出的。

自 1989 年至 1994 年，FDA 采集了 500 份以上的家畜饲料样品，分析有机氯和有机磷农药（Lovell et al.,1996），只有 16.1%的样品未检出农药残留。在检出农药残留的样品中发现了 804 种残留物（654 种可量化，150 种痕量），但均未超过规定限值。最常检出农药残留是 5 种有机磷化合物（马拉硫磷、甲基毒死蜱、二嗪磷、毒死蜱和甲基嘧啶磷），占检出种类的 93.4%。最常检出有机氯化合物是甲氧滴滴涕、DDE、多氯联苯（PCB）、狄氏剂、五氯硝基苯和林旦，但同时检出这 6 种化合物的样品只占 4.1%。

在家畜产品中以残留物形式出现的有机氯农药表明了某些已经禁用的产品仍持续残留，因为这些农药是脂质化合物，最终在食物链中浓缩并在脂肪中蓄积。Furusawa 和 Morita（2000）的发现证明了这一点，他们于 1998 年测定了日本一家普通家禽饲养场的蛋鸡基础日粮、鸡蛋、7 种组织（脂肪组织、血液、肾脏、肝脏、肌肉、卵巢和输卵管）及排泄物浸提出的脂肪中有机氯农药的污染和蓄积水平。日本大约于 1970 年在农业上停用有机氯农药，但在日粮脂肪中仍发现了狄氏剂及所检测的所有种类的 DDT。此外，在蛋黄脂肪和蛋鸡所有组织脂肪中检测出了狄氏剂和一些种类的 DDT，但在经干燥的排泄物中未检出。虽然对于所检测的有机氯农药，其长久残留是明显的，但其蓄积的水平仍远低于实际的残留物限值。

霉菌毒素

霉菌毒素是丝状真菌（霉菌）的次生代谢产物，动物摄入后，可引发一系列的有害生理反应。一些典型症状是：拒食、消化不良、神经系统紊乱如颤抖和无力、受胎率降低甚至流产的繁殖障碍、免疫抑制、器官损伤和癌症。虽然已经发现并鉴别了上百种霉菌毒素，但引起猪不良反应的霉菌毒素主要是黄曲霉毒素（B_1、B_2、G_1 和 G_2）、玉米赤霉烯酮、脱氧雪腐镰刀菌烯醇（DON 或呕吐毒素）、伏马毒素和赭曲霉毒素 A。产生这 5 种毒素的霉菌是曲霉菌属（产生黄曲霉毒素和赭曲霉毒素）、镰刀菌属（产生玉米赤霉烯酮、呕吐毒素和伏马毒素 B_1）和青霉菌属（产生赭曲霉毒素）。这些霉菌在田间和储藏期间均可产生。真菌生长很大程度上依赖于环境条件，尤其是植物生长和饲料储藏关键期的温度与湿度。虽然每种毒素可能引发几种非特异性反应，但已查明均有主反应。黄曲霉毒素是强效肝脏毒素，玉米赤霉烯酮具有高雌激素效应，DON 影响采食量和消化道，伏马毒素 B_1 引起猪的肺水肿，赭曲霉素是肾毒素。

Placinta 等（1998）发表了一篇关于世界范围内谷物和动物饲料遭受镰刀菌属霉菌毒素污染的综述。综述表明这些霉菌毒素普遍存在，但就其浓度而言，有明显的地域性。一项商业调查（BIOMIN，2010）也表明霉菌毒素不仅广泛存在于世界各地，也存在于不少农产品中。该调查对 2660 份样品做了 9030 次测试，对大量饲料样品（如谷物、副产物饲料和成品饲料）中的黄曲霉毒素、玉米赤霉烯酮、呕吐毒素和赭曲霉毒素 A 进行了分析。与该公司前些年的调查相同，玉米受到的污染是最广泛、最严重的，75%的样品至少被一种、40%的样品至少被一种以上的霉菌毒素污染。霉菌毒素除了存在于谷物中外，在由这些谷物制成的副产物中还被浓缩，如酒精糟及可溶物（DDGS）或浓缩酒精糟可溶物（CDS）。Schaafsma 等（2009）测定表明 CDS 以及最终副产物 DDGS 中 DON 的浓度高于最初的原料（玉米）中的浓度。以干重为基础，与玉米相比，DDGS 中毒素的浓度增加了 3 倍，而 CDS 中毒素的浓度增加了 4 倍。

FDA 已经发布了可能存在于原料谷物和成品饲料中的黄曲霉毒素和呕吐毒素及其污染物的法规指南。FDA 以三种形式之一发布政策指南或强制公告："忠告水平"（advisory levels）——监管机构确信食品或饲料中某物质的含量有足够的安全裕量保护人类和动物健康；"行动水平"（action levels）——希望界定污染物让监管机构采取执法行动的准确含量；"监管限值"（regulatory limits）——经过行政立法程序规定的公众通知和评论规则制定程序并发布有效法规后确立的毒素和污染物的含量。在国家谷物和饲料协会网站上可找到 FDA 毒素和污染物法规指南的概要，对于更详细的背景信息或最新信息可查阅 FDA 指南文件（2000，2001，2010b）。

NRC（1979）、CAST（1989，2003）及 Kanora 和 Maes（2009）提供了关于霉菌毒素的发生、它们对不同动物的影响或对猪的具体影响，以及被污染饲料可能的处理方法等更全面的信息。对于美国以外的国家，关于霉菌毒素和其他污染物以及应急行动水平的信息可从联合国粮农组织（FAO）的网站查阅。

重金属

猪料添加的矿物质可通过采矿或从产品中回收得到。由于矿物质来源和纯化或提取方法不同，各种不需要的元素可能存在于终产品中，相似地，当用于畜牧业的矿物质来源于回收原料时，回收方法将影响有害矿物质/金属的潜在含量。来源于这些工业过程的副产物及所采用的加工方法可对农牧业造成影响，影响的途径是通过空气传播或将副产物用作满足作物对氮、磷、钾需要的化肥。

Vreman 等（1986）进行了两项研究，评估奶牛直接或通过港口污泥和下水道排出的污泥采食金属元素后，镉、铅、汞和砷从饲料向牛奶和各种体组织的转移情况。在为期 3 个月的第一项研究中，奶牛直接摄入重金属，牛奶和体组织中的重金属水平是对照组的 4~75 倍。使用淤泥的第二项研究为期 28 个月，在饲喂期结束时，体组织检查表明肝脏和肾脏是金属蓄积的主要部位，骨骼中铅含量随摄入量增加而提高，但是重金属摄入量增加并不引起它们在牛奶、血液和肌肉中含量的显著增加。Timmons（2010）报道了一份亚太地区关于矿物质预混料和全价料重金属污染的行业调查，对样品进行分析，判定超过欧盟（EU）通过法规 2002/32/EC 确立的有害物质标准的比例。欧盟的这个法规列出了饲料添加剂中有害物质的最大限量：砷（15 ppm）、镉（10 ppm）、铅（100 ppm）、汞（0.05 ppm）。就至少一种重金属超过 EU 限量的污染样品的比例而言，来自 10 个接受调查的国家的样品中有 3%~43%受到了污染。在采集到的 25 个家禽预混料样品中，有 48%受到了至少一种重金属超过 EU 限量的污染，在 30

个添加了无机矿物质的全价料样品中，有 7%受到了至少一种重金属的污染。一项在美国进行的调查（Kerr et al.，2008）表明，某些矿物元素来源的铅含量超过欧盟标准。在美国，矿物质饲料原料允许的污染物水平指南由美国饲料管理协会制定（AAFCO，2010）。

除了使用下水道排污物淤泥作为化肥和不小心使用了受污染的矿物质预混料外，重金属污染最可能的来源是使用可能含有汞的鱼粉。众所周知，汞会在鱼体内蓄积，使用含汞的鱼粉会导致汞在家畜产品中蓄积。鱼粉中汞含量随用于生产鱼粉的鱼的种类以及捕捞鱼的水域而异（Johnston and Savage，1991）。Chang 等（1977）直接给猪补充汞、Stothers 等（1971）给猪和家禽饲喂鱼粉的早期研究确立了汞的添加剂量和添加形式与组织中汞水平的关系。两项研究也表明汞沉积最多的部位是被毛和肝脏。Stothers 等（1997）报道汞在动物体内的蓄积存在物种差异。在饲喂不同水平汞日粮时，家禽蓄积的汞较猪少。Dórea（2006）发表了关于对许多家畜使用鱼粉的可能性及其对人类健康影响的综述。

Lin 等（2004）发现向饲粮中添加 0.3%蒙脱石黏土纳米化合物显著降低（$P<0.05$）血液、肌肉、肾脏、肝脏组织中汞水平，这表明添加这种非营养性的吸附材料通过其特异性的吸附作用有效降低了消化道对汞的吸收。因此，任何重金属的潜在毒性不仅随其在成品饲料中的浓度，也随饲料中与其发生互作的其他成分而变化。

其他化学污染物

三聚氰胺（$C_3H_6N_6$；相对分子质量 126.12；Merk Index，2006）是一种含氮量高（66.64%）的化合物。当用测定的 N 含量乘以 6.25 计算饲料/食品中粗蛋白含量时，产品中掺入少量三聚氰胺就可很大程度地提高其粗蛋白含量，因为三聚氰胺本身似乎有高达 416.5%的粗蛋白含量（66.64%N×6.25）。正如引言所提到的，在中国三聚氰胺被人为地掺入谷物副产物和奶粉中以提高所谓的粗蛋白含量。Sharma 和 Paradakar（2010）总结了有关这次事件的报道和三聚氰胺在工业中的应用。简而言之，2007 年和 2008 年在北美检测出宠物食品中被人为地掺入了三聚氰胺，随后发现在中国三聚氰胺被蓄意加入到婴儿奶粉中，导致大约 30 万名儿童发病，至少 6 人死亡。

多氯联苯（PCB）和二噁英（多氯二苯并呋喃和多氯二苯并二噁英的统称）是强毒物并广受关注。它们能在消化道被迅速吸收，引起消化道、神经系统和繁殖系统的病理反应。多氯联苯和二噁英具有免疫抑制作用，有证据表明除了在母乳蓄积外，它们还能通过胎盘转运并在胎儿体内蓄积（Calamari，2002）。多氯联苯在工业中广泛应用的基础在于它们的理化特性，因为它们具有阻燃性、非常低的导电性、高的导热性、对化学降解极强的抗性，在正常的环境条件下化学性质非常稳定。二噁英是一系列含氯产品（工业化学试剂和农药）生产过程的污染物，燃烧含有氯化物的原料，特别是供氧不足和燃烧温度不够高时也会产生。Calamari（2002）就此方面的问题发表了一篇很好的综述。二噁英可通过受污染的脂肪（Feed Info News Service，2010b）或受污染的预混料（Feed Info News Service，2010c）进入饲料/食品供应链。当含量超过可接受的水平时，可导致大量的饲料召回事件及对饲料业和畜牧业的严重破坏（Feed Info News Service，2009，2011c,d；Feedstuffs，2011）。

业已证明，制定法规控制污染物有益于人类食品供应。Schwind 和 Jira（2008）调查了德国的肉与肉制品中二噁英和多氯联苯的水平，发现所有涉及肉类中上述两种有害物质的含量均显著低于欧盟所制定的最大残留水平。与大约十年前在德国进行的类似研究相比，二噁英的含量，特别是禽肉和和牛肉中的含量显著降低。

生物性污染物

生物性污染物主要有两小类：传染性海绵状脑病（TSE）和细菌。与动物有关的两种 TSE 在美国受到关注：牛海绵状脑病（BSE）和慢性消耗性疾病（CWD）。与家畜饲料污染有关而受到关注的细菌是：芽孢杆菌、梭状芽孢杆菌、大肠杆菌、分支杆菌、假单胞菌、沙门氏菌（各种血清型）和葡萄球菌，但并非所有的细菌都对猪有重要影响。

传染性海绵状脑病（TSE）是影响人和动物的一大类疾病，特征是脑组织退化呈海绵状，会导致死亡。TSE 包括牛 BSE、小型反刍动物（如绵羊和山羊）痒病和鹿科动物（如鹿和麋鹿）慢性消耗性疾病（CWD）。TSE 很大程度上归因于一种被称为感染性蛋白的颗粒，该物质是一种主要由蛋白质异构体组成的传染性病原体。BSE 于 1986 年首次在英国确诊，由于用受感染动物的胴体制成的肉骨粉配制饲料，造成该病流行。FAO（1998）的一份出版物详述了这个仍在发展中问题的大量观察史及其病因学发现过程。因为标准炼制过程并不能使 BSE 病原体完全失活或将其杀灭，所以由受感染的动物炼制的蛋白原料如肉骨粉可能含有传染性病原。一份 BSE 公告（USDA，2006）提到，美国农业部动植物卫生检验局（APHIS）联合食品和药物管理局（FDA）、美国农业部食品安全检查署已采取强力措施防止 BSE 进入美国和传播的可能。虽然已从在加拿大出口美国的牛中鉴别出了 BSE，动植物健康检查署自 1989 年来一直采取严格措施防止高风险动物和产品的进口。1997 年 FDA 禁止用大多数哺乳动物蛋白饲养包括牛在内的反刍动物的法规生效执行。对进口牛严格监管和禁用有风险的饲料是防止牛病传播的重要措施。虽然以上所述是饲料行业关注的重要领域，但目前还不是养猪业关注的问题。

饲料的细菌污染是一个争议颇多的领域，因为饲料是人类食品细菌污染发生的主要途径的观点并未得到广泛认同。正如 D'Mello（2000）所提到的，自从发现大肠杆菌中的 O157:H7 血清型与人类疾病的联系后，动物饲料中的大肠杆菌含量受到相当关注。毫无疑问，许多肉类大肠杆菌的潜在污染都与屠宰、零售过程及家中储藏环境有关。虽然 Lynn 等（1998）的调查发现从饲养场和商业过程（生产、销售）中采集的 209 个牛料样品无一呈 O157：H7 血清型大肠杆菌阳性，但是 30%的样品对大肠杆菌属呈阳性以及随后的试验证明混合日粮支持大肠杆菌复制，上述两项事实表明饲料可能是畜牧业大肠杆菌的来源。试验日粮支持大肠杆菌复制的能力与日粮中的青贮玉米的有机酸含量有关，从而提示饲料支持细菌复制的能力将随特定饲料来源和特定菌株生长所需的条件而变化。

Molla 等（2010）测定了从商业养猪场采集的饲料和粪样分离物中沙门氏菌的出现与遗传的相关性。在商业加工饲料和粪样中存在沙门氏菌的遗传关联菌株，某些情况下还存在无性繁殖菌株，包括对多种药物具有耐受性的菌株，作为沙门氏菌传播的可能载体，商业饲料对其有重大意义。Wales 等（2010）对一系列投放到饲料或饮水中为控制存在于饲料或家畜所处环境中沙门氏菌的不同化学试剂的作用模式和效价的资料进行了综述。综述阐述了使用化学试剂对饲料和饲料原料去除病菌污染必须考虑最初污染的程度、在储存和转运期间二次污染的可能性以及目标家畜对沙门氏菌的易感性，才能取得效果。FDA（2010a）最近向有关团体征求对动物饲料中沙门氏菌的执法政策指南草案的意见。FDA 要求对草案中提出的由于存在沙门氏菌而被判定不合格的动物饲料或饲料原料推荐采取执法行动的某些标准进行反馈。一旦完成，这份文件将为 FDA 关于受沙门氏菌污染的以及可直接接触到人的（如宠物食品）动物饲料或饲料原料的法规政策提供指南。这份政策指南草案聚焦于几种对人类健康有潜在影响的血清型而不是全面禁止，因此，并不是所有发现饲料中有沙门氏菌的事件都被认为是饲料不合格。

物理性污染物

在成品饲料中有时会发现塑料、玻璃和金属的物理污染物。这种潜在的污染大多可通过饲料厂清理和卫生措施控制。在谷物加工前要经过的设备中适当放置磁铁，可以清除谷物中的金属。其他污染，如虫害尸体，也与饲料厂的卫生措施和是否注意限制害虫入厂有关。Pedersen（1985）对卫生措施及害虫管理提出了指导建议。

将来潜在的问题

在美国和许多其他国家，转基因（GM）作物被广泛种植并用来喂猪。但是，一些国家禁用使用转基因技术生产的饲料，而且从国际贸易的角度，它们被认为是污染物。最近欧洲委员会的饲料链和食品安全常务委员会批准了法规 EC 619/2011，允许进口动物饲料中转基因物质最高含量为 0.1%（Europa，2011）。确立不被认为不合格的实际水平已被饲料行业充分接受（Feed Info News Service, 2011a）。因为将来分析技术的改进可能会发现当前检测不到的浓度，所以将"污染"设置在零水平会导致使用共同交通区域运输大宗产品极为困难，因为少量的溢出就能污染在该区域运输的许多其他产品。

Lynas 等（1998）调查了 400 多种饲料和预混料中抗菌剂的污染情况（样品中有 40%认为未添加药物，60%宣称添加了药物）。在添加了药物的饲料中，35%含有未标注的抗菌剂，而未添加药物的饲料 44%含有可检测水平的抗菌剂。最常见的检出抗菌剂是金霉素（15.2%）、磺胺类药物（6.9%）、青霉素（3.4%）和离子载体类药物（3.4%）。如果在屠宰前饲喂给动物，饲料中检出的任何污染浓度的磺胺二甲嘧啶均可导致组织中的残留。Lynas 等（1998）发现的问题可能与饲料厂对配料顺序和生产不同批次产品时混合机清洗的管理以及员工认识的不足有关。在维生素预混料中发现了氯霉素（一种在部分而非所有国家禁用的广谱抗生素）残留的例子，表明在国际经济中可能出现污染（Feed Info News Service, 2011b）。

由于抗生素用于许多工业过程，这些过程中产生的副产物的残留是一个潜在问题。在美国，FDA 对干酒精糟（DDG）中抗生素残留进行了一项全国调查，追踪和检测诸如维吉利亚霉素、青霉素和红霉素等抗生素残留，这些抗生素均用于控制发酵罐中细菌生长（FDA, 2009a）。此项调查检测了 60 个 DDG 样品，其中 40 个来自国内，20 个来自国外。因污染而成为潜在问题的程度依赖于每个酒精厂的生产过程，所以可能残留的多少随工厂而异，正如农业和贸易政策研究所 2009 年的一份报告所述，在美国将近 45%的酒精生产设施使用非抗生素方法控制发酵罐中的细菌（Geiver, 2010）。

美国动物饲料安全体系

2003 年，FDA 宣称有意使其动物饲料安全计划更具有风险控制和综合性。一旦完成，更适应现代需要的美国动物饲料安全体系（AFSS）将以控制风险为基础的预防控制措施确保动物饲料安全。在国家的帮助下，FDA 已拟定了一份框架性文件，明确了当前主要的过程、指南、法规、强调饲料安全的政策性文件，以及使管理机构的饲料安全计划具有综合性和风险可控性所需要的文件（FDA, 2011）。

此努力的一部分是提出了一种对动物饲料中潜在有毒或有害的生物、化学和物理物质进行相对风险排序的方法（FDA，2009b）。重要的是，要认识到运用风险排序方法的目的不是对与任何一种饲料污染物相关的风险进行估测，而是作为工具对饲料污染物相对风险进行排序，帮助FDA用一种以控制风险为基础的方式确定其资源配置的优先性。FDA（2009b）较为详细地解释了这种方法。FDA（2009c）提供了关于猪的具体案例。

其他信息来源

从根本上说，饲料安全涉及对原料来源和与原料相关的质量检查、原料及成品恰当的储存、饲料厂卫生和记录以及适宜的法规等一系列细节的关注。美国饲料工业在为养猪业提供安全饲料方面做了出色的工作。想进一步加强质量控制的公司可从几个方面获得信息。AAFCO（2010）提供了关于饲料安全计划制定指南模板方面的信息。由Miller Publishing Company每年更新的饲料添加剂纲要（Lundeen，2010）有几个很好的章节讨论了当前有助于制定和维护饲料安全计划的良好操作规范（GMP）。

美国饲料工业协会可提供一份具有极好前瞻性的食品安全领导计划，即安全饲料/安全食品认证计划（AFIA，2009）。这个计划很好地叙述了参与组织的认证检查、保持记录的责任、原料采购的说明或建议、成品的鉴定和追溯以及与本章提到的许多污染物有关的问题。

为了确保动物健康状况，从而最终实现食品安全，我们除了关注饲料生产控制污染物外，也要确保对水污染物的直接关注，本书第5章讨论水质量、污染物和猪的健康的问题。

参考文献

AAFCO (Association of American Feed Control Officials). 2010. *AAFCO Official Publication*. Atlanta: Georgia Department of Agriculture.

AFIA (American Feed Industry Association). 2009. Safe Feed/Safe Food Certification Program. Available online at http://www.afia.org/Afia/Files/SFSF%20files/Revised%20SFSFpacket4%2009.pdf. Accessed on February 2, 2011.

BIOMIN. 2010. BIOMIN Newsletter. 8(83), Special Edition. Herzogenburg, Austria: Biomin Holding GmbH.

Calamari, D. 2002. The fate of PCBs and dioxins in the environment and foodstuffs. Pp. 15-21 in *Proceedings of Aquaculture Europe 2002: Seafarming—Today and Tomorrow*, EAS Special Publication No. 32, B. Basurco and M. Saroglia, eds. Oostende, Belgium: European Aquaculture Society.

CAST (Council for Agricultural Science and Technology). 1989. *Mycotoxins: Economic and Health Risks*. Report No. 116. Ames, IA: CAST.

CAST. 2003. *Mycotoxins: Risks in Plant, Animal, and Human Systems*. Report No. 139. Ames, IA: CAST.

Chang, C. W. J., R. M. Nakamura, and C. C. Brooks. 1977. Effect of varied dietary levels and forms of mercury on swine. *Journal of Animal Science* 45:279-285.

D'Mello, J. P. F. 2000. Contaminants and toxins in animal feeds. Pp. 107-128 in *FAO Animal Production and Health: Assessing Quality and Safety of Animal Feeds*. Rome: FAO. Available online at ftp://ftp.fao.org/docrep/fao/007/y5159e/y5159e04.pdf. Accessed on January 27, 2011.

Dórea, J. G. 2006. Fish meal in animal feed and human exposure to persistent bioaccumulative and toxic substances. *Journal of Food Protection* 69:2777-2785.

Europa. 2011. Rules on GMOs in the EU—Harmonisation of controls. Available online at: http://ec.europa.eu/food/food/biotechnology/harmonisation_of_controls_en.htm Accessed on November 18, 2011.

FAO (Food and Agriculture Organization of the United Nations). 1998. *Manual on Bovine Spongiform Encephalopathy*, J. W. Willesmith. Rome: FAO. Available online at http://web.archive.org/wcb/20080302180353/www.fao.org/DOCREP/003/W8656E/W8656E00.htm. Accessed on January 28, 2011.

FDA (U.S. Food and Drug Administration). 2000. Guidance for Industry: Action Levels for Poisonous or Deleterious Substances in Human Food and Animal Feed. Available online at http://www.fda.gov/Food/GuidanceComplianceRegulatoryInformation/GuidanceDocuments/ChemicalContaminantsandPesticides/ucm077969.htm#afla. Accessed on January 27, 2011.

FDA. 2001. Guidance for Industry: Fumonisin Levels in Human Foods and Animal Feeds; Final Guidance. Available online at http://www.fda.gov/Food/GuidanceComplianceRegulatoryInformation/GuidanceDocuments/ChemicalContaminantsandPesticides/ucm109231.htm. Accessed on January 27, 2011.

FDA. 2009a. FY 2010 Nationwide Survey of Distillers Grains for Antibiotic Residues. Available online at http://www.fda.gov/AnimalVeterinary/Products/AnimalFoodFeeds/Contaminants/ucm190907.htm. Accessed on January 27, 2011.

FDA. 2009b. Determining Health Consequence Scoring for Feed Contaminants. Available online at http://www.fda.gov/AnimalVeterinary/SafetyHealth/AnimalFeedSafetySystemAFSS/ucm053716.htm. Accessed on January 27, 2011.

猪营养需要

FDA. 2009c. Exposure Scoring for Feed Contaminants—A Swine Feed Example. Available online at http://www.fda.gov/AnimalVeterinary/SafetyHealth/AnimalFeedSafetySystemAFSS/ucm053722.htm. Accessed on January 27, 2011.

FDA. 2010a. FDA Announces Draft Compliance Policy Guide: Salmonella in Animal Feed. Available online at http://www.fda.gov/AnimalVeterinary/NewsEvents/CVMUpdates/ucm220829.htm. Accessed on January 27, 2011.

FDA. 2010b. Guidance for Industry and FDA: Advisory Levels for Deoxynivalenol (DON) in Finished Wheat Products for Human Consumption and Grains and Grain By-Products Used for Animal Feed. Available online at http://www.fda.gov/Food/GuidanceComplianceRegulatoryInformation/GuidanceDocuments/NaturalToxins/ucm120184.htm. Accessed on January 27, 2011.

FDA. 2011. Animal Feed Safety System (AFSS). Available online at http://www.fda.gov/AnimalVeterinary/SafetyHealth/AnimalFeedSafetySystemAFSS/default.htm. Accessed on January 27, 2011.

Feed Info News Service. 2009. Farmers in Northern Ireland demand compensation from feed firm after pork dioxin contamination. December 5.

Feed Info News Service. 2010a. FEFAC Cautiously Welcomes Draft GM Feed Rules. November 29.

Feed Info News Service. 2010b. Germany to destroy dioxin-tainted vegetable feed fat. December 30.

Feed Info News Service. 2010c. Vitamin A: Authorities respond to dioxin case. March 23.

Feed Info News Service. 2011a. FEFAC welcomes EU adoption of "technical LLP solution" on GMO traces. June 27.

Feed Info News Service. 2011b. Germany: Chloramphenicol residues found in Chinese vitamin A/D3. January 18.

Feed Info News Service. 2011c. German pig prices collapse after dioxin alert. January 21.

Feed Info News Service. 2011d. Russia bans German pork for dioxin fears. January 24.

Feedstuffs. 2011. Dioxin rocks Europe's feed sector. 83(2):1.

Furusawa, N., and Y. Morita. 2000. Polluting profiles of dieldrin and DDTs in laying hens of Osaka. *Japanese Journal of Veterinary Medicine B* 47:511-515.

Geiver, L. 2010. DDG survey on antibiotic residue not finished. *Ethanol Producer Magazine*, May 21. Available online at http://www.ethanolproducer.com/articles/6678/ddg-survey-on-antibiotic-residue-not-finished/. Accessed on November 30, 2011.

Johnston, J. N., and G. P. Savage. 1991. Mercury consumption and toxicity with reference to fish and fish meal. *Nutrition Abstracts and Reviews, Series A, Human and Experimental* 61:73-116.

Kanora, A., and D. Maes. 2009. The role of mycotoxins in pig reproduction: A review. *Veterinarni Medicina* 54:565-576.

Kerr, B. J., C. J. Ziemer, T. E. Weber, S. L. Trabue, B. L. Bearson, G. C. Shurson, and M. H. Whitney. 2008. Comparative sulfur analysis using thermal combustion or inductively coupled plasma methodology and mineral composition of common livestock feedstuffs. *Journal of Animal Science* 86:2377-2384.

Lin, X. L., Z. R. Xu, X. T. Zou, F. Wang, X. H. Yan, and J. F. Jiang. 2004. Effects of montmorillonite nanocomposite on mercury residues in growing/finishing pigs. *Asian-Australasian Journal of Animal Science* 17:1434-1437.

Lovell, R. A., D. G. McChesney, and W. D. Price. 1996. Organohalogen and organophosphorus pesticides in mixed feed rations: Findings from FDA's domestic surveillance during fiscal years 1989-1994. *Journal of AOAC International* 79:544-548.

Lundeen, T., ed. 2010. Feed Additive Compendium. Minneapolis, MN: Miller. Available online at http://www.feedcompendium.com.

Lynas, L., D. Currie, W. J. McCaughey, J. D. G. McEvoy, and D. G. Kennedy. 1998. Contamination of animal feedingstuffs with undeclared antimicrobial additives. *Food Additives and Contaminants* 15:162-170.

Lynn, T. V., D. D. Hancock, T. E. Besser, J. H. Harrison, D. H. Rice, N. T. Stewart, and L. L. Rowan. 1998. The occurrence and replication of *Escherichia coli* in cattle feeds. *Journal of Dairy Science* 81:1102-1108.

Merck Index. 2006. 14th Ed.. Rahway, NJ: Merck and Co, Inc.

Molla, B., A. Sterman, J. Mathews, V. Artuso-Ponte, M. Abley, W. Farmer, P. Rajala-Schultz, W. E. M. Morrow, and W. A. Gebreyes. 2010. *Salmonella enterica* in commercial swine feed and subsequent isolation of phenotypically and genotypically related strains from fecal samples. *Applied and Environmental Microbiology* 76:7188-7193.

NRC (National Research Council). 1979. *Interactions of Mycotoxins in Animal Production*. Washington, DC: National Academy Press.

Pedersen, J. R. 1985. Sanitation and Pest Management. Pp. 379-389 in *Feed Manufacturing Technology III*, R. R. McEllhiney, ed. Arlington, VA: American Feed Industry Association.

Placinta, C. M., J. P. F. D'Mello, and A. M. C. Macdonald. 1998. A review of worldwide contamination of cereal grains and animal feed with *Fusarium* mycotoxins. *Animal Feed Science and Technology* 78:21-37.

Pollock, C. 2010. Moldy grain, vomitoxin contamination putting a damper on record Ohio corn yields. Ohio State University Extension News Article. Available online at http://www.ag.ohio-state.edu/~news/story.php?id=5515. Accessed on January 25, 2011.

Schaafsma, A. W., V. Limay-Rios, D. E. Paul, and J. D. Miller. 2009. Mycotoxins in fuel ethanol co-products derived from maize: A mass balance for deoxynivalenol. *Journal of the Science of Food and Agriculture* 89:1574-1580.

Schwind, K.-H., and W. Jira. 2008. Dioxins and PCBs in German meat and meat products—Results of a monitoring study. In *Proceedings of 54th International Congress of Meat Science and Technology*, August 10-15, 2008, Cape Town, South Africa. Eastern Cape, South Africa: Merino South Africa.

Sharma, K., and M. Paradakar. 2010. The melamine adulteration scandal. *Food Security* 2:97-107.

Stothers, S. C., L. D. Campbell, and F. A. Armstrong. 1971. Mercury levels in tissues of pigs and chicks fed mercury-contaminated fish-meal. *Canadian Journal of Animal Science*. 51:817 (Abstr.).

Timmons, R. A. 2010. Global trace mineral contamination and a review of EU legislation. Pp. 23-28 in the *Proceedings of the 71st Minnesota Nutrition Conference*, September 21-22, 2010, Owatonna, MN. St. Paul: University of Minnesota.

USDA (U.S. Department of Agriculture). 2006. Bovine spongiform encephalopathy: An overview. Available online at http://www.aphis.usda.gov/publications/animal_health/content/printable_version/BSEbrochure12-2006.pdf. Accessed on January 28, 2011.

Van Barneveld, R. J. 1999. Physical and chemical contaminants in grains used in livestock feeds. *Australian Journal of Agricultural Research* 50:807-823.

Vreman, K., N. G. Van Der Veen, E. J. Van Der Molen, and W. G. De Ruig. 1986. Transfer of cadmium, lead, mercury and arsenic from feed into milk and various tissues of dairy cows: Chemical and pathological data. *Netherlands Journal of Agricultural Science* 34:129-144.

Wales, A. D., V. M. Allen, and R. H. Davies. 2010. Chemical treatment of animal feed and water for the control of *Salmonella*. *Foodborne Pathogens and Disease* 7:3-15.

第12章 饲料加工

引言

植物碳水化合物通常分为：①单糖及其异构体（如葡萄糖和果糖）；②储存形式的化合物（如淀粉）；③结构形式的碳水化合物（如纤维素和半纤维素）。这种分类方法在第4章中有更加详细的描述。单糖通常容易在肠道前段被消化，因此不太可能通过饲料加工技术提高它们的消化率。淀粉也是首先在消化道的前段被消化（Svihus et al., 2005; Bach Knudsen et al., 2006; Wiseman, 2006），但其消化取决于淀粉酶:支链淀粉比例、淀粉颗粒大小及有无淀粉酶抑制剂的存在，加工过程可能会改善淀粉的消化率。结构形式的碳水化合物结构复杂并且组成变异大（Theander et al., 1989; Selvendran and Robertson, 1990; Bach Knudsen, 2001）；哺乳动物的消化酶不能完全消化，可以通过各种加工工艺改善其消化率。因此，开发饲料加工技术将有利于改善猪饲料中的能量及其他营养物质的消化率，这有助于以最低成本提供动物生长所需要的能量、矿物质和氨基酸。饲料加工（如膨化和膨胀处理、糊化、粉碎或微粉碎、调质和制粒）是改善猪日粮营养价值的技术之一。

饲料加工对养分利用率的影响

对饲料原料或日粮的加工可以提高营养物质的消化率，从而改善猪的生产性能（Hancock and Behnke, 2001; Lundbald, 2009）。粉碎可以有效增加日粮的表面积，从而增加饲料与消化酶的接触面积。Ohh 等（1983）、Healy 等（1994）和 Wondra 等（1995a）的研究资料显示，综合粉碎能耗、动物生产性能、胃形态学、营养物质的消化率等因素，粉碎粒度以 700 μm 最优。Hancock 和 Behnke（2001）的研究得出结论，玉米或高粱的粉碎平均粒径每降低 100 μm，饲料效率（增重:饲料）可以提高 1.3%。这相当于玉米或高粱的粉碎粒度每降低 100 μm，其表观消化能提高 0.86%，大约等于 30 kcal 消化能（Owsley et al., 1981; Giesmann et al., 1990; Healy et al., 1994; Wondra et al., 1995a,b,c,d）。尽管早已知道减小粉碎粒度可以提高燕麦的营养物质消化率（Crampton and Bell, 1946），但有关机械加工对纤维的消化率和能量利用率的影响方面的资料还是非常有限的。Nuzback 等（1984）研究报道，给妊娠母猪饲喂含 50% 苜蓿草粉的日粮，粉碎粒度从 646 μm 降低到 434 μm，改善了日粮干物质、中性洗涤纤维、酸性洗涤纤维、半纤维素和纤维素的消化率，粉碎粒度每降低 100 μm，能量的消化率改善 2.2%（相当于粉碎粒度每降低 100 μm，消化能大约增加 97 kcal）。最近的研究显示，多种来源的 DDGS 粉碎粒度从 716 μm 降低到 344 μm (Mendoza et al., 2010)，或单一来源的 DDGS 粉碎粒度从 818 μm 降低到 308 μm（Liu et al., 2011），都能改善能量的消化率，相当于每降低 100 μm 的粉碎粒度，消化能增加 45 kcal。

微粉碎也是一种通过利用水分、温度和机械力减少颗粒大小的加工方法。微粉碎在改善猪生产性能和营养物质的消化率方面，不同研究者的结论并不一致。一些研究人员的研究结果显示微粉碎可以改善猪的生产性能或营养物质的消化率（Lawrence, 1973; Thacker, 1999; Qwusu-Asiedu et al., 2002; Nyachoti et al., 2006），然而其他研究者（Zarkadas and Wiseman, 2001; Valencia et al., 2008）却未能得出相似的结论。

带压力或不带压力的热处理可能会影响营养物质的消化率和动物的生产性能（Lundbald, 2009）。

热处理的功效之一是改变淀粉的结构和使淀粉酶抑制剂变性，在有水条件下的加热处理，会导致晶体结构的破坏和凝胶化的膨胀过程，这已被证实可以改善淀粉的消化率（Sun et al., 2006; Vicente et al., 2009）。与此相反，如果凝胶淀粉不迅速冷却，而让凝胶淀粉慢慢冷却再结晶，就会变成无特定晶体结构的回生淀粉。有时把回生淀粉误称为抗性淀粉，然而回生淀粉与抗性淀粉之间存在许多差异（Bhandari et al., 2009）。抗性淀粉和回生淀粉都不能被小肠中的消化酶所消化，但可以被后肠微生物所分解而产生挥发性脂肪酸，所以粪便中也不会有淀粉存在（Heijnen and Beynen, 1997; Hedemann and Bach Knudsen, 2007）。热处理也可以破坏干扰蛋白质消化和利用的蛋白酶抑制因子。胰蛋白酶抑制因子和糜蛋白酶抑制因子是最有名的两种蛋白酶抑制因子，它们存在于豆科植物的种子（如大豆、豌豆、菜豆等）中。这两种蛋白酶抑制因子都能被适度的热加工处理所破坏（Liener, 2000）。

膨化和膨胀（高温和高压处理）加工方法被用于水产和宠物饲料加工，其优点已由 Hancock 和 Behnke（2001）进行了综述。最近，在猪上的研究显示膨化玉米改善了回肠干物质的消化率（Muley et al., 2007），以及改善了含有豌豆和豌豆+亚麻籽日粮的回肠和总肠道消化率（Stein and Bohlke, 2007; Htoo et al., 2008）。与此相反，膨胀加工对于豌豆+大豆+木薯的基础日粮，或小麦+大麦+豆粕+菜粕的基础日粮并没有改善总养分的消化率（van der Poel et al., 1997）或猪的生产性能（Callan et al., 2007）。造成这种差别的原因并没有找到。

制粒的日粮对猪的生产性能的影响是不确定的，但总体看来，能改善大约6%的饲料效率（Hancock and Behnke, 2001）。这种改善的原因是多方面的，包括理化特性的改变（如淀粉糊化）、容重增加、适口性改善、粉尘减少、病原微生物下降、营养物质消化率的改善和/或减少饲料浪费。对含有大量玉米纤维（玉米蛋白饲料）的日粮进行制粒，被证实能改善氮平衡，显然是因为改善了色氨酸的利用率（Yen et al., 1971）。膨化机和膨胀器也被应用于饲料业，以改善制粒效率和颗粒质量（Lundbald et al., 2009）；一些研究表明，膨胀调质改善增重和饲料采食量的程度要大于膨化处理，主要是改善了氨基酸的回肠消化率，而不是改善干物质、粗蛋白或磷的消化率（Lundbald et al., 2009）。

其他信息

应用各种加工技术提高以植物原料为基础的猪和禽日粮营养物质的消化率，已有数十年的研究历史。然而，由于各种饲料原料的物理化学成分复杂多样，能否获得能改善营养物质消化率和动物生产性能的日粮，取决于对这些物理化学特性以及加工方法对这些成分影响的认知。饲料加工的主要目的是减少抗营养因子对营养物质利用率的影响继而提高改善动物生产性能，同时勿使加工对有效营养成分产生负面影响。过热和水分会破坏一些营养成分，特别是氨基酸，这些已在第2章中讨论过。纤维素含量与能量消化率之间呈负相关，提高纤维消化率的加工技术，从而也可以提高能量的消化率，这是符合逻辑的，加工技术也许对代谢和经济效益都是有益的。这些饲料加工的实用信息可以在 Hancock 和 Behnke（2001）、Richert 和 DeRouchey（2010）的综述文献中找到。

参 考 文 献

Bach Knudsen, K. E. 2001. The nutritional significance of "dietary fibre" analysis. *Animal Feed Science and Technology* 90:3-20.

Bach Knudsen, K. E., H. N. Laerke, S. Steenfeldt, M. S. Hedemann, and H. Jorgensen. 2006. In vivo methods to study the digestion of starch in pigs and poultry. *Animal Feed Science and Technology* 130:114-135.

Bhandari, S. K., C. M. Nyachoti, and D. O. Krause. 2009. Raw potato starch in weaned pig diets and its influence on postweaning scours and the molecular microbial ecology of the digestive tract. *Journal of Animal Science* 87:984-993.

Callan, J. J., B. P. Garry, and U. J. V. O'Doherty. 2007. The effect of expander processing and screen size on nutrient digestibility, growth performance, selected faecal microbial populations and faecal volatile fatty acid concentrations in grower-finisher pigs. *Animal Feed Science and Technology* 134:223-234.

Crampton, E. W., and J. M. Bell. 1946. The effect of fineness of grinding on the utilization of oats by market pigs. *Journal of Animal Science* 5:200-210.

Giesemann, M. A., A. J. Lewis, J. D. Hancock, and E. R. Peo, Jr. 1990. Effect of particle size of corn and grain sorghum on growth and digestibility of growing pigs. *Journal of Animal Science* 68(Suppl. 1):104 (Abstr.).

Hancock, J. D., and K. C. Behnke. 2001. Use of ingredient and diet processing technologies (grinding, mixing, pelleting, and extruding) to produce quality feeds for pigs. Pp. 469-497 in *Swine Nutrition*, 2nd Ed., A. J. Lewis and L. L. Southern, eds. Boca Raton, FL: CRC Press.

Healy, B. J., J. D. Hancock, G. A. Kennedy, P. J. Bramel-Cox, K. C. Behnke, and R. H. Hines. 1994. Optimum particle size of corn and hard and soft sorghum for nursery pigs. *Journal of Animal Science* 72:2227-2236.

Hedemann, S. K., and K. E. Bach Knudsen. 2007. Resistant starch for weanling pigs-effects on concentration of short chain fatty acids in digesta and intestinal morphology. *Livestock Science* 108:175-177.

Heijnen, M-L. A., and A. C. Beynen. 1997. Consumption of retrograded (RS_3) but not uncooked (RS_2) resistant starch shifts nitrogen excretion from urine to feces in cannulated piglets. *Journal of Nutrition* 127:1828-1832.

Htoo, J. K., X. Meng, J. F. Patience, M. E. R. Dugan, and R. T. Zijlstra. 2008. Effects of coextrusion of flaxseed and field pea on the digestibility of energy, ether extract, fatty acids, protein, and amino acids in grower-finisher pigs. *Journal of Animal Science* 86:2942-2951.

Lawrence, T. L. J. 1973. An evaluation of the micronization process for preparing cereals for the growing pig. 1. Effects on digestibility and nitrogen retention. *Animal Production* 16:99-107.

Liener, I. E. 2000. Non-nutritive factors and bioactive compounds in soy. Pp. 13-45 in *Soy in Animal Nutrition*, J. K. Drackley, ed. Savoy, IL: Federation of Animal Science Societies.

Liu, P., L. W. O. Souza, S. K. Baidoo, and G., C. Shurson. 2011. Impact of DDGS particle size on nutrient digestibility, DE and ME content, and flowability in diets for growing pigs. *Journal of Animal Science* 89(E-Suppl. 2):96 (Abstr.).

Lundbald, K. K. 2009. Effect of diet conditioning on physical and nutritional quality of feed for pigs and chickens. Ph.D. Dissertation, Norwegian University of Life Sciences, Aas, Norway.

Lundbald, K. K., J. D. Hancock, K. C. Behnke, E. Prestlokken, L. J. McKinney, and M. Sorsensen. 2009. The effect of adding water into the mixer on pelleting efficiency and pellet quality in diets for finishing pigs without and with use of an expander. *Animal Feed Science and Technology* 150:295-302.

Mendoza, O. F., M. Ellis, A. M. Gaines, M. Kocher, T. Sauber, and D. Jones. 2010. Effect of particle size of corn distillers dried grains with solubles (DDGS) on digestible and metabolizable energy content for growing pigs. *Journal of Animal Science* 88(E-Suppl. 3):104 (Abstr.).

Muley, N. S., E. van Heugten, A. J. Moeser, K. D. Rausch, and T. A. T. G. van Kempen. 2007. Nutritional value for swine of extruded corn and corn fractions obtained after dry milling. *Journal of Animal Science* 85:1695-1701.

Nuzback, L. J., D. S. Pollmann, and K. C. Behnke. 1984. Effect of particle size and physical form on sun-cured alfalfa on digestibility for gravid swine. *Journal of Animal Science* 58:378-385.

Nyachoti, C. M., S. D. Arntfield, W. Guenter, S. Cenkowski, and F. O. Opapeju. 2006. Effect of micronized pea and enzyme supplementation on nutrient utilization and manure output in growing pigs. *Journal of Animal Science* 84:2150-2156.

Ohh, S. J., G. L. Allee, K. C. Behnke, and C. W. Deyoe. 1983. Effects of particle size of corn and sorghum grain on performance and digestibility of nutrients for weaned pigs. *Journal of Animal Science* 57(Suppl. 1):260 (Abstr.).

Owsley, W. F., D. A. Knabe, and T. D. Tanksley, Jr. 1981. Effect of sorghum particle size on digestibility of nutrients at the terminal ileum and over the total digestive tract of growing-finishing pigs. *Journal of Animal Science* 52:557-565.

Owusu-Asiedu, A., S. K. Baidoo, and C. M. Nyachoti. 2002. Effect of heat processing on nutrient digestibility in pea and supplementing amylase and xylanase to raw, extruded or micronized pea-based diets on performance of early-weaned pigs. *Canadian Journal of Animal Science* 82:367-374.

Richert, B. T., and J. M. DeRouchey. 2010. Swine feed processing and manufacturing. Pp. 245-250 in *National Swine Nutrition Guide*, D. J. Meisinger, ed. Ames, IA: U.S. Pork Center of Excellence.

Selvendran, R. R., and J. A. Robertson. 1990. The chemistry of dietary fibre: A holistic view of the cell wall matrix. Pp. 27-43 in *Dietary Fibre: Chemical and Biological Aspects*, Royal Society of Chemistry Special Publication No 83. D. A. T. Southgate, K. Waldron, I. T. Johnson, and G. R. Fenwick, eds. Cambridge, UK: Royal Society of Chemistry.

Stein, H. H., and R. A. Bohlke. 2007. The effects of thermal treatment of field peas (*Pisum sativum* L.) on nutrient and energy digestibility by growing pigs. *Journal of Animal Science* 85:1424-1431.

Sun, T., H. N. Laerke, H. Jorgensen, and K. E. Bach Knudsen. 2006. The effect of extrusion cooking of different starch sources on the *in vitro* and *in vivo* digestibility in growing pigs. *Animal Feed Science and Technology* 131:66-85.

Svihus, B., A. K. Uhlen, and O. M. Harstad. 2005. Effect of starch granule structure, associated components and processing on nutritive value of cereal starch: A review. *Animal Feed Science and Technology* 122:303-320.

Thacker, P. A. 1999. Effect of micronization on the performance of growing/finishing pigs fed diets based on hulled and hulless barley. *Animal Feed Science and Technology* 79:29-41.

Theander, O., E. Westerlund, P. Aman, and H. Graham. 1989. Plant cell walls and monogastric diets. *Animal Feed Science and Technology* 23:205-225.

Valencia, D. G., M. P. Serrano, R. Lazaro, M. A. Latorre, and G. G. Mateos. 2008. Influence of micronization (fine grinding) of soya bean meal and fullfat soya bean on productive performance and digestive traits in young pigs. *Animal Feed Science and Technology* 147:340-356.

van der Poel, A. F. B., H. M. P. Fransen, and M. W. Bosch. 1997. Effect of expander conditioning and/or pelleting of a diet containing tapioca, pea and soybean meal on the total tract digestibility in growing pigs. *Animal Feed Science and Technology* 66:289-295.

Vicente, B., D. G. Valencia, M. P. Serrano, R. Lazaro, and G. G. Mateos. 2009. Effects of feeding rice and the degree of starch gelatinization of rice on nutrient digestibility and ileal morphology of young pigs. *British Journal of Nutrition* 101:1278-1281.

Wiseman, J. 2006. Variations in starch digestibility in non-ruminants. *Animal Feed Science and Technology* 130:66-77.

Wondra, K. J., J. D. Hancock, K. C. Behnke, R. H. Hines, and C. R. Stark. 1995a. Effects of particle size and pelleting on growth performance, nutrient digestibility, and stomach morphology in finishing pigs. *Journal of Animal Science* 73:757-763.

Wondra, K. J., J. D. Hancock, K. C. Behnke, and C. R. Stark. 1995b. Effects of mill type and particle size uniformity on growth performance, nutrient digestibility, and stomach morphology in finishing pigs. *Journal of Animal Science* 73:2564-2573.

Wondra, K. J., J. D. Hancock, G. A. Kennedy, K. C. Behnke, and K. R. Wondra. 1995c. Effects of reducing particle size of corn in lactation diets on energy and nitrogen metabolism in second-parity sows. *Journal of Animal Science* 73:427-432.

Wondra, K. J., J. D. Hancock, G. A. Kennedy, R. H. Hines, and K. C. Behnke. 1995d. Reducing particle size of corn in lactation diets from 1,200 to 400 micrometers improved sow and litter performance. *Journal of Animal Science* 73:421-426.

Yen, J. T., D. H. Baker, B. G. Harmon, and A. H. Jensen. 1971. Corn gluten feed in swine diets and effect of pelleting on tryptophan availability to pigs and rats. *Journal of Animal Science* 33:987-991.

Zarkadas, L. N., and J. Wiseman. 2001. Influence of processing variables during micronization of wheat on starch structure and subsequent performance and digestibility in weaned piglets fed wheat-based diets. *Animal Feed Science and Technology* 93:93-107.

第 13 章 营养物质和能量的消化率

引言

猪日粮是由各种饲料原料组成以满足动物对营养素和能量需要的配合饲料,并通过单个原料的化学分析以计算配合饲料所设定的营养组成。但不同组分中营养物质的利用率并不相同,有些营养物质无法被动物利用而在粪中排出。因此,估算配合饲料中每种组分中各种营养成分被猪吸收的比例,使日粮营养和能量水平与猪需要量一致是非常重要的。

日粮营养成分的有效率可定义为在猪肠道内消化吸收并用于满足代谢或组织合成的那部分营养物质的比例(Batterham, 1992)。但日粮营养成分的有效性是一种"抽象的概念,无法测定,只能估算"(Sibbald, 1987)。营养成分的有效值可通过斜率法估算,但它提供的是一种相对有效性而非绝对有效性(Ammerman et al., 1995; Gabert et al., 2001)。斜率法过程繁琐且费用高昂,同时在配合饲料中需要各种组分的营养成分具有可加性,而斜率法所得有效率并非总是能达到该目标(Gabert et al., 2001)。

为了克服相对有效性的测定困难与不准确性,当前猪日粮配制中采用各种日粮组分的营养物质和能量消化率值,使各组分中被吸收的营养素和能量可以定量,并在配合饲料中具有可加性(Gabert et al., 2001; Stein et al., 2007)。营养物质和能量的消化率有多种表达方式,我们要选择在配合饲料中具有可加性的表达方式。本章描述了氨基酸、脂肪、碳水化合物、磷以及能量的消化率的测定方法。有关各种测定技术的方法及其优缺点可以参见 Adeola(2001)、Gabert 等(2001)及 Stein 等(2007)的详细综述。

粗蛋白和氨基酸

猪饲料原料中蛋白质的营养价值由蛋白质中必需氨基酸的组成和消化率决定。这些必需氨基酸每天都需由日粮中提供,通过小肠被猪吸收,并用于体蛋白的合成,因此有必要测定日粮中每种原料的氨基酸的消化率。当前大多数猪日粮配方是基于每种原料的可消化氨基酸来配制的。但也要认识到,饲料原料如若经热处理,某些可消化氨基酸因美拉德反应而发生结构变化,从而有可能不能用于体蛋白合成(Batterham, 1992; Moughan, 2003a, 2005; Finot, 2005; Pahm et al., 2009)。

日粮中氨基酸通过猪的小肠进行吸收,在回肠末端,尚未吸收的氨基酸将进入大肠并在大肠中微生物作用下发酵。发酵可导致氨基酸的分解代谢与合成,但大肠中氨基酸的吸收基本上可以忽略。未消化的氨基酸和微生物所合成的氨基酸在粪中排出。由于进入大肠的氨基酸会被微生物发酵,因而粪中的氨基酸含量不能精确代表小肠未吸收的氨基酸(Sauer and Ozimek, 1986)。因此,可通过回肠末端收取食糜,以估测氨基酸在小肠中的消失量。收取回肠末端食糜需要对猪进行外科手术,方法有多种(Sauer and de Lange, 1992; Moughan, 2003b)。在北美,应用最广泛的是 T 型瘘管法,即在距回-盲瓣 10~20 cm 处的回肠安装 T 型瘘管,用于收集回肠食糜。由于无法收集全部回肠食糜,需要在猪日粮中添加不消化的指示剂(通常为氧化铬或二氧化钛),即可计算氨基酸的消化率。

回肠氨基酸消化率可通过回肠表观氨基酸消化率(apparent ileal digestibility, AID)或标准回肠氨基酸消化率(standardized ileal digestibility, SID)来表示。AID 值可通过式 13-1 计算(Stein et al., 2007)。

$$\text{AID} (\%) = [1 - (\text{AA}_{digesta} / \text{AA}_{diet}) \times (\text{Marker}_{diet} / \text{Marker}_{digesta})] \times 100 \quad (式 13\text{-}1)$$

式中，$AA_{digesta}$ 和 AA_{diet} 分别为回肠食糜和日粮干物质（DM，g/kg）中氨基酸含量；$Marker_{diet}$ 和 $Marker_{digesta}$ 分别为日粮和回肠食糜干物质中指示剂含量（g/kg）。

与回肠末端真可消化率不同，日粮中氨基酸的 AID 值代表"表观的"回肠消化率，因为所收集的回肠食糜不仅含有未消化的日粮氨基酸，也有内源性的氨基酸。内源氨基酸源自分泌进入肠道的酶、脱落细胞、黏膜蛋白、血清白蛋白及其他组分（Nyachoti et al., 1997; Moughan et al., 1992; Jansman et al., 2002）。大多数的内源蛋白质在猪小肠中经消化后被重吸收，但一些进入大肠的内源性蛋白质没有被消化而排出体外，这些内源性氮所含的氨基酸称为内源性损失的氨基酸。内源性损失的比例与消化道内干物质相关。与氨基酸含量高的原料相比，氨基酸含量低的原料相应的内源性氨基酸占回肠末端氨基酸总量的比例更高（Fan et al., 1994; Mosenthin et al., 2000）。由此可知，AID 值明显会受到测试日粮氨基酸含量的影响（Donkoh and Moughan, 1994; Fan et al., 1994）。因而，饲料原料的氨基酸 AID 值在混合日粮中不一定总是具有可加性（Stein et al., 2005）。然而，如通过内源氨基酸来校正 AID 值，则可以将内源氨基酸的影响最小化。因此，有必要测定与干物质摄入量（DMI）相关的内源氨基酸损失（Mosenthin et al., 2000; Jansman et al., 2002）。这部分内源氨基酸损失称为基础性内源氨基酸损失，一般通过饲喂一无氮日粮进行测定，并按式 13-2 计算（Stein et al., 2007）：

$$IAA_{end} = AA_{digesta} \times (Marker_{diet} / Marker_{digesta}) \quad \text{（式 13-2）}$$

式中，IAA_{end} 为基础内源性氨基酸损失（g/kg 干物质进食量，g/kg DMI）；$AA_{digesta}$ 为回肠食糜干物质中氨基酸含量（g/kg DM）；$Marker_{diet}$ 和 $Marker_{digesta}$ 分别为日粮和回肠食糜中指示剂的浓度（g/kg DM）。

由于无氮日粮法会使试验猪处于非正常生理状态下，通过该方法测定基础性内源氨基酸损失存在一定争议（Low, 1980; Hodgkinson et al., 2000）。一些其他方法也用来测定基础性内源氨基酸损失，包括回归法、酶解酪蛋白法、合成晶体氨基酸日粮法等（Nyachoti et al., 1997; Moughan et al., 1992; Mariscal-Landin and Reis de Souza, 2006）。但通过与这些方法比较，无氮日粮法在测定基础性内源氨基酸损失上与之并无明显的差异（Jansman et al., 2002），因此，无氮日粮法仍为最常用的基础性内源氨基酸损失测定方法。通过基础性内源氨基酸损失来校正 AID 值可得到 SID 值，计算式 13-3 如下（Stein et al., 2007）：

$$SID (\%) = AID + [(basel\ IAA_{end} / AA_{diet}) \times 100] \quad \text{（式 13-3）}$$

式中，SID 为标准回肠氨基酸消化率（%）；$basel\ IAA_{end}$ 是基础性内源氨基酸损失（g/kg DMI）；AA_{diet} 为日粮干物质中待测氨基酸含量（g/kg DM）。

SID 值消除了基础性内源氨基酸损失的影响，因而 SID 值在配合日粮中应具有可加性，可用于实际日粮配合（Stein et al., 2005）。

测定每种饲料原料 SID 值的准确性基于一个假设，即从小肠吸收的氨基酸可用于蛋白质合成，而且在小肠内不存在微生物对氨基酸的代谢或净合成（Moughan, 2003a）。经过热处理的蛋白质会发生美拉德反应，这类蛋白质的氨基酸被吸收后不一定百分之百用于蛋白质合成。因此，在测定这类原料氨基酸的 SID 值时会有偏差（Moughan, 2005）。猪肠道微生物主要存在于大肠，但小肠内亦有一些微生物活动（Smiricky et al., 2002），且在小肠内亦可能存在微生物对氨基酸的分解代谢与合成。但当前尚无小肠微生物发酵所引起的氨基酸净合成或净消失的确切数据（Moughan, 2003a），因此，当前假定小肠微生物对日粮氨基酸的吸收与利用无影响。

脂类

大多数猪日粮配方都不以可消化脂肪来配制，而且在配方程序中也不包括脂肪消化率。然而，日粮脂肪在供能方面起着重要作用，因此有时会测定饲料原料的脂肪消化率。

脂肪在小肠的水性环境中几乎不溶解，所以它在小肠的消化和吸收需一系列步骤，包括乳化、酶解、形成乳糜微粒、通过不动水层并进入肠细胞（Bauer et al., 2005）。许多因素会影响脂肪的消化，猪全价日粮的脂肪全消化道表观消化率（Apparent total tract digestibility, ATTD）变异较大，在25%~77%范围内浮动（Noblet et al., 1994）。后肠微生物可以合成脂肪，导致粪中含有内源性脂肪。由于纤维可促进肠道微生物的增殖，高纤维日粮将会促进内源性脂肪的合成，增加内源性脂肪的损失（Kil et al., 2010）。因此，相比全消化道脂肪消化率，回肠脂肪消化率更准确。回肠表观脂肪消化率与回肠氨基酸消化率的测定方法相似，也需要在日粮中添加不消化的指示剂。

与日粮氨基酸含量影响氨基酸的 AID 值类似，日粮脂肪的含量也会影响脂肪的 AID 值（Kil et al., 2010），原因在于内源性脂肪对 AID 值的影响。为了最小化内源性脂肪的影响，有必要测定内源性脂肪损失。与氨基酸测定不同，当前并无合适的测定基础内源性脂肪损失的方法，因而脂肪的 SID 值无法计算。尽管通过回归法也无法测定基础内源性脂肪损失，但可以测定总的内源性脂肪损失，通过总内源性脂肪损失可以校正 AID 值，得到脂肪的回肠末端真消化率（true ileal digestibility, TID）（Jørgensen et al., 1993; Kil et al., 2010）如下：

$$TID(\%) = AID + [(total\ IL_{end} / L_{diet}) \times 100] \qquad （式13-4）$$

式中，total IL_{end} 是回肠总的内源性脂肪损失（g/kg DMI）；L_{diet} 为日粮干物质中脂肪含量（g/kg）。在回归法测定过程中，脂肪的 TID 值亦可直接通过线性回归方程的斜率得到（Jørgensen et al., 1993; Kil et al., 2010）。

饲料原料中的脂肪可通过醚提法或酸解醚提法进行测定。酸解醚提法的酸解过程可以将矿物质所结合的脂肪释放，通过该法测定的脂肪含量会高于醚提法（Sanderson, 1986）。在动物肠道内，脂肪可与矿物质形成复合物，所以酸解醚提法被认为在测定脂肪消化率方面准确性更高。

在测定脂肪消化率方面，脂肪的 TID 值由于更准确地反映了日粮脂肪的吸收率而优于 AID 值。TID 值不受日粮中脂肪含量的影响，同时与全消化道脂肪消化率的区别是，TID 值不受猪后肠微生物所合成脂肪的影响。因此，当前认为 TID 值在配合日粮中具有可加性。

碳水化合物

大多数猪日粮配方一般还按可消化碳水化合物来配制。与脂肪类似，碳水化合物起着重要的能量供应作用。为准确评估猪从日粮中摄取的能量，有必要准确测定日粮碳水化合物的消化率（Noblet et al., 1994）。碳水化合物包括糖与二糖、淀粉与糖原和日粮纤维，而且这三类成分的消化及发酵程度也不相同，需要分开考虑。

二糖

猪日粮中通常含有单糖和蔗糖，乳仔猪日粮中也可能含有乳糖。蔗糖和乳糖在小肠酶的作用下降解成单糖并最终通过主动与被动转运的方式被吸收利用（Englyst and Hudson, 2000）。该过程非常高效，可以认为二糖到达小肠末端前完全被消化吸收（van Beers et al., 1995），所以二糖的消化率一般不必测定。但如果要测定二糖的消化率，可以直接测定二糖的 AID 值，过程与氨基酸 AID 测定相似。并没有任何证据表明存在内源性二糖的分泌，所以二糖的 SID 或 TID 可以不予考虑。

淀粉和糖原

猪日粮通常含有大量淀粉，且当日粮中存在肉类副产物时，会存在糖原，但其含量基本上可以忽略。在胰淀粉酶、肠淀粉酶、肠麦芽糖酶及肠异麦芽糖酶（也称为α-糊精酶；Grofft and Gropper, 2000）作用下，日粮中绝大多数淀粉在猪小肠中被消化。大多数饲料原料的淀粉消化率均很高，在到达小肠末端前消化率为90%~95%（Bach Knudsen, 2001）。淀粉消化并生成葡萄糖被吸收供能。一些在小肠内没有被消化的淀粉（如抗性淀粉）进入大肠内发酵。通常粪中淀粉的含量非常低。猪全消化道内的淀粉消化率一般超过99%（Stein and Bohlke, 2007）。但也有例外，即日粮中的原料未粉碎至合适的粒径，使淀粉无法被酶解或微生物发酵。

未消化的淀粉会进入大肠内发酵，所以淀粉的消化率需在小肠末端测定。淀粉回肠表观消化率（AID）的测定与前文所提到的氨基酸、脂肪及二糖类似。与二糖相似，尚未发现淀粉以内源性方式分泌进入肠道，因而未用内源性损失来校正AID，也不计算淀粉的SID及TID值。

在小肠内无法被消化的淀粉称为抗性淀粉。可通过模拟小肠的消化环境，以酶解法估算饲料原料中抗性淀粉的含量。但如果已测定淀粉的体内AID值，抗性淀粉的含量等于100减去淀粉的AID。进入大肠内的抗性淀粉的发酵产物为短链脂肪酸。由于短链脂肪酸的能量利用效率低于葡萄糖，所以抗性淀粉的能值低于淀粉（Black, 1995）。

日粮纤维

日粮纤维是饲料中寡糖、抗性淀粉、非淀粉多糖及木质素的总称。根据日粮纤维的定义，它包括所有未在小肠中被酶解的碳水化合物。有些日粮纤维组分可在小肠中发酵，而其他日粮纤维组分则在大肠中发酵（Urriola et al., 2010）。无论在哪里发酵，唯一可吸收作为能量来源的发酵产物是短链脂肪酸。因而，从能量供应来看，日粮纤维在小肠还是大肠中发酵没有区别。为了准确评估日粮纤维的能量价值，需要评估全消化道日粮纤维的消失量。尽管现在已意识到日粮纤维的内源分泌量亦应该考虑，但当前日粮纤维的基础性或总的内源分泌量并没有被测定。因而，目前日粮纤维的能值贡献仍基于纤维的全消化道表观消失量。

磷

磷的吸收部位为小肠，内源磷也会分泌并进入小肠（Fan et al., 2001）。研究发现在大肠内既没有磷的净吸收，也没有内源磷的净分泌，大肠在维持体内磷平衡方面的作用基本上可以忽略（Bohlke et al., 2005）。因此，磷的回肠表观消化率（AID）与其全消化道表观消化率（ATTD）没有差别（Fan et al., 2001; Bohlke et al., 2005; Dilger and Adeola, 2006）。相比AID测定，ATTD测定更容易且费用也更低，因而磷的消化率一般采用ATTD，用式13-5计算（Almeida and Stein, 2010）：

$$P 的 ATTD (\%) = [(P_{intake} - P_{output}) / P_{intake}] \times 100 \quad （式 13\text{-}5）$$

式中，P_{intake}（进食量）和P_{output}（排出量）的单位为g/d，或g/全收集期。

尽管相对量较少，内源磷损失（EPL）仍会显著影响磷的ATTD，因此，磷的ATTD受到日粮中磷含量的影响，这种情况与氨基酸和脂肪的AID受日粮中氨基酸与脂肪含量的影响相似（Fan et al., 2001; Shen et al., 2002; Ajakaiye et al., 2003）。所以在配合日粮中磷的ATTD值并非总具有可加性，从而在实际配方中出现困难，因为假设饲料原料的消化率具有可加性。因此，对EPL的校正是必要的。汇总当前的研究报道，总 EPL 的测定结果差异较大（Shen et al., 2002; Dilger and Adeda, 2006; Pettey et al., 2006），无法得出有信服力的总 EPL 值。但与总 EPL 值不同，在不同的研究结果中，基础性 EPL 值则差异不大，均值为 190 mg 磷/kg 干物质摄入量（Traylor et al., 2001; Stein et al., 2006; Widmer et al., 2007; Almeida and Stein, 2010）。基础性 EPL 值可通过饲喂无磷日粮并根据粪中磷排出量计算，公式如下（Almeida and Stein, 2010）：

$$\text{Basal EPL (mg/kg DMI)} = [(P_{output}/DMI) \times 1000 \times 1000] \quad \text{（式 13-6）}$$

式中，Basal EPL 为基础性内源磷损失（mg/kg DMI）；P_{output} 为每日的粪磷排出量（g），DMI 为日干物质采食量（g）。

通过饲喂含磷日粮可得到粪磷排出量，并扣除基础性内源磷损失，可以得到磷的全消化道标准消化率（STTD），式 13-7 如下（Almeida and Stein, 2010）：

$$\text{STTD (\%)} = \{[P_{intake} - (P_{output} - \text{basal EPL})] / P_{intake}\} \times 100 \quad \text{（式 13-7）}$$

STTD 为磷的全消化道标准消化率(%)；P_{intake} 和 P_{output} 为每日磷摄入及排出量，单位为 g；basal EPL 为基础性内源磷损失（mg/kg DMI）与干物质日摄入量的乘积。

如果磷的 ATTD 值已测定，可以根据 ATTD 值及基础性内源磷损失计算磷的 STTD 值，式 13-8 如下：

$$\text{STTD (\%)} = \text{ATTD} + [(\text{baselEPL} / P_{diet}) \times 100] \quad \text{（式 13-8）}$$

式中，Basal EPL 为基础性内源磷损失（g/kg DMI），P_{diet} 为每千克日粮干物质中磷的含量（g）。

根据研究报道，基础性内源磷损失约为 190 mg/kg DMI。不同试验及动物体重差异对该值的影响较小（Baker, 2011）。因此，用猪测定某一原料中磷的 ATTD 值后，不需再用此同一批猪测定基础性内源磷损失，可用 190 mg/kg 干物质采食量作为基础性内源磷损失的常数值来校正 ATTD。如原料中 ATTD 值已知，可用此法计算磷的 STTD 值。磷的全消化道标准消化率（STTD）具有可加性，在配制实际日粮时使用磷的 STTD 值比 ATTD 值更为精确。

能量

猪从日粮中获取的能量是指蛋白质、脂肪和碳水化合物消化代谢所产生的能量总和。日粮总能（GE）通过弹式测热仪测定。日粮消化能（DE）为猪摄入总能与粪排出的总能之差。另外，日粮或饲料原料的消化能可通过计算日粮成分的 ATTD 得到。式 13-5 是用来计算磷的 ATTD，也可用来计算总能的 ATTD。根据日粮总能的 ATTD，可以得到 DE。因而，可通过全收粪法测定日粮或原料的消化能。具体过程为：将试验猪置于代谢笼，经过 5~10 天的预饲期，记录随后 5 天的采食量和粪排出量。为确保收集的粪源于 5 天收集期间所喂的日粮，收集期开始所喂日粮需添加起始指示剂，当该指示剂在粪中出现时即可开始收粪（Widmer et al., 2007）。同样的，在收集期结束时所喂的日粮中加入终止指示剂，当粪中出现该指示剂时便停止收粪（Adeola, 2001; Widmer et al., 2007）。

如果在收粪的同时也收集尿，测定消化能时亦测定尿能，从消化能中扣除尿能，即为动物代谢所需的能量，称为代谢能（ME）。大多数饲料原料的代谢能约为消化能的 92%~98%。尿中主要的能量物质是氮，尿氮随日粮蛋白质含量的变化而变化。需要特别提出的是，当待测日粮组分中蛋白质的氨基酸组成与猪所需的氨基酸组成存在显著差异时，该组分的代谢能将会被低估。为解决这个问题，代谢能有时被校正为 50%氮沉积值，该值基于一个假设，即在平衡日粮中，约 50%的可消化氮沉积在体内（Noblet et al., 2004）。该氮校正代谢能有时可通过计算获得（Cozannet et al., 2010）。

一些饲料原料的测定能值会受到猪日龄的影响，并且用特定体重猪所测的能值并不总是可以代表不同体重猪适用的能值（Le Goff and Noblet, 2001; Jørgensen et al., 2007; Cozannet et al., 2010），特别是饲料原料中非淀粉多糖含量较高时尤其如此（Le Goff and Noblet, 2001）。因此，对同一种饲料原料，有人建议将不同的能值赋予不同类型的猪（Noblet and van Milgen, 2004）。然而，尚缺乏有关不同类型猪利用各饲料原料的准确能值数据，因而在生长猪饲料评估体系中无法使用因年龄而异的能值。不过有人建议使用两种能值，一种适用于母猪，另一种适用于其他猪（Sauvant et al., 2004）。

用于评估饲料原料能量消化率所选用猪的品种也会影响测定结果，许多地方品种猪比典型的商品猪对纤维和能量具有更高的消化率（Kemp et al., 1991; Ndindana et al., 2002; Len et al., 2006; von Heimendahl et al., 2010）。但是有关不同品种的商品猪（如大白猪、长白猪、杜洛克和汉普夏猪）在能量消化率方面的差异尚无公开发表的数据，因而当前假定用一个品种的猪所测定的能值可用于其他品种的猪。

一些饲料原料的能量消化率也受用来测定消化率的颗粒大小的影响（Healy et al., 1994），这对母猪和生长猪都是如此（Wondra et al., 1995a,b）。一般来说，颗粒越小，能量消化率越高，从而使其经济价值也越高（Borg et al., 2008）。但过小的颗粒可能导致猪胃溃疡，并使肠道隐窝黏蛋白颗粒增大（Brunsgaard, 1998; Hedeman et al., 2005）。在实际养猪生产中最常用的粒径范围为 400~600 μm，建议在测定能量消化率时也使用此粒径范围。

参 考 文 献

Adeola, O. 2001. Digestion and balance techniques in pigs. Pp. 903-916 in *Swine Nutrition*, 2nd Ed., A. J. Lewis and L. L. Southern, eds. Boca Raton, FL: CRC Press.

Ajakaiye, A., M. Z. Fan, T. Archbold, R. R. Hacker, C. W. Forsberg, and J. P. Phillips. 2003. Determination of true digestive utilization of phosphorus and the endogenous phosphorus outputs associated with soybean meal for growing pigs. *Journal of Animal Science* 81:2766-2775.

Almeida, F. N., and H. H. Stein. 2010. Performance and phosphorus balance of pigs fed diets formulated on the basis of values for standardized total tract digestibility of phosphorus. *Journal of Animal Science* 88:2968-2977.

Ammerman, C. B., D. H. Baker, and A. J. Lewis. 1995. Introduction. Pp. 1-3 in *Bioavailability of Nutrients for Animals*, C. B. Ammerman, D. H. Baker, and A. J. Lewis, eds. San Diego, CA: Academic Press.

Bach Knudsen, K. E. 2001. The nutritional significance of "dietary fiber" analysis. *Animal Feed Science and Technology* 90:3-20.

Baker, S. R. 2011. Aspects of phosphorus nutrition in swine. MS Thesis, University of Illinois, Urbana.

Batterham, E. S. 1992. Availability and utilization of amino acids for growing pigs. *Nutrition Research Reviews* 5:1-18.

Bauer, E., S. Jakob, and R. Mosenthin. 2005. Principles of physiology of lipid digestion. *Asian-Australasian Journal of Animal Sciences* 18:282-295.

Black, J. L. 1995. Modelling energy metabolism in the pig—Critical evaluation of a simple reference model. Pp. 87-102 in *Modelling Growth in the Pig*, P. J. Moughan, M. W. A. Verstegen, and M. I. Visser-Reynevel, eds. Wageningen, The Netherlands: Wageningen Press.

Bohlke, R. A., R. C. Thaler, and H. H. Stein. 2005. Calcium, phosphorus, and amino acid digestibility in low-phytate corn, normal corn, and soybean meal by growing pigs. *Journal of Animal Science* 83:2396-2403.

Borg, B. S. 2008. Nutritional issues facing the swine industry—Squeezing the lemon. Pp. 3-7 in *Proceedings of the Swine Nutrition Conference*, September 4, 2008, Indianapolis, IN.

Brunsgaard, G. 1998. Effects of cereal type and feed particle size on morphological characteristics, epithelial cell proliferation, and lectin binding pattern in the large intestine of pigs. *Journal of Animal Science* 76:2787-2798.

Cervantes-Pahm, S. K. 2011. In vivo and in vitro disappearance of energy and nutrients in novel carbohydrates and cereal grains by pigs. PhD Dissertation, University of Illinois, Urbana.

Cozannet, P., Y. Primot, C. Gady, J. P. Métaver, M. Lessire, F. Skiba, and J. Noblet. 2010. Energy value of wheat distillers grains with solubles for growing pigs and adult sows. *Journal of Animal Science* 88:2382-2392.

Dilger, R. N., and O. Adeola. 2006. Estimation of true phosphorus digestibility and endogenous phosphorus loss in growing pigs fed conventional and low-phytate soybean meal. *Journal of Animal Science* 84:627-634.

Donkoh, A., and P. J. Moughan. 1994. The effect of dietary crude protein content on apparent and true ileal nitrogen and amino acid digestibilities. *British Journal of Nutrition* 72:59-68.

Englyst, K. N., and G. J. Hudson. 2000. Carbohydrates. Pp. 61-76 in *Human Nutrition and Dietetics*, 10th Ed., J. S. Garrow, W. P. T. James, and A. Ralph, eds. Edinburgh, UK: Churchill Livingston.

Fan, M. Z., W. C. Sauer, R. T. Hardin, and K. A. Lien. 1994. Determination of apparent ileal amino acid digestibility in pigs: Effect of dietary amino acid level. *Journal of Animal Science* 72:2851-2859.

Fan, M. Z., T. Archbold, W. C. Sauer, D. Lackeyram, T. Rideout, Y. Gao, C. F. M. de Lange, and R. R. Hacker. 2001. Novel methodology allows simultaneous measurement of true phosphorus digestibility and the gastrointestinal endogenous phosphorus outputs in studies with pigs. *Journal of Nutrition* 131:2388-2396.

Finot, P. A. 2005. The absorption and metabolism of modified amino acids in processed food. *Journal of AOAC International* 88:894-903.

Gabert, V. M., H. Jørgensen, and C. M. Nyachoti. 2001. Bioavailability of amino acids in feedstuffs for swine. Pp. 151-186 in *Swine Nutrition*, 2nd Ed., A. J. Lewis and L. L. Southern, eds. Boca Raton, FL: CRC Press.

Groff, J. L., and S. S. Gropper, eds. 2000. *Advanced Nutrition and Human Metabolism*, 3rd Ed. Belmont, CA: Wadsworth.

Healy, B. J., J. D. Hancock, G. A. Kennedy, P. J. Bramel-Cox, K. C. Behnke, and R. H. Hines. 1994. Optimum particle size of corn and hard and soft sorghum for nursery pigs. *Journal of Animal Science* 72:2227-2236.

Hedemann, M. S., L. L. Mikkelsen, P. J. Naughton, and B. B. Jhensen. 2005. Effects of feed particle size and feed processing on morphological characteristics in the small and large intestine of pigs and on adhesion of *Salmonella enterica* serovar Typhimurium DT12 in the ileum in vitro. *Journal of Animal Science* 83:1554-1562.

Hodgkinson, S. M., P. Moughan, G. W. Reynolds, and K. A. C. James. 2000. The effect of dietary peptide concentration on endogenous ileal amino acid loss in the growing pig. *British Journal of Nutrition* 83:421-430.

Jansman, A. J. M., W. Smink, P. van Leeuwen, and M. Rademacher. 2002. Evaluation through literature data of the amount and amino acid composition of basal endogenous crude protein at the terminal ileum of pigs. *Animal Feed Science and Technology* 98:49-60.

Jørgensen, H., K. Jakobsen, and B. O. Eggum. 1993. Determination of endogenous fat and fatty acids at the terminal ileum and on faeces in growing pigs. *Acta Agriculturae Scandinavica, Section A—Animal Science* 43:101-106.

Jørgensen, H., A. Serena, M. S. Hedemann, and K. E. Bach Knudsen. 2007. The fermentative capacity of growing pigs and adult sows fed diets with contrasting type and level of dietary fiber. *Livestock Science* 109:111-114.

Kemp, B., L. A. Hartog, J. J. den Klok, and T. Zandstra. 1991. The digestibility of nutrients, energy and nitrogen in the Meishan and Dutch Landrace pig. *Journal of Animal Physiology and Animal Nutrition* 65:263-266.

Kil, D. Y., T. E. Sauber, D. B. Jones, and H. H. Stein. 2010. Effect of the form of dietary fat and the concentration of dietary NDF on ileal and total tract endogenous losses and apparent and true digestibility of fat by growing pigs. *Journal of Animal Science* 88:2959-2967.

Le Goff, G., and J. Noblet. 2001. Comparative digestibility of dietary energy and nutrients in growing pigs and adult sows. *Journal of Animal Science* 79:2418-2427.

Len, N. T., J. E. Lindberg, and B. Ogle. 2006. Digestibility and nitrogen retention of diets containing different levels of fibre in local (Mong Cai), F1 (Mong Cai × Yorkshire) and exotic (Landrace × Yorkshire) growing pigs in Vietnam. *Journal of Animal Physiology and Animal Nutrition* 91:297-303.

Low, A. G. 1980. Nutrient absorption in pigs. *Journal of Food Science and Agriculture* 31:1087-1130.

Mariscal-Landin, G., and T. C. Reis de Souza. 2006. Endogenous ileal losses of nitrogen and amino acids in pigs and piglets fed graded levels of casein. *Archives of Animal Nutrition* 60:454-466.

Mosenthin, R., W. C. Sauer, R. Blank, J. Huisman, and M. Z. Fan. 2000. The concept of digestible amino acids in diet formulation for pigs. *Livestock Production Science* 64:265-280.

Moughan, P. J. 2003a. Amino acid availability: Aspects of chemical analysis and bioassay methodology. *Nutrition Research Reviews* 16:127-141.

Moughan, P. 2003b. Amino acid digestibility and availability in foods and feedstuffs. Pp. 199-221 in the *Proceedings of the 9th International Symposium on Digestive Physiology of Pigs*, May 14-17, 2003, Banff, Alberta, Canada.

Moughan, P. J. 2005. Absorption of chemically unmodified lysine from proteins in foods that have sustained damage during processing or storage. *Journal of AOAC International* 88:949-954.

Moughan, P. J., G. Schuttert, and M. Leenaars. 1992. Endogenous amino acid flow in the stomach and small intestine of the young growing pig. *Journal of the Science of Food and Agriculture* 60:437–442.

Ndindana, W., K. Dzama, P. N. B. Ndiweni, S. M. Maswaure, and M. Chimonyo. 2002. Digestibility of high fibre diets and performance of growing Zimbabwean indigenous Mukota pigs and exotic Large White pigs fed maize based diets with graded levels of maize cobs. *Animal Feed Science and Technology* 97:199-208.

Noblet, J., and J. van Milgen. 2004. Energy values of pig feeds: Effect of pig body weight and energy evaluation system. *Journal of Animal Science* 82 (E-Suppl.):E229-E238.

Noblet, J., H. Fortune, X. S. Shi, and S. Dubois. 1994. Prediction of net energy value of feeds for growing pigs. *Journal of Animal Science* 72:344-354.

Noblet, J., B. Seve, and C. Jondreville. 2004. Nutritional values for pigs. Pp. 25-35 in *Tables for Composition and Nutritional Values of Feed Materials*, 2nd Ed., D. Sauvant, J. M. Perez, and G. Tran, eds. Wageningen, The Netherlands: Wageningen Academic Publishers.

Nyachoti, C. M., C. F. M. de Lange, B. W. McBride, and H. Schulze. 1997. Significance of endogenous gut nitrogen losses in the nutrition of growing pigs: A review. *Canadian Journal of Animal Science* 77:149-163.

Pahm, A. A., C. Pedersen, and H. H. Stein. 2009. Standardized ileal digestibility of reactive lysine in distillers dried grains with solubles fed to growing pigs. *Journal of Agricultural and Food Chemistry* 57:535-539.

Pettey, L. A., G. L. Cromwell, and M. D. Lindemann. 2006. Estimation of endogenous phosphorus loss in growing and finishing pigs fed semi-purified diets. *Journal of Animal Science* 84:618-626.

Sanderson, P. 1986. A new method of analysis of feedingstuffs for the determination of crude oils and fats. Pp. 77-81 in *Recent Advances in Animal Nutrition*, W. Haresign and D. J. A. Cole, eds. London: Butterworths.

Sauer, W. C., and K. de Lange. 1992. Novel methods for determining protein and amino acid digestibilities in feed stuffs. Pp. 87-120 in *Modern Methods in Protein Nutrition and Metabolism*, S. Nissen, ed. San Diego, CA: Academic Press.

Sauer, W. C., and L. Ozimek. 1986. Digestibility of amino acids in swine: Results and their practical applications. A review. *Livestock Production Science* 15:367-388.

Sauvant, D., J. M. Perez, and G. Tran. 2004. *Tables of Composition and Nutritional Value of Feed Materials*, 2nd Ed. Wageningen, The Netherlands: Wageningen Academic.

Shen, Y., M. Z. Fan, A. Ajakaiye, and T. Archbold. 2002. Use of the regression analysis technique to determine the true phosphorus digestibility and the endogenous phosphorus output associated with corn in growing pigs. *Journal of Nutrition* 132:1199-1206.

Sibbald, I. R. 1987. Estimation of bioavailable amino acids in feedingstuffs for poultry and pigs: A review with emphasis on balance experiments. *Canadian Journal of Animal Science* 67:221-301.

Smiricky, M. R., C. M. Grieshop, D. M. Albin, J. E. Wubben, V. M. Gabert, and G. C. Fahey, Jr. 2002. The influence of soy oligosaccharides on apparent and true ileal amino acid digestibilities and fecal consistency in growing pigs. *Journal of Animal Science* 80:2433-2441.

Stein, H. H., and R. A. Bohlke. 2007. The effects of thermal treatment of field peas (*Pisum sativum* L.) on nutrient and energy digestibility by growing pigs. *Journal of Animal Science* 85:1424-1431.

Stein, H. H., C. Pedersen, A. R. Wirt, and R. A. Bohlke. 2005. Additivity of values for apparent and standardized ileal digestibility of amino acids in mixed diets fed to growing pigs. *Journal of Animal Science* 83:2387-2395.

Stein, H. H., M. G. Boersma, and C. Pedersen. 2006. Apparent and true total tract digestibility of phosphorus in field peas (*Pisum sativum* L.) by growing pigs. *Canadian Journal of Animal Science* 85:523-525.

Stein, H. H., B. Seve, M. F. Fuller, P. J. Moughan, and C. F. M. de Lange. 2007. Invited review: Amino acid bioavailability and digestibility in pig feed ingredients: Terminology and application. *Journal of Animal Science* 85:172-180.

Traylor, S. L., G. L. Cromwell, M. D. Lindemann, and D. A. Knabe. 2001. Effects of level of supplemental phytase on ileal digestibility of amino acids, calcium, and phosphorus in dehulled soybean meal for growing pigs. *Journal of Animal Science* 79:2634-2642.

Urriola, P. E., G. C. Shurson, and H. H. Stein. 2010. Digestibility of dietary fiber in distillers co-products fed to growing pigs. *Journal of Animal Science* 88:2373-2381.

van Beers, E. H., H. A. Büller, R. J. Grand, A. W. C. Einerhand, and J. Dekker. 1995. Intestinal brush border glycohydrolases: Structure, function, and development. *Critical Reviews in Biochemistry and Molecular Biology* 30:197-262.

von Heimendahl, E., G. Breves, and Hj. Abel. 2010. Fiber related digestive processes in three different breeds of pigs. *Journal of Animal Science* 88:972-981.

Widmer, M. R., L. M. McGinnis, and H. H. Stein. 2007. Energy, amino acid, and phosphorus digestibility of high protein distillers dried grain and corn germ fed to growing pigs. *Journal of Animal Science* 85:2994-3003.

Wondra, K. J., J. D. Hancock, K. C. Behnke, and C. R. Stark. 1995a. Effects of mill type and particle size uniformity on growth performance, nutrient digestibility, and stomach morphology in finishing pigs. *Journal of Animal Science* 73:2564-2573.

Wondra, K. J., J. D. Hancock, G. A. Kennedy, K. C. Behnke, and K. R. Wondra. 1995b. Effects of reducing particle size of corn in lactation diets on energy and nitrogen metabolism in second parity sows. *Journal of Animal Science* 73:427-432.

第 14 章　营养对养分排泄和环境的影响

引言

传统上，养猪生产者和营养师的目标是使猪的生产性能最大化，他们通常采用最低成本配方（用最低成本满足动物取得该生产性能的最低营养需要）达到上述目的，很少关注多种养分的过剩。配合日粮须考虑以下三个方面：①同一发育阶段动物营养需要的差异；②原料营养组分的变异；③同一饲料养分消化率和可利用率的变异，而上述因素都无一例外地引起日粮的多种养分过剩。因此，为获得最大生产性能而在日粮中过量添加养分会使粪、尿中的养分增加，这些养分最终会排泄到环境中。日粮养分（即蛋白质、各种矿物元素和电解质平衡）的高低会影响动物饮水量，从而影响粪、尿的排泄量。不过，研究表明动物自由饮水量变化中只有很小部分可以归因为营养素的摄入（Mroz et al., 1995; Shaw et al., 2006）。

随着体重增加，猪对大多数养分的需要量（作为日粮的百分数）逐渐降低。经常调整配方（阶段饲喂）与之匹配，将会缩小日粮养分与动物需要的差距，从而减少养分排泄（Boisen et al., 1991; Roth and Kirchgessner, 1993c）。结合阶段饲喂，公母分饲可以更加准确地满足不同性别猪的营养需要，从而降低养分的排泄（Campbell et al., 1985; Campbell and Taverner, 1988）。

在养猪生产中，与养分利用不断提高密切相关的是动物生产效率，随着日粮营养水平提高，遵循单位养分投入产出比递减的规律（Heady et al., 1954; Combs et al., 1991; Gahl et al., 1995）。因此，营养师在设计日粮时，其营养设置可能需要接近预设的应答水平。因为，当动物生产性能接近极限时，单位营养素提高带来的效益越来越小，而相应成本却越来越高。鉴于原料（养分）和养分浪费成本增加，日粮营养水平应该与猪持续改变的营养需要尽可能地接近。

猪用饲料原料各种各样，动物对其养分的可利用率存在一定变异，不能消化和不可利用的养分被排出体外。因此，采用高消化率、高可利用率的饲料原料，使大部分养分被吸收，进而可能被动物生产所利用，是减少养分排泄的手段之一。但是，在采用上述方法时，我们需要衡量原料成本、原料在配方中的使用限制，以及动物对某些原料的生理接受能力（如富含纤维原料引起肠道容积限制，酶如乳糖酶活性随着猪日龄的增加而降低）。提高饲料原料养分消化率的另一个方式是添加外源性饲料酶，它们可以分解饲料原料中某些特定复合物。在所用酶中，植酸酶效果最好，它可以分解和释放植酸磷；其次是其他酶，包括蛋白酶、脂肪酶和各式各样的碳水化合物酶。此外，使用不同来源的矿物元素（硫酸盐和氧化物）或者矿物络合物（即螯合物和蛋白盐）可能会提高动物对其的可利用率。最后，我们还要考虑原料中的抗营养因子诸如丹宁（Brand et al., 1990）、棉酚（Knabe et al., 1979; Mosenthin et al., 1993）、霉菌毒素（Goyarts and Danicke, 2005）及胰蛋白酶抑制剂（Herkelman et al., 1992; Barth et al., 1993）对养分消化带来的不利影响。

减少养分排泄的第二种途径是最大化地利用已吸收的养分。慎重地使用容易发生美拉德反应的原料就是例子（美拉德反应对赖氨酸的吸收影响微不足道，却很大程度上降低了其消化之后的利用率（Batterham, 1992）。此外，通过原料互补和添加合成氨基酸，为蛋白质合成提供比例最平衡的氨基酸组合将会提高氮的利用（Batterham and Bayley, 1989; Buraczewska and Swiech, 2000; Baker, 2004; Yen et al., 2004）。在矿物质领域，日粮中合适的钙磷比例有助于其吸收和利用（Selle et al., 2009）。而多种微量元素间潜在的相互作用对其消化吸收的影响也必须考虑（Davies, 1979; Underwood, 1981; Fairweather-Tait and Hurrell, 1996; Baker, 2008）。

所有降低养分排泄策略的成功最终取决于三个主要因素：①准确地评估同一发育阶段猪的营养需要；②精确地描述饲料原料养分的数据；③每种原料中各种养分的消化率和可利用率。

氮

猪的饲料氮沉积范围为30%~60%，远未达到100%（Kirchgessner et al., 1994; Otto et al., 2003a; van Kempen et al., 2003）。仅用天然原料配合日粮满足动物氨基酸需要，会导致很多必需和非必需氨基酸过剩。没有消化的氨基酸大多以微生物氮的形式，通过粪便排出体外。吸收而没有特定功用的氨基酸将参与代谢，最终大多以尿氮排出体外。使用多种原料与合成氨基酸，结合动物对氨基酸需要与理想蛋白理念可以在降低动物蛋白摄入量的基础上，满足它的氨基酸需要。降低日粮蛋白水平对猪生产性能和瘦肉组织的构成影响甚微。假如能以晶体形式补充任何缺乏的氨基酸，就会对氮的排泄产生巨大影响。33组猪代谢试验综合数据显示，在平衡氨基酸的基础上，日粮蛋白每降低一个百分点，无论猪体重如何，氮的排泄都会降低8%（Kerr, 2003）。该数据与Leek等（2005）发表的数据（8.7%）相吻合，却略高于Leek等（2007）发表的数据（6.7%）。减少氮排泄意义深远，粪氮含量降低将会影响农田的施肥（动物粪尿）量，从而改变地表水流或者地下渗流中氮的含量（Misselbrook et al., 1998）。此外，日粮蛋白摄入量会影响随后产生的粪尿中氨气的释放（Latimier et al., 1993; von Pfeiffer, 1993; Kreuzer et al., 1998; Otto et al., 2003b; Portejoie et al., 2004; Velthof et al., 2005; Panetta et al., 2006; Le et al., 2009）。氨气降低程度等于（Hayes et al., 2004; Leek et al., 2007）或者高于（Canh et al., 1998b; Panetta et al., 2006）Kerr（2003）报道的数据（8%）。Canh等（1998b）和Panetta等（2006）报道的数据更高。其原因是人们观察到，平衡试验回收氮时，如果操作不当（Just et al., 1982; van Kempen et al., 2003），将因粪尿收集过程中氨的挥发而高估氮的存留。氨挥发量的降低不仅改善环境，同时改善猪健康状况和生产力。即使在低水平通风时，畜舍氨气浓度也很少超过30 ppm（Sun and Hoff, 2010, 2011），但是人们观察到生活在氨气污染（>50 ppm）环境中的猪的肺比较重，与生活在过滤空气中的猪相比，肺中细菌数高50%，同时生长受阻（Drummond et al., 1978, 1980; Donham, 1991）。

氮排泄途径是通过粪还是尿也需要考虑。虽然氮净排泄量不变，但增加日粮中抗性淀粉、不可消化寡糖和非淀粉多糖含量会提高后肠发酵碳水化合物浓度，加快细菌繁殖，最终增加了以微生物蛋白存在的粪氮排出，减少了尿氮（Canh et al., 1997; Younes et al., 1997; Bakker and Dekker, 1998; Zervas and Zijlstra, 2002; Hansen et al., 2007），这也会降低氨气的释放（Canh et al., 1998c, d; Kreuzer et al., 1998）。

钙和磷

在常量矿物元素中，钙和磷的研究最为透彻。已知日粮中钙和磷只有20%~50%为机体存留（Kornegay and Harper, 1997），大部分随粪便排出。矿物元素来源（Combs and Wallace, 1962）、原料选择（Bohlke et al., 2005; Pedersen et al., 2007）、其他矿物元素水平（Stein et al., 2008）以及动物体重（Kemme et al., 1997a, b）诸多因素都会影响钙和磷的吸收。此外，钙磷比例不但影响其吸收（Vipperman et al., 1974），同时影响其存留（Crenshaw, 2001; Selle et al., 2009）。在很多植物源性饲料原料中，磷主要以植酸磷存在，大部分不能被单胃动物利用（Jongbloed and Kemme, 1990; Cromwell and Coffey, 1991; Pallauf and Rimbach, 1997），从而造成了猪饲料中大量的磷未被消化。不过，试验证明使用外源植酸酶可以释放植酸磷，提高了磷的消化率（Simons et al., 1990; Cromwell, 2002; Selle and Ravindran, 2008）。改善幅度与磷的来源和日粮中磷的水平、钙磷比例、动物体重、植酸酶添加水平和类型（Kornegay, 1996; Selle and Ravindran, 2008; Kerr et al., 2010）有关。因此，提高钙和磷消化率和消化之后的利用率，并尽可能满足动物特定生产需求将会减少钙和磷向环境的排泄。

铜、铁、锰、镁、钾和锌

各种猪用商业饲料的微量元素存留范围分别是铜5%~40%（Combs et al., 1966; Apgar and Kornegay, 1996）、铁5%~40%（Kornegay and Harper, 1997; Houdijk et al., 1999）、锰<10%（Kornegay and Harper, 1997）、镁15%~60%（Partridge, 1978; Dove, 1995）、钾5%~20%（Mroz et al., 2002）及锌5%~40%（Houdijk et al., 1999; Rincker et al., 2005）。此外，虽然高铜和高锌可以提高动物生产性能（Smith et al., 1997; Hill et al., 2000），但其中接近90%~95%会排出体外（Apgar and Kornegay, 1996; Veum et al., 2004; Buff et al., 2005）。因此，猪采食的矿物元素大部分随粪便排出体外。如果农作物和饲草只能利用其中的一小部分，这些土壤中的微量元素将有可能超过农作物的需要，结果是造成环境污染。不过由于土壤蓄积某些矿物元素的能力强，这不会对随后的农作物产量造成明显影响（Payne et al., 1988; Anderson et al., 1991）。

硫

对含硫氨基酸的代谢，人们进行了广泛研究（du Vigneaud, 1952; Shoveller et al., 2005; Baker, 2006），但很少关注动物的无机硫需要，只是认识到在某些特定营养状况下，动物需要无机硫（Lovett et al., 1986），或者担心动物饮水中硫酸盐含量过高（Anderson and Stothers, 1978; Paterson et al., 1979; Veehuizen et al., 1992; Anderson et al., 1994）。增加粪尿中硫的含量（通过日粮添加 $CaSO_4$）可以降低其pH，从而降低氨气挥发（Canh et al., 1998a; Mroz et al., 2000），尽管氨气挥发量也可以通过日粮蛋白水平进行调控（Velthof et al., 2005）。然而，鉴于各种饲料原料和矿物元素总硫含量比较高（Kerr et al., 2008）以及摄入总硫的存留近65%（Shurson et al., 1998），硫的排泄还是对土壤、水和空气造成影响。的确，众所周知，从动物粪便中散发的含硫毒气种类很多（Banwart and Bremner, 1975），硫含量高的日粮会增加粪便中含硫恶臭因子（Sutton et al., 1998; Whitney et al., 1999; Apgar et al., 2002; Eriksen et al., 2010; Li et al., 2011）。不像粪尿中氮含量和氨挥发量有明确关系（Latimier et al., 1993; von Pfeiffer, 1993; Panetta et al., 2006），粪尿中硫含量和散发量的关系不甚明了。

碳

虽然碳是能量原料（即淀粉、脂肪和非淀粉多糖）基本元素，同时在热量测定试验中被间接考虑，但在典型养分平衡试验中并不被考虑。动物平衡试验主要关注干物质、能量、脂肪和碳水化合物的利用。通过平衡试验，可测定动物消化饲料原料并释放出用于维持和生产目的的能量（用消化能、代谢能和净能衡量）的能力。迄今为止发表的几篇关于猪体蛋白、脂肪和矿物元素构成的文献（Mahan and Shields, 1998; Wiseman et al., 2009; Peters et al., 2010）都没有直接测定碳，然而，考虑到碳是能量代谢和排放气体的基本元素，碳平衡应该是估计养殖对环境影响的重要依据。

通常猪机体组分可分为灰分、脂肪、水分和蛋白质（Shields et al., 1983; Wagner et al., 1999）。将体蛋白（碳，53%；氢，7%；氧，23%；氮，16%）和体脂肪（碳，76%；氢，12%；氧，12%；氮，1%）

（Kleiber，1961）碳的估量应用到猪生长曲线和机体构成的估值上（Wagner et al.，1999），可以估计猪机体的含碳量。猪日粮通常含碳40%（Kerr et al.，2006），或者根据蛋白质、碳水化合物和脂肪的摄入量算出日粮的碳含量，加上估出的呼吸熵（根据生长速度调整瘦肉:脂肪比例）和碳消化率（从饲料、干物质或者能量消化率估算）可以得到碳摄入和沉积量，进而得出碳排泄量。最近，Kerr等（2006）研究发现碳在粪尿中含量接近0.9%，即6.5%的摄入日粮碳排泄在粪尿中。增加日粮纤维采食量，不但因为其低消化率而增加粪便排泄量（Graham et al.，1986；Canh et al.，1998d；Kreuzer et al.，1998；Burkhalter et al.，2001；Kerr et al.，2006），粪碳作为日粮碳的百分比也会进一步增加（Kerr et al.，2006），从而对农业种植产生不确定的影响（Unger and Kaspar，1994；Vitosh et al.，1997；Misselbrook et al.，1998；Sorensen and Fernandez，2003）。

日粮配方与气体排放

猪粪便散发的废气是微生物作用于未消化饲料、动物内源分泌物以及超出动物需要的养分产生的（Mackie et al.，1998；Zhu and Jacobson，1999；Le et al.，2005），包括"发臭"和"无嗅"气体。臭味气体虽然种类很多（Spoelstra，1980；Yasuhara et al.，1984；O'Neill and Phillips，1992），但大致可以分为4类：脂肪酸（即乙酸，$C_2H_4O_2$；丙酸，$C_3H_6O_2$；丁酸，$C_4H_8O_2$；异丁酸，$C_4H_8O_2$；异戊酸，$C_5H_{10}O_2$；正戊酸，$C_5H_{10}O_2$）、酚类（即苯酚，C_6H_6O；对甲基苯酚，C_7H_8O；4-乙基苯酚，$C_8H_{10}O$）、含硫化合物（即硫化氢，H_2S；二甲基三硫化物，$C_2H_6S_3$）和含氮化合物（即氨气，NH_3；吲哚，C_8H_7N；三甲基吲哚，C_9H_9N）。无嗅化合物大多归类为温室气体（即一氧化二氮，N_2O；甲烷，CH_4；二氧化碳，CO_2）。人类嗅觉天生非常复杂，只要当臭味物质的浓度达到一定阈值时，人类才能闻到臭味（Devos et al.，1990；Le et al.，2005），所以采样的地方就非常重要，比如在粪池的下风头（Wright et al.，2005），还是上方（Blanes-Vidal et al.，2009）。同理，要真正理解降低温室气体排放的潜在影响，一定要将其折算成二氧化碳当量（IPCC，2001）。

有关低蛋白日粮对非氨气体和臭味影响的研究很少，没有定论。Hobbs等（1996）、Shriver等（2003）和Le等（2008，2009）报道猪采食添加氨基酸的低蛋白日粮，其粪便中短链脂肪酸浓度降低；而Cromwell等（1999）和Otto等（2003b）报道猪采食添加氨基酸的低蛋白日粮，其粪便中总短链脂肪酸浓度升高；其他研究者（Obrock-Hegel，1997；Sutton et al.，1999；Leek et al.，2007）报道当猪采食不同蛋白水平日粮时，其粪便中挥发有机化合物浓度没有根本变化。在降低日粮蛋白的试验中，废气排放有的降低（Hayes et al.，2004；Le et al.，2007；Leek et al.，2007），有的升高（Cromwell et al.，1999；Otto et al.，2003b），有的没有变化（Obrock-Hegel，1997；Clark et al.，2005；Le et al.，2008，2009）。因此，降低日粮蛋白水平是否能影响粪便中挥发性有机化合物浓度和废气排放目前还没有一致性结论，关于饲喂添加氨基酸的低蛋白日粮对温室气体排放能否产生影响的研究也不充分。Velthof等（2005）观察到采食低蛋白日粮的猪甲烷释放量低，而一氧化二氮的量没有变化。相反，Clark等（2005）研究显示低蛋白日粮增加猪的二氧化碳和甲烷排泄，而一氧化二氮没有变化。Kerr等（2006）报道低蛋白日粮不影响储存容器中粪便的甲烷排放，但增加一氧化二氮排放。Le等（2009）报道对温室气体（甲烷、一氧化二氮或者二氧化碳）没有任何影响。

改变日粮中不可消化寡糖、非淀粉多糖和抗性淀粉含量会引发单胃动物盲肠和后肠的细菌增殖，产生短链脂肪酸（乙酸、丙酸和丁酸，并有微量的异丁酸、戊酸和异戊酸）和其他各种气体（二氧化碳、甲烷和氢气）（Eastwood，1992；Annison and Topping，1994；Jensne and Jorgensen，1994；van der Meulen et al.，1997）。据报道，添加含有上述成分的饲料原料改变粪便中短链脂肪酸浓度（Canh et al.，1997，

1998c, d; Sutton et al., 1999; Shriver et al., 2003; Lynch et al., 2007a; Le et al., 2008），对粪便和粪尿气味有着不确定的作用（DeCamp et al., 2001; Miller and Varel, 2003; Rideout et al., 2004; Willig et al., 2005; Garry et al., 2007; Le et al., 2008; O'Shea et al., 2010）。

有关日粮纤维对温室气体排放影响的研究结果是相互矛盾的。Galassi 等（2004）应用呼吸室研究表明，与对照日粮比较，小麦麸对甲烷排放没有影响，而添加甜菜渣却增加甲烷排放。Velthof 等（2005）报道甲烷排放随着日粮非淀粉多糖水平的提升而增加，但对一氧化二氮没有影响。Clark 等（2005）却报道日粮中添加 20% 的甜菜渣减少二氧化碳排放，但对甲烷和一氧化二氮没有影响。Kerr 等（2006）在日粮中添加大豆皮作为纤维素源提高了一氧化二氮浓度，但对甲烷没有影响。正如 Kirchgessner 等（1991）和 Jorgensen（2007）报道甲烷的产生与日粮发酵纤维水平密切相关。虽然与反刍动物相比，单胃动物甲烷产生量低（Jensen, 1996），但环境挑战将迫使我们在未来设计配方时必须考虑这个问题。

用以减少养猪产生的氨气、硫化氢及挥发臭气的饲料添加剂为数众多，包括植物提取物（Colina et al., 2001; Rideout et al., 2004; Panetta et al., 2005; Lynch et al., 2007b; Windisch et al., 2008; Biagi et al., 2010）、有机酸（Eriksen et al., 2010; Halas et al., 2010）、益生元和益生素（Wang et al., 2009; O'Shea et al., 2010）、植物精油（Varel, 2002; Michiels et al., 2009）、腐植酸（Ji et al., 2006）、酸化钙盐（Canh et al., 1998a）和微量元素（Armstrong et al., 2000）。上述添加剂的文献综述不在本书的讨论范围。

综合措施

通常，提高动物养分消化率和饲料（养分）利用效率将降低养分损失（Henry and Dourmad, 1992）。诸多因素可以提高饲料效率，包括品种改良（Campbell and Taverner, 1988; Bark et al., 1992）、改善环境条件（Verstegen et al., 1973）、合理搭配高品质原料的日粮、饲料加工诸如制粒和细粉碎工艺（Yen et al., 2004）、代谢调节剂（Quiniou et al., 1993; Caperna et al., 1995）、抗生素（Roth and Kirchgessner, 1993a, b）、免疫调控（Williams et al., 1997），以及正确地调整饲槽以减少饲料浪费。

随着单位面积养猪生产强度的增加，粪便回田一定与农业种植需要相平衡以防地表和地下水污染，并使土壤微量元素蓄积最小化。氮施用过多可以引起地表水中氮和地下水中亚硝酸盐含量上升。磷施用过多会使它在土壤中蓄积。虽然被土壤颗粒吸附而不会渗入地下水，但磷会随流失土壤进入溪水、湖泊和河流，而磷是调控水生植物生长的第一限制性营养素（Pierzynski et al., 1994; Sharpley et al., 1994），而其生长会导致水质恶化（Crenshaw and Johnason, 1995）。结合养分排泄最小化，将粪便组成［无论数据来自现成表格（ASAE, 2005）还是分析结果］、粪便储存的作用（Petersen et al., 1998）以及农田施肥方法（Hoff et al., 1981）综合在一起进行治理是养猪生产的目标。

参 考 文 献

Anderson, D. M., and S. C. Stothers. 1978. Effects of saline water high in sulfates, chlorides and nitrates on the performance of young weanling pigs. *Journal of Animal Science* 47:900-907.

Anderson, J. S., D. M. Anderson, and J. M. Murphy. 1994. The effect of water quality on nutrient availability for grower/finisher pigs. *Canadian Journal of Animal Science* 74:141-148.

Anderson, M. A., J. R. McKenna, D. C. Martens, S. J. Donohue, E. T. Kornegay, and M. D. Lindemann. 1991. Long-term effects of copper rich swine manure application on continuous corn production. *Communications in Soil Science and Plant Analysis* 22:993-1002.

Annison, G., and D. L. Topping. 1994. Nutritional role of resistant starch: Chemical structure vs. physiological function. *Annual Review of Nutrition* 14:297-320.

Apgar, G. A., and E. T. Kornegay. 1996. Mineral balance of finishing pigs fed copper sulfate or a copper lysine complex at growth stimulating levels. *Journal of Animal Science* 74:1594-1600.

Apgar, G., K. Griswold, B. Jacobson, and J. Salazar. 2002. Effects of elevated and reduced dietary N and S concentration upon growth and concentration of odor causing components in waste of finishing pigs. *Journal of Animal Science* 80(Suppl. 1):395 (Abstr.).

Armstrong, T. A., C. M. Williams, J. W. Spears, and S. S. Schiffman. 2000. High dietary copper improves odor characteristics of swine waste. *Journal of Animal Science* 78:859-864.

ASAE (American Society of Agricultural Engineers). 2005. Manure Production and Characteristics. *ASAE* D384.2, MAR2005. St. Joseph, MI: ASAE.

Baker, D. H. 2004. Animal models of human amino acid responses. *Journal of Nutrition* 134:1646S-1650S.

Baker, D. H. 2006. Comparative species utilization and toxicity of sulfur amino acids. *Journal of Nutrition* 136:1670S-1675S.

Baker, D. H. 2008. Animal models in nutrition research. *Journal of Nutrition* 138:391-396.

Bakker, G. C. M., and R. A. Dekker. 1998. Effect of source and amount of non-starch polysaccharides on the site of excretion of nitrogen in pigs. *Journal of Animal Science* 76(Suppl. 1):665 (Abstr.).

Banwart, W. L., and J. M. Bremner. 1975. Identification of sulfur gases evolved from animal manures. *Journal of Environmental Quality* 4:363-366.

Bark, L. J., T. S. Stahly, G. L. Cromwell, and J. Miyat. 1992. Influence of genetic capacity for lean tissue growth on rate and efficiency of tissue accretion in pigs fed ractopamine. *Journal of Animal Science* 70:3391-3400.

Barth, C., A. B. Lunding, M. Schmitx, and H. Hagemeister. 1993. Soybean trypsin inhibitor(s) reduce absorption of exogenous and increase loss of endogenous protein in miniature pigs. *Journal of Nutrition* 123:2195-2200.

Batterham, E. S. 1992. Availability and utilization of amino acids for growing pigs. *Nutrition Research Reviews* 5:1-18.

Batterham, E. S., and H. S. Bayley. 1989. Effect of frequency of feeding of diets containing free or protein-bound lysine on the oxidation of [^{14}C]lysine or [^{14}C]phenylalanine by growing pigs. *British Journal of Nutrition* 62:647-655.

Biagi, G., I. Cipollini, B. R. Paulicks, and F. X. Roth. 2010. Effect of tannins on growth performance and intestinal ecosystem in weaned piglets. *Archives of Animal Nutrition* 64:121-135.

Blanes-Vidal, V., M. N. Hansen, A. P. S. Adamsen, A. Feilberg, S. O. Petersen, and B. B. Jensen. 2009. Characterization of odor released during handling of swine slurry: Part I. Relationship between odorants and perceived odor concentrations. *Atmospheric Environment* 43:2997-3005.

Bohlke, R. A., R. C. Thaler, and H. H. Stein. 2005. Calcium, phosphorus and amino acid digestibility in low-phytase corn, normal corn, and soybean meal by growing pigs. *Journal of Animal Science* 83:2396-2403.

Boisen, S., J. A. Fernandez, and A. Madsen. 1991. Studies on ideal protein requirement of pigs from 20 to 95 kg live weight. Pp. 299-302 in *Proceedings of the 6th International Symposium on Protein Metabolism and Nutrition*, June 9-14, 1991, Herning, Denmark.

Brand, T. S., H. A. Badenhorst, F. K. Siebrits, and I. P. Hayes. 1990. The use of pigs both intact and with ileo-rectal anastomosis to estimate the apparent and true digestibility of amino acids in untreated, heat-treated and thermal-ammoniated high-tannin grain sorghum. *South African Journal of Animal Science* 20:223-228.

Buff, C. E., D. W. Bollinger, M. R. Ellersieck, W. A. Brommelsiek, and T. L. Veum. 2005. Comparison of growth performance and zinc absorption, retention, and excretion in weanling pigs fed diets supplemented with zinc-polysaccharide or zinc oxide. *Journal of Animal Science* 83:2380-2386.

Buraczewska, L., and E. Swiech. 2000. A note on absorption of crystalline threonine in pigs. *Journal of Animal and Feed Sciences* 9:489-492.

Burkhalter, T. M., N. R. Merchen, L. L. Bauer, S. M. Murray, A. R. Patil, J. L. Brent, Jr., and G. C. Fahey, Jr. 2001. The ratio of insoluble to soluble fiber components in soybean hulls affects ileal and total-tract nutrient digestibilities and fecal characteristics of dogs. *Journal of Nutrition* 131:1978-1985.

Campbell, R. G., and M. R. Taverner. 1988. Genotype and sex effect on the relationship between energy intake and protein deposition in growing pigs. *Journal of Animal Science* 66:676-686.

Campbell, R. G., M. R. Taverner, and D. M. Curic. 1985. Effects of sex and energy intake between 48 and 90 kg live weight on protein deposition in growing pigs. *Animal Production* 40:497-503.

Canh, T. T., M. W. A. Verstegen, A. J. A. Aarnink, and J. W. Schrama. 1997. Influence of dietary factors on nitrogen partitioning and composition of urine and feces of fattening pigs. *Journal of Animal Science* 75:700-706.

Canh, T. T., A. J. A. Aarnink, Z. Morz, A. W. Jongbloed, J. W. Schrama, and M. W. A. Verstegen. 1998a. Influence of electrolyte balance and acidifying calcium salts in the diet of growing-finishing pigs on urinary pH, slurry pH and ammonia volatilization from slurry. *Livestock Production Science* 56:1-13.

Canh, T. T., A. J. A. Aarnink, J. B. Schutte, A. Sutton, D. J. Langhout, and M. W. A. Verstegen. 1998b. Dietary protein affects nitrogen excretion and ammonia emission from slurry of growing-finishing pigs. *Livestock Production Science* 56:181-191.

Canh, T. T., A. J. A. Aarnink, M. W. A. Verstegen, and J. W. Schrama. 1998c. Influence of dietary factors on the pH and ammonia emission of slurry from growing-finishing pigs. *Journal of Animal Science* 76:1123-1130.

Canh, T. T., A. L. Sutton, A. J. A. Aarnink, M. W. A. Verstegen, J. W. Schrama, and G. C. M. Bakker. 1998d. Dietary carbohydrates alter the fecal composition and pH and the ammonia emission from slurry of growing pigs. *Journal of Animal Science* 76:1887-1895.

Caperna, T. J., R. G. Campbell, M. R. Ballard, and N. C. Steele. 1995. Somatotropin enhances the rate of amino acid deposition but has minimal impact on amino acid balance in growing pig. *Journal of Nutrition* 125:2104-2113.

Clark, O. G., S. Moehn, I. Edeogu, J. Price, and J. Leonard. 2005. Manipulation of dietary protein and nonstarch polysaccharide to control swine manure emissions. *Journal of Environmental Quality* 34:1461-1466.

Colina, J. J., A. J. Lewis, P. S. Miller, and R. L. Fisher. 2001. Dietary manipulation to reduce aerial ammonia concentrations in nursery pig facilities. *Journal of Animal Science* 79:3096-3103.

Combs, G. E., and H. D. Wallace. 1962. Growth and digestibility studies with young pigs fed various levels and sources of calcium. *Journal of Animal Science* 21:734-737.

Combs, G. E., C. B. Ammerman, R. L. Shirley, and H. D. Wallace. 1966. Effect of source and level of dietary protein on pigs fed high-copper rations. *Journal of Animal Science* 25:613-616.

Combs, N. R., E. T. Kornegay, M. D. Lindemann, and D. R. Notter. 1991. Calcium and phosphorus requirement of swine from weaning to market weight: 1. Development of response curves for performance. *Journal of Animal Science* 69:673-681.

Crenshaw, T. D. 2001. Calcium, phosphorus, vitamin D, and vitamin K in swine nutrition. Pp. 187-212 in *Swine Nutrition*, 2nd Ed., A. J. Lewis and L. L. Southern, eds. Washington, DC: CRC Press.

Crenshaw, T. D., and J. C. Johanson. 1995. Nutritional strategies for waste reduction management: Minerals. Pp. 69-78 in *New Horizons in Animal Nutrition and Health*, J. B. Longenecker and J. W. Spears, eds. Chapel Hill: Institute of Nutrition of the University of North Carolina.

Cromwell, G. L. 2002. Approaches to meeting the nonruminant's phosphorus requirements. Pp. 61-76 in *Proceedings of the 63rd Minnesota Nutrition Conference*, Egan, MN. St. Paul: University of Minnesota.

Cromwell, G. L., and R. D. Coffey. 1991. Phosphorus—a key essential nutrient, yet a possible major pollutant—its central role in animal nutrition. Pp. 133-145 in *Biotechnology in the Feed Industry*, T. P. Lyons, ed. Nicholasville, KY: Alltech Technical Publications.

Cromwell, G. L., L. W. Turner, R. S. Gates, J. L. Taraba, M. D. Lindemann, S. L. Traylor, W. A. Dozier, III, and H. J. Monegue. 1999. Manipulation of swine diets to reduce gaseous emissions from manure that contribute to odor. *Journal of Animal Science* 77(Suppl. 1):69 (Abstr.).

Davies, N. T. 1979. Anti-nutrient factors affecting mineral utilization. *Proceedings of the Nutrition Society* 38:121-128.

DeCamp, S. A., B. E. Hill, S. L. Hankins, D. C. Kendall, B. T. Richert, A. L. Sutton, D. T. Kelly, M. L. Cobb, D. W. Bundy, and W. J. Powers. 2001. Effects of soybean hulls in a commercial diet on pig performance, manure composition, and selected air quality parameters in swine facilities. *Journal of Animal Science* 79(Suppl. 1):252 (Abstr.).

Devos, M., F. Patte, J. Rouault, P. Laffort, and L. van Gemert. 1990. *Standardized Human Olfactory Thresholds*. New York: IRl Press at Oxford University Press.

Donham, K. J. 1991. Association of environmental air contaminants with disease and productivity in swine. *American Journal of Veterinary Research* 52:1723-1730.

Dove, C. R. 1995. The effect of copper level on nutrient utilization of weanling pigs. *Journal of Animal Science* 73:166-171.

Drummond, J. G., S. E. Curtis, and J. Simon. 1978. Effects of atmospheric ammonia on pulmonary bacterial clearance in the young pig. *American Journal of Veterinary Research* 39:211-212.

Drummond, J. G., S. E. Curtis, J. Simon, and H. W. Norton. 1980. Effects of aerial ammonia on growth and health of young pigs. *Journal of Animal Science* 50:1085-1091.

du Vigneaud, V. 1952. *A Trail of Research in Sulfur Chemistry and Metabolism*. Ithaca, NY: Cornell University Press.

Eastwood, M. A. 1992. The physiological effect of dietary fiber: An update. *Annual Review of Nutrition* 12:19-35.

Eriksen, J., A. P. S. Adamsen, J. V. Norgaard, H. D. Poulsen, B. B. Jensen, and S. O. Petersen. 2010. Emission of sulfur-containing odorants, ammonia, and methane from pig slurry: Effects of dietary methionine and benzoic acid. *Journal of Environmental Quality* 39:1097-1107.

Fairweather-Tait, S., and R. F. Hurrell. 1996. Bioavailability of minerals and trace elements. *Nutrition Research Reviews* 9:295-324.

Gahl, M. J., T. D. Crenshaw, and N. J. Benevenga. 1995. Diminishing returns in weight, nitrogen, and lysine gain of pigs fed six levels of lysine from three supplemental sources. *Journal of Animal Science* 72:3177-3187.

Galassi, G., G. M. Crovetto, L. Rapetti, and A. Tamburini. 2004. Energy and nitrogen balance in heavy pigs fed different fibre sources. *Livestock Production Science* 85:253-262.

Garry, B. P., M. Fogarty, T. P. Curran, M. J. O'Connell, and J. V. O'Doherty. 2007. The effect of cereal type and enzyme addition on pig performance, intestinal microflora, and ammonia and odour emissions. *Animal* 1:751-757.

Goyarts, T., and S. Danicke. 2005. Effects of deoxynivalenol (DON) on growth performance, nutrient digestibility and DON metabolism in pigs. *Mycotoxin Research* 21:139-142.

Graham, H., K. Hesselman, and P. Aman. 1986. The influence of wheat bran and sugar-beet pulp on the digestibility of dietary components in a cereal-based pig diet. *Journal of Nutrition* 116:242-251.

Halas, D., C. F. Hansen, D. J. Hampson, J-C. Kim, B. P. Mullan, R. H. Wilson, and J. R. Pluske. 2010. Effects of benzoic acid and inulin on ammonia-nitrogen excretion, plasma urea levels, and the pH in faeces and urine of weaner pigs. *Livestock Science* 134:243-245.

Hansen, M. J., A. Chwalibog, and A. H. Tauson. 2007. Influence of different fibre sources in diets for growing pigs on chemical composition of faeces and slurry and ammonia emission from slurry. *Animal Feed Science and Technology* 134:326-336.

Hayes, E. T., A. B. G. Leek, T. P. Curran, V. A. Dodd, O. T. Carton, V. E. Beattie, and J. V. O'Doherty. 2004. The influence of diet crude protein level on odour and ammonia emissions from finishing pig houses. *Bioresource Technology* 91:309-315.

Heady, E. O., R. Woodworth, D. R. Catron, and G. C. Ashton. 1954. New procedures in estimating feed substitution rates and in determining economic efficiency in pork production. *Ames Agriculture Experiment Station Research Bulletin* 893-976. Ames. Iowa State College.

Henry, Y., and J. Y. Dourmad. 1992. Protein nutrition and N pollution. *Feed Mix* (May):25-28.

Herkelman, K. L., G. L. Cromwell, T. S. Stahly, T. W. Pfeiffer, and D. A. Knabe. 1992. Apparent digestibility of amino acids in raw and heated conventional and low-trypsin-inhibitor soybeans for pigs. *Journal of Animal Science* 70:818-826.

Hill, G. M., G. L. Cromwell, T. D. Crenshaw, C. R. Dove, R. C. Ewan, D. A. Knabe, A. J. Lewis, G. W. Libal, D. C. Mahan, G. C. Shurson, L. L. Southern, and T. L. Veum. 2000. Growth promotion effects and plasma changes from feeding high dietary concentrations of zinc and copper to weanling pigs (regional study). *Journal of Animal Science* 78:1010-1016.

Hobbs, P. J., B. F. Pain, R. M. Kay, and P. A. Lee. 1996. Reduction of odorous compounds in fresh pig slurry by dietary control of crude protein. *Journal of the Science of Food and Agriculture* 71:508-514.

Hoff, J. D., D. W. Nelsen, and A. L. Sutton. 1981. Ammonia volatilization from liquid swine manure applied to cropland. *Journal of Environmental Quality* 10:87-90.

Houdijk, J. G., M. W. Bosch, S. Tamminga, M. W. Verstgen, E. B. Berenpas, and H. Knoop. 1999. Apparent ileal and total-tract nutrient digestion by pigs as affected by dietary nondigestible oligosaccharides. *Journal of Animal Science* 77:148-158.

IPCC (Intergovernmental Panel on Climate Change). 2001. *Intergovernmental Panel on Climate Change: Technical Summary of the 3rd Assessment Report of Working Group 1, The Scientific Basis*. Geneva, Switzerland: IPCC. Available online at http://www.grida.no/climate/ipcc_tar/wg1/010.htm. Accessed on March 29, 2010.

Jensen, B. B. 1996. Methanogenesis in monogastric animals. *Environmental Monitoring and Assessment* 42:99-112.

Jensen, B. B., and H. Jorgensen. 1994. Effect of dietary fiber on microbial activity and microbial gas production in various regions of the gastrointestinal tract of pigs. *Applied and Environmental Microbiology* 60:1897-1904.

Ji, F., J. McGlone, and S. W. Kim. 2006. Effects of dietary humic substances on pig growth performance, carcass characteristics, and ammonia emission. *Journal of Animal Science* 84:2482-2490.

Jongbloed, A. W., and P. A. Kemme. 1990. Apparent digestible phosphorus in the feeding of pigs in relation to availability, requirement and environment. 1. Digestible phosphorus in feedstuffs from plant and animal origin. *Netherlands Journal of Agricultural Science* 38:567-575.

Jorgensen, H. 2007. Methane emission by growing pigs and adult sows as influenced by fermentation. *Livestock Science* 109:216-219.

Just, A., J. A. Fernandez, and H. Horgensen. 1982. Nitrogen balance studies and nitrogen retention. *Physiologie Digestive chez le Porc* 12:111-122.

Kemme, P. A., A. W. Jongbloed, Z. Mroz, and A. C. Beynen. 1997a. The efficacy of *Aspergillus niger* phytase in rendering phytate phosphorus available for absorption in pigs is influenced by pig physiological status. *Journal of Animal Science* 75:2129-2138.

Kemme, P. A., J. S. Radcliffe, A. W. Jongbloed, and Z. Mroz. 1997b. Factors affecting phosphorus and calcium digestibility in diets for growing-finishing pigs. *Journal of Animal Science* 75:2139-2146.

Kerr, B. J. 2003. Dietary manipulation to reduce environmental impact. Pp. 139-158 in *9th International Symposium on Digestive Physiology in Pigs*, May 14-17, 2003, Banff, Alberta, Canada.

Kerr, B. J., C. J. Ziemer, S. L. Trabue, J. D. Crouse, and T. B. Parkin. 2006. Manure composition of swine as affected by dietary protein and cellulose concentrations. *Journal of Animal Science* 84:1584-1592.

Kerr, B. J., C. J. Ziemer, T. E. Weber, S. L. Trabue, B. L. Bearson, G. C. Shurson, and M. H. Whitney. 2008. Comparative sulfur analysis using thermal combustion on inductively coupled plasma methodology and mineral composition of common livestock feeds. *Journal of Animal Science* 86:2377-2384.

Kerr, B. J., T. E. Weber, P. S. Miller, and L. L. Southern. 2010. Effect of phytase on apparent total tract digestibility of phosphorus in corn-soybean meal diets fed to finishing pigs. *Journal of Animal Science* 88:238-247.

Kirchgessner, M., M. Kreuzer, H. L. Muller, and W. Windisch. 1991. Release of methane and of carbon dioxide by the pig. *Agribiological Research-Zeitschrift fur Agrarbiologie Agrikulturchemie Okologie* 44:103-113.

Kirchgessner, M., W. Windisch, and F. X. Roth. 1994. The efficiency of nitrogen conversion in animal nutrition. *Nova Acta Leopoldina* 70:393-412.

Kleiber, M. 1961. *The Fire of Life, an Introduction to Animal Energetics.* Malabar, FL: R. E. Krieger.

Knabe, D. A., T. D. Tanksley, Jr., and J. H. Hesby. 1979. Effect of lysine, crude fiber and free gossypol in cottonseed meal on the performance of growing pigs. *Journal of Animal Science* 49:134-142.

Kornegay, E. T. 1996. Nutritional, environmental and economic considerations for using phytase in pig diets. Pp. 279-304 in *Nutrient Management of Food Animals to Enhance and Protect the Environment*, E. T. Kornegay, ed. Boca Raton, FL: CRC Press, Inc.

Kornegay, E. T., and A. F. Harper. 1997. Environmental nutrition: Nutrient management strategies to reduce nutrient excretion of swine. *The Professional Animal Scientist* 13:99-111.

Kreuzer, M., A. Machmuller, M. M. Gerdemann, H. Hanneken, and M. Whittmann. 1998. Reduction of gaseous nitrogen loss from pig manure using feed rich in easily-fermentable non-starch polysaccharides. *Animal Feed Science and Technology* 73:1-19.

Latimier, P., J. Y. Dourmad, A. Corlouer, J. Chauvel, J. le Pan, M. Gautier, and D. Lesaicherre. 1993. Effect of three protein feeding strategies, for growing-finishing pigs, on growth performance and nitrogen output in the slurry. *Journées de la Recherche Porcine en France* 25:295-300.

Le, P. D., A. J. A. Aarnink, N. W. M. Ogink, P. M. Becker, and M. W. A. Verstegen. 2005. Odour from animal production facilities: Its relationship to diet. *Nutrition Research Reviews* 18:3-30.

Le, P. D., A. J. A. Aarnink, A. W. Jongbloed, C. M. C. van der Peet-Schwering, N. W. M. Ogink, and M. W. A. Verstegen. 2007. Effects of dietary crude protein level on odour from pig manure. *Animal* 1:734-744.

Le., P. D., A. J. A. Aarnink, A. W. Jongbloed, C. M. C. van der Peet-Schwering, N. W. M. Ogink, and M. W. A. Verstegen. 2008. Interactive effects of dietary crude protein and fermentable carbohydrate levels on odour from pig manure. *Livestock Science* 114:48-61.

Le, P. D., A. J. A. Aarnink, and A. W. Jongbloed. 2009. Odour and ammonia emission from pig manure as affected by dietary crude protein level. *Livestock Science* 121:267-274.

Leek, A. B. G., J. J. Callan, R. W. Henry, and J. B. O'Doherty. 2005. The application of low crude protein wheat-soyabean diets to growing and finishing pigs. 2. The effects on nutrient digestibility, nitrogen excretion, faecal volatile fatty acid concentration and ammonia emission from boars. *Irish Journal of Agricultural and Food Research* 44:247-260.

Leek, A. B. G., E. T. Hayes, T. P. Curran, J. J. Callan, V. E. Beattie, V. A. Dodd, and J. V. O'Doherty. 2007. The influence of manure composition on emissions of odour and ammonia from finishing pigs fed different concentrations of dietary crude protein. *Bioresource Technology* 98:3431-3439.

Li, W. T., W. J. Powers, and G. M. Hill. 2011. Feeding distillers dried grains with solubles and organic trace mineral sources to swine and the resulting effect on gaseous emissions. *Journal of Animal Science* 89:3286-3299.

Lovett, T. D., M. T. Coffey, R. D. Miles, and G. E. Combs. 1986. Methionine, choline and sulfate interrelationships in the diet of weanling swine. *Journal of Animal Science* 63:467-471.

Lynch, M. B., T. Sweeney, J. J. Callan, and J. V. O'Doherty. 2007a. Effects on increasing the intake of dietary β-glucans by exchanging wheat for barley on nutrient digestibility, nitrogen excretion, intestinal microflora, volatile fatty acid concentration and manure ammonia emissions in finishing pigs. *Animal* 1:812-819.

Lynch, M. B., T. Sweeney, J. J. Callan, and J. V. O'Doherty. 2007b. The effect of high and low dietary crude protein and inulin supplementation on nutrient digestibility, nitrogen excretion, intestinal microflora and manure ammonia emissions from finisher pigs. *Animal* 1:1112-1121.

Mackie, R. I., P. G. Stroot, and V. H. Varel. 1998. Biochemical identification and biological origin of key odor components in livestock waste. *Journal of Animal Science* 76:1331-1342.

Mahan, D. C., and R. G. Shields, Jr. 1998. Macro- and micromineral composition of pigs from birth to 145 kilograms of body weight. *Journal of Animal Science* 76:506-512.

Michiels, J., J. A. M. Missotten, D. Fremaut, S. De Smet, and N. A. Dierick. 2009. In vitro characterization of the antimicrobial activity of selected essential oil compounds and binary combinations against the pig gut flora. *Animal Feed Science and Technology* 151:111-127.

Miller, D. N., and V. H. Varel. 2003. Swine manure composition affects the biochemical origins, composition, and accumulation of odorous compounds. *Journal of Animal Science* 81:2131-2138.

Misselbrook, T. H., D. R. Chadwick, B. F. Pain, and D. M. Headon. 1998. Dietary manipulation as a means of decreasing N losses and methane emissions and improving herbage N uptake following application of pig slurry to grassland. *Journal of Agricultural Science* 130:183-191.

Mosenthin, R., W. C. Sauer, K. A. Lien, and C. F. M. de Lange. 1993. Apparent, true and real ileal protein and amino acid digestibilities in growing pigs fed two varieties of fababeans (*Vicia faba* L.) different in tannin content. *Journal of Animal Physiology and Animal Nutrition* 70:253-265.

Mroz, Z., A. W. Jongbloed, N. P. Lenis, and K. Vreman. 1995. Water in pig nutrition: Physiology, allowance and environmental implications. *Nutrition Research Reviews* 8:137-164.

Mroz, Z., A. J. Moeser, K. Vreman, J. T. M. van Diepen, T. van Kempen, T. T. Canh, and A. W. Jongbloed. 2000. Effects of dietary carbohydrates and buffering capacity on nutrient digestibility and manure characteristics in finishing pigs. *Journal of Animal Science* 78:3096-3106.

Mroz, Z., D. E. Reese, M. Overland, J. T. van Diepen, and J. Kogus. 2002. The effects of potassium diformate and its molecular constituents on the apparent ileal and fecal digestibility and retention of nutrients in growing-finishing pigs. *Journal of Animal Science* 80:681-690.

Obrock-Hegel, C. E. 1997. The effects of reducing dietary crude protein concentration on odor in swine facilities. M.S. Thesis, University of Nebraska, Lincoln.

O'Neill, D. H., and V. R. Phillips. 1992. A review of the control of odour nuisance from livestock buildings: Part 3, Properties of the odorous substances which have been identified in livestock wastes or in the air around them. *Journal of Agricultural Engineering and Resources* 53:23-50.

O'Shea, C. J., T. Sweeney, M. B. Lynch, D. A. Gahn, J. J. Callan, and J. V. O'Doherty. 2010. Effect of β-glucans contained in barley- and oat-based diets and exogenous enzyme supplementation on gastrointestinal fermentation of finisher pigs and subsequent manure odor and ammonia emissions. *Journal of Animal Science* 88:1411-1420.

Otto, E. R., M. Yokoyama, P. K. Ku, N. K. Ames, and N. L. Trottier. 2003a. Nitrogen balance and ileal amino acid digestibility in growing pigs fed diets reduced in protein concentration. *Journal of Animal Science* 81:1743-1753.

Otto, E. R., M. Yokoyama, R. D. von Bermuth, T. van Kempen, and N. L. Trottier. 2003b. Ammonia, volatile fatty acids, phenolics and odor offensiveness in manure from growing pigs fed diets reduced in protein concentration. *Journal of Animal Science* 81:1754-1763.

Panetta, D. M., W. J. Powers, and J. C. Lorimor. 2005. Management strategy impacts on ammonia volatilization from swine manure. *Journal of Environmental Quality* 34:1119-1130.

Panetta, D. M., W. J. Powers, H. Xin, B. J. Kerr, and K. J. Stalder. 2006. Nitrogen excretion and ammonia emissions from pigs fed modified diets. *Journal of Environmental Quality* 35:1297-1308.

Pallauf, J., and G. Rimbach. 1997. Nutritional significance of phytic acid and phytase. *Archives of Animal Nutrition* 50:301-319.

Partridge, I. G. 1978. Studies on digestion and absorption in the intestines of growing pigs. 3. Net movements of mineral nutrients in the digestive tract. *British Journal of Nutrition* 39:527-537.

Paterson, D. W., R. C. Wahlstrom, G. W. Libal, and O. E. Olson. 1979. Effects of sulfate in water on swine reproduction and young pig performance. *Journal of Animal Science* 49:664-667.

Payne, G. G., D. C. Martens, E. T. Kornegay, and M. D. Lindemann. 1988. Availability and form of copper in three soils following eight annual applications of Cu-enriched swine manure. *Journal of Environmental Quality* 17:740-746.

Pedersen, C., M. G. Boersma, and H. H. Stein. 2007. Digestibility of energy and phosphorus in ten samples of distillers dried grains with solubles fed to growing pigs. *Journal of Animal Science* 85:1168-1176.

Peters, J. C., D. C. Mahan, T. G. Wiseman, and N. D. Fastinger. 2010. Effect of dietary organic and inorganic micromineral source and level on sow body, liver, colostrums, mature milk, and progeny mineral compositions over six parities. *Journal of Animal Science* 88:626-637.

Petersen, S. O., A. M. Lind, and S. G. Sommer. 1998. Nitrogen and organic matter losses during storage of cattle and pig manure. *Journal of Agricultural Science* 130:69-79.

Pierzynski, G. M., J. T. Sims, and G. F. Vance. 1994. *Soils and Environmental Quality*. Boca Raton, FL: Lewis Publishers, CRC Press.

Portejoie, S., J. Y. Dourmad, J. Martinez, and Y. Lebreton. 2004. Effect of lowering dietary crude protein on nitrogen excretion, manure composition and ammonia emission from fattening pigs. *Livestock Production Science* 91:45-55.

Quiniou, N., J. Noblet, and J. Y. Dounnad. 1993. Effect of porcine somatotropin and dietary protein level on the nitrogen and phosphorus losses of pigs. *Journées de la Recherche Porcine en France* 25:287-294.

Rideout, T. C., M. Z. Fan, J. P. Cant, C. Wagner-Riddle, and P. Stonehouse. 2004. Excretion of major odor-causing and acidifying compounds in response to dietary supplementation of chicory inulin in growing pigs. *Journal of Animal Science* 82:1678-1684.

Rincker, M. J., G. M. Hill, J. E. Link, A. M. Meyer, and J. E. Rowntree. 2005. Effects of dietary zinc and iron supplementation on mineral excretion, body composition, and mineral status of nursery pigs. *Journal of Animal Science* 83:2762-2774.

Roth, F. X., and M. Kirchgessner. 1993a. Influence of avilamycin and tylosin on retention and excretion of nitrogen in finishing pigs. *Journal of Animal Physiology and Animal Nutrition* 69:245-250.

Roth, F. X., and M. Kirchgessner. 1993b. Influence of avilamycin and tylosin on retention and excretion of nitrogen in growing pigs. *Journal of Animal Physiology and Animal Nutrition* 69:175-185.

Roth, F. X., and M. Kirchgessner. 1993c. Reducing nitrogen excretion in pigs by optimum dietary protein and amino acid supply. *Zuchtungskunde* 65:420-429.

Selle, P. H., and V. Ravindran. 2008. Phytate degrading enzymes in pig nutrition. *Livestock Science* 113:99-122.

Selle, P. H., A. J. Cowieson, and V. Ravindran. 2009. Consequences of calcium interactions with phytate and phytase for poultry and pigs. *Livestock Science* 124:126-141.

Sharpley, A. N., S. C. Chapra, R. Wedepohl, J. T. Sims, T. C. Daniel, and K. R. Reddy. 1994. Managing agricultural phosphorus for protection of surface waters: Issues and options. *Journal of Environmental Quality* 23:437-451.

Shaw, M. I., A. D. Beaulieu, and J. F. Patience. 2006. Effect of diet composition on water consumption in growing pigs. *Journal of Animal Science* 84:3123-3132.

Shields, R. G., Jr., D. C. Mahan, and P. L. Graham. 1983. Changes in swine body composition from birth to 145 kg. *Journal of Animal Science* 57:43-54.

Shoveller, A. K., B. Stoll, R. O. Ball, and D. G. Burrin. 2005. Nutritional and functional importance of intestinal sulfur amino acid metabolism. *Journal of Nutrition* 135:1609-1612.

Shriver, J. A., S. D. Carter, A. L. Sutton, B. T. Richert, B. W. Senne, and L. A. Pettey. 2003. Effects of adding fiber sources to reduced crude protein, amino acid-supplemented diets on nitrogen excretion, growth performance, and carcass traits of finishing pigs. *Journal of Animal Science* 81:492-502.

Shurson, J., M. Whitney, and R. Nicolai. 1998. Nutritional manipulation of swine diets to reduce hydrogen sulfide emissions. Pp. 219-240 in *Proceedings of the 59th Minnesota Nutrition Conference IPC Technical Symposium*, September 21-23, 1998, Bloomington, MN.

Simons, P. C. M., H. A. J. Versteegh, A. W. Jongbloed, P. A. Kemme, P. Slump, K. D. Bos, M. G. E. Wolters, R. F. Beudeker, and G. J. Verschoor. 1990. Improvement of phosphorus availability by microbial phytase in broilers and pigs. *British Journal of Nutrition* 64:525-540.

Smith, J. W., II, M. D. Tokach, R. D. Goodband, J. L. Nelssen, and B. T. Richert. 1997. Effects of the interrelationship between zinc oxide and copper sulfate on growth performance of early-weaned pigs. *Journal of Animal Science* 75:1861-1866.

Sorensen, P., and J. A. Fernandez. 2003. Dietary effects on the composition of pig slurry and on the plant utilization of pig slurry nitrogen. *Journal of Agricultural Science* 140:343-355.

Spoelstra, S. F. 1980. Origin of objectionable odorous components in piggery wastes and the possibility of applying indicator components for studying odour development. *Agriculture and Environment* 5:241-260.

Stein, H. H., C. T. Kadzere, S. W. Kim, and P. S. Miller. 2008. Influence of dietary phosphorus concentration on the digestibility of phosphorus in monocalcium phosphate by growing pigs. *Journal of Animal Science* 86:1861-1867.

Sun, G., and S. J. Hoff. 2010. Prediction of indoor climate and long-term air quality using the BTA-AQP model: Part II. Overall model evaluation and application. *Transactions of the American Society of Agricultural and Biological Engineers* 53:871-881.

Sun, G., and S. J. Hoff. 2011. Simulation of impacts of different animal management practices and geographic area on long-term air quality. *Transactions of the American Society of Agricultural and Biological Engineers* 54:1465-1477.

Sutton, A. L., J. A. Patterson, O. L. Adeola, B. A. Richert, D. T. Kelly, A. J. Heber, K. B. Kephart, R. Mumma, and E. Bogus. 1998. Reducing sulfur-containing odors through diet manipulation. Pp. 125-130 in *Animal Production Systems and the Environment*. Ames: Iowa State University.

Sutton, A. L., K. B. Kephart, M. W. A. Verstegen, T. T. Canh, and P. J. Hobbs. 1999. Potential for reduction of odorous compounds in swine manure through diet modification. *Journal of Animal Science* 77:430-439.

Underwood, E. J. 1981. *The Mineral Nutrition of Livestock*, 2nd Ed. Farnham Royal, UK: Commonwealth Agricultural Bureaux.

Unger, P. W., and T. C. Kaspar. 1994. Soil compaction and root growth: A review. *Agronomy Journal* 86:759-766.

van der Meulen, J., G. C. M. Bakker, J. G. M. Bakker, H. de Visser, A. W. Jongbloed, and H. Everts. 1997. Effect of resistant starch on net portal-drained viscera flux on glucose, volatile fatty acids, urea, and ammonia in growing pigs. *Journal of Animal Science* 75:2697-2704.

van Kempen, T. A. T. G., D. H. Baker, and E. van Heugten. 2003. Nitrogen losses in metabolism trials. *Journal of Animal Science* 81:2649-2650.

Varel, V. H. 2002. Livestock manure odor abatement with plant-derived oils and nitrogen conservation with urease inhibitors: A review. *Journal of Animal Science* 80(E. Suppl. 2):E1-E7.

Veenhuizen, M. F., G. C. Shruson, and E. M. Kohler. 1992. Effect of concentration and source of sulfate on nursery pig performance and health. *Journal of the American Veterinary Medical Association* 201:1203-1208.

Velthof, G. L., J. A. Nelemans, O. Oenema, and P. J. Kuikman. 2005. Gaseous nitrogen and carbon losses from pig manure derived from different diets. *Journal of Environmental Quality* 34:698-706.

Verstegen, M. W. A., W. H. Close, I. B. Start, and L. E. Mount. 1973. The effects of environmental temperature and plane of nutrition on heat loss, energy retention and deposition of protein and fat in groups of growing pigs. *British Journal of Nutrition* 30:21-35.

Veum, T. L., M. S. Carlson, C. W. Wu, D. W. Bollinger, and M. R. Ellersieck. 2004. Copper proteinate in weanling pig diets for enhancing growth performance and reducing fecal copper excretion compared with copper sulfate. *Journal of Animal Science* 82:1062-1070.

Vipperman, Jr., P. E., E. R. Peo, Jr., and P. J. Cunningham. 1974. Effect of dietary calcium and phosphorus level upon calcium, phosphorus and nitrogen balance in swine. *Journal of Animal Science* 38:758-765.

Vitosh, M. L., R. E. Lucas, and G. H. Silva. 1997. Long-term effects of fertilizer and manure on corn yield, soil carbon, and other soil chemical properties in Michigan. Pp. 129-169 in *Soil Organic Matter in Temperate Agroecosystems*. Boca Raton, FL: CRC Press.

von Pfeiffer, A. 1993. Protein reduced feeding concepts, a contribution to reduced ammoniac emissions in pig fattening. *Zuchtungskunde*

65:431-443.

Wagner, J. R., A. P. Schinckel, W. Chen, J. C. Forrest, and B. L. Coe. 1999. Analysis of body composition changes of swine during growth and development. *Journal of Animal Science* 77:1442-1466.

Wang, Y., J. H. Cho, Y. J. Chen, J. S. Yoo, Y. Huang, H. J. Kim, and I. H. Kim. 2009. The effect of probiotic BioPlus 2B® on growth performance, dry matter and nitrogen digestibility and slurry noxious gas emission in growing pigs. *Livestock Science* 120:35-42.

Whitney, M. H., R. Nicolai, and G. C. Shurson. 1999. Effects of feeding low sulfur starter diets on growth performance of early weaned pigs and odor, hydrogen sulfide, and ammonia emissions in nursery rooms. *Journal of Animal Science* 77(Suppl. 1):70 (Abstr.).

Williams, N. H., T. S. Stahly, and D. R. Zimmerman. 1997. Effect of level of chronic immune system activation on the growth and dietary lysine needs of pigs fed from 6 to 112 kg. *Journal of Animal Science* 75:2481-2496.

Willig, S., D. Losel, and R. Claus. 2005. Effects of resistant potato starch on odor emission from feces in swine production units. *Journal of Agricultural and Food Chemistry* 53:1173-1178.

Windisch, W., K. Schedle, C. Plitzner, and A. Kroismayr. 2008. Use of phytogentic products as feed additives for swine and poultry. *Journal of Animal Science* 86(E. Suppl.):E140-E148.

Wiseman, T. G., D. C. Mahan, and N. R. St-Pierre. 2009. Mineral composition of two genetic lines of barrows and gilts from twenty to one hundred twenty-five kilograms of body weight. *Journal of Animal Science* 87:2306-2314.

Wright, D., D. Eaton, L. Nielsen, F. Kuhurt, J. Kozier, J. Spinhime, and D. Parker. 2005. Multidimensional gas chromatography-olfactometry for identification and prioritization of malodors from confined animal feeding operations. *Journal of Agricultural and Food Chemistry* 53:8663-8671.

Yasuhara, A., K. Fuwa, and M. Jimbu. 1984. Identification of odorous compounds in fresh and rotten swine manure. *Agricultural and Biological Chemistry* 48:3001-3010.

Yen, J. T., B. J. Kerr, R. A. Easter, and A. M. Parkhurst. 2004. Difference in rates of net portal absorption between crystalline and protein-bound lysine and threonine in growing pigs fed once daily. *Journal of Animal Science* 82:1079-1090.

Younes, H., C. Remesy, S. Behr, and C. Demigne. 1997. Fermentable carbohydrate exerts a urea-lowering effect in normal and nephrectomized rats. *American Journal of Physiology* 272:G515-G521.

Zervas, S., and R. T. Zijlstra. 2002. Effects of dietary protein and fermentable fiber on nitrogen excretion patterns and plasma urea in grower pigs. *Journal of Animal Science* 80:3247-3256.

Zhu, J., and L. D. Jacobson. 1999. Correlating microbes to major odorous compounds in swine manure. *Journal of Environmental Quality* 28:737-744.

第15章 未来研究方向

引言

本书前言中曾提出，要介绍"未来需进一步研究的问题"，那么本章节将作详细探讨。那些目前研究较少或诸多领域未解的难题，以及当今主流的猪品种所对应的营养需要参数，亟待研究。同时，饲料原料组成也需更多的探索。其中一些未知信息将在提高猪生产效率中有着更重要的经济价值。

营养需要的评估方法

要评估营养需要，选择、分析和定量试验日粮中的可利用营养素，决定营养需要量的主要因素，严格和标准的试验设计，以及实验室流程是至关重要的。如果在试验中，既能测定猪表观生产性能数据，又能辅以代谢试验量化营养素的利用情况，这样的试验结果就非常有价值了。后者能为我们提供建立数学模型的参数，以便能预测动物在不同的营养素摄入量下不同的反应，并预测特定猪群的营养需要。将来建立营养需要量的数学模型必须包括测试模型预测的营养需要量与在明确说明试验条件下得到的营养需要量之间的吻合度。

群体内动物个体间的差异是影响评估最佳营养需要的关键因素之一。因此，在开展营养需要的试验时，需要评估群体内动物个体差异。同时，日粮营养水平对动物间变异的影响也是一个重要因素。

养分的利用和饲料采食量

需要进一步研究的领域有：整个机体和可食产品对氮/氨基酸的利用率有待进一步探讨；主要机体功能（如体蛋白和脂肪的沉积、奶中营养素的产出）对可消化营养素和能量的利用效率有待进一步探讨；估测机体用于维持功能的营养损失（如体蛋白周转中氨基酸的分解代谢以及导致的最低尿氮损失）。因为没有足够的数据证明这些因素如何影响营养物质和能量吸收后用于机体各种功能的效率，日粮因素（如日粮中可发酵纤维与抗营养因子水平及饲料加工）、动物本身（如生长阶段、品种、健康状况和应激）和代谢调控剂（免疫去势和β-受体激动剂）对营养物质和能量利用的影响需要进一步研究。

要精确定量肉品中蛋白质和脂肪沉积及动物对环境的排放，需要进一步研究机体各部分的化学组成（蛋白质、脂肪、水分、灰分、钙和磷）和物理组成（内脏和肌肉）之间的相关参数。而且动物在生长早期所摄入的营养素对后期生长、营养物质利用和体组成的影响需要更深入的研究。

妊娠期、哺乳期及生长早期的营养摄入对繁殖性能的影响是非常重要的。在哺乳母猪中，需要更深入地研究营养物质吸收后的利用规律，这将有助于了解能量、氨基酸和矿物质的摄入量对产奶量、奶组成的影响，以及与机体营养物质的沉积和动员之间的关系。同时，也要研究不同品种、胎次和体况的母猪的营养物质需要量之间的差异。

在准确预测不同品种、健康状况、日粮组成和环境变化（如温度和其他物理性因素）情况下动物的采食量方面，还需要更多的研究。

能量

在能量体系中，净能（net energy, NE）值可以通过消化能（digestible energy, DE）值或代谢能（metabolizable energy, ME）值，或通过日粮中全消化道营养素消化率系数（如干物质、氮、粗脂肪和无氮浸出物），或饲料原料营养组成来测算。当前饲料数据库中原料营养组成、全消化道营养消化率系数及多数原料的经验能量值等信息仍然较缺乏。因此，迫切需要检测这些原料的化学组成，测定营养素的生物学利用率或者通过回肠校正后的消化能和营养素消化率评估原料的生物学利用率，再通过标准化的试验程序评估其净能含量，然后以生长性能试验和体组成屠宰试验验证。此外，不同来源脂肪中脂肪酸组成、消化率和能值，以及脂肪形式（细胞内和浸提的）对能量消化率的影响并未被充分地研究。因此，为了更透彻地了解能量代谢，今后需要更多地探讨这些因素对能量的消化和利用。此外，用数学模型来估测非常规性饲料原料（如干法或湿法碾磨工艺的副产物）的净能值，将比传统代谢能体系和净能体系更具有优势。这是因为这些非常规饲料含有非常高的某营养成分，这些成分超过用来预测 DE/ME/NE 的营养浓度范围。

能量利用效率的表达通常认为是单一数值；但是针对某些能量利用效率不受日粮组分影响，而且不能通过当前预测方法（模型）推导的猪群的特定能量（例如，维持、能量用于脂类和蛋白沉积的效率）利用效率，则存在变异。未来的研究有必要考虑设定维持能量需要的变异值，并开发适宜的预测公式。

为获得更准确的能量需要的预测模型，需要进一步探讨育肥猪阶段能量的摄入量与蛋白质/脂肪沉积之间的关系；需要测定妊娠阶段胎儿/母猪组织中营养物质沉积/动员情况；需要测定哺乳阶段中不同繁殖能力的母猪母乳产量、母乳组成和仔猪生长性能。最后，关于免疫去势和外源促生长剂对能量的摄入，对维持和生长所需能量利用的数据仍较缺乏。

氨基酸

相对于任何其他营养物质，各种猪对氨基酸需要量的研究较多，赖氨酸需要量的确定已较合理，但是还有些相关信息不明确。今后需要进一步研究蛋白质沉积和体增重所需的可消化色氨酸、苏氨酸、缬氨酸、异亮氨酸和蛋氨酸的需要量。需要更多地研究某些因素（如猪健康状况和日粮可发酵纤维）对一些特殊氨基酸（如胱氨酸、色氨酸和苏氨酸）需要量的影响，因为这些氨基酸与动物免疫和其他非生产功能有关。而且，当日粮中添加了大量人工合成的氨基酸时，要进一步研究关于合成非必需氨基酸对氮的需要量。

在妊娠阶段，需要另评估母猪对赖氨酸、苏氨酸、色氨酸、蛋氨酸和精氨酸的需要量；测定妊娠后期（最后三分之一）各种蛋白质库中的氨基酸组成；记录整个妊娠期的体重变化；评估母猪从妊娠 30 d 至 110 d 中以上各种氨基酸沉积为氮的效率，还要评估各个不同妊娠阶段母猪乳腺、胎儿和胎盘生长的氨基酸组成及母体沉积蛋白质的氨基酸组成。这些信息对于建立需要量数学模型来说非常重要，建立的模型可以预测所有必需氨基酸、条件性必需氨基酸和总氮的需要量。

在哺乳阶段，需要更多地研究氨基酸参与合成乳汁的效率，以及乳汁中蛋白转化为仔猪增重的效率。赖氨酸、苏氨酸、蛋氨酸、色氨酸、缬氨酸和异亮氨酸的需要量也需要进一步研究。

关于后备公猪和种公猪的氨基酸及其他营养素需要量的数据非常少，相关的评估公猪活力的标准和指标尚有待确定。

矿物质

测定整个机体中钙和磷的存留量，并把它与相关变量（体蛋白沉积或某重要生理指标）联系起来，这些信息对于测定钙和磷需要量非常重要。相对于以前测定钙和磷需要量的试验来说，由于品种改良，日粮和原料的变异，各种猪用于维持生长和骨骼强度的钙和磷需要量需要重新评估。同样，针对后备母猪发育、经产母猪生产性能和使用年限等进展，钙和磷需要量也有待重新评估。

电解质平衡和钠、氯的需要量有待重新评估，尤其是在即将出栏猪的不同日粮（如不同纤维类型和含量）中和添加有植酸酶的情况下。在农业中，水的利用将越来越受到重视，过多的氯化钠会影响水的摄入和排泄，但从乳猪和小猪研究中发现，这两种矿物质还明显影响营养素的消化率。

锌是日粮中继钙、磷、钠和氯之后最易缺乏的一种矿物质。锌的功能与蛋白质的合成有关，随着育肥猪的肌肉量越来越高，关于猪生命周期内锌的需要量有待重新评估。

植酸酶是研究最多的酶制剂之一，除了提高磷的利用，还可以影响日粮中除磷以外的其他数种矿物质的利用。植酸酶在较高的添加水平下还有可能影响能量利用率，但这一推测有待更多的研究证实。

脂类

脂肪的营养价值已经被公认，而且脂肪在生命周转代谢中的利用情况也研究得比较透彻。过去10年，关于脂肪在猪饲料中的应用研究较多，主要是因为人们对有活性功能脂肪逐渐了解和高脂肪含量的农业副产品的出现。目前，仍需要进一步测定不同来源脂肪的校正回肠可消化率，尤其是针对保育猪；要研究不同来源脂肪对不同猪品种的净能值，研究抗氧化剂对脂肪的保护作用，研究 n-6 和 n-3 生物活性脂肪酸在育肥猪和母猪健康及繁殖过程中的生物学功能；研究脂肪的品质对饲料品质、猪群健康和猪肉品质的影响，以及夏季哺乳母猪热应激时脂肪的饲用价值。由于需要测定含有大量的多不饱和脂肪酸的油脂生物利用率，以前用来预测胴体中和日粮中碘值的公式需要重新建立。

维生素

许多关于维生素的研究结果已过时，无法采用这些数据来修改需要量，因为这些研究更多的是定性研究（对高剂量的反应），而不是定量研究（梯度添加时的反应）。最值得研究的领域是关于母猪繁殖对维生素的需要量，而且要更多地关注母猪整个生命周期（最少 2 个胎次，最好 4 个胎次）的营养，因为这影响着母猪一生的繁殖性能、健康和福利状况，而不是仅仅关注母猪一个胎次内的产仔数和窝重。从一些母猪的研究中，我们发现添加维生素 D 能改善骨骼健康，进而延长母猪使用年限，这表明维生素 D 在骨骼的钙和磷代谢中起着重要作用。因此，需要更多研究来确定维生素的最佳水平。几乎没有人研究母猪维生素 K 的需要量，母猪繁殖对烟酸、泛酸和硫胺素需要量的研究也缺乏。而且，从目前研究看，维生素 B_6 和维生素 B_{12} 在母猪营养研究中最有意义，所以有必要通过试验来确定添加时间和最佳添加水平。

饲料原料组成

本版《猪营养需要》回顾和总结了过去 10~20 年间最新的关于饲料原料组成的研究成果。在列出的这 122 种原料中，每一种都列出了所含有的 130 种营养素指标，但只有少数原料的营养素可消化率和生物学利用率数据较全。

正是这些缺乏的数据对于某些营养素更具经济价值。例如，农产品加工副产物中的维生素组成几乎是空白，目前，关于任何一种原料中维生素的最新数据也非常少，这样导致营养师在猪的日粮中添加多种维生素预混剂往往超过了猪对维生素的需要。由于原料中维生素检测费用较高，如果期望获得每种原料中的各种维生素数据，这样做可能非常不经济。

首先我们可以把工作重点放在分析有重要经济价值的营养素的校正回肠表观消化率或者全消化道消化率或生物学利用率；同时积累一些常用原料的校正氨基酸回肠消化率、校正全消化道磷消化率和全消化道钙表观消化率的数值和变异情况，这将非常有意义。

其他领域和重点

今后需要进一步研究日粮中氮、硫和纤维素（来源和水平）对氨、挥发性脂肪酸、温室效应气体释放的影响，同样也包括如何测定臭气方法的研究。当日粮中有复杂的碳水化合物时，要研究复合酶是否能提高碳水化合物的消化率，进而研究提高能量消化率的机理。对于饲料添加剂对肠道健康的影响及其导致猪生长性能的变化方面的研究比较缺乏。为更好地了解肠道微生物对整个动物机体生产性能的影响，不仅要研究微生物对肠道特定部位引起的变化或者某个特定的免疫反应，而且需要从整体角度来研究微生物的功效。今后，需要研究饲料加工工艺、饲料粉碎粒度和酶制剂之间的互作效应。

尽管看完本章内容，好像我们对猪的营养需要知之甚少，但相比其他动物，我们还是取得了很多成果。如果资源允许的情况下，我们可以在本章概述领域进行研究，但是在资源有限的情况下，研究重点应放在各种类型猪对氨基酸、钙和磷的需要量评估，尤其重要的是针对母猪的研究。

营养需要、饲料组成及其他表格

第 16 章 营养需要列表

引言

  本章表格主要介绍仔猪、生长猪和育肥猪、妊娠母猪和泌乳母猪，以及种公猪的营养需要。表中提供的数据是猪在相对无应激的环境下所获得的，包括环境的温度、与致病菌的接触和活动空间。表中给出了能量、氨基酸、氮、维生素、矿物质和亚油酸的需要量。氨基酸的需要量以回肠标准可消化氨基酸、回肠表观可消化氨基酸和总氨基酸三种形式表述，其中前两者适用于所有类型的日粮，后者仅适用于玉米-豆粕型日粮。同样地，磷的需要量也以全消化道标准可消化磷、全消化道表观可消化磷和总磷三种形式表述。所有营养素的需要量都包括了饲料原料所能提供的营养素的量。

  生长-育肥猪（25~135 kg 体重）、妊娠母猪和泌乳母猪氨基酸、氮、钙和磷的需要量由第 8 章的模型计算所得。断奶仔猪（5~25 kg 体重）的赖氨酸需要量主要是根据以往需要量研究的经验，其他氨基酸和氮的需要量通过第 8 章描述的模型计算得到。除此之外的营养成分需要量是根据以往研究经验和编委会对于猪日粮营养需要平均水平估算得到。

  表 16-1~表 16-4 给出了评估的 5~25 kg 体重断奶仔猪和 25~135 kg 体重生长-育肥猪的营养需要。表 16-1 中氨基酸需要量适用于公、母各半的高、中度瘦肉型猪（25~125 kg 体重，平均蛋白沉积量 135 g/d）。表 16-2 分别给出了高、中度瘦肉型的阉公猪、母猪及公猪在体重 50~75 kg、75~100 kg、100~135 kg 时的营养需要量。表 16-3 介绍了蛋白沉积率分别为 115 g/d、135 g/d 和 155 g/d（公、母各半）的生长育肥猪营养需要量。表 16-4 给出了针对促性腺激素释放激素（GnRH）免疫接种或者饲喂莱克多巴胺的未阉公猪，以及饲喂莱克多巴胺的阉公猪和小母猪的营养需要量。总消化道标准可消化的钙和磷、全消化道表观可消化钙和磷及总的钙和磷需要量也列在表 16-1~表 16-4 中。矿物质、维生素和亚油酸的需要量在表 16-5 中进行介绍。

  表 16-6 和表 16-7 介绍了不同配种体重、怀孕期增重和预计窝产仔数的妊娠母猪氨基酸需要量，以及不同分娩后体重、泌乳期体重变化和仔猪增重的泌乳母猪的氨基酸需要量。表 16-8 中列出了妊娠母猪和泌乳母猪每天矿物质、维生素和亚油酸的营养需要量及日粮中的添加浓度。表 16-9 列出了公猪的营养需要量。

  表中所列各种类型猪对氨基酸、氮、钙和磷的需要量仅为示例。读者可以根据模型，结合实际情况（如不同的瘦肉生长速度、采食量、日粮能量浓度、环境温度和饲养密度）确定适合的需要量。用模型进行推算断奶仔猪和生长-育肥猪的矿物质和维生素需要量所得结果可能会与表中所列情况略有差别，因为模型中用指数方程估算不同体重猪的需要量。同样，用模型推算的断奶仔猪的氨基酸需要量也可能会与表中所列情况略有差别。

  具有优良遗传背景和健康状况的高瘦肉率猪对于某种特定矿物质或维生素的需要量可能会比表中列出的平均水平要高，但是目前没有准确的信息预估更高生产性能的猪的定量需要。生长公猪或后备母猪的钙和磷需要量比表中所列 50~135 kg 体重猪的高大约 15%（第 7 章）。

  表中所列的需要量不包括任何有意识的增加使用量的情况。这些都是编委会推荐的最低需要量的最佳估测值。但是在生产实践中经常会在表中所列需要量的基础上加一个安全裕量，这样形成的营养水平通常称做营养素供给量。营养素供给量一般由营养专家设定，主要在考虑到饲料营养组成与其生物利用率的差异、饲料中的抑制剂或毒素、饲料的加工或混合不充分，以及储存过程中的部分损失等因素的基础上设定的。例如，饲料中微量元素和维生素的含量及生物学效价可能变异很大，而且经常未经测定。微量元素或维生素的实际添加量可能等于或高于估算的推荐量，而由饲料原料来源的这些营养素就作为安全剂量。由于这些因素，饲料产品标签上"符合或超过 NRC 营养需要"的描述本身并不一定是一个全价和平衡日粮的证明。关于营养约束与限制知识对于正确使用以下营养需要表格很重要。

表 16-1A 仔猪和生长育肥猪日粮钙、磷和氨基酸需要量（自由采食、日粮含 90% 干物质）[a]

指标	体重/kg						
	5~7	7~11	11~25	25~50	50~75	75~100	100~135
日粮净能[b]/（kcal/kg）	2448	2448	2412	2475	2475	2475	2475
日粮有效消化能[b]/（kcal/kg）	3542	3542	3490	3402	3402	3402	3402
日粮有效代谢能[b]/（kcal/kg）	3400	3400	3350	3300	3300	3300	3300
估算有效代谢能摄入量/（kcal/d）	904	1592	3033	4959	6989	8265	9196
估算采食量+浪费[c]/（g/d）	280	493	953	1582	2229	2636	2933
体增重/（g/d）	210	335	585	758	900	917	867
体蛋白沉积/（g/d）	—	—	—	128	147	141	122
钙和磷/%							
总钙	0.85	0.80	0.70	0.66	0.59	0.52	0.46
STTD 磷[d]	0.45	0.40	0.33	0.31	0.27	0.24	0.21
ATTD 磷[e,f]	0.41	0.36	0.29	0.26	0.23	0.21	0.18
总磷[f]	0.70	0.65	0.60	0.56	0.52	0.47	0.43
氨基酸[g,h]							
回肠标准可消化基础/%							
精氨酸	0.68	0.61	0.56	0.45	0.39	0.33	0.28
组氨酸	0.52	0.46	0.42	0.34	0.29	0.25	0.21
异亮氨酸	0.77	0.69	0.63	0.51	0.45	0.39	0.33
亮氨酸	1.50	1.35	1.23	0.99	0.85	0.74	0.62
赖氨酸	**1.50**	**1.35**	**1.23**	**0.98**	**0.85**	**0.73**	**0.61**
蛋氨酸	0.43	0.39	0.36	0.28	0.24	0.21	0.18
蛋氨酸+半胱氨酸	0.82	0.74	0.68	0.55	0.48	0.42	0.36
苯丙氨酸	0.88	0.79	0.72	0.59	0.51	0.44	0.37
苯丙氨酸+酪氨酸	1.38	1.25	1.14	0.92	0.80	0.69	0.58
苏氨酸	0.88	0.79	0.73	0.59	0.52	0.46	0.40
色氨酸	0.25	0.22	0.20	0.17	0.15	0.13	0.11
缬氨酸	0.95	0.86	0.78	0.64	0.55	0.48	0.41
总氮	3.10	2.80	2.56	2.11	1.84	1.61	1.37
回肠表观可消化基础/%							
精氨酸	0.64	0.57	0.51	0.41	0.34	0.29	0.24
组氨酸	0.49	0.44	0.40	0.32	0.27	0.24	0.19
异亮氨酸	0.74	0.66	0.60	0.49	0.42	0.36	0.30
亮氨酸	1.45	1.30	1.18	0.94	0.81	0.69	0.57
赖氨酸	**1.45**	**1.31**	**1.19**	**0.94**	**0.81**	**0.69**	**0.57**
蛋氨酸	0.42	0.38	0.34	0.27	0.23	0.20	0.16
蛋氨酸+半胱氨酸	0.79	0.71	0.65	0.53	0.46	0.40	0.33
苯丙氨酸	0.85	0.76	0.69	0.56	0.48	0.41	0.34
苯丙氨酸+酪氨酸	1.32	1.19	1.08	0.87	0.75	0.65	0.54
苏氨酸	0.81	0.73	0.67	0.54	0.47	0.41	0.35
色氨酸	0.23	0.21	0.19	0.16	0.13	0.12	0.10
缬氨酸	0.89	0.80	0.73	0.59	0.51	0.44	0.36
总氮	2.84	2.55	2.32	1.88	1.62	1.40	1.16

续表

指标	体重/kg						
	5~7	7~11	11~25	25~50	50~75	75~100	100~135
	总氨基酸和总氮基础/%						
精氨酸	0.75	0.68	0.62	0.50	0.44	0.38	0.32
组氨酸	0.58	0.53	0.48	0.39	0.34	0.30	0.25
异亮氨酸	0.88	0.79	0.73	0.59	0.52	0.45	0.39
亮氨酸	1.71	1.54	1.41	1.13	0.98	0.85	0.71
赖氨酸	**1.70**	**1.53**	**1.40**	**1.12**	**0.97**	**0.84**	**0.71**
蛋氨酸	0.49	0.44	0.40	0.32	0.28	0.25	0.21
蛋氨酸+半胱氨酸	0.96	0.87	0.79	0.65	0.57	0.50	0.43
苯丙氨酸	1.01	0.91	0.83	0.68	0.59	0.51	0.43
苯丙氨酸+酪氨酸	1.60	1.44	1.32	1.08	0.94	0.82	0.70
苏氨酸	1.05	0.95	0.87	0.72	0.64	0.56	0.49
色氨酸	0.28	0.25	0.23	0.19	0.17	0.15	0.13
缬氨酸	1.10	1.00	0.91	0.75	0.65	0.57	0.49
总氮	3.63	3.29	3.02	2.51	2.20	1.94	1.67

a 25~125 kg 体重阶段，公母 1:1 混养，高-中度瘦肉生长速度（每日平均体蛋白沉积 135 g）。

b 玉米-豆粕型日粮的能量含量。有效消化能和有效代谢能是根据体重 25 kg 左右猪的固定净能转化率计算得到。对于玉米-豆粕型日粮，有效消化能和有效代谢能与实际消化能和代谢能相似。最适宜的日粮能量水平根据原料的可利用率及当地原料成本不同而改变。当使用替代原料时，建议基于净能含量设计日粮配方，调整营养需要量确保营养素含量与净能比率保持不变。

c 假设有 5%的饲料浪费。

d 全消化道标准可消化。

e 全消化道表观可消化。

f 全消化道表观可消化磷和总磷需要量只适用于玉米-豆粕型的日粮，是依据全消化道标准可消化磷的需要量和玉米、去皮豆粕及磷酸氢钙的营养成分计算得出。假设日粮中添加 0.1%的赖氨酸盐酸盐及 3%的维生素和矿物质。玉米和豆粕添加水平以满足回肠标准可消化赖氨酸需要量为前提计算得到，而磷酸氢钙的添加量以满足全消化道标准可消化磷水平为前提计算得到。

g 5~25 kg 的猪赖氨酸需要量（%）根据经验计算所得，其他氨基酸的需要量都参照满足维持和生长需要的氨基酸与赖氨酸的比率计算所得。25~135 kg 猪的需要量由生长模型估算得到。

h 回肠表观可消化和总氨基酸需要量只适用于玉米-豆粕型日粮，主要根据回肠标准可消化氨基酸需要量和玉米、去皮豆粕为基础的日粮（日粮添加 0.1%的赖氨酸盐酸盐和 3%的维生素和矿物质）中氨基酸含量计算得出。对每一种氨基酸而言，日粮中的玉米和豆粕水平及营养素需要量都以满足回肠标准可消化氨基酸需要量为前提计算得到。

表 16-1B 仔猪和生长育肥猪每日钙、磷和氨基酸需要量（自由采食、日粮含90%干物质）[a]

指标	体重/kg						
	5~7	7~11	11~25	25~50	50~75	75~100	100~135
日粮净能[b]/（kcal/kg）	2448	2448	2412	2475	2475	2475	2475
日粮有效消化能[b]/（kcal/kg）	3542	3542	3490	3402	3402	3402	3402
日粮有效代谢能[b]/（kcal/kg）	3400	3400	3350	3300	3300	3300	3300
估算有效代谢能摄入量/（kcal/d）	904	1592	3033	4959	6989	8265	9196
估算采食量+浪费[c]/（g/d）	280	493	953	1582	2229	2636	2933
增重/（g/d）	210	335	585	758	900	917	867
蛋白沉积/（g/d）	—	—	—	128	147	141	122
钙和磷/（g/d）							
总钙	2.26	3.75	6.34	9.87	12.43	13.14	12.80
STTD 磷[d]	1.20	1.87	2.99	4.59	5.78	6.11	5.95
ATTD 磷[e,f]	1.09	1.69	2.63	3.90	4.89	5.15	4.98
总磷[f]	1.86	3.04	5.43	8.47	10.92	11.86	11.97
氨基酸[g,h]							
回肠标准可消化基础/（g/d）							
精氨酸	1.8	2.9	5.1	6.8	8.2	8.4	7.8
组氨酸	1.4	2.2	3.8	5.1	6.2	6.3	5.8
异亮氨酸	2.0	3.2	5.7	7.7	9.4	9.7	9.1
亮氨酸	4.0	6.3	11.1	14.9	18.1	18.5	17.2
赖氨酸	**4.0**	**6.3**	**11.1**	**14.8**	**17.9**	**18.3**	**16.9**
蛋氨酸	1.2	1.8	3.2	4.3	5.2	5.3	4.9
蛋氨酸+半胱氨酸	2.2	3.5	6.1	8.3	10.2	10.5	9.9
苯丙氨酸	2.3	3.7	6.6	8.8	10.8	11.0	10.3
苯丙氨酸+酪氨酸	3.7	5.8	10.3	13.8	16.9	17.3	16.3
苏氨酸	2.3	3.7	6.6	8.9	11.1	11.6	11.1
色氨酸	0.7	1.0	1.8	2.5	3.1	3.2	3.0
缬氨酸	2.5	4.0	7.1	9.6	11.7	12.1	11.4
总氮	8.3	13.1	23.2	31.7	39.0	40.2	38.1
回肠表观可消化基础/（g/d）							
精氨酸	1.7	2.7	4.7	6.1	7.3	7.3	6.6
组氨酸	1.3	2.1	3.6	4.8	5.8	5.9	5.4
异亮氨酸	2.0	3.1	5.5	7.3	8.9	9.0	8.4
亮氨酸	3.8	6.1	10.7	14.1	17.1	17.3	16.0
赖氨酸	**3.9**	**6.1**	**10.7**	**14.1**	**17.1**	**17.3**	**15.9**
蛋氨酸	1.1	1.8	3.1	4.1	4.9	5.0	4.6
蛋氨酸+半胱氨酸	2.1	3.3	5.9	7.9	9.7	9.9	9.3
苯丙氨酸	2.3	3.6	6.3	8.4	10.1	10.3	9.6
苯丙氨酸+酪氨酸	3.5	5.6	9.8	13.1	15.9	16.3	15.1
苏氨酸	2.2	3.4	6.0	8.1	9.9	10.3	9.7
色氨酸	0.6	1.0	1.7	2.3	2.8	2.9	2.7
缬氨酸	2.4	3.7	6.6	8.8	10.7	10.9	10.2
总氮	7.6	12.0	21.0	28.3	34.3	35.0	32.5

续表

指标	体重/kg						
	5~7	7~11	11~25	25~50	50~75	75~100	100~135
	总氨基酸和氮基础/(g/d)						
精氨酸	2.0	3.2	5.6	7.6	9.3	9.6	9.0
组氨酸	1.6	2.5	4.4	5.9	7.2	7.4	7.0
异亮氨酸	2.3	3.7	6.6	8.9	11.0	11.4	10.8
亮氨酸	4.6	7.2	12.7	17.0	20.8	21.3	19.9
赖氨酸	**4.5**	**7.2**	**12.6**	**16.9**	**20.6**	**21.1**	**19.7**
蛋氨酸	1.3	2.1	3.6	4.9	6.0	6.1	5.8
蛋氨酸+半胱氨酸	2.5	4.1	7.2	9.8	12.1	12.6	12.0
苯丙氨酸	2.7	4.3	7.5	10.2	12.5	12.8	12.1
苯丙氨酸+酪氨酸	4.2	6.8	12.0	16.2	20.0	20.6	19.5
苏氨酸	2.8	4.4	7.9	10.8	13.4	14.1	13.7
色氨酸	0.7	1.2	2.1	2.9	3.5	3.7	3.5
缬氨酸	2.9	4.7	8.3	11.3	13.9	14.4	13.6
总氮	9.7	15.4	27.3	37.7	46.6	48.6	46.5

a 5~135 kg体重阶段，公母1:1混养，高-中度瘦肉生长速度（平均每天体蛋白沉积135 g）。

b 玉米-豆粕型日粮的能量含量。有效消化能和有效代谢能是根据体重在25 kg左右猪的净能固定转化率计算得到。对于玉米-豆粕型日粮，有效消化能和有效代谢能接近实际消化能和代谢能。最适宜的日粮能量水平根据原料的可利用率与当地原料成本不同而改变。当使用替代原料时，建议基于净能含量设计日粮配方，调整营养需要量确保营养含量与净能比率保持不变。

c 假设有5%的饲料浪费。

d 全消化道标准可消化。

e 全消化道表观可消化。

f 全消化道表观可消化磷和总磷需要量只适用于玉米-豆粕型的日粮，是依据全消化道标准可消化磷的需要量和玉米、去皮豆粕以及磷酸氢钙的营养成分计算得出。假设日粮中添加0.1%的赖氨酸盐酸盐及3%的维生素和矿物质。玉米和豆粕添加水平以满足回肠标准可消化赖氨酸需要量为前提计算得到，而磷酸氢钙的添加量以满足全消化道标准可消化磷水平而变动。

g 5~25 kg猪的赖氨酸需要量（%）根据经验计算所得，其他氨基酸的需要量都参照满足维持和生长需要的氨基酸与赖氨酸的比率计算所得。25~135 kg猪的需要量由生长模型估算得到。

h 回肠表观可消化和总氨基酸需要量只适用于玉米-豆粕型日粮，主要根据回肠标准可消化氨基酸需要量和玉米、去皮豆粕为基础的日粮（日粮添加0.1%的赖氨酸盐酸盐和3%的维生素和矿物质）中氨基酸含量计算得出。对每一种氨基酸，日粮中的玉米和豆粕水平以及营养素需要量都以满足回肠标准可消化氨基酸需要量为前提而计算得到。

表 16-2A 不同体重阉公猪、母猪与公猪日粮钙、磷和氨基酸需要量（自由采食、日粮含90%干物质）

体重范围/kg	50~75			75~100			100~135		
性别	阉公猪	母猪	公猪	阉公猪	母猪	公猪	阉公猪	母猪	公猪
日粮净能 [a]/（kcal/kg）	2475	2475	2475	2475	2475	2475	2475	2475	2475
日粮有效消化能 [a]/（kcal/kg）	3402	3402	3402	3402	3402	3402	3402	3402	3402
日粮有效代谢能 [a]/（kcal/kg）	3300	3300	3300	3300	3300	3300	3300	3300	3300
估算有效代谢能摄入量/（kcal/d）	7282	6658	6466	8603	7913	7657	9495	8910	8633
估算采食量+浪费 [b]/（g/d）	2323	2124	2062	2744	2524	2442	3029	2842	2754
体增重/（g/d）	917	866	872	936	897	922	879	853	906
体蛋白沉积/（g/d）	145	145	150	139	144	156	119	126	148
钙和磷/%									
总钙	0.56	0.61	0.64	0.50	0.56	0.61	0.43	0.49	0.57
STTD 磷 [c]	0.26	0.28	0.30	0.23	0.26	0.29	0.20	0.23	0.27
ATTD 磷 [d,e]	0.22	0.24	0.25	0.19	0.22	0.24	0.17	0.19	0.23
总磷 [e]	0.50	0.53	0.55	0.45	0.49	0.53	0.41	0.45	0.50
氨基酸 [f,g]									
回肠标准可消化基础/%									
精氨酸	0.37	0.40	0.40	0.32	0.35	0.37	0.27	0.29	0.33
组氨酸	0.28	0.30	0.30	0.24	0.26	0.28	0.20	0.22	0.25
异亮氨酸	0.43	0.46	0.46	0.37	0.41	0.43	0.31	0.34	0.39
亮氨酸	0.82	0.88	0.89	0.70	0.78	0.83	0.59	0.65	0.74
赖氨酸	**0.81**	**0.87**	**0.88**	**0.69**	**0.77**	**0.82**	**0.58**	**0.64**	**0.73**
蛋氨酸	0.23	0.25	0.26	0.20	0.22	0.24	0.17	0.18	0.21
蛋氨酸+半胱氨酸	0.46	0.49	0.50	0.40	0.44	0.47	0.34	0.37	0.42
苯丙氨酸	0.49	0.52	0.53	0.42	0.46	0.49	0.35	0.39	0.44
苯丙氨酸+酪氨酸	0.76	0.82	0.83	0.66	0.73	0.77	0.56	0.61	0.69
苏氨酸	0.50	0.53	0.54	0.44	0.48	0.51	0.38	0.42	0.46
色氨酸	0.14	0.15	0.15	0.12	0.13	0.14	0.10	0.11	0.13
缬氨酸	0.53	0.57	0.58	0.46	0.51	0.54	0.39	0.43	0.48
总氮	1.76	1.88	1.91	1.54	1.69	1.78	1.31	1.43	1.61
回肠表观可消化基础/%									
精氨酸	0.33	0.35	0.36	0.28	0.31	0.33	0.22	0.25	0.29
组氨酸	0.26	0.28	0.29	0.22	0.25	0.26	0.18	0.20	0.24
异亮氨酸	0.40	0.43	0.44	0.34	0.38	0.40	0.29	0.32	0.36
亮氨酸	0.77	0.83	0.84	0.66	0.73	0.78	0.54	0.60	0.70
赖氨酸	**0.77**	**0.83**	**0.84**	**0.65**	**0.73**	**0.78**	**0.54**	**0.60**	**0.69**
蛋氨酸	0.22	0.24	0.24	0.19	0.21	0.22	0.16	0.17	0.20
蛋氨酸+半胱氨酸	0.44	0.47	0.47	0.38	0.42	0.44	0.32	0.35	0.40
苯丙氨酸	0.46	0.49	0.50	0.39	0.44	0.46	0.33	0.36	0.41
苯丙氨酸+酪氨酸	0.72	0.77	0.78	0.62	0.68	0.73	0.52	0.57	0.65
苏氨酸	0.45	0.48	0.49	0.39	0.43	0.45	0.33	0.36	0.41
色氨酸	0.13	0.14	0.14	0.11	0.12	0.13	0.09	0.10	0.12
缬氨酸	0.48	0.52	0.53	0.42	0.46	0.49	0.35	0.38	0.44
总氮	1.55	1.66	1.69	1.33	1.47	1.56	1.11	1.22	1.40

续表

体重范围/kg	50~75			75~100			100~135		
性别	阉公猪	母猪	公猪	阉公猪	母猪	公猪	阉公猪	母猪	公猪
	总氨基酸和总氮基础/%								
精氨酸	0.42	0.45	0.46	0.37	0.40	0.42	0.31	0.34	0.38
组氨酸	0.32	0.35	0.35	0.28	0.31	0.33	0.24	0.26	0.30
异亮氨酸	0.50	0.53	0.54	0.43	0.48	0.50	0.37	0.40	0.45
亮氨酸	0.94	1.00	1.02	0.81	0.89	0.95	0.68	0.75	0.85
赖氨酸	**0.93**	**0.99**	**1.01**	**0.80**	**0.89**	**0.94**	**0.67**	**0.74**	**0.85**
蛋氨酸	0.27	0.29	0.29	0.23	0.26	0.27	0.20	0.22	0.25
蛋氨酸+半胱氨酸	0.55	0.58	0.59	0.48	0.53	0.55	0.41	0.45	0.50
苯丙氨酸	0.56	0.60	0.61	0.49	0.54	0.57	0.41	0.45	0.51
苯丙氨酸+酪氨酸	0.90	0.96	0.98	0.79	0.86	0.91	0.67	0.73	0.83
苏氨酸	0.61	0.65	0.66	0.54	0.59	0.62	0.47	0.51	0.56
色氨酸	0.16	0.17	0.17	0.14	0.15	0.16	0.12	0.13	0.15
缬氨酸	0.63	0.67	0.68	0.55	0.60	0.63	0.47	0.51	0.57
总氮	2.12	2.25	2.28	1.86	2.03	2.13	1.60	1.74	1.94

a 玉米-豆粕型日粮的能量含量。有效消化能和有效代谢能是根据体重在 25 kg 左右猪的净能固定转化率计算所得。对于玉米-豆粕型日粮，有效消化能和有效代谢能接近实际消化能和代谢能。最适宜的日粮能量水平根据原料的可利用率及当地原料成本不同而改变。当使用替代原料时，建议按净能含量设计日粮配方，调整营养需要量确保营养含量与净能比率保持不变。

b 假设有 5%的饲料浪费。

c 全消化道标准可消化。

d 全消化道表观可消化。

e 全消化道表观可消化磷和总磷需要量只适用于玉米-豆粕型的日粮，是依据全消化道标准可消化磷的需要量和玉米、去皮豆粕及磷酸氢钙的营养成分计算而得。假设日粮中添加 0.1%的赖氨酸盐酸盐及 3%的维生素和矿物质。玉米和豆粕添加水平以满足回肠标准可消化赖氨酸需要量为前提计算得到，而磷酸氢钙的添加量以满足全消化道标准可消化磷水平为前提而计算得到。

f 需要量从生长模型估算得到。

g 回肠表观可消化和总氨基酸需要量只适用于玉米-豆粕型日粮，主要根据回肠标准可消化氨基酸需要量和玉米、去皮豆粕为基础的日粮（日粮添加 0.1%的赖氨酸盐酸盐和 3%的维生素和矿物质）中氨基酸含量计算所得。对每一种氨基酸而言，日粮中的玉米和豆粕水平以及营养素需要量都以满足回肠标准可消化氨基酸需要量为前提而计算得到。

表 16-2B 不同体重阉公猪、小母猪与公猪每日钙、磷和氨基酸需要量（自由采食、日粮含 90% 干物质）

体重范围/kg	50~75			75~100			100~135		
性别	阉公猪	小母猪	公猪	阉公猪	小母猪	公猪	阉公猪	小母猪	公猪
日粮净能[a]/(kcal/kg)	2475	2475	2475	2475	2475	2475	2475	2475	2475
日粮有效消化能[a]/(kcal/kg)	3402	3402	3402	3402	3402	3402	3402	3402	3402
日粮有效代谢能[a]/(kcal/kg)	3300	3300	3300	3300	3300	3300	3300	3300	3300
估算有效代谢能摄入量/(kcal/d)	7282	6658	6466	8603	7913	7657	9495	8910	8633
估算采食量+浪费[b]/(g/d)	2323	2124	2062	2744	2524	2442	3029	2842	2754
体增重/(g/d)	917	866	872	936	897	922	879	853	906
体蛋白沉积/(g/d)	145	145	150	139	144	156	119	126	148
钙和磷/(g/d)									
总钙	12.27	12.22	12.59	12.91	13.36	14.26	12.47	13.11	15.01
STTD 磷[c]	5.71	5.68	5.85	6.00	6.21	6.63	5.80	6.10	6.98
ATTD 磷[d,e]	4.81	4.81	4.97	5.04	5.25	5.63	4.84	5.12	5.91
总磷[e]	10.95	10.65	10.77	11.85	11.86	12.30	11.88	12.05	13.13
氨基酸[f,g]									
回肠标准可消化基础/(g/d)									
精氨酸	8.2	8.0	7.9	8.3	8.4	8.7	7.6	7.9	8.8
组氨酸	6.1	6.0	6.0	6.2	6.3	6.5	5.7	5.9	6.6
异亮氨酸	9.4	9.2	9.1	9.6	9.7	10.0	9.0	9.2	10.1
亮氨酸	18.0	17.7	17.5	18.3	18.7	19.2	16.9	17.5	19.4
赖氨酸	**17.8**	**17.5**	**17.3**	**18.1**	**18.4**	**19.0**	**16.6**	**17.2**	**19.2**
蛋氨酸	5.1	5.0	5.0	5.2	5.3	5.5	4.8	5.0	5.5
蛋氨酸+半胱氨酸	10.2	9.9	9.8	10.4	10.6	10.8	9.8	10.1	11.0
苯丙氨酸	10.7	10.5	10.4	10.9	11.1	11.4	10.2	10.5	11.5
苯丙氨酸+酪氨酸	16.8	16.5	16.3	17.2	17.5	17.9	16.0	16.5	18.2
苏氨酸	11.1	10.8	10.6	11.6	11.6	11.8	11.1	11.2	12.1
色氨酸	3.1	3.0	3.0	3.2	3.2	3.3	3.0	3.1	3.3
缬氨酸	11.7	11.4	11.3	12.0	12.2	12.4	11.2	11.5	12.6
总氮	38.9	37.9	37.4	40.1	40.4	41.3	37.6	38.6	42.1
回肠表观可消化基础/(g/d)									
精氨酸	7.2	7.1	7.1	7.2	7.4	7.7	6.4	6.7	7.6
组氨酸	5.8	5.7	5.6	5.8	6.0	6.1	5.3	5.5	6.2
异亮氨酸	8.8	8.6	8.5	8.9	9.1	9.4	8.2	8.5	9.4
亮氨酸	17.0	16.7	16.5	17.1	17.5	18.1	15.7	16.3	18.2
赖氨酸	**16.9**	**16.7**	**16.5**	**17.1**	**17.5**	**18.1**	**15.6**	**16.2**	**18.1**
蛋氨酸	4.9	4.8	4.8	4.9	5.1	5.2	4.5	4.7	5.2
蛋氨酸+半胱氨酸	9.6	9.4	9.3	9.8	10.0	10.2	9.2	9.5	10.4
苯丙氨酸	10.1	9.9	9.8	10.2	10.4	10.7	9.4	9.7	10.8
苯丙氨酸+酪氨酸	15.9	15.6	15.4	16.1	16.4	16.9	14.9	15.4	17.0
苏氨酸	9.9	9.7	9.5	10.2	10.3	10.5	9.6	9.8	10.7
色氨酸	2.8	2.8	2.7	2.9	2.9	3.0	2.7	2.8	3.0
缬氨酸	10.7	10.5	10.3	10.8	11.1	11.3	10.0	10.3	11.4
总氮	34.1	33.5	33.1	34.6	35.3	36.2	31.9	33.0	36.5

续表

体重范围/kg	50~75			75~100			100~135		
性别	阉公猪	小母猪	公猪	阉公猪	小母猪	公猪	阉公猪	小母猪	公猪
	总氨基酸和总氮基础/(g/d)								
精氨酸	9.3	9.0	8.9	9.5	9.6	9.8	8.9	9.1	10.0
组氨酸	7.2	7.0	6.9	7.3	7.4	7.6	6.9	7.1	7.8
异亮氨酸	11.0	10.7	10.5	11.3	11.4	11.6	10.6	10.9	11.9
亮氨酸	20.7	20.3	20.0	21.1	21.5	22.0	19.6	20.2	22.3
赖氨酸	**20.5**	**20.1**	**19.9**	**20.9**	**21.3**	**21.8**	**19.4**	**20.0**	**22.1**
蛋氨酸	5.9	5.8	5.8	6.1	6.2	6.3	5.7	5.9	6.4
蛋氨酸+半胱氨酸	12.1	11.8	11.6	12.5	12.6	12.9	11.9	12.1	13.2
苯丙氨酸	12.4	12.1	12.0	12.7	12.9	13.2	11.9	12.2	13.4
苯丙氨酸+酪氨酸	19.9	19.4	19.2	20.5	20.7	21.2	19.3	19.8	21.6
苏氨酸	13.5	13.1	12.8	14.2	14.1	14.3	13.6	13.8	14.8
色氨酸	3.5	3.4	3.4	3.7	3.7	3.7	3.5	3.5	3.8
缬氨酸	13.9	13.5	13.3	14.3	14.4	14.7	13.5	13.8	15.0
总氮	46.7	45.4	44.7	48.5	48.7	49.5	46.1	46.9	50.8

  a 玉米-豆粕型日粮的能量含量。有效消化能和有效代谢能是根据体重在 25 kg 以上猪的净能的固定转化率计算得到。对于玉米-豆粕型日粮，有效消化能和有效代谢能接近实际消化能和代谢能。最适宜的日粮能量水平根据原料的可利用率及当地原料成本不同而改变。当使用替代原料时，建议按净能含量设计日粮配方，调整营养需要量确保营养含量与净能比率保持不变。

  b 假设有 5%的饲料浪费。

  c 全消化道标准可消化。

  d 全消化道表观可消化。

  e 全消化道表观可消化磷和总磷需要量只适用于玉米-豆粕型的日粮，是依据全消化道标准可消化磷的需要量和玉米、去皮豆粕与磷酸氢钙的营养成分计算所得。假设日粮中添加 0.1%的赖氨酸盐酸盐及 3%的维生素和矿物质。玉米和豆粕添加水平以满足回肠标准可消化赖氨酸需要量而计算得到，而磷酸氢钙的添加量以满足全消化道标准可消化磷水平而变动。

  f 需要量从生长模型估算得到。

  g 回肠表观可消化和总氨基酸需要量只适用于玉米-豆粕型日粮，主要根据回肠标准可消化氨基酸需要量和玉米、去皮豆粕为基础的日粮（日粮添加 0.1%的赖氨酸盐酸盐和 3%的维生素和矿物质）中氨基酸含量计算所得。对每一种氨基酸，日粮中的玉米和豆粕水平以及营养素需要量以满足回肠标准可消化氨基酸需要量为前提计算得到。

表16-3A 不同平均体蛋白沉积水平、体重25~125 kg猪的日粮钙、磷和氨基酸需要量
（自由采食，日粮含90%干物质）

体重范围/kg	50~75			75~100			100~135		
平均蛋白沉积量/(g/d)	115	135	155	115	135	155	115	135	155
日粮净能 a/(kcal/kg)	2475	2475	2475	2475	2475	2475	2475	2475	2475
日粮有效消化能 a/(kcal/kg)	3402	3402	3402	3402	3402	3402	3402	3402	3402
日粮有效代谢能 a/(kcal/kg)	3300	3300	3300	3300	3300	3300	3300	3300	3300
估算有效代谢能摄入量/(kcal/d)	6980	6989	6982	8254	8265	8250	9204	9196	9197
估算采食量+浪费 b/(g/d)	2226	2229	2227	2633	2636	2632	2936	2933	2934
体增重/(g/d)	817	900	982	842	917	994	804	867	930
体蛋白沉积/(g/d)	125	147	168	121	141	163	104	122	140
钙和磷/%									
总钙	0.51	0.59	0.66	0.46	0.52	0.59	0.40	0.46	0.52
STTD 磷 c	0.24	0.27	0.31	0.21	0.24	0.28	0.19	0.21	0.24
ATTD 磷 d,e	0.20	0.23	0.26	0.18	0.21	0.23	0.15	0.18	0.20
总磷 e	0.47	0.52	0.56	0.43	0.47	0.52	0.39	0.43	0.46
氨基酸 f,g									
回肠标准可消化基础/%									
精氨酸	0.36	0.39	0.41	0.31	0.33	0.36	0.26	0.28	0.30
组氨酸	0.27	0.29	0.31	0.23	0.25	0.27	0.19	0.21	0.22
异亮氨酸	0.41	0.45	0.47	0.36	0.39	0.41	0.30	0.33	0.35
亮氨酸	0.79	0.85	0.91	0.68	0.74	0.79	0.57	0.62	0.66
赖氨酸	**0.78**	**0.85**	**0.91**	**0.67**	**0.73**	**0.78**	**0.56**	**0.61**	**0.65**
蛋氨酸	0.22	0.24	0.26	0.19	0.21	0.23	0.16	0.18	0.19
蛋氨酸+半胱氨酸	0.45	0.48	0.51	0.39	0.42	0.45	0.33	0.36	0.38
苯丙氨酸	0.47	0.51	0.54	0.41	0.44	0.47	0.34	0.37	0.39
苯丙氨酸+酪氨酸	0.74	0.80	0.85	0.64	0.69	0.74	0.54	0.58	0.62
苏氨酸	0.49	0.52	0.55	0.43	0.46	0.49	0.38	0.40	0.42
色氨酸	0.14	0.15	0.16	0.12	0.13	0.14	0.10	0.11	0.12
缬氨酸	0.51	0.55	0.59	0.45	0.48	0.51	0.38	0.41	0.43
总氮	1.71	1.84	1.95	1.50	1.61	1.71	1.28	1.37	1.44
回肠表观可消化基础/%									
精氨酸	0.31	0.34	0.37	0.26	0.29	0.32	0.21	0.24	0.26
组氨酸	0.25	0.27	0.29	0.22	0.24	0.25	0.18	0.19	0.21
异亮氨酸	0.38	0.42	0.45	0.33	0.36	0.39	0.28	0.30	0.32
亮氨酸	0.74	0.81	0.87	0.64	0.69	0.75	0.53	0.57	0.62
赖氨酸	**0.74**	**0.81**	**0.87**	**0.63**	**0.69**	**0.74**	**0.52**	**0.57**	**0.61**
蛋氨酸	0.21	0.23	0.25	0.18	0.20	0.22	0.15	0.16	0.18
蛋氨酸+半胱氨酸	0.42	0.46	0.49	0.37	0.40	0.42	0.31	0.33	0.35
苯丙氨酸	0.44	0.48	0.51	0.38	0.41	0.44	0.32	0.34	0.37
苯丙氨酸+酪氨酸	0.69	0.75	0.80	0.60	0.65	0.70	0.50	0.54	0.58
苏氨酸	0.44	0.47	0.50	0.38	0.41	0.43	0.33	0.35	0.37
色氨酸	0.12	0.13	0.14	0.11	0.12	0.13	0.09	0.10	0.10
缬氨酸	0.47	0.51	0.54	0.40	0.44	0.47	0.34	0.36	0.39
总氮	1.50	1.62	1.73	1.29	1.40	1.49	1.08	1.16	1.24

续表

体重范围/kg	50~75			75~100			100~135		
平均蛋白沉积量/(g/d)	115	135	155	115	135	155	115	135	155
总氨基酸和总氮基础/%									
精氨酸	0.41	0.44	0.47	0.35	0.38	0.41	0.30	0.32	0.34
组氨酸	0.31	0.34	0.36	0.27	0.30	0.32	0.23	0.25	0.27
异亮氨酸	0.48	0.52	0.55	0.42	0.45	0.48	0.36	0.39	0.41
亮氨酸	0.90	0.98	1.05	0.78	0.85	0.91	0.66	0.71	0.76
赖氨酸	**0.89**	**0.97**	**1.04**	**0.78**	**0.84**	**0.90**	**0.65**	**0.71**	**0.76**
蛋氨酸	0.26	0.28	0.30	0.23	0.25	0.26	0.19	0.21	0.22
蛋氨酸+半胱氨酸	0.53	0.57	0.61	0.47	0.50	0.53	0.40	0.43	0.45
苯丙氨酸	0.54	0.59	0.63	0.48	0.51	0.55	0.40	0.43	0.46
苯丙氨酸+酪氨酸	0.87	0.94	1.00	0.77	0.82	0.88	0.65	0.70	0.74
苏氨酸	0.60	0.64	0.67	0.53	0.56	0.59	0.47	0.49	0.51
色氨酸	0.16	0.17	0.18	0.14	0.15	0.16	0.12	0.13	0.13
缬氨酸	0.61	0.65	0.69	0.53	0.57	0.61	0.46	0.49	0.52
总氮	2.05	2.20	2.33	1.82	1.94	2.05	1.57	1.67	1.75

a 玉米-豆粕型日粮的能量含量。有效消化能和有效代谢能是根据体重在 25 kg 以上猪的净能的固定转化率计算得到。对于玉米-豆粕型日粮，有效消化能和有效代谢能接近实际消化能和代谢能。最适宜的日粮能量水平随当地饲料原料的可利用性和成本不同而变动。当使用替代原料时，建议按净能含量设计日粮配方，调整营养需要量确保营养含量与净能比率保持不变。

b 假设有 5%的饲料浪费。

c 全消化道标准可消化。

d 全消化道表观可消化。

e 全消化道表观可消化磷和总磷需要量只适用于玉米-豆粕型的日粮，是依据全消化道标准可消化磷的需要量和玉米、去皮豆粕与磷酸氢钙的营养成分计算所得。假设日粮中添加 0.1%的赖氨酸盐酸盐及 3%的维生素和矿物质。玉米和豆粕添加水平以满足回肠标准可消化赖氨酸需要量而计算得到，而磷酸氢钙的添加量以满足全消化道标准可消化磷水平而变。

f 需要量从生长模型估算得到。

g 回肠表观可消化和总氨基酸需要量只适用于玉米-豆粕型日粮，主要根据回肠标准可消化氨基酸需要量和玉米、去皮豆粕为基础的日粮（日粮添加 0.1%的赖氨酸盐酸盐和 3%的维生素和矿物质）中氨基酸含量计算所得。对于每一种氨基酸，日粮中的玉米和豆粕水平以及营养素需要量都以满足回肠标准可消化氨基酸需要量为前提而计算得到。

表 16-3B 不同平均体蛋白沉积水平下的体重在 25~125 kg 猪的每日钙、磷和氨基酸需要量
（自由采食，日粮含 90%干物质）

体重范围/kg	50~75			75~100			100~135		
平均蛋白沉积量/（g/d）	115	135	155	115	135	155	115	135	155
日粮净能[a]/（kcal/kg）	2475	2475	2475	2475	2475	2475	2475	2475	2475
日粮有效消化能[a]/（kcal/kg）	3402	3402	3402	3402	3402	3402	3402	3402	3402
日粮有效代谢能[a]/（kcal/kg）	3300	3300	3300	3300	3300	3300	3300	3300	3300
估算有效代谢能摄入量/（kcal/d）	6980	6989	6982	8254	8265	8250	9204	9196	9197
估算采食量+浪费[b]/（g/d）	2226	2229	2227	2633	2636	2632	2936	2933	2934
体增重/（g/d）	817	900	982	842	917	994	804	867	930
体蛋白沉积/（g/d）	125	147	168	121	141	163	104	122	140
钙和磷/（g/d）									
总钙	10.80	12.43	13.99	11.45	13.14	14.83	11.21	12.8	14.39
STTD磷[c]	5.02	5.78	6.51	5.33	6.11	6.90	5.21	5.95	6.69
ATTD磷[d,e]	4.21	4.89	5.54	4.44	5.15	5.85	4.32	4.98	5.64
总磷[e]	9.91	10.92	11.88	10.8	11.86	12.90	10.98	11.97	12.94
氨基酸[f,g]									
回肠标准可消化基础/（g/d）									
精氨酸	7.5	8.2	8.8	7.7	8.4	9	7.2	7.8	8.3
组氨酸	5.6	6.2	6.6	5.8	6.3	6.7	5.4	5.8	6.2
异亮氨酸	8.7	9.4	10.0	9.0	9.7	10.3	8.4	9.1	9.7
亮氨酸	16.6	18.1	19.3	17	18.5	19.8	15.9	17.2	18.4
赖氨酸	**16.4**	**17.9**	**19.2**	**16.8**	**18.3**	**19.6**	**15.6**	**16.9**	**18.1**
蛋氨酸	4.7	5.2	5.5	4.8	5.3	5.7	4.5	4.9	5.2
蛋氨酸+半胱氨酸	9.4	10.2	10.8	9.8	10.5	11.2	9.2	9.9	10.5
苯丙氨酸	9.9	10.8	11.5	10.2	11.0	11.8	9.6	10.3	11.0
苯丙氨酸+酪氨酸	15.6	16.9	18	16	17.3	18.5	15.1	16.3	17.3
苏氨酸	10.4	11.1	11.7	10.9	11.6	12.2	10.5	11.1	11.7
色氨酸	2.9	3.1	3.3	3.0	3.2	3.4	2.8	3	3.2
缬氨酸	10.9	11.7	12.5	11.2	12.1	12.9	10.6	11.4	12.1
总氮	36.2	39.0	41.3	37.5	40.3	42.7	35.7	38.1	40.3
回肠表观可消化基础/（g/d）									
精氨酸	6.6	7.3	7.8	6.6	7.3	7.9	6	6.6	7.1
组氨酸	5.3	5.8	6.2	5.4	5.9	6.3	5	5.4	5.8
异亮氨酸	8.1	8.9	9.5	8.3	9.0	9.7	7.7	8.4	8.9
亮氨酸	15.6	17.1	18.3	15.9	17.3	18.6	14.7	16	17.1
赖氨酸	**15.6**	**17.1**	**18.3**	**15.8**	**17.3**	**18.6**	**14.6**	**15.9**	**17.1**
蛋氨酸	4.5	4.9	5.3	4.6	5	5.4	4.2	4.6	4.9
蛋氨酸+半胱氨酸	8.9	9.7	10.3	9.2	9.9	10.6	8.7	9.3	9.9
苯丙氨酸	9.3	10.1	10.8	9.5	10.3	11.1	8.8	9.6	10.2
苯丙氨酸+酪氨酸	14.7	15.9	17	15	16.3	17.4	14.0	15.1	16.1
苏氨酸	9.2	9.9	10.5	9.6	10.3	10.9	9.1	9.7	10.3
色氨酸	2.6	2.8	3	2.7	2.9	3.1	2.5	2.7	2.9
缬氨酸	9.9	10.7	11.4	10.1	10.9	11.7	9.4	10.2	10.8
总氮	31.6	34.3	36.6	32.3	35	37.3	30.1	32.5	34.5

续表

体重范围/kg	50~75			75~100			100~135		
平均蛋白沉积量/（g/d）	115	135	155	115	135	155	115	135	155
	总氨基酸和总氮基础/（g/d）								
精氨酸	8.6	9.3	9.9	8.9	9.6	10.2	8.4	9	9.6
组氨酸	6.6	7.2	7.7	6.8	7.4	7.9	6.5	7	7.4
异亮氨酸	10.2	11	11.6	10.6	11.4	12.1	10.1	10.8	11.4
亮氨酸	19.1	20.8	22.2	19.6	21.3	22.8	18.4	19.9	21.2
赖氨酸	**18.9**	**20.6**	**22**	**19.4**	**21.1**	**22.6**	**18.2**	**19.7**	**21.1**
蛋氨酸	5.5	6	6.4	5.7	6.1	6.6	5.3	5.8	6.2
蛋氨酸+半胱氨酸	11.2	12.1	12.8	11.7	12.6	13.3	11.2	12	12.7
苯丙氨酸	11.5	12.5	13.3	11.9	12.8	13.7	11.2	12.1	12.8
苯丙氨酸+酪氨酸	18.5	20	21.2	19.2	20.6	21.9	18.2	19.5	20.7
苏氨酸	12.6	13.4	14.1	13.3	14.1	14.9	13	13.7	14.3
色氨酸	3.3	3.5	3.7	3.4	3.7	3.9	3.3	3.5	3.7
缬氨酸	12.9	13.9	14.7	13.4	14.4	15.2	12.7	13.6	14.4
总氮	43.5	46.6	49.2	45.5	48.6	51.3	43.8	46.5	48.9

  a 玉米-豆粕型日粮的能量含量。有效消化能和有效代谢能是根据体重在 25 kg 左右猪的净能的固定转化率计算得到。对于玉米-豆粕型日粮，有效消化能和有效代谢能接近实际消化能和代谢能。最适宜的日粮能量水平根据当地饲料原料的可利用性和成本不同而改变。当使用替代原料时，建议按净能含量设计日粮配方，调整营养需要量确保营养含量与净能比率保持不变。

  b 假设有 5%的饲料浪费。

  c 全消化道标准可消化。

  d 全消化道表观可消化。

  e 全消化道表观可消化磷和总磷需要量只适用于玉米-豆粕型的日粮，是依据全消化道标准可消化磷的需要量和玉米、去皮豆粕与磷酸氢钙的营养成分计算所得。假设日粮中添加 0.1%的赖氨酸盐酸盐及 3%的维生素和矿物质。玉米和豆粕添加水平以满足回肠标准可消化赖氨酸需要量而计算得到，而磷酸氢钙的添加量以满足全消化道标准可消化磷水平而变动。

  f 需要量从生长模型估算得到。

  g 回肠表观可消化和总氨基酸需要量只适用于玉米-豆粕型日粮，主要根据回肠标准可消化氨基酸需要量和玉米、去皮豆粕为基础的日粮（日粮添加 0.1%的赖氨酸盐酸盐和 3%的维生素和矿物质）中氨基酸含量计算所得。对于每一种氨基酸，日粮中的玉米和豆粕水平以及营养素需要量都以满足回肠标准可消化氨基酸需要量为前提而计算得到。

表16-4A 针对促性腺激素释放激素（GnRH）免疫接种或者饲喂莱克多巴胺的未阉公猪，以及饲喂莱克多巴胺的阉公猪和小母猪的钙、磷与氨基酸营养需要量（自由采食，日粮含90%干物质）

体重范围/kg	免疫的公猪	公猪（5 ppm莱克多巴胺）	公猪（10 ppm莱克多巴胺）	阉公猪（5 ppm莱克多巴胺）	阉公猪（10 ppm莱克多巴胺）
	105~135	115~135	115~135	115~135	115~135
日粮净能[a]/（kcal/kg）	2 475	2 475	2475	2 475	2 475
日粮有效消化能[a]/（kcal/kg）	3 402	3 402	3 402	3 402	3 402
日粮有效代谢能[a]/（kcal/kg）	3 300	3 300	3 300	3 300	3 300
估算有效代谢能摄入量/（kcal/d）	10 203	8 722	8 647	9 262	9 181
估算采食量+浪费[b]/（g/d）	3 255	2 782	2 758	2 954	2 929
增重/（g/d）	1 023	1 029	1 064	957	983
蛋白沉积/（g/d）	137	187	199	152	161
钙和磷/%					
总钙	0.47	0.71	0.75	0.56	0.59
STTD磷[c]	0.22	0.33	0.35	0.26	0.27
ATTD磷[d,e]	0.18	0.28	0.30	0.22	0.23
总磷[e]	0.43	0.59	0.62	0.49	0.52
氨基酸[f,g]					
回肠标准可消化基础/%					
精氨酸	0.27	0.42	0.45	0.34	0.37
组氨酸	0.20	0.31	0.33	0.25	0.27
异亮氨酸	0.32	0.51	0.54	0.42	0.45
亮氨酸	0.60	0.93	1.00	0.77	0.82
赖氨酸	**0.59**	**0.94**	**1.01**	**0.77**	**0.83**
蛋氨酸	0.17	0.28	0.30	0.23	0.24
蛋氨酸+半胱氨酸	0.35	0.54	0.58	0.45	0.48
苯丙氨酸	0.36	0.56	0.60	0.46	0.49
苯丙氨酸+酪氨酸	0.57	0.88	0.95	0.73	0.78
苏氨酸	0.39	0.57	0.61	0.49	0.52
色氨酸	0.11	0.17	0.18	0.14	0.15
缬氨酸	0.40	0.61	0.65	0.50	0.54
总氮	1.33	1.96	2.08	1.64	1.74
回肠表观可消化基础/%					
精氨酸	0.23	0.37	0.40	0.30	0.32
组氨酸	0.19	0.29	0.31	0.24	0.25
异亮氨酸	0.29	0.48	0.52	0.39	0.42
亮氨酸	0.56	0.89	0.95	0.72	0.77
赖氨酸	**0.56**	**0.90**	**0.97**	**0.73**	**0.79**
蛋氨酸	0.16	0.27	0.29	0.21	0.23
蛋氨酸+半胱氨酸	0.32	0.51	0.55	0.42	0.45
苯丙氨酸	0.33	0.53	0.57	0.43	0.46
苯丙氨酸+酪氨酸	0.53	0.84	0.90	0.68	0.73
苏氨酸	0.34	0.52	0.55	0.43	0.46
色氨酸	0.09	0.15	0.16	0.13	0.14
缬氨酸	0.35	0.56	0.60	0.46	0.49
总氮	1.13	1.74	1.86	1.42	1.52

续表

体重范围/kg	免疫的公猪	公猪（5ppm莱克多巴胺）	公猪（10ppm莱克多巴胺）	阉公猪（5ppm莱克多巴胺）	阉公猪（10ppm莱克多巴胺）
	105~135	115~135	115~135	115~135	115~135
	总氨基酸和总氮基础/%				
精氨酸	0.32	0.47	0.50	0.39	0.41
组氨酸	0.24	0.36	0.38	0.30	0.32
异亮氨酸	0.38	0.59	0.63	0.49	0.52
亮氨酸	0.70	1.07	1.15	0.88	0.94
赖氨酸	**0.69**	**1.08**	**1.16**	**0.89**	**0.95**
蛋氨酸	0.20	0.32	0.34	0.26	0.28
蛋氨酸+半胱氨酸	0.42	0.63	0.68	0.53	0.57
苯丙氨酸	0.42	0.64	0.69	0.53	0.57
苯丙氨酸+酪氨酸	0.68	1.04	1.11	0.86	0.92
苏氨酸	0.48	0.69	0.74	0.59	0.63
色氨酸	0.12	0.19	0.20	0.16	0.17
缬氨酸	0.48	0.72	0.76	0.60	0.64
总氮	1.62	2.34	2.48	1.97	2.08

a 玉米-豆粕型日粮的能量含量。有效消化能和有效代谢能是根据体重在 25 kg 以上猪的净能固定转化率计算得到。对于玉米-豆粕型日粮，有效消化能和有效代谢能接近实际消化能和代谢能。最适宜的日粮能量水平随当地饲料原料的可利用性和成本不同而变动。当使用替代原料时，建议按净能含量设计日粮配方，调整营养需要量确保营养含量与净能比率保持不变。

b 假设有 5%的饲料浪费。

c 全消化道标准可消化。

d 全消化道表观可消化。

e 全消化道表观可消化磷和总磷需要量只适用于玉米-豆粕型的日粮，是依据全消化道标准可消化磷的需要量和玉米、去皮豆粕以及磷酸氢钙的营养成分计算所得。假设日粮中添加 0.1%的赖氨酸盐酸盐及 3%的维生素和矿物质。玉米和豆粕的添加水平以满足标准回肠可消化赖氨酸需要量而计算得到，而磷酸氢钙的添加量以满足全消化道标准可消化磷水平而有所变动。

f 需要量从生长模型估算得到。

g 回肠表观可消化与总氨基酸需要量只适用于玉米-豆粕型日粮，主要根据回肠标准可消化氨基酸需要量和玉米、去皮豆粕为基础的日粮（日粮添加 0.1%的赖氨酸盐酸盐和 3%的维生素和矿物质）中氨基酸含量计算所得。对于每一种氨基酸，日粮中的玉米和豆粕水平以及营养素需要量都以满足回肠标准可消化氨基酸需要量为前提而计算得到。

表16-4B 针对促性腺激素释放激素（GnRH）免疫接种或者饲喂莱克多巴胺的未阉公猪，以及饲喂莱克多巴胺的阉公猪和小母猪每日的钙、磷与氨基酸每日需要量（自由采食，日粮含90%干物质）

体重范围/kg	免疫公猪	公猪（5 ppm 莱克多巴胺）	公猪（10 ppm 莱克多巴胺）	阉公猪和小母猪（5 ppm莱克多巴胺）	阉公猪和小母猪（10 ppm莱克多巴胺）
	105~135	115~135	115~135	115~135	115~135
日粮净能[a]/（kcal/kg）	2 475	2 475	2 475	2 475	2 475
日粮有效消化能[a]/（kcal/kg）	3 402	3 402	3 402	3 402	3 402
日粮有效代谢能[a]/（kcal/kg）	3 300	3 300	3 300	3 300	3 300
估算有效代谢能摄入量/（kcal/d）	10 203	8 722	8 647	9 262	9 181
估算采食量+浪费[b]/（g/d）	3 255	2 782	2 758	2 954	2 929
增重/（g/d）	1 023	1 029	1 064	957	983
蛋白沉积/（g/d）	137	187	199	152	161
钙和磷/（g/d）					
总钙	14.44	18.73	19.76	15.60	16.43
STTD磷[c]	6.72	8.71	9.19	7.26	7.64
ATTD磷[d,e]	5.63	7.44	7.86	6.13	6.47
总磷[e]	13.38	15.61	16.27	13.85	14.38
氨基酸[f,g]					
回肠标准可消化基础/（g/d）					
精氨酸	8.4	11.0	11.7	9.6	10.2
组氨酸	6.3	8.2	8.6	7.1	7.5
异亮氨酸	9.8	13.4	14.3	11.7	12.5
亮氨酸	18.6	24.7	26.3	21.5	22.8
赖氨酸	**18.3**	**24.9**	**26.5**	**21.6**	**23.0**
蛋氨酸	5.3	7.3	7.8	6.3	6.8
蛋氨酸+半胱氨酸	10.7	14.2	15.1	12.5	13.3
苯丙氨酸	11.2	14.7	15.6	12.9	13.6
苯丙氨酸+酪氨酸	17.6	23.3	24.8	20.4	21.7
苏氨酸	12.0	15.2	16.0	13.6	14.4
色氨酸	3.3	4.4	4.7	3.9	4.1
缬氨酸	12.3	16.1	17.1	14.1	15.0
总氮	41.1	51.8	54.6	45.9	48.3
回肠表观可消化基础/（g/d）					
精氨酸	7.1	9.9	10.5	8.4	9.0
组氨酸	5.9	7.7	8.2	6.7	7.0
异亮氨酸	9.0	12.6	13.5	11.0	11.7
亮氨酸	17.3	23.4	25.0	20.2	21.5
赖氨酸	**17.2**	**23.8**	**25.4**	**20.5**	**21.9**
蛋氨酸	5	7.0	7.5	6.0	6.5
蛋氨酸+半胱氨酸	10.0	13.5	14.4	11.8	12.6
苯丙氨酸	10.3	13.9	14.8	12.1	12.8
苯丙氨酸+酪氨酸	16.3	22.1	23.5	19.2	20.4
苏氨酸	10.5	13.7	14.5	12.1	12.9
色氨酸	2.9	4.0	4.3	3.5	3.8
缬氨酸	10.9	14.8	15.7	12.8	13.6
总氮	34.9	45.9	48.7	40.0	42.3

续表

体重范围/kg	免疫公猪	公猪（5 ppm莱克多巴胺）	公猪（10 ppm莱克多巴胺）	阉公猪和小母猪（5 ppm莱克多巴胺）	阉公猪和小母猪（10 ppm莱克多巴胺）
	105~135	115~135	115~135	115~135	115~135
	总氨基酸和总氮基础/（g/d）				
精氨酸	9.8	12.4	13.1	11.0	11.5
组氨酸	7.6	9.5	10.0	8.4	8.8
异亮氨酸	11.6	15.5	16.5	13.7	14.5
亮氨酸	21.5	18.3	30.1	24.7	26.2
赖氨酸	**21.4**	**28.5**	**30.3**	**24.8**	**26.4**
蛋氨酸	6.3	8.4	8.9	7.3	7.8
蛋氨酸+半胱氨酸	12.9	16.8	17.8	14.9	15.8
苯丙氨酸	13.1	17.0	18.0	14.9	15.8
苯丙氨酸+酪氨酸	21.1	27.4	29.1	24.2	25.6
苏氨酸	14.8	18.3	19.3	16.6	17.4
色氨酸	3.8	4.9	5.3	4.4	4.7
缬氨酸	14.7	18.9	20.0	16.8	17.7
总氮	50.2	61.8	65.0	55.3	58.0

a 玉米-豆粕型日粮的能量含量。有效消化能和有效代谢能是根据体重在 25 kg 以上猪的净能的固定转化率计算得到。对于玉米-豆粕型日粮，有效消化能和有效代谢能接近实际消化能和代谢能。最适宜的日粮能量水平根据原料的可利用性与当地原料成本不同而变动。当使用替代原料时，建议按净能含量设计日粮配方，调整营养需要量确保营养含量与净能比率保持不变。

b 假设有 5%的饲料浪费。

c 全消化道标准可消化。

d 全消化道表观可消化。

e 全消化道表观可消化磷和总磷需要量只适用于玉米-豆粕型的日粮，是依据全消化道标准可消化磷的需要量和玉米、去皮豆粕以及磷酸氢钙的营养成分计算所得。假设日粮中添加 0.1%的赖氨酸盐酸盐及 3%的维生素和矿物质。玉米和豆粕添加水平以满足标准回肠可消化赖氨酸需要量而计算所得，而磷酸氢钙的添加量以满足全消化道标准可消化磷水平而有所变动。

f 需要量从生长模型估算得到。

g 回肠表观可消化与总氨基酸需要量只适用于玉米-豆粕型日粮，主要根据回肠标准可消化氨基酸需要量和玉米、去皮豆粕为基础的日粮（日粮添加 0.1%的赖氨酸盐酸盐和 3%的维生素和矿物质）中氨基酸含量计算所得。对于每一种氨基酸，日粮中的玉米和豆粕水平以及营养素需要量都以满足回肠标准可消化氨基酸需要量为前提而计算得到。

猪营养需要

表 16-5A　仔猪和生长育肥猪日粮矿物质、维生素和脂肪酸需要量（自由采食，日粮含 90%干物质）

指标	体重范围/kg						
	5~7	7~11	11~25	25~50	50~75	75~100	100~135
日粮净能[a]/（kcal/kg）	2448	2448	2412	2475	2475	2475	2475
日粮有效消化能[a]/（kcal/kg）	3542	3542	3490	3402	3402	3402	3402
日粮有效代谢能[a]/（kcal/kg）	3400	3400	3350	3300	3300	3300	3300
估算有效代谢能摄入量/（kcal/d）	904	1592	3033	4959	6989	8265	9196
估算采食量+浪费[b]/（g/d）	280	493	953	1582	2229	2636	2933
增重/（g/d）	210	335	585	758	900	917	867
蛋白沉积/（g/d）	—	—	—	128	147	141	122
	需要量（%或每 kg 日粮需要量）						
矿物质							
钠/%	0.40	0.35	0.28	0.10	0.10	0.10	0.10
氯/%	0.50	0.45	0.32	0.08	0.08	0.08	0.08
镁/%	0.04	0.04	0.04	0.04	0.04	0.04	0.04
钾/%	0.30	0.28	0.26	0.23	0.19	0.17	0.17
铜/（mg/kg）	6.00	6.00	5.00	4.00	3.50	3.00	3.00
碘/（mg/kg）	0.14	0.14	0.14	0.14	0.14	0.14	0.14
铁/（mg/kg）	100	100	100	60	50	40	40
锰/（mg/kg）	4.00	4.00	3.00	2.00	2.00	2.00	2.00
硒/（mg/kg）	0.30	0.30	0.25	0.20	0.15	0.15	0.15
锌/（mg/kg）	100	100	80	60	50	50	50
维生素							
维生素 A[c]/（IU/kg）	2200	2200	1750	1300	1300	1300	1300
维生素 D[d]/（IU/kg）	220	220	200	150	150	150	150
维生素 E[e]/（IU/kg）	16	16	11	11	11	11	11
维生素 K/（mg/kg）	0.50	0.50	0.50	0.50	0.50	0.50	0.50
生物素/（mg/kg）	0.08	0.05	0.05	0.05	0.05	0.05	0.05
胆碱/（g/kg）	0.60	0.50	0.40	0.30	0.30	0.30	0.30
叶酸/（mg/kg）	0.30	0.30	0.30	0.30	0.30	0.30	0.30
烟酸，可利用[f]/（mg/kg）	30.00	30.00	30.00	30.00	30.00	30.00	30.00
泛酸/（mg/kg）	12.00	10.00	9.00	8.00	7.00	7.00	7.00
核黄素/（mg/kg）	4.00	3.50	3.00	2.50	2.00	2.00	2.00
硫胺素/（mg/kg）	1.50	1.00	1.00	1.00	1.00	1.00	1.00
维生素 B_6/（mg/kg）	7.00	7.00	3.00	1.00	1.00	1.00	1.00
维生素 B_{12}/（μg/kg）	20.00	17.50	15.00	10.00	5.00	5.00	5.00
亚油酸/%	0.10	0.10	0.10	0.10	0.10	0.10	0.10

　　a 玉米-豆粕型日粮的能量含量。有效消化能和有效代谢能是根据体重在 25 kg 左右猪的净能的固定转化率计算得到。对于玉米-豆粕型日粮，有效消化能和有效代谢能接近实际消化能和代谢能。最适宜的日粮能量水平根据原料的可利用性与当地原料成本不同而改变。当使用替代原料时，建议按净能含量设计日粮配方，调整营养需要量确保营养含量与净能比率保持不变。

　　b 假设有 5%的饲料浪费。

　　c 1 IU 维生素 A=0.30 μg 视黄醇或者 0.344 μg 视黄醇乙酸酯。维生素 A 的活性（视黄醇当量）也由 β-胡萝卜素提供（见维生素章节）。

　　d 1 IU 维生素 D_2 或 D_3=0.025 μg。

　　e 1 IU 维生素 E=0.67 mg D-α-生育酚或者 1 mg DL-α-醋酸生育酚。近来在猪上的研究表明，天然的和人工合成的 α-醋酸生育酚有实质性的差别（见维生素章节）。

　　f 玉米、高粱、小麦和大麦中的烟酸不可被利用。同样，这些谷物来源的副产物中的烟酸利用率也很低，除非对这些副产物进行了湿磨工艺的发酵。

表 16-5B　仔猪和生长育肥猪每日矿物质、维生素和脂肪酸需要量（自由采食，日粮含90%干物质）

指标	体重范围/kg						
	5~7	7~11	11~25	25~50	50~75	75~100	100~135
日粮净能[a]/（kcal/kg）	2448	2448	2412	2475	2475	2475	2475
日粮有效消化能[a]/（kcal/kg）	3542	3542	3490	3402	3402	3402	3402
日粮有效代谢能[a]/（kcal/kg）	3400	3400	3350	3300	3300	3300	3300
估算有效代谢能摄入量/（kcal/d）	904	1592	3033	4959	6989	8265	9196
估算采食量+浪费[b]/（g/d）	280	493	953	1582	2229	2636	2933
体增重/（g/d）	210	335	585	758	900	917	867
体蛋白沉积/（g/d）	—	—	—	128	147	141	122
	每日需要量						
矿物质							
钠/g	1.06	1.64	2.53	1.50	2.12	2.51	2.79
氯/g	1.33	2.11	2.90	1.20	1.69	2.00	2.23
镁/g	0.11	0.19	0.36	0.60	0.85	1.00	1.11
钾/g	0.80	1.31	2.35	3.46	4.02	4.26	4.74
铜/mg	1.60	2.81	4.53	6.01	7.41	7.52	8.36
碘/mg	0.04	0.07	0.13	0.21	0.30	0.35	0.39
铁/mg	26.6	46.8	90.5	90.2	105.9	100.2	111.5
锰/mg	1.06	1.87	2.72	3.01	4.24	5.01	5.57
硒/mg	0.08	0.14	0.23	0.30	0.32	0.38	0.42
锌/mg	26.6	46.8	72.4	90.2	105.9	125.3	139.4
维生素							
维生素 A[c]/IU	585	1030	1584	1954	2753	3257	3623
维生素 D[d]/I	59	103	181	225	318	376	418
维生素 E[e]/I	4.3	7.5	10.0	16.5	23.3	27.6	30.7
维生素 K（甲萘醌）/mg	0.13	0.23	0.45	0.75	1.06	1.25	1.39
生物素/mg	0.02	0.02	0.05	0.08	0.11	0.13	0.14
胆碱/g	0.16	0.23	0.36	0.45	0.64	0.75	0.84
叶酸/mg	0.08	0.14	0.27	0.45	0.64	0.75	0.84
烟酸，可利用[f]/mg	7.98	14.05	27.16	45.09	63.53	75.15	83.62
泛酸/mg	3.19	4.68	8.15	12.02	14.82	17.54	19.51
核黄素/mg	1.06	1.64	2.72	3.76	4.24	5.01	5.57
硫胺素/mg	0.40	0.47	0.91	1.50	2.12	2.51	2.79
维生素 B_6/mg	1.86	3.28	2.72	1.50	2.12	2.51	2.79
维生素 B_{12}/μg	5.32	8.20	13.58	15.03	10.59	12.53	13.94
亚油酸/g	0.3	0.5	0.9	1.5	2.1	2.5	2.8

　　a 玉米-豆粕型日粮的能量含量。有效消化能和有效代谢能是根据体重在25 kg左右猪的净能的固定转化率计算得到。对于玉米-豆粕型日粮，有效消化能和有效代谢能接近实际消化能和代谢能。最适宜的日粮能量水平根据原料的可利用率及当地原料成本不同而改变。当使用替代原料时，建议以净能含量为基础设计日粮配方，调整营养需要量确保营养含量与净能比率保持不变。

　　b 假设有5%的饲料浪费。

　　c 1 IU 维生素 A=0.30 μg 视黄醇或者 0.344 μg 视黄醇乙酸酯。维生素A的活性（视黄醇当量）也由β-胡萝卜素提供（见维生素章节）。

　　d 1 IU 维生素 D_2 或 D_3=0.025 μg。

　　e 1 IU 维生素 E=0.67 mg D-α-生育酚或者 1 mg DL-α-醋酸生育酚。近来在猪上的研究表明，天然的和人工合成的α-醋酸生育酚有实质性的差别（见维生素章节）。

　　f 玉米、高粱、小麦和大麦中的烟酸不可被利用。同样，这些谷物来源的副产物中的烟酸利用率也很低，除非对这些副产物进行了湿磨工艺的发酵。

表 16-6A 怀孕母猪日粮钙、磷和氨基酸需要量（按 90%干物质计算）[a]

胎次（配种时体重，kg）	1(140)		2(165)		3(185)		4+(205)					
预期孕期体重增重/kg	65		60		52.2		45		40		45	
预期窝产仔数[b]	12.5		13.5		13.5		13.5		13.5		15.5	
妊娠天数	<90	>90	<90	>90	<90	>90	<90	>90	<90	>90	<90	>90
日粮净能[a]/(kcal/kg)	2518	2518	2518	2518	2518	2518	2518	2518	2518	2518	2518	2518
日粮有效消化能[a]/(kcal/kg)	3388	3388	3388	3388	3388	3388	3388	3388	3388	3388	3388	3388
日粮有效代谢能[a]/(kcal/kg)	3300	3300	3300	3300	3300	3300	3300	3300	3300	3300	3300	3300
估算有效代谢能摄入量/(kcal/d)	6678	7932	6928	8182	6928	8182	6897	8151	6427	7681	6521	7775
估算采食量+浪费[c]	2130	2530	2210	2610	2210	2610	2200	2600	2050	2450	2080	2480
增重/(g/d)	578	543	539	481	472	408	410	340	364	298	416	313
钙和磷/%												
总钙	0.61	0.83	0.54	0.78	0.49	0.72	0.43	0.67	0.46	0.71	0.46	0.75
STTD 磷[d]	0.27	0.36	0.24	0.34	0.21	0.31	0.19	0.29	0.20	0.31	0.20	0.33
ATTD 磷[e,f]	0.23	0.31	0.20	0.29	0.18	0.27	0.16	0.25	0.17	0.26	0.17	0.28
总磷[f]	0.49	0.62	0.45	0.58	0.41	0.55	0.38	0.52	0.40	0.54	0.40	0.56
氨基酸[g,h]												
回肠标准可消化基础/%												
精氨酸	0.28	0.37	0.23	0.32	0.19	0.28	0.17	0.24	0.17	0.25	0.17	0.26
组氨酸	0.18	0.22	0.15	0.19	0.13	0.16	0.11	0.14	0.11	0.14	0.11	0.15
异亮氨酸	0.30	0.36	0.25	0.32	0.22	0.27	0.19	0.24	0.19	0.24	0.20	0.26
亮氨酸	0.47	0.65	0.40	0.57	0.35	0.51	0.30	0.45	0.31	0.47	0.32	0.49
赖氨酸	**0.52**	**0.69**	**0.44**	**0.61**	**0.37**	**0.53**	**0.32**	**0.46**	**0.32**	**0.48**	**0.33**	**0.50**
蛋氨酸	0.15	0.20	0.12	0.17	0.10	0.15	0.09	0.13	0.09	0.13	0.09	0.14
蛋氨酸+半胱氨酸	0.34	0.45	0.29	0.40	0.26	0.36	0.23	0.33	0.23	0.33	0.24	0.35
苯丙氨酸	0.29	0.38	0.25	0.34	0.21	0.30	0.19	0.27	0.19	0.27	0.19	0.29
苯丙氨酸+酪氨酸	0.50	0.66	0.43	0.58	0.37	0.51	0.32	0.46	0.33	0.47	0.33	0.49
苏氨酸	0.37	0.48	0.33	0.43	0.29	0.39	0.27	0.36	0.27	0.36	0.28	0.38
色氨酸	0.09	0.13	0.08	0.12	0.07	0.11	0.07	0.10	0.07	0.10	0.07	0.11
缬氨酸	0.37	0.49	0.32	0.43	0.28	0.39	0.25	0.35	0.25	0.36	0.26	0.37
总氮	1.32	1.79	1.15	1.61	1.01	1.45	0.90	1.32	0.91	1.35	0.94	1.43
回肠表观可消化基础/%												
精氨酸	0.23	0.32	0.19	0.28	0.15	0.23	0.12	0.20	0.12	0.21	0.13	0.22
组氨酸	0.17	0.21	0.14	0.18	0.11	0.15	0.10	0.13	0.10	0.13	0.10	0.14
异亮氨酸	0.27	0.34	0.23	0.29	0.19	0.25	0.17	0.22	0.17	0.22	0.17	0.23
亮氨酸	0.43	0.60	0.36	0.53	0.30	0.46	0.26	0.41	0.27	0.42	0.28	0.45
赖氨酸	**0.49**	**0.66**	**0.40**	**0.57**	**0.34**	**0.49**	**0.29**	**0.43**	**0.29**	**0.44**	**0.30**	**0.47**
蛋氨酸	0.14	0.19	0.11	0.16	0.09	0.14	0.08	0.12	0.08	0.12	0.08	0.13
蛋氨酸+半胱氨酸	0.32	0.43	0.27	0.38	0.24	0.34	0.21	0.31	0.21	0.31	0.22	0.33
苯丙氨酸	0.26	0.35	0.22	0.31	0.19	0.27	0.16	0.24	0.16	0.25	0.17	0.26
苯丙氨酸+酪氨酸	0.46	0.62	0.39	0.54	0.33	0.47	0.29	0.42	0.29	0.43	0.30	0.45
苏氨酸	0.32	0.43	0.28	0.38	0.25	0.34	0.22	0.32	0.22	0.32	0.23	0.33
色氨酸	0.08	0.12	0.07	0.11	0.06	0.10	0.05	0.09	0.06	0.09	0.06	0.10
缬氨酸	0.33	0.44	0.28	0.39	0.24	0.34	0.21	0.31	0.21	0.31	0.22	0.33
总氮	1.12	1.58	0.95	1.41	0.82	1.25	0.72	1.12	0.73	1.15	0.75	1.23

续表

胎次（配种时体重，kg）	1(140)		2(165)		3(185)		4+(205)					
预期孕期体增重/kg	65		60		52.2		45		40		45	
预期窝产仔数[b]	12.5		13.5		13.5		13.5		13.5		15.5	
妊娠天数	<90	>90	<90	>90	<90	>90	<90	>90	<90	>90	<90	>90
总氨基酸和总氮基础/%												
精氨酸	0.32	0.42	0.27	0.37	0.23	0.32	0.20	0.29	0.21	0.29	0.21	0.31
组氨酸	0.22	0.27	0.19	0.23	0.16	0.20	0.14	0.18	0.14	0.18	0.14	0.19
异亮氨酸	0.36	0.43	0.31	0.38	0.27	0.33	0.24	0.29	0.24	0.30	0.24	0.31
亮氨酸	0.55	0.75	0.47	0.66	0.41	0.59	0.36	0.53	0.36	0.54	0.37	0.57
赖氨酸	**0.61**	**0.80**	**0.52**	**0.71**	**0.45**	**0.62**	**0.39**	**0.55**	**0.39**	**0.56**	**0.40**	**0.59**
蛋氨酸	0.18	0.23	0.15	0.20	0.13	0.18	0.11	0.16	0.11	0.16	0.12	0.17
蛋氨酸+半胱氨酸	0.41	0.54	0.36	0.48	0.32	0.44	0.29	0.40	0.29	0.41	0.30	0.43
苯丙氨酸	0.34	0.44	0.29	0.40	0.25	0.35	0.23	0.31	0.23	0.32	0.23	0.34
苯丙氨酸+酪氨酸	0.61	0.79	0.53	0.70	0.46	0.62	0.41	0.56	0.41	0.57	0.42	0.60
苏氨酸	0.46	0.58	0.41	0.53	0.37	0.48	0.34	0.44	0.34	0.45	0.35	0.47
色氨酸	0.11	0.15	0.10	0.14	0.09	0.13	0.08	0.12	0.08	0.12	0.08	0.13
缬氨酸	0.45	0.58	0.39	0.52	0.34	0.46	0.31	0.42	0.31	0.43	0.32	0.45
总氮	1.62	2.15	1.42	1.95	1.26	1.77	1.14	1.62	1.15	1.65	1.18	1.74

a 玉米-豆粕型日粮的能量含量。有效消化能和有效代谢能是根据母猪的净能固定转化率计算得到。对于玉米-豆粕型日粮，有效消化能和有效代谢能接近实际消化能和代谢能。最适宜的日粮能量水平依原料的可利用性和当地原料成本不同而改变。当使用替代原料时，建议按净能含量设计日粮配方，调整营养需要量确保营养含量与净能比率保持不变。

b 预期平均出生体重为 1.40 kg。

c 假定饲料浪费为 5%。

d 全消化道标准可消化。

e 全消化道表观可消化。

f 全消化道表观可消化磷和总磷需要量只适用于玉米-豆粕型的日粮，是依据全消化道标准可消化磷的需要量和玉米、去皮豆粕及磷酸氢钙的营养成分计算得出。假设日粮中添加 0.1%的赖氨酸盐酸盐及 3%的维生素和矿物质。玉米和豆粕添加水平以满足回肠标准可消化赖氨酸需要量计算得到，而磷酸氢钙的添加量以满足全消化道标准可消化磷水平计算得到。

g 需要量以生长模型为基础估算得到。

h 回肠表观可消化和总氨基酸需要量只适用于玉米-豆粕型日粮，主要根据回肠标准可消化氨基酸需要量和玉米、去皮豆粕为基础的日粮（日粮添加 0.1%的赖氨酸盐酸盐和 3%的维生素和矿物质）中氨基酸含量计算得出。对每一种氨基酸而言，日粮中的玉米和豆粕水平及营养素需要量都以满足回肠标准可消化氨基酸需要量计算得到。

表16-6B 怀孕母猪钙、磷、氨基酸每日需要量（90%干物质基础）[a]

胎次（配种时体重, kg）	1(140)		2(165)		3(185)		4+(205)					
预期孕期体重增重/kg	65		60		52.2		45		40		45	
预期窝产仔数[b]	12.5		13.5		13.5		13.5		13.5		15.5	
怀孕期	<90	>90	<90	>90	<90	>90	<90	>90	<90	>90	<90	>90
日粮净能[a]/(kcal/kg)	2518	2518	2518	2518	2518	2518	2518	2518	2518	2518	2518	2518
日粮有效消化能[a]/(kcal/kg)	3388	3388	3388	3388	3388	3388	3388	3388	3388	3388	3388	3388
日粮有效代谢能[a]/(kcal/kg)	3300	3300	3300	3300	3300	3300	3300	3300	3300	3300	3300	3300
估算有效代谢能摄入量/(kcal/d)	6678	7932	6928	8182	6928	8182	6897	8151	6427	7681	6521	7775
估算采食量+浪费[c]	2130	2530	2210	2610	2210	2610	2200	2600	2050	2450	2080	2480
增重/(g/d)	578	543	539	481	472	408	410	340	364	298	416	313
钙和磷/(g/d)												
总钙	12.42	19.94	11.42	19.31	10.20	17.91	9.05	16.55	8.89	16.40	9.18	17.77
STTD 磷[d]	5.40	8.67	4.96	8.39	4.43	7.79	3.93	7.20	3.87	7.13	3.99	7.73
ATTD 磷[e,f]	4.61	7.49	4.22	7.25	3.75	6.71	3.30	6.19	3.26	6.15	3.37	6.68
总磷[f]	9.91	14.78	9.40	14.45	8.67	13.59	7.98	12.75	7.69	12.47	7.89	13.29
氨基酸[g,h]												
回肠标准可消化基础/(g/d)												
精氨酸	5.6	8.8	4.8	7.9	4.1	6.9	3.5	6.0	3.2	5.8	3.4	6.2
组氨酸	3.7	5.4	3.2	4.8	2.6	4.1	2.2	3.5	2.1	3.3	2.2	3.5
异亮氨酸	6.1	8.8	5.3	7.9	4.6	6.9	4.0	5.9	3.7	5.7	3.9	6.1
亮氨酸	9.6	15.6	8.5	14.2	7.3	12.6	6.4	11.2	6.0	10.8	6.3	11.6
赖氨酸	**10.6**	**16.7**	**9.2**	**15.1**	**7.8**	**13.1**	**6.7**	**11.5**	**6.3**	**11.1**	**6.6**	**11.9**
蛋氨酸	3.0	4.7	2.6	4.3	2.2	3.7	1.8	3.2	1.7	3.1	1.8	3.4
蛋氨酸+半胱氨酸	6.8	10.8	6.1	10.0	5.4	8.9	4.8	8.1	4.5	7.8	4.7	8.3
苯丙氨酸	5.8	9.1	5.1	8.4	4.4	7.4	3.9	6.6	3.7	6.3	3.8	6.8
苯丙氨酸+酪氨酸	10.1	15.9	9.0	14.5	7.7	12.7	6.7	11.3	6.3	10.9	6.6	11.6
苏氨酸	7.6	11.5	6.9	10.7	6.2	9.7	5.6	8.8	5.3	8.5	5.4	9.0
色氨酸	1.9	3.2	1.7	3.0	1.5	2.7	1.4	2.5	1.3	2.4	1.3	2.6
缬氨酸	7.5	11.8	6.7	10.8	5.8	9.5	5.2	8.6	4.9	8.3	5.0	8.8
总氮	26.8	43.1	24.1	40.1	21.2	36.0	18.9	32.6	17.8	31.5	18.5	33.8
回肠表观可消化基础/(g/d)												
精氨酸	4.7	7.8	3.9	6.9	3.2	5.8	2.6	4.9	2.4	4.8	2.6	5.2
组氨酸	3.4	5.0	2.9	4.4	2.4	3.7	2.0	3.1	1.9	3.0	1.9	3.2
异亮氨酸	5.5	8.1	4.8	7.3	4.1	6.2	3.5	5.3	3.3	5.1	3.4	5.5
亮氨酸	8.7	14.5	7.6	13.1	6.4	11.5	5.5	10.1	5.2	9.8	5.4	10.6
赖氨酸	**9.9**	**15.8**	**8.5**	**14.1**	**7.1**	**12.21**	**6.0**	**10.6**	**5.6**	**10.2**	**5.9**	**11.0**
蛋氨酸	2.7	4.5	2.3	4.0	1.9	3.4	1.6	3.0	1.5	2.9	1.6	3.1
蛋氨酸+半胱氨酸	6.4	10.2	5.7	9.4	5.0	8.4	4.4	7.6	4.2	7.3	4.3	7.8
苯丙氨酸	5.3	8.5	4.6	7.7	3.9	6.7	3.4	5.9	3.2	5.7	3.3	6.2
苯丙氨酸+酪氨酸	9.4	14.9	8.2	13.5	7.0	11.8	6.0	10.4	5.7	10.0	5.9	10.7
苏氨酸	6.6	10.3	5.9	9.4	5.2	8.3	4.6	7.6	4.4	7.4	4.5	7.8
色氨酸	1.6	2.9	1.5	2.7	1.3	2.4	1.1	2.2	1.1	2.2	1.1	2.3
缬氨酸	6.6	10.7	5.8	9.6	5.0	8.5	4.3	7.6	4.1	7.3	4.3	7.8
总氮	22.7	37.9	20.0	34.9	17.1	30.9	15.0	27.6	14.1	26.8	14.8	28.9

续表

胎次（配种时体重，kg）	1(140)		2(165)		3(185)		4+(205)					
预期孕期体增重/kg	65		60		52.2		45		40		45	
预期窝产仔数[b]	12.5		13.5		13.5		13.5		13.5		15.5	
怀孕期	<90	>90	<90	>90	<90	>90	<90	>90	<90	>90	<90	>90
总氨基酸和总氮基础/(g/d)												
精氨酸	6.5	10.0	5.7	9.1	4.9	8.0	4.3	7.1	4.0	6.8	4.2	7.3
组氨酸	4.4	6.4	3.9	5.7	3.3	5.0	2.9	4.3	2.7	4.1	2.8	4.4
异亮氨酸	7.2	10.3	6.4	9.4	5.6	8.2	4.9	7.2	4.6	6.9	4.8	7.4
亮氨酸	11.1	17.9	9.9	16.5	8.5	14.6	7.5	13.0	7.1	12.6	7.4	13.5
赖氨酸	12.4	19.3	11.0	17.5	9.4	15.4	8.2	13.6	7.7	13.1	8.0	14.0
蛋氨酸	3.6	5.6	3.1	5.1	2.7	4.5	2.4	3.9	2.2	3.8	2.3	4.1
蛋氨酸+半胱氨酸	8.3	12.9	7.5	12.0	6.7	10.8	6.0	9.8	5.7	9.5	5.9	10.1
苯丙氨酸	6.9	10.7	6.1	9.8	5.3	8.7	4.7	7.8	4.5	7.5	4.6	8.0
苯丙氨酸+酪氨酸	12.3	18.9	11.0	17.4	9.6	15.4	8.5	13.8	8.0	13.3	8.3	14.1
苏氨酸	9.4	14.0	8.6	13.2	7.8	12.0	7.1	10.9	6.7	10.5	6.9	11.1
色氨酸	2.2	3.6	2.0	3.4	1.8	3.1	1.6	2.9	1.6	2.8	1.6	3.0
缬氨酸	9.0	14.0	8.1	12.9	7.2	11.5	6.4	10.4	6.0	10.0	6.2	10.65
总氮	32.7	51.7	29.8	48.4	26.5	43.8	23.9	39.9	22.5	38.5	23.3	41.1

a 玉米-豆粕型日粮的能量含量。有效消化能和有效代谢能是根据母猪的净能固定转化率计算得到。对于玉米-豆粕型日粮，有效消化能和有效代谢能接近实际消化能和代谢能。最适宜的日粮能量水平根据原料的可利用率及当地原料成本不同而改变。当使用替代原料时，建议按照净能含量设计日粮配方，调整营养需要量确保营养含量与净能比率保持不变。

b 预期平均出生体重为1.40 kg。

c 假定饲料浪费为5%。

d 全消化道标准可消化。

e 全消化道表观可消化。

f 全消化道表观可消化磷和总磷需要量只适用于玉米-豆粕型的日粮，是依据全消化道标准可消化磷的需要量和玉米、去皮豆粕及磷酸氢钙的营养成分计算得出。假设日粮中添加0.1%的赖氨酸盐酸盐及3%的维生素和矿物质。玉米和豆粕添加水平以满足标准回肠可消化赖氨酸需要量为前提计算得到，而磷酸氢钙的添加量以满足全消化道标准可消化磷水平为前提计算得到。

g 需要量以生长模型为基础估算得到。

h 回肠表观可消化和总氨基酸需要量只适用于玉米-豆粕型日粮，主要根据回肠标准可消化氨基酸需要量和玉米、去皮豆粕为基础的日粮（日粮添加0.1%的赖氨酸盐酸盐和3%的维生素和矿物质）中氨基酸含量计算得出。对每一种氨基酸而言，日粮中的玉米和豆粕水平及营养素需要量都以满足回肠标准可消化氨基酸需要量为前提计算得到。

表 16-7A 泌乳母猪日粮钙、磷和氨基酸需要量（自由采食，日粮含 90%干物质）[a]

胎次	1			2+		
产仔后体重/kg	175	175	175	210	210	210
窝产仔数	11	11	11	11.5	11.5	11.5
泌乳期长度/d	21	21	21	21	21	21
泌乳仔猪平均日增重/g	190	230	270	190	230	270
日粮净能[a]	2518	2518	2518	2518	2518	2518
日粮有效消化能[a]/(kcal/kg)	3388	3388	3388	3388	3388	3388
日粮有效代谢能[a]/(kcal/kg)	3300	3300	3300	3300	3300	3300
估算有效代谢能摄入量/(Mcal/d)	18.7	18.7	18.7	20.7	20.7	20.7
估算采食量+浪费[b]/(g/d)	5.95	5.95	5.93	6.61	6.61	6.61
母猪预计体重变化/kg	1.5	−7.7	−17.4	3.7	−5.8	−15.9
钙和磷/%						
总钙	0.63	0.71	0.80	0.60	0.68	0.76
STTD磷[c]	0.31	0.36	0.40	0.30	0.34	0.38
ATTD磷[d,e]	0.27	0.31	0.35	0.26	0.29	0.33
总磷[e]	0.56	0.62	0.67	0.54	0.60	0.65
氨基酸[f,g]						
回肠标准可消化基础/%						
精氨酸	0.43	0.44	0.46	0.42	0.43	0.45
组氨酸	0.30	0.32	0.34	0.29	0.31	0.33
异亮氨酸	0.41	0.45	0.49	0.40	0.43	0.47
亮氨酸	0.83	0.92	1.00	0.80	0.88	0.96
赖氨酸	**0.75**	**0.81**	**0.87**	**0.72**	**0.78**	**0.84**
蛋氨酸	0.20	0.21	0.23	0.19	0.21	0.22
蛋氨酸+半胱氨酸	0.39	0.43	0.47	0.38	0.41	0.45
苯丙氨酸	0.41	0.44	0.48	0.39	0.42	0.46
苯丙氨酸+酪氨酸	0.83	0.91	0.99	0.80	0.87	0.95
苏氨酸	0.47	0.51	0.55	0.46	0.49	0.53
色氨酸	0.14	0.15	0.17	0.13	0.15	0.16
缬氨酸	0.64	0.69	0.74	0.61	0.66	0.71
总氮	1.62	1.73	1.86	1.56	1.67	1.79
回肠表观可消化基础/%						
精氨酸	0.39	0.40	0.41	0.38	0.39	0.40
组氨酸	0.28	0.30	0.33	0.27	0.29	0.31
异亮氨酸	0.39	0.42	0.46	0.37	0.41	0.44
亮氨酸	0.79	0.87	0.95	0.76	0.83	0.91
赖氨酸	**0.71**	**0.77**	**0.83**	**0.68**	**0.74**	**0.80**
蛋氨酸	0.19	0.20	0.22	0.18	0.20	0.21
蛋氨酸+半胱氨酸	0.37	0.41	0.44	0.36	0.39	0.42
苯丙氨酸	0.38	0.41	0.45	0.36	0.40	0.43
苯丙氨酸+酪氨酸	0.78	0.86	0.95	0.75	0.83	0.90
苏氨酸	0.42	0.46	0.50	0.41	0.44	0.48
色氨酸	0.13	0.14	0.16	0.12	0.14	0.15
缬氨酸	0.58	0.64	0.69	0.56	0.61	0.66
总氮	1.40	1.52	1.64	1.35	1.46	1.57

续表

胎次	1			2 +		
产仔后体重/kg	175	175	175	210	210	175
窝产仔数	11	11	11	11.5	11.5	11.5
泌乳期长度/d	21	21	21	21	21	21
泌乳仔猪平均日增重/g	190	230	270	190	230	270
	总氨基酸和总氮基础/%					
精氨酸	0.48	0.50	0.51	0.47	0.48	0.50
组氨酸	0.35	0.37	0.40	0.34	0.36	0.38
异亮氨酸	0.49	0.52	0.56	0.47	0.50	0.54
亮氨酸	0.96	1.05	1.15	0.92	1.01	1.10
赖氨酸	**0.86**	**0.93**	**1.00**	**0.83**	**0.90**	**0.96**
蛋氨酸	0.23	0.25	0.27	0.23	0.24	0.26
蛋氨酸+半胱氨酸	0.47	0.51	0.55	0.46	0.49	0.53
苯丙氨酸	0.47	0.51	0.55	0.46	0.49	0.53
苯丙氨酸+酪氨酸	0.98	1.07	1.16	0.94	1.03	1.12
苏氨酸	0.58	0.62	0.67	0.56	0.60	0.65
色氨酸	0.16	0.18	0.19	0.15	0.17	0.18
缬氨酸	0.75	0.81	0.87	0.72	0.78	0.84
总氮	1.95	2.08	2.22	1.89	2.01	2.15

a 玉米-豆粕型日粮的能量含量。有效消化能和有效代谢能是根据母猪的净能固定转化率计算得到。对于玉米-豆粕型日粮，有效消化能和有效代谢能接近实际消化能和代谢能。最适宜的日粮能量水平根据原料的可利用性及当地原料成本不同而改变。当使用替代原料时，建议按照净能含量设计日粮配方，调整营养需要量确保营养含量与净能比率保持不变。

b 假定饲料浪费为5%。

c 全消化道标准可消化。

d 全消化道表观可消化。

e 全消化道表观可消化磷和总磷需要量只适用于玉米-豆粕型的日粮，是依据全消化道标准可消化磷的需要量和玉米、去皮豆粕及磷酸氢钙的营养成分计算得出。假设日粮中添加0.1%的赖氨酸盐酸盐及3%的维生素和矿物质。玉米和豆粕添加水平以满足标准回肠可消化赖氨酸需要量为前提计算得到，而磷酸氢钙的添加量以满足全消化道标准可消化磷水平为前提计算得到。

f 需要量以生长模型为基础估算得到。

g 回肠表观可消化和总氨基酸需要量只适用于玉米-豆粕型日粮，主要根据回肠标准可消化氨基酸需要量和玉米、去皮豆粕为基础的日粮（日粮添加0.1%的赖氨酸盐酸盐和3%的维生素和矿物质）中氨基酸含量计算得出。对每一种氨基酸而言，日粮中的玉米和豆粕水平及营养素需要量都以满足回肠标准可消化氨基酸需要量为前提计算得到。

表 16-7B　泌乳母猪每日钙、磷以及氨基酸需要量（90%干物质基础）

胎次	1			2+		
产仔后体重/kg	175	175	175	210	210	210
窝产仔数	11	11	11	11.5	11.5	11.5
哺乳期长度/d	21	21	21	21	21	21
哺乳仔猪平均日增重/(g)	190	230	270	190	230	270
日粮净能[a]/(kcal/kg)	2518	2518	2518	2518	2518	2518
日粮有效消化能[a]/(kcal/kg)	3388	3388	3388	3388	3388	3388
日粮有效代谢能[a]/(kcal/kg)	3300	3300	3300	3300	3300	3300
估算有效代谢能摄入量/(Mcal/d)	18.7	18.7	18.7	20.7	20.7	20.7
估算采食量+浪费[b]/(g/d)	5.95	5.95	5.93	6.61	6.61	6.61
母猪预计体重变化/kg	1.5	-7.7	-17.4	3.7	-5.8	-15.9
钙和磷/(g/d)						
总钙	35.3	40.3	45.0	37.7	42.9	48.1
STTD磷[c]	17.7	20.1	22.6	18.9	21.4	24.0
ATTD磷[d,e]	15.1	17.3	19.6	16.1	18.4	20.8
总磷[e]	31.6	34.8	38.1	34.1	37.4	40.8
氨基酸[f,g]						
回肠标准可消化基础/(g/d)						
精氨酸	24.3	25.1	26.0	26.3	27.1	28.0
组氨酸	16.9	18.2	19.5	18.1	19.4	20.8
异亮氨酸	23.4	25.5	27.5	25.1	27.2	29.4
亮氨酸	47.1	51.9	56.7	50.3	55.2	60.3
赖氨酸	**42.2**	**45.7**	**49.3**	**45.3**	**48.9**	**52.6**
蛋氨酸	11.3	12.2	13.1	12.1	13.0	14.0
蛋氨酸+半胱氨酸	22.3	24.3	26.4	23.8	26.0	28.1
苯丙氨酸	22.9	24.9	27.0	24.5	26.6	28.8
苯丙氨酸+酪氨酸	46.9	51.6	56.3	50.1	55.0	59.9
苏氨酸	26.8	29.0	31.3	28.8	31.1	33.5
色氨酸	7.9	8.7	9.6	8.4	9.3	10.2
缬氨酸	35.9	38.9	42.0	38.5	41.6	44.9
总氮	91.1	98.1	105.2	97.9	105.1	112.5
回肠表观可消化基础/(g/d)						
精氨酸	21.8	22.6	23.5	23.6	24.4	25.2
组氨酸	15.9	17.2	18.5	17.1	18.4	19.7
异亮氨酸	21.9	23.9	26.0	23.4	25.5	27.7
亮氨酸	44.5	49.2	54.0	47.4	52.3	57.3
赖氨酸	**40.0**	**43.5**	**47.0**	**42.9**	**46.5**	**50.1**
蛋氨酸	10.7	11.6	12.5	11.4	12.3	13.3
蛋氨酸+半胱氨酸	21.0	22.9	24.9	22.4	24.5	26.6
苯丙氨酸	21.3	23.3	25.4	22.8	24.9	27.0
苯丙氨酸+酪氨酸	44.3	48.9	53.5	47.2	52.0	56.8
苏氨酸	23.8	26.0	28.1	25.5	27.7	30.0
色氨酸	7.2	8.1	8.9	7.7	8.5	9.4
缬氨酸	33.0	36.0	39.0	35.4	38.4	41.6
总氮	79.2	85.9	92.8	84.8	91.7	98.9

续表

胎次	1			2 +		
产仔后体重/kg	175	175	175	210	210	210
窝产仔数	11	11	11	11.5	11.5	11.5
哺乳期长度/d	21	21	21	21	21	21
哺乳仔猪平均日增重/(g)	190	230	270	190	230	270
	总氨基酸和氮基础/(g/d)					
精氨酸	27.3	28.2	29.1	29.6	30.5	31.4
组氨酸	19.7	21.1	22.5	21.1	22.6	24.1
异亮氨酸	27.4	29.6	31.9	29.4	31.7	34.1
亮氨酸	54.1	59.5	65.0	57.8	63.4	69.1
赖氨酸	**48.7**	**52.6**	**56.5**	**52.4**	**56.4**	**60.5**
蛋氨酸	13.2	14.2	15.1	14.2	15.2	16.2
蛋氨酸+半胱氨酸	26.7	29.0	31.3	28.7	31.1	33.5
苯丙氨酸	26.7	29.0	31.3	28.6	31.0	33.4
苯丙氨酸+酪氨酸	55.3	60.5	65.8	59.1	64.6	70.2
苏氨酸	32.7	35.3	37.9	35.2	37.9	40.6
色氨酸	9.0	9.9	10.9	9.6	10.6	11.6
缬氨酸	42.2	45.7	49.2	45.3	48.9	52.5
总氮	109.9	117.8	125.8	118.4	126.5	134.9

a 玉米-豆粕型日粮的能量含量。有效消化能和有效代谢能是根据母猪的净能固定转化率计算得到。对于玉米-豆粕型日粮，有效消化能和有效代谢能接近实际消化能和代谢能。最适宜的日粮能量水平依据原料的可利用性及当地原料成本不同而改变。当使用替代原料时，建议按照净能含量设计日粮配方，调整营养需要量确保营养含量与净能比率保持不变。

b 假定饲料浪费为5%。

c 全消化道标准可消化。

d 全消化道表观可消化。

e 全消化道表观可消化磷和总磷需要量只适用于玉米-豆粕型的日粮，是依据全消化道标准可消化磷的需要量和玉米、去皮豆粕及磷酸氢钙的营养成分计算得到。假设日粮中添加0.1%的赖氨酸盐酸盐及3%的维生素和矿物质。玉米和豆粕添加水平以满足标准回肠可消化赖氨酸需要量为前提计算得到，而磷酸氢钙的添加量以满足全消化道标准可消化磷水平为前提计算得到。

f 需要量以生长模型为基础估算得到。

g 回肠表观可消化和总氨基酸需要量只适用于玉米-豆粕型日粮，主要根据回肠标准可消化氨基酸需要量和玉米、去皮豆粕为基础的日粮（日粮添加0.1%的赖氨酸盐酸盐和3%的维生素和矿物质）中氨基酸含量计算得出。对每一种氨基酸而言，日粮中的玉米和豆粕水平及营养素需要量都以满足回肠标准可消化氨基酸需要为前提计算得到。

表16-8A 妊娠和泌乳母猪日粮矿物质、维生素和脂肪酸的需要量（日粮含90%干物质）

指标	妊娠母猪	泌乳母猪
日粮净能[a]/（kcal/kg）	2 518	2 518
日粮有效消化能[a]/（kcal/kg）	3 388	3 388
日粮有效代谢能[a]/（kcal/kg）	3 300	3 300
估算有效代谢能摄入量/（kcal/d）	6 928	19 700
估算采食量+浪费[b]/（g/d）	2 210	6 280
	需要量（%或每千克含量）	
矿物质元素		
钠/%	0.15	0.20
氯/%	0.12	0.16
镁/%	0.06	0.06
钾/%	0.20	0.20
铜/（mg/kg）	10	20
碘/（mg/kg）	0.14	0.14
铁/（mg/kg）	80	80
锰/（mg/kg）	25	25
硒/（mg/kg）	0.15	0.15
锌/（mg/kg）	100	100
维生素		
维生素A[c]/（IU/kg）	4 000	2 000
维生素D_3[d]/（IU/kg）	800	800
维生素E[e]/（IU/kg）	44	44
维生素K（甲萘醌）/（mg/kg）	0.50	0.50
生物素/（mg/kg）	0.20	0.20
胆碱（g/kg）	1.25	1.00
叶酸/（mg/kg）	1.30	1.30
烟酸，可利用[f]/（mg/kg）	10	10
泛酸/（mg/kg）	12	12
核黄素/（mg/kg）	3.75	3.75
硫胺素/（mg/kg）	1.00	1.00
维生素B_6/（mg/kg）	1.00	1.00
维生素B_{12}/（μg/kg）	15	15
亚油酸/%	0.10	0.10

  a 玉米-豆粕型日粮的能量含量。有效消化能和有效代谢能是根据母猪的净能固定转化率计算得到。对于玉米-豆粕型日粮，有效消化能和有效代谢能接近实际消化能和代谢能。最适宜的日粮能量水平依据原料的可利用性与当地原料成本的不同而改变。当使用替代原料时，建议基于净能含量设计日粮配方，调整营养需要量确保营养含量与净能比率保持不变。
  b 假设有5%的饲料浪费。
  c 1 IU 维生素A=0.30 μg 视黄醇或者 0.344 μg 视黄醇乙酸酯。维生素A的活性（视黄醇当量）也由β-胡萝卜素提供（见维生素章节）。
  d 1 IU 维生素D_2或D_3=0.025 μg。
  e 1 IU 维生素E=0.67 mg D-α-生育酚或者1 mg DL-α-醋酸生育酚。近来在猪上的研究表明，天然的和人工合成的α-醋酸生育酚有实质性的差别（见维生素章节）。
  f 玉米、高粱、小麦和大麦中的烟酸不可被利用。同样这些谷物来源的副产物中的烟酸利用率也很低，除非对这些副产物进行了湿磨工艺的发酵。

表16-8B 妊娠和泌乳母猪每日矿物质、维生素和脂肪酸的需要量（日粮含90%干物质）

指标	妊娠母猪	泌乳母猪
日粮净能 [a]/（kcal/kg）	2 518	2 518
日粮有效消化能 [a]/（kcal/kg）	3 388	3 388
日粮有效代谢能 [a]/（kcal/kg）	3 300	3 300
估算有效代谢能摄入量/（kcal/d）	6 928	19 700
估算采食量+浪费 [b]/（g/d）	2 210	6 280
	日需要量	
矿物质元素		
钠/g	3.15	11.93
氯/g	2.52	9.55
镁/g	1.26	3.58
钾/g	4.20	11.93
铜/mg	21.00	119.32
碘/mg	0.29	0.84
铁/mg	168.0	477.3
锰/mg	52.49	149.15
硒/mg	0.31	0.89
锌/mg	210.0	596.6
维生素		
维生素 A[c]/IU	8 398	11 932
维生素 D_3[d]/IU	1 680	4 773
维生素 E[e]/IU	92.4	262.5
维生素 K（甲萘醌）/mg	1.05	2.98
生物素/mg	0.42	1.19
胆碱/g	2.62	5.97
叶酸/mg	2.73	7.76
烟酸，可利用 [f]/mg	21.00	59.66
泛酸/mg	25.19	71.59
核黄素/mg	7.87	22.37
硫胺素/mg	2.10	5.97
维生素 B_6/mg	2.10	5.97
维生素 B_{12}/μg	31.49	89.49
亚油酸/g	2.1	6.0

a 玉米-豆粕型日粮的能量含量。有效消化能和有效代谢能是根据母猪的净能固定转化率计算得到。对于玉米-豆粕型日粮，有效消化能和有效代谢能接近实际消化能和代谢能。最适宜的日粮能量水平依据原料的可利用性与当地原料成本的不同而改变。当使用替代原料时，建议按照净能含量设计日粮配方，调整营养需要量确保营养含量与净能比率保持不变。

b 假设有5%的饲料浪费。

c 1 IU 维生素 A=0.30 μg 视黄醇或者 0.344 μg 视黄醇乙酸酯。维生素 A 的活性（视黄醇当量）也由 β-胡萝卜素提供（见维生素章节）。

d 1 IU 维生素 D_2 或 D_3=0.025 μg。

e 1 IU 维生素 E=0.67 mg D-α-生育酚或者 1 mg DL-α-醋酸生育酚。近来在猪上的研究表明，天然的和人工合成的 α-醋酸生育酚有实质性的差别（见维生素章节）。

f 玉米、高粱、小麦和大麦中的烟酸不可被利用。同样，这些谷物来源的副产物中的烟酸利用率也很低，除非对这些副产物进行了湿磨工艺的发酵。

表 16-9　种公猪配种期日粮和每日氨基酸、矿物质、维生素和脂肪酸需要量（日粮含 90% 干物质）[a]

日粮净能 [b] /（kcal/kg）		2475
日粮有效消化能 [b] /（kcal/kg）		3402
日粮有效代谢能 [b] /（kcal/kg）		3300
估算有效代谢能摄入量 [b] /（kcal/d）		7838
估算采食量+浪费 [c] /（g/d）		2500
	需要量（%或每千克含量）	日需要量
氨基酸（回肠标准可消化）		
精氨酸	0.20%	4.86 g
组氨酸	0.15%	3.46 g
异亮氨酸	0.31%	7.41 g
亮氨酸	0.33%	7.83 g
赖氨酸	**0.51%**	**11.99 g**
蛋氨酸	0.08%	1.96 g
蛋氨酸+半胱氨酸	0.25%	5.98 g
苯丙氨酸	0.36%	8.50 g
苯丙氨酸+酪氨酸	0.58%	13.77 g
苏氨酸	0.22%	5.19 g
色氨酸	0.20%	4.82 g
缬氨酸	0.27%	6.52 g
总氮	1.14%	27.04 g
氨基酸（回肠表观可消化基础）[d]		
精氨酸	0.16%	3.86 g
组氨酸	0.13%	3.16 g
异亮氨酸	0.29%	6.81 g
亮氨酸	0.29%	6.84 g
赖氨酸	**0.47%**	**11.13 g**
蛋氨酸	0.07%	1.72 g
蛋氨酸+半胱氨酸	0.23%	5.55 g
苯丙氨酸	0.33%	7.86 g
苯丙氨酸+酪氨酸	0.54%	12.81 g
苏氨酸	0.17%	4.15 g
色氨酸	0.19%	4.52 g
缬氨酸	0.23%	5.58 g
总氮	0.94%	22.40 g
总氨基酸基础 [d]		
精氨酸	0.25%	5.83 g
组氨酸	0.18%	4.30 g
异亮氨酸	0.37%	8.81 g
亮氨酸	0.39%	9.20 g
赖氨酸	**0.60%**	**14.25 g**
蛋氨酸	0.11%	2.55 g
蛋氨酸+半胱氨酸	0.31%	7.44 g
苯丙氨酸	0.42%	9.96 g
苯丙氨酸+酪氨酸	0.70%	16.55 g
苏氨酸	0.28%	6.70 g
色氨酸	0.23%	5.42 g
缬氨酸	0.34%	8.01 g
总氮	1.41%	33.48 g
矿物质		
总钙	0.75%	17.81 g

续表

日粮净能 [b]/（kcal/kg）		2475
日粮有效消化能 [b]/（kcal/kg）		3402
日粮有效代谢能 [b]/（kcal/kg）		3300
估算有效代谢能摄入量 [b]/（kcal/d）		7838
估算采食量+浪费 [c]/（g/d）		2500
矿物质	需要量（%或每千克含量）	日需要量
STTD 磷 [e]	0.33%	7.84 g
ATTD 磷 [f,g]	0.31%	7.36 g
总磷 [g]	0.75%	17.81 g
钠	0.15%	3.56 g
氯	0.12%	2.85 g
镁	0.04%	0.95 g
钾	0.20%	4.75 g
铜	5 mg	11.88 mg
碘	0.14 mg	0.33 mg
铁	80 mg	190 mg
锰	20 mg	47.5 mg
硒	0.30 mg	0.71 mg
锌	50 mg	118.75 mg
维生素		
维生素 A [h]	4000 IU	9500 IU
维生素 D_3 [i]	200 IU	475 IU
维生素 E [j]	44 IU	104.5 IU
维生素 K（甲萘醌）	0.50 mg	1.19 mg
生物素	0.20 mg	0.48 mg
胆碱	1.25 g	2.97 g
叶酸	1.30 mg	3.09 mg
烟酸，可利用 [k]	10 mg	23.75 mg
泛酸	12 mg	28.50 mg
核黄素	3.75 mg	8.91 mg
硫胺素	1.0 mg	2.38 mg
维生素 B_6	1.0 mg	2.38 mg
维生素 B_{12}	15 μg	35.63 μg
亚油酸	0.1%	2.38%

a 需要量为日采食量加上浪费为 2.5 kg 饲料。采食量可能需要根据公猪体重和预期增重进行调整。

b 玉米-豆粕型日粮的能量含量。有效消化能和有效代谢能是根据体重在 25kg 左右猪的净能的固定转化率计算得到。对于玉米-豆粕型日粮，有效消化能和有效代谢能接近实际消化能和代谢能。最佳日粮能量水平依据当地饲料原料的可利用性及当地原料成本的不同而改变。当使用替代原料时，建议按照净能含量设计日粮配方，调整营养需要量确保营养含量与净能比率保持不变。

c 假设有 5%的饲料浪费率。

d 表观回肠可消化和总氨基酸需要量只适用于玉米-豆粕型日粮，主要根据回肠标准可消化氨基酸需要量和玉米、去皮豆粕为基础的日粮（日粮添加 0.1%的赖氨酸盐酸盐和 3%的维生素和矿物质）中氨基酸含量计算得出。对每一种氨基酸而言，日粮中的玉米和豆粕水平及营养素需要量都以满足回肠标准可消化氨基酸需要量为前提计算得到。

e 全消化道标准可消化。

f 全消化道表观可消化。

g 全消化道表观可消化磷和总磷需要量只适用于玉米-豆粕型的日粮，是依据全消化道标准可消化磷的需要量和玉米、去皮豆粕及磷酸氢钙的营养成分计算得出。假设日粮中添加 0.1%的赖氨酸盐酸盐及 3%的维生素和矿物质。玉米和豆粕添加水平满足回肠标准可消化赖氨酸需要，而磷酸氢钙的添加量满足全消化道标准可消化磷水平。

h 1 IU 维生素 A=0.30 μg 视黄醇或者 0.344 μg 视黄醇乙酸酯。维生素 A 的活性（视黄醇当量）也由 β-胡萝卜素提供（见维生素章节）。

i 1 IU 维生素 D_2 或 D_3=0.025 μg。

j 1 IU 维生素 E=0.67 mg D-α-生育酚或者 1mg DL-α-醋酸生育酚。近来在猪上的研究表明，天然的和人工合成的 α-醋酸生育酚有实质性的差别（见维生素章节）。

k 玉米、高粱、小麦和大麦中的烟酸不可被利用。同样，这些谷物来源的副产物中的烟酸利用率也很低，除非对这些副产物进行了湿磨工艺的发酵。

第 17 章 饲料原料成分

引言

饲料原料成分见表 17-1。表中所有数据都以饲喂状态为基础。本版本中原料的营养和常规成分呈现形式与以前版本的《猪营养需要》不同。在此版本中，每一种原料都单独列为一页。之所以采用这种形式是为了方便使用。因为在大多数配方软件上一种原料的所有营养和常规成分都是一次性录入，而不是单个或一组相同成分一次性录入。原料名称、美国饲料管理协会（AAFCO，2010）设计的原料编号及原料在 AAFCO（2010）的页码都列在该页中，某些原料具有国际饲料号码（IFN），也一并列在该原料所在页上。如果这种原料的描述与 AAFCO（2010）有不同或 AAFCO（2010）没有对这种原料的描述，本版本提供一个简要的描述。

编委会对每种原料的营养/常规成分的相关文献进行了详尽的检索。对于总的营养/常规成分，文献的检索集中在最近 15 年。对于氨基酸回肠表观和标准消化率、磷的全消化道表观和标准消化率的文献检索没有考虑文献的发表时间，而是尽量全部囊括。下文对表 17-1 中每一种原料成分的组成进行了简单的解释。

对于所有的营养素成分，如果列出了所参照的文献的数量和标准差（如果文献数量大于 1），那么相关信息就是来源于编委会对文献的综述。如果没有列出文献的数量，那么内容就是来源于其他的综述（NRC，1998，2007；Sauvant et al.，2004；CVB，2008；AminoDat，2010）。

虽然编委会考虑到某些谷物的营养成分由于产地原因差异很大，但是并没有足够的数据可以在表格中做出区分。其他数据库，如国际生命科学研究所（http://www.cropcomposition.org）编译的数据库，含有大量产地对一些主要谷物影响的资料。

常规成分和碳水化合物

本段落中表 17-1 的内容几乎完全来源于编委会对于文献的综述，除了以下说明的例外情况。淀粉和酸性洗涤纤维的内容或来自编委会的综述或从其他文献的综述，其他综述的文献用于计算净能（NE，第 1 章）。乙醚浸出物值在本节的表中给出，如下所示，它也在"脂肪酸"部分给出。虽然在已发表的文献中实验室方法并不总是清楚的，但我们还是假设乙醚浸出物是采取石油乙醚提取的方法，酸乙醚浸出物采用酸水解的方法。粗纤维包括在"常规成分"部分。虽然普遍认为粗纤维值无论在理论上还是实践中对猪的营养没有价值，但它仍然在世界各地被采用，也标注在美国饲料标签内容里。

氨基酸

表 17-1 中总氨基酸含量完全来源于编委会的综述或 NRC（1998）。氨基酸的表观消化率来自编委会对文献的综述和其他综述性文献。如果检索到的关于氨基酸表观消化率文献为 3 个或更少，则将这些数据与 NRC（1998）或 CVB（2008）对比。如果编委会检索到文献的数据，不管文献的数量多少，与 NRC（1998）或 CVB（2008）的值比较接近，就采用这些文献的数据。如果检索到的文献只有 3 个

或更少，而且数据与 NRC（1998）或 CVB（2008）差别较大，那么就采用 NRC（1998）的数据。如果 NRC（1998）没有这些数据，则采用 CVB（2008）的数据。回肠标准消化率的数据也采用同样的方法，或者是对 NRC（1998）、Sauvant 等（2004）、CVB（2008）和 AminoDat（2010）的数据综合对比后得出定论。如果没有检索到相关文献，则采用综述中数据的平均值。对于某些特定的原料系列如玉米副产物（第9章），系列里的所有原料都采用这个系列的平均值，这在表中很容易看到。

矿物质

矿物元素的含量数据来自于编委会对文献的综述或 NRC（1998）。微量元素的数据基本上来源于 NRC（1998）。磷的全消化道表观或标准消化率全部来自编委会的文献综述，并且按照第13章描述的方法计算得到。几种常量元素的含量在表17-2中列出，数据来源于 NRC（1998），做了少量的编辑。表17-3列出了微量元素的来源和生物利用率，数据也来源于 NRC（1998）。

维生素

维生素含量几乎完全来源于 NRC（1998）。

脂肪酸

脂肪酸含量数据来源于 Sauvant 等（2004）或美国农业部（USDA，2010）。脂肪酸以乙醚浸出物的百分比的形式给出。乙醚浸出物的值与脂肪酸值来源相同，这个值与"常规成分"部分编委会综述的乙醚浸出物值不全部相同。碘价和碘价物按照第3章中描述的方法进行计算。不同来源的脂肪和油的特性和能值列在表17-4中。

能量

总能数据来源于编委会综述或 NRC（1998）。消化能数据来源于编委会综述、NRC（1998）或 Sauvant 等（2004）。净能数据按照第1章描述的方法进行计算。

原料列表

以下原料列在表17-1中。
1 苜蓿干草
2 苜蓿粉
3 面包渣
4 大麦
5 裸大麦
6 蚕豆
7 菜豆
8 血细胞

9 血粉
10 血浆
11 啤酒糟
12 亚麻荞粕
13 全脂双低菜籽
14 压榨双低菜籽粕
15 浸提双低菜籽粕
16 木薯粉
17 柑橘渣
18 压榨椰子粕
19 椰子粕
20 普通黄玉米
21 高营养玉米
22 玉米麸
23 玉米酒精糟
24 玉米酒精糟及可溶物，脂肪含量 >10%
25 玉米酒精糟及可溶物，6%<脂肪含量< 9%
26 玉米酒精糟及可溶物，脂肪含量< 4%
27 玉米高蛋白酒精糟
28 玉米酒精糟可溶物
29 玉米胚芽
30 玉米胚芽粕
31 玉米蛋白饲料
32 玉米蛋白粉
33 玉米糁，玉米渣
34 全脂棉籽
35 棉粕
36 喷干全鸡蛋粉
37 羽毛粉
38 混合鱼粉
39 亚麻籽
40 亚麻籽粕
41 明胶
42 膨化芸豆
43 生芸豆
44 扁豆
45 羽扇豆
46 肉骨粉，磷>4%
47 肉粉
48 酪蛋白（来源于牛奶）
49 乳糖（来源于牛奶）
50 脱脂奶粉
51 低蛋白乳清粉，乳糖 80%
52 低蛋白乳清粉，乳糖 85%
53 乳清粉
54 乳清浓缩蛋白

55 黍
56 甜菜糖蜜
57 甘蔗糖蜜
58 去壳燕麦
59 燕麦
60 裸燕麦
61 去壳压片燕麦
62 压榨棕榈仁
63 棕榈仁粕
64 压榨花生粕
65 花生粕
66 豌豆浓缩蛋白
67 鹰嘴豆
68 豇豆
69 紫花豌豆
70 裂荚紫花豌豆
71 宠物食品的副产物
72 猪肠膜蛋白
73 马铃薯浓缩蛋白
74 禽副产物
75 禽肉粉
76 大米
77 米糠
78 脱脂米糠
79 碎大米
80 抛光大米
81 大米浓缩蛋白
82 黑麦
83 红花粕
84 去壳红花粕
85 鲑鱼蛋白水解物
86 芝麻粕
87 高粱
88 高粱酒精糟及可溶物
89 大豆皮
90 压榨去皮豆粕
91 溶剂浸提去皮豆粕
92 酶处理豆粕
93 压榨豆粕
94 发酵豆粕
95 溶剂浸提去皮高蛋白豆粕
96 压榨高蛋白豆粕
97 溶剂浸提去皮低寡聚糖豆粕
98 压榨低寡聚糖豆粕
99 溶剂浸提豆粕
100 全脂大豆

101 高蛋白全脂大豆
102 低寡聚糖全脂大豆
103 大豆浓缩蛋白
104 大豆分离蛋白
105 甜菜渣
106 全脂葵花籽
107 溶剂浸提去壳葵花粕
108 溶剂浸提葵花粕
109 黑小麦
110 黑小麦酒精糟及可溶物
111 硬红小麦
112 软红小麦
113 小麦麸
114 小麦酒精糟及可溶物
115 小麦蛋白
116 小麦细麸
117 小麦加工筛下物
118 小麦次粉
119 啤酒酵母
120 白酒酵母
121 酵母单细胞蛋白
122 圆酵母

参 考 文 献

AAFCO (Association of American Feed Control Officials). 2010. *Official Publication.* Oxford, IN: AAFCO.

AminoDat 4.0. 2010. Evonik Industries, Hanau, Germany.

Cera, K. R., D. C. Mahan, and G. A. Reinhart. 1989. Apparent fat digestibilities and performance responses of postweaning swine fed diets supplemented with coconut oil, corn oil or tallow. *Journal of Animal Science* 67:2040-2047.

CVB (Dutch PDV [Product Board Animal Feed]). 2008. CVB Feedstuff Database. Available online at http://www.pdv.nl/english/Voederwaardering/about_cvb/index.php. Accessed on June 9, 2011.

NRC (National Research Council). 1998. *Nutrient Requirements of Swine,* 10th Rev. Ed. Washington, DC: National Academy Press.

NRC. 2007. *Nutrient Requirements of Horses,* 6th Rev. Ed. Washington, DC: The National Academies Press.

Powles, J., J. Wiseman, D. J. A. Cole, and S. Jagger. 1995. Prediction of the apparent digestible energy value of fats given to pigs. *Animal Science* 61:149-154.

Sauvant, D., J. M. Perez, and G. Tran. 2004. *Tables of Composition and Nutritional Value of Feed Materials: Pigs, Poultry, Sheep, Goats, Rabbits, Horses, Fish,* INRA, Paris, France, ed. Wageningen, the Netherlands: Wageningen Academic.

USDA (U.S. Department of Agriculture), Agricultural Research Service. 2010. USDA National Nutrient Database for Standard Reference, Release 23. Nutrient Data Laboratory Home Page. Available online at http://www.ars.usda.gov/ba/bhnrc/ndl. Accessed on August 10, 2011.

van Milgen, J., J. Noblet, and S. Dubois. 2001. Energetic efficiency of starch, protein, and lipid utilization in growing pigs. *Journal of Nutrition* 131:1309-1318.

表 17-1 猪日粮中原料组成成分(数据以饲喂状态为基础)

原料名称:苜蓿干草(Alfalfa Hay)
AAFCO#: 3.1, AAFCO 2010, p.324
IFN#: 1-30-293

常规成分, %				氨基酸, %								
				总氨基酸				可消化氨基酸				
								回肠表观消化率			回肠标准消化率		
	$\bar{x}$	n	SD		$\bar{x}$	n	SD	$\bar{x}$	n	SD	$\bar{x}$	n	SD
干物质 Dry matter	90.33	3	0.61	必需氨基酸 Essential									
粗蛋白 Crude protein	19.32	7	3.47	粗蛋白 CP	19.32	7	3.47						
粗纤维 Crude fiber				精氨酸 Arg									
乙醚浸出物/粗脂肪 Ether extract	2.30	6	0.63	组氨酸 His									
酸乙醚浸出物 Acid ether extract				异亮氨酸 Ile									
粗灰分 Ash	11.00	6	2.33	亮氨酸 Leu									
碳水化合物成分, %				赖氨酸 Lys									
				蛋氨酸 Met									
乳糖 Lactose				苯丙氨酸 Phe									
蔗糖 Sucrose				苏氨酸 Thr									
棉子糖 Raffinose				色氨酸 Trp									
水苏糖 Stachyose				缬氨酸 Val									
毛蕊花糖 Verbascose				非必需氨基酸 Nonessential									
低聚糖 Oligosaccharides				丙氨酸 Ala									
淀粉 Starch	1.02	1		天冬氨酸 Asp									
中性洗涤纤维 Neutral detergent fiber	37.00	7	7.50	半胱氨酸 Cys									
酸性洗涤纤维 Acid detergent fiber	31.01	7	7.95	谷氨酸 Glu									
半纤维素 Hemicellulose				甘氨酸 Gly									
酸性洗涤木质素 Acid detergent lignin	6.65	1		脯氨酸 Pro									
总膳食纤维 Total dietary fiber				丝氨酸 Ser									
不溶性膳食纤维 Insoluble dietary fiber				酪氨酸 Tyr									
可溶性膳食纤维 Soluble dietary fiber													

矿物质				维生素, mg/kg(标注单位者除外)				脂肪酸, 占乙醚提取物, %			
	$\bar{x}$	n	SD		$\bar{x}$	n	SD		$\bar{x}$	n	SD
常量元素 Macro,%				脂溶性维生素 Fat soluble				粗脂肪			
钙 Ca	1.46	8	0.29	β-胡萝卜素 β-Carotene				C-12:0			
氯 Cl				维生素 E Vitamin E				C-14:0			
钾 K	2.48	4	0.75	水溶性维生素 Water soluble				C-16:0			
镁 Mg	0.27	5	0.04	维生素 B_6 Vitamin B_6				C-16:1			
钠 Na	0.02	1		维生素 B_{12} Vitamin B_{12}, μg/kg				C-18:0			
磷 P	0.26	8	0.07	生物素 Biotin				C-18:1			
硫 S	0.28	2	0.01	叶酸 Folacin				C-18:2			
微量元素 Micro,ppm				烟酸 Niacin				C-18:3			
铬 Cr				泛酸 Pantothenic acid				C-18:4			
铜 Cu	5.50	2	0.01	核黄素 Riboflavin				C-20:0			
铁 Fe	587	1		硫胺素 Thiamin				C-20:1			
碘 I				胆碱 Choline				C-20:4			
锰 Mn	41.32	2	2.81					C-20:5			
硒 Se	0.24	1						C-22:0			
锌 Zn	25.33	2	3.49	能量, kcal/kg				C-22:1			
								C-22:5			
植酸磷 Phytate P,%				总能 GE	4077			C-22:6			
磷全消化道表观消化率 ATTD of P,%				消化能 DE	1830			C-24:0			
磷全消化道标准消化率 STTD of P,%				代谢能 ME	1699			饱和脂肪酸			
				净能 NE	878			单不饱和脂肪酸			
								多不饱和脂肪酸			
								碘价			
								碘价物			

续表

原料名称：苜蓿粉(Alfalfa Meal)
AAFCO#:3.2, AAFCO 2010, p.324
IFN#:1-00-022

常规成分, %				氨基酸, %							
				总氨基酸				可消化氨基酸			
								回肠表观消化率		回肠标准消化率	
	$\bar{x}$	n	SD		$\bar{x}$	n	SD	$\bar{x}$	n	SD	$\bar{x}$	n	SD
干物质 Dry matter	92.30	1		必需氨基酸 Essential									
粗蛋白 Crude protein	16.25	2	3.04	粗蛋白 CP	16.25	2	3.04	39					
粗纤维 Crude fiber				精氨酸 Arg	0.71			64			74		
乙醚浸出物/粗脂肪 Ether extract	1.70			组氨酸 His	0.37			50			59		
酸乙醚浸出物 Acid ether extract	1.70	2	0.99	异亮氨酸 Ile	0.68			59			68		
				亮氨酸 Leu	1.21			63			71		
粗灰分 Ash	10.10	2	1.41	赖氨酸 Lys	0.74			50			56		
碳水化合物成分, %				蛋氨酸 Met	0.25			64			71		
乳糖 Lactose				苯丙氨酸 Phe	0.84			62			70		
蔗糖 Sucrose				苏氨酸 Thr	0.70			51			63		
棉子糖 Raffinose				色氨酸 Trp	0.24			39			46		
水苏糖 Stachyose				缬氨酸 Val	0.86			55			64		
毛蕊花糖 Verbascose				非必需氨基酸 Nonessential									
低聚糖 Oligosaccharides				丙氨酸 Ala	0.87			53			59		
淀粉 Starch	3.40	1		天冬氨酸 Asp	1.93			64			68		
中性洗涤纤维 Neutral detergent fiber	42.00	2	4.95	半胱氨酸 Cys	0.18			20			37		
酸性洗涤纤维 Acid detergent fiber	32.15	2	1.91	谷氨酸 Glu	1.61			51			58		
半纤维素 Hemicellulose	14.70	1		甘氨酸 Gly	0.81			41			51		
酸性洗涤木质素 Acid detergent lignin	8.30	1		脯氨酸 Pro	0.89			61			74		
总膳食纤维 Total dietary fiber				丝氨酸 Ser	0.73			50			59		
不溶性膳食纤维 Insoluble dietary fiber				酪氨酸 Tyr	0.55			59			66		
可溶性膳食纤维 Soluble dietary fiber													

矿物质				维生素, mg/kg(标注单位者除外)				脂肪酸, 占乙醚提取物, %		
	$\bar{x}$	n	SD		$\bar{x}$	n	SD		$\bar{x}$	n	SD
常量元素 Macro,%				脂溶性维生素 Fat soluble				粗脂肪	2.60		
钙 Ca	1.14	2	0.56	β-胡萝卜素 β-Carotene	94.60			C-12:0	1.00		
氯 Cl	0.47			维生素 E Vitamin E	49.80			C-14:0	0.95		
钾 K	2.30			水溶性维生素 Water Soluble				C-16:0	12.80		
镁 Mg	0.23			维生素 B_6 Vitamin B_6	6.50			C-16:1	0.70		
钠 Na	0.09			维生素 B_{12} Vitamin B_{12}, μg/kg	0			C-18:0	1.90		
磷 P	0.30	2	0.06	生物素 Biotin	0.54			C-18:1	2.20		
硫 S	0.29			叶酸 Folacin	4.36			C-18:2	9.65		
微量元素 Micro,ppm				烟酸 Niacin	38.00			C-18:3	18.50		
铬 Cr				泛酸 Pantothenic acid	29.00			C-18:4	0.00		
铜 Cu	10.00			核黄素 Riboflavin	13.60			C-20:0	1.80		
铁 Fe	333			硫胺素 Thiamin	3.40			C-20:1	0.00		
碘 I				胆碱 Choline	1401			C-20:4	0.00		
锰 Mn	32.00							C-20:5	0.00		
硒 Se	0.34							C-22:0	1.45		
锌 Zn	24.00			能量, kcal/kg				C-22:1	0.00		
								C-22:5	0.00		
植酸磷 Phytate P,%				总能 GE	4038			C-22:6	0.00		
磷全消化道表观消化率 ATTD of P,%	50			消化能 DE	1830			C-24:0	0.70		
磷全消化道标准消化率 STTD of P,%	55			代谢能 ME	1720			饱和脂肪酸	20.60		
				净能 NE	897			单不饱和脂肪酸	2.90		
								多不饱和脂肪酸	28.15		
								碘价	70.73		
								碘价物	18.39		

续表

原料名称：面包渣(Bakery Meal)
AAFCO#:60.15, AAFCO 2010, p.375
IFN#: 4-00-466

常规成分, %				氨基酸, %								
				总氨基酸				可消化氨基酸				
	$\bar{x}$	n	SD		$\bar{x}$	n	SD	回肠表观消化率			回肠标准消化率	
								$\bar{x}$	n	SD	$\bar{x}$	n	SD
干物质 Dry matter	90.8	1		必需氨基酸 Essential									
粗蛋白 Crude protein	12.30	1		粗蛋白 CP	12.30	1							
粗纤维 Crude fiber				精氨酸 Arg	0.58	1							
乙醚浸出物/粗脂肪 Ether extract	8.05			组氨酸 His	0.22	1							
酸乙醚浸出物 Acid ether extract				异亮氨酸 Ile	0.51	1					94		
粗灰分 Ash				亮氨酸 Leu	0.88	1					90		
碳水化合物成分, %				赖氨酸 Lys	0.41	1					77		
				蛋氨酸 Met	0.19	1					90		
乳糖 Lactose				苯丙氨酸 Phe	0.50	1							
蔗糖 Sucrose				苏氨酸 Thr	0.42	1					69		
棉子糖 Raffinose				色氨酸 Trp	0.15	1					91		
水苏糖 Stachyose				缬氨酸 Val	0.53	1					93		
毛蕊花糖 Verbascose				非必需氨基酸 Nonessential									
低聚糖 Oligosaccharides				丙氨酸 Ala	0.52	1							
淀粉 Starch	52.80			天冬氨酸 Asp	0.45	1							
中性洗涤纤维 Neutral detergent fiber	2.00			半胱氨酸 Cys	0.18	1					91		
酸性洗涤纤维 Acid detergent fiber	5.51			谷氨酸 Glu	1.92	1							
半纤维素 Hemicellulose				甘氨酸 Gly	0.78	1							
酸性洗涤木质素 Acid detergent lignin				脯氨酸 Pro	0.98	1							
总膳食纤维 Total dietary fiber				丝氨酸 Ser	0.56	1							
不溶性膳食纤维 Insoluble dietary fiber				酪氨酸 Tyr	0.55	1							
可溶性膳食纤维 Soluble dietary fiber													

矿物质				维生素, mg/kg(标注单位者除外)				脂肪酸, 占乙醚提取物, %			
	$\bar{x}$	n	SD		$\bar{x}$	n	SD		$\bar{x}$	n	SD
常量元素 Macro,%				脂溶性维生素 Fat soluble				粗脂肪			
钙 Ca	0.13			β-胡萝卜素 β-Carotene	4.2			C-12:0			
氯 Cl	1.48			维生素 E Vitamin E				C-14:0			
钾 K	0.39			水溶性维生素 Water soluble				C-16:0			
镁 Mg	0.24			维生素 B_6 Vitamin B_6	4.3			C-16:1			
钠 Na	1.14			维生素 B_{12} Vitamin B_{12}, μg/kg	0			C-18:0			
磷 P	0.25			生物素 Biotin	0.07			C-18:1			
硫 S	0.02			叶酸 Folacin	0.20			C-18:2			
微量元素 Micro,ppm				烟酸 Niacin	26			C-18:3			
铬 Cr				泛酸 Pantothenic acid	8.3			C-18:4			
铜 Cu	5.00			核黄素 Riboflavin	1.4			C-20:0			
铁 Fe	28.00			硫胺素 Thiamin	2.9			C-20:1			
碘 I				胆碱 Choline	923			C-20:4			
锰 Mn	65.00							C-20:5			
硒 Se								C-22:0			
锌 Zn	15.00							C-22:1			
				能量, kcal/kg				C-22:5			
植酸磷 Phytate P,%				总能 GE	4558			C-22:6			
磷全消化道表观消化率 ATTD of P, %				消化能 DE	3940			C-24:0			
磷全消化道标准消化率 STTD of P, %				代谢能 ME	3856			饱和脂肪酸			
				净能 NE	2981			单不饱和脂肪酸			
								多不饱和脂肪酸			
								碘价			
								碘价物			

续表

原料名称：大麦(Barley)
AAFCO#：无官方定义
IFN#：4-00-572

常规成分, %				氨基酸, %							
				总氨基酸			可消化氨基酸				
							回肠表观消化率			回肠标准消化率	
	$\bar{x}$	n	SD		$\bar{x}$	n	SD	$\bar{x}$	n	SD	$\bar{x}$	n	SD

常规成分	$\bar{x}$	n	SD	氨基酸	$\bar{x}$	n	SD	$\bar{x}$(AID)	n	SD	$\bar{x}$(SID)	n	SD
干物质 Dry matter	89.90	52	2.65	必需氨基酸 Essential									
粗蛋白 Crude protein	11.33	76	1.54	粗蛋白 CP	11.33	76	1.54	66	20	7.41	79	18	6.03
粗纤维 Crude fiber	3.90	12	0.70	精氨酸 Arg	0.53	31	0.09	75	22	6.13	85	22	5.92
乙醚浸出物/粗脂肪 Ether extract	2.11	33	0.65	组氨酸 His	0.27	28	0.07	73	21	7.14	81	21	4.89
酸乙醚浸出物 Acid ether extract	2.10	4	0.48	异亮氨酸 Ile	0.37	37	0.07	72	22	5.37	79	22	9.00
粗灰分 Ash	2.38	38	0.42	亮氨酸 Leu	0.72	30	0.11	74	22	4.90	81	22	4.71
碳水化合物成分, %				赖氨酸 Lys	0.40	38	0.05	66	21	8.77	75	21	8.70
				蛋氨酸 Met	0.20	35	0.03	76	19	5.69	82	19	5.62
乳糖 Lactose				苯丙氨酸 Phe	0.53	28	0.11	76	21	5.71	81	21	5.31
蔗糖 Sucrose				苏氨酸 Thr	0.36	37	0.05	60	22	9.43	76	22	9.56
棉子糖 Raffinose				色氨酸 Trp	0.13	23	0.02	73	8	6.98	82	8	7.03
水苏糖 Stachyose				缬氨酸 Val	0.52	37	0.08	71	22	6.88	80	22	7.16
毛蕊花糖 Verbascose				非必需氨基酸 Nonessential									
低聚糖 Oligosaccharides				丙氨酸 Ala	0.44	25	0.06	60	21	9.09	73	21	8.36
淀粉 Starch	50.21	17	5.20	天冬氨酸 Asp	0.65	25	0.10	63	21	9.97	75	21	9.78
中性洗涤纤维 Neutral detergent fiber	18.29	32	3.38	半胱氨酸 Cys	0.26	34	0.06	73	17	7.77	81	17	7.54
酸性洗涤纤维 Acid detergent fiber	5.78	33	1.32	谷氨酸 Glu	2.50	25	0.60	82	21	8.77	87	21	5.64
半纤维素 Hemicellulose	14.1	1		甘氨酸 Gly	0.45	26	0.07	53	21	15.62	82	21	15.75
酸性洗涤木质素 Acid detergent lignin	2.28	9	0.67	脯氨酸 Pro	1.11	23	0.32	60	16	19.77	88	16	29.25
总膳食纤维 Total dietary fiber	15.35	1		丝氨酸 Ser	0.45	27	0.08	68	21	8.71	80	20	8.38
不溶性膳食纤维 Insoluble dietary fiber				酪氨酸 Tyr	0.28	28	0.06	68	18	14.54	78	17	12.65
可溶性膳食纤维 Soluble dietary fiber													

矿物质	$\bar{x}$	n	SD	维生素, mg/kg(标注单位者除外)	$\bar{x}$	n	SD	脂肪酸, 占乙醚提取物, %	$\bar{x}$	n	SD
常量元素 Macro,%				脂溶性维生素 Fat soluble				粗脂肪	1.60		
钙 Ca	0.06	32	0.02	β-胡萝卜素 β-Carotene	4.1			C-12:0	0.25		
氯 Cl	0.12			维生素 E Vitamin E	7.4			C-14:0	0.50		
钾 K	0.38	3	0.17	水溶性维生素 Water soluble				C-16:0	17.88		
镁 Mg	0.14	5	0.01	维生素 B_6 Vitamin B_6	5.0			C-16:1	0.25		
钠 Na	0.02	1		维生素 B_{12} Vitamin B_{12}, μg/kg	0			C-18:0	0.75		
磷 P	0.35	39	0.04	生物素 Biotin	0.14			C-18:1	10.50		
硫 S	0.13	3	0.05	叶酸 Folacin	0.31			C-18:2	43.44		
微量元素 Micro,ppm				烟酸 Niacin	55			C-18:3	4.81		
铬 Cr				泛酸 Pantothenic acid	8.0			C-18:4			
铜 Cu	5.43	4	1.94	核黄素 Riboflavin	1.8			C-20:0	0.00		
铁 Fe	75.70	2	19.80	硫胺素 Thiamin	4.5			C-20:1	0.00		
碘 I				胆碱 Choline	1034			C-20:4			
锰 Mn	16.29	3	0.77					C-20:5			
硒 Se	0.10	1						C-22:0			
锌 Zn	28.09	4	6.95	能量, kcal/kg				C-22:1			
								C-22:5			
植酸磷 Phytate P,%	0.22	17	0.04	总能 GE	3939	24	87	C-22:6			
磷全消化道表观消化率 ATTD of P, %	39	11	5.31	消化能 DE	3150	8	350	C-24:0			
磷全消化道标准消化率 STTD of P, %	45	11	5.84	代谢能 ME	3073			饱和脂肪酸	19.38		
				净能 NE	2327			单不饱和脂肪酸	10.75		
								多不饱和脂肪酸	48.25		
								碘价	101.46		
								碘价物	16.23		

续表

原料名称：裸大麦(Barley, Hulless)
AAFCO#：无官方定义
IFN#：4-00-552

常规成分, %				氨基酸, %									
				总氨基酸				可消化氨基酸					
								回肠表观消化率			回肠标准消化率		
	$\bar{x}$	n	SD		$\bar{x}$	n	SD	$\bar{x}$	n	SD	$\bar{x}$	n	SD
干物质 Dry matter	89.58	13	1.80	必需氨基酸 Essential									
粗蛋白 Crude protein	12.77	20	0.91	粗蛋白 CP	12.77	20	0.91	63	9	3.25	69	10	20.64
粗纤维 Crude fiber	1.1	1		精氨酸 Arg	0.68	15	0.22	68	10	6.72	77	10	7.79
乙醚浸出物/粗脂肪 Ether extract	3.17	9	0.59	组氨酸 His	0.40	14	0.14	71	9	7.11	77	9	8.81
酸乙醚浸出物 Acid ether extract				异亮氨酸 Ile	0.35	16	0.12	65	10	7.32	75	10	5.36
粗灰分 Ash	1.94	3	0.39	亮氨酸 Leu	0.74	15	0.16	68	10	5.99	75	10	5.27
碳水化合物成分, %				赖氨酸 Lys	0.51	16	0.14	56	10	5.01	65	10	5.48
				蛋氨酸 Met	0.20	14	0.03	68	8	4.39	73	8	4.21
乳糖 Lactose				苯丙氨酸 Phe	0.54	14	0.14	70	9	5.11	75	9	5.21
蔗糖 Sucrose				苏氨酸 Thr	0.37	16	0.05	56	10	4.15	70	10	5.13
棉子糖 Raffinose				色氨酸 Trp	0.13	2	0.03						
水苏糖 Stachyose				缬氨酸 Val	0.55	14	0.08	66	10	6.95	75	10	6.51
毛蕊花糖 Verbascose				非必需氨基酸 Nonessential									
低聚糖 Oligosaccharides				丙氨酸 Ala	0.58	14	0.15	54	10	7.65	66	10	8.81
淀粉 Starch	54.56	2	1.73	天冬氨酸 Asp	0.64	14	0.15	58	10	4.81	70	10	5.57
中性洗涤纤维 Neutral detergent fiber	12.55	11	1.84	半胱氨酸 Cys	0.23	14	0.06	64	8	6.17	72	8	6.06
酸性洗涤纤维 Acid detergent fiber	2.18	3	0.55	谷氨酸 Glu	3.61	14	1.04	77	10	4.33	80	10	4.56
半纤维素 Hemicellulose				甘氨酸 Gly	0.71	14	0.38	47	10	8.43	77	10	15.25
酸性洗涤木质素 Acid detergent lignin				脯氨酸 Pro	0.97	10	0.54	67	6	6.86	112	6	18.91
总膳食纤维 Total dietary fiber				丝氨酸 Ser	0.63	14	0.14	63	10	5.98	73	10	7.42
不溶性膳食纤维 Insoluble dietary fiber				酪氨酸 Tyr	0.25	14	0.12	65	9	8.61	74	9	9.17
可溶性膳食纤维 Soluble dietary fiber													

矿物质				维生素, mg/kg(标注单位者除外)				脂肪酸, 占乙醚提取物, %			
	$\bar{x}$	n	SD		$\bar{x}$	n	SD		$\bar{x}$	n	SD
常量元素 Macro,%				脂溶性维生素 Fat soluble				粗脂肪			
钙 Ca	0.06	5	0.03	β-胡萝卜素 β-Carotene				C-12:0			
氯 Cl	0.10			维生素 E Vitamin E	6.0			C-14:0			
钾 K	0.44			水溶性维生素 Water soluble				C-16:0			
镁 Mg	0.12			维生素 B_6 Vitamin B_6	5.6			C-16:1			
钠 Na	0.02			维生素 B_{12} Vitamin B_{12}, μg/kg	0			C-18:0			
磷 P	0.36	9	0.06	生物素 Biotin	0.07			C-18:1			
硫 S				叶酸 Folacin	0.62			C-18:2			
微量元素 Micro,ppm				烟酸 Niacin	48			C-18:3			
铬 Cr				泛酸 Pantothenic acid	6.8			C-18:4			
铜 Cu	5			核黄素 Riboflavin	1.8			C-20:0			
铁 Fe	56			硫胺素 Thiamin	4.3			C-20:1			
碘 I				胆碱 Choline				C-20:4			
锰 Mn	16							C-20:5			
硒 Se								C-22:0			
锌 Zn	27			能量, kcal/kg				C-22:1			
								C-22:5			
植酸磷 Phytate P,%	0.26	3	0.03	总能 GE	3959	5	71	C-22:6			
磷全消化道表观消化率 ATTD of P,%	31	1		消化能 DE	3266			C-24:0			
磷全消化道标准消化率 STTD of P,%	36	1		代谢能 ME	3179			饱和脂肪酸			
				净能 NE	2464			单不饱和脂肪酸			
								多不饱和脂肪酸			
								碘价			
								碘价物			

续表

原料名称：蚕豆(Beans, Faba)
AAFCO#: 无官方定义

常规成分, %				氨基酸, %									
				总氨基酸				可消化氨基酸					
								回肠表观消化率			回肠标准消化率		
	$\bar{x}$	n	SD		$\bar{x}$	n	SD	$\bar{x}$	n	SD	$\bar{x}$	n	SD
干物质 Dry matter	88.12	15	0.84	必需氨基酸 Essential									
粗蛋白 Crude protein	27.16	26	1.83	粗蛋白 CP	27.16	26	1.83	73	24	5.88	79	24	5.67
粗纤维 Crude fiber	8.55	3	0.82	精氨酸 Arg	2.43	19	0.31	88	18	3.28	90	18	3.09
乙醚浸出物/粗脂肪 Ether extract	1.30	13	0.14	组氨酸 His	0.72	21	0.05	76	20	7.83	79	20	8.10
酸乙醚浸出物 Acid ether extract				异亮氨酸 Ile	1.13	25	0.10	77	25	6.03	81	25	5.25
粗灰分 Ash	3.43	15	0.38	亮氨酸 Leu	1.94	25	0.20	79	25	4.80	82	25	4.94
碳水化合物成分, %				赖氨酸 Lys	1.65	25	0.20	82	25	4.16	85	25	4.26
				蛋氨酸 Met	0.19	25	0.02	65	23	8.22	73	23	11.69
乳糖 Lactose	0.00	6	0	苯丙氨酸 Phe	1.19	21	0.11	77	20	5.37	80	20	5.93
蔗糖 Sucrose	0.00	6	0	苏氨酸 Thr	0.91	25	0.13	70	25	6.37	78	25	6.34
棉子糖 Raffinose	0.00	6	0	色氨酸 Trp	0.22	16	0.06	61	16	11.40	64	14	11.22
水苏糖 Stachyose	0.00	6	0	缬氨酸 Val	1.22	25	0.13	73	25	5.84	78	25	4.95
毛蕊花糖 Verbascose	0.00	6	0	非必需氨基酸 Nonessential									
低聚糖 Oligosaccharides				丙氨酸 Ala	1.05	19	0.12	72	18	5.45	78	18	5.60
淀粉 Starch	39.22	14	2.38	天冬氨酸 Asp	2.80	19	0.34	81	18	4.39	85	18	4.18
中性洗涤纤维 Neutral detergent fiber	13.29	16	2.39	半胱氨酸 Cys	0.34	23	0.03	56	22	9.80	62	22	10.87
酸性洗涤纤维 Acid detergent fiber	10.33	16	1.07	谷氨酸 Glu	4.40	19	0.65	85	18	4.10	88	18	3.14
半纤维素 Hemicellulose	1.86	6	0.43	甘氨酸 Gly	1.09	19	0.15	62	18	9.85	76	18	9.24
酸性洗涤木质素 Acid detergent lignin	0.48	8	0.41	脯氨酸 Pro	0.99	13	0.34	50	11	23.39	87	11	20.89
总膳食纤维 Total dietary fiber				丝氨酸 Ser	1.22	19	0.24	77	18	7.47	83	18	5.50
不溶性膳食纤维 Insoluble dietary fiber				酪氨酸 Tyr	0.84	7	0.14	74	10	5.65	82	9	6.80
可溶性膳食纤维 Soluble dietary fiber													

矿物质				维生素, mg/kg (标注单位者除外)				脂肪酸, 占乙醚提取物, %		
	$\bar{x}$	n	SD		$\bar{x}$	n	SD		$\bar{x}$	SD
常量元素 Macro, %				脂溶性维生素 Fat soluble				粗脂肪	1.30	
钙 Ca	0.14	3	0.04	β-胡萝卜素 β-Carotene				C-12:0	0.00	
氯 Cl	0.07			维生素 E Vitamin E	0.8			C-14:0	0.32	
钾 K	1.20			水溶性维生素 Water soluble				C-16:0	13.52	
镁 Mg	0.15			维生素 B_6 Vitamin B_6				C-16:1	0.00	
钠 Na	0.03			维生素 B_{12} Vitamin B_{12}, μg/kg	0			C-18:0	2.08	
磷 P	0.42	3	0.01	生物素 Biotin	0.09			C-18:1	20.80	
硫 S	0.29			叶酸 Folacin				C-18:2	39.68	
微量元素 Micro, ppm				烟酸 Niacin	26			C-18:3	2.80	
铬 Cr				泛酸 Pantothenic acid	3.0			C-18:4	0.00	
铜 Cu	11			核黄素 Riboflavin	2.9			C-20:0		
铁 Fe	75			硫胺素 Thiamin	5.5			C-20:1	0.00	
碘 I				胆碱 Choline	1670			C-20:4	0.00	
锰 Mn	15							C-20:5	0.00	
硒 Se	0.02							C-22:0	0.00	
锌 Zn	42			能量, kcal/kg				C-22:1	0.00	
								C-22:5	0.00	
植酸磷 Phytate P, %	0.23	1		总能 GE	4473			C-22:6	0.00	
磷全消化道表观消化率 ATTD of P, %	32	1		消化能 DE	3245			C-24:0	0.00	
磷全消化道标准消化率 STTD of P, %	36	1		代谢能 ME	3060			饱和脂肪酸	15.92	
				净能 NE	2143			单不饱和脂肪酸	20.80	
								多不饱和脂肪酸	42.48	
								碘价	98.16	
								碘价物	12.76	

续表

原料名称: 菜豆(Beans, Phaselous Beans)
AAFCO#: 无官方定义

常规成分, %				氨基酸, %								
				总氨基酸				可消化氨基酸				
	$\bar{x}$	n	SD		$\bar{x}$	n	SD	回肠表观消化率			回肠标准消化率		
干物质 Dry matter				必需氨基酸 Essential				$\bar{x}$	n	SD	$\bar{x}$	n	SD
粗蛋白 Crude protein	22.90			粗蛋白 CP	22.90			49					
粗纤维 Crude fiber				精氨酸 Arg	1.91			70			72		
乙醚浸出物/粗脂肪 Ether extract				组氨酸 His	0.74						58		
酸乙醚浸出物 Acid ether extract				异亮氨酸 Ile	1.17			50			54		
粗灰分 Ash				亮氨酸 Leu	2.05			52			55		
碳水化合物成分, %				赖氨酸 Lys	1.67			65			68		
				蛋氨酸 Met	0.29			52			55		
乳糖 Lactose				苯丙氨酸 Phe	1.41			41			44		
蔗糖 Sucrose				苏氨酸 Thr	1.12			50			55		
棉子糖 Raffinose				色氨酸 Trp	0.27			50			55		
水苏糖 Stachyose				缬氨酸 Val	1.33			49			53		
毛蕊花糖 Verbascose				非必需氨基酸 Nonessential									
低聚糖 Oligosaccharides				丙氨酸 Ala	1.12			50			55		
淀粉 Starch	34.40			天冬氨酸 Asp	3.06			45			47		
中性洗涤纤维 Neutral detergent fiber				半胱氨酸 Cys	0.29			38			45		
酸性洗涤纤维 Acid detergent fiber				谷氨酸 Glu	4.17			53			56		
半纤维素 Hemicellulose				甘氨酸 Gly	1.06			41			50		
酸性洗涤木质素 Acid detergent lignin				脯氨酸 Pro	1.04						60		
总膳食纤维 Total dietary fiber				丝氨酸 Ser	1.54			53			57		
不溶性膳食纤维 Insoluble dietary fiber				酪氨酸 Tyr	0.85			52			56		
可溶性膳食纤维 Soluble dietary fiber													

矿物质				维生素, mg/kg(标注单位者除外)				脂肪酸, 占乙醚提取物, %			
	$\bar{x}$	n	SD		$\bar{x}$	n	SD		$\bar{x}$	n	SD
常量元素 Macro, %				脂溶性维生素 Fat soluble				粗脂肪			
钙 Ca	0.21	1		β-胡萝卜素 β-Carotene				C-12:0			
氯 Cl				维生素 E Vitamin E				C-14:0			
钾 K				水溶性维生素 Water soluble				C-16:0			
镁 Mg				维生素 B_6 Vitamin B_6				C-16:1			
钠 Na				维生素 B_{12} Vitamin B_{12}, μg/kg				C-18:0			
磷 P	0.52	1		生物素 Biotin				C-18:1			
硫 S				叶酸 Folacin				C-18:2			
微量元素 Micro, ppm				烟酸 Niacin				C-18:3			
铬 Cr				泛酸 Pantothenic acid				C-18:4			
铜 Cu				核黄素 Riboflavin				C-20:0			
铁 Fe				硫胺素 Thiamin				C-20:1			
碘 I				胆碱 Choline				C-20:4			
锰 Mn								C-20:5			
硒 Se								C-22:0			
锌 Zn								C-22:1			
				能量, kcal/kg				C-22:5			
植酸磷 Phytate P, %	0.17	1		总能 GE				C-22:6			
磷全消化道表观消化率 ATTD of P, %	38	1		消化能 DE				C-24:0			
磷全消化道标准消化率 STTD of P, %	43	1		代谢能 ME				饱和脂肪酸			
				净能 NE				单不饱和脂肪酸			
								多不饱和脂肪酸			
								碘价			
								碘价物			

续表

原料名称：血细胞(Blood Cells)
AAFCO#: 9.24, AAFCO 2010, p.328

常规成分, %				氨基酸, %							
				总氨基酸				可消化氨基酸			
								回肠表观消化率		回肠标准消化率	
	$\bar{x}$	n	SD		$\bar{x}$	n	SD	$\bar{x}$	n	SD	$\bar{x}$	n	SD
干物质 Dry matter	93.43	3	2.75	必需氨基酸 Essential							
粗蛋白 Crude protein	92.83	3	1.27	粗蛋白 CP	92.83	3	1.27				
粗纤维 Crude fiber				精氨酸 Arg	3.37	3	0.19				
乙醚浸出物/粗脂肪 Ether extract	1.50			组氨酸 His	5.84	2	0.20				
酸乙醚浸出物 Acid ether extract	1.50	1		异亮氨酸 Ile	0.31	3	0.06				
粗灰分 Ash	7.00	1		亮氨酸 Leu	12.72	3	0.25				
碳水化合物成分, %				赖氨酸 Lys	7.75	3	1.49				
乳糖 Lactose				蛋氨酸 Met	0.97	3	0.32				
蔗糖 Sucrose				苯丙氨酸 Phe	6.66	3	0.40				
棉子糖 Raffinose				苏氨酸 Thr	3.43	3	0.78				
水苏糖 Stachyose				色氨酸 Trp	1.72	2	0.09				
毛蕊花糖 Verbascose				缬氨酸 Val	8.44	3	0.17				
低聚糖 Oligosaccharides				非必需氨基酸 Nonessential							
淀粉 Starch	0.00			丙氨酸 Ala							
中性洗涤纤维 Neutral detergent fiber				天冬氨酸 Asp							
酸性洗涤纤维 Acid detergent fiber	0.00			半胱氨酸 Cys	0.58	2	0.08				
半纤维素 Hemicellulose				谷氨酸 Glu							
酸性洗涤木质素 Acid detergent lignin				甘氨酸 Gly							
总膳食纤维 Total dietary fiber				脯氨酸 Pro							
不溶性膳食纤维 Insoluble dietary fiber				丝氨酸 Ser							
可溶性膳食纤维 Soluble dietary fiber				酪氨酸 Tyr	2.32	2	0.01				

矿物质				维生素, mg/kg(标注单位者除外)				脂肪酸, 占乙醚提取物, %			
	$\bar{x}$	n	SD		$\bar{x}$	n	SD		$\bar{x}$	n	SD
常量元素 Macro,%				脂溶性维生素 Fat soluble				粗脂肪			
钙 Ca	0.02	2	0.01	β-胡萝卜素 β-Carotene				C-12:0			
氯 Cl	0.96	2	0.49	维生素 E Vitamin E				C-14:0			
钾 K	0.80	1		水溶性维生素 Water soluble				C-16:0			
镁 Mg	0.02	1		维生素 B_6 Vitamin B_6				C-16:1			
钠 Na	0.84	2	0.40	维生素 B_{12} Vitamin B_{12}, μg/kg				C-18:0			
磷 P	0.34	2	0.00	生物素 Biotin				C-18:1			
硫 S	0.49	1		叶酸 Folacin				C-18:2			
微量元素 Micro,ppm				烟酸 Niacin				C-18:3			
铬 Cr				泛酸 Pantothenic acid				C-18:4			
铜 Cu	2.55	2	0.64	核黄素 Riboflavin				C-20:0			
铁 Fe	2675	2	80.04	硫胺素 Thiamin				C-20:1			
碘 I				胆碱 Choline				C-20:4			
锰 Mn	0.4	1						C-20:5			
硒 Se	1.00	1						C-22:0			
锌 Zn	15.75	2	0.35	能量, kcal/kg				C-22:1			
								C-22:5			
植酸磷 Phytate P,%				总能 GE	5216	1		C-22:6			
磷全消化道表观消化率 ATTD of P,%	80	1		消化能 DE				C-24:0			
磷全消化道标准消化率 STTD of P,%	93	1		代谢能 ME				饱和脂肪酸			
				净能 NE				单不饱和脂肪酸			
								多不饱和脂肪酸			
								碘价			
								碘价物			

续表

原料名称：血粉 (Blood Meal)
AAFCO#: 9.61, AAFCO 2010, p.330
IFN#: 5-26-005

常规成分, %				氨基酸, %								
					总氨基酸			可消化氨基酸				
								回肠表观消化率			回肠标准消化率		
	$\bar{x}$	n	SD		$\bar{x}$	n	SD	$\bar{x}$	n	SD	$\bar{x}$	n	SD
干物质 Dry matter	93.23	7	1.97	必需氨基酸 Essential									
粗蛋白 Crude protein	88.65	13	2.74	粗蛋白 CP	88.65	13	2.74	87	3	1.73	89	3	1.84
粗纤维 Crude fiber				精氨酸 Arg	3.83	9	0.43	91	6	5.99	92	5	6.02
乙醚浸出物/粗脂肪 Ether extract	1.45	4	0.06	组氨酸 His	5.39	9	0.33	90	6	6.09	91	5	6.09
酸乙醚浸出物 Acid ether extract	2.00	1		异亮氨酸 Ile	0.97	9	0.63	68	4	9.25	73	4	9.19
粗灰分 Ash	5.82	5	0.76	亮氨酸 Leu	11.45	9	1.10	85	6	1.68	93	5	1.67
碳水化合物成分, %				赖氨酸 Lys	8.60	8	0.57	93	6	1.71	93	5	1.71
				蛋氨酸 Met	1.18	6	0.20	82	4	1.46	88		
乳糖 Lactose				苯丙氨酸 Phe	6.15	9	0.82	91	6	1.64	92	5	1.56
蔗糖 Sucrose				苏氨酸 Thr	4.36	9	0.32	86	6	2.39	87	5	2.36
棉子糖 Raffinose				色氨酸 Trp	1.34	8	0.35	89	4	3.51	91	3	3.55
水苏糖 Stachyose				缬氨酸 Val	7.96	9	0.66	91	6	2.64	92	5	2.62
毛蕊花糖 Verbascose				非必需氨基酸 Nonessential									
低聚糖 Oligosaccharides				丙氨酸 Ala	7.29	2	0.75	89	2	1.77	90	2	1.57
淀粉 Starch	0.00			天冬氨酸 Asp	7.78	2	2.23	87	2	1.77	88	2	1.64
中性洗涤纤维 Neutral detergent fiber				半胱氨酸 Cys	1.26	4	0.44	81			86		
酸性洗涤纤维 Acid detergent fiber	0.00			谷氨酸 Glu	7.18	2	1.13	86	2	0.40	87	2	0.51
半纤维素 Hemicellulose				甘氨酸 Gly	3.69	2	0.24	86			88		
酸性洗涤木质素 Acid detergent lignin				脯氨酸 Pro	5.03	2	1.78	85			88		
总膳食纤维 Total dietary fiber				丝氨酸 Ser	4.64	2	0.21	88	2	1.10	89	2	1.15
不溶性膳食纤维 Insoluble dietary fiber				酪氨酸 Tyr	2.66	5	0.25	82	4	7.56	88		
可溶性膳食纤维 Soluble dietary fiber													

矿物质				维生素, mg/kg (标注单位者除外)				脂肪酸，占乙醚提取物, %			
	$\bar{x}$	n	SD		$\bar{x}$	n	SD		$\bar{x}$	n	SD
常量元素 Macro, %				脂溶性维生素 Fat soluble				粗脂肪			
钙 Ca	0.05	2	0.01	β-胡萝卜素 β-Carotene				C-12:0			
氯 Cl	0.63	1		维生素 E Vitamin E	1.0			C-14:0			
钾 K	0.15			水溶性维生素 Water soluble				C-16:0			
镁 Mg	0.11			维生素 B_6 Vitamin B_6	4.4			C-16:1			
钠 Na	0.63	1		维生素 B_{12} Vitamin B_{12}, μg/kg	44			C-18:0			
磷 P	0.21	2	0.15	生物素 Biotin	0.03			C-18:1			
硫 S	0.47			叶酸 Folacin	0.10			C-18:2			
微量元素 Micro, ppm				烟酸 Niacin	31			C-18:3			
铬 Cr				泛酸 Pantothenic acid	2.0			C-18:4			
铜 Cu	7.60	1		核黄素 Riboflavin	2.4			C-20:0			
铁 Fe	1494	1		硫胺素 Thiamin	0.4			C-20:1			
碘 I				胆碱 Choline	852			C-20:4			
锰 Mn	0.00	1						C-20:5			
硒 Se								C-22:0			
锌 Zn	49.10	1		能量, kcal/kg				C-22:1			
								C-22:5			
植酸磷 Phytate P, %				总能 GE	5330	1		C-22:6			
磷全消化道表观消化率 ATTD of P, %	67	2	13.29	消化能 DE	4376			C-24:0			
磷全消化道标准消化率 STTD of P, %	88	2	2.55	代谢能 ME	3773			饱和脂肪酸			
				净能 NE	2279			单不饱和脂肪酸			
								多不饱和脂肪酸			
								碘价			
								碘价物			

续表

原料名称：血浆(Blood Plasma)
AAFCO#: 9.72, AAFCO 2010, p.331

常规成分, %				氨基酸, %								
				总氨基酸				可消化氨基酸				
								回肠表观消化率			回肠标准消化率		
	$\bar{x}$	n	SD		$\bar{x}$	n	SD	$\bar{x}$	n	SD	$\bar{x}$	n	SD
干物质 Dry matter	91.97	6	1.1	必需氨基酸 Essential									
粗蛋白 Crude protein	77.84	12	2.12	粗蛋白 CP	77.84	12	2.12	76	2	3.96	81	2	7.90
粗纤维 Crude fiber				精氨酸 Arg	4.39	13	0.29	88	4	6.11	91	4	6.98
乙醚浸出物/粗脂肪 Ether extract	2.00	2	0	组氨酸 His	2.53	13	0.18	85	4	4.62	87	4	5.30
酸乙醚浸出物 Acid ether extract	2.7	1		异亮氨酸 Ile	2.69	13	0.36	81	4	10.41	85	4	12.32
粗灰分 Ash	8.68	4	0.22	亮氨酸 Leu	7.39	13	0.64	84	4	5.57	87	4	6.89
碳水化合物成分, %				赖氨酸 Lys	6.90	12	0.30	85	4	5.57	87	4	6.31
				蛋氨酸 Met	0.79	13	0.19	80	4	5.74	84	4	9.00
乳糖 Lactose				苯丙氨酸 Phe	4.25	13	0.33	83	4	5.55	86	4	7.08
蔗糖 Sucrose				苏氨酸 Thr	4.47	13	0.31	77	4	11.32	80	4	12.86
棉子糖 Raffinose				色氨酸 Trp	1.41	10	0.13	85	2	8.56	92	2	12.23
水苏糖 Stachyose				缬氨酸 Val	5.12	12	0.27	79	4	9.84	82	4	11.31
毛蕊花糖 Verbascose				非必需氨基酸 Nonessential									
低聚糖 Oligosaccharides				丙氨酸 Ala	4.01	7	0.33	81	3	7.89	85	3	7.34
淀粉 Starch	0.00			天冬氨酸 Asp	7.39	7	0.33	83	3	7.62	86	3	6.48
中性洗涤纤维 Neutral detergent fiber				半胱氨酸 Cys	2.60	9	0.33	82	2	12.23	85	2	7.78
酸性洗涤纤维 Acid detergent fiber	0.00			谷氨酸 Glu	10.92	7	0.65	85	3	7.72	87	3	6.38
半纤维素 Hemicellulose				甘氨酸 Gly	2.75	7	0.18	73	3	5.96	85	3	2.16
酸性洗涤木质素 Acid detergent lignin				脯氨酸 Pro	4.30	7	0.38	87	3	4.42	99	3	5.03
总膳食纤维 Total dietary fiber				丝氨酸 Ser	4.15	7	0.33	84	3	5.49	87	3	5.46
不溶性膳食纤维 Insoluble dietary fiber				酪氨酸 Tyr	3.89	10	0.32	74	3	25.71	76	3	27.25
可溶性膳食纤维 Soluble dietary fiber													

矿物质				维生素, mg/kg(标注单位者除外)				脂肪酸, 占乙醚提取物, %			
	$\bar{x}$	n	SD		$\bar{x}$	n	SD		$\bar{x}$	n	SD
常量元素 Macro,%				脂溶性维生素 Fat soluble				粗脂肪			
钙 Ca	0.13	3	0.04	β-胡萝卜素 β-Carotene				C-12:0			
氯 Cl	1.19	1		维生素 E Vitamin E				C-14:0			
钾 K	0.02	1		水溶性维生素 Water soluble				C-16:0			
镁 Mg	0.03	1		维生素 B_6 Vitamin B_6				C-16:1			
钠 Na	2.76	1		维生素 B_{12} Vitamin B_{12}, μg/kg				C-18:0			
磷 P	1.28	3	0.55	生物素 Biotin				C-18:1			
硫 S	1.02	1		叶酸 Folacin				C-18:2			
微量元素 Micro,ppm				烟酸 Niacin				C-18:3			
铬 Cr				泛酸 Pantothenic acid				C-18:4			
铜 Cu	14.75	2	4.60	核黄素 Riboflavin				C-20:0			
铁 Fe	81	2	5.59	硫胺素 Thiamin				C-20:1			
碘 I				胆碱 Choline				C-20:4			
锰 Mn	2.50	1						C-20:5			
硒 Se	1.60	1						C-22:0			
锌 Zn	13.45	2	0.64	能量, kcal/kg				C-22:1			
								C-22:5			
植酸磷 Phytate P,%				总能 GE	4733	3	98	C-22:6			
磷全消化道表观消化率 ATTD of P,%	92	1		消化能 DE	4546	1		C-24:0			
磷全消化道标准消化率 STTD of P,%	98	1		代谢能 ME	4017			饱和脂肪酸			
				净能 NE	2506			单不饱和脂肪酸			
								多不饱和脂肪酸			
								碘价			
								碘价物			

续表

原料名称：啤酒糟(Brewers Grains)
AAFCO#:15.1, AAFCO 2010, p.333
IFN#: 5-00-516

常规成分, %				氨基酸, %								
				总氨基酸				可消化氨基酸				
								回肠表观消化率		回肠标准消化率		
	$\bar{x}$	n	SD		$\bar{x}$	n	SD	$\bar{x}$	n	SD	$\bar{x}$	n	SD
干物质 Dry matter	92.00			必需氨基酸 Essential									
粗蛋白 Crude protein	26.50			粗蛋白 CP	26.50			70					
粗纤维 Crude fiber				精氨酸 Arg	1.53			81			93		
乙醚浸出物/粗脂肪 Ether extract	4.72			组氨酸 His	0.53			70			83		
酸乙醚浸出物 Acid ether extract	7.30			异亮氨酸 Ile	1.02			81			87		
粗灰分 Ash				亮氨酸 Leu	2.08			73			86		
碳水化合物成分, %				赖氨酸 Lys	1.08			69			80		
				蛋氨酸 Met	0.45			74			87		
乳糖 Lactose				苯丙氨酸 Phe	1.22			81			90		
蔗糖 Sucrose				苏氨酸 Thr	0.95			70			80		
棉子糖 Raffinose				色氨酸 Trp	0.26			73			81		
水苏糖 Stachyose				缬氨酸 Val	1.26			73			84		
毛蕊花糖 Verbascose				非必需氨基酸 Nonessential									
低聚糖 Oligosaccharides				丙氨酸 Ala	1.43			71			74		
淀粉 Starch	5.30			天冬氨酸 Asp	1.94			70			74		
中性洗涤纤维 Neutral detergent fiber	48.70			半胱氨酸 Cys	0.49			67			76		
酸性洗涤纤维 Acid detergent fiber	20.14			谷氨酸 Glu	5.13			71			74		
半纤维素 Hemicellulose				甘氨酸 Gly	1.10			66			74		
酸性洗涤木质素 Acid detergent lignin				脯氨酸 Pro	2.36			69			74		
总膳食纤维 Total dietary fiber				丝氨酸 Ser	1.20			68			74		
不溶性膳食纤维 Insoluble dietary fiber				酪氨酸 Tyr	0.88			91			93		
可溶性膳食纤维 Soluble dietary fiber													

矿物质				维生素, mg/kg(标注单位者除外)				脂肪酸, 占乙醚提取物, %			
	$\bar{x}$	n	SD		$\bar{x}$	n	SD		$\bar{x}$	n	SD
常量元素 Macro,%				脂溶性维生素 Fat soluble				粗脂肪	6.70		
钙 Ca	0.21			β-胡萝卜素 β-Carotene	0.2			C-12:0	0.00		
氯 Cl	0.15			维生素 E Vitamin E				C-14:0	0.54		
钾 K	0.08			水溶性维生素 Water soluble				C-16:0	9.99		
镁 Mg	0.16			维生素 B_6 Vitamin B_6	0.7			C-16:1	0.00		
钠 Na	0.26			维生素 B_{12} Vitamin B_{12}, μg/kg	0			C-18:0	0.68		
磷 P	0.58			生物素 Biotin	0.06			C-18:1	5.40		
硫 S	0.31			叶酸 Folacin	7.10			C-18:2	24.93		
微量元素 Micro,ppm				烟酸 Niacin	43			C-18:3	2.52		
铬 Cr				泛酸 Pantothenic acid	8.0			C-18:4	0.00		
铜 Cu	21			核黄素 Riboflavin	1.4			C-20:0	0.00		
铁 Fe	250			硫胺素 Thiamin	0.6			C-20:1	0.00		
碘 I				胆碱 Choline	1723			C-20:4	0.00		
锰 Mn	38							C-20:5	0.00		
硒 Se	0.70							C-22:0	0.00		
锌 Zn	62			能量, kcal/kg				C-22:1	0.00		
								C-22:5	0.00		
植酸磷 Phytate P,%	0.35			总能 GE	4805			C-22:6	0.00		
磷全消化道表观消化率 ATTD of P,%	32			消化能 DE	2100			C-24:0	0.00		
磷全消化道标准消化率 STTD of P,%	39			代谢能 ME	1920			饱和脂肪酸	11.21		
				净能 NE	1155			单不饱和脂肪酸	5.40		
								多不饱和脂肪酸	27.45		
								碘价	56.86		
								碘价物	38.10		

续表

原料名称：亚麻荠粕(Camelina Meal)
AAFCO#：无官方定义

常规成分, %				氨基酸, %								
				总氨基酸				可消化氨基酸				
								回肠表观消化率			回肠标准消化率		
	$\bar{x}$	n	SD		$\bar{x}$	n	SD	$\bar{x}$	n	SD	$\bar{x}$	n	SD
干物质 Dry matter				必需氨基酸 Essential									
粗蛋白 Crude protein	35.15	2	2.90	粗蛋白 CP	35.15	2	2.90						
粗纤维 Crude fiber	11.9	1		精氨酸 Arg	2.11	2	0.96						
乙醚浸出物/粗脂肪 Ether extract	18.5	1		组氨酸 His	0.80	2	0.06						
酸乙醚浸出物 Acid ether extract				异亮氨酸 Ile	1.32	2	0.09						
粗灰分 Ash				亮氨酸 Leu	2.21	2	0.12						
碳水化合物成分, %				赖氨酸 Lys	1.62	2	0.16						
				蛋氨酸 Met	0.87	2	0.06						
乳糖 Lactose				苯丙氨酸 Phe	1.40	2	0.09						
蔗糖 Sucrose				苏氨酸 Thr	1.30	2	0.10						
棉子糖 Raffinose				色氨酸 Trp	0.42	2	0.11						
水苏糖 Stachyose				缬氨酸 Val	1.81	2	0.14						
毛蕊花糖 Verbascose				非必需氨基酸 Nonessential									
低聚糖 Oligosaccharides				丙氨酸 Ala	1.55	2	0.08						
淀粉 Starch	6.50			天冬氨酸 Asp	2.75	2	0.16						
中性洗涤纤维 Neutral detergent fiber				半胱氨酸 Cys	0.95	1							
酸性洗涤纤维 Acid detergent fiber				谷氨酸 Glu	5.77	2	0.05						
半纤维素 Hemicellulose				甘氨酸 Gly	1.75	2	0.09						
酸性洗涤木质素 Acid detergent lignin				脯氨酸 Pro									
总膳食纤维 Total dietary fiber				丝氨酸 Ser	1.34	2	0.07						
不溶性膳食纤维 Insoluble dietary fiber				酪氨酸 Tyr	0.77	2	0.06						
可溶性膳食纤维 Soluble dietary fiber													

矿物质				维生素, mg/kg(标注单位者除外)				脂肪酸, 占乙醚提取物, %			
	$\bar{x}$	n	SD		$\bar{x}$	n	SD		$\bar{x}$	n	SD
常量元素 Macro,%				脂溶性维生素 Fat soluble				粗脂肪			
钙 Ca	0.21	2	0.00	β-胡萝卜素 β-Carotene				C-12:0			
氯 Cl				维生素 E Vitamin E				C-14:0			
钾 K	1.23	2	0.11	水溶性维生素 Water soluble				C-16:0			
镁 Mg	0.40	2	0.02	维生素 B_6 Vitamin B_6				C-16:1			
钠 Na	0.01	2	0.00	维生素 B_{12} Vitamin B_{12}, μg/kg				C-18:0			
磷 P	0.77	2	0.03	生物素 Biotin				C-18:1			
硫 S	0.72	2	0.12	叶酸 Folacin				C-18:2			
微量元素 Micro,ppm				烟酸 Niacin				C-18:3			
铬 Cr				泛酸 Pantothenic acid				C-18:4			
铜 Cu	6.80	2	0.35	核黄素 Riboflavin				C-20:0			
铁 Fe	137	2	16.26	硫胺素 Thiamin				C-20:1			
碘 I				胆碱 Choline				C-20:4			
锰 Mn	23.85	2	1.91					C-20:5			
硒 Se								C-22:0			
锌 Zn	47.95	2	4.17	能量, kcal/kg				C-22:1			
								C-22:5			
植酸磷 Phytate P,%				总能 GE	4931	1		C-22:6			
磷全消化道表观消化率 ATTD of P,%				消化能 DE				C-24:0			
磷全消化道标准消化率 STTD of P,%				代谢能 ME				饱和脂肪酸			
				净能 NE				单不饱和脂肪酸			
								多不饱和脂肪酸			
								碘价			
								碘价物			

续表

原料名称：全脂双低菜籽(Canola, Full Fat)
AAFCO#：无官方定义

常规成分, %				氨基酸, %								
				总氨基酸				可消化氨基酸				
								回肠表观消化率			回肠标准消化率		
	$\bar{x}$	n	SD		$\bar{x}$	n	SD	$\bar{x}$	n	SD	$\bar{x}$	n	SD
干物质 Dry matter	94.57	5	1.56	必需氨基酸 Essential									
粗蛋白 Crude protein	22.06	6	4.67	粗蛋白 CP	22.06	6	4.67	66			64	1	
粗纤维 Crude fiber	6.10	2	0.13	精氨酸 Arg	1.00	3	0.06	81			84		
乙醚浸出物/粗脂肪 Ether extract	43.61	5	3.51	组氨酸 His	0.60	3	0.06	77			80		
酸乙醚浸出物 Acid ether extract				异亮氨酸 Ile	0.60	3	0.12	73			74		
粗灰分 Ash	3.71	2	0.06	亮氨酸 Leu	1.14	3	0.06	73			76		
碳水化合物成分, %				赖氨酸 Lys	1.01	3	0.05	70			73		
				蛋氨酸 Met	0.38	3	0.04	78			81		
乳糖 Lactose				苯丙氨酸 Phe	0.73	3	0.10	73			77		
蔗糖 Sucrose				苏氨酸 Thr	0.83	3	0.07	64			70		
棉子糖 Raffinose				色氨酸 Trp	0.23	2	0.01	66			71		
水苏糖 Stachyose				缬氨酸 Val	0.83	3	0.17	66			71		
毛蕊花糖 Verbascose				非必需氨基酸 Nonessential									
低聚糖 Oligosaccharides				丙氨酸 Ala	0.84	1		70			73	1	
淀粉 Starch	0.70			天冬氨酸 Asp	1.48	1		66			71		
中性洗涤纤维 Neutral detergent fiber	16.71	5	3.07	半胱氨酸 Cys	0.46	3	0.02	66			70		
酸性洗涤纤维 Acid detergent fiber	12.57	5	0.94	谷氨酸 Glu	3.66	1		77	1		84		
半纤维素 Hemicellulose	1.46	2	0.60	甘氨酸 Gly	0.74	1		62			73		
酸性洗涤木质素 Acid detergent lignin	5.40	2	0.33	脯氨酸 Pro	0.60			70			79		
总膳食纤维 Total dietary fiber				丝氨酸 Ser	0.85	1		69			76		
不溶性膳食纤维 Insoluble dietary fiber				酪氨酸 Tyr	0.55	1		70			75		
可溶性膳食纤维 Soluble dietary fiber													

矿物质				维生素, mg/kg(标注单位者除外)				脂肪酸, 占乙醚提取物, %			
	$\bar{x}$	n	SD		$\bar{x}$	n	SD		$\bar{x}$	n	SD
常量元素 Macro, %				脂溶性维生素 Fat soluble				粗脂肪	42.00		
钙 Ca	0.36	3	0.05	β-胡萝卜素 β-Carotene				C-12:0	0.00		
氯 Cl				维生素 E Vitamin E				C-14:0	0.10		
钾 K	1.02	1		水溶性维生素 Water soluble				C-16:0	3.99		
镁 Mg	0.19	1		维生素 B_6 Vitamin B_6				C-16:1	0.38		
钠 Na				维生素 B_{12} Vitamin B_{12}, μg/kg				C-18:0	1.71		
磷 P	0.70	3	0.14	生物素 Biotin				C-18:1	55.10		
硫 S				叶酸 Folacin				C-18:2	19.48		
微量元素 Micro, ppm				烟酸 Niacin				C-18:3	9.31		
铬 Cr				泛酸 Pantothenic acid				C-18:4	0.00		
铜 Cu	2.50	1		核黄素 Riboflavin				C-20:0	0.00		
铁 Fe	51.60	1		硫胺素 Thiamin				C-20:1	0.00		
碘 I				胆碱 Choline				C-20:4	0.00		
锰 Mn	38.10	1						C-20:5	0.00		
硒 Se								C-22:0	0.00		
锌 Zn	27.23	1		能量, kcal/kg				C-22:1	0.00		
								C-22:5	0.00		
植酸磷 Phytate P, %	0.79	1		总能 GE	6371	1		C-22:6	0.00		
磷全消化道表观消化率 ATTD of P, %	28			消化能 DE	5234			C-24:0	0.00		
磷全消化道标准消化率 STTD of P, %	32			代谢能 ME	5084			饱和脂肪酸	5.80		
				净能 NE	4059			单不饱和脂肪酸	55.48		
								多不饱和脂肪酸	28.79		
								碘价	110.59		
								碘价物	464.49		

原料名称：压榨双低菜籽粕(Canola Meal, Expelled)
AAFCO#: 71.25, AAFCO 2010, p.385
IFN#: 5-03-870

常规成分, %				氨基酸, %								
				总氨基酸			可消化氨基酸					
							回肠表观消化率			回肠标准消化率		
	$\bar{x}$	n	SD		$\bar{x}$	n	SD	$\bar{x}$	n	SD	$\bar{x}$	n	SD
干物质 Dry matter	93.11	3	2.21	必需氨基酸 Essential									
粗蛋白 Crude protein	35.19	14	4.08	粗蛋白 CP	35.19	14	4.08	70	6	5.05	75	6	4.26
粗纤维 Crude fiber	9.77	3	2.66	精氨酸 Arg	1.76	12	0.26	80	13	6.37	83	13	6.79
乙醚浸出物/粗脂肪 Ether extract	9.97	4	3.34	组氨酸 His	0.82	12	0.25	76	13	11.46	78	13	11.30
酸乙醚浸出物 Acid ether extract				异亮氨酸 Ile	1.67	12	0.54	76	13	8.17	78	13	7.94
粗灰分 Ash	6.39	3	0.19	亮氨酸 Leu	1.95	12	0.30	77	13	7.39	78	13	7.60
碳水化合物成分, %				赖氨酸 Lys	1.58	12	0.58	70	13	13.25	71	13	13.18
				蛋氨酸 Met	0.61	12	0.16	82	13	4.23	83	13	4.30
乳糖 Lactose				苯丙氨酸 Phe	1.48	12	0.49	79	12	8.37	80	12	8.51
蔗糖 Sucrose				苏氨酸 Thr	1.22	12	0.20	67	13	11.84	70	13	12.09
棉子糖 Raffinose				色氨酸 Trp	0.32	4	0.21	72	4	13.09	73		
水苏糖 Stachyose				缬氨酸 Val	1.63	12	0.36	71	13	11.00	73	13	10.74
毛蕊花糖 Verbascose				非必需氨基酸 Nonessential									
低聚糖 Oligosaccharides				丙氨酸 Ala	1.36	9	0.11	73	12	8.45	76	12	8.66
淀粉 Starch	3.80			天冬氨酸 Asp	2.17	9	0.33	71	12	10.61	73	12	10.76
中性洗涤纤维 Neutral detergent fiber	23.77	4	2.22	半胱氨酸 Cys	0.79	11	0.29	74	10	8.38	76	10	8.19
酸性洗涤纤维 Acid detergent fiber	17.57	3	0.72	谷氨酸 Glu	5.82	9	0.97	82	12	6.78	84	12	7.24
半纤维素 Hemicellulose	5.48	1		甘氨酸 Gly	1.67	9	0.24	64	12	14.96	70	12	17.39
酸性洗涤木质素 Acid detergent lignin	7.31	1		脯氨酸 Pro	0.99	4	0.85	66	7	9.87	132	7	71.15
总膳食纤维 Total dietary fiber	25.81	1		丝氨酸 Ser	0.99	9	0.27	68	12	14.53	71	12	15.55
不溶性膳食纤维 Insoluble dietary fiber				酪氨酸 Tyr	0.78	10	0.16	72	12	11.11	74	12	11.65
可溶性膳食纤维 Soluble dietary fiber													

矿物质				维生素, mg/kg(标注单位者除外)				脂肪酸, 占乙醚提取物, %			
	$\bar{x}$	n	SD		$\bar{x}$	n	SD		$\bar{x}$	n	SD
常量元素 Macro, %				脂溶性维生素 Fat soluble				粗脂肪	2.30		
钙 Ca	0.69	9	0.11	β-胡萝卜素 β-Carotene				C-12:0	0.00		
氯 Cl				维生素 E Vitamin E				C-14:0	0.08		
钾 K				水溶性维生素 Water soluble				C-16:0	3.36		
镁 Mg	0.52	1		维生素 B_6 Vitamin B_6				C-16:1	0.32		
钠 Na				维生素 B_{12} Vitamin B_{12}, µg/kg				C-18:0	1.44		
磷 P	1.15	10	0.16	生物素 Biotin				C-18:1	46.40		
硫 S				叶酸 Folacin				C-18:2	16.40		
微量元素 Micro, ppm				烟酸 Niacin				C-18:3	7.84		
铬 Cr				泛酸 Pantothenic acid				C-18:4	0.00		
铜 Cu	5.40	1		核黄素 Riboflavin				C-20:0			
铁 Fe	232	1		硫胺素 Thiamin				C-20:1			
碘 I				胆碱 Choline				C-20:4	0.00		
锰 Mn	60.30	1						C-20:5	0.00		
硒 Se								C-22:0	0.00		
锌 Zn	72.00	1		能量, kcal/kg				C-22:1	0.00		
								C-22:5	0.00		
植酸磷 Phytate P, %	0.87	2	0.06	总能 GE	4873	3	120	C-22:6	0.00		
磷全消化道表观消化率 ATTD of P, %	28			消化能 DE	3779	2	17	C-24:0	0.00		
磷全消化道标准消化率 STTD of P, %	32			代谢能 ME	3540			饱和脂肪酸	4.88		
				净能 NE	2351			单不饱和脂肪酸	46.72		
								多不饱和脂肪酸	24.24		
								碘价	93.13		
								碘价物	21.42		

续表

原料名称：浸提双低菜籽粕(Canola Meal, Solvent Extracted)
AAFCO#: 71.77, AAFCO 2010, p.384
IFN#: 5-05-146

常规成分, %				氨基酸, %								
				总氨基酸				可消化氨基酸				
								回肠表观消化率			回肠标准消化率		
	$\bar{x}$	n	SD		$\bar{x}$	n	SD	$\bar{x}$	n	SD	$\bar{x}$	n	SD
干物质 Dry matter	91.33	46	2.40	必需氨基酸 Essential									
粗蛋白 Crude protein	37.50	96	3.01	粗蛋白 CP	37.50	96	3.01	68	42	9.49	74	39	8.24
粗纤维 Crude fiber	10.50	16	1.59	精氨酸 Arg	2.28	78	0.57	82	44	6.42	85	41	5.56
乙醚浸出物/粗脂肪 Ether extract	3.22	34	1.23	组氨酸 His	1.07	71	0.25	75	39	10.89	78	36	10.24
酸乙醚浸出物 Acid ether extract				异亮氨酸 Ile	1.42	78	0.14	72	44	9.22	76	41	8.34
粗灰分 Ash	6.89	22	0.84	亮氨酸 Leu	2.45	78	0.27	74	44	7.88	78	41	6.44
碳水化合物成分, %				赖氨酸 Lys	2.07	78	0.33	71	44	10.43	74	41	9.65
				蛋氨酸 Met	0.71	55	0.18	82	41	7.77	85	39	4.06
乳糖 Lactose				苯丙氨酸 Phe	1.48	72	0.24	74	39	8.60	77	36	8.42
蔗糖 Sucrose				苏氨酸 Thr	1.55	78	0.38	65	44	9.67	70	41	9.64
棉子糖 Raffinose				色氨酸 Trp	0.43	35	0.10	66	22	9.49	71		
水苏糖 Stachyose				缬氨酸 Val	1.78	78	0.21	69	44	10.95	74	41	9.78
毛蕊花糖 Verbascose				非必需氨基酸 Nonessential									
低聚糖 Oligosaccharides	26.77	1		丙氨酸 Ala	1.61	50	0.22	72	29	7.99	77	27	7.25
淀粉 Starch	6.07	2	1.37	天冬氨酸 Asp	2.56	48	0.22	72	29	7.73	76	27	7.11
中性洗涤纤维 Neutral detergent fiber	22.64	33	4.51	半胱氨酸 Cys	0.86	49	0.12	70	33	8.46	74	31	7.44
酸性洗涤纤维 Acid detergent fiber	15.42	24	3.18	谷氨酸 Glu	6.35	48	0.94	81	29	7.37	84	27	3.94
半纤维素 Hemicellulose	5.29	1		甘氨酸 Gly	1.80	50	0.25	69	29	8.71	78	27	8.60
酸性洗涤木质素 Acid detergent lignin	3.36	7	2.49	脯氨酸 Pro	2.02	48	0.78	66	27	13.91	92	25	11.82
总膳食纤维 Total dietary fiber	26.6	3	1.63	丝氨酸 Ser	1.49	50	0.24	69	29	8.50	75	27	6.51
不溶性膳食纤维 Insoluble dietary fiber				酪氨酸 Tyr	1.06	48	0.22	72	27	9.75	77	22	8.33
可溶性膳食纤维 Soluble dietary fiber													

矿物质				维生素, mg/kg(标注单位者除外)				脂肪酸, 占乙醚提取物, %			
	$\bar{x}$	n	SD		$\bar{x}$	n	SD		$\bar{x}$	n	SD
常量元素 Macro,%				脂溶性维生素 Fat soluble				粗脂肪	2.30		
钙 Ca	0.69	19	0.10	β-胡萝卜素 β-Carotene				C-12:0	0.00		
氯 Cl	0.11			维生素 E Vitamin E	13.4			C-14:0	0.08		
钾 K	1.69	1		水溶性维生素 Water soluble				C-16:0	3.36		
镁 Mg	0.28	1		维生素 B_6 Vitamin B_6	7.2			C-16:1	0.32		
钠 Na	0.07			维生素 B_{12} Vitamin B_{12}, μg/kg	0			C-18:0	1.44		
磷 P	1.08	19	0.07	生物素 Biotin	0.98			C-18:1	46.40		
硫 S	0.85			叶酸 Folacin	0.83			C-18:2	16.40		
微量元素 Micro,ppm				烟酸 Niacin	160			C-18:3	7.84		
铬 Cr				泛酸 Pantothenic acid	9.5			C-18:4	0.00		
铜 Cu	4.90	1		核黄素 Riboflavin	5.8			C-20:0	0.00		
铁 Fe	163	1		硫胺素 Thiamin	5.2			C-20:1	0.00		
碘 I				胆碱 Choline	6700			C-20:4	0.00		
锰 Mn	76.90	1						C-20:5	0.00		
硒 Se	1.10							C-22:0	0.00		
锌 Zn	49.73	1		能量, kcal/kg				C-22:1	0.00		
								C-22:5	0.00		
植酸磷 Phytate P,%	0.65	5	0.30	总能 GE	4332	19	112	C-22:6	0.00		
磷全消化道表观消化率 ATTD of P, %	28	7	4.02	消化能 DE	3273	20	361	C-24:0	0.00		
磷全消化道标准消化率 STTD of P, %	32	7	5.73	代谢能 ME	3013			饱和脂肪酸	4.88		
				净能 NE	1890			单不饱和脂肪酸	46.72		
								多不饱和脂肪酸	24.24		
								碘价	93.13		
								碘价物	21.42		

续表

原料名称：木薯粉(Cassava Meal)
AAFCO#：无官方定义
IFN#：4-01-152

常规成分, %				氨基酸, %							
				总氨基酸			可消化氨基酸				
	$\bar{x}$	n	SD		$\bar{x}$	n	SD	回肠表观消化率		回肠标准消化率	
干物质 Dry matter	88.09	7	1.03	必需氨基酸 Essential				$\bar{x}$	n	SD	$\bar{x}$	n	SD
粗蛋白 Crude protein	2.88	7	0.75	粗蛋白 CP	2.88	7	0.75				
粗纤维 Crude fiber	4.18	6	1.08	精氨酸 Arg	0.18					90	
乙醚浸出物/粗脂肪 Ether extract	0.94	7	0.20	组氨酸 His	0.08					80	
酸乙醚浸出物 Acid ether extract				异亮氨酸 Ile	0.11					81	
粗灰分 Ash	5.70	7	0.75	亮氨酸 Leu	0.19					79	
碳水化合物成分, %				赖氨酸 Lys	0.12					71	
				蛋氨酸 Met	0.04					84	
乳糖 Lactose	0.00	5	0.00	苯丙氨酸 Phe	0.15					80	
蔗糖 Sucrose	0.00	5	0.00	苏氨酸 Thr	0.11					73	
棉子糖 Raffinose	0.00	5	0.00	色氨酸 Trp	0.04					77	
水苏糖 Stachyose	0.00	5	0.00	缬氨酸 Val	0.14					76	
毛蕊花糖 Verbascose	0.00	5	0.00	非必需氨基酸 Nonessential							
低聚糖 Oligosaccharides				丙氨酸 Ala							
淀粉 Starch	67.85	4	4.90	天冬氨酸 Asp							
中性洗涤纤维 Neutral detergent fiber	6.55	3	1.21	半胱氨酸 Cys	0.05					68	
酸性洗涤纤维 Acid detergent fiber	5.99	5	1.95	谷氨酸 Glu							
半纤维素 Hemicellulose				甘氨酸 Gly							
酸性洗涤木质素 Acid detergent lignin				脯氨酸 Pro							
总膳食纤维 Total dietary fiber				丝氨酸 Ser							
不溶性膳食纤维 Insoluble dietary fiber				酪氨酸 Tyr	0.04					76	
可溶性膳食纤维 Soluble dietary fiber											

矿物质				维生素, mg/kg(标注单位者除外)				脂肪酸, 占乙醚提取物, %			
	$\bar{x}$	n	SD		$\bar{x}$	n	SD		$\bar{x}$	n	SD
常量元素 Macro,%				脂溶性维生素 Fat soluble				粗脂肪	0.70		
钙 Ca	0.28	1		β-胡萝卜素 β-Carotene				C-12:0	3.12		
氯 Cl	0.07			维生素 E Vitamin E	0.2			C-14:0	1.36		
钾 K	0.49			水溶性维生素 Water soluble				C-16:0	25.52		
镁 Mg	0.11			维生素 B_6 Vitamin B_6	0.7			C-16:1	0.56		
钠 Na	0.03			维生素 B_{12} Vitamin B_{12}, µg/kg	0			C-18:0	2.32		
磷 P	0.12	2	0.03	生物素 Biotin	0.05			C-18:1	28.16		
硫 S	0.50			叶酸 Folacin				C-18:2	13.12		
微量元素 Micro,ppm				烟酸 Niacin	3			C-18:3	6.08		
铬 Cr				泛酸 Pantothenic acid	0.3			C-18:4	0.00		
铜 Cu	4			核黄素 Riboflavin	0.8			C-20:0	0.00		
铁 Fe	18			硫胺素 Thiamin	1.6			C-20:1	0.00		
碘 I				胆碱 Choline				C-20:4	0.00		
锰 Mn	28							C-20:5	0.00		
硒 Se	0.10							C-22:0	0.00		
锌 Zn	10			能量, kcal/kg				C-22:1	0.00		
								C-22:5	0.00		
植酸磷 Phytate P,%	0.04	1		总能 GE	3451	5	83	C-22:6	0.00		
磷全消化道表观消化率 ATTD of P,%	10	1		消化能 DE	3407	1		C-24:0	0.00		
磷全消化道标准消化率 STTD of P,%	24	1		代谢能 ME	3387			饱和脂肪酸	32.80		
				净能 NE	2647			单不饱和脂肪酸	28.72		
								多不饱和脂肪酸	19.20		
								碘价	66.23		
								碘价物	4.64		

续表

原料名称：柑橘渣(Citrus Pulp)
AAFCO#: 21.1, AAFCO 2010, p.337
IFN#: 4-01-237

常规成分, %				氨基酸, %							
				总氨基酸				可消化氨基酸			
	$\bar{x}$	n	SD		$\bar{x}$	n	SD	回肠表观消化率		回肠标准消化率	
								$\bar{x}$	n	SD	$\bar{x}$	n	SD
干物质 Dry matter	90.9	2	1.7	必需氨基酸 Essential							
粗蛋白 Crude protein	6.64	3	1.29	粗蛋白 CP	6.64	3	1.29				
粗纤维 Crude fiber				精氨酸 Arg	0.26			89			
乙醚浸出物/粗脂肪 Ether extract	2.49	3	0.42	组氨酸 His	0.12			84			
酸乙醚浸出物 Acid ether extract				异亮氨酸 Ile	0.18			81			
粗灰分 Ash	7.73	1		亮氨酸 Leu	0.32			83			
碳水化合物成分, %				赖氨酸 Lys	0.19			77			
				蛋氨酸 Met	0.07			85			
乳糖 Lactose				苯丙氨酸 Phe	0.24			84			
蔗糖 Sucrose				苏氨酸 Thr	0.18			76			
棉子糖 Raffinose				色氨酸 Trp	0.05			77			
水苏糖 Stachyose				缬氨酸 Val	0.24			78			
毛蕊花糖 Verbascose				非必需氨基酸 Nonessential							
低聚糖 Oligosaccharides				丙氨酸 Ala	0.25						
淀粉 Starch	2.53	2	0.66	天冬氨酸 Asp	0.60						
中性洗涤纤维 Neutral detergent fiber	21.23	2	2.11	半胱氨酸 Cys	0.08			73			
酸性洗涤纤维 Acid detergent fiber	20.2	2	3.57	谷氨酸 Glu	0.52						
半纤维素 Hemicellulose				甘氨酸 Gly	0.24						
酸性洗涤木质素 Acid detergent lignin				脯氨酸 Pro	0.53						
总膳食纤维 Total dietary fiber				丝氨酸 Ser	0.23						
不溶性膳食纤维 Insoluble dietary fiber				酪氨酸 Tyr	0.16			86			
可溶性膳食纤维 Soluble dietary fiber											

矿物质				维生素, mg/kg(标注单位者除外)				脂肪酸, 占乙醚提取物, %			
	$\bar{x}$	n	SD		$\bar{x}$	n	SD		$\bar{x}$	n	SD
常量元素 Macro,%				脂溶性维生素 Fat soluble				粗脂肪	2.20		
钙 Ca	1.71	3	0.41	β-胡萝卜素 β-Carotene				C-12:0	0.48		
氯 Cl				维生素 E Vitamin E				C-14:0	0.42		
钾 K	0.74	1		水溶性维生素 Water soluble				C-16:0	15.54		
镁 Mg	0.11	1		维生素 B_6 Vitamin B_6				C-16:1	0.00		
钠 Na	0.52	1		维生素 B_{12} Vitamin B_{12}, μg/kg				C-18:0	2.88		
磷 P	0.09	3	0.01	生物素 Biotin				C-18:1	15.54		
硫 S	0.07	1		叶酸 Folacin				C-18:2	21.54		
微量元素 Micro,ppm				烟酸 Niacin				C-18:3	3.84		
铬 Cr				泛酸 Pantothenic acid				C-18:4	0.00		
铜 Cu	2.69	1		核黄素 Riboflavin				C-20:0	0.00		
铁 Fe	76.87	1		硫胺素 Thiamin				C-20:1	0.00		
碘 I				胆碱 Choline				C-20:4	0.00		
锰 Mn	8.52	1						C-20:5	0.00		
硒 Se								C-22:0	0.00		
锌 Zn	30.59	1		能量, kcal/kg				C-22:1	0.00		
								C-22:5	0.00		
植酸磷 Phytate P,%	0.04	1		总能 GE	3828	1		C-22:6	0.00		
磷全消化道表观消化率 ATTD of P, %				消化能 DE	2773			C-24:0	0.00		
磷全消化道标准消化率 STTD of P, %				代谢能 ME	2728			饱和脂肪酸	19.32		
				净能 NE	1757			单不饱和脂肪酸	15.54		
								多不饱和脂肪酸	25.38		
								碘价	63.45		
								碘价物	13.96		

续表

原料名称：压榨椰子粕(Copra Expelled)
AAFCO#：无官方定义

常规成分, %				氨基酸, %							
				总氨基酸				可消化氨基酸			
								回肠表观消化率		回肠标准消化率	
	$\bar{x}$	n	SD		$\bar{x}$	n	SD	$\bar{x}$	n	SD	$\bar{x}$	n	SD
干物质 Dry matter				必需氨基酸 Essential							
粗蛋白 Crude protein	20.40			粗蛋白 CP	20.40			52			
粗纤维 Crude fiber				精氨酸 Arg	2.45			56			58	
乙醚浸出物/粗脂肪 Ether extract				组氨酸 His	0.40			53			58	
酸乙醚浸出物 Acid ether extract				异亮氨酸 Ile	0.72			53			58	
粗灰分 Ash				亮氨酸 Leu	1.39			54			58	
碳水化合物成分, %				赖氨酸 Lys	0.56			51			58	
				蛋氨酸 Met	0.34			55			58	
乳糖 Lactose				苯丙氨酸 Phe	0.94			54			58	
蔗糖 Sucrose				苏氨酸 Thr	0.67			49			58	
棉子糖 Raffinose				色氨酸 Trp	0.16			49			58	
水苏糖 Stachyose				缬氨酸 Val	1.08			53			58	
毛蕊花糖 Verbascose				非必需氨基酸 Nonessential							
低聚糖 Oligosaccharides				丙氨酸 Ala	0.94			53			58	
淀粉 Starch	0.60			天冬氨酸 Asp	1.77			53			58	
中性洗涤纤维 Neutral detergent fiber				半胱氨酸 Cys	0.34			52			58	
酸性洗涤纤维 Acid detergent fiber				谷氨酸 Glu	4.08			55			58	
半纤维素 Hemicellulose				甘氨酸 Gly	0.94			48			58	
酸性洗涤木质素 Acid detergent lignin				脯氨酸 Pro	0.79			43			58	
总膳食纤维 Total dietary fiber				丝氨酸 Ser	0.94			51			58	
不溶性膳食纤维 Insoluble dietary fiber				酪氨酸 Tyr	0.54			52			58	
可溶性膳食纤维 Soluble dietary fiber											

矿物质				维生素, mg/kg(标注单位者除外)				脂肪酸, 占乙醚提取物, %			
	$\bar{x}$	n	SD		$\bar{x}$	n	SD		$\bar{x}$	n	SD
常量元素 Macro,%				脂溶性维生素 Fat soluble				粗脂肪	8.20		
钙 Ca	0.04	1		β-胡萝卜素 β-Carotene				C-12:0	41.76		
氯 Cl				维生素 E Vitamin E				C-14:0	15.93		
钾 K				水溶性维生素 Water soluble				C-16:0	8.01		
镁 Mg				维生素 B_6 Vitamin B_6				C-16:1	0.36		
钠 Na				维生素 B_{12} Vitamin B_{12}, μg/kg				C-18:0	2.70		
磷 P	0.52	1		生物素 Biotin				C-18:1	5.85		
硫 S				叶酸 Folacin				C-18:2	1.62		
微量元素 Micro,ppm				烟酸 Niacin				C-18:3	0.09		
铬 Cr				泛酸 Pantothenic acid				C-18:4	0.00		
铜 Cu				核黄素 Riboflavin				C-20:0	0.45		
铁 Fe				硫胺素 Thiamin				C-20:1	0.00		
碘 I				胆碱 Choline				C-20:4	0.00		
锰 Mn								C-20:5	0.00		
硒 Se								C-22:0	0.00		
锌 Zn				能量, kcal/kg				C-22:1	0.00		
								C-22:5	0.00		
植酸磷 Phytate P,%	0.22	1		总能 GE	4308	1		C-22:6	0.00		
磷全消化道表观消化率 ATTD of P,%	61	1		消化能 DE	3756	1		C-24:0	0.00		
磷全消化道标准消化率 STTD of P,%	72	1		代谢能 ME	3617			饱和脂肪酸	80.64		
				净能 NE				单不饱和脂肪酸	6.21		
								多不饱和脂肪酸	1.71		
								碘价	8.79		
								碘价物	7.21		

续表

原料名称：椰子粕(Copra Meal)
AAFCO#: 71.61, AAFCO 2010, p.384
IFN#: 5-01-573

常规成分, %				氨基酸, %							
				总氨基酸				可消化氨基酸			
								回肠表观消化率		回肠标准消化率	
	$\bar{x}$	n	SD		$\bar{x}$	n	SD	$\bar{x}$	n	SD	$\bar{x}$	n	SD
干物质 Dry matter	92.00			必需氨基酸 Essential								
粗蛋白 Crude protein	21.90			粗蛋白 CP	21.90			52				
粗纤维 Crude fiber				精氨酸 Arg	2.38			81			88	
乙醚浸出物/粗脂肪 Ether extract	3.00			组氨酸 His	0.39			63			70	
酸乙醚浸出物 Acid ether extract				异亮氨酸 Ile	0.75			64			72	
粗灰分 Ash				亮氨酸 Leu	1.36			68			73	
碳水化合物成分, %				赖氨酸 Lys	0.58			51			64	
				蛋氨酸 Met	0.35			67			77	
乳糖 Lactose				苯丙氨酸 Phe	0.84			71			75	
蔗糖 Sucrose				苏氨酸 Thr	0.67			51			67	
棉子糖 Raffinose				色氨酸 Trp	0.19			63			69	
水苏糖 Stachyose				缬氨酸 Val	1.07			68			71	
毛蕊花糖 Verbascose				非必需氨基酸 Nonessential								
低聚糖 Oligosaccharides				丙氨酸 Ala	0.83			53			58	
淀粉 Starch	2.60			天冬氨酸 Asp	1.58			54			58	
中性洗涤纤维 Neutral detergent fiber	51.30			半胱氨酸 Cys	0.29			54			65	
酸性洗涤纤维 Acid detergent fiber	25.50			谷氨酸 Glu	3.71			55			58	
半纤维素 Hemicellulose				甘氨酸 Gly	0.83			49			58	
酸性洗涤木质素 Acid detergent lignin				脯氨酸 Pro	0.69			44			58	
总膳食纤维 Total dietary fiber				丝氨酸 Ser	0.85			51			58	
不溶性膳食纤维 Insoluble dietary fiber				酪氨酸 Tyr	0.58			53			72	
可溶性膳食纤维 Soluble dietary fiber												

矿物质	$\bar{x}$	n	SD	维生素, mg/kg(标注单位者除外)	$\bar{x}$	n	SD	脂肪酸, 占乙醚提取物, %	$\bar{x}$	n	SD
常量元素 Macro,%				脂溶性维生素 Fat soluble				粗脂肪			
钙 Ca	0.13	1		β-胡萝卜素 β-Carotene				C-12:0			
氯 Cl	0.37			维生素 E Vitamin E	7.7			C-14:0			
钾 K	1.83			水溶性维生素 Water soluble				C-16:0			
镁 Mg	0.31			维生素 B_6 Vitamin B_6	4.4			C-16:1			
钠 Na	0.04			维生素 B_{12} Vitamin B_{12}, μg/kg				C-18:0			
磷 P	0.58	1		生物素 Biotin	0.25			C-18:1			
硫 S	0.31			叶酸 Folacin	0.30			C-18:2			
微量元素 Micro,ppm				烟酸 Niacin	28			C-18:3			
铬 Cr				泛酸 Pantothenic acid	6.5			C-18:4			
铜 Cu	25			核黄素 Riboflavin	3.5			C-20:0			
铁 Fe	486			硫胺素 Thiamin	0.70			C-20:1			
碘 I				胆碱 Choline	1089			C-20:4			
锰 Mn	69							C-20:5			
硒 Se								C-22:0			
锌 Zn	49			能量, kcal/kg				C-22:1			
								C-22:5			
植酸磷 Phytate P,%	0.26	1		总能 GE	4199			C-22:6			
磷全消化道表观消化率 ATTD of P, %	34	1		消化能 DE	3010			C-24:0			
磷全消化道标准消化率 STTD of P, %	44	1		代谢能 ME	2861			饱和脂肪酸			
				净能 NE	1747			单不饱和脂肪酸			
								多不饱和脂肪酸			
								碘价			
								碘价物			

续表

原料名称：普通黄玉米(Corn,Yellow Dent)
AAFCO#:48.4, AAFCO 2010, p.355
IFN#:4-02-861

常规成分, %				氨基酸, %								
				总氨基酸				可消化氨基酸				
								回肠表观消化率			回肠标准消化率	
	$\bar{x}$	n	SD		$\bar{x}$	n	SD	$\bar{x}$	n	SD	$\bar{x}$	n	SD
干物质 Dry matter	88.31	133	2.41	必需氨基酸 Essential									
粗蛋白 Crude protein	8.24	163	0.93	粗蛋白 CP	8.24	163	0.93	65	19	10.34	80	19	9.18
粗纤维 Crude fiber	1.98	78	0.61	精氨酸 Arg	0.37	127	0.05	75	27	7.98	87	27	7.62
乙醚浸出物/粗脂肪 Ether extract	3.48	115	0.78	组氨酸 His	0.24	121	0.05	77	27	5.75	83	27	5.42
酸乙醚浸出物 Acid ether extract	3.68	7	1.26	异亮氨酸 Ile	0.28	128	0.06	73	27	6.70	82	27	6.26
粗灰分 Ash	1.30	76	0.32	亮氨酸 Leu	0.96	121	0.15	82	27	7.47	87	27	7.37
碳水化合物成分, %				赖氨酸 Lys	0.25	132	0.04	60	27	11.63	74	27	10.62
				蛋氨酸 Met	0.18	130	0.03	77	25	11.15	83	25	10.12
乳糖 Lactose	0.00	8	0.00	苯丙氨酸 Phe	0.39	120	0.05	78	27	6.89	85	27	6.58
蔗糖 Sucrose	0.09	9	0.28	苏氨酸 Thr	0.28	129	0.04	61	27	10.29	77	27	10.70
棉子糖 Raffinose	0.01	9	0.04	色氨酸 Trp	0.06	111	0.01	62	13	10.01	80	13	9.54
水苏糖 Stachyose	0.01	9	0.02	缬氨酸 Val	0.38	128	0.05	71	27	8.23	82	27	7.38
毛蕊花糖 Verbascose	0.01	9	0.02	非必需氨基酸 Nonessential									
低聚糖 Oligosaccharides				丙氨酸 Ala	0.60	87	0.08	77	22	5.78	81	21	16.94
淀粉 Starch	62.55	37	4.61	天冬氨酸 Asp	0.54	87	0.09	71	22	9.21	79	21	16.81
中性洗涤纤维 Neutral detergent fiber	9.11	54	1.97	半胱氨酸 Cys	0.19	112	0.02	75	19	6.37	80	20	17.60
酸性洗涤纤维 Acid detergent fiber	2.88	45	0.83	谷氨酸 Glu	1.48	79	0.26	80	22	11.31	84	21	18.99
半纤维素 Hemicellulose				甘氨酸 Gly	0.31	85	0.04	50	22	24.33	84	21	22.06
酸性洗涤木质素 Acid detergent lignin	0.32	2	0.12	脯氨酸 Pro	0.71	83	0.12	50	18	24.62	93	17	18.98
总膳食纤维 Total dietary fiber	13.73	2	4.65	丝氨酸 Ser	0.38	81	0.06	74	22	7.18	82	21	17.20
不溶性膳食纤维 Insoluble dietary fiber				酪氨酸 Tyr	0.26	101	0.07	74	22	7.17	79	20	17.83
可溶性膳食纤维 Soluble dietary fiber													

矿物质				维生素, mg/kg(标注单位者除外)				脂肪酸, 占乙醚提取物, %		
	$\bar{x}$	n	SD		$\bar{x}$	n	SD		$\bar{x}$	SD
常量元素 Macro,%				脂溶性维生素 Fat soluble				粗脂肪	4.74	
钙 Ca	0.02	61	0.01	β-胡萝卜素 β-Carotene	0.8			C-12:0	0.00	
氯 Cl	0.05			维生素 E Vitamin E	11.65	1		C-14:0	0.00	
钾 K	0.32	6	0.01	水溶性维生素 Water soluble				C-16:0	12.00	
镁 Mg	0.12	9	0.07	维生素 B_6 Vitamin B_6	5.0			C-16:1	0.08	
钠 Na	0.02	2	0.00	维生素 B_{12} Vitamin B_{12}, μg/kg	0			C-18:0	1.58	
磷 P	0.26	76	0.05	生物素 Biotin	0.06			C-18:1	26.31	
硫 S				叶酸 Folacin	0.15			C-18:2	44.24	
微量元素 Micro,ppm				烟酸 Niacin	24			C-18:3	1.37	
铬 Cr				泛酸 Pantothenic acid	6.0			C-18:4		
铜 Cu	3.41	5	2.02	核黄素 Riboflavin	1.2			C-20:0	0.00	
铁 Fe	18.38	3	10.86	硫胺素 Thiamin	3.5			C-20:1	0.00	
碘 I				胆碱 Choline	620			C-20:4		
锰 Mn	4.31	5	2.50					C-20:5		
硒 Se	0.07							C-22:0		
锌 Zn	16.51	5	4.96	能量, kcal/kg				C-22:1		
								C-22:5		
植酸磷 Phytate P,%	0.21	10	0.04	总能 GE	3933	48	86	C-22:6		
磷全消化道表观消化率 ATTD of P,%	26	17	7.11	消化能 DE	3451	11	111	C-24:0		
磷全消化道标准消化率 STTD of P,%	34	17	7.22	代谢能 ME	3395			饱和脂肪酸	13.59	
				净能 NE	2672			单不饱和脂肪酸	26.39	
								多不饱和脂肪酸	45.61	
								碘价	107.54	
								碘价物	50.98	

续表

原料名称：高营养玉米(Corn, Nutridense)
AAFCO#: 48.4, AAFCO 2010, p.355
IFN#: 4-02-861

常规成分, %				氨基酸, %								
				总氨基酸				可消化氨基酸				
	$\bar{x}$	n	SD		$\bar{x}$	n	SD	回肠表观消化率			回肠标准消化率		
干物质 Dry matter	87.93	8	2.55	必需氨基酸 Essential				$\bar{x}$	n	SD	$\bar{x}$	n	SD
粗蛋白 Crude protein	9.02	12	1.12	粗蛋白 CP	9.02	12	1.12	74	1		83	1	
粗纤维 Crude fiber	2.22	2	0.06	精氨酸 Arg	0.44	9	0.05	75	3	4.61	83	3	4.56
乙醚浸出物/粗脂肪 Ether extract	4.85	6	1.08	组氨酸 His	0.26	9	0.03	77	3	4.80	82	3	3.93
酸乙醚浸出物 Acid ether extract	5.01	3	0.48	异亮氨酸 Ile	0.32	10	0.04	76	3	2.43	85	3	3.04
粗灰分 Ash	1.44	8	0.26	亮氨酸 Leu	1.09	10	0.15	83	3	2.50	87	3	2.52
碳水化合物成分, %				赖氨酸 Lys	0.27	10	0.05	65	3	6.20	79	3	5.06
				蛋氨酸 Met	0.20	10	0.01	79	3	5.57	83	3	4.10
乳糖 Lactose				苯丙氨酸 Phe	0.43	7	0.05	80	3	3.26	86	3	4.12
蔗糖 Sucrose				苏氨酸 Thr	0.31	10	0.03	62	3	9.61	78	3	8.03
棉子糖 Raffinose				色氨酸 Trp	0.07	4	0.01	65	1		76	1	
水苏糖 Stachyose				缬氨酸 Val	0.44	10	0.05	72	3	5.75	81	3	5.04
毛蕊花糖 Verbascose				非必需氨基酸 Nonessential									
低聚糖 Oligosaccharides				丙氨酸 Ala	0.66	7	0.08	76	3	6.38	85	1	
淀粉 Starch	67.44	4	3.07	天冬氨酸 Asp	0.60	7	0.08	75	3	2.00	82	1	
中性洗涤纤维 Neutral detergent fiber	6.98	2	0.96	半胱氨酸 Cys	0.22	8	0.02	78	1		82	1	
酸性洗涤纤维 Acid detergent fiber	2.33	1		谷氨酸 Glu	1.66	7	0.21	68	3	18.83	75	1	
半纤维素 Hemicellulose				甘氨酸 Gly	0.32	5	0.01	51	3	29.65	88	1	
酸性洗涤木质素 Acid detergent lignin				脯氨酸 Pro	0.77	7	0.09	45	3	3.82	85	1	
总膳食纤维 Total dietary fiber	9.6	2	0.33	丝氨酸 Ser	0.42	7	0.05	74	3	3.91	85	1	
不溶性膳食纤维 Insoluble dietary fiber				酪氨酸 Tyr	0.28	7	0.04	70	3	7.98	80	1	
可溶性膳食纤维 Soluble dietary fiber													

矿物质				维生素, mg/kg(标注单位者除外)				脂肪酸, 占乙醚提取物, %			
	$\bar{x}$	n	SD		$\bar{x}$	n	SD		$\bar{x}$	n	SD
常量元素 Macro, %				脂溶性维生素 Fat soluble				粗脂肪			
钙 Ca	0.04	3	0.02	β-胡萝卜素 β-Carotene				C-12:0			
氯 Cl				维生素 E Vitamin E				C-14:0			
钾 K	0.30	2	0.03	水溶性维生素 Water soluble				C-16:0			
镁 Mg	0.11	2	0.01	维生素 B_6 Vitamin B_6				C-16:1			
钠 Na				维生素 B_{12} Vitamin B_{12}, μg/kg				C-18:0			
磷 P	0.27	7	0.02	生物素 Biotin				C-18:1			
硫 S				叶酸 Folacin				C-18:2			
微量元素 Micro, ppm				烟酸 Niacin				C-18:3			
铬 Cr				泛酸 Pantothenic acid				C-18:4			
铜 Cu				核黄素 Riboflavin				C-20:0			
铁 Fe				硫胺素 Thiamin				C-20:1			
碘 I				胆碱 Choline				C-20:4			
锰 Mn								C-20:5			
硒 Se								C-22:0			
锌 Zn				能量, kcal/kg				C-22:1			
								C-22:5			
植酸磷 Phytate P, %	0.16	2	0.11	总能 GE	3987	6	140	C-22:6			
磷全消化道表观消化率 ATTD of P, %	26			消化能 DE	3455	1		C-24:0			
磷全消化道标准消化率 STTD of P, %	34			代谢能 ME	3394			饱和脂肪酸			
				净能 NE	2718			单不饱和脂肪酸			
								多不饱和脂肪酸			
								碘价			
								碘价物			

续表

原料名称：玉米麸(Corn Bran)
AAFCO#: 48.2, AAFCO 2010, p.355
IFN#: 4-02-841

常规成分, %				氨基酸, %							
				总氨基酸				可消化氨基酸			
								回肠表观消化率		回肠标准消化率	
	$\bar{x}$	n	SD		$\bar{x}$	n	SD	$\bar{x}$	n	SD	$\bar{x}$	n	SD
干物质 Dry matter	88.50	2	1.41	必需氨基酸 Essential							
粗蛋白 Crude protein	9.53	2	0.19	粗蛋白 CP	9.53	2	0.19	63			
粗纤维 Crude fiber	6.61	2	0.35	精氨酸 Arg	0.56						89
乙醚浸出物/粗脂肪 Ether extract	8.52	2	1.36	组氨酸 His	0.29						83
酸乙醚浸出物 Acid ether extract				异亮氨酸 Ile	0.30			70			81
				亮氨酸 Leu	0.97			80			84
粗灰分 Ash	2.53	2	0.03	赖氨酸 Lys	0.35			59			74
碳水化合物成分, %				蛋氨酸 Met	0.19			82			86
乳糖 Lactose	0.00	2	0.00	苯丙氨酸 Phe	0.37			74			83
蔗糖 Sucrose	0.00	2	0.00	苏氨酸 Thr	0.35			55			74
棉子糖 Raffinose	0.00	2	0.00	色氨酸 Trp	0.08			54			75
水苏糖 Stachyose	0.00	2	0.00	缬氨酸 Val	0.46			69			79
毛蕊花糖 Verbascose	0.00	2	0.00	非必需氨基酸 Nonessential							
低聚糖 Oligosaccharides				丙氨酸 Ala	0.67			74			80
淀粉 Starch	31.73	1		天冬氨酸 Asp	0.65			62			73
中性洗涤纤维 Neutral detergent fiber	32.96	1		半胱氨酸 Cys	0.20			64			73
酸性洗涤纤维 Acid detergent fiber	9.23	1		谷氨酸 Glu	1.49			73			80
半纤维素 Hemicellulose				甘氨酸 Gly	0.41			50			70
酸性洗涤木质素 Acid detergent lignin				脯氨酸 Pro	0.76			65			77
总膳食纤维 Total dietary fiber				丝氨酸 Ser	0.43			68			81
不溶性膳食纤维 Insoluble dietary fiber				酪氨酸 Tyr	0.30			76			85
可溶性膳食纤维 Soluble dietary fiber											

矿物质				维生素, mg/kg(标注单位者除外)				脂肪酸, 占乙醚提取物, %			
	$\bar{x}$	n	SD		$\bar{x}$	n	SD		$\bar{x}$	n	SD
常量元素 Macro,%				脂溶性维生素 Fat soluble				粗脂肪	0.92		
钙 Ca	0.47			β-胡萝卜素 β-Carotene				C-12:0	0.00		
氯 Cl				维生素 E Vitamin E				C-14:0	0.00		
钾 K				水溶性维生素 Water soluble				C-16:0	12.07		
镁 Mg				维生素 B_6 Vitamin B_6				C-16:1	0.11		
钠 Na				维生素 B_{12} Vitamin B_{12}, μg/kg				C-18:0	1.63		
磷 P	0.29			生物素 Biotin				C-18:1	26.41		
硫 S				叶酸 Folacin				C-18:2	44.35		
微量元素 Micro,ppm				烟酸 Niacin				C-18:3	1.41		
铬 Cr				泛酸 Pantothenic acid				C-18:4			
铜 Cu				核黄素 Riboflavin				C-20:0	0.00		
铁 Fe				硫胺素 Thiamin				C-20:1	0.00		
碘 I				胆碱 Choline				C-20:4			
锰 Mn								C-20:5			
硒 Se								C-22:0			
锌 Zn				能量, kcal/kg				C-22:1			
								C-22:5			
植酸磷 Phytate P,%				总能 GE	4652			C-22:6			
磷全消化道表观消化率 ATTD of P,%	20			消化能 DE	2649	3	55	C-24:0			
磷全消化道标准消化率 STTD of P,%	27			代谢能 ME	2584			饱和脂肪酸	13.70		
				净能 NE	1977			单不饱和脂肪酸	26.52		
								多不饱和脂肪酸	45.76		
								碘价	107.97		
								碘价物	9.93		

续表

原料名称：玉米酒精糟 (Corn DDG)
AAFCO#: 27.5, AAFCO 2010, p.343
IFN#: 5-02-842

常规成分,%				氨基酸,%								
					总氨基酸			可消化氨基酸				
								回肠表观消化率			回肠标准消化率		
	$\bar{x}$	n	SD		$\bar{x}$	n	SD	$\bar{x}$	n	SD	$\bar{x}$	n	SD
干物质 Dry matter	90.82	2	3.98	必需氨基酸 Essential									
粗蛋白 Crude protein	28.89	3	2.56	粗蛋白 CP	28.89	3	2.56	67	1		76	1	
粗纤维 Crude fiber	9.48	1		精氨酸 Arg	1.22	2	0.10	75	1		83	1	
乙醚浸出物/粗脂肪 Ether extract	8.69	2	1.09	组氨酸 His	0.78	2	0.14	81	1		84	1	
酸乙醚浸出物 Acid ether extract				异亮氨酸 Ile	1.19	2	0.15	80	1		83	1	
粗灰分 Ash	3.04	2	1.67	亮氨酸 Leu	4.03	2	0.47	84	1		86	1	
碳水化合物成分,%				赖氨酸 Lys	0.87	2	0.08	73	1		78	1	
				蛋氨酸 Met	0.62	2	0.08	88	1		89	1	
乳糖 Lactose				苯丙氨酸 Phe	1.62	2	0.14	83	1		87	1	
蔗糖 Sucrose				苏氨酸 Thr	1.13	2	0.04	71	1		78	1	
棉子糖 Raffinose				色氨酸 Trp	0.21	2	0.01	63	1		71	1	
水苏糖 Stachyose				缬氨酸 Val	1.56	2	0.23	78	1		81	1	
毛蕊花糖 Verbascose				非必需氨基酸 Nonessential									
低聚糖 Oligosaccharides				丙氨酸 Ala	2.33	2	0.24	78	1		82	1	
淀粉 Starch	3.83	1		天冬氨酸 Asp	1.94	2	0.11	69	1		74	1	
中性洗涤纤维 Neutral detergent fiber	41.86	3	6.71	半胱氨酸 Cys	0.57	2	0.04	77	1		81	1	
酸性洗涤纤维 Acid detergent fiber	15.55	3	4.33	谷氨酸 Glu	5.14	2	0.11	85	1		87	1	
半纤维素 Hemicellulose				甘氨酸 Gly	1.09	2	0.12	40	1		66	1	
酸性洗涤木质素 Acid detergent lignin				脯氨酸 Pro	2.54	2	0.05	12	1		55	1	
总膳食纤维 Total dietary fiber	43.90	1		丝氨酸 Ser	1.39	2	0.09	76	1		82	1	
不溶性膳食纤维 Insoluble dietary fiber				酪氨酸 Tyr	1.31	1					80		
可溶性膳食纤维 Soluble dietary fiber													

矿物质	$\bar{x}$	n	SD	维生素,mg/kg(标注单位者除外)	$\bar{x}$	n	SD	脂肪酸,占乙醚提取物,%	$\bar{x}$	n	SD
常量元素 Macro,%				脂溶性维生素 Fat soluble				粗脂肪			
钙 Ca	0.08	2	0.09	β-胡萝卜素 β-Carotene	3.0			C-12:0			
氯 Cl	0.08			维生素 E Vitamin E	12.9			C-14:0			
钾 K	0.17			水溶性维生素 Water soluble				C-16:0			
镁 Mg	0.25			维生素 B_6 Vitamin B_6	4.4			C-16:1			
钠 Na	0.09			维生素 B_{12} Vitamin B_{12}, μg/kg	0			C-18:0			
磷 P	0.56	2	0.11	生物素 Biotin	0.49			C-18:1			
硫 S				叶酸 Folacin	0.90			C-18:2			
微量元素 Micro,ppm				烟酸 Niacin	37			C-18:3			
铬 Cr				泛酸 Pantothenic acid	11.7			C-18:4			
铜 Cu	45			核黄素 Riboflavin	5.2			C-20:0			
铁 Fe	220			硫胺素 Thiamin	1.7			C-20:1			
碘 I				胆碱 Choline	1180			C-20:4			
锰 Mn	22							C-20:5			
硒 Se	0.40							C-22:0			
锌 Zn	55			能量, kcal/kg				C-22:1			
								C-22:5			
植酸磷 Phytate P,%				总能 GE	4919	5	342	C-22:6			
磷全消化道表观消化率 ATTD of P,%				消化能 DE	3355	4	173	C-24:0			
磷全消化道标准消化率 STTD of P,%				代谢能 ME	3158			饱和脂肪酸			
				净能 NE	2109			单不饱和脂肪酸			
								多不饱和脂肪酸			
								碘价			
								碘价物			

续表

原料名称：玉米酒精糟及可溶物，脂肪含量>10%(Corn DDGS, >10% Oil)
AAFCO#: 27.6, AAFCO 2010, p.343
IFN#: 5-02-843

常规成分, %				氨基酸, %								
				总氨基酸				可消化氨基酸				
								回肠表观消化率			回肠标准消化率		
	$\bar{x}$	n	SD		$\bar{x}$	n	SD	$\bar{x}$	n	SD	$\bar{x}$	n	SD
干物质 Dry matter	89.31	59	1.91	必需氨基酸 Essential									
粗蛋白 Crude protein	27.33	81	1.53	粗蛋白 CP	27.33	81	1.53	64	40	5.19	74	35	5.83
粗纤维 Crude fiber	7.06	12	1.24	精氨酸 Arg	1.16	67	0.17	74	40	5.02	81	40	5.25
乙醚浸出物/粗脂肪 Ether extract	10.43	34	1.03	组氨酸 His	0.71	67	0.07	74	40	4.97	78	40	4.75
酸乙醚浸出物 Acid ether extract	11.27	8	1.36	异亮氨酸 Ile	1.02	77	0.09	72	40	5.03	76	40	4.87
粗灰分 Ash	4.11	39	0.91	亮氨酸 Leu	3.13	67	0.46	82	40	4.09	84	40	4.00
碳水化合物成分, %				赖氨酸 Lys	0.77	68	0.12	55	40	10.76	61	40	8.75
				蛋氨酸 Met	0.55	68	0.09	80	40	4.30	82	40	4.13
乳糖 Lactose				苯丙氨酸 Phe	1.34	67	0.10	78	40	3.87	81	40	3.96
蔗糖 Sucrose				苏氨酸 Thr	0.99	64	0.08	64	40	6.51	71	40	5.73
棉子糖 Raffinose				色氨酸 Trp	0.21	67	0.03	63	40	8.34	71	40	8.16
水苏糖 Stachyose				缬氨酸 Val	1.35	67	0.12	71	40	5.16	75	40	4.95
毛蕊花糖 Verbascose				非必需氨基酸 Nonessential									
低聚糖 Oligosaccharides				丙氨酸 Ala	1.93	58	0.16	74	40	4.72	79	40	4.64
淀粉 Starch	6.73	32	1.70	天冬氨酸 Asp	1.82	58	0.18	63	40	5.73	69	40	5.52
中性洗涤纤维 Neutral detergent fiber	32.50	76	5.42	半胱氨酸 Cys	0.51	60	0.11	69	40	5.97	73	40	5.70
酸性洗涤纤维 Acid detergent fiber	11.75			谷氨酸 Glu	4.35	58	0.69	76	40	7.81	81	40	5.63
半纤维素 Hemicellulose				甘氨酸 Gly	1.04	56	0.09	42	40	10.79	64	40	11.16
酸性洗涤木质素 Acid detergent lignin	2.61	1		脯氨酸 Pro	2.09	58	0.18	34	40	19.40	74	40	21.54
总膳食纤维 Total dietary fiber	31.35	8	3.28	丝氨酸 Ser	1.18	58	0.16	70	40	5.36	77	40	5.48
不溶性膳食纤维 Insoluble dietary fiber				酪氨酸 Tyr	1.04	38	0.14	78	20	4.48	81	20	3.98
可溶性膳食纤维 Soluble dietary fiber													

矿物质				维生素, mg/kg(标注单位者除外)				脂肪酸, 占乙醚提取物, %			
	$\bar{x}$	n	SD		$\bar{x}$	n	SD		$\bar{x}$	n	SD
常量元素 Macro,%				脂溶性维生素 Fat soluble				粗脂肪			
钙 Ca	0.12	38	0.19	β-胡萝卜素 β-Carotene				C-12:0			
氯 Cl				维生素 E Vitamin E				C-14:0			
钾 K	0.90	22	0.12	水溶性维生素 Water soluble				C-16:0			
镁 Mg	0.29	25	0.04	维生素 B_6 Vitamin B_6				C-16:1			
钠 Na	0.22	23	0.13	维生素 B_{12} Vitamin B_{12}, µg/kg				C-18:0			
磷 P	0.73	66	0.10	生物素 Biotin				C-18:1			
硫 S	0.66	19	0.28	叶酸 Folacin				C-18:2			
微量元素 Micro,ppm				烟酸 Niacin				C-18:3			
铬 Cr				泛酸 Pantothenic acid				C-18:4			
铜 Cu	7.65	22	4.14	核黄素 Riboflavin				C-20:0			
铁 Fe	1.26	21	73.07	硫胺素 Thiamin				C-20:1			
碘 I				胆碱 Choline				C-20:4			
锰 Mn	17.92	22	10.05					C-20:5			
硒 Se								C-22:0			
锌 Zn	65.05	21	19.62	能量, kcal/kg				C-22:1			
								C-22:5			
植酸磷 Phytate P,%	0.26	1		总能 GE	4849	41	113	C-22:6			
磷全消化道表观消化率 ATTD of P, %	60	17	6.49	消化能 DE	3620	16	166	C-24:0			
磷全消化道标准消化率 STTD of P, %	65	17	6.54	代谢能 ME	3434			饱和脂肪酸			
				净能 NE	2384			单不饱和脂肪酸			
								多不饱和脂肪酸			
								碘价			
								碘价物			

续表

原料名称：玉米酒精糟及可溶物，6%<脂肪含量< 9% (Corn DDGS, >6% and <9% Oil)
该原料是将酒精糟与脂肪离心分离后的可溶物混和所得。
AAFCO#: 27.6, AAFCO 2010, p.343
IFN#: 5-02-843

常规成分, %				氨基酸, %								
					总氨基酸			可消化氨基酸				
	$\bar{x}$	n	SD		$\bar{x}$	n	SD	回肠表观消化率			回肠标准消化率	
干物质 Dry matter	89.35	13	1.55	必需氨基酸 Essential				$\bar{x}$	n	SD	$\bar{x}$	n	SD
粗蛋白 Crude protein	27.36	13	2.00	粗蛋白 CP	27.36	13	2.00	64	40	5.19	74	35	5.83
粗纤维 Crude fiber	8.92	4	1.38	精氨酸 Arg	1.23	6	0.16	74	40	5.02	81	40	5.25
乙醚浸出物/粗脂肪 Ether extract	8.90	8	0.46	组氨酸 His	0.74	6	0.08	74	40	4.97	78	40	4.75
酸乙醚浸出物 Acid ether extract	8.71	4	0.16	异亮氨酸 Ile	1.06	9	0.09	72	40	5.03	76	40	4.87
粗灰分 Ash	4.04	9	1	亮氨酸 Leu	3.25	9	0.44	82	40	4.09	84	40	4.00
碳水化合物成分, %				赖氨酸 Lys	0.90	9	0.13	55	40	10.76	61	40	8.75
				蛋氨酸 Met	0.57	9	0.11	80	40	4.30	82	40	4.13
乳糖 Lactose				苯丙氨酸 Phe	1.37	6	0.16	78	40	3.87	81	40	3.96
蔗糖 Sucrose				苏氨酸 Thr	0.99	9	0.06	64	40	6.51	71	40	5.73
棉子糖 Raffinose				色氨酸 Trp	0.20	9	0.03	63	40	8.34	71	40	8.16
水苏糖 Stachyose				缬氨酸 Val	1.39	9	0.12	71	40	5.16	75	40	4.95
毛蕊花糖 Verbascose				非必需氨基酸 Nonessential									
低聚糖 Oligosaccharides				丙氨酸 Ala	2.13	4	0.30	74	40	4.72	79	40	4.64
淀粉 Starch	9.63	4	2.95	天冬氨酸 Asp	2.01	4	0.26	63	40	5.73	69	40	5.52
中性洗涤纤维 Neutral detergent fiber	30.46	11	5.68	半胱氨酸 Cys	0.44	7	0.06	69	40	5.97	73	40	5.70
酸性洗涤纤维 Acid detergent fiber	12.02	9	2.47	谷氨酸 Glu	5.35	4	0.83	76	40	7.81	81	40	5.63
半纤维素 Hemicellulose				甘氨酸 Gly	1.13	4	0.09	42	40	10.79	64	40	11.16
酸性洗涤木质素 Acid detergent lignin				脯氨酸 Pro	2.36	4	0.31	34	40	19.40	74	40	21.54
总膳食纤维 Total dietary fiber				丝氨酸 Ser	1.40	4	0.20	70	40	5.36	77	40	5.48
不溶性膳食纤维 Insoluble dietary fiber				酪氨酸 Tyr	1.22	3	0.16	78	20	4.48	81	20	3.98
可溶性膳食纤维 Soluble dietary fiber													

矿物质				维生素, mg/kg(标注单位者除外)				脂肪酸, 占乙醚提取物, %			
	$\bar{x}$	n	SD		$\bar{x}$	n	SD		$\bar{x}$	n	SD
常量元素 Macro,%				脂溶性维生素 Fat soluble				粗脂肪			
钙 Ca	0.08	9	0.07	β-胡萝卜素 β-Carotene	3.5			C-12:0			
氯 Cl	0.20			维生素 E Vitamin E				C-14:0			
钾 K	0.88	4	0.11	水溶性维生素 Water soluble				C-16:0			
镁 Mg	0.49	4	0.24	维生素 B_6 Vitamin B_6	8.0			C-16:1			
钠 Na	0.30	2	0.23	维生素 B_{12} Vitamin B_{12}, μg/kg	0			C-18:0			
磷 P	0.60	9	0.20	生物素 Biotin	0.78			C-18:1			
硫 S	0.48	2	0.27	叶酸 Folacin	0.90			C-18:2			
微量元素 Micro,ppm				烟酸 Niacin	75			C-18:3			
铬 Cr				泛酸 Pantothenic acid	14.0			C-18:4			
铜 Cu	6.04	2	1.13	核黄素 Riboflavin	8.6			C-20:0			
铁 Fe	147	2	8.68	硫胺素 Thiamin	2.9			C-20:1			
碘 I				胆碱 Choline	2637			C-20:4			
锰 Mn	16.51	2	2.98					C-20:5			
硒 Se	0.39							C-22:0			
锌 Zn	51.62	2	16.11					C-22:1			
				能量, kcal/kg				C-22:5			
植酸磷 Phytate P,%				总能 GE	4710	3	120	C-22:6			
磷全消化道表观消化率 ATTD of P, %	60			消化能 DE	3582	3	161	C-24:0			
磷全消化道标准消化率 STTD of P, %	65			代谢能 ME	3396			饱和脂肪酸			
				净能 NE	2343			单不饱和脂肪酸			
								多不饱和脂肪酸			
								碘价			
								碘价物			

续表

原料名称：玉米酒精糟及可溶物，脂肪含量< 4% (Corn DDGS, <4% Oil)
该原料是将玉米酒精糟及内溶物中的脂肪通过溶剂浸提后所得。
AAFCO#: 27.6, AAFCO 2010, p.343
IFN#: 5-02-843

常规成分, %				氨基酸, %									
					总氨基酸			可消化氨基酸					
								回肠表观消化率			回肠标准消化率		
	$\bar{x}$	n	SD	必需氨基酸 Essential	$\bar{x}$	n	SD	$\bar{x}$	n	SD	$\bar{x}$	n	SD
干物质 Dry matter	89.25	2	2.20										
粗蛋白 Crude protein	27.86	2	4.73	粗蛋白 CP	27.86	2	4.73	64	40	5.19	74	35	5.83
粗纤维 Crude fiber	6.19	1		精氨酸 Arg	1.31	1		74	40	5.02	81	40	5.25
乙醚浸出物/粗脂肪 Ether extract	3.57	2	0.62	组氨酸 His	0.82	1		74	40	4.97	78	40	4.75
酸乙醚浸出物 Acid ether extract				异亮氨酸 Ile	1.02	2	0.28	72	40	5.03	76	40	4.87
粗灰分 Ash	4.64	1		亮氨酸 Leu	3.64	1		82	40	4.09	84	40	4.00
碳水化合物成分, %				赖氨酸 Lys	0.68	2	0.28	55	40	10.76	61	40	8.75
				蛋氨酸 Met	0.50	2	0.12	80	40	4.30	82	40	4.13
乳糖 Lactose				苯丙氨酸 Phe	1.69	1		78	40	3.87	81	40	3.96
蔗糖 Sucrose				苏氨酸 Thr	0.97	2	0.18	64	40	6.51	71	40	5.73
棉子糖 Raffinose				色氨酸 Trp	0.18	2	0.01	63	40	8.34	71	40	8.16
水苏糖 Stachyose				缬氨酸 Val	1.34	2	0.28	71	40	5.16	75	40	4.95
毛蕊花糖 Verbascose				非必需氨基酸 Nonessential									
低聚糖 Oligosaccharides				丙氨酸 Ala	2.13	1		74	40	4.72	79	40	4.64
淀粉 Starch	10.00			天冬氨酸 Asp	1.84	1		63	40	5.73	69	40	5.52
中性洗涤纤维 Neutral detergent fiber	33.75	2	1.20	半胱氨酸 Cys	0.51	2	0.04	69	40	5.97	73	40	5.70
酸性洗涤纤维 Acid detergent fiber	16.91	1		谷氨酸 Glu	4.26	1		76	40	7.81	81	40	5.63
半纤维素 Hemicellulose				甘氨酸 Gly	1.18	1		42	40	10.79	64	40	11.16
酸性洗涤木质素 Acid detergent lignin				脯氨酸 Pro	2.11	1		34	40	19.40	74	40	21.54
总膳食纤维 Total dietary fiber				丝氨酸 Ser	1.30	1		70	40	5.36	77	40	5.48
不溶性膳食纤维 Insoluble dietary fiber				酪氨酸 Tyr	1.13	1		78	20	4.48	81	20	3.98
可溶性膳食纤维 Soluble dietary fiber													
矿物质				维生素, mg/kg(标注单位者除外)				脂肪酸, 占乙醚提取物, %					
	$\bar{x}$	n	SD		$\bar{x}$	n	SD		$\bar{x}$	n	SD		
常量元素 Macro,%				脂溶性维生素 Fat soluble				粗脂肪	3.90				
钙 Ca	0.05	1		β-胡萝卜素 β-Carotene				C-12:0	0.00				
氯 Cl				维生素 E Vitamin E				C-14:0	0.08				
钾 K				水溶性维生素 Water soluble				C-16:0	8.33				
镁 Mg				维生素 B_6 Vitamin B_6				C-16:1	0.30				
钠 Na				维生素 B_{12} Vitamin B_{12}, μg/kg				C-18:0	1.35				
磷 P	0.76	1		生物素 Biotin				C-18:1	20.18				
硫 S				叶酸 Folacin				C-18:2	42.38				
微量元素 Micro,ppm				烟酸 Niacin				C-18:3	0.75				
铬 Cr				泛酸 Pantothenic acid				C-18:4	0.00				
铜 Cu				核黄素 Riboflavin				C-20:0	0.00				
铁 Fe				硫胺素 Thiamin				C-20:1	0.00				
碘 I				胆碱 Choline				C-20:4	0.00				
锰 Mn								C-20:5	0.00				
硒 Se								C-22:0	0.00				
锌 Zn				能量, kcal/kg				C-22:1	0.00				
								C-22:5	0.00				
植酸磷 Phytate P,%				总能 GE	5098	1		C-22:6	0.00				
磷全消化道表观消化率 ATTD of P, %	60			消化能 DE	3291	2	269	C-24:0	0.00				
磷全消化道标准消化率 STTD of P, %	65			代谢能 ME	3102			饱和脂肪酸	9.75				
				净能 NE	2009			单不饱和脂肪酸	20.48				
								多不饱和脂肪酸	43.13				
								碘价	97.17				
								碘价物	37.90				

续表

原料名称：玉米高蛋白酒精糟(Corn HP DDG)
玉米在发酵和蒸馏前进行脱皮和脱胚处理。可溶物不添加到酒精糟中。如果可溶物添加到酒精糟中就是玉米高蛋白酒精糟及可溶物。
AAFCO#: 27.5, AAFCO 2010, p.343
IFN#: 5-02-842

常规成分, %				氨基酸, %								
					总氨基酸			可消化氨基酸				
	$\bar{x}$	n	SD		$\bar{x}$	n	SD	回肠表观消化率			回肠标准消化率	
干物质 Dry matter	91.20	7	2.04	必需氨基酸 Essential				$\bar{x}$	n	SD	$\bar{x}$	n	SD
粗蛋白 Crude protein	45.35	6	4.32	粗蛋白 CP	45.35	6	4.32	70	2	2.76	76	2	5.37
粗纤维 Crude fiber	7.30	1		精氨酸 Arg	1.62	3	0.18	81	3	5.20	85	3	2.25
乙醚浸出物/粗脂肪 Ether extract	3.54	5	0.69	组氨酸 His	1.07	3	0.07	77	3	2.19	79	3	2.25
酸乙醚浸出物 Acid ether extract	3.70	1		异亮氨酸 Ile	1.83	3	0.18	77	3	2.91	80	3	2.78
粗灰分 Ash	2.39	3	0.89	亮氨酸 Leu	6.18	3	0.38	85	3	6.85	86	3	7.09
碳水化合物成分, %				赖氨酸 Lys	1.22	3	0.11	65	3	7.59	69	3	5.80
				蛋氨酸 Met	0.93	3	0.12	85	3	2.70	86	3	2.87
乳糖 Lactose				苯丙氨酸 Phe	2.42	3	0.12	82	3	4.97	84	3	5.01
蔗糖 Sucrose				苏氨酸 Thr	1.59	3	0.09	70	3	2.27	75	3	2.25
棉子糖 Raffinose				色氨酸 Trp	0.24	3	0.03	76	3	4.95	82	3	3.29
水苏糖 Stachyose				缬氨酸 Val	2.12	3	0.02	75	3	3.54	78	3	3.78
毛蕊花糖 Verbascose				非必需氨基酸 Nonessential									
低聚糖 Oligosaccharides				丙氨酸 Ala	3.32	3	0.2	80	3	5.43	82	3	5.43
淀粉 Starch	10.15	2	1.48	天冬氨酸 Asp	2.75	3	0.26	71	3	1.34	74	3	1.72
中性洗涤纤维 Neutral detergent fiber	33.63	3	7.06	半胱氨酸 Cys	0.82	3	0.04	75	3	2.86	78	3	3.72
酸性洗涤纤维 Acid detergent fiber	20.63	3	6.02	谷氨酸 Glu	7.52	3	0.58	82	3	5.94	83	3	6.03
半纤维素 Hemicellulose				甘氨酸 Gly	1.39	3	0.07	55	3	9.22	70	3	4.27
酸性洗涤木质素 Acid detergent lignin	3.77	1		脯氨酸 Pro	3.65	3	0.06	64	3	15.64	79	3	5.51
总膳食纤维 Total dietary fiber				丝氨酸 Ser	1.96	3	0.12	79	3	2.87	82	3	2.85
不溶性膳食纤维 Insoluble dietary fiber				酪氨酸 Tyr	1.92	3	0.1	83	3	3.72	85	3	4.01
可溶性膳食纤维 Soluble dietary fiber													

矿物质				维生素, mg/kg(标注单位者除外)				脂肪酸, 占乙醚提取物, %			
	$\bar{x}$	n	SD		$\bar{x}$	n	SD		$\bar{x}$	n	SD
常量元素 Macro,%				脂溶性维生素 Fat soluble				粗脂肪			
钙 Ca	0.02	6	0.01	β-胡萝卜素 β-Carotene				C-12:0			
氯 Cl				维生素 E Vitamin E				C-14:0			
钾 K	0.37	1		水溶性维生素 Water soluble				C-16:0			
镁 Mg	0.09	1		维生素 B_6 Vitamin B_6				C-16:1			
钠 Na	0.06	2	0.05	维生素 B_{12} Vitamin B_{12}, μg/kg				C-18:0			
磷 P	0.36	7	0.03	生物素 Biotin				C-18:1			
硫 S	0.75	1		叶酸 Folacin				C-18:2			
微量元素 Micro,ppm				烟酸 Niacin				C-18:3			
铬 Cr				泛酸 Pantothenic acid				C-18:4			
铜 Cu	2.03	1		核黄素 Riboflavin				C-20:0			
铁 Fe	65.30	1		硫胺素 Thiamin				C-20:1			
碘 I				胆碱 Choline				C-20:4			
锰 Mn	7.00	1						C-20:5			
硒 Se								C-22:0			
锌 Zn	27.30	1		能量, kcal/kg				C-22:1			
								C-22:5			
植酸磷 Phytate P,%	0.11	1		总能 GE	5173	3	162	C-22:6			
磷全消化道表观消化率 ATTD of P,%	64	2	6.36	消化能 DE	4040	3	351	C-24:0			
磷全消化道标准消化率 STTD of P,%	73	2	5.45	代谢能 ME	3732			饱和脂肪酸			
				净能 NE	2342			单不饱和脂肪酸			
								多不饱和脂肪酸			
								碘价			
								碘价物			

续表

原料名称：玉米酒精糟可溶物(Corn Distillers Solubles)
AAFCO#:27.4, AAFCO 2010, p.342
IFN#: 5-02-844

常规成分, %				氨基酸, %								
				总氨基酸				可消化氨基酸				
								回肠表观消化率			回肠标准消化率	
	$\bar{x}$	n	SD	必需氨基酸 Essential	$\bar{x}$	n	SD	$\bar{x}$	n	SD	$\bar{x}$	n	SD
干物质 Dry matter	87.80	1		粗蛋白 CP	18.70	1							
粗蛋白 Crude protein	18.70	1		精氨酸 Arg	0.9	1							
粗纤维 Crude fiber				组氨酸 His	0.6	1							
乙醚浸出物/粗脂肪 Ether extract	12.07	1		异亮氨酸 Ile	0.7	1							
酸乙醚浸出物 Acid ether extract				亮氨酸 Leu	1.8	1							
粗灰分 Ash	8.70	1		赖氨酸 Lys	0.8	1							
碳水化合物成分, %				蛋氨酸 Met	0.4	1							
乳糖 Lactose				苯丙氨酸 Phe	0.8	1							
蔗糖 Sucrose				苏氨酸 Thr	0.8	1							
棉子糖 Raffinose				色氨酸 Trp	0.2	1							
水苏糖 Stachyose				缬氨酸 Val	1.1	1							
毛蕊花糖 Verbascose				非必需氨基酸 Nonessential									
低聚糖 Oligosaccharides				丙氨酸 Ala	1.3	1							
淀粉 Starch	5.27			天冬氨酸 Asp	1.3	1							
中性洗涤纤维 Neutral detergent fiber	24.80			半胱氨酸 Cys	0.4	1							
酸性洗涤纤维 Acid detergent fiber	7.50			谷氨酸 Glu	2.3	1							
半纤维素 Hemicellulose				甘氨酸 Gly									
酸性洗涤木质素 Acid detergent lignin				脯氨酸 Pro	1.3	1							
总膳食纤维 Total dietary fiber				丝氨酸 Ser	0.8	1							
不溶性膳食纤维 Insoluble dietary fiber				酪氨酸 Tyr	0.6	1							
可溶性膳食纤维 Soluble dietary fiber													

矿物质				维生素, mg/kg(标注单位者除外)				脂肪酸, 占乙醚提取物, %			
	$\bar{x}$	n	SD		$\bar{x}$	n	SD		$\bar{x}$	n	SD
常量元素 Macro,%				脂溶性维生素 Fat soluble				粗脂肪			
钙 Ca	0.29			β-胡萝卜素 β-Carotene				C-12:0			
氯 Cl	0.25			维生素 E Vitamin E				C-14:0			
钾 K	1.50			水溶性维生素 Water soluble				C-16:0			
镁 Mg	0.64			维生素 B_6 Vitamin B_6	8.8			C-16:1			
钠 Na	0.26			维生素 B_{12} Vitamin B_{12}, μg/kg	3			C-18:0			
磷 P	1.24	1		生物素 Biotin	1.66			C-18:1			
硫 S	0.37			叶酸 Folacin	1.10			C-18:2			
微量元素 Micro,ppm				烟酸 Niacin	116			C-18:3			
铬 Cr				泛酸 Pantothenic acid	21.0			C-18:4			
铜 Cu	83			核黄素 Riboflavin	17.0			C-20:0			
铁 Fe	560			硫胺素 Thiamin	6.9			C-20:1			
碘 I				胆碱 Choline	4842			C-20:4			
锰 Mn	74							C-20:5			
硒 Se	0.33							C-22:0			
锌 Zn	85			能量, kcal/kg				C-22:1			
								C-22:5			
植酸磷 Phytate P,%				总能 GE	4717			C-22:6			
磷全消化道表观消化率 ATTD of P, %				消化能 DE	3325			C-24:0			
磷全消化道标准消化率 STTD of P, %				代谢能 ME	3198			饱和脂肪酸			
				净能 NE	2312			单不饱和脂肪酸			
								多不饱和脂肪酸			
								碘价			
								碘价物			

续表

原料名称：玉米胚芽(Corn Germ)
AAFCO#: 48.32, AAFCO 2010, p.357

常规成分, %	$\bar{x}$	n	SD	氨基酸, % 总氨基酸	$\bar{x}$	n	SD	可消化氨基酸 回肠表观消化率 $\bar{x}$	n	SD	回肠标准消化率 $\bar{x}$	n	SD
干物质 Dry matter	90.87	7	2.97	必需氨基酸 Essential									
粗蛋白 Crude protein	14.79	8	1.03	粗蛋白 CP	14.79	8	1.03	33	1		56	1	
粗纤维 Crude fiber				精氨酸 Arg	1.11	3	0.03	79	2	8.73	87	2	5.02
乙醚浸出物/粗脂肪 Ether extract	19.74	6	2.41	组氨酸 His	0.42	3	0.01	65	2	7.55	72	2	4.03
酸乙醚浸出物 Acid ether extract	17.6	1		异亮氨酸 Ile	0.43	3	0.03	51	2	9.46	61	2	6.22
粗灰分 Ash	5.54	5	1.30	亮氨酸 Leu	1.05	3	0.07	61	2	3.93	69	2	0.64
碳水化合物成分, %				赖氨酸 Lys	0.78	3	0.02	56	2	12.76	64	2	8.63
				蛋氨酸 Met	0.26	3	0.01	67	2	8.24	72	2	6.15
乳糖 Lactose	0.00	3	0.00	苯丙氨酸 Phe	0.57	3	0.03	57	2	5.96	66	2	2.47
蔗糖 Sucrose	0.00	3	0.00	苏氨酸 Thr	0.52	3	0.01	42	2	11.41	57	2	5.16
棉子糖 Raffinose	0.00	3	0.00	色氨酸 Trp	0.10	3	0.02	50	2	4.69	63	2	6.01
水苏糖 Stachyose	0.00	3	0.00	缬氨酸 Val	0.72	3	0.02	57	2	11.16	67	2	6.93
毛蕊花糖 Verbascose	0.00	3	0.00	非必需氨基酸 Nonessential									
低聚糖 Oligosaccharides				丙氨酸 Ala	0.91	2	0	53	1		64	1	
淀粉 Starch	23.51	4	2.58	天冬氨酸 Asp	1.10	2	0.06	47	1		60	1	
中性洗涤纤维 Neutral detergent fiber	18.27	5	4.33	半胱氨酸 Cys	0.32	3	0.03	58	2	8.06	66	2	3.25
酸性洗涤纤维 Acid detergent fiber	6.67	4	2.11	谷氨酸 Glu	1.94	2	0.16	63	1		72	1	
半纤维素 Hemicellulose				甘氨酸 Gly	0.77	2	0.01	14	1		76	1	
酸性洗涤木质素 Acid detergent lignin	2.37	1		脯氨酸 Pro	0.95	2	0.04	34	1		84	1	
总膳食纤维 Total dietary fiber				丝氨酸 Ser	0.59	2	0.04	48	1		65	1	
不溶性膳食纤维 Insoluble dietary fiber				酪氨酸 Tyr	0.41	3	0.02	51	2	7.52	61	2	3.11
可溶性膳食纤维 Soluble dietary fiber													

矿物质	$\bar{x}$	n	SD	维生素, mg/kg(标注单位者除外)	$\bar{x}$	n	SD	脂肪酸, 占乙醚提取物, %	$\bar{x}$	n	SD
常量元素 Macro,%				脂溶性维生素 Fat soluble				粗脂肪			
钙 Ca	0.02	4	0.01	β-胡萝卜素 β-Carotene				C-12:0			
氯 Cl				维生素 E Vitamin E				C-14:0			
钾 K	1.53	1		水溶性维生素 Water soluble				C-16:0			
镁 Mg	0.52	1		维生素 B_6 Vitamin B_6				C-16:1			
钠 Na	0.01	1		维生素 B_{12} Vitamin B_{12}, μg/kg				C-18:0			
磷 P	1.27	5	0.13	生物素 Biotin				C-18:1			
硫 S	0.17	1		叶酸 Folacin				C-18:2			
微量元素 Micro,ppm				烟酸 Niacin				C-18:3			
铬 Cr				泛酸 Pantothenic acid				C-18:4			
铜 Cu	5.30	1		核黄素 Riboflavin				C-20:0			
铁 Fe	96.7	1		硫胺素 Thiamin				C-20:1			
碘 I				胆碱 Choline				C-20:4			
锰 Mn	22.30	1						C-20:5			
硒 Se								C-22:0			
锌 Zn	83.70	1		能量, kcal/kg				C-22:1			
								C-22:5			
植酸磷 Phytate P,%	1.07	1		总能 GE	4919	1		C-22:6			
磷全消化道表观消化率 ATTD of P, %	33	2	6.15	消化能 DE	3670	1		C-24:0			
磷全消化道标准消化率 STTD of P, %	37	2	4.95	代谢能 ME	3569			饱和脂肪酸			
				净能 NE	2807			单不饱和脂肪酸			
								多不饱和脂肪酸			
								碘价			
								碘价物			

续表

原料名称：玉米胚芽粕(Corn Germ Meal)
AAFCO#: 48.22, AAFCO 2010, p.357
IFN#: 5-02-894

常规成分,%	$\bar{x}$	n	SD	氨基酸,% 总氨基酸	$\bar{x}$	n	SD	可消化氨基酸 回肠表观消化率 $\bar{x}$	n	SD	回肠标准消化率 $\bar{x}$	n	SD
干物质 Dry matter	90.10	1		必需氨基酸 Essential									
粗蛋白 Crude protein	23.33	2	3.19	粗蛋白 CP	23.33	2	3.19	60					
粗纤维 Crude fiber	9.53	1		精氨酸 Arg	1.49	1		76			83		
乙醚浸出物/粗脂肪 Ether extract	2.12	1		组氨酸 His	1.17	1		71			78		
酸乙醚浸出物 Acid ether extract	5.41	1		异亮氨酸 Ile	0.64	1		66			75		
粗灰分 Ash	2.96	2	0.78	亮氨酸 Leu	0.75	1		72			78		
碳水化合物成分,%				赖氨酸 Lys	1.70	1		53			62		
				蛋氨酸 Met	1.04	1		77			80		
乳糖 Lactose				苯丙氨酸 Phe	0.37	1		75			81		
蔗糖 Sucrose				苏氨酸 Thr	0.89	1		59			70		
棉子糖 Raffinose				色氨酸 Trp	0.78	1		53			66		
水苏糖 Stachyose				缬氨酸 Val	0.63	1		64			73		
毛蕊花糖 Verbascose				非必需氨基酸 Nonessential									
低聚糖 Oligosaccharides				丙氨酸 Ala	1.26	1		62			65		
淀粉 Starch	14.20	1		天冬氨酸 Asp	1.50	1		60			65		
中性洗涤纤维 Neutral detergent fiber	44.46	2	14.07	半胱氨酸 Cys	0.25			59			63		
酸性洗涤纤维 Acid detergent fiber	10.75	2	0.54	谷氨酸 Glu	0.33	1		62			65		
半纤维素 Hemicellulose	43.28	1		甘氨酸 Gly	2.87	1		55			65		
酸性洗涤木质素 Acid detergent lignin	1.09	1		脯氨酸 Pro	0.91	1		59			65		
总膳食纤维 Total dietary fiber	41.56	2	1.43	丝氨酸 Ser	1.07	1		59			65		
不溶性膳食纤维 Insoluble dietary fiber				酪氨酸 Tyr	0.63	1		75			79		
可溶性膳食纤维 Soluble dietary fiber													

矿物质	$\bar{x}$	n	SD	维生素, mg/kg(标注单位者除外)	$\bar{x}$	n	SD	脂肪酸,占乙醚提取物,%	$\bar{x}$	n	SD
常量元素 Macro,%				脂溶性维生素 Fat soluble				粗脂肪	2.50		
钙 Ca	0.03	1		β-胡萝卜素 β-Carotene				C-12:0	0.00		
氯 Cl				维生素 E Vitamin E				C-14:0	0.08		
钾 K	0.54	1		水溶性维生素 Water soluble				C-16:0	8.33		
镁 Mg	0.36	1		维生素 B_6 Vitamin B_6				C-16:1	0.30		
钠 Na				维生素 B_{12} Vitamin B_{12}, μg/kg				C-18:0	1.35		
磷 P	0.90	1		生物素 Biotin				C-18:1	20.18		
硫 S	0.36	1		叶酸 Folacin				C-18:2	42.38		
微量元素 Micro,ppm				烟酸 Niacin				C-18:3	0.75		
铬 Cr				泛酸 Pantothenic acid				C-18:4	0.00		
铜 Cu	7.03	1		核黄素 Riboflavin				C-20:0	0.00		
铁 Fe				硫胺素 Thiamin				C-20:1	0.00		
碘 I				胆碱 Choline				C-20:4	0.00		
锰 Mn	20.99	1						C-20:5	0.00		
硒 Se								C-22:0	0.00		
锌 Zn	133	1		能量, kcal/kg				C-22:1	0.00		
								C-22:5	0.00		
植酸磷 Phytate P,%				总能 GE	4178	2	100	C-22:6	0.00		
磷全消化道表观消化率 ATTD of P,%	33			消化能 DE	2988			C-24:0	0.00		
磷全消化道标准消化率 STTD of P,%	37			代谢能 ME	2830			饱和脂肪酸	9.75		
				净能 NE	1888			单不饱和脂肪酸	20.48		
								多不饱和脂肪酸	43.13		
								碘价	97.17		
								碘价物	24.29		

续表

原料名称：玉米蛋白饲料(Corn Gluten Feed)
AAFCO#: 48.13, AAFCO 2010, p.356
IFN#: 5-02-903

常规成分, %				氨基酸, %								
				总氨基酸				可消化氨基酸				
								回肠表观消化率			回肠标准消化率	
	$\bar{x}$	n	SD		$\bar{x}$	n	SD	$\bar{x}$	n	SD	$\bar{x}$	n	SD
干物质 Dry matter	87.13	4	2.89	必需氨基酸 Essential									
粗蛋白 Crude protein	17.39	4	3.82	粗蛋白 CP	17.39	4	3.82	64					
粗纤维 Crude fiber	7.08	3	0.75	精氨酸 Arg	1.04			79			86		
乙醚浸出物/粗脂肪 Ether extract	4.21	3	0.21	组氨酸 His	0.67			69			75		
酸乙醚浸出物 Acid ether extract				异亮氨酸 Ile	0.66			68			80		
粗灰分 Ash	5.14	4	0.72	亮氨酸 Leu	1.96			81			85		
碳水化合物成分, %				赖氨酸 Lys	0.63			51			66		
				蛋氨酸 Met	0.35			79			82		
乳糖 Lactose	0.00	3	0.00	苯丙氨酸 Phe	0.76			80			85		
蔗糖 Sucrose	0.00	3	0.00	苏氨酸 Thr	0.74			57			71		
棉子糖 Raffinose	0.00	3	0.00	色氨酸 Trp	0.07			47			66		
水苏糖 Stachyose	0.00	3	0.00	缬氨酸 Val	1.01			71			77		
毛蕊花糖 Verbascose	0.00	3	0.00	非必需氨基酸 Nonessential									
低聚糖 Oligosaccharides				丙氨酸 Ala	1.28			80			84		
淀粉 Starch	23.67	3	9.39	天冬氨酸 Asp	1.05			66			72		
中性洗涤纤维 Neutral detergent fiber	27.50	4	3.06	半胱氨酸 Cys	0.46			53			62		
酸性洗涤纤维 Acid detergent fiber	8.43	4	2.22	谷氨酸 Glu	3.11			78			82		
半纤维素 Hemicellulose				甘氨酸 Gly	0.79			52			62		
酸性洗涤木质素 Acid detergent lignin				脯氨酸 Pro	1.56			71			78		
总膳食纤维 Total dietary fiber	26.8	1		丝氨酸 Ser	0.78			68			76		
不溶性膳食纤维 Insoluble dietary fiber				酪氨酸 Tyr	0.58			80			84		
可溶性膳食纤维 Soluble dietary fiber													

矿物质				维生素, mg/kg(标注单位者除外)				脂肪酸, 占乙醚提取物, %			
	$\bar{x}$	n	SD		$\bar{x}$	n	SD		$\bar{x}$	n	SD
常量元素 Macro,%				脂溶性维生素 Fat soluble				粗脂肪	2.70		
钙 Ca	0.09	4	0.04	β-胡萝卜素 β-Carotene	1.0			C-12:0	0.00		
氯 Cl	0.22			维生素 E Vitamin E	8.5			C-14:0	0.09		
钾 K	0.98			水溶性维生素 Water soluble				C-16:0	9.99		
镁 Mg	0.33			维生素 B_6 Vitamin B_6	13.0			C-16:1	0.36		
钠 Na	0.15			维生素 B_{12} Vitamin B_{12}, μg/kg	0			C-18:0	1.62		
磷 P	0.78	4	0.15	生物素 Biotin	0.14			C-18:1	24.21		
硫 S	0.22			叶酸 Folacin	0.28			C-18:2	50.85		
微量元素 Micro,ppm				烟酸 Niacin	66			C-18:3	0.90		
铬 Cr				泛酸 Pantothenic acid	17.0			C-18:4	0.00		
铜 Cu	48			核黄素 Riboflavin	2.4			C-20:0	0.00		
铁 Fe	460			硫胺素 Thiamin	2.0			C-20:1	0.00		
碘 I				胆碱 Choline	1518			C-20:4	0.00		
锰 Mn	24							C-20:5	0.00		
硒 Se	0.27							C-22:0	0.00		
锌 Zn	70			能量, kcal/kg				C-22:1	0.00		
								C-22:5	0.00		
植酸磷 Phytate P,%	0.62	2	0.01	总能 GE	3989	2	294	C-22:6	0.00		
磷全消化道表观消化率 ATTD of P, %	26	4	8.21	消化能 DE	2990			C-24:0	0.00		
磷全消化道标准消化率 STTD of P, %	32	4	8.85	代谢能 ME	2872			饱和脂肪酸	11.70		
				净能 NE	2043			单不饱和脂肪酸	24.57		
								多不饱和脂肪酸	51.75		
								碘价	116.61		
								碘价物	31.48		

续表

原料名称：玉米蛋白粉(Corn Gluten Meal)
AAFCO#: 48.22, AAFCO 2010, p.356
IFN#: 5-02-900

常规成分, %				氨基酸, %							
				总氨基酸				可消化氨基酸			
								回肠表观消化率		回肠标准消化率	
	$\bar{x}$	n	SD		$\bar{x}$	n	SD	$\bar{x}$	n	SD	$\bar{x}$	n	SD
干物质 Dry matter	90.04	6	0.72	必需氨基酸 Essential									
粗蛋白 Crude protein	58.25	10	5.97	粗蛋白 CP	58.25	10	5.97	72	6	20.43	75	6	20.11
粗纤维 Crude fiber	0.70	4	0.13	精氨酸 Arg	1.66	8	0.46	88	6	2.10	91	6	2.00
乙醚浸出物/粗脂肪 Ether extract	4.74	5	1.97	组氨酸 His	1.32	8	0.33	86	6	2.94	87	6	3.07
酸乙醚浸出物 Acid ether extract	0.63	1		异亮氨酸 Ile	2.23	8	0.33	91	6	1.48	93	6	1.55
粗灰分 Ash	1.46	5	0.56	亮氨酸 Leu	9.82	7	0.98	96	6	1.17	96	6	1.22
碳水化合物成分, %				赖氨酸 Lys	0.93	8	0.18	77	6	4.79	81	6	4.78
				蛋氨酸 Met	1.21	7	0.44	92	5	8.82	93	5	8.85
乳糖 Lactose	0.00	3	0.00	苯丙氨酸 Phe	3.25	8	0.57	93	6	2.76	94	6	2.76
蔗糖 Sucrose	0.00	3	0.00	苏氨酸 Thr	1.81	8	0.47	84	6	5.02	87	6	6.14
棉子糖 Raffinose	0.00	3	0.00	色氨酸 Trp	0.27	6	0.07	61	5	10.15	77		
水苏糖 Stachyose	0.00	3	0.00	缬氨酸 Val	2.42	8	0.53	89	6	1.51	91	6	1.18
毛蕊花糖 Verbascose	0.00	3	0.00	非必需氨基酸 Nonessential									
低聚糖 Oligosaccharides				丙氨酸 Ala	4.33	5	1.11	92	5	3.59	93	5	3.24
淀粉 Starch	17.93	2	1.21	天冬氨酸 Asp	2.97	5	0.82	86	5	2.05	89	5	2.70
中性洗涤纤维 Neutral detergent fiber	1.57	2	0.05	半胱氨酸 Cys	1.01	6	0.29	86	4	2.30	88	4	3.13
酸性洗涤纤维 Acid detergent fiber	7.08			谷氨酸 Glu	11.20	5	2.99	93	5	2.73	94	5	2.74
半纤维素 Hemicellulose				甘氨酸 Gly	1.28	5	0.37	78	5	13.22	89	5	14.31
酸性洗涤木质素 Acid detergent lignin				脯氨酸 Pro	4.93	5	1.25	78	5	15.81	86	5	14.54
总膳食纤维 Total dietary fiber				丝氨酸 Ser	2.29	5	0.87	91	5	2.23	93	5	3.51
不溶性膳食纤维 Insoluble dietary fiber				酪氨酸 Tyr	2.86	5	0.28	93	3	2.02	94	3	2.00
可溶性膳食纤维 Soluble dietary fiber													

矿物质				维生素, mg/kg(标注单位者除外)				脂肪酸, 占乙醚提取物, %			
	$\bar{x}$	n	SD		$\bar{x}$	n	SD		$\bar{x}$	n	SD
常量元素 Macro, %				脂溶性维生素 Fat soluble				粗脂肪	2.50		
钙 Ca	0.03	2	0.00	β-胡萝卜素 β-Carotene				C-12:0	0.00		
氯 Cl	0.06			维生素 E Vitamin E	6.7			C-14:0	0.08		
钾 K	0.18	1		水溶性维生素 Water soluble				C-16:0	8.88		
镁 Mg	0.09	1		维生素 B_6 Vitamin B_6	6.9			C-16:1	0.32		
钠 Na	0.02			维生素 B_{12} Vitamin B_{12}, μg/kg	0			C-18:0	1.44		
磷 P	0.49	3	0.04	生物素 Biotin	0.15			C-18:1	21.52		
硫 S	1.00	1		叶酸 Folacin	0.13			C-18:2	45.20		
微量元素 Micro, ppm				烟酸 Niacin	55			C-18:3	0.80		
铬 Cr				泛酸 Pantothenic acid	3.5			C-18:4	0.00		
铜 Cu	11.04	1		核黄素 Riboflavin	2.2			C-20:0	0.00		
铁 Fe	282			硫胺素 Thiamin	0.3			C-20:1	0.00		
碘 I				胆碱 Choline	330			C-20:4	0.00		
锰 Mn	3.98	1						C-20:5	0.00		
硒 Se	1.00							C-22:0	0.00		
锌 Zn	25.97	1		能量, kcal/kg				C-22:1	0.00		
								C-22:5	0.00		
植酸磷 Phytate P, %				总能 GE	4865	5	324	C-22:6	0.00		
磷全消化道表观消化率 ATTD of P, %	38	2	2.40	消化能 DE	4133	4	124	C-24:0	0.00		
磷全消化道标准消化率 STTD of P, %	47	2	2.40	代谢能 ME	3737			饱和脂肪酸	10.40		
				净能 NE	2464			单不饱和脂肪酸	21.84		
								多不饱和脂肪酸	46.00		
								碘价	103.65		
								碘价物	25.91		

续表

原料名称：玉米糁，玉米渣 (Corn Grits, Hominy Feed)
AAFCO#：无官方定义
IFN#：4-03-011

常规成分, %				氨基酸, %							
				总氨基酸				可消化氨基酸			
								回肠表观消化率		回肠标准消化率	
	$\bar{x}$	n	SD		$\bar{x}$	n	SD	$\bar{x}$	n	SD	$\bar{x}$	n	SD
干物质 Dry matter	87.47	6	0.98	必需氨基酸 Essential								
粗蛋白 Crude protein	9.12	6	0.91	粗蛋白 CP	9.12	6	0.91					
粗纤维 Crude fiber	3.19	5	0.52	精氨酸 Arg	0.56						87	
乙醚浸出物/粗脂肪 Ether extract	7.40	6	3.34	组氨酸 His	0.28						80	
酸乙醚浸出物 Acid ether extract				异亮氨酸 Ile	0.36						81	
粗灰分 Ash	2.34	5	0.58	亮氨酸 Leu	0.98						86	
碳水化合物成分, %				赖氨酸 Lys	0.38						71	
				蛋氨酸 Met	0.18						87	
乳糖 Lactose	0.00	5	0.00	苯丙氨酸 Phe	0.43						86	
蔗糖 Sucrose	0.00	5	0.00	苏氨酸 Thr	0.40						73	
棉子糖 Raffinose	0.00	5	0.00	色氨酸 Trp	0.10						68	
水苏糖 Stachyose	0.00	5	0.00	缬氨酸 Val	0.52						80	
毛蕊花糖 Verbascose	0.00	5	0.00	非必需氨基酸 Nonessential								
低聚糖 Oligosaccharides				丙氨酸 Ala								
淀粉 Starch	47.58	5	7.96	天冬氨酸 Asp								
中性洗涤纤维 Neutral detergent fiber	14.30	4	0.22	半胱氨酸 Cys	0.18						74	
酸性洗涤纤维 Acid detergent fiber	4.51	4	0.34	谷氨酸 Glu								
半纤维素 Hemicellulose				甘氨酸 Gly								
酸性洗涤木质素 Acid detergent lignin				脯氨酸 Pro								
总膳食纤维 Total dietary fiber	10.11	1		丝氨酸 Ser								
不溶性膳食纤维 Insoluble dietary fiber				酪氨酸 Tyr	0.40						88	
可溶性膳食纤维 Soluble dietary fiber												

矿物质				维生素, mg/kg(标注单位者除外)				脂肪酸, 占乙醚提取物, %			
	$\bar{x}$	n	SD		$\bar{x}$	n	SD		$\bar{x}$	n	SD
常量元素 Macro,%				脂溶性维生素 Fat soluble				粗脂肪	1.20		
钙 Ca	0.29	1		β-胡萝卜素 β-Carotene	9.0			C-12:0	0.08		
氯 Cl	0.07			维生素 E Vitamin E	6.5			C-14:0	0.08		
钾 K	0.61			水溶性维生素 Water soluble				C-16:0	11.25		
镁 Mg	0.24			维生素 B_6 Vitamin B_6	11.0			C-16:1	0.33		
钠 Na	0.08			维生素 B_{12} Vitamin B_{12}, μg/kg	0			C-18:0	1.50		
磷 P	0.73	1		生物素 Biotin	0.13			C-18:1	24.67		
硫 S	0.03			叶酸 Folacin	0.21			C-18:2	41.83		
微量元素 Micro,ppm				烟酸 Niacin	47			C-18:3	1.25		
铬 Cr				泛酸 Pantothenic acid	8.2			C-18:4			
铜 Cu	13.00			核黄素 Riboflavin	2.1			C-20:0	0.00		
铁 Fe	67			硫胺素 Thiamin	8.1			C-20:1	0.00		
碘 I				胆碱 Choline	1155			C-20:4			
锰 Mn	15.00							C-20:5			
硒 Se	0.10							C-22:0			
锌 Zn	30.00			能量, kcal/kg				C-22:1			
								C-22:5			
植酸磷 Phytate P,%	0.49	1		总能 GE	4145	5	179	C-22:6			
磷全消化道表观消化率 ATTD of P,%	26			消化能 DE	3355			C-24:0			
磷全消化道标准消化率 STTD of P,%	34			代谢能 ME	3293			饱和脂肪酸	12.92		
				净能 NE	2574			单不饱和脂肪酸	25.00		
								多不饱和脂肪酸	43.08		
								碘价	101.63		
								碘价物	12.20		

续表

原料名称：全脂棉籽(Cottonseed, Full Fat)
AAFCO#: 24.4, AAFCO 2010, p.341
IFN#: 5-01-609

常规成分, %				氨基酸, %								
				总氨基酸				可消化氨基酸				
								回肠表观消化率			回肠标准消化率	
	$\bar{x}$	n	SD		$\bar{x}$	n	SD	$\bar{x}$	n	SD	$\bar{x}$	n	SD
干物质 Dry matter	92.56	8	0.38	必需氨基酸 Essential									
粗蛋白 Crude protein	23.77	9	1.88	粗蛋白 CP	23.77	9	1.88	73	17	5.18	77	17	5.16
粗纤维 Crude fiber				精氨酸 Arg	2.41			87	20	3.73	88	20	3.77
乙醚浸出物/粗脂肪 Ether extract	16.51	9	1.26	组氨酸 His	0.61			72	20	8.93	74	20	8.36
酸乙醚浸出物 Acid ether extract				异亮氨酸 Ile	0.7			67	20	9.32	70	20	9.46
粗灰分 Ash	4	9	0.28	亮氨酸 Leu	1.18			70	20	7.90	73	20	7.76
碳水化合物成分, %				赖氨酸 Lys	0.87			59	20	10.87	63	20	10.85
				蛋氨酸 Met	0.33			70	16	14.08	73	16	13.63
乳糖 Lactose				苯丙氨酸 Phe	1.17			79	20	5.61	81	20	5.56
蔗糖 Sucrose				苏氨酸 Thr	0.67			64	20	9.89	68	20	9.61
棉子糖 Raffinose				色氨酸 Trp	0.25			68	11	8.81	71	11	8.95
水苏糖 Stachyose				缬氨酸 Val	0.98			69	20	8.45	73	20	8.23
毛蕊花糖 Verbascose				非必需氨基酸 Nonessential									
低聚糖 Oligosaccharides				丙氨酸 Ala	0.78			66	17	8.91	70	17	8.86
淀粉 Starch	2.30			天冬氨酸 Asp	1.87			74	17	6.40	76	17	6.27
中性洗涤纤维 Neutral detergent fiber	51.04	9	3.77	半胱氨酸 Cys	0.33			73	7	9.92	76	7	9.83
酸性洗涤纤维 Acid detergent fiber	38.59	9	2.9	谷氨酸 Glu	4.24			83	16	4.96	84	16	4.87
半纤维素 Hemicellulose				甘氨酸 Gly	0.80			67	17	9.66	77	17	9.71
酸性洗涤木质素 Acid detergent lignin	10.75	4	0.49	脯氨酸 Pro	0.79			58	15	17.42	84	14	16.06
总膳食纤维 Total dietary fiber				丝氨酸 Ser	0.90			72	17	7.49	75	17	7.12
不溶性膳食纤维 Insoluble dietary fiber				酪氨酸 Tyr	0.56			73	16	6.39	76	14	6.24
可溶性膳食纤维 Soluble dietary fiber													

矿物质				维生素, mg/kg(标注单位者除外)				脂肪酸, 占乙醚提取物, %			
	$\bar{x}$	n	SD		$\bar{x}$	n	SD		$\bar{x}$	n	SD
常量元素 Macro,%				脂溶性维生素 Fat soluble				粗脂肪			
钙 Ca	0.15	1		β-胡萝卜素 β-Carotene				C-12:0	0.00		
氯 Cl				维生素 E Vitamin E				C-14:0	0.94		
钾 K				水溶性维生素 Water soluble				C-16:0	23.24		
镁 Mg				维生素 B_6 Vitamin B_6				C-16:1	0.71		
钠 Na				维生素 B_{12} Vitamin B_{12}, μg/kg				C-18:0	2.35		
磷 P	0.65	1		生物素 Biotin				C-18:1	18.22		
硫 S				叶酸 Folacin				C-18:2	49.23		
微量元素 Micro,ppm				烟酸 Niacin				C-18:3	0.19		
铬 Cr				泛酸 Pantothenic acid				C-18:4			
铜 Cu				核黄素 Riboflavin				C-20:0	0.00		
铁 Fe				硫胺素 Thiamin				C-20:1	0.00		
碘 I				胆碱 Choline				C-20:4			
锰 Mn								C-20:5			
硒 Se								C-22:0			
锌 Zn				能量, kcal/kg				C-22:1			
								C-22:5			
植酸磷 Phytate P,%				总能 GE	5248			C-22:6			
磷全消化道表观消化率 ATTD of P, %	31			消化能 DE	3207			C-24:0			
磷全消化道标准消化率 STTD of P, %	36			代谢能 ME	3045			饱和脂肪酸	26.53		
				净能 NE	1970			单不饱和脂肪酸	18.93		
								多不饱和脂肪酸	49.42		
								碘价	106.69		
								碘价物	176.15		

续表

原料名称：棉粕(Cottonseed Meal)
AAFCO#: 24.12, AAFCO 2010, p.341
IFN#: 5-01-632

常规成分, %				氨基酸, %							
					总氨基酸			可消化氨基酸			
	$\bar{x}$	n	SD		$\bar{x}$	n	SD	回肠表观消化率		回肠标准消化率	
干物质 Dry matter	90.69	8	1.87	必需氨基酸 Essential				$\bar{x}$	n	SD	$\bar{x}$	n	SD
粗蛋白 Crude protein	39.22	25	3.59	粗蛋白 CP	39.22	25	3.59	73	17	5.18	77	17	5.16
粗纤维 Crude fiber	13.96	5	1.68	精氨酸 Arg	4.04	16	0.68	87	20	3.73	88	20	3.77
乙醚浸出物/粗脂肪 Ether extract	5.50	6	2.50	组氨酸 His	1.11	19	0.16	72	20	8.93	74	20	8.36
酸乙醚浸出物 Acid ether extract				异亮氨酸 Ile	1.21	19	0.24	67	20	9.32	70	20	9.46
粗灰分 Ash	6.39	5	0.46	亮氨酸 Leu	2.18	19	0.39	70	20	7.90	73	20	7.76
碳水化合物成分, %				赖氨酸 Lys	1.50	19	0.28	59	20	10.87	63	20	10.85
				蛋氨酸 Met	0.51	15	0.14	70	16	14.08	73	16	13.63
乳糖 Lactose	0.00	5	0.00	苯丙氨酸 Phe	1.98	19	0.32	79	20	5.61	81	20	5.56
蔗糖 Sucrose	0.00	5	0.00	苏氨酸 Thr	1.36	19	0.18	64	20	9.89	68	20	9.61
棉子糖 Raffinose	0.00	5	0.00	色氨酸 Trp	0.53	13	0.16	68	11	8.81	71	11	8.95
水苏糖 Stachyose	0.00	5	0.00	缬氨酸 Val	1.86	19	0.42	69	20	8.45	73	20	8.23
毛蕊花糖 Verbascose	0.00	5	0.00	非必需氨基酸 Nonessential									
低聚糖 Oligosaccharides				丙氨酸 Ala	1.51	13	0.31	66	17	8.91	70	17	8.86
淀粉 Starch	1.95	4	0.48	天冬氨酸 Asp	3.28	14	0.78	74	17	6.40	76	17	6.27
中性洗涤纤维 Neutral detergent fiber	25.15	4	4.07	半胱氨酸 Cys	0.82	12	0.31	73	7	9.92	76	7	9.83
酸性洗涤纤维 Acid detergent fiber	17.92	5	1.99	谷氨酸 Glu	6.93	14	1.56	83	16	4.96	84	16	4.87
半纤维素 Hemicellulose				甘氨酸 Gly	1.58	14	0.32	67	17	9.66	77	17	9.71
酸性洗涤木质素 Acid detergent lignin				脯氨酸 Pro	1.50	11	0.43	58	15	17.42	84	14	16.06
总膳食纤维 Total dietary fiber				丝氨酸 Ser	1.80	14	0.61	72	17	7.49	75	17	7.12
不溶性膳食纤维 Insoluble dietary fiber				酪氨酸 Tyr	0.98	15	0.19	73	16	6.39	76	14	6.24
可溶性膳食纤维 Soluble dietary fiber													

矿物质				维生素, mg/kg(标注单位者除外)				脂肪酸, 占乙醚提取物, %			
	$\bar{x}$	n	SD		$\bar{x}$	n	SD		$\bar{x}$	n	SD
常量元素 Macro,%				脂溶性维生素 Fat soluble				粗脂肪	4.77		
钙 Ca	0.25	4	0.03	β-胡萝卜素 β-Carotene	0.2			C-12:0	0.00		
氯 Cl	0.05			维生素 E Vitamin E	14.0			C-14:0	0.88		
钾 K	1.40			水溶性维生素 Water soluble				C-16:0	21.99		
镁 Mg	0.50			维生素 B_6 Vitamin B_6	5.1			C-16:1	0.67		
钠 Na	0.04			维生素 B_{12} Vitamin B_{12}, μg/kg	0			C-18:0	2.22		
磷 P	0.98	6	0.09	生物素 Biotin	0.30			C-18:1	17.25		
硫 S	0.31			叶酸 Folacin	1.65			C-18:2	46.60		
微量元素 Micro,ppm				烟酸 Niacin	40			C-18:3	0.19		
铬 Cr				泛酸 Pantothenic acid	12.0			C-18:4			
铜 Cu	18.00			核黄素 Riboflavin	5.9			C-20:0	0.00		
铁 Fe	184			硫胺素 Thiamin	7.0			C-20:1	0.00		
碘 I				胆碱 Choline	2933			C-20:4			
锰 Mn	20.00							C-20:5			
硒 Se	0.80							C-22:0			
锌 Zn	70.00			能量, kcal/kg				C-22:1			
								C-22:5			
植酸磷 Phytate P,%				总能 GE	4383	5	148	C-22:6			
磷全消化道表观消化率 ATTD of P, %	31	5	9.98	消化能 DE	2912			C-24:0			
磷全消化道标准消化率 STTD of P, %	36	5	8.99	代谢能 ME	2645			饱和脂肪酸	25.09		
				净能 NE	1624			单不饱和脂肪酸	17.92		
								多不饱和脂肪酸	46.79		
								碘价	101.04		
								碘价物	48.19		

续表

原料名称：喷干全鸡蛋粉 (Egg, Whole, Spray Dried)
AAFCO#: 9.74, AAFCO 2010, p.331

常规成分, %				氨基酸, %								
				总氨基酸				可消化氨基酸				
								回肠表观消化率			回肠标准消化率		
	$\bar{x}$	n	SD		$\bar{x}$	n	SD	$\bar{x}$	n	SD	$\bar{x}$	n	SD
干物质 Dry matter	95.16	3	0.98	必需氨基酸 Essential									
粗蛋白 Crude protein	50.97	4	2.25	粗蛋白 CP	50.97	4	2.25						
粗纤维 Crude fiber				精氨酸 Arg	3.01	3	0.01						
乙醚浸出物/粗脂肪 Ether extract	34.26	2	4.6	组氨酸 His	1.20	3	0.12						
酸乙醚浸出物 Acid ether extract	35.40	1		异亮氨酸 Ile	2.81	3	0.21						
粗灰分 Ash	5.75	1		亮氨酸 Leu	4.41	3	0.28						
碳水化合物成分, %				赖氨酸 Lys	3.54	3	0.15						
				蛋氨酸 Met	1.62	3	0.09						
乳糖 Lactose				苯丙氨酸 Phe	2.68	3	0.1						
蔗糖 Sucrose				苏氨酸 Thr	2.13	3	0.03						
棉子糖 Raffinose				色氨酸 Trp	0.94	3	0.14						
水苏糖 Stachyose				缬氨酸 Val	3.34	3	0.16						
毛蕊花糖 Verbascose				非必需氨基酸 Nonessential									
低聚糖 Oligosaccharides				丙氨酸 Ala	2.63	1							
淀粉 Starch				天冬氨酸 Asp	4.65	1							
中性洗涤纤维 Neutral detergent fiber				半胱氨酸 Cys	1.19	2	0.06						
酸性洗涤纤维 Acid detergent fiber				谷氨酸 Glu	5.92	1							
半纤维素 Hemicellulose				甘氨酸 Gly	1.54	1							
酸性洗涤木质素 Acid detergent lignin				脯氨酸 Pro	1.57	1							
总膳食纤维 Total dietary fiber				丝氨酸 Ser	2.72	1							
不溶性膳食纤维 Insoluble dietary fiber				酪氨酸 Tyr	1.95	2	0.06						
可溶性膳食纤维 Soluble dietary fiber													

矿物质				维生素, mg/kg(标注单位者除外)				脂肪酸, 占乙醚提取物, %		
	$\bar{x}$	n	SD		$\bar{x}$	n	SD		$\bar{x}$	n	SD
常量元素 Macro, %				脂溶性维生素 Fat soluble				粗脂肪			
钙 Ca	0.29	2	0.11	β-胡萝卜素 β-Carotene				C-12:0			
氯 Cl				维生素 E Vitamin E				C-14:0			
钾 K				水溶性维生素 Water soluble				C-16:0			
镁 Mg				维生素 B_6 Vitamin B_6				C-16:1			
钠 Na				维生素 B_{12} Vitamin B_{12}, μg/kg				C-18:0			
磷 P	0.69	2	0.03	生物素 Biotin				C-18:1			
硫 S				叶酸 Folacin				C-18:2			
微量元素 Micro, ppm				烟酸 Niacin				C-18:3			
铬 Cr				泛酸 Pantothenic acid				C-18:4			
铜 Cu	1.80	1		核黄素 Riboflavin				C-20:0			
铁 Fe	61	1		硫胺素 Thiamin				C-20:1			
碘 I				胆碱 Choline				C-20:4			
锰 Mn	0.00	1						C-20:5			
硒 Se								C-22:0			
锌 Zn	43.70	1		能量, kcal/kg				C-22:1			
								C-22:5			
植酸磷 Phytate P, %				总能 GE	6283	2	202	C-22:6			
磷全消化道表观消化率 ATTD of P, %	50	1		消化能 DE				C-24:0			
磷全消化道标准消化率 STTD of P, %	55	1		代谢能 ME				饱和脂肪酸			
				净能 NE				单不饱和脂肪酸			
								多不饱和脂肪酸			
								碘价			
								碘价物			

续表

原料名称：羽毛粉 (Feather Meal)
AAFCO#: 9.15, AAFCO 2010, p.327
IFN#: 5-03-795

常规成分, %				氨基酸, %								
				总氨基酸				可消化氨基酸				
								回肠表观消化率			回肠标准消化率	
	$\bar{x}$	n	SD		$\bar{x}$	n	SD	$\bar{x}$	n	SD	$\bar{x}$	n	SD
干物质 Dry matter	94.24	4	1.44	必需氨基酸 Essential									
粗蛋白 Crude protein	80.90	6	6.58	粗蛋白 CP	80.90	6	6.58	75	2	16.97	68	2	4.31
粗纤维 Crude fiber	0.32	1		精氨酸 Arg	5.63	7	0.58	81	2	2.83	81	2	2.17
乙醚浸出物/粗脂肪 Ether extract	5.97	1		组氨酸 His	0.82	8	0.18	54	2	26.87	56	2	24.77
酸乙醚浸出物 Acid ether extract				异亮氨酸 Ile	3.63	8	0.91	75	2	0.71	76	2	0.06
粗灰分 Ash	5.08	1		亮氨酸 Leu	6.59	7	1.24	77	2	2.12	77	2	2.28
碳水化合物成分, %				赖氨酸 Lys	2.00	8	0.36	54	2	19.80	56	2	18.61
				蛋氨酸 Met	0.59	5	0.13	65			73	1	
乳糖 Lactose				苯丙氨酸 Phe	3.95	7	0.99	78	2	2.83	79	2	2.64
蔗糖 Sucrose				苏氨酸 Thr	3.72	8	0.40	69	2	4.24	71	2	4.62
棉子糖 Raffinose				色氨酸 Trp	0.60	6	0.16	60	1		63	1	
水苏糖 Stachyose				缬氨酸 Val	5.75	8	1.28	75	2	3.54	75	2	3.34
毛蕊花糖 Verbascose				非必需氨基酸 Nonessential									
低聚糖 Oligosaccharides				丙氨酸 Ala	3.90	4	0.44	70			71		
淀粉 Starch	0.00			天冬氨酸 Asp	4.95	4	1.41	47			48		
中性洗涤纤维 Neutral detergent fiber				半胱氨酸 Cys	4.32	4	0.44	71			73		
酸性洗涤纤维 Acid detergent fiber	0.00			谷氨酸 Glu	8.40	4	2.61	75	1		76	1	
半纤维素 Hemicellulose				甘氨酸 Gly	7.08	4	1.50	78	1		80	1	
酸性洗涤木质素 Acid detergent lignin				脯氨酸 Pro	10.16	4	1.61	86			87		
总膳食纤维 Total dietary fiber				丝氨酸 Ser	8.18	4	2.66	76	1		77	1	
不溶性膳食纤维 Insoluble dietary fiber				酪氨酸 Tyr	2.12	6	0.55	73			79		
可溶性膳食纤维 Soluble dietary fiber													

矿物质				维生素, mg/kg (标注单位者除外)				脂肪酸, 占乙醚提取物, %			
	$\bar{x}$	n	SD		$\bar{x}$	n	SD		$\bar{x}$	n	SD
常量元素 Macro, %				脂溶性维生素 Fat soluble				粗脂肪	6.80		
钙 Ca	0.41	2	0.06	β-胡萝卜素 β-Carotene				C-12:0	0.00		
氯 Cl	0.26			维生素 E Vitamin E	7.3			C-14:0	1.00		
钾 K	0.19			水溶性维生素 Water soluble				C-16:0	17.40		
镁 Mg	0.20			维生素 B_6 Vitamin B_6	3.0			C-16:1	3.10		
钠 Na	0.34			维生素 B_{12} Vitamin B_{12}, μg/kg	78			C-18:0	6.90		
磷 P	0.28	3	0.10	生物素 Biotin	0.13			C-18:1	19.95		
硫 S	1.39			叶酸 Folacin	0.20			C-18:2	1.65		
微量元素 Micro, ppm				烟酸 Niacin	21			C-18:3	0.00		
铬 Cr				泛酸 Pantothenic acid	10.0			C-18:4	0.00		
铜 Cu	10.00			核黄素 Riboflavin	2.1			C-20:0	0.00		
铁 Fe	76			硫胺素 Thiamin	0.1			C-20:1	0.00		
碘 I				胆碱 Choline	891			C-20:4	0.00		
锰 Mn	10.00							C-20:5	0.00		
硒 Se	0.69							C-22:0	0.00		
锌 Zn	111			能量, kcal/kg				C-22:1	0.00		
								C-22:5	0.00		
植酸磷 Phytate P, %				总能 GE	5467			C-22:6	0.00		
磷全消化道表观消化率 ATTD of P, %	74	2	1.91	消化能 DE	3400			C-24:0	0.00		
磷全消化道标准消化率 STTD of P, %	89	2	2.33	代谢能 ME	2850			饱和脂肪酸	25.30		
				净能 NE	1740			单不饱和脂肪酸	23.05		
								多不饱和脂肪酸	1.65		
								碘价	24.00		
								碘价物	16.32		

续表

原料名称：混合鱼粉(Fish Meal, Combined)
所有鱼粉的数据综合分析，因为大多数文献并没有区分鱼的种类。
AAFCO#: 51.14, AAFCO 2010, p.358
IFN#: 5-01-977

常规成分, %				氨基酸, %									
					总氨基酸			可消化氨基酸					
								回肠表观消化率			回肠标准消化率		
	$\bar{x}$	n	SD		$\bar{x}$	n	SD	$\bar{x}$	n	SD	$\bar{x}$	n	SD
干物质 Dry matter	93.70	8	2.42	必需氨基酸 Essential									
粗蛋白 Crude protein	63.28	23	4.66	粗蛋白 CP	63.28	23	4.66	82	16	7.04	85	16	6.16
粗纤维 Crude fiber	0.24	4	0.22	精氨酸 Arg	3.84	24	0.48	85	22	9.98	86	22	10.11
乙醚浸出物/粗脂肪 Ether extract	9.71	5	1.28	组氨酸 His	1.44	21	0.29	82	22	10.56	84	22	10.55
酸乙醚浸出物 Acid ether extract	8.73	1		异亮氨酸 Ile	2.56	25	0.31	82	22	12.03	83	22	12.06
粗灰分 Ash	16.07	5	3.16	亮氨酸 Leu	4.47	25	0.50	82	22	11.64	83	22	11.71
碳水化合物成分, %				赖氨酸 Lys	4.56	24	0.90	85	22	8.35	86	22	8.37
				蛋氨酸 Met	1.73	22	0.45	86	18	7.53	87	18	7.57
乳糖 Lactose				苯丙氨酸 Phe	2.47	24	0.22	80	22	12.37	82	22	12.43
蔗糖 Sucrose				苏氨酸 Thr	2.58	25	0.33	78	21	14.37	81	22	14.49
棉子糖 Raffinose				色氨酸 Trp	0.63	16	0.10	73	10	9.43	76	10	9.97
水苏糖 Stachyose				缬氨酸 Val	3.06	25	0.45	81	22	10.16	83	22	10.22
毛蕊花糖 Verbascose				非必需氨基酸 Nonessential									
低聚糖 Oligosaccharides				丙氨酸 Ala	3.93	18	0.54	79	15	14.67	80	15	14.65
淀粉 Starch	0.00			天冬氨酸 Asp	5.41	17	1.18	71	15	22.27	73	15	22.53
中性洗涤纤维 Neutral detergent fiber				半胱氨酸 Cys	0.61	16	0.20	62	11	18.94	64	11	17.71
酸性洗涤纤维 Acid detergent fiber	0.00			谷氨酸 Glu	7.88	17	1.18	79	15	14.48	80	15	14.54
半纤维素 Hemicellulose				甘氨酸 Gly	4.71	18	0.98	71	15	20.64	75	15	20.63
酸性洗涤木质素 Acid detergent lignin				脯氨酸 Pro	2.89	18	1.07	65	14	25.52	86	14	21.49
总膳食纤维 Total dietary fiber				丝氨酸 Ser	2.43	18	0.59	72	15	20.75	75	15	20.96
不溶性膳食纤维 Insoluble dietary fiber				酪氨酸 Tyr	1.88	15	0.38	73	13	17.12	74	12	17.65
可溶性膳食纤维 Soluble dietary fiber													

矿物质				维生素, mg/kg(标注单位者除外)				脂肪酸, 占乙醚提取物, %			
	$\bar{x}$	n	SD		$\bar{x}$	n	SD		$\bar{x}$	n	SD
常量元素 Macro,%				脂溶性维生素 Fat soluble				粗脂肪			
钙 Ca	4.28	11	1.14	β-胡萝卜素 β-Carotene				C-12:0			
氯 Cl				维生素 E Vitamin E				C-14:0			
钾 K	0.62	2	0.10	水溶性维生素 Water soluble				C-16:0			
镁 Mg	0.13	1		维生素 B_6 Vitamin B_6				C-16:1			
钠 Na				维生素 B_{12} Vitamin B_{12}, μg/kg				C-18:0			
磷 P	2.93	14	0.51	生物素 Biotin				C-18:1			
硫 S				叶酸 Folacin				C-18:2			
微量元素 Micro,ppm				烟酸 Niacin				C-18:3			
铬 Cr				泛酸 Pantothenic acid				C-18:4			
铜 Cu	8.00	1		核黄素 Riboflavin				C-20:0			
铁 Fe	411	2	416	硫胺素 Thiamin				C-20:1			
碘 I				胆碱 Choline				C-20:4			
锰 Mn	38.90	1						C-20:5			
硒 Se								C-22:0			
锌 Zn	88.98	2	27.61	能量, kcal/kg				C-22:1			
								C-22:5			
植酸磷 Phytate P,%				总能 GE	4496	4	84.4	C-22:6			
磷全消化道表观消化率 ATTD of P,%	79	7	11.53	消化能 DE	3958	3	392	C-24:0			
磷全消化道标准消化率 STTD of P,%	82	7	11.44	代谢能 ME	3528			饱和脂肪酸			
				净能 NE	2351			单不饱和脂肪酸			
								多不饱和脂肪酸			
								碘价			
								碘价物			

续表

原料名称：亚麻籽(Flaxseed)
AAFCO#：无官方定义

常规成分, %				氨基酸, %							
				总氨基酸				可消化氨基酸			
								回肠表观消化率		回肠标准消化率	
	$\bar{x}$	n	SD		$\bar{x}$	n	SD	$\bar{x}$	SD	$\bar{x}$	n	SD
干物质 Dry matter	92.13	3	1.49	必需氨基酸 Essential							
粗蛋白 Crude protein	22.53	7	1.53	粗蛋白 CP	22.53	7	1.53				
粗纤维 Crude fiber	6.00	1		精氨酸 Arg	2.2	1					
乙醚浸出物/粗脂肪 Ether extract	33.77	5	4.41	组氨酸 His	0.51	1					
酸乙醚浸出物 Acid ether extract				异亮氨酸 Ile	0.95	1				77	
粗灰分 Ash	3.33	4	0.20	亮氨酸 Leu	1.35	1					
碳水化合物成分, %				赖氨酸 Lys	0.91	1				84	
				蛋氨酸 Met	0.43	1				85	
乳糖 Lactose				苯丙氨酸 Phe	1.08	1					
蔗糖 Sucrose				苏氨酸 Thr	0.85	1				82	
棉子糖 Raffinose				色氨酸 Trp						86	
水苏糖 Stachyose				缬氨酸 Val	1.16	1				77	
毛蕊花糖 Verbascose				非必需氨基酸 Nonessential							
低聚糖 Oligosaccharides				丙氨酸 Ala	1.05	1				77	
淀粉 Starch				天冬氨酸 Asp	2.18	1				77	
中性洗涤纤维 Neutral detergent fiber	39.65	2	11.38	半胱氨酸 Cys	0.41	1					
酸性洗涤纤维 Acid detergent fiber	24.85	2	6.86	谷氨酸 Glu	4.46	1				77	
半纤维素 Hemicellulose				甘氨酸 Gly	1.38	1				77	
酸性洗涤木质素 Acid detergent lignin				脯氨酸 Pro	0.84	1				77	
总膳食纤维 Total dietary fiber				丝氨酸 Ser	1.06	1				77	
不溶性膳食纤维 Insoluble dietary fiber				酪氨酸 Tyr							
可溶性膳食纤维 Soluble dietary fiber											

矿物质				维生素, mg/kg(标注单位者除外)				脂肪酸, 占乙醚提取物, %			
	$\bar{x}$	n	SD		$\bar{x}$	n	SD		$\bar{x}$	n	SD
常量元素 Macro,%				脂溶性维生素 Fat soluble				粗脂肪	42.16		
钙 Ca	0.38			β-胡萝卜素 β-Carotene				C-12:0	0.00		
氯 Cl				维生素 E Vitamin E				C-14:0	0.02		
钾 K				水溶性维生素 Water soluble				C-16:0	5.14		
镁 Mg				维生素 B_6 Vitamin B_6				C-16:1	0.06		
钠 Na				维生素 B_{12} Vitamin B_{12}, μg/kg				C-18:0	3.15		
磷 P	0.61			生物素 Biotin				C-18:1	17.45		
硫 S				叶酸 Folacin				C-18:2	14.00		
微量元素 Micro,ppm				烟酸 Niacin				C-18:3	54.11		
铬 Cr				泛酸 Pantothenic acid				C-18:4			
铜 Cu				核黄素 Riboflavin				C-20:0	0.12		
铁 Fe				硫胺素 Thiamin				C-20:1	0.16		
碘 I				胆碱 Choline				C-20:4	0.00		
锰 Mn								C-20:5	0.00		
硒 Se								C-22:0	0.12		
锌 Zn				能量, kcal/kg				C-22:1	0.03		
								C-22:5	0.00		
植酸磷 Phytate P,%				总能 GE	6117	5	72	C-22:6	0.00		
磷全消化道表观消化率 ATTD of P, %	21			消化能 DE				C-24:0	0.07		
磷全消化道标准消化率 STTD of P, %	28			代谢能 ME				饱和脂肪酸	8.63		
				净能 NE				单不饱和脂肪酸	17.70		
								多不饱和脂肪酸	68.11		
								碘价	189.20		
								碘价物	797.66		

续表

原料名称：亚麻籽粕(Flaxseed Meal)
AAFCO#: 71.11, AAFCO 2010, p.385
IFN#: 5-30-288

常规成分, %				氨基酸, %								
				总氨基酸				可消化氨基酸				
								回肠表观消化率			回肠标准消化率	
	$\bar{x}$	n	SD		$\bar{x}$	n	SD	$\bar{x}$	n	SD	$\bar{x}$	n	SD
干物质 Dry matter	90.18	4	1.43	必需氨基酸 Essential									
粗蛋白 Crude protein	33.28	8	1.79	粗蛋白 CP	33.28	8	1.79	61	1		78	1	
粗纤维 Crude fiber	9.18	4	1.10	精氨酸 Arg	3.00	3	0.58	80	3	8.01	82	3	9.23
乙醚浸出物/粗脂肪 Ether extract	6.45	7	3.20	组氨酸 His	0.67	3	0.03	69	3	4.02	74	3	4.48
酸乙醚浸出物 Acid ether extract				异亮氨酸 Ile	1.33	3	0.02	74	3	10.15	79	3	10.59
粗灰分 Ash	5.23	5	0.55	亮氨酸 Leu	1.91	3	0.08	73	3	8.19	78	3	8.47
碳水化合物成分, %				赖氨酸 Lys	1.19	3	0.07	65	3	6.14	77		
乳糖 Lactose				蛋氨酸 Met	0.77	3	0.27	74	3	1.41	82		
蔗糖 Sucrose	4.67	1		苯丙氨酸 Phe	1.49	3	0.08	75	3	4.48	79	3	6.08
棉子糖 Raffinose				苏氨酸 Thr	1.13	3	0.05	59	3	1.59	74		
水苏糖 Stachyose				色氨酸 Trp	0.51	3	0.14	57	3	27.71	78		
毛蕊花糖 Verbascose				缬氨酸 Val	1.55	3	0.05	68	3	3.50	75		
低聚糖 Oligosaccharides				非必需氨基酸 Nonessential									
淀粉 Starch	5.17	2	5.48	丙氨酸 Ala	1.45	2	0.01	72			75		
中性洗涤纤维 Neutral detergent fiber	24.93	6	2.44	天冬氨酸 Asp	2.80	2	0.17	73			75	2	12.22
酸性洗涤纤维 Acid detergent fiber	15.87	6	2.07	半胱氨酸 Cys	0.59	2	0.07	60	2	12.37	77		
半纤维素 Hemicellulose				谷氨酸 Glu	6.15	2	0.04	73			75		
酸性洗涤木质素 Acid detergent lignin	5.89	1		甘氨酸 Gly	1.84	2	0.05	71			77	2	2.22
总膳食纤维 Total dietary fiber				脯氨酸 Pro	1.45	2	0.33	64	2	16.62	75		
不溶性膳食纤维 Insoluble dietary fiber				丝氨酸 Ser	1.39	2	0.04	71			76	2	5.86
可溶性膳食纤维 Soluble dietary fiber				酪氨酸 Tyr	0.72	2	0.11	65	2	2.90	78		

矿物质				维生素, mg/kg(标注单位者除外)				脂肪酸, 占乙醚提取物, %			
	$\bar{x}$	n	SD		$\bar{x}$	n	SD		$\bar{x}$	n	SD
常量元素 Macro,%				脂溶性维生素 Fat soluble				粗脂肪	3.00		
钙 Ca	0.37	4	0.03	β-胡萝卜素 β-Carotene	0.2			C-12:0	0.00		
氯 Cl	0.06			维生素 E Vitamin E	2.0			C-14:0	0.08		
钾 K	1.26			水溶性维生素 Water soluble				C-16:0	4.80		
镁 Mg	0.50	3	0.04	维生素 B_6 Vitamin B_6	6.0			C-16:1	0.08		
钠 Na	0.13			维生素 B_{12} Vitamin B_{12}, μg/kg	0			C-18:0	2.55		
磷 P	0.87	5	0.05	生物素 Biotin	0.41			C-18:1	14.03		
硫 S	0.39			叶酸 Folacin	1.30			C-18:2	11.03		
微量元素 Micro,ppm				烟酸 Niacin	33			C-18:3	40.65		
铬 Cr				泛酸 Pantothenic acid	14.7			C-18:4	0.00		
铜 Cu	16.20	3	1.80	核黄素 Riboflavin	2.9			C-20:0	0.00		
铁 Fe	111	2	32.46	硫胺素 Thiamin	7.5			C-20:1	0.00		
碘 I				胆碱 Choline	1512			C-20:4	0.00		
锰 Mn	45.90	3	0.90					C-20:5	0.00		
硒 Se	0.63							C-22:0	0.00		
锌 Zn	57.90	3	6.32	能量, kcal/kg				C-22:1	0.00		
								C-22:5	0.00		
植酸磷 Phytate P,%				总能 GE	4887	2	175	C-22:6	0.00		
磷全消化道表观消化率 ATTD of P, %	21	1		消化能 DE	3060			C-24:0	0.00		
磷全消化道标准消化率 STTD of P, %	28	1		代谢能 ME	2834			饱和脂肪酸	7.43		
				净能 NE	1830			单不饱和脂肪酸	14.10		
								多不饱和脂肪酸	51.68		
								碘价	143.79		
								碘价物	43.14		

续表

原料名称：明胶(Gelatin)
AAFCO#: 60.29, AAFCO 2010, p.376
IFN#: 5-14-503

常规成分, %				氨基酸, %							
				总氨基酸				可消化氨基酸			
								回肠表观消化率		回肠标准消化率	
	$\bar{x}$	n	SD		$\bar{x}$	n	SD	$\bar{x}$	n	SD	$\bar{x}$	n	SD
干物质 Dry matter	87.54	3	2.53	必需氨基酸 Essential								
粗蛋白 Crude protein	100.1	1		粗蛋白 CP	100.1	1		82	2	1.41	85	2	0.71
粗纤维 Crude fiber				精氨酸 Arg	7.91	4	0.23	85			95	2	0.00
乙醚浸出物/粗脂肪 Ether extract	0.00			组氨酸 His	0.76	4	0.12	88	2	2.83	91	2	2.12
酸乙醚浸出物 Acid ether extract				异亮氨酸 Ile	1.25	4	0.15	93	2	0.00	96	2	0.71
粗灰分 Ash				亮氨酸 Leu	2.79	4	0.22	85	2	1.41	88	2	1.41
碳水化合物成分, %				赖氨酸 Lys	3.87	4	0.29	90	2	0.71	92	2	0.71
				蛋氨酸 Met	0.97	4	0.06	91	2	0.71	92	2	0.71
乳糖 Lactose				苯丙氨酸 Phe	1.89	4	0.16	91	2	0.71	93	2	1.41
蔗糖 Sucrose				苏氨酸 Thr	2.17	4	0.85	78	2	1.41	81	2	1.41
棉子糖 Raffinose				色氨酸 Trp	0.09	4	0.09	93	2	0.71	98	2	1.41
水苏糖 Stachyose				缬氨酸 Val	2.27	4	0.22	86	2	0.00	90	2	0.00
毛蕊花糖 Verbascose				非必需氨基酸 Nonessential								
低聚糖 Oligosaccharides				丙氨酸 Ala	8.99	3	0.37	87	2	0.71	88	2	0.71
淀粉 Starch	0.00			天冬氨酸 Asp	4.73	3	1.62	63	2	11.31	66	2	11.31
中性洗涤纤维 Neutral detergent fiber				半胱氨酸 Cys	0.11	4	0.06						
酸性洗涤纤维 Acid detergent fiber	0.00			谷氨酸 Glu	8.73	3	3.02	80	2	2.83	82	2	2.12
半纤维素 Hemicellulose				甘氨酸 Gly	25.39	3	7.11	82	2	0.71	83	2	0.71
酸性洗涤木质素 Acid detergent lignin				脯氨酸 Pro	15.25	3	4.63	79	2	1.41	83	2	1.41
总膳食纤维 Total dietary fiber				丝氨酸 Ser	2.95	3	0.42	79	2	0.00	86		
不溶性膳食纤维 Insoluble dietary fiber				酪氨酸 Tyr	0.65	4	0.27	62	2	26.87	76	2	19.09
可溶性膳食纤维 Soluble dietary fiber												

矿物质				维生素, mg/kg(标注单位者除外)				脂肪酸, 占乙醚提取物, %			
	$\bar{x}$	n	SD		$\bar{x}$	n	SD		$\bar{x}$	n	SD
常量元素 Macro,%				脂溶性维生素 Fat soluble				粗脂肪			
钙 Ca				β-胡萝卜素 β-Carotene				C-12:0			
氯 Cl				维生素 E Vitamin E				C-14:0			
钾 K				水溶性维生素 Water soluble				C-16:0			
镁 Mg				维生素 B_6 Vitamin B_6				C-16:1			
钠 Na				维生素 B_{12} Vitamin B_{12}, μg/kg				C-18:0			
磷 P				生物素 Biotin				C-18:1			
硫 S				叶酸 Folacin				C-18:2			
微量元素 Micro,ppm				烟酸 Niacin				C-18:3			
铬 Cr				泛酸 Pantothenic acid				C-18:4			
铜 Cu				核黄素 Riboflavin				C-20:0			
铁 Fe				硫胺素 Thiamin				C-20:1			
碘 I				胆碱 Choline				C-20:4			
锰 Mn								C-20:5			
硒 Se								C-22:0			
锌 Zn								C-22:1			
				能量, kcal/kg				C-22:5			
植酸磷 Phytate P,%				总能 GE	5645			C-22:6			
磷全消化道表观消化率 ATTD of P,%				消化能 DE	4900			C-24:0			
磷全消化道标准消化率 STTD of P,%				代谢能 ME	4219			饱和脂肪酸			
				净能 NE	2519			单不饱和脂肪酸			
								多不饱和脂肪酸			
								碘价			
								碘价物			

续表

原料名称：膨化芸豆(Kidney Beans, Extruded)
AAFCO#：
IFN#：

常规成分, %				氨基酸, %								
				总氨基酸				可消化氨基酸				
								回肠表观消化率			回肠标准消化率	
	$\bar{x}$	n	SD		$\bar{x}$	n	SD	$\bar{x}$	n	SD	$\bar{x}$	n	SD
干物质 Dry matter	91.45	2	2.47	必需氨基酸 Essential									
粗蛋白 Crude protein	20.03	3	5.89	粗蛋白 CP	20.03	3	5.89	64	1		76	1	
粗纤维 Crude fiber	4.42	2	2.95	精氨酸 Arg	1.28	1		84	1		94	1	
乙醚浸出物/粗脂肪 Ether extract	1.10	2	0.59	组氨酸 His	0.19	1		58	1		66	1	
酸乙醚浸出物 Acid ether extract				异亮氨酸 Ile	0.94	1		66	1		72	1	
粗灰分 Ash	2.65	2	1.34	亮氨酸 Leu	1.9	1		65	1		71	1	
碳水化合物成分, %				赖氨酸 Lys	1.51	1		82	1		85	1	
				蛋氨酸 Met	0.25	1							
乳糖 Lactose				苯丙氨酸 Phe	1.35	1		70	1		74	1	
蔗糖 Sucrose				苏氨酸 Thr	0.94	1		67	1		76	1	
棉子糖 Raffinose				色氨酸 Trp									
水苏糖 Stachyose				缬氨酸 Val	1.13	1		58	1		65	1	
毛蕊花糖 Verbascose				非必需氨基酸 Nonessential									
低聚糖 Oligosaccharides				丙氨酸 Ala	1	1		68	1		82	1	
淀粉 Starch				天冬氨酸 Asp	2.08	1		86	1		89	1	
中性洗涤纤维 Neutral detergent fiber				半胱氨酸 Cys	0.21	1							
酸性洗涤纤维 Acid detergent fiber				谷氨酸 Glu	2	1		83	1		87	1	
半纤维素 Hemicellulose				甘氨酸 Gly	1.16	1		47	1		101	1	
酸性洗涤木质素 Acid detergent lignin				脯氨酸 Pro	0.77	1		45	1				
总膳食纤维 Total dietary fiber				丝氨酸 Ser	1.35	1		68	1		77	1	
不溶性膳食纤维 Insoluble dietary fiber				酪氨酸 Tyr	0.81	1		61	1		67	1	
可溶性膳食纤维 Soluble dietary fiber													

矿物质				维生素, mg/kg(标注单位者除外)				脂肪酸, 占乙醚提取物, %			
	$\bar{x}$	n	SD		$\bar{x}$	n	SD		$\bar{x}$	n	SD
常量元素 Macro,%				脂溶性维生素 Fat soluble				粗脂肪			
钙 Ca				β-胡萝卜素 β-Carotene				C-12:0			
氯 Cl				维生素 E Vitamin E				C-14:0			
钾 K				水溶性维生素 Water soluble				C-16:0			
镁 Mg				维生素 B_6 Vitamin B_6				C-16:1			
钠 Na				维生素 B_{12} Vitamin B_{12}, μg/kg				C-18:0			
磷 P				生物素 Biotin				C-18:1			
硫 S				叶酸 Folacin				C-18:2			
微量元素 Micro,ppm				烟酸 Niacin				C-18:3			
铬 Cr				泛酸 Pantothenic acid				C-18:4			
铜 Cu				核黄素 Riboflavin				C-20:0			
铁 Fe				硫胺素 Thiamin				C-20:1			
碘 I				胆碱 Choline				C-20:4			
锰 Mn								C-20:5			
硒 Se								C-22:0			
锌 Zn				能量, kcal/kg				C-22:1			
								C-22:5			
植酸磷 Phytate P,%				总能 GE				C-22:6			
磷全消化道表观消化率 ATTD of P, %				消化能 DE				C-24:0			
磷全消化道标准消化率 STTD of P, %				代谢能 ME				饱和脂肪酸			
				净能 NE				单不饱和脂肪酸			
								多不饱和脂肪酸			
								碘价			
								碘价物			

续表

原料名称：生芸豆(Kidney Beans, Raw)
AAFCO#:
IFN#:

常规成分, %				氨基酸, %							
				总氨基酸				可消化氨基酸			
	$\bar{x}$	n	SD		$\bar{x}$	n	SD	回肠表观消化率		回肠标准消化率	
								$\bar{x}$	n	SD	$\bar{x}$	n	SD
干物质 Dry matter	86.6	1		必需氨基酸 Essential							
粗蛋白 Crude protein	20.00	1		粗蛋白 CP	20.00	1					
粗纤维 Crude fiber	6.4	1		精氨酸 Arg	1.27	1					
乙醚浸出物/粗脂肪 Ether extract	1.35	1		组氨酸 His	0.2	1					
酸乙醚浸出物 Acid ether extract				异亮氨酸 Ile	0.96	1					
粗灰分 Ash	3.5	1		亮氨酸 Leu	1.9	1					
碳水化合物成分, %				赖氨酸 Lys	1.53	1					
				蛋氨酸 Met	0.28	1					
乳糖 Lactose				苯丙氨酸 Phe	1.31	1					
蔗糖 Sucrose				苏氨酸 Thr	0.93	1					
棉子糖 Raffinose				色氨酸 Trp							
水苏糖 Stachyose				缬氨酸 Val	1.15	1					
毛蕊花糖 Verbascose				非必需氨基酸 Nonessential							
低聚糖 Oligosaccharides				丙氨酸 Ala	1.02	1					
淀粉 Starch				天冬氨酸 Asp	2.04	1					
中性洗涤纤维 Neutral detergent fiber				半胱氨酸 Cys	0.24	1					
酸性洗涤纤维 Acid detergent fiber				谷氨酸 Glu	1.94	1					
半纤维素 Hemicellulose				甘氨酸 Gly	1.12	1					
酸性洗涤木质素 Acid detergent lignin				脯氨酸 Pro	0.76	1					
总膳食纤维 Total dietary fiber				丝氨酸 Ser	1.36	1					
不溶性膳食纤维 Insoluble dietary fiber				酪氨酸 Tyr	0.8	1					
可溶性膳食纤维 Soluble dietary fiber											

矿物质				维生素, mg/kg(标注单位者除外)				脂肪酸, 占乙醚提取物, %			
	$\bar{x}$	n	SD		$\bar{x}$	n	SD		$\bar{x}$	n	SD
常量元素 Macro,%				脂溶性维生素 Fat soluble				粗脂肪	0.83		
钙 Ca				β-胡萝卜素 β-Carotene				C-12:0	0.00		
氯 Cl				维生素 E Vitamin E				C-14:0	0.00		
钾 K				水溶性维生素 Water soluble				C-16:0	12.77		
镁 Mg				维生素 B_6 Vitamin B_6				C-16:1	0.00		
钠 Na				维生素 B_{12} Vitamin B_{12}, μg/kg				C-18:0	1.69		
磷 P				生物素 Biotin				C-18:1	7.71		
硫 S				叶酸 Folacin				C-18:2	21.45		
微量元素 Micro,ppm				烟酸 Niacin				C-18:3	33.61		
铬 Cr				泛酸 Pantothenic acid				C-18:4	0.00		
铜 Cu				核黄素 Riboflavin				C-20:0	0.00		
铁 Fe				硫胺素 Thiamin				C-20:1	0.00		
碘 I				胆碱 Choline				C-20:4	0.00		
锰 Mn								C-20:5	0.00		
硒 Se								C-22:0	0.00		
锌 Zn				能量, kcal/kg				C-22:1	0.00		
								C-22:5	0.00		
植酸磷 Phytate P,%				总能 GE				C-22:6	0.00		
磷全消化道表观消化率 ATTD of P, %				消化能 DE				C-24:0	0.00		
磷全消化道标准消化率 STTD of P, %				代谢能 ME				饱和脂肪酸	14.46		
				净能 NE				单不饱和脂肪酸	7.71		
								多不饱和脂肪酸	55.06		
								碘价	137.66		
								碘价物	11.43		

续表

原料名称：扁豆(Lentils)
AAFCO#：无官方定义
IFN#：5-02-506

常规成分, %				氨基酸, %							
					总氨基酸			可消化氨基酸			
								回肠表观消化率		回肠标准消化率	
	$\bar{x}$	n	SD		$\bar{x}$	n	SD	$\bar{x}$	n	SD	$\bar{x}$	n	SD
干物质 Dry matter	90.00	1		必需氨基酸 Essential									
粗蛋白 Crude protein	26.00	1		粗蛋白 CP	26.00	1		73					
粗纤维 Crude fiber				精氨酸 Arg	2.05			84			86		
乙醚浸出物/粗脂肪 Ether extract	1.30			组氨酸 His	0.78			76			79		
酸乙醚浸出物 Acid ether extract				异亮氨酸 Ile	1.00			73			77		
粗灰分 Ash	2.79	1		亮氨酸 Leu	1.84			73			76		
碳水化合物成分, %				赖氨酸 Lys	1.71			77			79		
				蛋氨酸 Met	0.18			66			71		
乳糖 Lactose				苯丙氨酸 Phe	1.29			72			75		
蔗糖 Sucrose				苏氨酸 Thr	0.84			66			73		
棉子糖 Raffinose				色氨酸 Trp	0.21			62			68		
水苏糖 Stachyose				缬氨酸 Val	1.27			70			75		
毛蕊花糖 Verbascose				非必需氨基酸 Nonessential									
低聚糖 Oligosaccharides				丙氨酸 Ala	1.24			69			73		
淀粉 Starch	41.75			天冬氨酸 Asp	2.82			76			79		
中性洗涤纤维 Neutral detergent fiber	17.37	1		半胱氨酸 Cys	0.27			57			66		
酸性洗涤纤维 Acid detergent fiber	2.97	1		谷氨酸 Glu	4.03			79			82		
半纤维素 Hemicellulose				甘氨酸 Gly	1.11			67			75		
酸性洗涤木质素 Acid detergent lignin				脯氨酸 Pro	1.05			73			84		
总膳食纤维 Total dietary fiber				丝氨酸 Ser	1.13			72			78		
不溶性膳食纤维 Insoluble dietary fiber				酪氨酸 Tyr	0.70			73			77		
可溶性膳食纤维 Soluble dietary fiber													

矿物质				维生素, mg/kg(标注单位者除外)				脂肪酸, 占乙醚提取物, %			
	$\bar{x}$	n	SD		$\bar{x}$	n	SD		$\bar{x}$	n	SD
常量元素 Macro,%				脂溶性维生素 Fat soluble				粗脂肪			
钙 Ca	0.10			β-胡萝卜素 β-Carotene	1.0			C-12:0			
氯 Cl	0.03			维生素 E Vitamin E	0			C-14:0			
钾 K	0.89			水溶性维生素 Water soluble				C-16:0			
镁 Mg	0.12			维生素 B_6 Vitamin B_6	5.5			C-16:1			
钠 Na	0.02			维生素 B_{12} Vitamin B_{12}, µg/kg	0			C-18:0			
磷 P	0.38			生物素 Biotin	0.13			C-18:1			
硫 S	0.20			叶酸 Folacin	0.70			C-18:2			
微量元素 Micro,ppm				烟酸 Niacin	22			C-18:3			
铬 Cr				泛酸 Pantothenic acid	14.9			C-18:4			
铜 Cu	10.00			核黄素 Riboflavin	2.4			C-20:0			
铁 Fe	85			硫胺素 Thiamin	3.9			C-20:1			
碘 I				胆碱 Choline				C-20:4			
锰 Mn	13.00							C-20:5			
硒 Se	0.10							C-22:0			
锌 Zn	25.00			能量, kcal/kg				C-22:1			
								C-22:5			
植酸磷 Phytate P,%				总能 GE	4483			C-22:6			
磷全消化道表观消化率 ATTD of P, %				消化能 DE	3540			C-24:0			
磷全消化道标准消化率 STTD of P, %				代谢能 ME	3363			饱和脂肪酸			
				净能 NE	2437			单不饱和脂肪酸			
								多不饱和脂肪酸			
								碘价			
								碘价物			

续表

原料名称: 羽扇豆(Lupins)
AAFCO#: 无官方定义
IFN#: 5-27-717

常规成分, %				氨基酸, %								
					总氨基酸			可消化氨基酸				
								回肠表观消化率			回肠标准消化率		
	$\bar{x}$	n	SD		$\bar{x}$	n	SD	$\bar{x}$	n	SD	$\bar{x}$	n	SD
干物质 Dry matter	91.13	23	1.34	必需氨基酸 Essential									
粗蛋白 Crude protein	32.44	31	4.63	粗蛋白 CP	32.44	31	4.63	80	18	4.23	86	16	3.78
粗纤维 Crude fiber	14.25	2	2.91	精氨酸 Arg	3.61	13	0.73	92	11	4.40	93	11	4.47
乙醚浸出物/粗脂肪 Ether extract	6.08	20	1.14	组氨酸 His	0.92	19	0.24	83	18	5.15	86	18	5.37
酸乙醚浸出物 Acid ether extract				异亮氨酸 Ile	1.39	19	0.24	83	18	5.87	85	18	5.93
粗灰分 Ash	3.67	15	0.38	亮氨酸 Leu	2.31	19	0.40	82	18	5.71	85	18	5.79
碳水化合物成分, %				赖氨酸 Lys	1.58	19	0.25	82	18	4.07	85	18	4.00
				蛋氨酸 Met	0.21	19	0.07	75	18	8.95	81	18	7.95
乳糖 Lactose				苯丙氨酸 Phe	1.34	19	0.24	82	18	5.99	84	18	6.12
蔗糖 Sucrose				苏氨酸 Thr	1.20	18	0.15	76	18	5.78	82	18	6.00
棉子糖 Raffinose				色氨酸 Trp	0.26	14	0.08	78	13	3.94	82	11	4.40
水苏糖 Stachyose				缬氨酸 Val	1.32	19	0.19	77	18	6.01	81	18	5.82
毛蕊花糖 Verbascose				非必需氨基酸 Nonessential									
低聚糖 Oligosaccharides				丙氨酸 Ala	1.14	13	0.18	72	11	7.35	78	11	7.31
淀粉 Starch	7.44	9	1.77	天冬氨酸 Asp	3.26	13	0.65	81	11	9.36	85	11	9.42
中性洗涤纤维 Neutral detergent fiber	24.11	22	2.88	半胱氨酸 Cys	0.46	17	0.09	78	16	7.93	83	18	7.91
酸性洗涤纤维 Acid detergent fiber	19.90	22	3.08	谷氨酸 Glu	7.00	13	1.57	86	11	7.97	88	11	7.83
半纤维素 Hemicellulose	3.70	5	0.72	甘氨酸 Gly	1.38	13	0.23	70	11	7.81	80	11	8.25
酸性洗涤木质素 Acid detergent lignin	1.52	14	0.75	脯氨酸 Pro	1.37	13	0.21	67	11	14.04	93	11	6.91
总膳食纤维 Total dietary fiber				丝氨酸 Ser	1.61	13	0.33	80	11	8.16	84	11	8.13
不溶性膳食纤维 Insoluble dietary fiber	30.03	1		酪氨酸 Tyr	1.16	8	0.42	79	7	6.59	82	8	5.68
可溶性膳食纤维 Soluble dietary fiber	1.61	1											

矿物质				维生素, mg/kg(标注单位者除外)				脂肪酸, 占乙醚提取物, %			
	$\bar{x}$	n	SD		$\bar{x}$	n	SD		$\bar{x}$	n	SD
常量元素 Macro,%				脂溶性维生素 Fat soluble				粗脂肪	9.74		
钙 Ca	0.37	2	0.04	β-胡萝卜素 β-Carotene				C-12:0	0.08		
氯 Cl	0.03			维生素 E Vitamin E	7.5			C-14:0	0.13		
钾 K	1.10			水溶性维生素 Water soluble				C-16:0	7.62		
镁 Mg	0.19			维生素 B_6 Vitamin B_6				C-16:1	0.35		
钠 Na	0.02			维生素 B_{12} Vitamin B_{12}, μg/kg				C-18:0	3.24		
磷 P	0.31	9	0.05	生物素 Biotin	0.05			C-18:1	36.53		
硫 S	0.24			叶酸 Folacin				C-18:2	20.48		
微量元素 Micro,ppm				烟酸 Niacin				C-18:3	4.58		
铬 Cr				泛酸 Pantothenic acid				C-18:4	0.00		
铜 Cu	6.00			核黄素 Riboflavin				C-20:0	0.00		
铁 Fe	54			硫胺素 Thiamin				C-20:1	2.62		
碘 I				胆碱 Choline				C-20:4			
锰 Mn	1390							C-20:5			
硒 Se	0.07							C-22:0			
锌 Zn	32.00			能量, kcal/kg				C-22:1	0.95		
								C-22:5			
植酸磷 Phytate P,%	0.21	9	0.05	总能 GE	4366	9	70	C-22:6			
磷全消化道表观消化率 ATTD of P,%	50			消化能 DE	3397	8	183	C-24:0			
磷全消化道标准消化率 STTD of P,%	57			代谢能 ME	3176			饱和脂肪酸	11.08		
				净能 NE	2043			单不饱和脂肪酸	40.45		
								多不饱和脂肪酸	25.06		
								碘价	85.62		
								碘价物	83.39		

续表

原料名称：肉骨粉，磷>4% (Meat and Bone Meal, P >4%)
磷水平大于4%的产品被归类到肉骨粉，但是许多该类产品并不符合AAFCO的定义，即钙的水平要低于2.2倍磷的水平。
AAFCO#: 9.41, AAFCO 2010, p.328
IFN#: 5-00-388

常规成分, %				氨基酸, %								
					总氨基酸			可消化氨基酸				
								回肠表观消化率			回肠标准消化率	
	$\bar{x}$	n	SD		$\bar{x}$	n	SD	$\bar{x}$	n	SD	$\bar{x}$	n	SD
干物质 Dry matter	95.16	16	1.55	必需氨基酸 Essential									
粗蛋白 Crude protein	50.05	33	4.33	粗蛋白 CP	50.05	33	4.33	68	11	5.49	72	11	6.62
粗纤维 Crude fiber				精氨酸 Arg	3.53	27	0.30	80	12	3.84	83	12	5.14
乙醚浸出物/粗脂肪 Ether extract	9.21	16	1.54	组氨酸 His	0.91	27	0.17	68	12	7.75	71	12	8.76
酸乙醚浸出物 Acid ether extract				异亮氨酸 Ile	1.47	27	0.26	69	12	7.86	73	12	8.56
粗灰分 Ash	31.95	20	5.59	亮氨酸 Leu	3.06	27	0.42	72	12	5.75	76	12	5.87
碳水化合物成分, %				赖氨酸 Lys	2.59	27	0.38	70	12	7.42	73	12	8.17
				蛋氨酸 Met	0.69	21	0.18	81	4	4.35	84	4	2.90
乳糖 Lactose				苯丙氨酸 Phe	1.65	27	0.22	76	12	5.91	79	12	5.98
蔗糖 Sucrose				苏氨酸 Thr	1.63	27	0.28	64	12	7.07	69	12	8.00
棉子糖 Raffinose				色氨酸 Trp	0.30	26	0.06	52	10	10.82	62	10	13.17
水苏糖 Stachyose				缬氨酸 Val	2.19	27	0.35	72	12	5.87	76	12	6.19
毛蕊花糖 Verbascose				非必需氨基酸 Nonessential									
低聚糖 Oligosaccharides				丙氨酸 Ala	3.87	13	0.44	76	6	1.94	79	6	3.66
淀粉 Starch	0.00			天冬氨酸 Asp	3.74	13	0.64	61	6	4.82	65	6	6.49
中性洗涤纤维 Neutral detergent fiber	32.50			半胱氨酸 Cys	0.46	20	0.15	46	4	28.55	56	4	24.15
酸性洗涤纤维 Acid detergent fiber	5.05	2	0.95	谷氨酸 Glu	6.09	13	0.89	71	6	3.39	75	6	5.23
半纤维素 Hemicellulose				甘氨酸 Gly	7.06	13	0.68	74	6	4.91	78	6	5.88
酸性洗涤木质素 Acid detergent lignin				脯氨酸 Pro	4.38	13	0.62	70	4	6.43	81	4	3.87
总膳食纤维 Total dietary fiber				丝氨酸 Ser	1.89	13	0.32	66	6	4.22	71	6	6.84
不溶性膳食纤维 Insoluble dietary fiber				酪氨酸 Tyr	1.08	20	0.19	59	6	15.12	68	6	11.10
可溶性膳食纤维 Soluble dietary fiber													

矿物质				维生素, mg/kg (标注单位者除外)				脂肪酸, 占乙醚提取物, %			
	$\bar{x}$	n	SD		$\bar{x}$	n	SD		$\bar{x}$	n	SD
常量元素 Macro, %				脂溶性维生素 Fat soluble				粗脂肪	10.60		
钙 Ca	10.94	28	1.79	β-胡萝卜素 β-Carotene				C-12:0	0.14		
氯 Cl	0.69			维生素 E Vitamin E	1.6			C-14:0	1.89		
钾 K	0.65			水溶性维生素 Water soluble				C-16:0	19.25		
镁 Mg	0.41			维生素 B_6 Vitamin B_6	4.6			C-16:1	2.59		
钠 Na	0.63			维生素 B_{12} Vitamin B_{12}, μg/kg	90			C-18:0	13.44		
磷 P	5.26	30	0.88	生物素 Biotin	0.08			C-18:1	28.49		
硫 S	0.38			叶酸 Folacin	0.41			C-18:2	2.52		
微量元素 Micro, ppm				烟酸 Niacin	49			C-18:3	0.63		
铬 Cr				泛酸 Pantothenic acid	4.1			C-18:4	0.00		
铜 Cu	11.00			核黄素 Riboflavin	4.7			C-20:0	1.05		
铁 Fe	606			硫胺素 Thiamin	0.4			C-20:1	0.00		
碘 I				胆碱 Choline	1996			C-20:4	0.00		
锰 Mn	17.00							C-20:5	0.00		
硒 Se	0.31							C-22:0	0.00		
锌 Zn	96.00			能量, kcal/kg				C-22:1	0.00		
								C-22:5	0.00		
植酸磷 Phytate P, %				总能 GE	3806	13	481	C-22:6	0.00		
磷全消化道表观消化率 ATTD of P, %	68	3	10.40	消化能 DE	3303	7	405	C-24:0	0.00		
磷全消化道标准消化率 STTD of P, %	70	3	10.38	代谢能 ME	2963			饱和脂肪酸	35.77		
				净能 NE	1961			单不饱和脂肪酸	31.08		
								多不饱和脂肪酸	3.15		
								碘价	34.47		
								碘价物	36.53		

续表

原料名称：肉粉(Meat Meal)

磷水平小于4%的产品被归类到肉粉,但是许多该类产品并不符合AAFCO的定义,即钙的水平要低于2.2倍磷的水平。
AAFCO#: 9.40, AAFCO 2010, p.328
IFN#: 5-00-385

常规成分, %				氨基酸, %							
				总氨基酸				可消化氨基酸			
								回肠表观消化率		回肠标准消化率	
	$\bar{x}$	n	SD	必需氨基酸 Essential	$\bar{x}$	n	SD	$\bar{x}$	n	SD	$\bar{x}$	n	SD
干物质 Dry matter	96.12	28	1.38										
粗蛋白 Crude protein	56.40	35	3.33	粗蛋白 CP	56.40	35	3.33	73	9	6.87	76	9	7.39
粗纤维 Crude fiber				精氨酸 Arg	3.65	33	0.28	83	9	4.69	84	9	4.57
乙醚浸出物/粗脂肪 Ether extract	11.09	25	1.33	组氨酸 His	1.24	33	0.21	73	9	9.83	75	9	9.95
酸乙醚浸出物 Acid ether extract				异亮氨酸 Ile	1.82	33	0.23	75	9	6.94	78	9	6.53
粗灰分 Ash	21.59	29	3.6	亮氨酸 Leu	3.70	33	0.40	75	9	8.59	77	9	8.54
碳水化合物成分, %				赖氨酸 Lys	3.20	33	0.40	76	9	7.61	78	9	7.62
				蛋氨酸 Met	0.83	30	0.13	80	6	7.19	82	6	6.73
乳糖 Lactose				苯丙氨酸 Phe	1.98	33	0.27	77	9	7.12	79	9	7.12
蔗糖 Sucrose				苏氨酸 Thr	1.89	33	0.20	71	9	7.32	74	9	7.49
棉子糖 Raffinose				色氨酸 Trp	0.40	30	0.06	67	4	12.47	76		
水苏糖 Stachyose				缬氨酸 Val	2.61	33	0.31	74	9	8.69	76	9	8.51
毛蕊花糖 Verbascose				非必需氨基酸 Nonessential									
低聚糖 Oligosaccharides				丙氨酸 Ala	3.82	28	0.38	78	5	8.06	80	5	7.92
淀粉 Starch	0.00			天冬氨酸 Asp	4.28	28	0.39	68	5	7.90	71	5	7.99
中性洗涤纤维 Neutral detergent fiber	31.6			半胱氨酸 Cys	0.56	30	0.15	59	5	9.45	62	5	9.60
酸性洗涤纤维 Acid detergent fiber	8.30			谷氨酸 Glu	7.03	28	0.48	75	5	8.74	77	5	8.26
半纤维素 Hemicellulose				甘氨酸 Gly	5.98	28	0.69	77	5	6.18	79	5	5.94
酸性洗涤木质素 Acid detergent lignin				脯氨酸 Pro	3.92	28	0.56	77	5	6.61	86	5	8.20
总膳食纤维 Total dietary fiber				丝氨酸 Ser	1.99	28	0.35	73	5	8.81	76	5	8.53
不溶性膳食纤维 Insoluble dietary fiber				酪氨酸 Tyr	1.35	30	0.13	77	6	7.38	78	6	8.18
可溶性膳食纤维 Soluble dietary fiber													

矿物质				维生素, mg/kg(标注单位者除外)				脂肪酸,占乙醚提取物, %			
	$\bar{x}$	n	SD		$\bar{x}$	n	SD		$\bar{x}$	n	SD
常量元素 Macro,%				脂溶性维生素 Fat soluble				粗脂肪			
钙 Ca	6.37	37	1.43	β-胡萝卜素 β-Carotene				C-12:0			
氯 Cl	0.97			维生素 E Vitamin E	1.2			C-14:0			
钾 K	0.57			水溶性维生素 Water soluble				C-16:0			
镁 Mg	0.35			维生素 B_6 Vitamin B_6	2.4			C-16:1			
钠 Na	0.80			维生素 B_{12} Vitamin B_{12}, μg/kg	80			C-18:0			
磷 P	3.16	37	0.62	生物素 Biotin	0.08			C-18:1			
硫 S	0.45			叶酸 Folacin	0.50			C-18:2			
微量元素 Micro,ppm				烟酸 Niacin	57			C-18:3			
铬 Cr				泛酸 Pantothenic acid	5.0			C-18:4			
铜 Cu	10.00			核黄素 Riboflavin	4.7			C-20:0			
铁 Fe	440			硫胺素 Thiamin	0.6			C-20:1			
碘 I				胆碱 Choline	2077			C-20:4			
锰 Mn	10.00							C-20:5			
硒 Se	0.37							C-22:0			
锌 Zn	94.00							C-22:1			
				能量, kcal/kg				C-22:5			
植酸磷 Phytate P,%				总能 GE	4497	26	251	C-22:6			
磷全消化道表观消化率 ATTD of P,%	82	6	4.15	消化能 DE	3452	14	424	C-24:0			
磷全消化道标准消化率 STTD of P,%	86	6	3.48	代谢能 ME	3068			饱和脂肪酸			
				净能 NE	2010			单不饱和脂肪酸			
								多不饱和脂肪酸			
								碘价			
								碘价物			

续表

原料名称：酪蛋白(来源于牛奶) (Milk, Casein)
AAFCO#: 54.16, AAFCO 2010, p. 361
IFN#: 5-01-162

常规成分, %				氨基酸, %								
					总氨基酸			可消化氨基酸				
								回肠表观消化率			回肠标准消化率		
	$\bar{x}$	n	SD		$\bar{x}$	n	SD	$\bar{x}$	n	SD	$\bar{x}$	n	SD
干物质 Dry matter	91.72	7	2.41	必需氨基酸 Essential									
粗蛋白 Crude protein	88.95	15	4.94	粗蛋白 CP	88.95	15	4.94	87	13	8.96	94	13	6.11
粗纤维 Crude fiber	0	1		精氨酸 Arg	3.13	17	0.20	88	14	11.52	95	14	5.45
乙醚浸出物/粗脂肪 Ether extract	0.17	2	0.11	组氨酸 His	2.57	17	0.32	93	15	4.89	97	15	3.42
酸乙醚浸出物 Acid ether extract				异亮氨酸 Ile	4.49	17	0.48	91	15	5.37	95	15	3.40
粗灰分 Ash				亮氨酸 Leu	8.24	17	0.51	94	15	4.09	97	15	3.01
碳水化合物成分, %				赖氨酸 Lys	6.87	17	0.57	95	15	4.08	97	15	2.76
				蛋氨酸 Met	2.52	17	0.28	96	14	2.88	98	14	2.17
乳糖 Lactose				苯丙氨酸 Phe	4.49	17	0.30	93	15	5.48	96	15	5.22
蔗糖 Sucrose				苏氨酸 Thr	3.77	17	0.44	86	15	10.60	93	15	6.47
棉子糖 Raffinose				色氨酸 Trp	1.33	13	0.59	92	9	7.17	96	10	4.56
水苏糖 Stachyose				缬氨酸 Val	5.81	17	0.53	92	15	5.01	96	15	3.42
毛蕊花糖 Verbascose				非必需氨基酸 Nonessential									
低聚糖 Oligosaccharides				丙氨酸 Ala	2.58	14	0.19	83	15	10.91	92	15	6.47
淀粉 Starch	0.00			天冬氨酸 Asp	5.93	14	0.69	88	15	6.92	94	15	4.74
中性洗涤纤维 Neutral detergent fiber				半胱氨酸 Cys	0.45	15	0.16	67	13	25.29	85	13	18.21
酸性洗涤纤维 Acid detergent fiber	0.00			谷氨酸 Glu	18.06	14	2.87	93	15	3.64	96	15	2.70
半纤维素 Hemicellulose				甘氨酸 Gly	1.60	14	0.16	63	14	30.14	87	14	20.85
酸性洗涤木质素 Acid detergent lignin				脯氨酸 Pro	9.82	13	0.74	80	14	27.87	99	14	7.64
总膳食纤维 Total dietary fiber				丝氨酸 Ser	4.55	14	0.62	86	15	8.46	92	15	4.39
不溶性膳食纤维 Insoluble dietary fiber				酪氨酸 Tyr	4.87	12	0.36	94	14	4.26	97	14	3.64
可溶性膳食纤维 Soluble dietary fiber													

矿物质				维生素, mg/kg (标注单位者除外)				脂肪酸, 占乙醚提取物, %			
	$\bar{x}$	n	SD		$\bar{x}$	n	SD		$\bar{x}$	n	SD
常量元素 Macro, %				脂溶性维生素 Fat soluble				粗脂肪			
钙 Ca	0.20	3	0.17	β-胡萝卜素 β-Carotene				C-12:0			
氯 Cl	0.04			维生素 E Vitamin E				C-14:0			
钾 K	0.01			水溶性维生素 Water soluble				C-16:0			
镁 Mg	0.01			维生素 B_6 Vitamin B_6	0.4			C-16:1			
钠 Na	0.01			维生素 B_{12} Vitamin B_{12}, μg/kg				C-18:0			
磷 P	0.68	3	0.01	生物素 Biotin	0.04			C-18:1			
硫 S	0.60			叶酸 Folacin	0.51			C-18:2			
微量元素 Micro, ppm				烟酸 Niacin	1			C-18:3			
铬 Cr				泛酸 Pantothenic acid	2.7			C-18:4			
铜 Cu	4.00			核黄素 Riboflavin	1.5			C-20:0			
铁 Fe	14			硫胺素 Thiamin	0.4			C-20:1			
碘 I				胆碱 Choline	205			C-20:4			
锰 Mn	4.00							C-20:5			
硒 Se	0.16							C-22:0			
锌 Zn	30.00			能量, kcal/kg				C-22:1			
								C-22:5			
植酸磷 Phytate P, %				总能 GE	5670	1		C-22:6			
磷全消化道表观消化率 ATTD of P, %	87	10	7.05	消化能 DE	4135			C-24:0			
磷全消化道标准消化率 STTD of P, %	98			代谢能 ME	3530			饱和脂肪酸			
				净能 NE	2088			单不饱和脂肪酸			
								多不饱和脂肪酸			
								碘价			
								碘价物			

续表

原料名称：乳糖(来源于牛奶) (Milk, Lactose)
AAFCO#：无官方定义
在计算净能的公式里将乳糖等同于淀粉。

常规成分, %				氨基酸, %							
				总氨基酸			可消化氨基酸				
							回肠表观消化率		回肠标准消化率		
	$\bar{x}$	n	SD		$\bar{x}$	n	SD				
干物质 Dry matter	95.00			必需氨基酸 Essential				$\bar{x}$	n	SD	$\bar{x}$	n	SD
粗蛋白 Crude protein	0.00			粗蛋白 CP									
粗纤维 Crude fiber				精氨酸 Arg									
乙醚浸出物/粗脂肪 Ether extract	0.00			组氨酸 His									
酸乙醚浸出物 Acid ether extract				异亮氨酸 Ile									
粗灰分 Ash				亮氨酸 Leu									
碳水化合物成分, %				赖氨酸 Lys									
				蛋氨酸 Met									
乳糖 Lactose	95.00			苯丙氨酸 Phe									
蔗糖 Sucrose				苏氨酸 Thr									
棉子糖 Raffinose				色氨酸 Trp									
水苏糖 Stachyose				缬氨酸 Val									
毛蕊花糖 Verbascose				非必需氨基酸 Nonessential									
低聚糖 Oligosaccharides				丙氨酸 Ala									
淀粉 Starch				天冬氨酸 Asp									
中性洗涤纤维 Neutral detergent fiber				半胱氨酸 Cys									
酸性洗涤纤维 Acid detergent fiber	0.00			谷氨酸 Glu									
半纤维素 Hemicellulose				甘氨酸 Gly									
酸性洗涤木质素 Acid detergent lignin				脯氨酸 Pro									
总膳食纤维 Total dietary fiber				丝氨酸 Ser									
不溶性膳食纤维 Insoluble dietary fiber				酪氨酸 Tyr									
可溶性膳食纤维 Soluble dietary fiber													

矿物质				维生素, mg/kg(标注单位者除外)				脂肪酸, 占乙醚提取物, %			
	$\bar{x}$	n	SD		$\bar{x}$	n	SD		$\bar{x}$	n	SD
常量元素 Macro,%				脂溶性维生素 Fat soluble				粗脂肪			
钙 Ca				β-胡萝卜素 β-Carotene				C-12:0			
氯 Cl				维生素 E Vitamin E				C-14:0			
钾 K				水溶性维生素 Water soluble				C-16:0			
镁 Mg				维生素 B_6 Vitamin B_6				C-16:1			
钠 Na				维生素 B_{12} Vitamin B_{12}, μg/kg				C-18:0			
磷 P				生物素 Biotin				C-18:1			
硫 S				叶酸 Folacin				C-18:2			
微量元素 Micro,ppm				烟酸 Niacin				C-18:3			
铬 Cr				泛酸 Pantothenic acid				C-18:4			
铜 Cu	0.00	1		核黄素 Riboflavin				C-20:0			
铁 Fe	5.80	1		硫胺素 Thiamin				C-20:1			
碘 I				胆碱 Choline				C-20:4			
锰 Mn	0.00	1						C-20:5			
硒 Se								C-22:0			
锌 Zn	0.20	1		能量, kcal/kg				C-22:1			
								C-22:5			
植酸磷 Phytate P,%				总能 GE	4143			C-22:6			
磷全消化道表观消化率 ATTD of P,%				消化能 DE	3525			C-24:0			
磷全消化道标准消化率 STTD of P,%				代谢能 ME	3525			饱和脂肪酸			
				净能 NE	2923			单不饱和脂肪酸			
								多不饱和脂肪酸			
								碘价			
								碘价物			

续表

原料名称：脱脂奶粉(Milk, Skim Milk Powder)
AAFCO#: 54.3, AAFCO 2010, p. 360
IFN#: 5-01-175
在计算净能的公式里将乳糖等同于淀粉。

常规成分, %				氨基酸, %								
					总氨基酸			可消化氨基酸				
	$\bar{x}$	n	SD		$\bar{x}$	n	SD	回肠表观消化率			回肠标准消化率	
								$\bar{x}$	n	SD	$\bar{x}$	n	SD
干物质 Dry matter	94.60			必需氨基酸 Essential									
粗蛋白 Crude protein	36.77	5	5.85	粗蛋白 CP	36.77	5	5.85	86	6	3.33	90	6	3.00
粗纤维 Crude fiber				精氨酸 Arg	1.17	5	0.21	90	7	5.41	95	7	5.05
乙醚浸出物/粗脂肪 Ether extract	0.90			组氨酸 His	0.94	5	0.14	91	7	5.48	93	7	5.42
酸乙醚浸出物 Acid ether extract				异亮氨酸 Ile	1.45	5	0.34	89	7	7.05	91	7	6.77
粗灰分 Ash				亮氨酸 Leu	3.02	5	0.29	92	7	3.01	94	7	2.94
碳水化合物成分, %				赖氨酸 Lys	2.42	4	0.28	92	7	4.35	94	7	4.53
				蛋氨酸 Met	0.82	5	0.08	91	6	3.74	92	6	3.78
乳糖 Lactose	47.82			苯丙氨酸 Phe	1.51	5	0.12	93	7	3.67	95	7	3.43
蔗糖 Sucrose				苏氨酸 Thr	1.44	5	0.15	88	7	5.82	92	7	5.48
棉子糖 Raffinose				色氨酸 Trp	0.44			90			88		
水苏糖 Stachyose				缬氨酸 Val	1.85	5	0.45	89	7	5.88	92	7	5.68
毛蕊花糖 Verbascose				非必需氨基酸 Nonessential									
低聚糖 Oligosaccharides				丙氨酸 Ala	1.19	3	0.15	85	3	4.36	90	3	4.17
淀粉 Starch				天冬氨酸 Asp	2.67	3	0.16	88	3	0.54	91	3	0.47
中性洗涤纤维 Neutral detergent fiber				半胱氨酸 Cys	0.33	2	0.10	81			86		
酸性洗涤纤维 Acid detergent fiber	0.00			谷氨酸 Glu	7.05	3	0.42	89	3	4.29	90	3	4.32
半纤维素 Hemicellulose				甘氨酸 Gly	0.76	3	0.20	76	3	7.79	99	3	5.63
酸性洗涤木质素 Acid detergent lignin				脯氨酸 Pro	3.17	3	0.19	91	3	4.42	100		
总膳食纤维 Total dietary fiber				丝氨酸 Ser	1.81	3	0.02	82	3	10.74	85	3	10.73
不溶性膳食纤维 Insoluble dietary fiber				酪氨酸 Tyr	1.48	3	0.25	91	4	5.96	93	4	5.86
可溶性膳食纤维 Soluble dietary fiber													

矿物质				维生素, mg/kg(标注单位者除外)				脂肪酸, 占乙醚提取物, %			
	$\bar{x}$	n	SD		$\bar{x}$	n	SD		$\bar{x}$	n	SD
常量元素 Macro,%				脂溶性维生素 Fat soluble				粗脂肪	0.77		
钙 Ca	1.27	1		β-胡萝卜素 β-Carotene				C-12:0	1.82		
氯 Cl	1.00			维生素 E Vitamin E	4.1			C-14:0	10.78		
钾 K	1.60			水溶性维生素 Water soluble				C-16:0	30.52		
镁 Mg	0.12			维生素 B_6 Vitamin B_6	4.1			C-16:1	2.86		
钠 Na	0.48			维生素 B_{12} Vitamin B_{12}, μg/kg	36			C-18:0	11.04		
磷 P	1.06	1		生物素 Biotin	0.25			C-18:1	21.69		
硫 S	0.32			叶酸 Folacin	0.47			C-18:2	2.47		
微量元素 Micro,ppm				烟酸 Niacin	12			C-18:3	1.43		
铬 Cr				泛酸 Pantothenic acid	36.4			C-18:4			
铜 Cu	0.10	1		核黄素 Riboflavin	19.1			C-20:0	0.00		
铁 Fe	0.00	1		硫胺素 Thiamin	3.7			C-20:1	0.00		
碘 I				胆碱 Choline	1393			C-20:4			
锰 Mn	0.00	1						C-20:5			
硒 Se	0.12							C-22:0			
锌 Zn	43.10	1		能量, kcal/kg				C-22:1			
								C-22:5			
植酸磷 Phytate P,%				总能 GE	4437			C-22:6			
磷全消化道表观消化率 ATTD of P, %	91	1		消化能 DE	3980			C-24:0			
磷全消化道标准消化率 STTD of P, %	98			代谢能 ME	3730			饱和脂肪酸	56.49		
				净能 NE	2695			单不饱和脂肪酸	24.55		
								多不饱和脂肪酸	3.90		
								碘价	30.71		
								碘价物	2.36		

续表

原料名称：低蛋白乳清粉，乳糖 80% (Milk, Whey Permeate, 80% Lactose)
脱水前将蛋白从乳清粉中分离出去。本产品是一种低蛋白产品，主要含来源于乳清粉的乳糖和灰分。
AAFCO#:无官方定义
在计算净能的公式里将乳糖等同于淀粉。

常规成分, %				氨基酸, %							
				总氨基酸			可消化氨基酸				
	$\bar{x}$	n	SD		$\bar{x}$	n	SD	回肠表观消化率		回肠标准消化率	
干物质 Dry matter	96.00			必需氨基酸 Essential				$\bar{x}$	n	SD	$\bar{x}$	n	SD
粗蛋白 Crude protein	3.50			粗蛋白 CP						1			
粗纤维 Crude fiber				精氨酸 Arg									
乙醚浸出物/粗脂肪 Ether extract	0.20			组氨酸 His									
酸乙醚浸出物 Acid ether extract				异亮氨酸 Ile									
粗灰分 Ash				亮氨酸 Leu									
碳水化合物成分, %				赖氨酸 Lys									
				蛋氨酸 Met									
乳糖 Lactose	80.00			苯丙氨酸 Phe									
蔗糖 Sucrose				苏氨酸 Thr									
棉子糖 Raffinose				色氨酸 Trp									
水苏糖 Stachyose				缬氨酸 Val									
毛蕊花糖 Verbascose				非必需氨基酸 Nonessential									
低聚糖 Oligosaccharides				丙氨酸 Ala									
淀粉 Starch				天冬氨酸 Asp									
中性洗涤纤维 Neutral detergent fiber				半胱氨酸 Cys									
酸性洗涤纤维 Acid detergent fiber	0.00			谷氨酸 Glu									
半纤维素 Hemicellulose				甘氨酸 Gly									
酸性洗涤木质素 Acid detergent lignin				脯氨酸 Pro									
总膳食纤维 Total dietary fiber				丝氨酸 Ser									
不溶性膳食纤维 Insoluble dietary fiber				酪氨酸 Tyr									
可溶性膳食纤维 Soluble dietary fiber													

矿物质				维生素, mg/kg(标注单位者除外)				脂肪酸, 占乙醚提取物, %			
	$\bar{x}$	n	SD		$\bar{x}$	n	SD		$\bar{x}$	n	SD
常量元素 Macro,%				脂溶性维生素 Fat soluble				粗脂肪			
钙 Ca				β-胡萝卜素 β-Carotene				C-12:0			
氯 Cl				维生素 E Vitamin E				C-14:0			
钾 K				水溶性维生素 Water soluble				C-16:0			
镁 Mg				维生素 B_6 Vitamin B_6				C-16:1			
钠 Na				维生素 B_{12} Vitamin B_{12}, μg/kg				C-18:0			
磷 P				生物素 Biotin				C-18:1			
硫 S				叶酸 Folacin				C-18:2			
微量元素 Micro,ppm				烟酸 Niacin				C-18:3			
铬 Cr				泛酸 Pantothenic acid				C-18:4			
铜 Cu				核黄素 Riboflavin				C-20:0			
铁 Fe				硫胺素 Thiamin				C-20:1			
碘 I				胆碱 Choline				C-20:4			
锰 Mn								C-20:5			
硒 Se								C-22:0			
锌 Zn								C-22:1			
				能量, kcal/kg				C-22:5			
植酸磷 Phytate P,%				总能 GE	3426	1		C-22:6			
磷全消化道表观消化率 ATTD of P,%				消化能 DE	3177	1		C-24:0			
磷全消化道标准消化率 STTD of P,%				代谢能 ME	3153			饱和脂肪酸			
				净能 NE	2579			单不饱和脂肪酸			
								多不饱和脂肪酸			
								碘价			
								碘价物			

续表

原料名称：低蛋白乳清粉，乳糖 85% (Milk, Whey Permeate, 85% Lactose)

脱水前将蛋白从乳清粉中分离出去。本产品是一种低蛋白产品，主要含乳糖，大多数灰分被去除。

AAFCO#：无官方定义

在计算净能的公式里将乳糖等同于淀粉。

常规成分, %				氨基酸, %							
				总氨基酸				可消化氨基酸			
								回肠表观消化率			回肠标准消化率
	$\bar{x}$	n	SD		$\bar{x}$	n	SD	$\bar{x}$	n	SD	$\bar{x}$	n	SD
干物质 Dry matter	98.00			必需氨基酸 Essential							
粗蛋白 Crude protein	3.00			粗蛋白 CP							
粗纤维 Crude fiber				精氨酸 Arg							
乙醚浸出物/粗脂肪 Ether extract	0.20			组氨酸 His							
酸乙醚浸出物 Acid ether extract				异亮氨酸 Ile							
粗灰分 Ash				亮氨酸 Leu							
碳水化合物成分, %				赖氨酸 Lys							
				蛋氨酸 Met							
乳糖 Lactose	85.00			苯丙氨酸 Phe							
蔗糖 Sucrose				苏氨酸 Thr							
棉子糖 Raffinose				色氨酸 Trp							
水苏糖 Stachyose				缬氨酸 Val							
毛蕊花糖 Verbascose				非必需氨基酸 Nonessential							
低聚糖 Oligosaccharides				丙氨酸 Ala							
淀粉 Starch				天冬氨酸 Asp							
中性洗涤纤维 Neutral detergent fiber				半胱氨酸 Cys							
酸性洗涤纤维 Acid detergent fiber	0.00			谷氨酸 Glu							
半纤维素 Hemicellulose				甘氨酸 Gly							
酸性洗涤木质素 Acid detergent lignin				脯氨酸 Pro							
总膳食纤维 Total dietary fiber				丝氨酸 Ser							
不溶性膳食纤维 Insoluble dietary fiber				酪氨酸 Tyr							
可溶性膳食纤维 Soluble dietary fiber											

矿物质				维生素, mg/kg(标注单位者除外)				脂肪酸, 占乙醚提取物, %			
	$\bar{x}$	n	SD		$\bar{x}$	n	SD		$\bar{x}$	n	SD
常量元素 Macro,%				脂溶性维生素 Fat soluble				粗脂肪			
钙 Ca	0.27	2	0.22	β-胡萝卜素 β-Carotene				C-12:0			
氯 Cl				维生素 E Vitamin E				C-14:0			
钾 K				水溶性维生素 Water soluble				C-16:0			
镁 Mg				维生素 B_6 Vitamin B_6				C-16:1			
钠 Na				维生素 B_{12} Vitamin B_{12}, μg/kg				C-18:0			
磷 P	0.34	2	0.33	生物素 Biotin				C-18:1			
硫 S				叶酸 Folacin				C-18:2			
微量元素 Micro,ppm				烟酸 Niacin				C-18:3			
铬 Cr				泛酸 Pantothenic acid				C-18:4			
铜 Cu				核黄素 Riboflavin				C-20:0			
铁 Fe				硫胺素 Thiamin				C-20:1			
碘 I				胆碱 Choline				C-20:4			
锰 Mn								C-20:5			
硒 Se								C-22:0			
锌 Zn				能量, kcal/kg				C-22:1			
								C-22:5			
植酸磷 Phytate P,%				总能 GE	3657	1		C-22:6			
磷全消化道表观消化率 ATTD of P, %	82			消化能 DE	3626	1		C-24:0			
磷全消化道标准消化率 STTD of P, %	92			代谢能 ME	3606			饱和脂肪酸			
				净能 NE	2922			单不饱和脂肪酸			
								多不饱和脂肪酸			
								碘价			
								碘价物			

原料名称：乳清粉(Milk, Whey Powder)
AAFCO#: 54.7, AAFCO 2010, p.360
IFN#: 4-01-182
在计算净能的公式里将乳糖等同于淀粉。

常规成分, %				氨基酸, %							
				总氨基酸				可消化氨基酸			
								回肠表观消化率		回肠标准消化率	
	$\bar{x}$	n	SD		$\bar{x}$	n	SD	$\bar{x}$	n	SD	$\bar{x}$	n	SD
干物质 Dry matter	97.15	4	0.82	必需氨基酸 Essential									
粗蛋白 Crude protein	11.55	6	0.93	粗蛋白 CP	11.55	6	0.93	87	1		102	1	
粗纤维 Crude fiber	0.08	2	0.05	精氨酸 Arg	0.26	7	0.02	83	2	13.57	98		
乙醚浸出物/粗脂肪 Ether extract	0.83	3	0.79	组氨酸 His	0.21	6	0.04	90	2	3.80	96		
酸乙醚浸出物 Acid ether extract				异亮氨酸 Ile	0.64	8	0.04	94	2	3.26	96		
粗灰分 Ash	8.00	2	0.44	亮氨酸 Leu	1.11	7	0.08	94	2	2.82	98		
碳水化合物成分, %				赖氨酸 Lys	0.88	8	0.09	94	2	2.61	97		
				蛋氨酸 Met	0.17	8	0.00	95	2	6.26	98		
乳糖 Lactose	72.88			苯丙氨酸 Phe	0.35	7	0.03	78	2	13.65	90		
蔗糖 Sucrose				苏氨酸 Thr	0.71	8	0.04	85	2	2.35	89		
棉子糖 Raffinose				色氨酸 Trp	0.20	7	0.03	78			97		
水苏糖 Stachyose				缬氨酸 Val	0.61	8	0.03	91	2	5.27	96		
毛蕊花糖 Verbascose				非必需氨基酸 Nonessential									
低聚糖 Oligosaccharides				丙氨酸 Ala	0.54	3	0.07	81			90		
淀粉 Starch				天冬氨酸 Asp	1.16	3	0.10	83			91		
中性洗涤纤维 Neutral detergent fiber				半胱氨酸 Cys	0.26	6	0.03	86			93		
酸性洗涤纤维 Acid detergent fiber	0.00			谷氨酸 Glu	1.95	3	0.14	85	2	10.91	90		
半纤维素 Hemicellulose				甘氨酸 Gly	0.20	3	0.04	55			99		
酸性洗涤木质素 Acid detergent lignin				脯氨酸 Pro	0.66	2	0.07	74			100		
总膳食纤维 Total dietary fiber				丝氨酸 Ser	0.54	3	0.06	78			85		
不溶性膳食纤维 Insoluble dietary fiber				酪氨酸 Tyr	0.27	5	0.02	86	1		97	1	
可溶性膳食纤维 Soluble dietary fiber													

矿物质				维生素, mg/kg(标注单位者除外)				脂肪酸, 占乙醚提取物, %			
	$\bar{x}$	n	SD		$\bar{x}$	n	SD		$\bar{x}$	n	SD
常量元素 Macro,%				脂溶性维生素 Fat soluble				粗脂肪	1.07		
钙 Ca	0.62	2	0.18	β-胡萝卜素 β-Carotene				C-12:0	1.12		
氯 Cl	1.40			维生素 E Vitamin E	0.3			C-14:0	9.72		
钾 K	1.96			水溶性维生素 Water soluble				C-16:0	30.47		
镁 Mg	0.13			维生素 B_6 Vitamin B_6	4.0			C-16:1	3.08		
钠 Na	0.94			维生素 B_{12} Vitamin B_{12}, μg/kg	23			C-18:0	9.07		
磷 P	0.69	4	0.04	生物素 Biotin	0.27			C-18:1	23.46		
硫 S	0.72			叶酸 Folacin	0.85			C-18:2	2.34		
微量元素 Micro,ppm				烟酸 Niacin	10			C-18:3	0.84		
铬 Cr				泛酸 Pantothenic acid	47.0			C-18:4			
铜 Cu	6.60	1		核黄素 Riboflavin	27.1			C-20:0	0.00		
铁 Fe	57	1		硫胺素 Thiamin	4.1			C-20:1	0.00		
碘 I				胆碱 Choline	1820			C-20:4			
锰 Mn	3.00							C-20:5			
硒 Se	0.12							C-22:0			
锌 Zn	9.90	1		能量, kcal/kg				C-22:1			
								C-22:5			
植酸磷 Phytate P,%				总能 GE	3647	1		C-22:6			
磷全消化表观消化率 ATTD of P, %	82	4	1.80	消化能 DE	3494	1		C-24:0			
磷全消化道标准消化率 STTD of P, %	92	4	1.56	代谢能 ME	3415			饱和脂肪酸	52.06		
				净能 NE	2704			单不饱和脂肪酸	26.54		
								多不饱和脂肪酸	3.18		
								碘价	30.68		
								碘价物	3.28		

续表

原料名称：乳清浓缩蛋白(Milk, Whey Protein Concentrate)
AAFCO#: 54.25, AAFCO 2010, p.361
IFN#: 5-06-836
在计算净能的公式里将乳糖等同于淀粉。

常规成分, %				氨基酸, %								
					总氨基酸			可消化氨基酸				
								回肠表观消化率			回肠标准消化率	
	$\bar{x}$	n	SD		$\bar{x}$	n	SD	$\bar{x}$	n	SD	$\bar{x}$	n	SD
干物质 Dry matter	94.40	8	1.72	必需氨基酸 Essential									
粗蛋白 Crude protein	76.32	7	4.61	粗蛋白 CP	76.32	7	4.61	84	2	4.95	86	2	4.78
粗纤维 Crude fiber	1.33	1		精氨酸 Arg	2.01	8	0.21	88	3	4.36	93	3	4.11
乙醚浸出物/粗脂肪 Ether extract	0.20			组氨酸 His	1.46	8	0.21	86	3	3.79	92		
酸乙醚浸出物 Acid ether extract				异亮氨酸 Ile	4.74	8	0.63	89	3	6.21	95		
粗灰分 Ash	2.63	8	0.44	亮氨酸 Leu	8.43	8	1.15	90	3	5.73	95		
碳水化合物成分, %				赖氨酸 Lys	6.85	8	0.86	92	3	1.66	93	3	2.54
				蛋氨酸 Met	1.65	8	0.28	91	3	4.59	96		
乳糖 Lactose	5.00			苯丙氨酸 Phe	2.70	8	0.27	82	3	4.26	87		
蔗糖 Sucrose				苏氨酸 Thr	4.82	8	0.69	83	3	1.63	85	3	3.00
棉子糖 Raffinose				色氨酸 Trp	1.59	8	0.21	88	3	4.71	95		
水苏糖 Stachyose				缬氨酸 Val	4.54	8	0.50	90	3	2.42	95		
毛蕊花糖 Verbascose				非必需氨基酸 Nonessential									
低聚糖 Oligosaccharides				丙氨酸 Ala	3.77	8	0.37	86	3	3.17	90	3	3.37
淀粉 Starch				天冬氨酸 Asp	7.80	8	0.88	89	3	3.59	91	3	4.02
中性洗涤纤维 Neutral detergent fiber				半胱氨酸 Cys	1.79	8	0.35	84	3	4.80	85	3	4.68
酸性洗涤纤维 Acid detergent fiber	0.00			谷氨酸 Glu	12.29	8	1.79	88	3	4.41	89	3	4.56
半纤维素 Hemicellulose				甘氨酸 Gly	1.45	8	0.11	72	3	17.70	87	3	10.37
酸性洗涤木质素 Acid detergent lignin				脯氨酸 Pro	4.29	8	0.83	81	3	3.75	93	3	2.63
总膳食纤维 Total dietary fiber				丝氨酸 Ser	3.28	8	0.50	85	3	2.01	88	3	2.62
不溶性膳食纤维 Insoluble dietary fiber				酪氨酸 Tyr	2.34	6	0.22	81	1		86	1	
可溶性膳食纤维 Soluble dietary fiber													

矿物质				维生素, mg/kg(标注单位者除外)				脂肪酸, 占乙醚提取物, %			
	$\bar{x}$	n	SD		$\bar{x}$	n	SD		$\bar{x}$	n	SD
常量元素 Macro,%				脂溶性维生素 Fat soluble				粗脂肪			
钙 Ca	0.63	1		β-胡萝卜素 β-Carotene				C-12:0			
氯 Cl				维生素 E Vitamin E				C-14:0			
钾 K				水溶性维生素 Water soluble				C-16:0			
镁 Mg				维生素 B_6 Vitamin B_6				C-16:1			
钠 Na				维生素 B_{12} Vitamin B_{12}, μg/kg				C-18:0			
磷 P	0.38	1		生物素 Biotin				C-18:1			
硫 S				叶酸 Folacin				C-18:2			
微量元素 Micro,ppm				烟酸 Niacin				C-18:3			
铬 Cr				泛酸 Pantothenic acid				C-18:4			
铜 Cu				核黄素 Riboflavin				C-20:0			
铁 Fe				硫胺素 Thiamin				C-20:1			
碘 I				胆碱 Choline				C-20:4			
锰 Mn								C-20:5			
硒 Se								C-22:0			
锌 Zn				能量, kcal/kg				C-22:1			
								C-22:5			
植酸磷 Phytate P,%				总能 GE	5245	1		C-22:6			
磷全消化道表观消化率 ATTD of P,%	82			消化能 DE	4949	1		C-24:0			
磷全消化道标准消化率 STTD of P,%	92			代谢能 ME	4430			饱和脂肪酸			
				净能 NE	2797			单不饱和脂肪酸			
								多不饱和脂肪酸			
								碘价			
								碘价物			

续表

原料名称：黍(Millet)
AAFCO#：无官方定义
IFN#：4-03-120

常规成分, %				氨基酸, %							
				总氨基酸				可消化氨基酸			
	$\bar{x}$	n	SD		$\bar{x}$	n	SD	回肠表观消化率		回肠标准消化率	
干物质 Dry matter	88.50	1		必需氨基酸 Essential				$\bar{x}$	n	$\bar{x}$	n	SD
粗蛋白 Crude protein	11.90	5	2.85	粗蛋白 CP	11.90	5	2.85	79	1	88	1
粗纤维 Crude fiber				精氨酸 Arg	0.57	4	0.09	82	1	89	
乙醚浸出物/粗脂肪 Ether extract	4.25	1		组氨酸 His	0.29	4	0.05	85	1	90	1
酸乙醚浸出物 Acid ether extract				异亮氨酸 Ile	0.49	4	0.08	83	1	89	1
粗灰分 Ash				亮氨酸 Leu	1.22	4	0.20	87	1	91	1
碳水化合物成分, %				赖氨酸 Lys	0.37	4	0.05	74	1	83	1
				蛋氨酸 Met	0.28	4	0.05	72	1	75	1
乳糖 Lactose				苯丙氨酸 Phe	0.55	4	0.05	85	1	91	1
蔗糖 Sucrose				苏氨酸 Thr	0.45	4	0.07	75	1	86	
棉子糖 Raffinose				色氨酸 Trp	0.17	4	0.06	84	1	97	
水苏糖 Stachyose				缬氨酸 Val	0.66	4	0.10	81	1	87	1
毛蕊花糖 Verbascose				非必需氨基酸 Nonessential							
低聚糖 Oligosaccharides				丙氨酸 Ala	1.07	1		85	1	91	1
淀粉 Starch	54.95			天冬氨酸 Asp	1.09	1		79	1	86	1
中性洗涤纤维 Neutral detergent fiber	15.80			半胱氨酸 Cys	0.32	1		82	1	88	
酸性洗涤纤维 Acid detergent fiber	13.80			谷氨酸 Glu	2.84	1		89	1	92	1
半纤维素 Hemicellulose				甘氨酸 Gly	0.42	1		55		84	
酸性洗涤木质素 Acid detergent lignin				脯氨酸 Pro	0.80			81		95	
总膳食纤维 Total dietary fiber	4.78	1		丝氨酸 Ser	0.64	1		81	1	90	
不溶性膳食纤维 Insoluble dietary fiber				酪氨酸 Tyr	0.58	1		81	1	86	
可溶性膳食纤维 Soluble dietary fiber											

矿物质				维生素, mg/kg(标注单位者除外)				脂肪酸, 占乙醚提取物, %			
	$\bar{x}$	n	SD		$\bar{x}$	n	SD		$\bar{x}$	n	SD
常量元素 Macro,%				脂溶性维生素 Fat soluble				粗脂肪	4.22		
钙 Ca	0.03			β-胡萝卜素 β-Carotene				C-12:0	0.07		
氯 Cl	0.03			维生素 E Vitamin E				C-14:0	0.00		
钾 K	0.43			水溶性维生素 Water soluble				C-16:0	12.51		
镁 Mg	0.16			维生素 B_6 Vitamin B_6	5.8			C-16:1	0.33		
钠 Na	0.04			维生素 B_{12} Vitamin B_{12}, μg/kg	0			C-18:0	3.65		
磷 P	0.31			生物素 Biotin	0.16			C-18:1	17.51		
硫 S	0.14			叶酸 Folacin	0.23			C-18:2	47.75		
微量元素 Micro,ppm				烟酸 Niacin	23			C-18:3	2.80		
铬 Cr				泛酸 Pantothenic acid	11.0			C-18:4			
铜 Cu	26.00			核黄素 Riboflavin	3.8			C-20:0	0.00		
铁 Fe	71			硫胺素 Thiamin	7.3			C-20:1	0.47		
碘 I				胆碱 Choline	440			C-20:4			
锰 Mn	30.00							C-20:5			
硒 Se	0.70							C-22:0			
锌 Zn	18.00			能量, kcal/kg				C-22:1			
								C-22:5			
植酸磷 Phytate P,%				总能 GE	4472			C-22:6			
磷全消化道表观消化率 ATTD of P,%				消化能 DE	3020			C-24:0			
磷全消化道标准消化率 STTD of P,%				代谢能 ME	2939			饱和脂肪酸	16.23		
				净能 NE	2218			单不饱和脂肪酸	18.32		
								多不饱和脂肪酸	50.55		
								碘价	110.52		
								碘价物	46.64		

续表

原料名称：甜菜糖蜜(Molasses, Sugar Beets)
AAFCO#: 63.1, AAFCO 2010, p.380
IFN#: 4-30-289
在计算净能的公式里将蔗糖等同于淀粉。

常规成分, %				氨基酸, %								
				总氨基酸				可消化氨基酸				
								回肠表观消化率			回肠标准消化率	
	$\bar{x}$	n	SD		$\bar{x}$	n	SD	$\bar{x}$	n	SD	$\bar{x}$	n	SD
干物质 Dry matter	72.20			必需氨基酸 Essential									
粗蛋白 Crude protein	10.00			粗蛋白 CP	10.00			86					
粗纤维 Crude fiber				精氨酸 Arg	0.06						92		
乙醚浸出物/粗脂肪 Ether extract	0.16			组氨酸 His	0.04						90		
酸乙醚浸出物 Acid ether extract				异亮氨酸 Ile	0.24			79			88		
粗灰分 Ash				亮氨酸 Leu	0.24			74			89		
碳水化合物成分, %				赖氨酸 Lys	0.10			37			86		
				蛋氨酸 Met	0.03			68			90		
乳糖 Lactose				苯丙氨酸 Phe	0.06			46			90		
蔗糖 Sucrose	47.50			苏氨酸 Thr	0.08			32			86		
棉子糖 Raffinose				色氨酸 Trp	0.05			44			86		
水苏糖 Stachyose				缬氨酸 Val	0.15			59			87		
毛蕊花糖 Verbascose				非必需氨基酸 Nonessential									
低聚糖 Oligosaccharides				丙氨酸 Ala	0.23			79			95		
淀粉 Starch				天冬氨酸 Asp	0.62			84			95		
中性洗涤纤维 Neutral detergent fiber				半胱氨酸 Cys	0.05			44			84		
酸性洗涤纤维 Acid detergent fiber	0.08			谷氨酸 Glu	4.75			92			95		
半纤维素 Hemicellulose				甘氨酸 Gly	0.20			58			95		
酸性洗涤木质素 Acid detergent lignin				脯氨酸 Pro	0.10						95		
总膳食纤维 Total dietary fiber				丝氨酸 Ser	0.21			66			95		
不溶性膳食纤维 Insoluble dietary fiber				酪氨酸 Tyr	0.24			81			91		
可溶性膳食纤维 Soluble dietary fiber													
矿物质				维生素, mg/kg(标注单位者除外)				脂肪酸, 占乙醚提取物, %					
	$\bar{x}$	n	SD		$\bar{x}$	n	SD		$\bar{x}$	n	SD		
常量元素 Macro,%				脂溶性维生素 Fat soluble				粗脂肪					
钙 Ca	0.25			β-胡萝卜素 β-Carotene				C-12:0					
氯 Cl				维生素 E Vitamin E				C-14:0					
钾 K				水溶性维生素 Water soluble				C-16:0					
镁 Mg				维生素 B_6 Vitamin B_6				C-16:1					
钠 Na				维生素 B_{12} Vitamin B_{12}, μg/kg				C-18:0					
磷 P	0.16			生物素 Biotin				C-18:1					
硫 S				叶酸 Folacin				C-18:2					
微量元素 Micro,ppm				烟酸 Niacin				C-18:3					
铬 Cr				泛酸 Pantothenic acid				C-18:4					
铜 Cu				核黄素 Riboflavin				C-20:0					
铁 Fe				硫胺素 Thiamin				C-20:1					
碘 I				胆碱 Choline				C-20:4					
锰 Mn								C-20:5					
硒 Se								C-22:0					
锌 Zn				能量, kcal/kg				C-22:1					
								C-22:5					
植酸磷 Phytate P,%				总能 GE	3045	1		C-22:6					
磷全消化道表观消化率 ATTD of P, %	50			消化能 DE	2366	1		C-24:0					
磷全消化道标准消化率 STTD of P, %	63			代谢能 ME	2298			饱和脂肪酸					
				净能 NE	1795			单不饱和脂肪酸					
								多不饱和脂肪酸					
								碘价					
								碘价物					

续表

原料名称：甘蔗糖蜜 (Molasses, Sugar Cane)
AAFCO#: 63.7, AAFCO 2010, p.380
IFN#: 4-13-251
在计算净能的公式里将蔗糖等同于淀粉。

常规成分, %				氨基酸, %							
				总氨基酸				可消化氨基酸			
								回肠表观消化率		回肠标准消化率	
	$\bar{x}$	n	SD		$\bar{x}$	n	SD	$\bar{x}$	n	SD	$\bar{x}$	n	SD
干物质 Dry matter	74.10			必需氨基酸 Essential									
粗蛋白 Crude protein	4.80			粗蛋白 CP	4.80			77					
粗纤维 Crude fiber				精氨酸 Arg	0.02						92		
乙醚浸出物/粗脂肪 Ether extract	0.15			组氨酸 His	0.01						90		
酸乙醚浸出物 Acid ether extract				异亮氨酸 Ile	0.04			29			88		
粗灰分 Ash				亮氨酸 Leu	0.06			25			89		
碳水化合物成分, %				赖氨酸 Lys	0.02						86		
				蛋氨酸 Met	0.02			52			90		
乳糖 Lactose				苯丙氨酸 Phe	0.03						90		
蔗糖 Sucrose	47.50			苏氨酸 Thr	0.05						86		
棉子糖 Raffinose				色氨酸 Trp	0.01						86		
水苏糖 Stachyose				缬氨酸 Val	0.11			51			87		
毛蕊花糖 Verbascose				非必需氨基酸 Nonessential									
低聚糖 Oligosaccharides				丙氨酸 Ala	0.20			72			95		
淀粉 Starch				天冬氨酸 Asp	0.89			88			95		
中性洗涤纤维 Neutral detergent fiber				半胱氨酸 Cys	0.04			40			84		
酸性洗涤纤维 Acid detergent fiber	0.15			谷氨酸 Glu	0.41			69			95		
半纤维素 Hemicellulose				甘氨酸 Gly	0.07						95		
酸性洗涤木质素 Acid detergent lignin				脯氨酸 Pro	0.05						95		
总膳食纤维 Total dietary fiber				丝氨酸 Ser	0.07						95		
不溶性膳食纤维 Insoluble dietary fiber				酪氨酸 Tyr	0.03						91		
可溶性膳食纤维 Soluble dietary fiber													

矿物质				维生素, mg/kg (标注单位者除外)				脂肪酸, 占乙醚提取物, %			
	$\bar{x}$	n	SD		$\bar{x}$	n	SD		$\bar{x}$	n	SD
常量元素 Macro, %				脂溶性维生素 Fat soluble				粗脂肪	0.10		
钙 Ca	0.82	2	0.11	β-胡萝卜素 β-Carotene				C-12:0	0.00		
氯 Cl				维生素 E Vitamin E				C-14:0	0.00		
钾 K				水溶性维生素 Water soluble				C-16:0	18.00		
镁 Mg				维生素 B_6 Vitamin B_6				C-16:1	0.00		
钠 Na				维生素 B_{12} Vitamin B_{12}, μg/kg				C-18:0	2.00		
磷 P	0.08	2	0.02	生物素 Biotin				C-18:1	32.00		
硫 S				叶酸 Folacin				C-18:2	50.00		
微量元素 Micro, ppm				烟酸 Niacin				C-18:3	0.00		
铬 Cr				泛酸 Pantothenic acid				C-18:4			
铜 Cu				核黄素 Riboflavin				C-20:0	0.00		
铁 Fe				硫胺素 Thiamin				C-20:1	0.00		
碘 I				胆碱 Choline				C-20:4			
锰 Mn								C-20:5			
硒 Se								C-22:0			
锌 Zn								C-22:1			
				能量, kcal/kg				C-22:5			
植酸磷 Phytate P, %	0.01			总能 GE	4223			C-22:6			
磷全消化道表观消化率 ATTD of P, %	50			消化能 DE	2366			C-24:0			
磷全消化道标准消化率 STTD of P, %	63			代谢能 ME	2333			饱和脂肪酸	20.00		
				净能 NE	1842			单不饱和脂肪酸	32.00		
								多不饱和脂肪酸	50.00		
								碘价	119.25		
								碘价物	1.19		

续表

原料名称：去壳燕麦(Oat Groats)
AAFCO#: 69.1, AAFCO 2010, p. 383

常规成分, %				氨基酸, %							
				总氨基酸				可消化氨基酸			
	$\bar{x}$	n	SD		$\bar{x}$	n	SD	回肠表观消化率		回肠标准消化率	
								$\bar{x}$	n	SD	$\bar{x}$	n	SD
干物质 Dry matter	87.10	1		必需氨基酸 Essential							
粗蛋白 Crude protein	13.90	1		粗蛋白 CP	13.90	1					
粗纤维 Crude fiber				精氨酸 Arg	0.85			86		86	
乙醚浸出物/粗脂肪 Ether extract	5.90	1		组氨酸 His	0.24			83		83	
酸乙醚浸出物 Acid ether extract				异亮氨酸 Ile	0.55			83		83	
粗灰分 Ash	2.40	1		亮氨酸 Leu	0.98			83		83	
碳水化合物成分, %				赖氨酸 Lys	0.48			79		79	
				蛋氨酸 Met	0.20			85		86	
乳糖 Lactose				苯丙氨酸 Phe	0.66			86		86	
蔗糖 Sucrose				苏氨酸 Thr	0.44			76		80	
棉子糖 Raffinose				色氨酸 Trp	0.18			80		82	
水苏糖 Stachyose				缬氨酸 Val	0.72			82		82	
毛蕊花糖 Verbascose				非必需氨基酸 Nonessential							
低聚糖 Oligosaccharides				丙氨酸 Ala	0.60						
淀粉 Starch	46.80	1		天冬氨酸 Asp	1.04						
中性洗涤纤维 Neutral detergent fiber	9.70	1		半胱氨酸 Cys	0.22			80		85	
酸性洗涤纤维 Acid detergent fiber	6.50	1		谷氨酸 Glu	2.59						
半纤维素 Hemicellulose				甘氨酸 Gly	0.64						
酸性洗涤木质素 Acid detergent lignin				脯氨酸 Pro	0.69						
总膳食纤维 Total dietary fiber				丝氨酸 Ser	0.62						
不溶性膳食纤维 Insoluble dietary fiber				酪氨酸 Tyr	0.51			82		84	
可溶性膳食纤维 Soluble dietary fiber											

矿物质				维生素, mg/kg(标注单位者除外)				脂肪酸, 占乙醚提取物, %			
	$\bar{x}$	n	SD		$\bar{x}$	n	SD		$\bar{x}$	n	SD
常量元素 Macro,%				脂溶性维生素 Fat soluble				粗脂肪			
钙 Ca	0.08			β-胡萝卜素 β-Carotene				C-12:0			
氯 Cl	0.09			维生素 E Vitamin E				C-14:0			
钾 K	0.38			水溶性维生素 Water soluble				C-16:0			
镁 Mg	0.11			维生素 B_6 Vitamin B_6	1.1			C-16:1			
钠 Na	0.05			维生素 B_{12} Vitamin B_{12}, μg/kg	0			C-18:0			
磷 P	0.41			生物素 Biotin	0.20			C-18:1			
硫 S	0.20			叶酸 Folacin	0.50			C-18:2			
微量元素 Micro,ppm				烟酸 Niacin	14			C-18:3			
铬 Cr				泛酸 Pantothenic acid	13.4			C-18:4			
铜 Cu	6.00			核黄素 Riboflavin	1.5			C-20:0			
铁 Fe	49			硫胺素 Thiamin	6.5			C-20:1			
碘 I				胆碱 Choline	1139			C-20:4			
锰 Mn	32.00							C-20:5			
硒 Se								C-22:0			
锌 Zn				能量, kcal/kg				C-22:1			
								C-22:5			
植酸磷 Phytate P,%				总能 GE	4576			C-22:6			
磷全消化道表观消化率 ATTD of P, %				消化能 DE	3690			C-24:0			
磷全消化道标准消化率 STTD of P, %				代谢能 ME	3595			饱和脂肪酸			
				净能 NE	2720			单不饱和脂肪酸			
								多不饱和脂肪酸			
								碘价			
								碘价物			

续表

原料名称：燕麦(Oats)
AAFCO#：无官方定义
IFN#: 4-03-309

常规成分, %				氨基酸, %								
					总氨基酸			可消化氨基酸				
								回肠表观消化率			回肠标准消化率	
	$\bar{x}$	n	SD	必需氨基酸 Essential	$\bar{x}$	n	SD	$\bar{x}$	n	SD	$\bar{x}$	n	SD
干物质 Dry matter	89.88	5	1.75										
粗蛋白 Crude protein	11.16	5	1.44	粗蛋白 CP	11.16	5	1.44	62	1				
粗纤维 Crude fiber				精氨酸 Arg	0.73	2	0.12	85	1		90		
乙醚浸出物/粗脂肪 Ether extract	5.42	4	0.84	组氨酸 His	0.24	2	0.04	81			85		
酸乙醚浸出物 Acid ether extract	4.20	1		异亮氨酸 Ile	0.41	2	0.11	73	1		81	1	
粗灰分 Ash	2.64	4	1.15	亮氨酸 Leu	0.79	2	0.16	75	1		83		
				赖氨酸 Lys	0.49	2	0.06	70			76		
碳水化合物成分, %				蛋氨酸 Met	0.68	2	0.01	79			83		
乳糖 Lactose				苯丙氨酸 Phe	0.52	2	0.12	81			84		
蔗糖 Sucrose				苏氨酸 Thr	0.42	2	0.03	59			71		
棉子糖 Raffinose				色氨酸 Trp	0.14			59	1		75		
水苏糖 Stachyose				缬氨酸 Val	0.63	2	0.13	72	1		80	1	
毛蕊花糖 Verbascose				非必需氨基酸 Nonessential									
低聚糖 Oligosaccharides				丙氨酸 Ala	0.46			67			76		
淀粉 Starch	39.06	1		天冬氨酸 Asp	0.81			67			76		
中性洗涤纤维 Neutral detergent fiber	25.30	1		半胱氨酸 Cys	0.36			69			75		
酸性洗涤纤维 Acid detergent fiber	13.73	4	1.21	谷氨酸 Glu	2.14			78			84		
半纤维素 Hemicellulose				甘氨酸 Gly	0.48			61			77		
酸性洗涤木质素 Acid detergent lignin				脯氨酸 Pro	0.54			68			86		
总膳食纤维 Total dietary fiber	33.93	1		丝氨酸 Ser	0.47			69			81		
不溶性膳食纤维 Insoluble dietary fiber				酪氨酸 Tyr	0.41			76			82		
可溶性膳食纤维 Soluble dietary fiber													

矿物质				维生素, mg/kg(标注单位者除外)				脂肪酸, 占乙醚提取物, %			
	$\bar{x}$	n	SD		$\bar{x}$	n	SD		$\bar{x}$	n	SD
常量元素 Macro,%				脂溶性维生素 Fat soluble				粗脂肪	6.90		
钙 Ca	0.03	1		β-胡萝卜素 β-Carotene	3.7			C-12:0	0.35		
氯 Cl	0.10			维生素 E Vitamin E	7.8			C-14:0	0.22		
钾 K	0.42			水溶性维生素 Water soluble				C-16:0	14.99		
镁 Mg	0.16			维生素 B_6 Vitamin B_6	2.0			C-16:1	0.19		
钠 Na	0.08			维生素 B_{12} Vitamin B_{12}, μg/kg	0			C-18:0	0.94		
磷 P	0.35	2	0.04	生物素 Biotin	0.24			C-18:1	31.38		
硫 S	0.21			叶酸 Folacin	0.30			C-18:2	35.13		
微量元素 Micro,ppm				烟酸 Niacin	19			C-18:3	1.61		
铬 Cr				泛酸 Pantothenic acid	13.0			C-18:4			
铜 Cu	6.00			核黄素 Riboflavin	1.7			C-20:0	0.00		
铁 Fe	85			硫胺素 Thiamin	6.0			C-20:1	0.00		
碘 I				胆碱 Choline	946			C-20:4			
锰 Mn	43.00							C-20:5			
硒 Se	0.30							C-22:0			
锌 Zn	38.00			能量, kcal/kg				C-22:1			
								C-22:5			
植酸磷 Phytate P,%	0.19	2	0	总能 GE	4272	1		C-22:6			
磷全消化道表观消化率 ATTD of P,%	33	3	3.10	消化能 DE	2627			C-24:0			
磷全消化道标准消化率 STTD of P,%	39	3	3.53	代谢能 ME	2551			饱和脂肪酸	16.49		
				净能 NE	1893			单不饱和脂肪酸	31.57		
								多不饱和脂肪酸	36.74		
								碘价	96.36		
								碘价物	66.49		

续表

原料名称：裸燕麦(Oats, Naked)
AAFCO#：无官方定义
IFN#：4-25-101

常规成分, %				氨基酸, %							
				总氨基酸				可消化氨基酸			
								回肠表观消化率		回肠标准消化率	
	$\bar{x}$	n	SD		$\bar{x}$	n	SD	$\bar{x}$	n	SD	$\bar{x}$	n	SD
干物质 Dry matter	91.80	3	0.26	必需氨基酸 Essential									
粗蛋白 Crude protein	14.70	3	3.48	粗蛋白 CP	14.70	3	3.48	73			81		
粗纤维 Crude fiber	2.20	2	0.14	精氨酸 Arg	0.89	2	0.23	89	2	0.00	95	2	1.68
乙醚浸出物/粗脂肪 Ether extract	10.65	2	1.34	组氨酸 His	0.27	2	0.08	84	2	0.00	93		
酸乙醚浸出物 Acid ether extract	7.20	1		异亮氨酸 Ile	0.54	2	0.16	83	2	1.41	89		
粗灰分 Ash	1.73	3	0.29	亮氨酸 Leu	0.96	2	0.25	85	2	0.00	91	2	1.50
碳水化合物成分, %				赖氨酸 Lys	0.56	2	0.11	86	2	4.95	90		
				蛋氨酸 Met	0.22	2	0.05	83	2	2.83	89		
乳糖 Lactose				苯丙氨酸 Phe	0.65	2	0.19	87	2	0.00	92	2	1.59
蔗糖 Sucrose				苏氨酸 Thr	0.48	2	0.12	78	2	0.71	85		
棉子糖 Raffinose				色氨酸 Trp	0.15			75			83		
水苏糖 Stachyose				缬氨酸 Val	0.70	2	0.20	85	2	0.00	90		
毛蕊花糖 Verbascose				非必需氨基酸 Nonessential									
低聚糖 Oligosaccharides				丙氨酸 Ala	0.65	2	0.15	75			82		
淀粉 Starch	56.35			天冬氨酸 Asp	1.09	2	0.31	75			82		
中性洗涤纤维 Neutral detergent fiber	11.07	3	2.73	半胱氨酸 Cys	0.41	2	0.06	76	2	2.83	81	2	3.45
酸性洗涤纤维 Acid detergent fiber	3.70			谷氨酸 Glu	3.02	2	0.84	90	2	0.00	90		
半纤维素 Hemicellulose				甘氨酸 Gly	0.63	2	0.13	68	2	2.83	83		
酸性洗涤木质素 Acid detergent lignin				脯氨酸 Pro	0.65	2	0.16	77			92		
总膳食纤维 Total dietary fiber				丝氨酸 Ser	0.70	2	0.18	77			88	2	1.28
不溶性膳食纤维 Insoluble dietary fiber				酪氨酸 Tyr	0.32	2	0.10	82	2	0.71	91		
可溶性膳食纤维 Soluble dietary fiber													

矿物质				维生素, mg/kg(标注单位者除外)				脂肪酸, 占乙醚提取物, %			
	$\bar{x}$	n	SD		$\bar{x}$	n	SD		$\bar{x}$	n	SD
常量元素 Macro,%				脂溶性维生素 Fat soluble				粗脂肪			
钙 Ca	0.08			β-胡萝卜素 β-Carotene				C-12:0			
氯 Cl	0.11			维生素 E Vitamin E	2.0			C-14:0			
钾 K	0.36			水溶性维生素 Water soluble				C-16:0			
镁 Mg	0.12			维生素 B_6 Vitamin B_6	9.6			C-16:1			
钠 Na	0.02			维生素 B_{12} Vitamin B_{12}, μg/kg	0			C-18:0			
磷 P	0.38			生物素 Biotin	0.12			C-18:1			
硫 S	0.14			叶酸 Folacin	0.50			C-18:2			
微量元素 Micro,ppm				烟酸 Niacin	20			C-18:3			
铬 Cr				泛酸 Pantothenic acid	7.1			C-18:4			
铜 Cu	4.00			核黄素 Riboflavin	1.3			C-20:0			
铁 Fe	58			硫胺素 Thiamin	5.2			C-20:1			
碘 I				胆碱 Choline	1240			C-20:4			
锰 Mn	37.00							C-20:5			
硒 Se	0.09							C-22:0			
锌 Zn	34.00			能量, kcal/kg				C-22:1			
								C-22:5			
植酸磷 Phytate P,%				总能 GE	4422	2	34	C-22:6			
磷全消化道表观消化率 ATTD of P,%				消化能 DE	4126	2	69	C-24:0			
磷全消化道标准消化率 STTD of P,%				代谢能 ME	4026			饱和脂肪酸			
				净能 NE	3164			单不饱和脂肪酸			
								多不饱和脂肪酸			
								碘价			
								碘价物			

续表

原料名称：去壳压片燕麦(Oats, Rolled, Dehulled)
AAFCO#：无官方定义

常规成分, %				氨基酸, %							
				总氨基酸				可消化氨基酸			
								回肠表观消化率		回肠标准消化率	
	$\bar{x}$	n	SD		$\bar{x}$	n	SD	$\bar{x}$	n	SD	$\bar{x}$	n	SD
干物质 Dry matter	91.10	1		必需氨基酸 Essential									
粗蛋白 Crude protein	12.94	1		粗蛋白 CP	12.94	1							
粗纤维 Crude fiber				精氨酸 Arg									
乙醚浸出物/粗脂肪 Ether extract	8.29	1		组氨酸 His									
酸乙醚浸出物 Acid ether extract				异亮氨酸 Ile									
粗灰分 Ash				亮氨酸 Leu									
碳水化合物成分, %				赖氨酸 Lys									
				蛋氨酸 Met									
乳糖 Lactose				苯丙氨酸 Phe									
蔗糖 Sucrose				苏氨酸 Thr									
棉子糖 Raffinose				色氨酸 Trp									
水苏糖 Stachyose				缬氨酸 Val									
毛蕊花糖 Verbascose				非必需氨基酸 Nonessential									
低聚糖 Oligosaccharides				丙氨酸 Ala									
淀粉 Starch	51.02	1		天冬氨酸 Asp									
中性洗涤纤维 Neutral detergent fiber				半胱氨酸 Cys									
酸性洗涤纤维 Acid detergent fiber				谷氨酸 Glu									
半纤维素 Hemicellulose				甘氨酸 Gly									
酸性洗涤木质素 Acid detergent lignin				脯氨酸 Pro									
总膳食纤维 Total dietary fiber	9.11	1		丝氨酸 Ser									
不溶性膳食纤维 Insoluble dietary fiber				酪氨酸 Tyr									
可溶性膳食纤维 Soluble dietary fiber													

矿物质				维生素, mg/kg(标注单位者除外)				脂肪酸, 占乙醚提取物, %			
	$\bar{x}$	n	SD		$\bar{x}$	n	SD		$\bar{x}$	n	SD
常量元素 Macro,%				脂溶性维生素 Fat soluble				粗脂肪			
钙 Ca				β-胡萝卜素 β-Carotene				C-12:0			
氯 Cl				维生素 E Vitamin E				C-14:0			
钾 K				水溶性维生素 Water soluble				C-16:0			
镁 Mg				维生素 B_6 Vitamin B_6				C-16:1			
钠 Na				维生素 B_{12} Vitamin B_{12}, μg/kg				C-18:0			
磷 P				生物素 Biotin				C-18:1			
硫 S				叶酸 Folacin				C-18:2			
微量元素 Micro,ppm				烟酸 Niacin				C-18:3			
铬 Cr				泛酸 Pantothenic acid				C-18:4			
铜 Cu				核黄素 Riboflavin				C-20:0			
铁 Fe				硫胺素 Thiamin				C-20:1			
碘 I				胆碱 Choline				C-20:4			
锰 Mn								C-20:5			
硒 Se								C-22:0			
锌 Zn				能量, kcal/kg				C-22:1			
								C-22:5			
植酸磷 Phytate P,%				总能 GE				C-22:6			
磷全消化道表观消化率 ATTD of P, %				消化能 DE				C-24:0			
磷全消化道标准消化率 STTD of P, %				代谢能 ME				饱和脂肪酸			
				净能 NE				单不饱和脂肪酸			
								多不饱和脂肪酸			
								碘价			
								碘价物			

续表

原料名称：压榨棕榈仁(Palm Kernel Expelled)
通过螺旋压榨棕榈果的机械法提取棕榈油所得。
AAFCO#：无官方定义

常规成分, %				氨基酸, %							
				总氨基酸				可消化氨基酸			
								回肠表观消化率		回肠标准消化率	
	$\bar{x}$	n	SD		$\bar{x}$	n	SD	$\bar{x}$	n	SD	$\bar{x}$	n	SD
干物质 Dry matter	92.00	2	3.39	必需氨基酸 Essential									
粗蛋白 Crude protein	16.64	2	0.22	粗蛋白 CP	16.64	2	0.22						
粗纤维 Crude fiber	16.71	2	1.46	精氨酸 Arg									
乙醚浸出物/粗脂肪 Ether extract	11.24	2	3.49	组氨酸 His									
酸乙醚浸出物 Acid ether extract				异亮氨酸 Ile									
粗灰分 Ash	3.82	2	0.08	亮氨酸 Leu									
碳水化合物成分, %				赖氨酸 Lys									
				蛋氨酸 Met									
乳糖 Lactose	0.00	2	0.00	苯丙氨酸 Phe									
蔗糖 Sucrose	0.00	2	0.00	苏氨酸 Thr									
棉子糖 Raffinose	0.00	2	0.00	色氨酸 Trp									
水苏糖 Stachyose	0.00	2	0.00	缬氨酸 Val									
毛蕊花糖 Verbascose	0.00	2	0.00	非必需氨基酸 Nonessential									
低聚糖 Oligosaccharides				丙氨酸 Ala									
淀粉 Starch	2.58	2	0.49	天冬氨酸 Asp									
中性洗涤纤维 Neutral detergent fiber	56.48	2	9.04	半胱氨酸 Cys									
酸性洗涤纤维 Acid detergent fiber	37.31	2	4.24	谷氨酸 Glu									
半纤维素 Hemicellulose				甘氨酸 Gly									
酸性洗涤木质素 Acid detergent lignin				脯氨酸 Pro									
总膳食纤维 Total dietary fiber				丝氨酸 Ser									
不溶性膳食纤维 Insoluble dietary fiber				酪氨酸 Tyr									
可溶性膳食纤维 Soluble dietary fiber													

矿物质				维生素, mg/kg(标注单位者除外)				脂肪酸, 占乙醚提取物, %		
	$\bar{x}$	n	SD		$\bar{x}$	n	SD		$\bar{x}$	
常量元素 Macro,%				脂溶性维生素 Fat soluble				粗脂肪	8.50	
钙 Ca	0.31	2	0.07	β-胡萝卜素 β-Carotene				C-12:0	42.21	
氯 Cl				维生素 E Vitamin E				C-14:0	14.13	
钾 K				水溶性维生素 Water soluble				C-16:0	7.65	
镁 Mg				维生素 B_6 Vitamin B_6				C-16:1	0.00	
钠 Na				维生素 B_{12} Vitamin B_{12}, μg/kg				C-18:0	2.34	
磷 P	0.52	2	0.01	生物素 Biotin				C-18:1	13.41	
硫 S				叶酸 Folacin				C-18:2	1.98	
微量元素 Micro,ppm				烟酸 Niacin				C-18:3	0.36	
铬 Cr				泛酸 Pantothenic acid				C-18:4	0.00	
铜 Cu				核黄素 Riboflavin				C-20:0	0.00	
铁 Fe				硫胺素 Thiamin				C-20:1	0.00	
碘 I				胆碱 Choline				C-20:4	0.00	
锰 Mn								C-20:5	0.00	
硒 Se								C-22:0	0.00	
锌 Zn				能量, kcal/kg				C-22:1	0.00	
								C-22:5	0.00	
植酸磷 Phytate P,%	0.37	2	0.02	总能 GE	3981	3	206	C-22:6	0.00	
磷全消化道表观消化率 ATTD of P,%	39	2	0.64	消化能 DE	3176	3	107	C-24:0	0.00	
磷全消化道标准消化率 STTD of P,%	49	2	0.50	代谢能 ME	3063			饱和脂肪酸	73.35	
				净能 NE	1941			单不饱和脂肪酸	13.41	
								多不饱和脂肪酸	2.34	
								碘价	16.62	
								碘价物	14.12	

续表

原料名称：棕榈仁粕(Palm Kernel Meal)
用溶剂浸提法从棕榈果中提取棕榈油所得。
AAFCO#：无官方定义

常规成分, %				氨基酸, %								
				总氨基酸				可消化氨基酸				
								回肠表观消化率			回肠标准消化率		
	$\bar{x}$	n	SD		$\bar{x}$	n	SD	$\bar{x}$	n	SD	$\bar{x}$	n	SD
干物质 Dry matter				必需氨基酸 Essential									
粗蛋白 Crude protein	14.39	3	0.51	粗蛋白 CP	14.39	3	0.51	51	2	7.24	63	2	7.78
粗纤维 Crude fiber				精氨酸 Arg	1.41	3	0.31	80	2	4.62	84	2	4.24
乙醚浸出物/粗脂肪 Ether extract				组氨酸 His	0.26	3	0.02	58	2	7.23	65	2	7.07
酸乙醚浸出物 Acid ether extract				异亮氨酸 Ile	0.55	3	0.03	57	2	4.37	63	2	4.24
粗灰分 Ash				亮氨酸 Leu	0.90	3	0.04	67	2	4.88	73	2	4.95
碳水化合物成分, %				赖氨酸 Lys	0.36	3	0.08	35	2	11.65	48	2	9.90
				蛋氨酸 Met	0.19	3	0.04	63	2	3.70	70	2	2.12
乳糖 Lactose				苯丙氨酸 Phe	0.56	3	0.03	69	2	2.81	75	2	2.83
蔗糖 Sucrose				苏氨酸 Thr	0.47	3	0.04	56	2	5.40	68	2	6.36
棉子糖 Raffinose				色氨酸 Trp	0.11	3	0.03	48			58		
水苏糖 Stachyose				缬氨酸 Val	0.83	3	0.03	63	2	8.04	70	2	7.78
毛蕊花糖 Verbascose				非必需氨基酸 Nonessential									
低聚糖 Oligosaccharides				丙氨酸 Ala	0.60	3	0.03	57	2	4.08	68	2	4.24
淀粉 Starch				天冬氨酸 Asp	1.22	3	0.10	59			65		
中性洗涤纤维 Neutral detergent fiber				半胱氨酸 Cys	0.18	3	0.06	33	2	10.63	46	2	5.66
酸性洗涤纤维 Acid detergent fiber	35.0			谷氨酸 Glu	2.69	3	0.11	63	2	4.22	67	2	4.24
半纤维素 Hemicellulose				甘氨酸 Gly	0.65	3	0.04	53			65	2	7.78
酸性洗涤木质素 Acid detergent lignin				脯氨酸 Pro	0.39	3	0.04	45			65		
总膳食纤维 Total dietary fiber				丝氨酸 Ser	0.85	3	0.32	55			65		
不溶性膳食纤维 Insoluble dietary fiber				酪氨酸 Tyr	0.34	3	0.02	49	2	1.44	57	2	1.41
可溶性膳食纤维 Soluble dietary fiber													

矿物质				维生素, mg/kg(标注单位者除外)				脂肪酸, 占乙醚提取物, %			
	$\bar{x}$	n	SD		$\bar{x}$	n	SD		$\bar{x}$	n	SD
常量元素 Macro,%				脂溶性维生素 Fat soluble				粗脂肪			
钙 Ca	0.20	1		β-胡萝卜素 β-Carotene				C-12:0			
氯 Cl				维生素 E Vitamin E				C-14:0			
钾 K				水溶性维生素 Water soluble				C-16:0			
镁 Mg				维生素 B_6 Vitamin B_6				C-16:1			
钠 Na				维生素 B_{12} Vitamin B_{12}, μg/kg				C-18:0			
磷 P	0.54	1		生物素 Biotin				C-18:1			
硫 S				叶酸 Folacin				C-18:2			
微量元素 Micro,ppm				烟酸 Niacin				C-18:3			
铬 Cr				泛酸 Pantothenic acid				C-18:4			
铜 Cu				核黄素 Riboflavin				C-20:0			
铁 Fe				硫胺素 Thiamin				C-20:1			
碘 I				胆碱 Choline				C-20:4			
锰 Mn								C-20:5			
硒 Se								C-22:0			
锌 Zn				能量, kcal/kg				C-22:1			
								C-22:5			
植酸磷 Phytate P,%	0.31	1		总能 GE	3640	1		C-22:6			
磷全消化道表观消化率 ATTD of P, %	49	1		消化能 DE	2970			C-24:0			
磷全消化道标准消化率 STTD of P, %	58	1		代谢能 ME	2868			饱和脂肪酸			
				净能 NE	1641			单不饱和脂肪酸			
								多不饱和脂肪酸			
								碘价			
								碘价物			

续表

原料名称：压榨花生粕(Peanut Meal, Expelled)
AAFCO#: 71.9, AAFCO 2010, p.385
IFN#: 5-03-649

常规成分, %				氨基酸, %								
				总氨基酸				可消化氨基酸				
								回肠表观消化率			回肠标准消化率	
	$\bar{x}$	n	SD	必需氨基酸 Essential	$\bar{x}$	n	SD	$\bar{x}$	n	SD	$\bar{x}$	n	SD
干物质 Dry matter	92.00												
粗蛋白 Crude protein	44.23	3	3.89	粗蛋白 CP	44.23	3	3.89	79			87	3	3.78
粗纤维 Crude fiber				精氨酸 Arg	5.20	3	0.37	93			93		
乙醚浸出物/粗脂肪 Ether extract	6.50			组氨酸 His	1.04	3	0.09	79			81		
酸乙醚浸出物 Acid ether extract				异亮氨酸 Ile	1.46	3	0.08	78			81		
粗灰分 Ash				亮氨酸 Leu	2.65	3	0.17	79			81		
碳水化合物成分, %				赖氨酸 Lys	1.55	3	0.10	73			76		
				蛋氨酸 Met	0.50			80	4	4.08	83	4	4.44
乳糖 Lactose				苯丙氨酸 Phe	2.12	3	0.17	86	4	4.56	88	4	4.85
蔗糖 Sucrose				苏氨酸 Thr	1.16	3	0.06	70			74		
棉子糖 Raffinose				色氨酸 Trp	0.33	3	0.03	73			76		
水苏糖 Stachyose				缬氨酸 Val	1.75	3	0.09	75			78	4	10.38
毛蕊花糖 Verbascose				非必需氨基酸 Nonessential									
低聚糖 Oligosaccharides				丙氨酸 Ala									
淀粉 Starch	6.65			天冬氨酸 Asp									
中性洗涤纤维 Neutral detergent fiber	14.6			半胱氨酸 Cys	0.60			78	1		81	1	
酸性洗涤纤维 Acid detergent fiber	9.1			谷氨酸 Glu									
半纤维素 Hemicellulose				甘氨酸 Gly									
酸性洗涤木质素 Acid detergent lignin				脯氨酸 Pro									
总膳食纤维 Total dietary fiber				丝氨酸 Ser									
不溶性膳食纤维 Insoluble dietary fiber				酪氨酸 Tyr	1.74			88	2	3.11	92	1	
可溶性膳食纤维 Soluble dietary fiber													

矿物质				维生素, mg/kg(标注单位者除外)				脂肪酸, 占乙醚提取物, %	
	$\bar{x}$	n	SD		$\bar{x}$	n	SD		$\bar{x}$
常量元素 Macro,%				脂溶性维生素 Fat soluble				粗脂肪	
钙 Ca	0.17			β-胡萝卜素 β-Carotene				C-12:0	0.00
氯 Cl	0.03			维生素 E Vitamin E	2.7			C-14:0	0.00
钾 K	1.20			水溶性维生素 Water soluble				C-16:0	8.73
镁 Mg	0.33			维生素 B_6 Vitamin B_6	7.4			C-16:1	0.00
钠 Na	0.06			维生素 B_{12} Vitamin B_{12}, μg/kg	0			C-18:0	1.82
磷 P	0.63			生物素 Biotin	0.35			C-18:1	39.82
硫 S	0.29			叶酸 Folacin	0.70			C-18:2	26.00
微量元素 Micro,ppm				烟酸 Niacin	166			C-18:3	0.00
铬 Cr				泛酸 Pantothenic acid	47.0			C-18:4	
铜 Cu	15			核黄素 Riboflavin	5.2			C-20:0	0.00
铁 Fe	285			硫胺素 Thiamin	7.1			C-20:1	1.09
碘 I				胆碱 Choline	1848			C-20:4	
锰 Mn	39							C-20:5	
硒 Se	0.28							C-22:0	
锌 Zn	47			能量, kcal/kg				C-22:1	
								C-22:5	
植酸磷 Phytate P,%				总能 GE	4906			C-22:6	
磷全消化道表观消化率 ATTD of P, %				消化能 DE	3895			C-24:0	
磷全消化道标准消化率 STTD of P, %				代谢能 ME	3594			饱和脂肪酸	10.55
				净能 NE	2381			单不饱和脂肪酸	40.91
								多不饱和脂肪酸	26.00
								碘价	83.73
								碘价物	54.42

续表

原料名称：花生粕(Peanut Meal, Extracted)
AAFCO#: 71.9, AAFCO 2010, p.385
IFNF#: 5-03-650

常规成分, %				氨基酸, %								
				总氨基酸				可消化氨基酸				
								回肠表观消化率			回肠标准消化率		
	$\bar{x}$	n	SD		$\bar{x}$	n	SD	$\bar{x}$	n	SD	$\bar{x}$	n	SD
干物质 Dry matter	91.80	1		必需氨基酸 Essential									
粗蛋白 Crude protein	45.03	5	4.24	粗蛋白 CP	45.03	5	4.24	79			87	3	3.78
粗纤维 Crude fiber				精氨酸 Arg	5.27	6	0.63	93			93		
乙醚浸出物/粗脂肪 Ether extract	1.20			组氨酸 His	0.98	6	0.17	79			81		
酸乙醚浸出物 Acid ether extract				异亮氨酸 Ile	1.42	6	0.17	78			81		
粗灰分 Ash				亮氨酸 Leu	2.61	6	0.25	79			81		
碳水化合物成分, %				赖氨酸 Lys	1.44	6	0.13	73			76		
				蛋氨酸 Met	0.50	6	0.16	80	4	4.08	83	4	4.44
乳糖 Lactose				苯丙氨酸 Phe	1.97	6	0.17	86	4	4.56	88	4	4.85
蔗糖 Sucrose				苏氨酸 Thr	1.26	6	0.23	70			74		
棉子糖 Raffinose				色氨酸 Trp	0.40	4	0.05	73			76		
水苏糖 Stachyose				缬氨酸 Val	1.58	6	0.27	75			78	4	10.38
毛蕊花糖 Verbascose				非必需氨基酸 Nonessential									
低聚糖 Oligosaccharides				丙氨酸 Ala	1.87	4	0.30	81			84		
淀粉 Starch	6.70			天冬氨酸 Asp	4.49	4	1.40	86			87		
中性洗涤纤维 Neutral detergent fiber	16.20			半胱氨酸 Cys	0.54	4	0.05	78	1		81	1	
酸性洗涤纤维 Acid detergent fiber	12.46			谷氨酸 Glu	7.51	4	2.42	88			89		
半纤维素 Hemicellulose				甘氨酸 Gly	2.73	4	0.40	73			76		
酸性洗涤木质素 Acid detergent lignin				脯氨酸 Pro	1.52	4	0.82	87			92		
总膳食纤维 Total dietary fiber				丝氨酸 Ser	2.13	4	0.26	83			86		
不溶性膳食纤维 Insoluble dietary fiber				酪氨酸 Tyr	1.42	5	0.13	88	2	3.11	92	1	
可溶性膳食纤维 Soluble dietary fiber													

矿物质				维生素, mg/kg(标注单位者除外)				脂肪酸, 占乙醚提取物, %			
	$\bar{x}$	n	SD		$\bar{x}$	n	SD		$\bar{x}$	n	SD
常量元素 Macro,%				脂溶性维生素 Fat soluble				粗脂肪			
钙 Ca	0.39	2	0.16	β-胡萝卜素 β-Carotene				C-12:0			
氯 Cl	0.04			维生素 E Vitamin E	2.7			C-14:0			
钾 K	1.25			水溶性维生素 Water soluble				C-16:0			
镁 Mg	0.31			维生素 B_6 Vitamin B_6	7.4			C-16:1			
钠 Na	0.07			维生素 B_{12} Vitamin B_{12}, μg/kg	0			C-18:0			
磷 P	0.58	2	0.03	生物素 Biotin	0.35			C-18:1			
硫 S	0.30			叶酸 Folacin	0.70			C-18:2			
微量元素 Micro,ppm				烟酸 Niacin	166			C-18:3			
铬 Cr				泛酸 Pantothenic acid	47.0			C-18:4			
铜 Cu	15.00			核黄素 Riboflavin	5.2			C-20:0			
铁 Fe	260			硫胺素 Thiamin	7.1			C-20:1			
碘 I				胆碱 Choline	1848			C-20:4			
锰 Mn	40.00							C-20:5			
硒 Se	0.21							C-22:0			
锌 Zn	41.00			能量, kcal/kg				C-22:1			
								C-22:5			
植酸磷 Phytate P,%				总能 GE	4622			C-22:6			
磷全消化道表观消化率 ATTD of P, %				消化能 DE	3415			C-24:0			
磷全消化道标准消化率 STTD of P, %				代谢能 ME	3109			饱和脂肪酸			
				净能 NE	1924			单不饱和脂肪酸			
								多不饱和脂肪酸			
								碘价			
								碘价物			

续表

原料名称：豌豆浓缩蛋白(Pea Protein Concentrate)
通过空气分级处理技术将轻质粒从豆粉中分离所得。
AAFCO#：无官方定义

常规成分, %				氨基酸, %								
					总氨基酸			可消化氨基酸				
								回肠表观消化率			回肠标准消化率		
	$\bar{x}$	n	SD		$\bar{x}$	n	SD	$\bar{x}$	n	SD	$\bar{x}$	n	SD
干物质 Dry matter	94.31	1		必需氨基酸 Essential									
粗蛋白 Crude protein	82.82	1		粗蛋白 CP	82.82	1		73	41	4.02	80	41	3.48
粗纤维 Crude fiber				精氨酸 Arg	6.46	1		87	39	4.35	90	39	3.28
乙醚浸出物/粗脂肪 Ether extract	8.04	1		组氨酸 His	1.96	1		78	45	4.37	82	45	3.65
酸乙醚浸出物 Acid ether extract				异亮氨酸 Ile	3.73	1		76	45	5.10	81	45	3.62
粗灰分 Ash	6.22	1		亮氨酸 Leu	6.57	1		77	45	4.39	81	45	4.16
碳水化合物成分, %				赖氨酸 Lys	5.78	1		82	45	2.91	85	45	2.72
				蛋氨酸 Met	0.80	1		72	39	4.08	77	39	3.78
乳糖 Lactose				苯丙氨酸 Phe	4.48	1		77	45	3.98	80	45	3.84
蔗糖 Sucrose				苏氨酸 Thr	3.01	1		68	45	6.13	76	45	5.92
棉子糖 Raffinose				色氨酸 Trp	0.83	1		63	29	4.65	69	25	5.13
水苏糖 Stachyose				缬氨酸 Val	4.06	1		72	45	5.60	78	45	4.60
毛蕊花糖 Verbascose				非必需氨基酸 Nonessential									
低聚糖 Oligosaccharides				丙氨酸 Ala	3.39	1		70	39	5.15	77	39	4.25
淀粉 Starch				天冬氨酸 Asp	9.36	1		78	39	3.53	82	39	3.41
中性洗涤纤维 Neutral detergent fiber				半胱氨酸 Cys	0.80	1		61	37	4.17	68	37	4.02
酸性洗涤纤维 Acid detergent fiber	0.00			谷氨酸 Glu	12.94	1		83	39	3.61	86	39	3.51
半纤维素 Hemicellulose				甘氨酸 Gly	3.21	1		64	39	6.22	79	39	5.97
酸性洗涤木质素 Acid detergent lignin				脯氨酸 Pro	3.27	1		59	31	14.05	97	31	18.47
总膳食纤维 Total dietary fiber				丝氨酸 Ser	4.06	1		73	39	5.55	79	39	4.64
不溶性膳食纤维 Insoluble dietary fiber				酪氨酸 Tyr									
可溶性膳食纤维 Soluble dietary fiber													

矿物质				维生素, mg/kg(标注单位者除外)				脂肪酸, 占乙醚提取物, %	
	$\bar{x}$	n	SD		$\bar{x}$	n	SD		
常量元素 Macro,%				脂溶性维生素 Fat soluble				粗脂肪	
钙 Ca				β-胡萝卜素 β-Carotene				C-12:0	
氯 Cl				维生素 E Vitamin E				C-14:0	
钾 K				水溶性维生素 Water soluble				C-16:0	
镁 Mg				维生素 B_6 Vitamin B_6				C-16:1	
钠 Na				维生素 B_{12} Vitamin B_{12}, μg/kg				C-18:0	
磷 P				生物素 Biotin				C-18:1	
硫 S				叶酸 Folacin				C-18:2	
微量元素 Micro,ppm				烟酸 Niacin				C-18:3	
铬 Cr				泛酸 Pantothenic acid				C-18:4	
铜 Cu				核黄素 Riboflavin				C-20:0	
铁 Fe				硫胺素 Thiamin				C-20:1	
碘 I				胆碱 Choline				C-20:4	
锰 Mn								C-20:5	
硒 Se								C-22:0	
锌 Zn				能量, kcal/kg				C-22:1	
								C-22:5	
植酸磷 Phytate P,%				总能 GE	5562			C-22:6	
磷全消化道表观消化率 ATTD of P,%				消化能 DE	4620			C-24:0	
磷全消化道标准消化率 STTD of P,%				代谢能 ME	4057			饱和脂肪酸	
				净能 NE	2610			单不饱和脂肪酸	
								多不饱和脂肪酸	
								碘价	
								碘价物	

续表

原料名称：鹰嘴豆(Peas, Chick Peas)
AAFCO#：无官方定义

常规成分, %				氨基酸, %							
				总氨基酸				可消化氨基酸			
								回肠表观消化率		回肠标准消化率	
	$\bar{x}$	n	SD		$\bar{x}$	n	SD	$\bar{x}$	n	SD	$\bar{x}$	n	SD
干物质 Dry matter	89.74	3	1.15	必需氨基酸 Essential									
粗蛋白 Crude protein	20.33	3	0.89	粗蛋白 CP	20.33	3	0.89	73	41	4.02	80	41	3.48
粗纤维 Crude fiber				精氨酸 Arg	2.25	2	0.52	87	39	4.35	90	39	3.28
乙醚浸出物/粗脂肪 Ether extract	4.14	2	0.23	组氨酸 His	0.84	2	0.01	78	45	4.37	82	45	3.65
酸乙醚浸出物 Acid ether extract				异亮氨酸 Ile	0.91	2	0.17	76	45	5.10	81	45	3.62
粗灰分 Ash	2.86	3	0.04	亮氨酸 Leu	1.61	2	0.06	77	45	4.39	81	45	4.16
碳水化合物成分, %				赖氨酸 Lys	1.41	2	0.22	82	45	2.91	85	45	2.72
				蛋氨酸 Met	0.30	2	0.00	72	39	4.08	77	39	3.78
乳糖 Lactose				苯丙氨酸 Phe	1.23	2	0.08	77	45	3.98	80	45	3.84
蔗糖 Sucrose				苏氨酸 Thr	0.91	2	0.05	68	45	6.13	76	45	5.92
棉子糖 Raffinose				色氨酸 Trp									
水苏糖 Stachyose				缬氨酸 Val	1.02	2	0.08	72	45	5.60	78	45	4.60
毛蕊花糖 Verbascose				非必需氨基酸 Nonessential									
低聚糖 Oligosaccharides				丙氨酸 Ala	0.59	2	0.00	70	39	5.15	77	39	4.25
淀粉 Starch	44.80			天冬氨酸 Asp	2.50	2	0.05	78	39	3.53	82	39	3.41
中性洗涤纤维 Neutral detergent fiber	15.82	3	4.96	半胱氨酸 Cys	0.44	2	0.00	61	37	4.17	68	37	4.02
酸性洗涤纤维 Acid detergent fiber	6.75	3	3.49	谷氨酸 Glu	3.12	2	0.08	83	39	3.61	86	39	3.51
半纤维素 Hemicellulose	7.84	2	1.39	甘氨酸 Gly	0.99	2	0.05	64	39	6.22	79	39	5.97
酸性洗涤木质素 Acid detergent lignin	0.57	2	0.79	脯氨酸 Pro									
总膳食纤维 Total dietary fiber				丝氨酸 Ser	1.06	2	0.02	73	39	5.55	79	39	4.64
不溶性膳食纤维 Insoluble dietary fiber				酪氨酸 Tyr	0.82	2	0.10	74	32	5.62	78	31	4.96
可溶性膳食纤维 Soluble dietary fiber													

矿物质				维生素, mg/kg(标注单位者除外)				脂肪酸, 占乙醚提取物, %			
	$\bar{x}$	n	SD		$\bar{x}$	n	SD		$\bar{x}$	n	SD
常量元素 Macro,%				脂溶性维生素 Fat soluble				粗脂肪	6.04		
钙 Ca				β-胡萝卜素 β-Carotene				C-12:0	0.00		
氯 Cl				维生素 E Vitamin E				C-14:0	0.15		
钾 K				水溶性维生素 Water soluble				C-16:0	8.29		
镁 Mg				维生素 B_6 Vitamin B_6				C-16:1	0.20		
钠 Na				维生素 B_{12} Vitamin B_{12}, μg/kg				C-18:0	1.41		
磷 P				生物素 Biotin				C-18:1	22.28		
硫 S				叶酸 Folacin				C-18:2	42.93		
微量元素 Micro,ppm				烟酸 Niacin				C-18:3	1.67		
铬 Cr				泛酸 Pantothenic acid				C-18:4			
铜 Cu				核黄素 Riboflavin				C-20:0	0.00		
铁 Fe				硫胺素 Thiamin				C-20:1	0.00		
碘 I				胆碱 Choline				C-20:4			
锰 Mn								C-20:5			
硒 Se								C-22:0			
锌 Zn				能量, kcal/kg				C-22:1			
								C-22:5			
植酸磷 Phytate P,%				总能 GE	4554			C-22:6			
磷全消化道表观消化率 ATTD of P, %				消化能 DE	3504			C-24:0			
磷全消化道标准消化率 STTD of P, %				代谢能 ME	3366			饱和脂肪酸	9.85		
				净能 NE	2491			单不饱和脂肪酸	22.48		
								多不饱和脂肪酸	44.60		
								碘价	102.49		
								碘价物	61.91		

续表

原料名称：豇豆 (Peas, Cow Peas)
AAFCO#：无官方定义

常规成分, %				氨基酸, %								
				总氨基酸				可消化氨基酸				
								回肠表观消化率			回肠标准消化率	
	$\bar{x}$	n	SD		$\bar{x}$	n	SD	$\bar{x}$	n	SD	$\bar{x}$	n	SD
干物质 Dry matter	88.10			必需氨基酸 Essential									
粗蛋白 Crude protein	22.19			粗蛋白 CP	22.19			73	41	4.02	80	41	3.48
粗纤维 Crude fiber				精氨酸 Arg									
乙醚浸出物/粗脂肪 Ether extract	1.20			组氨酸 His									
酸乙醚浸出物 Acid ether extract				异亮氨酸 Ile									
粗灰分 Ash				亮氨酸 Leu									
碳水化合物成分, %				赖氨酸 Lys									
				蛋氨酸 Met									
乳糖 Lactose				苯丙氨酸 Phe									
蔗糖 Sucrose				苏氨酸 Thr									
棉子糖 Raffinose				色氨酸 Trp									
水苏糖 Stachyose				缬氨酸 Val									
毛蕊花糖 Verbascose				非必需氨基酸 Nonessential									
低聚糖 Oligosaccharides				丙氨酸 Ala									
淀粉 Starch	43.46			天冬氨酸 Asp									
中性洗涤纤维 Neutral detergent fiber				半胱氨酸 Cys									
酸性洗涤纤维 Acid detergent fiber	6.75			谷氨酸 Glu									
半纤维素 Hemicellulose				甘氨酸 Gly									
酸性洗涤木质素 Acid detergent lignin				脯氨酸 Pro									
总膳食纤维 Total dietary fiber				丝氨酸 Ser									
不溶性膳食纤维 Insoluble dietary fiber				酪氨酸 Tyr									
可溶性膳食纤维 Soluble dietary fiber													

矿物质				维生素, mg/kg (标注单位者除外)				脂肪酸，占乙醚提取物, %			
	$\bar{x}$	n	SD		$\bar{x}$	n	SD		$\bar{x}$	n	SD
常量元素 Macro,%				脂溶性维生素 Fat soluble				粗脂肪	1.26		
钙 Ca				β-胡萝卜素 β-Carotene				C-12:0	0.08		
氯 Cl				维生素 E Vitamin E				C-14:0	0.24		
钾 K				水溶性维生素 Water soluble				C-16:0	20.16		
镁 Mg				维生素 B_6 Vitamin B_6				C-16:1	0.32		
钠 Na				维生素 B_{12} Vitamin B_{12}, μg/kg				C-18:0	4.21		
磷 P				生物素 Biotin				C-18:1	6.98		
硫 S				叶酸 Folacin				C-18:2	27.22		
微量元素 Micro,ppm				烟酸 Niacin				C-18:3	15.79		
铬 Cr				泛酸 Pantothenic acid				C-18:4			
铜 Cu				核黄素 Riboflavin				C-20:0	0.00		
铁 Fe				硫胺素 Thiamin				C-20:1	0.08		
碘 I				胆碱 Choline				C-20:4			
锰 Mn								C-20:5			
硒 Se								C-22:0			
锌 Zn				能量, kcal/kg				C-22:1			
								C-22:5			
植酸磷 Phytate P,%				总能 GE	4417			C-22:6			
磷全消化道表观消化率 ATTD of P, %				消化能 DE	3504			C-24:0			
磷全消化道标准消化率 STTD of P, %				代谢能 ME	3353			饱和脂肪酸	24.68		
				净能 NE	2420			单不饱和脂肪酸	7.38		
								多不饱和脂肪酸	43.02		
								碘价	99.83		
								碘价物	12.58		

续表

原料名称：紫花豌豆(Peas, Field Peas)
AAFCO#：无官方定义

常规成分, %				氨基酸, %							
				总氨基酸				可消化氨基酸			
	$\bar{x}$	n	SD		$\bar{x}$	n	SD	回肠表观消化率		回肠标准消化率	
干物质 Dry matter	88.10	28	2.67	必需氨基酸 Essential				$\bar{x}$	n	SD	$\bar{x}$	n	SD
粗蛋白 Crude protein	22.17	61	1.51	粗蛋白 CP	22.17	61	1.51	73	41	4.02	80	41	3.48
粗纤维 Crude fiber	6.16	20	0.92	精氨酸 Arg	1.91	53	0.36	87	39	4.35	90	39	3.28
乙醚浸出物/粗脂肪 Ether extract	1.20	35	0.48	组氨酸 His	0.53	59	0.05	78	45	4.37	82	45	3.65
酸乙醚浸出物 Acid ether extract				异亮氨酸 Ile	0.94	59	0.12	76	45	5.10	81	45	3.62
粗灰分 Ash	2.86	34	0.18	亮氨酸 Leu	1.56	59	0.14	77	45	4.39	81	45	4.16
碳水化合物成分, %				赖氨酸 Lys	1.63	61	0.18	82	45	2.91	85	45	2.72
				蛋氨酸 Met	0.21	59	0.03	72	39	4.08	77	39	3.78
乳糖 Lactose				苯丙氨酸 Phe	1.02	58	0.10	77	45	3.98	80	45	3.84
蔗糖 Sucrose	0.19	9	0.58	苏氨酸 Thr	0.83	59	0.10	68	45	6.13	76	45	5.92
棉子糖 Raffinose	0.04	9	0.13	色氨酸 Trp	0.21	47	0.08	63	29	4.65	69	25	5.13
水苏糖 Stachyose	0.23	9	0.68	缬氨酸 Val	1.03	59	0.10	72	45	5.60	78	45	4.60
毛蕊花糖 Verbascose	0.32	9	0.96	非必需氨基酸 Nonessential									
低聚糖 Oligosaccharides				丙氨酸 Ala	0.95	49	0.11	70	39	5.15	77	39	4.25
淀粉 Starch	43.46	30	3.72	天冬氨酸 Asp	2.56	49	0.27	78	39	3.53	82	39	3.41
中性洗涤纤维 Neutral detergent fiber	12.84	30	3.90	半胱氨酸 Cys	0.31	57	0.04	61	37	4.17	68	37	4.02
酸性洗涤纤维 Acid detergent fiber	6.90	24	1.50	谷氨酸 Glu	3.87	49	0.54	83	39	3.61	86	39	3.51
半纤维素 Hemicellulose	2.79	6	0.84	甘氨酸 Gly	0.95	49	0.11	64	39	6.22	79	39	5.97
酸性洗涤木质素 Acid detergent lignin	0.45	10	0.51	脯氨酸 Pro	0.94	29	0.19	59	31	14.05	97	31	18.47
总膳食纤维 Total dietary fiber				丝氨酸 Ser	1.05	48	0.15	73	39	5.55	79	39	4.64
不溶性膳食纤维 Insoluble dietary fiber				酪氨酸 Tyr	0.59	46	0.13	74	32	5.62	78	31	4.96
可溶性膳食纤维 Soluble dietary fiber													

矿物质				维生素, mg/kg(标注单位者除外)				脂肪酸, 占乙醚提取物, %			
	$\bar{x}$	n	SD		$\bar{x}$	n	SD		$\bar{x}$	n	SD
常量元素 Macro,%				脂溶性维生素 Fat soluble				粗脂肪			
钙 Ca	0.09	10	0.04	β-胡萝卜素 β-Carotene				C-12:0			
氯 Cl				维生素 E Vitamin E				C-14:0			
钾 K				水溶性维生素 Water soluble				C-16:0			
镁 Mg				维生素 B_6 Vitamin B_6				C-16:1			
钠 Na				维生素 B_{12} Vitamin B_{12}, μg/kg				C-18:0			
磷 P	0.42	13	0.06	生物素 Biotin				C-18:1			
硫 S				叶酸 Folacin				C-18:2			
微量元素 Micro,ppm				烟酸 Niacin				C-18:3			
铬 Cr				泛酸 Pantothenic acid				C-18:4			
铜 Cu				核黄素 Riboflavin				C-20:0			
铁 Fe				硫胺素 Thiamin				C-20:1			
碘 I				胆碱 Choline				C-20:4			
锰 Mn								C-20:5			
硒 Se								C-22:0			
锌 Zn								C-22:1			
				能量, kcal/kg				C-22:5			
植酸磷 Phytate P,%	0.17	7	0.07	总能 GE	4035	4	54	C-22:6			
磷全消化道表观消化率 ATTD of P, %	49	8	6.13	消化能 DE	3504	2	21	C-24:0			
磷全消化道标准消化率 STTD of P, %	56	7	5.65	代谢能 ME	3353			饱和脂肪酸			
				净能 NE	2419			单不饱和脂肪酸			
								多不饱和脂肪酸			
								碘价			
								碘价物			

续表

原料名称：裂荚紫花豌豆(Peas, Field Pea Splits)
AAFCO#：无官方定义

常规成分, %				氨基酸, %									
				总氨基酸				可消化氨基酸					
								回肠表观消化率			回肠标准消化率		
	$\bar{x}$	n	SD		$\bar{x}$	n	SD	$\bar{x}$	n	SD	$\bar{x}$	n	SD
干物质 Dry matter	88.10			必需氨基酸 Essential									
粗蛋白 Crude protein	22.19			粗蛋白 CP	22.19			73	41	4.02	80	41	3.48
粗纤维 Crude fiber				精氨酸 Arg									
乙醚浸出物/粗脂肪 Ether extract	1.20			组氨酸 His									
酸乙醚浸出物 Acid ether extract				异亮氨酸 Ile									
粗灰分 Ash				亮氨酸 Leu									
碳水化合物成分, %				赖氨酸 Lys									
				蛋氨酸 Met									
乳糖 Lactose				苯丙氨酸 Phe									
蔗糖 Sucrose				苏氨酸 Thr									
棉子糖 Raffinose				色氨酸 Trp									
水苏糖 Stachyose				缬氨酸 Val									
毛蕊花糖 Verbascose				非必需氨基酸 Nonessential									
低聚糖 Oligosaccharides				丙氨酸 Ala									
淀粉 Starch	43.36			天冬氨酸 Asp									
中性洗涤纤维 Neutral detergent fiber				半胱氨酸 Cys									
酸性洗涤纤维 Acid detergent fiber	6.90			谷氨酸 Glu									
半纤维素 Hemicellulose				甘氨酸 Gly									
酸性洗涤木质素 Acid detergent lignin				脯氨酸 Pro									
总膳食纤维 Total dietary fiber				丝氨酸 Ser									
不溶性膳食纤维 Insoluble dietary fiber				酪氨酸 Tyr									
可溶性膳食纤维 Soluble dietary fiber													

矿物质				维生素, mg/kg(标注单位者除外)				脂肪酸, 占乙醚提取物, %			
	$\bar{x}$	n	SD		$\bar{x}$	n	SD		$\bar{x}$	n	SD
常量元素 Macro,%				脂溶性维生素 Fat soluble				粗脂肪			
钙 Ca	0.10	1		β-胡萝卜素 β-Carotene				C-12:0			
氯 Cl				维生素 E Vitamin E				C-14:0			
钾 K				水溶性维生素 Water soluble				C-16:0			
镁 Mg				维生素 B_6 Vitamin B_6				C-16:1			
钠 Na				维生素 B_{12} Vitamin B_{12}, μg/kg				C-18:0			
磷 P	0.43	1		生物素 Biotin				C-18:1			
硫 S				叶酸 Folacin				C-18:2			
微量元素 Micro,ppm				烟酸 Niacin				C-18:3			
铬 Cr				泛酸 Pantothenic acid				C-18:4			
铜 Cu				核黄素 Riboflavin				C-20:0			
铁 Fe				硫胺素 Thiamin				C-20:1			
碘 I				胆碱 Choline				C-20:4			
锰 Mn								C-20:5			
硒 Se								C-22:0			
锌 Zn								C-22:1			
				能量, kcal/kg				C-22:5			
植酸磷 Phytate P,%				总能 GE	4417			C-22:6			
磷全消化道表观消化率 ATTD of P, %	49			消化能 DE	3504			C-24:0			
磷全消化道标准消化率 STTD of P, %	56			代谢能 ME	3353			饱和脂肪酸			
				净能 NE	2419			单不饱和脂肪酸			
								多不饱和脂肪酸			
								碘价			
								碘价物			

续表

原料名称：宠物食品的副产物(Pet Food Byproduct)
AAFCO#: T60, 108, AAFCO 2010, p.379

常规成分, %				氨基酸, %				可消化氨基酸					
				总氨基酸				回肠表观消化率			回肠标准消化率		
	$\bar{x}$	n	SD		$\bar{x}$	n	SD	$\bar{x}$	n	SD	$\bar{x}$	n	SD
干物质 Dry matter	92.80	1		必需氨基酸 Essential									
粗蛋白 Crude protein	20.94	1		粗蛋白 CP	20.94	1							
粗纤维 Crude fiber	1.70	1		精氨酸 Arg	1.60	1							
乙醚浸出物/粗脂肪 Ether extract	8.29	1		组氨酸 His	0.53	1							
酸乙醚浸出物 Acid ether extract				异亮氨酸 Ile	0.90	1							
粗灰分 Ash	5.65	1		亮氨酸 Leu	1.59	1							
碳水化合物成分, %				赖氨酸 Lys	1.25	1							
				蛋氨酸 Met	0.45	1							
乳糖 Lactose				苯丙氨酸 Phe	0.97	1							
蔗糖 Sucrose				苏氨酸 Thr	0.82	1							
棉子糖 Raffinose				色氨酸 Trp									
水苏糖 Stachyose				缬氨酸 Val	1.05	1							
毛蕊花糖 Verbascose				非必需氨基酸 Nonessential									
低聚糖 Oligosaccharides				丙氨酸 Ala	1.28	1							
淀粉 Starch				天冬氨酸 Asp	1.89	1							
中性洗涤纤维 Neutral detergent fiber				半胱氨酸 Cys	0.09	1							
酸性洗涤纤维 Acid detergent fiber				谷氨酸 Glu	3.66	1							
半纤维素 Hemicellulose				甘氨酸 Gly	1.60	1							
酸性洗涤木质素 Acid detergent lignin				脯氨酸 Pro	1.20	1							
总膳食纤维 Total dietary fiber				丝氨酸 Ser	0.89	1							
不溶性膳食纤维 Insoluble dietary fiber				酪氨酸 Tyr									
可溶性膳食纤维 Soluble dietary fiber													

矿物质				维生素, mg/kg(标注单位者除外)				脂肪酸, 占乙醚提取物, %			
	$\bar{x}$	n	SD		$\bar{x}$	n	SD		$\bar{x}$	n	SD
常量元素 Macro, %				脂溶性维生素 Fat soluble				粗脂肪			
钙 Ca	0.82	1		β-胡萝卜素 β-Carotene				C-12:0			
氯 Cl	0.32	1		维生素 E Vitamin E				C-14:0			
钾 K	0.74	1		水溶性维生素 Water soluble				C-16:0			
镁 Mg	0.15	1		维生素 B_6 Vitamin B_6				C-16:1			
钠 Na	0.22	1		维生素 B_{12} Vitamin B_{12}, μg/kg				C-18:0			
磷 P	0.84	1		生物素 Biotin				C-18:1			
硫 S				叶酸 Folacin				C-18:2			
微量元素 Micro, ppm				烟酸 Niacin				C-18:3			
铬 Cr				泛酸 Pantothenic acid				C-18:4			
铜 Cu	4.40	1		核黄素 Riboflavin				C-20:0			
铁 Fe	152	1		硫胺素 Thiamin				C-20:1			
碘 I				胆碱 Choline				C-20:4			
锰 Mn	85.80	1						C-20:5			
硒 Se								C-22:0			
锌 Zn	293	1						C-22:1			
				能量, kcal/kg				C-22:5			
植酸磷 Phytate P, %				总能 GE	4601	1		C-22:6			
磷全消化道表观消化率 ATTD of P, %				消化能 DE				C-24:0			
磷全消化道标准消化率 STTD of P, %				代谢能 ME				饱和脂肪酸			
				净能 NE				单不饱和脂肪酸			
								多不饱和脂肪酸			
								碘价			
								碘价物			

续表

原料名称：猪肠膜蛋白 (Porcine Solubles, Dried)
AAFCO#: 9.12, AFFCO 2010, p.327
IFN#: 5-00-393

常规成分, %				氨基酸, %								
				总氨基酸				可消化氨基酸				
								回肠表观消化率			回肠标准消化率		
	$\bar{x}$	n	SD		$\bar{x}$	n	SD	$\bar{x}$	n	SD	$\bar{x}$	n	SD
干物质 Dry matter				必需氨基酸 Essential									
粗蛋白 Crude protein	51.01	1		粗蛋白 CP	51.01	1							
粗纤维 Crude fiber				精氨酸 Arg	2.72	1							
乙醚浸出物/粗脂肪 Ether extract				组氨酸 His	1.06	1							
酸乙醚浸出物 Acid ether extract				异亮氨酸 Ile	2.06	1							
粗灰分 Ash				亮氨酸 Leu	3.94	1							
碳水化合物成分, %				赖氨酸 Lys	3.81	1							
				蛋氨酸 Met	0.96	1							
乳糖 Lactose				苯丙氨酸 Phe	2.23	1							
蔗糖 Sucrose				苏氨酸 Thr	2.10	1							
棉子糖 Raffinose				色氨酸 Trp	0.25	1							
水苏糖 Stachyose				缬氨酸 Val	2.60	1							
毛蕊花糖 Verbascose				非必需氨基酸 Nonessential									
低聚糖 Oligosaccharides				丙氨酸 Ala	2.95	1							
淀粉 Starch				天冬氨酸 Asp									
中性洗涤纤维 Neutral detergent fiber				半胱氨酸 Cys	0.78	1							
酸性洗涤纤维 Acid detergent fiber				谷氨酸 Glu									
半纤维素 Hemicellulose				甘氨酸 Gly	3.65	1							
酸性洗涤木质素 Acid detergent lignin				脯氨酸 Pro	2.83	1							
总膳食纤维 Total dietary fiber				丝氨酸 Ser	1.86	1							
不溶性膳食纤维 Insoluble dietary fiber				酪氨酸 Tyr	1.86	1							
可溶性膳食纤维 Soluble dietary fiber													

矿物质				维生素, mg/kg(标注单位者除外)				脂肪酸, 占乙醚提取物, %			
	$\bar{x}$	n	SD		$\bar{x}$	n	SD		$\bar{x}$	n	SD
常量元素 Macro,%				脂溶性维生素 Fat soluble				粗脂肪			
钙 Ca				β-胡萝卜素 β-Carotene				C-12:0			
氯 Cl				维生素 E Vitamin E				C-14:0			
钾 K				水溶性维生素 Water soluble				C-16:0			
镁 Mg				维生素 B_6 Vitamin B_6				C-16:1			
钠 Na				维生素 B_{12} Vitamin B_{12}, µg/kg				C-18:0			
磷 P				生物素 Biotin				C-18:1			
硫 S				叶酸 Folacin				C-18:2			
微量元素 Micro,ppm				烟酸 Niacin				C-18:3			
铬 Cr				泛酸 Pantothenic acid				C-18:4			
铜 Cu				核黄素 Riboflavin				C-20:0			
铁 Fe				硫胺素 Thiamin				C-20:1			
碘 I				胆碱 Choline				C-20:4			
锰 Mn								C-20:5			
硒 Se								C-22:0			
锌 Zn				能量, kcal/kg				C-22:1			
								C-22:5			
植酸磷 Phytate P,%				总能 GE				C-22:6			
磷全消化道表观消化率 ATTD of P, %				消化能 DE				C-24:0			
磷全消化道标准消化率 STTD of P, %				代谢能 ME				饱和脂肪酸			
				净能 NE				单不饱和脂肪酸			
								多不饱和脂肪酸			
								碘价			
								碘价物			

第17章 饲料原料成分

续表

原料名称：马铃薯浓缩蛋白(Potato Protein Concentrate)
AAFCO#: 60.94, AAFCO 2010, p.378
IFN#: 5-25-392

常规成分, %				氨基酸, %								
					总氨基酸			可消化氨基酸				
								回肠表观消化率			回肠标准消化率	
	$\bar{x}$	n	SD	必需氨基酸 Essential	$\bar{x}$	n	SD	$\bar{x}$	n	SD	$\bar{x}$	n	SD
干物质 Dry matter	93.39	2	1.84	粗蛋白 CP	79.80	2	1.91	85	2	2.83	87	2	2.78
粗蛋白 Crude protein	79.80	2	1.91	精氨酸 Arg	4.14	2	0.11	91	2	1.41	92	2	1.38
粗纤维 Crude fiber	1.43	2	1.40	组氨酸 His	1.76	2	0.07	87	2	2.83	88	2	2.79
乙醚浸出物/粗脂肪 Ether extract	2.78	2	1.74	异亮氨酸 Ile	4.18	2	0.52	87	2	2.12	87	2	2.02
酸乙醚浸出物 Acid ether extract				亮氨酸 Leu	8.14	2	0.03	89	2	2.12	89	2	2.12
粗灰分 Ash	1.28	2	1.07	赖氨酸 Lys	6.18	2	0.13	88	2	2.12	88	2	2.11
碳水化合物成分, %				蛋氨酸 Met	1.74	2	0.05	90	2	1.41	91	2	1.39
乳糖 Lactose				苯丙氨酸 Phe	5.10	2	0.01	82	2	2.12	82	2	2.12
蔗糖 Sucrose				苏氨酸 Thr	4.61	2	0.04	84	2	2.83	85	2	2.82
棉子糖 Raffinose				色氨酸 Trp	1.10	2	0.00	78	2	3.54	79	2	3.54
水苏糖 Stachyose				缬氨酸 Val	5.36	2	0.10	88	2	2.12	88	2	2.10
毛蕊花糖 Verbascose				非必需氨基酸 Nonessential								
低聚糖 Oligosaccharides				丙氨酸 Ala	4.02	2	0.18	86	2	2.12	87	2	2.05
淀粉 Starch				天冬氨酸 Asp	9.99	2	0.28	84	2	4.24	85	2	4.22
中性洗涤纤维 Neutral detergent fiber				半胱氨酸 Cys	1.13	2	0.03	65	2	4.24	67	2	4.29
酸性洗涤纤维 Acid detergent fiber				谷氨酸 Glu	8.65	2	0.26	86	2	3.54	87	2	3.50
半纤维素 Hemicellulose				甘氨酸 Gly	4.08	2	0.01	85	2	4.24	89	2	4.23
酸性洗涤木质素 Acid detergent lignin				脯氨酸 Pro	4.06	2	0.01	88	2	1.41	100	2	1.37
总膳食纤维 Total dietary fiber				丝氨酸 Ser	4.35	2	0.08	86	2	2.83	87	2	2.85
不溶性膳食纤维 Insoluble dietary fiber				酪氨酸 Tyr	3.93			78			85		
可溶性膳食纤维 Soluble dietary fiber													
矿物质				维生素, mg/kg(标注单位者除外)				脂肪酸, 占乙醚提取物, %					
	$\bar{x}$	n	SD		$\bar{x}$	n	SD		$\bar{x}$	n	SD		
常量元素 Macro,%				脂溶性维生素 Fat soluble				粗脂肪	0.90				
钙 Ca				β-胡萝卜素 β-Carotene				C-12:0	0.00				
氯 Cl				维生素 E Vitamin E				C-14:0	0.32				
钾 K				水溶性维生素 Water soluble				C-16:0	13.76				
镁 Mg				维生素 B_6 Vitamin B_6				C-16:1	0.40				
钠 Na				维生素 B_{12} Vitamin B_{12}, μg/kg				C-18:0	3.12				
磷 P				生物素 Biotin				C-18:1	1.28				
硫 S				叶酸 Folacin				C-18:2	23.36				
微量元素 Micro,ppm				烟酸 Niacin				C-18:3	16.56				
铬 Cr				泛酸 Pantothenic acid				C-18:4	0.00				
铜 Cu	38.50	1		核黄素 Riboflavin				C-20:0	0.80				
铁 Fe	128	1		硫胺素 Thiamin				C-20:1	0.00				
碘 I				胆碱 Choline				C-20:4	0.00				
锰 Mn	0.10	1						C-20:5	0.00				
硒 Se								C-22:0	0.48				
锌 Zn	14.30	1		能量, kcal/kg				C-22:1	0.00				
								C-22:5	0.00				
植酸磷 Phytate P,%				总能 GE	5439			C-22:6	0.00				
磷全消化道表观消化率 ATTD of P, %				消化能 DE	4140			C-24:0	0.00				
磷全消化道标准消化率 STTD of P, %				代谢能 ME	3597			饱和脂肪酸	18.48				
				净能 NE				单不饱和脂肪酸	1.68				
								多不饱和脂肪酸	39.92				
								碘价	89.11				
								碘价物	8.02				

续表

原料名称：禽副产物(Poultry Byproduct)
AAFCO#: 9.14, AAFCO 2010, p.327
IFN#: 5-03-800

常规成分, %				氨基酸, %									
					总氨基酸			可消化氨基酸					
								回肠表观消化率			回肠标准消化率		
	$\bar{x}$	n	SD		$\bar{x}$	n	SD	$\bar{x}$	n	SD	$\bar{x}$	n	SD
干物质 Dry matter	92.08	7	3.69	必需氨基酸 Essential									
粗蛋白 Crude protein	64.03	11	2.35	粗蛋白 CP	64.03	11	2.35	75	5	2.71	78	5	2.14
粗纤维 Crude fiber	0.35	2	0.21	精氨酸 Arg	4.35	11	0.26	87	5	2.48	89	5	1.91
乙醚浸出物/粗脂肪 Ether extract	12.02	3	1.08	组氨酸 His	1.28	11	0.09	80	5	4.97	82	5	3.92
酸乙醚浸出物 Acid ether extract	18.30	2	1.27	异亮氨酸 Ile	2.38	11	0.13	79	5	1.96	81	5	1.49
粗灰分 Ash	13.32	5	2.25	亮氨酸 Leu	4.42	11	0.21	80	5	4.41	82	5	3.05
碳水化合物成分, %				赖氨酸 Lys	3.69	11	0.31	84	5	5.08	85	2	4.07
				蛋氨酸 Met	1.25	7	0.12	74			77		
乳糖 Lactose				苯丙氨酸 Phe	2.23	11	0.19	82	5	3.79	84	5	3.15
蔗糖 Sucrose				苏氨酸 Thr	2.35	11	0.25	74	5	3.95	77	5	2.59
棉子糖 Raffinose				色氨酸 Trp	0.46	9	0.12	74	5	13.02	78	5	9.46
水苏糖 Stachyose				缬氨酸 Val	2.91	11	0.39	78	5	2.45	80	5	2.17
毛蕊花糖 Verbascose				非必需氨基酸 Nonessential									
低聚糖 Oligosaccharides				丙氨酸 Ala	3.75	3	1.02	78	1		81	1	
淀粉 Starch	0.00			天冬氨酸 Asp	4.11	3	0.09	59	1		63	1	
中性洗涤纤维 Neutral detergent fiber				半胱氨酸 Cys	0.63	5	0.38	70			72		
酸性洗涤纤维 Acid detergent fiber	0.00			谷氨酸 Glu	6.41	3	1.33	75	1		78	1	
半纤维素 Hemicellulose				甘氨酸 Gly	6.17	4	2.25	75	1		79	1	
酸性洗涤木质素 Acid detergent lignin				脯氨酸 Pro	3.91	4	1.98	75	1		81	1	
总膳食纤维 Total dietary fiber				丝氨酸 Ser	2.27	4	0.28	72	1		76	1	
不溶性膳食纤维 Insoluble dietary fiber				酪氨酸 Tyr	1.93	5	0.24	51	1		69	1	
可溶性膳食纤维 Soluble dietary fiber													

矿物质				维生素, mg/kg (标注单位者除外)				脂肪酸, 占乙醚提取物, %			
	$\bar{x}$	n	SD		$\bar{x}$	n	SD		$\bar{x}$	n	SD
常量元素 Macro,%				脂溶性维生素 Fat soluble				粗脂肪			
钙 Ca	4.54	8	0.41	β-胡萝卜素 β-Carotene				C-12:0			
氯 Cl	0.49			维生素 E Vitamin E				C-14:0			
钾 K	0.53			水溶性维生素 Water soluble				C-16:0			
镁 Mg	0.18			维生素 B_6 Vitamin B_6	4.4			C-16:1			
钠 Na	0.49			维生素 B_{12} Vitamin B_{12}, μg/kg				C-18:0			
磷 P	2.51	8	0.18	生物素 Biotin	0.09			C-18:1			
硫 S	0.52			叶酸 Folacin	0.50			C-18:2			
微量元素 Micro,ppm				烟酸 Niacin	47			C-18:3			
铬 Cr				泛酸 Pantothenic acid	11.1			C-18:4			
铜 Cu	10.00			核黄素 Riboflavin	10.5			C-20:0			
铁 Fe	442			硫胺素 Thiamin	0.2			C-20:1			
碘 I				胆碱 Choline	6029			C-20:4			
锰 Mn	9.00							C-20:5			
硒 Se	0.88							C-22:0			
锌 Zn	94.00			能量, kcal/kg				C-22:1			
								C-22:5			
植酸磷 Phytate P,%				总能 GE	5300	2	112	C-22:6			
磷全消化道表观消化率 ATTD of P, %	48			消化能 DE	3090			C-24:0			
磷全消化道标准消化率 STTD of P, %	53			代谢能 ME	2655			饱和脂肪酸			
				净能 NE	1774			单不饱和脂肪酸			
								多不饱和脂肪酸			
								碘价			
								碘价物			

续表

原料名称：禽肉粉(Poultry Meal)
AAFCO#: 9.71, AAFCO 2010, p.331
IFN#: 5-03-798

常规成分, %				氨基酸, %							
				总氨基酸				可消化氨基酸			
								回肠表观消化率		回肠标准消化率	
	$\bar{x}$	n	SD		$\bar{x}$	n	SD	$\bar{x}$	n	SD	$\bar{x}$	n	SD
干物质 Dry matter	96.20	1		必需氨基酸 Essential							
粗蛋白 Crude protein	64.72	5	4.54	粗蛋白 CP	64.72	5	4.54				
粗纤维 Crude fiber				精氨酸 Arg	4.46	2	0.36				
乙醚浸出物/粗脂肪 Ether extract				组氨酸 His	1.69	2	0.06				
酸乙醚浸出物 Acid ether extract	14.40	1		异亮氨酸 Ile	2.50	5	0.16				
粗灰分 Ash	12.06	3	1.32	亮氨酸 Leu	4.63	5	0.23				
碳水化合物成分, %				赖氨酸 Lys	3.99	5	0.6				
				蛋氨酸 Met	1.15	5	0.23				
乳糖 Lactose				苯丙氨酸 Phe	2.64	2	0.07				
蔗糖 Sucrose				苏氨酸 Thr	2.55	5	0.25				
棉子糖 Raffinose				色氨酸 Trp	0.62	4	0.1				
水苏糖 Stachyose				缬氨酸 Val	3.07	5	0.13				
毛蕊花糖 Verbascose				非必需氨基酸 Nonessential							
低聚糖 Oligosaccharides				丙氨酸 Ala	4.18	2	0.07				
淀粉 Starch				天冬氨酸 Asp	5.71	2	0.49				
中性洗涤纤维 Neutral detergent fiber				半胱氨酸 Cys	0.87	5	0.25				
酸性洗涤纤维 Acid detergent fiber				谷氨酸 Glu	8.80	2	0.75				
半纤维素 Hemicellulose				甘氨酸 Gly	5.79	2	0.7				
酸性洗涤木质素 Acid detergent lignin				脯氨酸 Pro	4.23	1					
总膳食纤维 Total dietary fiber	2.60	1		丝氨酸 Ser	3.67	2	0.83				
不溶性膳食纤维 Insoluble dietary fiber				酪氨酸 Tyr	1.84	2	0.26				
可溶性膳食纤维 Soluble dietary fiber											

矿物质				维生素, mg/kg(标注单位者除外)				脂肪酸, 占乙醚提取物, %			
	$\bar{x}$	n	SD		$\bar{x}$	n	SD		$\bar{x}$	n	SD
常量元素 Macro,%				脂溶性维生素 Fat soluble				粗脂肪			
钙 Ca	2.82	3	0.28	β-胡萝卜素 β-Carotene				C-12:0			
氯 Cl				维生素 E Vitamin E				C-14:0			
钾 K				水溶性维生素 Water soluble				C-16:0			
镁 Mg				维生素 B_6 Vitamin B_6				C-16:1			
钠 Na				维生素 B_{12} Vitamin B_{12}, μg/kg				C-18:0			
磷 P	1.94	3	0.14	生物素 Biotin				C-18:1			
硫 S				叶酸 Folacin				C-18:2			
微量元素 Micro,ppm				烟酸 Niacin				C-18:3			
铬 Cr				泛酸 Pantothenic acid				C-18:4			
铜 Cu	35.70	1		核黄素 Riboflavin				C-20:0			
铁 Fe	230	1		硫胺素 Thiamin				C-20:1			
碘 I				胆碱 Choline				C-20:4			
锰 Mn	5.20	1						C-20:5			
硒 Se								C-22:0			
锌 Zn	99.40	1		能量, kcal/kg				C-22:1			
								C-22:5			
植酸磷 Phytate P,%				总能 GE				C-22:6			
磷全消化道表观消化率 ATTD of P, %	49	1		消化能 DE				C-24:0			
磷全消化道标准消化率 STTD of P, %	62	1		代谢能 ME				饱和脂肪酸			
				净能 NE				单不饱和脂肪酸			
								多不饱和脂肪酸			
								碘价			
								碘价物			

续表

原料名称：大米(Rice)
AAFCO#：无官方定义
IFN#：4-03-932

常规成分, %				氨基酸, %								
					总氨基酸			可消化氨基酸				
								回肠表观消化率			回肠标准消化率	
	$\bar{x}$	n	SD		$\bar{x}$	n	SD	$\bar{x}$	n	SD	$\bar{x}$	n	SD
干物质 Dry matter	87.78	6	2.20	必需氨基酸 Essential									
粗蛋白 Crude protein	7.87	9	1.04	粗蛋白 CP	7.87	9	1.04	80	2	5.66	94	1	
粗纤维 Crude fiber	0.52	5	0.25	精氨酸 Arg	0.44	3	0.05	88	3	1.00	93	3	1.15
乙醚浸出物/粗脂肪 Ether extract	1.30	4	0.47	组氨酸 His	0.33	3	0.17	80	3	4.51	85	3	2.65
酸乙醚浸出物 Acid ether extract	1.71	2	0.57	异亮氨酸 Ile	0.32	3	0.03	73	3	11.68	81	3	11.24
粗灰分 Ash	0.81	6	0.52	亮氨酸 Leu	0.56	3	0.06	77	3	6.56	83	3	6.03
碳水化合物成分, %				赖氨酸 Lys	0.35	3	0.12	80	3	3.00	89	3	3.79
				蛋氨酸 Met	0.25	3	0.19	85	2	9.19	87	2	9.90
乳糖 Lactose	0.00	4	0.00	苯丙氨酸 Phe	0.44	3	0.01	75	3	3.61	80	3	3.06
蔗糖 Sucrose	0.19	5	0.42	苏氨酸 Thr	0.23	3	0.04	72	3	2.52	85	3	6.66
棉子糖 Raffinose	0.00	4	0.00	色氨酸 Trp	0.11			63			77		
水苏糖 Stachyose	0.00	4	0.00	缬氨酸 Val	0.42	3	0.04	73	3	5.20	86	3	3.21
毛蕊花糖 Verbascose	0.00	4	0.00	非必需氨基酸 Nonessential									
低聚糖 Oligosaccharides				丙氨酸 Ala	0.34	3	0.05	72	3	6.03	74	3	6.03
淀粉 Starch	75.19	5	3.60	天冬氨酸 Asp	0.59	3	0.09	77	3	5.69	88	3	7.00
中性洗涤纤维 Neutral detergent fiber	1.28	4	0.95	半胱氨酸 Cys	0.18			63			77		
酸性洗涤纤维 Acid detergent fiber	0.64	3	0.14	谷氨酸 Glu	1.12	3	0.09	82	3	5.29	89	3	5.86
半纤维素 Hemicellulose				甘氨酸 Gly	0.31	3	0.05	73	3	4.73	77		
酸性洗涤木质素 Acid detergent lignin				脯氨酸 Pro	0.15	3	0.21	73	2	4.24	86		
总膳食纤维 Total dietary fiber				丝氨酸 Ser	0.28	3	0.06	74	3	7.21	92	3	10.00
不溶性膳食纤维 Insoluble dietary fiber				酪氨酸 Tyr	0.18	3	0.03	67	3	5.51	84		
可溶性膳食纤维 Soluble dietary fiber													

矿物质				维生素, mg/kg(标注单位者除外)				脂肪酸, 占乙醚提取物, %	
	$\bar{x}$	n	SD		$\bar{x}$	n	SD		
常量元素 Macro,%				脂溶性维生素 Fat soluble				粗脂肪	2.78
钙 Ca	0.09	1		β-胡萝卜素 β-Carotene				C-12:0	0.11
氯 Cl				维生素 E Vitamin E				C-14:0	0.36
钾 K				水溶性维生素 Water soluble				C-16:0	17.09
镁 Mg				维生素 B_6 Vitamin B_6				C-16:1	0.36
钠 Na				维生素 B_{12} Vitamin B_{12}, μg/kg				C-18:0	1.80
磷 P	0.34	2	0.19	生物素 Biotin				C-18:1	35.90
硫 S				叶酸 Folacin				C-18:2	34.42
微量元素 Micro,ppm				烟酸 Niacin				C-18:3	1.51
铬 Cr				泛酸 Pantothenic acid				C-18:4	
铜 Cu				核黄素 Riboflavin				C-20:0	0.00
铁 Fe				硫胺素 Thiamin				C-20:1	0.00
碘 I				胆碱 Choline				C-20:4	
锰 Mn								C-20:5	
硒 Se								C-22:0	
锌 Zn				能量, kcal/kg				C-22:1	
								C-22:5	
植酸磷 Phytate P,%	0.18	1		总能 GE	3723	4	49	C-22:6	
磷全消化道表观消化率 ATTD of P,%	29	1		消化能 DE	3681			C-24:0	
磷全消化道标准消化率 STTD of P,%	33	1		代谢能 ME	3627			饱和脂肪酸	19.35
				净能 NE	2881			单不饱和脂肪酸	36.26
								多不饱和脂肪酸	35.83
								碘价	98.86
								碘价物	27.48

续表

原料名称：米糠(Rice Bran)
AAFCO#: 75.7, AAFCO 2010, p.388
IFN#: 4-03-928

常规成分, %				氨基酸, %								
					总氨基酸			可消化氨基酸				
								回肠表观消化率			回肠标准消化率		
	$\bar{x}$	n	SD		$\bar{x}$	n	SD	$\bar{x}$	n	SD	$\bar{x}$	n	SD
干物质 Dry matter	91.60	3	1.5	必需氨基酸 Essential									
粗蛋白 Crude protein	15.11	3	1.28	粗蛋白 CP	15.11	3	1.28	57					
粗纤维 Crude fiber				精氨酸 Arg	1.24	3	0.08	85			89		
乙醚浸出物/粗脂肪 Ether extract	13.77			组氨酸 His	0.42	3	0.03	78			87		
酸乙醚浸出物 Acid ether extract				异亮氨酸 Ile	0.51	3	0.03	64			69		
粗灰分 Ash	14.80	3	4.82	亮氨酸 Leu	1.04	3	0.07	65			70		
碳水化合物成分, %				赖氨酸 Lys	0.67	3	0.03	72			78		
				蛋氨酸 Met	0.30	3	0.02	74			77		
乳糖 Lactose				苯丙氨酸 Phe	0.65	3	0.05	68			73		
蔗糖 Sucrose				苏氨酸 Thr	0.56	3	0.04	61			71		
棉子糖 Raffinose				色氨酸 Trp	0.19	3	0.01	64			73		
水苏糖 Stachyose				缬氨酸 Val	0.78	3	0.04	66			69		
毛蕊花糖 Verbascose				非必需氨基酸 Nonessential									
低聚糖 Oligosaccharides				丙氨酸 Ala	0.89	3	0.05	61			66		
淀粉 Starch	27.00			天冬氨酸 Asp	1.23	3	0.09	58			64		
中性洗涤纤维 Neutral detergent fiber	26.28	3	4.05	半胱氨酸 Cys	0.27	3	0.02	66			68		
酸性洗涤纤维 Acid detergent fiber	11.87			谷氨酸 Glu	1.95	3	0.22	66			71		
半纤维素 Hemicellulose				甘氨酸 Gly	0.81	3	0.05	48			59		
酸性洗涤木质素 Acid detergent lignin				脯氨酸 Pro	0.69	3	0.06	51			67		
总膳食纤维 Total dietary fiber				丝氨酸 Ser	0.69	3	0.05	60			69		
不溶性膳食纤维 Insoluble dietary fiber				酪氨酸 Tyr	0.40			77			81		
可溶性膳食纤维 Soluble dietary fiber													

矿物质				维生素, mg/kg(标注单位者除外)				脂肪酸, 占乙醚提取物, %			
	$\bar{x}$	n	SD		$\bar{x}$	n	SD		$\bar{x}$	n	SD
常量元素 Macro,%				脂溶性维生素 Fat soluble				粗脂肪	20.85		
钙 Ca	0.22	3	0.05	β-胡萝卜素 β-Carotene				C-12:0	0.09		
氯 Cl	0.07			维生素 E Vitamin E	9.7			C-14:0	0.37		
钾 K	1.56			水溶性维生素 Water soluble				C-16:0	17.06		
镁 Mg	0.90			维生素 B_6 Vitamin B_6	26.0			C-16:1	0.36		
钠 Na	0.03			维生素 B_{12} Vitamin B_{12}, μg/kg	0			C-18:0	1.79		
磷 P	2.16	4	0.32	生物素 Biotin	0.35			C-18:1	35.85		
硫 S	0.18			叶酸 Folacin	2.20			C-18:2	34.26		
微量元素 Micro,ppm				烟酸 Niacin	293			C-18:3	1.52		
铬 Cr				泛酸 Pantothenic acid	23.0			C-18:4			
铜 Cu	9.00			核黄素 Riboflavin	2.5			C-20:0	0.00		
铁 Fe	190			硫胺素 Thiamin	22.5			C-20:1	0.00		
碘 I				胆碱 Choline	1135			C-20:4			
锰 Mn	228							C-20:5			
硒 Se	0.40							C-22:0			
锌 Zn	30.00			能量, kcal/kg				C-22:1			
								C-22:5			
植酸磷 Phytate P,%	1.74	3	0.32	总能 GE	4772	3	299	C-22:6			
磷全消化道表观消化率 ATTD of P, %	13	4	1.24	消化能 DE	3100			C-24:0			
磷全消化道标准消化率 STTD of P, %	23	4	1.41	代谢能 ME	2997			饱和脂肪酸	19.31		
				净能 NE	2281			单不饱和脂肪酸	36.21		
								多不饱和脂肪酸	35.77		
								碘价	98.72		
								碘价物	205.83		

续表

原料名称：脱脂米糠(Rice Bran, Defatted)
AAFCO#:75.3, AAFCO 2010, p.388
IFN#: 4-03-930

常规成分, %				氨基酸, %								
				总氨基酸				可消化氨基酸				
								回肠表观消化率			回肠标准消化率	
	$\bar{x}$	n	SD		$\bar{x}$	n	SD	$\bar{x}$	n	SD	$\bar{x}$	n	SD
干物质 Dry matter	91.35	2	1.06	必需氨基酸 Essential									
粗蛋白 Crude protein	17.30	2	2.45	粗蛋白 CP	17.30	2	2.45						
粗纤维 Crude fiber				精氨酸 Arg	1.57	1					83		
乙醚浸出物/粗脂肪 Ether extract	3.52	2	0.28	组氨酸 His	0.55	1					75		
酸乙醚浸出物 Acid ether extract				异亮氨酸 Ile	0.62	1					75		
粗灰分 Ash	11.51	1		亮氨酸 Leu	1.25	1					75		
碳水化合物成分, %				赖氨酸 Lys	0.80	1					70		
				蛋氨酸 Met	0.36	1					78		
乳糖 Lactose				苯丙氨酸 Phe	0.78	1					74		
蔗糖 Sucrose				苏氨酸 Thr	0.68	1					69		
棉子糖 Raffinose				色氨酸 Trp	0.25	1					76		
水苏糖 Stachyose				缬氨酸 Val	0.94	1					73		
毛蕊花糖 Verbascose				非必需氨基酸 Nonessential									
低聚糖 Oligosaccharides				丙氨酸 Ala	1.11	1							
淀粉 Starch	26.25	1		天冬氨酸 Asp	1.59	1							
中性洗涤纤维 Neutral detergent fiber	23.56	1		半胱氨酸 Cys	0.36	1					63		
酸性洗涤纤维 Acid detergent fiber	1.31			谷氨酸 Glu	2.55	1							
半纤维素 Hemicellulose				甘氨酸 Gly	0.99	1							
酸性洗涤木质素 Acid detergent lignin				脯氨酸 Pro	0.81	1							
总膳食纤维 Total dietary fiber	25.79	1		丝氨酸 Ser	0.84	1							
不溶性膳食纤维 Insoluble dietary fiber				酪氨酸 Tyr	0.31						86		
可溶性膳食纤维 Soluble dietary fiber													

矿物质				维生素, mg/kg(标注单位者除外)				脂肪酸, 占乙醚提取物, %			
	$\bar{x}$	n	SD		$\bar{x}$	n	SD		$\bar{x}$	n	SD
常量元素 Macro,%				脂溶性维生素 Fat soluble				粗脂肪			
钙 Ca	0.1			β-胡萝卜素 β-Carotene				C-12:0			
氯 Cl				维生素 E Vitamin E				C-14:0			
钾 K				水溶性维生素 Water soluble				C-16:0			
镁 Mg				维生素 B_6 Vitamin B_6				C-16:1			
钠 Na				维生素 B_{12} Vitamin B_{12}, μg/kg				C-18:0			
磷 P	1.89			生物素 Biotin				C-18:1			
硫 S				叶酸 Folacin				C-18:2			
微量元素 Micro,ppm				烟酸 Niacin				C-18:3			
铬 Cr				泛酸 Pantothenic acid				C-18:4			
铜 Cu				核黄素 Riboflavin				C-20:0			
铁 Fe				硫胺素 Thiamin				C-20:1			
碘 I				胆碱 Choline				C-20:4			
锰 Mn								C-20:5			
硒 Se								C-22:0			
锌 Zn				能量, kcal/kg				C-22:1			
								C-22:5			
植酸磷 Phytate P,%				总能 GE	4056	1		C-22:6			
磷全消化道表观消化率 ATTD of P,%				消化能 DE	2199			C-24:0			
磷全消化道标准消化率 STTD of P,%	28			代谢能 ME	2081			饱和脂肪酸			
				净能 NE	1553			单不饱和脂肪酸			
								多不饱和脂肪酸			
								碘价			
								碘价物			

续表

原料名称：碎大米(Rice, Broken)
AAFCO#: 75.4, AAFCO 2010, p.388
IFN#: 4-03-932

常规成分, %				氨基酸, %								
				总氨基酸				可消化氨基酸				
								回肠表观消化率			回肠标准消化率		
	$\bar{x}$	n	SD		$\bar{x}$	n	SD	$\bar{x}$	n	SD	$\bar{x}$	n	SD
干物质 Dry matter	89.00			必需氨基酸 Essential									
粗蛋白 Crude protein	7.90			粗蛋白 CP	7.90								
粗纤维 Crude fiber				精氨酸 Arg	0.52						89		
乙醚浸出物/粗脂肪 Ether extract	1.30			组氨酸 His	0.18						84		
酸乙醚浸出物 Acid ether extract				异亮氨酸 Ile	0.34						81		
粗灰分 Ash				亮氨酸 Leu	0.67						83		
碳水化合物成分, %				赖氨酸 Lys	0.30						77		
				蛋氨酸 Met	0.18						85		
乳糖 Lactose				苯丙氨酸 Phe	0.39						84		
蔗糖 Sucrose				苏氨酸 Thr	0.26						76		
棉子糖 Raffinose				色氨酸 Trp	0.10						77		
水苏糖 Stachyose				缬氨酸 Val	0.49						78		
毛蕊花糖 Verbascose				非必需氨基酸 Nonessential									
低聚糖 Oligosaccharides				丙氨酸 Ala									
淀粉 Starch	75.19			天冬氨酸 Asp									
中性洗涤纤维 Neutral detergent fiber	12.20			半胱氨酸 Cys	0.11						73		
酸性洗涤纤维 Acid detergent fiber	6.40			谷氨酸 Glu									
半纤维素 Hemicellulose				甘氨酸 Gly									
酸性洗涤木质素 Acid detergent lignin				脯氨酸 Pro									
总膳食纤维 Total dietary fiber				丝氨酸 Ser									
不溶性膳食纤维 Insoluble dietary fiber				酪氨酸 Tyr	0.38						86		
可溶性膳食纤维 Soluble dietary fiber													

矿物质				维生素, mg/kg(标注单位者除外)				脂肪酸, 占乙醚提取物, %			
	$\bar{x}$	n	SD		$\bar{x}$	n	SD		$\bar{x}$	n	SD
常量元素 Macro,%				脂溶性维生素 Fat soluble				粗脂肪	1.20		
钙 Ca	0.04			β-胡萝卜素 β-Carotene				C-12:0	0.09		
氯 Cl	0.07			维生素 E Vitamin E	2.00			C-14:0	0.63		
钾 K	0.13			水溶性维生素 Water soluble				C-16:0	16.29		
镁 Mg	0.11			维生素 B_6 Vitamin B_6	28.00			C-16:1	0.27		
钠 Na	0.04			维生素 B_{12} Vitamin B_{12}, μg/kg	0.00			C-18:0	1.71		
磷 P	0.21	2	0.06	生物素 Biotin	0.08			C-18:1	36.18		
硫 S	0.06			叶酸 Folacin	0.20			C-18:2	32.31		
微量元素 Micro,ppm				烟酸 Niacin	25			C-18:3	1.35		
铬 Cr				泛酸 Pantothenic acid	3.30			C-18:4	0.00		
铜 Cu	21			核黄素 Riboflavin	0.40			C-20:0	0.18		
铁 Fe	18			硫胺素 Thiamin	1.40			C-20:1	0.00		
碘 I				胆碱 Choline	1003			C-20:4	0.00		
锰 Mn	12							C-20:5	0.00		
硒 Se	0.27							C-22:0	0.00		
锌 Zn	17			能量, kcal/kg				C-22:1	0.00		
								C-22:5	0.00		
植酸磷 Phytate P,%	0.14	1		总能 GE	4290			C-22:6	0.00		
磷全消化道表观消化率 ATTD of P,%	31	2	1.34	消化能 DE	3565			C-24:0	0.00		
磷全消化道标准消化率 STTD of P,%	38	2	2.76	代谢能 ME	3511			饱和脂肪酸	18.90		
				净能 NE	2778			单不饱和脂肪酸	36.45		
								多不饱和脂肪酸	33.66		
								碘价	94.95		
								碘价物	11.39		

续表

原料名称：抛光大米 (Rice, Polished)
AAFCO#：无官方定义
IFN#：4-03-932

常规成分, %				氨基酸, %								
				总氨基酸				可消化氨基酸				
								回肠表观消化率			回肠标准消化率		
	$\bar{x}$	n	SD		$\bar{x}$	n	SD	$\bar{x}$	n	SD	$\bar{x}$	n	SD
干物质 Dry matter	87.90	1		必需氨基酸 Essential									
粗蛋白 Crude protein	8.00	1		粗蛋白 CP	8.00	1							
粗纤维 Crude fiber				精氨酸 Arg	0.52								
乙醚浸出物/粗脂肪 Ether extract	1.41	1		组氨酸 His	0.18								
酸乙醚浸出物 Acid ether extract				异亮氨酸 Ile	0.34								
粗灰分 Ash				亮氨酸 Leu	0.67								
碳水化合物成分, %				赖氨酸 Lys	0.30								
				蛋氨酸 Met	0.18								
乳糖 Lactose				苯丙氨酸 Phe	0.39								
蔗糖 Sucrose				苏氨酸 Thr	0.26								
棉子糖 Raffinose				色氨酸 Trp	0.10								
水苏糖 Stachyose				缬氨酸 Val	0.49								
毛蕊花糖 Verbascose				非必需氨基酸 Nonessential									
低聚糖 Oligosaccharides				丙氨酸 Ala									
淀粉 Starch	83.59	1		天冬氨酸 Asp									
中性洗涤纤维 Neutral detergent fiber	12.2			半胱氨酸 Cys	0.11								
酸性洗涤纤维 Acid detergent fiber	3.10			谷氨酸 Glu									
半纤维素 Hemicellulose				甘氨酸 Gly									
酸性洗涤木质素 Acid detergent lignin				脯氨酸 Pro									
总膳食纤维 Total dietary fiber	1.32	1		丝氨酸 Ser									
不溶性膳食纤维 Insoluble dietary fiber				酪氨酸 Tyr	0.38								
可溶性膳食纤维 Soluble dietary fiber													

矿物质				维生素, mg/kg (标注单位者除外)				脂肪酸, 占乙醚提取物, %			
	$\bar{x}$	n	SD		$\bar{x}$	n	SD		$\bar{x}$	n	SD
常量元素 Macro, %				脂溶性维生素 Fat soluble				粗脂肪	1.42		
钙 Ca	0.04			β-胡萝卜素 β-Carotene				C-12:0	0.00		
氯 Cl	0.07			维生素 E Vitamin E	2.0			C-14:0	0.56		
钾 K	0.13			水溶性维生素 Water soluble				C-16:0	24.30		
镁 Mg	0.11			维生素 B_6 Vitamin B_6	28.0			C-16:1	0.35		
钠 Na	0.04			维生素 B_{12} Vitamin B_{12}, μg/kg	0			C-18:0	1.83		
磷 P	0.18			生物素 Biotin	0.08			C-18:1	30.70		
硫 S	0.06			叶酸 Folacin	0.20			C-18:2	22.04		
微量元素 Micro, ppm				烟酸 Niacin	25			C-18:3	4.72		
铬 Cr				泛酸 Pantothenic acid	3.3			C-18:4			
铜 Cu	21			核黄素 Riboflavin	0.4			C-20:0	0.00		
铁 Fe	18			硫胺素 Thiamin	1.4			C-20:1	0.00		
碘 I				胆碱 Choline	1003			C-20:4			
锰 Mn	12							C-20:5			
硒 Se	0.27							C-22:0			
锌 Zn	17			能量, kcal/kg				C-22:1			
								C-22:5			
植酸磷 Phytate P, %				总能 GE	4298			C-22:6			
磷全消化道表观消化率 ATTD of P, %				消化能 DE	3565			C-24:0			
磷全消化道标准消化率 STTD of P, %				代谢能 ME	3511			饱和脂肪酸	26.69		
				净能 NE	2847			单不饱和脂肪酸	31.06		
								多不饱和脂肪酸	26.76		
								碘价	80.74		
								碘价物	11.46		

续表

原料名称：大米浓缩蛋白 (Rice Protein Concentrate)

大米蛋白是生产大米淀粉的副产物，生产工艺类似优质小麦蛋白生产工艺。

AAFCO#：无官方定义

常规成分, %				氨基酸, %								
					总氨基酸			可消化氨基酸				
	$\bar{x}$	n	SD		$\bar{x}$	n	SD	回肠表观消化率			回肠标准消化率	
								$\bar{x}$	n	SD	$\bar{x}$	n	SD
干物质 Dry matter	92.68	1		必需氨基酸 Essential								
粗蛋白 Crude protein	67.51	1		粗蛋白 CP	67.51	1						
粗纤维 Crude fiber				精氨酸 Arg	5.26	1						
乙醚浸出物/粗脂肪 Ether extract	0.00			组氨酸 His	1.65	1						
酸乙醚浸出物 Acid ether extract				异亮氨酸 Ile	2.91	1						
粗灰分 Ash	3.41	1		亮氨酸 Leu	5.31	1						
碳水化合物成分, %				赖氨酸 Lys	2.21	1						
				蛋氨酸 Met	1.77	1						
乳糖 Lactose				苯丙氨酸 Phe	3.52	1						
蔗糖 Sucrose				苏氨酸 Thr	2.12	1						
棉子糖 Raffinose				色氨酸 Trp	0.81	1						
水苏糖 Stachyose				缬氨酸 Val	4.13	1						
毛蕊花糖 Verbascose				非必需氨基酸 Nonessential								
低聚糖 Oligosaccharides				丙氨酸 Ala	3.47	1						
淀粉 Starch	0.00			天冬氨酸 Asp	5.39	1						
中性洗涤纤维 Neutral detergent fiber				半胱氨酸 Cys	1.45	1						
酸性洗涤纤维 Acid detergent fiber	0.00			谷氨酸 Glu	10.87	1						
半纤维素 Hemicellulose				甘氨酸 Gly	2.77	1						
酸性洗涤木质素 Acid detergent lignin				脯氨酸 Pro	2.94	1						
总膳食纤维 Total dietary fiber				丝氨酸 Ser	2.36	1						
不溶性膳食纤维 Insoluble dietary fiber				酪氨酸 Tyr	3.32	1						
可溶性膳食纤维 Soluble dietary fiber												

矿物质				维生素, mg/kg (标注单位者除外)				脂肪酸, 占乙醚提取物, %			
	$\bar{x}$	n	SD		$\bar{x}$	n	SD		$\bar{x}$	n	SD
常量元素 Macro, %				脂溶性维生素 Fat soluble				粗脂肪			
钙 Ca	0.10	1		β-胡萝卜素 β-Carotene				C-12:0			
氯 Cl				维生素 E Vitamin E				C-14:0			
钾 K				水溶性维生素 Water soluble				C-16:0			
镁 Mg				维生素 B_6 Vitamin B_6				C-16:1			
钠 Na				维生素 B_{12} Vitamin B_{12}, μg/kg				C-18:0			
磷 P	0.75	1		生物素 Biotin				C-18:1			
硫 S				叶酸 Folacin				C-18:2			
微量元素 Micro, ppm				烟酸 Niacin				C-18:3			
铬 Cr				泛酸 Pantothenic acid				C-18:4			
铜 Cu				核黄素 Riboflavin				C-20:0			
铁 Fe				硫胺素 Thiamin				C-20:1			
碘 I				胆碱 Choline				C-20:4			
锰 Mn								C-20:5			
硒 Se								C-22:0			
锌 Zn								C-22:1			
				能量, kcal/kg				C-22:5			
植酸磷 Phytate P, %				总能 GE	4954	1		C-22:6			
磷全消化道表观消化率 ATTD of P, %				消化能 DE	4724	1		C-24:0			
磷全消化道标准消化率 STTD of P, %				代谢能 ME	4265			饱和脂肪酸			
				净能 NE	2692			单不饱和脂肪酸			
								多不饱和脂肪酸			
								碘价			
								碘价物			

续表

原料名称：黑麦(Rye)
AAFCO#：无官方定义
IFN#：4-04-047

常规成分, %				氨基酸, %								
				总氨基酸				可消化氨基酸				
								回肠表观消化率			回肠标准消化率	
	$\bar{x}$	n	SD		$\bar{x}$	n	SD	$\bar{x}$	n	SD	$\bar{x}$	n	SD
干物质 Dry matter	89.40	1		必需氨基酸 Essential									
粗蛋白 Crude protein	11.66	3	2.67	粗蛋白 CP	11.66	3	2.67	69	4	6.87	83	4	9.58
粗纤维 Crude fiber	2.71	2	1.12	精氨酸 Arg	0.70	2	0.27	73			79		
乙醚浸出物/粗脂肪 Ether extract	1.98	2	0.74	组氨酸 His	0.25	2	0.08	71			79		
酸乙醚浸出物 Acid ether extract				异亮氨酸 Ile	0.34	2	0.07	68			78		
粗灰分 Ash	1.78	1		亮氨酸 Leu	0.70	2	0.26	71			79		
碳水化合物成分, %				赖氨酸 Lys	0.43	2	0.10	64			74		
				蛋氨酸 Met	0.16	2	0.00	76			81		
乳糖 Lactose	0.00	1	0.00	苯丙氨酸 Phe	0.50	2	0.16	76			82		
蔗糖 Sucrose	0.00	1	0.00	苏氨酸 Thr	0.37	2	0.12	59			74		
棉子糖 Raffinose	0.00	1	0.00	色氨酸 Trp	0.10	1		67			76		
水苏糖 Stachyose	0.00	1	0.00	缬氨酸 Val	0.49	2	0.13	67			77		
毛蕊花糖 Verbascose	0.00	1	0.00	非必需氨基酸 Nonessential									
低聚糖 Oligosaccharides				丙氨酸 Ala	0.44	2	0.05	60			70		
淀粉 Starch	59.34	2	1.36	天冬氨酸 Asp	0.77	2	0.13	68			79		
中性洗涤纤维 Neutral detergent fiber	12.26	1		半胱氨酸 Cys	0.19	2	0.01	74			83		
酸性洗涤纤维 Acid detergent fiber	4.60			谷氨酸 Glu	2.63	2	0.74	89	2	0.21	93	2	0.25
半纤维素 Hemicellulose				甘氨酸 Gly	0.48	2	0.11	60			79		
酸性洗涤木质素 Acid detergent lignin	0.77	1		脯氨酸 Pro	1.57	1		86	1		98		
总膳食纤维 Total dietary fiber				丝氨酸 Ser	0.44	2	0.08	73			84		
不溶性膳食纤维 Insoluble dietary fiber				酪氨酸 Tyr	0.25	2	0.10	65			76		
可溶性膳食纤维 Soluble dietary fiber													

矿物质				维生素, mg/kg(标注单位者除外)				脂肪酸, 占乙醚提取物, %			
	$\bar{x}$	n	SD		$\bar{x}$	n	SD		$\bar{x}$	n	SD
常量元素 Macro,%				脂溶性维生素 Fat soluble				粗脂肪	2.50		
钙 Ca	0.08			β-胡萝卜素 β-Carotene				C-12:0	0.00		
氯 Cl	0.03			维生素 E Vitamin E	9.0			C-14:0	0.12		
钾 K	0.48			水溶性维生素 Water soluble				C-16:0	10.84		
镁 Mg	0.12			维生素 B_6 Vitamin B_6	2.6			C-16:1	0.40		
钠 Na	0.02			维生素 B_{12} Vitamin B_{12}, μg/kg	0			C-18:0	0.36		
磷 P	0.30			生物素 Biotin	0.08			C-18:1	11.20		
硫 S	0.15			叶酸 Folacin	0.60			C-18:2	38.32		
微量元素 Micro,ppm				烟酸 Niacin	19			C-18:3	6.28		
铬 Cr				泛酸 Pantothenic acid	8.0			C-18:4			
铜 Cu	7			核黄素 Riboflavin	1.6			C-20:0	0.00		
铁 Fe	60			硫胺素 Thiamin	3.6			C-20:1	0.52		
碘 I				胆碱 Choline	419			C-20:4			
锰 Mn	58							C-20:5			
硒 Se	0.38							C-22:0			
锌 Zn	31			能量, kcal/kg				C-22:1			
								C-22:5			
植酸磷 Phytate P,%	0.2			总能 GE	4350			C-22:6			
磷全消化道表观消化率 ATTD of P,%	43			消化能 DE	3270			C-24:0			
磷全消化道标准消化率 STTD of P,%	50			代谢能 ME	3191			饱和脂肪酸	11.32		
				净能 NE	2460			单不饱和脂肪酸	12.12		
								多不饱和脂肪酸	44.60		
								碘价	97.42		
								碘价物	24.35		

续表

原料名称：红花粕(Safflower Meal)
AAFCO#: 71.131, AAFCO 2010, p.386
IFN#: 5-04-110

常规成分, %				氨基酸, %								
					总氨基酸			可消化氨基酸				
								回肠表观消化率			回肠标准消化率	
	$\bar{x}$	n	SD		$\bar{x}$	n	SD	$\bar{x}$	n	SD	$\bar{x}$	n	SD
干物质 Dry matter	92.00			必需氨基酸 Essential									
粗蛋白 Crude protein	23.40			粗蛋白 CP	23.4								
粗纤维 Crude fiber				精氨酸 Arg	2.04						84		
乙醚浸出物/粗脂肪 Ether extract	2.24			组氨酸 His	0.59						84		
酸乙醚浸出物 Acid ether extract				异亮氨酸 Ile	0.67						87		
粗灰分 Ash				亮氨酸 Leu	1.52						87		
碳水化合物成分, %				赖氨酸 Lys	0.74						82		
				蛋氨酸 Met	0.34						84		
乳糖 Lactose				苯丙氨酸 Phe	1.07						90		
蔗糖 Sucrose				苏氨酸 Thr	0.65						79		
棉子糖 Raffinose				色氨酸 Trp	0.33						84		
水苏糖 Stachyose				缬氨酸 Val	1.18						88		
毛蕊花糖 Verbascose				非必需氨基酸 Nonessential									
低聚糖 Oligosaccharides				丙氨酸 Ala									
淀粉 Starch	0.90			天冬氨酸 Asp									
中性洗涤纤维 Neutral detergent fiber	55.9			半胱氨酸 Cys	0.38						84		
酸性洗涤纤维 Acid detergent fiber	36.56			谷氨酸 Glu									
半纤维素 Hemicellulose				甘氨酸 Gly									
酸性洗涤木质素 Acid detergent lignin				脯氨酸 Pro									
总膳食纤维 Total dietary fiber				丝氨酸 Ser									
不溶性膳食纤维 Insoluble dietary fiber				酪氨酸 Tyr	0.77								
可溶性膳食纤维 Soluble dietary fiber													

矿物质				维生素, mg/kg(标注单位者除外)				脂肪酸, 占乙醚提取物, %			
	$\bar{x}$	n	SD		$\bar{x}$	n	SD		$\bar{x}$	n	SD
常量元素 Macro,%				脂溶性维生素 Fat soluble				粗脂肪	2.39		
钙 Ca	0.34			β-胡萝卜素 β-Carotene				C-12:0	0.00		
氯 Cl	0.08			维生素 E Vitamin E	16.0			C-14:0	0.08		
钾 K	0.76			水溶性维生素 Water soluble				C-16:0	6.03		
镁 Mg	0.35			维生素 B_6 Vitamin B_6	12.0			C-16:1	0.08		
钠 Na	0.05			维生素 B_{12} Vitamin B_{12}, μg/kg	0			C-18:0	2.18		
磷 P	0.75			生物素 Biotin	1.03			C-18:1	11.30		
硫 S	0.13			叶酸 Folacin	0.50			C-18:2	65.94		
微量元素 Micro,ppm				烟酸 Niacin	11			C-18:3	0.25		
铬 Cr				泛酸 Pantothenic acid	33.9			C-18:4			
铜 Cu	10			核黄素 Riboflavin	3.3			C-20:0	0.00		
铁 Fe	495			硫胺素 Thiamin	4.6			C-20:1	0.00		
碘 I				胆碱 Choline	820			C-20:4			
锰 Mn	18							C-20:5			
硒 Se								C-22:0			
锌 Zn	41							C-22:1			
								C-22:5			
植酸磷 Phytate P,%				能量, kcal/kg				C-22:6			
磷全消化道表观消化率 ATTD of P,%				总能 GE	4589			C-24:0			
磷全消化道标准消化率 STTD of P,%				消化能 DE	2840			饱和脂肪酸	8.28		
				代谢能 ME	2681			单不饱和脂肪酸	11.38		
				净能 NE	1497			多不饱和脂肪酸	66.19		
								碘价	130.27		
								碘价物	31.13		

续表

原料名称：去壳红花粕(Safflower Meal, Dehulled)
AAFCO#：无官方定义
IFN#：5-07-959

常规成分, %				氨基酸, %							
				总氨基酸			可消化氨基酸				
							回肠表观消化率			回肠标准消化率		
	$\bar{x}$	n	SD		$\bar{x}$	n	SD	$\bar{x}$	n	SD	$\bar{x}$	n	SD
干物质 Dry matter	92.00			必需氨基酸 Essential								
粗蛋白 Crude protein	42.50			粗蛋白 CP	42.50							
粗纤维 Crude fiber				精氨酸 Arg	3.59							
乙醚浸出物/粗脂肪 Ether extract	1.30			组氨酸 His	1.07							
酸乙醚浸出物 Acid ether extract				异亮氨酸 Ile	1.69							
粗灰分 Ash				亮氨酸 Leu	2.57							
碳水化合物成分, %				赖氨酸 Lys	1.17							
				蛋氨酸 Met	0.66							
乳糖 Lactose				苯丙氨酸 Phe	2.00							
蔗糖 Sucrose				苏氨酸 Thr	1.28							
棉子糖 Raffinose				色氨酸 Trp	0.54							
水苏糖 Stachyose				缬氨酸 Val	2.33							
毛蕊花糖 Verbascose				非必需氨基酸 Nonessential								
低聚糖 Oligosaccharides				丙氨酸 Ala								
淀粉 Starch	1.40			天冬氨酸 Asp								
中性洗涤纤维 Neutral detergent fiber	25.9			半胱氨酸 Cys	0.69							
酸性洗涤纤维 Acid detergent fiber	18.0			谷氨酸 Glu								
半纤维素 Hemicellulose				甘氨酸 Gly								
酸性洗涤木质素 Acid detergent lignin				脯氨酸 Pro								
总膳食纤维 Total dietary fiber				丝氨酸 Ser								
不溶性膳食纤维 Insoluble dietary fiber				酪氨酸 Tyr	1.08							
可溶性膳食纤维 Soluble dietary fiber												

矿物质				维生素, mg/kg(标注单位者除外)				脂肪酸, 占乙醚提取物, %			
	$\bar{x}$	n	SD		$\bar{x}$	n	SD		$\bar{x}$	n	SD
常量元素 Macro,%				脂溶性维生素 Fat soluble				粗脂肪			
钙 Ca	0.37			β-胡萝卜素 β-Carotene				C-12:0			
氯 Cl	0.16			维生素 E Vitamin E	16.0			C-14:0			
钾 K	1.00			水溶性维生素 Water soluble				C-16:0			
镁 Mg	1.02			维生素 B_6 Vitamin B_6	11.3			C-16:1			
钠 Na	0.04			维生素 B_{12} Vitamin B_{12}, μg/kg	0			C-18:0			
磷 P	1.31			生物素 Biotin	1.03			C-18:1			
硫 S	0.20			叶酸 Folacin	1.60			C-18:2			
微量元素 Micro,ppm				烟酸 Niacin	22			C-18:3			
铬 Cr				泛酸 Pantothenic acid	39.1			C-18:4			
铜 Cu	9			核黄素 Riboflavin	2.4			C-20:0			
铁 Fe	484			硫胺素 Thiamin	4.5			C-20:1			
碘 I				胆碱 Choline	3248			C-20:4			
锰 Mn	39							C-20:5			
硒 Se								C-22:0			
锌 Zn	33			能量, kcal/kg				C-22:1			
								C-22:5			
植酸磷 Phytate P,%				总能 GE	4823			C-22:6			
磷全消化道表观消化率 ATTD of P,%				消化能 DE	3055			C-24:0			
磷全消化道标准消化率 STTD of P,%				代谢能 ME	2766			饱和脂肪酸			
				净能 NE	1623			单不饱和脂肪酸			
								多不饱和脂肪酸			
								碘价			
								碘价物			

续表

原料名称: 鲑鱼蛋白水解物(Salmon Protein Hydrolysate)
AAFCO#: 51.11, AAFCO 2010, p.359
IFN#: 5-18-778

常规成分, %				氨基酸, %								
				总氨基酸				可消化氨基酸				
								回肠表观消化率			回肠标准消化率		
	$\bar{x}$	n	SD		$\bar{x}$	n	SD	$\bar{x}$	n	SD	$\bar{x}$	n	SD
干物质 Dry matter	91.99	2	0.78	必需氨基酸 Essential									
粗蛋白 Crude protein	90.79	2	2.69	粗蛋白 CP	90.79	2	2.69						
粗纤维 Crude fiber				精氨酸 Arg	5.33	2	0.19						
乙醚浸出物/粗脂肪 Ether extract	2.12	1		组氨酸 His	1.55	2	0.05						
酸乙醚浸出物 Acid ether extract				异亮氨酸 Ile	2.11	2	0.06						
粗灰分 Ash	4.77	2	2.93	亮氨酸 Leu	3.97	2	0						
碳水化合物成分, %				赖氨酸 Lys	4.96	2	0.12						
				蛋氨酸 Met	1.84	2	0.06						
乳糖 Lactose				苯丙氨酸 Phe	2.08	2	0.02						
蔗糖 Sucrose				苏氨酸 Thr	2.68	2	0.08						
棉子糖 Raffinose				色氨酸 Trp	0.44	2	0.06						
水苏糖 Stachyose				缬氨酸 Val	2.69	2	0.12						
毛蕊花糖 Verbascose				非必需氨基酸 Nonessential									
低聚糖 Oligosaccharides				丙氨酸 Ala	5.77	2	0.22						
淀粉 Starch				天冬氨酸 Asp	6.05	2	0.18						
中性洗涤纤维 Neutral detergent fiber				半胱氨酸 Cys	0.41	2	0.01						
酸性洗涤纤维 Acid detergent fiber				谷氨酸 Glu	9.82	2	0.26						
半纤维素 Hemicellulose				甘氨酸 Gly	11.18	2	1.14						
酸性洗涤木质素 Acid detergent lignin				脯氨酸 Pro	5.74	2	0.61						
总膳食纤维 Total dietary fiber				丝氨酸 Ser	2.85	2	0.35						
不溶性膳食纤维 Insoluble dietary fiber				酪氨酸 Tyr	1.34	2	0.03						
可溶性膳食纤维 Soluble dietary fiber													

矿物质				维生素, mg/kg (标注单位者除外)				脂肪酸, 占乙醚提取物, %			
	$\bar{x}$	n	SD		$\bar{x}$	n	SD		$\bar{x}$	n	SD
常量元素 Macro,%				脂溶性维生素 Fat soluble				粗脂肪			
钙 Ca	0.09	2	0.05	β-胡萝卜素 β-Carotene				C-12:0			
氯 Cl				维生素 E Vitamin E				C-14:0			
钾 K	1.79	1		水溶性维生素 Water soluble				C-16:0			
镁 Mg	0.07	1		维生素 B_6 Vitamin B_6				C-16:1			
钠 Na				维生素 B_{12} Vitamin B_{12}, μg/kg				C-18:0			
磷 P	0.84	2	0.27	生物素 Biotin				C-18:1			
硫 S				叶酸 Folacin				C-18:2			
微量元素 Micro,ppm				烟酸 Niacin				C-18:3			
铬 Cr				泛酸 Pantothenic acid				C-18:4			
铜 Cu				核黄素 Riboflavin				C-20:0			
铁 Fe	6.29	1		硫胺素 Thiamin				C-20:1			
碘 I				胆碱 Choline				C-20:4			
锰 Mn								C-20:5			
硒 Se								C-22:0			
锌 Zn	54.13	1						C-22:1			
				能量, kcal/kg				C-22:5			
植酸磷 Phytate P,%				总能 GE	4713	2	135	C-22:6			
磷全消化道表观消化率 ATTD of P, %				消化能 DE	4173	1		C-24:0			
磷全消化道标准消化率 STTD of P, %				代谢能 ME	3556			饱和脂肪酸			
				净能 NE	2129			单不饱和脂肪酸			
								多不饱和脂肪酸			
								碘价			
								碘价物			

续表

原料名称：芝麻粕(Sesame Meal)
AAFCO#：无官方定义
IFN#:5-04-220

常规成分, %				氨基酸, %								
				总氨基酸				可消化氨基酸				
								回肠表观消化率		回肠标准消化率		
	$\bar{x}$	n	SD		$\bar{x}$	n	SD	$\bar{x}$	n	SD	$\bar{x}$	n	SD
干物质 Dry matter	93.00			必需氨基酸 Essential									
粗蛋白 Crude protein	42.60			粗蛋白 CP	42.60			81					
粗纤维 Crude fiber				精氨酸 Arg	4.86			94			96		
乙醚浸出物/粗脂肪 Ether extract	7.50			组氨酸 His	0.98			76			84		
酸乙醚浸出物 Acid ether extract				异亮氨酸 Ile	1.47			85			87		
粗灰分 Ash				亮氨酸 Leu	2.74			85			92		
碳水化合物成分, %				赖氨酸 Lys	1.01			76			85		
				蛋氨酸 Met	1.15			90			92		
乳糖 Lactose				苯丙氨酸 Phe	1.77			89			93		
蔗糖 Sucrose				苏氨酸 Thr	1.44			78			90		
棉子糖 Raffinose				色氨酸 Trp	0.54			85			85		
水苏糖 Stachyose				缬氨酸 Val	1.87			84			89		
毛蕊花糖 Verbascose				非必需氨基酸 Nonessential									
低聚糖 Oligosaccharides				丙氨酸 Ala	1.62			82			84		
淀粉 Starch	1.80			天冬氨酸 Asp	2.30			82			84		
中性洗涤纤维 Neutral detergent fiber	18.00			半胱氨酸 Cys	0.82			86			92		
酸性洗涤纤维 Acid detergent fiber	13.20			谷氨酸 Glu	6.53			83			84		
半纤维素 Hemicellulose				甘氨酸 Gly	1.65			80			84		
酸性洗涤木质素 Acid detergent lignin				脯氨酸 Pro	1.23			78			84		
总膳食纤维 Total dietary fiber				丝氨酸 Ser	1.50			81			84		
不溶性膳食纤维 Insoluble dietary fiber				酪氨酸 Tyr	1.52			87			91		
可溶性膳食纤维 Soluble dietary fiber													

矿物质				维生素, mg/kg(标注单位者除外)				脂肪酸,占乙醚提取物, %			
	$\bar{x}$	n	SD		$\bar{x}$	n	SD		$\bar{x}$	n	SD
常量元素 Macro,%				脂溶性维生素 Fat soluble				粗脂肪			
钙 Ca	1.70			β-胡萝卜素 β-Carotene	0.2			C-12:0	0.00		
氯 Cl	0.07			维生素 E Vitamin E	1.0			C-14:0	0.25		
钾 K	1.10			水溶性维生素 Water soluble				C-16:0	8.94		
镁 Mg	0.54			维生素 B_6 Vitamin B_6	12.5			C-16:1	0.30		
钠 Na	0.04			维生素 B_{12} Vitamin B_{12}, μg/kg	0			C-18:0	4.21		
磷 P	1.18			生物素 Biotin	0.24			C-18:1	37.29		
硫 S	0.56			叶酸 Folacin				C-18:2	43.03		
微量元素 Micro,ppm				烟酸 Niacin	30			C-18:3	0.76		
铬 Cr				泛酸 Pantothenic acid	6.0			C-18:4			
铜 Cu	34			核黄素 Riboflavin	3.6			C-20:0	0.00		
铁 Fe	93			硫胺素 Thiamin	2.8			C-20:1	0.14		
碘 I				胆碱 Choline	1536			C-20:4			
锰 Mn	53							C-20:5			
硒 Se	0.21							C-22:0			
锌 Zn	100							C-22:1			
				能量, kcal/kg				C-22:5			
植酸磷 Phytate P,%	0.89			总能 GE	4702			C-22:6			
磷全消化道表观消化率 ATTD of P,%	29			消化能 DE	3350			C-24:0			
磷全消化道标准消化率 STTD of P,%	42			代谢能 ME	3060			饱和脂肪酸	13.40		
				净能 NE	1972			单不饱和脂肪酸	37.73		
								多不饱和脂肪酸	43.79		
								碘价	113.86		
								碘价物	85.40		

第17章 饲料原料成分

续表

原料名称：高粱(Sorghum)
AAFCO#: 42.1, AAFCO 2010, p.354
IFN#: 4-04-379

常规成分, %				氨基酸, %								
				总氨基酸				可消化氨基酸				
								回肠表观消化率			回肠标准消化率		
	$\bar{x}$	n	SD	必需氨基酸 Essential	$\bar{x}$	n	SD	$\bar{x}$	n	SD	$\bar{x}$	n	SD
干物质 Dry matter	89.39	26	2.63										
粗蛋白 Crude protein	9.36	29	1.10	粗蛋白 CP	9.36	29	1.10	63	16	6.96	77	16	7.29
粗纤维 Crude fiber	2.14	4	0.16	精氨酸 Arg	0.36	22	0.05	68	16	10.02	80	16	10.39
乙醚浸出物/粗脂肪 Ether extract	3.42	6	0.43	组氨酸 His	0.21	21	0.03	64	15	8.32	74	15	8.19
酸乙醚浸出物 Acid ether extract	2.12	3	1.18	异亮氨酸 Ile	0.36	22	0.05	69	16	6.19	78	16	6.15
粗灰分 Ash	1.64	17	0.34	亮氨酸 Leu	1.21	22	0.15	78	16	4.77	83	16	5.06
碳水化合物成分, %				赖氨酸 Lys	0.20	22	0.05	53	16	11.87	74	16	12.44
				蛋氨酸 Met	0.16	20	0.03	74	16	6.99	79	16	7.16
乳糖 Lactose	0.00	4	0.00	苯丙氨酸 Phe	0.48	19	0.06	76	15	5.51	83	15	6.14
蔗糖 Sucrose	0.00	4	0.00	苏氨酸 Thr	0.30	22	0.04	54	16	9.35	75	16	8.48
棉子糖 Raffinose	0.00	4	0.00	色氨酸 Trp	0.07	18	0.02	65	14	8.88	74	2	24.75
水苏糖 Stachyose	0.00	4	0.00	缬氨酸 Val	0.46	22	0.06	66	16	7.08	77	16	7.38
毛蕊花糖 Verbascose	0.00	4	0.00	非必需氨基酸 Nonessential									
低聚糖 Oligosaccharides				丙氨酸 Ala	0.84	20	0.10	73	16	5.02	79	16	5.07
淀粉 Starch	70.05	5	8.71	天冬氨酸 Asp	0.60	20	0.10	66	16	6.43	79	16	7.08
中性洗涤纤维 Neutral detergent fiber	10.63	16	3.28	半胱氨酸 Cys	0.18	20	0.02	56	16	9.26	67	16	9.06
酸性洗涤纤维 Acid detergent fiber	4.93	16	1.48	谷氨酸 Glu	1.84	20	0.27	74	16	15.45	81	16	8.70
半纤维素 Hemicellulose				甘氨酸 Gly	0.31	20	0.04	34	16	17.67	67	16	19.01
酸性洗涤木质素 Acid detergent lignin	0.44	1		脯氨酸 Pro	0.74	19	0.10	46	15	22.86	74	15	29.54
总膳食纤维 Total dietary fiber	4.35	1		丝氨酸 Ser	0.39	20	0.05	66	16	6.02	81	16	6.23
不溶性膳食纤维 Insoluble dietary fiber				酪氨酸 Tyr	0.32	19	0.05	69	15	6.56	75	15	7.71
可溶性膳食纤维 Soluble dietary fiber													

矿物质				维生素, mg/kg(标注单位者除外)				脂肪酸, 占乙醚提取物, %			
	$\bar{x}$	n	SD		$\bar{x}$	n	SD		$\bar{x}$	n	SD
常量元素 Macro,%				脂溶性维生素 Fat soluble				粗脂肪	3.30		
钙 Ca	0.02	9	0.01	β-胡萝卜素 β-Carotene				C-12:0	0.21		
氯 Cl	0.09			维生素 E Vitamin E	5.0			C-14:0	0.27		
钾 K	0.35			水溶性维生素 Water soluble				C-16:0	12.33		
镁 Mg	0.15			维生素 B_6 Vitamin B_6	5.2			C-16:1	0.88		
钠 Na	0.01			维生素 B_{12} Vitamin B_{12}, μg/kg	0			C-18:0	1.06		
磷 P	0.27	10	0.06	生物素 Biotin	0.26			C-18:1	29.21		
硫 S	0.08			叶酸 Folacin	0.17			C-18:2	39.55		
微量元素 Micro,ppm				烟酸 Niacin	41			C-18:3	1.97		
铬 Cr				泛酸 Pantothenic acid	12.4			C-18:4			
铜 Cu	5.00			核黄素 Riboflavin	1.3			C-20:0	0.00		
铁 Fe	45			硫胺素 Thiamin	3.0			C-20:1	0.00		
碘 I				胆碱 Choline	668			C-20:4			
锰 Mn	15.00							C-20:5			
硒 Se	0.20							C-22:0			
锌 Zn	15.00			能量, kcal/kg				C-22:1			
								C-22:5			
植酸磷 Phytate P,%	0.18	2	0.05	总能 GE	3881	4	49	C-22:6			
磷全消化道表观消化率 ATTD of P, %	30	4	7.24	消化能 DE	3596	3	17	C-24:0			
磷全消化道标准消化率 STTD of P, %	40	4	7.33	代谢能 ME	3532			饱和脂肪酸	13.88		
				净能 NE	2780			单不饱和脂肪酸	30.09		
								多不饱和脂肪酸	41.52		
								碘价	104.08		
								碘价物	34.35		

续表

原料名称：高粱酒精糟及可溶物(Sorghum, DDGS)
AAFCO#: 27.6, AAFCO 2010, p.343
IFN#: 5-04-375

常规成分, %				氨基酸, %								
				总氨基酸				可消化氨基酸				
								回肠表观消化率			回肠标准消化率	
	$\bar{x}$	n	SD		$\bar{x}$	n	SD	$\bar{x}$	n	SD	$\bar{x}$	n	SD
干物质 Dry matter	89.84	4	1.69	必需氨基酸 Essential									
粗蛋白 Crude protein	30.80	4	1.34	粗蛋白 CP	30.80	4	1.34	65	1		73	1	
粗纤维 Crude fiber	7.06	2	0.22	精氨酸 Arg	1.10	1		70	1		79	1	
乙醚浸出物/粗脂肪 Ether extract	9.75	4	1.69	组氨酸 His	0.71	1		69	1		72	1	
酸乙醚浸出物 Acid ether extract				异亮氨酸 Ile	1.29	3	0.06	72	1		74	1	
粗灰分 Ash	6.62	3	4.57	亮氨酸 Leu	4.01	3	0.14	76	1		77	1	
碳水化合物成分, %				赖氨酸 Lys	0.82	3	0.14	59	1		64	1	
				蛋氨酸 Met	0.54	3	0.04	75	1		77	1	
乳糖 Lactose				苯丙氨酸 Phe	1.68	1		74	1		77	1	
蔗糖 Sucrose				苏氨酸 Thr	1.06	3	0.03	64	1		70	1	
棉子糖 Raffinose				色氨酸 Trp	0.25	3	0.09	67	1		72	1	
水苏糖 Stachyose				缬氨酸 Val	1.65	3	0.03	71	1		74	1	
毛蕊花糖 Verbascose				非必需氨基酸 Nonessential									
低聚糖 Oligosaccharides				丙氨酸 Ala	2.90	1		72	1		75	1	
淀粉 Starch	0.00			天冬氨酸 Asp	2.17	1		65	1		69	1	
中性洗涤纤维 Neutral detergent fiber	33.60	4	6.17	半胱氨酸 Cys	0.53	3	0.05	63	1		67	1	
酸性洗涤纤维 Acid detergent fiber	22.68	4	3.44	谷氨酸 Glu	6.31	1		75	1		77	1	
半纤维素 Hemicellulose				甘氨酸 Gly	1.03	1		41	1		69	1	
酸性洗涤木质素 Acid detergent lignin				脯氨酸 Pro	2.50	1		35	1		74	1	
总膳食纤维 Total dietary fiber				丝氨酸 Ser	1.40	1		68	1		78	1	
不溶性膳食纤维 Insoluble dietary fiber				酪氨酸 Tyr									
可溶性膳食纤维 Soluble dietary fiber													

矿物质				维生素, mg/kg(标注单位者除外)				脂肪酸, 占乙醚提取物, %			
	$\bar{x}$	n	SD		$\bar{x}$	n	SD		$\bar{x}$	n	SD
常量元素 Macro,%				脂溶性维生素 Fat soluble				粗脂肪			
钙 Ca	0.12	1		β-胡萝卜素 β-Carotene				C-12:0			
氯 Cl				维生素 E Vitamin E				C-14:0			
钾 K				水溶性维生素 Water soluble				C-16:0			
镁 Mg				维生素 B_6 Vitamin B_6				C-16:1			
钠 Na				维生素 B_{12} Vitamin B_{12}, μg/kg				C-18:0			
磷 P	0.76	1		生物素 Biotin				C-18:1			
硫 S				叶酸 Folacin				C-18:2			
微量元素 Micro,ppm				烟酸 Niacin				C-18:3			
铬 Cr				泛酸 Pantothenic acid				C-18:4			
铜 Cu				核黄素 Riboflavin				C-20:0			
铁 Fe				硫胺素 Thiamin				C-20:1			
碘 I				胆碱 Choline				C-20:4			
锰 Mn								C-20:5			
硒 Se								C-22:0			
锌 Zn								C-22:1			
				能量, kcal/kg				C-22:5			
植酸磷 Phytate P,%				总能 GE	4860			C-22:6			
磷全消化道表观消化率 ATTD of P,%				消化能 DE	3878			C-24:0			
磷全消化道标准消化率 STTD of P,%				代谢能 ME	3669			饱和脂肪酸			
				净能 NE	2394			单不饱和脂肪酸			
								多不饱和脂肪酸			
								碘价			
								碘价物			

续表

原料名称：大豆皮 (Soybean Hulls)
AAFCO#: 84.3, AAFCO 2010, p.390
IFN#: 1-04-560

常规成分, %	$\bar{x}$	n	SD	氨基酸, %								
				总氨基酸				可消化氨基酸				
								回肠表观消化率			回肠标准消化率		
					$\bar{x}$	n	SD	$\bar{x}$	n	SD	$\bar{x}$	n	SD
干物质 Dry matter	90.59	4	1.23	必需氨基酸 Essential									
粗蛋白 Crude protein	10.27	7	1.45	粗蛋白 CP	10.27	7	1.45	44					
粗纤维 Crude fiber	35.75	2	5.15	精氨酸 Arg	0.60	1		74			82		
乙醚浸出物/粗脂肪 Ether extract	1.29	2	0.71	组氨酸 His	0.29	1		47			56		
酸乙醚浸出物 Acid ether extract	1.71	1		异亮氨酸 Ile	0.38	2	0.09	56			67		
粗灰分 Ash	4.46	4	0.31	亮氨酸 Leu	0.76	1		59			68		
碳水化合物成分, %				赖氨酸 Lys	0.66	2	0.08	51			58		
				蛋氨酸 Met	0.14	2	0.04	60			70		
乳糖 Lactose				苯丙氨酸 Phe	0.46	1		62			71		
蔗糖 Sucrose				苏氨酸 Thr	0.39	2	0.09	47			62		
棉子糖 Raffinose				色氨酸 Trp	0.09	2	0.01	49			62		
水苏糖 Stachyose				缬氨酸 Val	0.51	2	0.09	50			61		
毛蕊花糖 Verbascose				非必需氨基酸 Nonessential									
低聚糖 Oligosaccharides				丙氨酸 Ala	0.48	1		44			54		
淀粉 Starch	3.65			天冬氨酸 Asp	1.20	1		47			54		
中性洗涤纤维 Neutral detergent fiber	59.39	7	4.7	半胱氨酸 Cys	0.20	2	0.05	51			63		
酸性洗涤纤维 Acid detergent fiber	41.55	6	1.93	谷氨酸 Glu	1.30	1		45			54		
半纤维素 Hemicellulose				甘氨酸 Gly	0.82	1		43			54		
酸性洗涤木质素 Acid detergent lignin				脯氨酸 Pro	0.47	1		34			54		
总膳食纤维 Total dietary fiber				丝氨酸 Ser	0.62	1		43			54		
不溶性膳食纤维 Insoluble dietary fiber				酪氨酸 Tyr	0.51	1		56			63		
可溶性膳食纤维 Soluble dietary fiber													

矿物质	$\bar{x}$	n	SD	维生素, mg/kg (标注单位者除外)	$\bar{x}$	n	SD	脂肪酸, 占乙醚提取物, %	$\bar{x}$	n	SD
常量元素 Macro, %				脂溶性维生素 Fat soluble				粗脂肪	2.20		
钙 Ca	0.54	2	0.07	β-胡萝卜素 β-Carotene				C-12:0	0.00		
氯 Cl				维生素 E Vitamin E				C-14:0	0.10		
钾 K				水溶性维生素 Water soluble				C-16:0	9.98		
镁 Mg				维生素 B_6 Vitamin B_6				C-16:1	0.19		
钠 Na				维生素 B_{12} Vitamin B_{12}, μg/kg				C-18:0	3.61		
磷 P	0.12	2	0.04	生物素 Biotin				C-18:1	20.62		
硫 S				叶酸 Folacin				C-18:2	50.45		
微量元素 Micro, ppm				烟酸 Niacin				C-18:3	7.03		
铬 Cr				泛酸 Pantothenic acid				C-18:4	0.00		
铜 Cu				核黄素 Riboflavin				C-20:0	0.00		
铁 Fe				硫胺素 Thiamin				C-20:1	0.00		
碘 I				胆碱 Choline				C-20:4	0.00		
锰 Mn								C-20:5	0.00		
硒 Se								C-22:0	0.00		
锌 Zn				能量, kcal/kg				C-22:1	0.00		
								C-22:5	0.00		
植酸磷 Phytate P, %	0.08			总能 GE	4210	1		C-22:6	0.00		
磷全消化道表观消化率 ATTD of P, %	20			消化能 DE	2008			C-24:0	0.00		
磷全消化道标准消化率 STTD of P, %	33			代谢能 ME	1938			饱和脂肪酸	13.68		
				净能 NE	989			单不饱和脂肪酸	20.81		
								多不饱和脂肪酸	57.48		
								碘价	129.24		
								碘价物	28.43		

续表

原料名称：压榨去皮豆粕(Soybean Meal, Dehulled, Expelled)
AAFCO#: 84.71, AAFCO 2010, p.392

常规成分, %				氨基酸, %								
				总氨基酸				可消化氨基酸				
	$\bar{x}$	n	SD		$\bar{x}$	n	SD	回肠表观消化率			回肠标准消化率	
干物质 Dry matter	95.57	4	1.56	必需氨基酸 Essential				$\bar{x}$	n	SD	$\bar{x}$	n	SD
粗蛋白 Crude protein	45.13	4	3.60	粗蛋白 CP	45.13	4	3.60	81	3	4.20	89	3	0.55
粗纤维 Crude fiber	3.30	1		精氨酸 Arg	3.02	4	0.4	90	3	3.07	95	3	0.39
乙醚浸出物/粗脂肪 Ether extract	6.64	2	1.10	组氨酸 His	1.14	4	0.15	86	3	3.39	90	3	1.49
酸乙醚浸出物 Acid ether extract				异亮氨酸 Ile	1.90	4	0.33	85	3	3.94	89	3	1.81
粗灰分 Ash	6.24	1		亮氨酸 Leu	3.21	4	0.51	85	3	3.35	88	3	1.77
碳水化合物成分, %				赖氨酸 Lys	2.79	4	0.22	86	3	4.05	90	3	2.45
				蛋氨酸 Met	0.60	4	0.07	80	3	7.82	85	3	4.84
乳糖 Lactose				苯丙氨酸 Phe	2.15	4	0.31	86	3	3.52	89	3	2.51
蔗糖 Sucrose				苏氨酸 Thr	1.73	4	0.14	76	3	3.62	84	3	1.82
棉子糖 Raffinose				色氨酸 Trp	0.69	2	0.04	87	1		89	1	
水苏糖 Stachyose				缬氨酸 Val	2.01	4	0.36	83	3	4.00	88	3	1.49
毛蕊花糖 Verbascose				非必需氨基酸 Nonessential									
低聚糖 Oligosaccharides				丙氨酸 Ala	1.88	3	0.29	81	3	3.59	88	3	0.62
淀粉 Starch	1.89			天冬氨酸 Asp	4.73	3	0.75	85	3	4.16	88	3	2.62
中性洗涤纤维 Neutral detergent fiber				半胱氨酸 Cys	0.72	3	0.1	79	3	3.48	87	3	1.84
酸性洗涤纤维 Acid detergent fiber	6.33			谷氨酸 Glu	7.35	3	1.19	88	3	3.83	91	3	2.33
半纤维素 Hemicellulose				甘氨酸 Gly	1.82	3	0.22	72	3	4.17	91	3	6.31
酸性洗涤木质素 Acid detergent lignin				脯氨酸 Pro	2.16	3	0.25	70	3	11.60	131	3	23.54
总膳食纤维 Total dietary fiber				丝氨酸 Ser	2.11	3	0.07	82	3	2.75	88	3	1.10
不溶性膳食纤维 Insoluble dietary fiber				酪氨酸 Tyr	1.47	3	0.27	84	3	2.63	87	3	1.26
可溶性膳食纤维 Soluble dietary fiber													

矿物质				维生素, mg/kg(标注单位者除外)				脂肪酸, 占乙醚提取物, %			
	$\bar{x}$	n	SD		$\bar{x}$	n	SD		$\bar{x}$	n	SD
常量元素 Macro, %				脂溶性维生素 Fat soluble				粗脂肪	1.90		
钙 Ca				β-胡萝卜素 β-Carotene				C-12:0	0.00		
氯 Cl				维生素 E Vitamin E				C-14:0	0.08		
钾 K				水溶性维生素 Water soluble				C-16:0	7.88		
镁 Mg				维生素 B_6 Vitamin B_6				C-16:1	0.15		
钠 Na				维生素 B_{12} Vitamin B_{12}, μg/kg				C-18:0	2.85		
磷 P				生物素 Biotin				C-18:1	16.28		
硫 S				叶酸 Folacin				C-18:2	39.83		
微量元素 Micro, ppm				烟酸 Niacin				C-18:3	5.55		
铬 Cr				泛酸 Pantothenic acid				C-18:4	0.00		
铜 Cu				核黄素 Riboflavin				C-20:0	0.00		
铁 Fe				硫胺素 Thiamin				C-20:1	0.00		
碘 I				胆碱 Choline				C-20:4	0.00		
锰 Mn								C-20:5	0.00		
硒 Se								C-22:0	0.00		
锌 Zn				能量, kcal/kg				C-22:1	0.00		
								C-22:5	0.00		
植酸磷 Phytate P, %				总能 GE	4710	1		C-22:6	0.00		
磷全消化道表观消化率 ATTD of P, %				消化能 DE	4210	1		C-24:0	0.00		
磷全消化道标准消化率 STTD of P, %				代谢能 ME	3903			饱和脂肪酸	10.80		
				净能 NE	2598			单不饱和脂肪酸	16.43		
								多不饱和脂肪酸	45.38		
								碘价	102.03		
								碘价物	19.39		

续表

原料名称：溶剂浸提去皮豆粕(Soybean Meal, Dehulled, Solvent Extracted)
AAFCO#: 84.7, AAFCO 2010, p.391
IFN#: 5-04-612

常规成分, %	$\bar{x}$	n	SD	氨基酸, %				可消化氨基酸				
				总氨基酸				回肠表观消化率			回肠标准消化率	
					$\bar{x}$	n	SD	$\bar{x}$	n	SD	$\bar{x}$	n	SD
干物质 Dry matter	89.98	101	2.62	必需氨基酸 Essential									
粗蛋白 Crude protein	47.73	154	2.30	粗蛋白 CP	47.73	154	2.30	82	69	5.03	87	68	4.48
粗纤维 Crude fiber	3.89	38	1.60	精氨酸 Arg	3.45	107	0.26	92	83	3.22	94	83	3.12
乙醚浸出物/粗脂肪 Ether extract	1.52	70	0.91	组氨酸 His	1.28	104	0.10	87	82	4.30	90	82	4.15
酸乙醚浸出物 Acid ether extract	2.86	6	0.96	异亮氨酸 Ile	2.14	113	0.18	87	83	3.96	89	82	3.79
粗灰分 Ash	6.27	56	0.51	亮氨酸 Leu	3.62	107	0.27	86	83	3.58	88	83	3.45
碳水化合物成分, %				赖氨酸 Lys	2.96	118	0.19	87	83	3.38	89	83	3.44
				蛋氨酸 Met	0.66	112	0.08	88	77	4.82	90	77	4.70
乳糖 Lactose	0.00	7	0.00	苯丙氨酸 Phe	2.40	105	0.19	86	82	3.85	88	82	3.65
蔗糖 Sucrose	4.30	19	3.60	苏氨酸 Thr	1.86	117	0.11	80	83	4.59	85	83	4.47
棉子糖 Raffinose	3.78	19	14.25	色氨酸 Trp	0.66	87	0.08	88	59	4.23	91	59	3.32
水苏糖 Stachyose	7.33	19	19.54	缬氨酸 Val	2.23	115	0.19	83	83	4.53	87	83	4.16
毛蕊花糖 Verbascose	0.00	7	0.00	非必需氨基酸 Nonessential									
低聚糖 Oligosaccharides	3.81	3	0.16	丙氨酸 Ala	2.06	80	0.16	80	61	5.37	85	61	5.94
淀粉 Starch	1.89			天冬氨酸 Asp	5.41	81	0.46	85	60	3.61	87	60	3.42
中性洗涤纤维 Neutral detergent fiber	8.21	32	2.90	半胱氨酸 Cys	0.70	98	0.08	79	74	4.64	84	74	4.55
酸性洗涤纤维 Acid detergent fiber	5.28	30	2.43	谷氨酸 Glu	8.54	80	1.19	87	61	4.01	89	61	4.24
半纤维素 Hemicellulose	3.90	6	0.48	甘氨酸 Gly	1.99	78	0.20	75	61	7.41	84	61	6.38
酸性洗涤木质素 Acid detergent lignin	1.10	1		脯氨酸 Pro	2.53	63	0.41	79	51	10.99	113	51	85.14
总膳食纤维 Total dietary fiber	16.71	8	3.47	丝氨酸 Ser	2.36	81	0.23	84	61	4.64	89	61	5.62
不溶性膳食纤维 Insoluble dietary fiber				酪氨酸 Tyr	1.59	86	0.20	84	59	5.15	88	56	4.70
可溶性膳食纤维 Soluble dietary fiber													

矿物质	$\bar{x}$	n	SD	维生素, mg/kg(标注单位者除外)	$\bar{x}$	n	SD	脂肪酸, 占乙醚提取物, %	$\bar{x}$	n	SD
常量元素 Macro,%				脂溶性维生素 Fat soluble				粗脂肪	1.50		
钙 Ca	0.33	65	0.10	β-胡萝卜素 β-Carotene	0.2			C-12:0	0.00		
氯 Cl	0.49	9	0.12	维生素 E Vitamin E	0.07	3	0.01	C-14:0	0.08		
钾 K	2.24	15	0.12	水溶性维生素 Water soluble				C-16:0	7.88		
镁 Mg	0.27	13	0.01	维生素 B_6 Vitamin B_6	6.4			C-16:1	0.15		
钠 Na	0.08	5	0.05	维生素 B_{12} Vitamin B_{12}, μg/kg	0			C-18:0	2.85		
磷 P	0.71	73	0.09	生物素 Biotin	0.26			C-18:1	16.28		
硫 S	0.40	10	0.04	叶酸 Folacin	1.37			C-18:2	39.83		
微量元素 Micro,ppm				烟酸 Niacin	22			C-18:3	5.55		
铬 Cr				泛酸 Pantothenic acid	15.0			C-18:4	0.00		
铜 Cu	15.13	15	1.30	核黄素 Riboflavin	3.1			C-20:0	0.00		
铁 Fe	98.19	11	42.43	硫胺素 Thiamin	3.2			C-20:1	0.00		
碘 I				胆碱 Choline	2731			C-20:4	0.00		
锰 Mn	35.49	14	5.56					C-20:5	0.00		
硒 Se	0.27							C-22:0	0.00		
锌 Zn	48.81	15	9.39	能量, kcal/kg				C-22:1	0.00		
								C-22:5	0.00		
植酸磷 Phytate P,%	0.38	20	0.07	总能 GE	4256	42	192	C-22:6	0.00		
磷全消化道表观消化率 ATTD of P, %	39	20	6.24	消化能 DE	3619	3	184	C-24:0	0.00		
磷全消化道标准消化率 STTD of P, %	48	20	7.62	代谢能 ME	3294			饱和脂肪酸	10.80		
				净能 NE	2087			单不饱和脂肪酸	16.43		
								多不饱和脂肪酸	45.38		
								碘价	102.03		
								碘价物	15.30		

续表

原料名称：酶处理豆粕(Soybean Meal, Enzyme Treated)
AAFCO#: 84.63, AAFCO 2010, p.392

常规成分, %				氨基酸, %							
				总氨基酸				可消化氨基酸			
	$\bar{x}$	n	SD		$\bar{x}$	n	SD	回肠表观消化率		回肠标准消化率	
								$\bar{x}$	n	SD	$\bar{x}$	n	SD
干物质 Dry matter	92.70	4	0.84	必需氨基酸 Essential							
粗蛋白 Crude protein	55.62	4	2.11	粗蛋白 CP	55.62	4	2.11	82	4	3.95	88	4	2.94
粗纤维 Crude fiber	4.06	4	0.94	精氨酸 Arg	3.95	4	0.19	92	5	1.81	96	5	2.94
乙醚浸出物/粗脂肪 Ether extract	1.82	4	0.48	组氨酸 His	1.41	4	0.06	87	5	3.13	90	5	4.71
酸乙醚浸出物 Acid ether extract				异亮氨酸 Ile	2.48	4	0.15	86	5	3.17	89	5	3.65
粗灰分 Ash	7.05	3	0.06	亮氨酸 Leu	4.09	4	0.19	86	5	3.75	89	5	4.42
碳水化合物成分, %				赖氨酸 Lys	3.20	4	0.13	83	5	3.30	86	5	3.78
				蛋氨酸 Met	0.71	4	0.03	88	4	2.10	91	4	1.89
乳糖 Lactose				苯丙氨酸 Phe	2.78	4	0.15	83	5	6.07	86	5	7.81
蔗糖 Sucrose				苏氨酸 Thr	2.13	4	0.13	78	5	5.24	83	5	5.93
棉子糖 Raffinose				色氨酸 Trp	0.72	4	0.04	80	4	3.32	83	4	4.41
水苏糖 Stachyose				缬氨酸 Val	2.57	4	0.17	84	5	5.06	89	5	5.33
毛蕊花糖 Verbascose				非必需氨基酸 Nonessential							
低聚糖 Oligosaccharides				丙氨酸 Ala	2.41	4	0.14	82	4	3.58	86	4	2.61
淀粉 Starch				天冬氨酸 Asp	6.14	4	0.40	83	4	2.57	86	4	2.91
中性洗涤纤维 Neutral detergent fiber				半胱氨酸 Cys	0.78	4	0.04	68	4	7.51	73	4	10.43
酸性洗涤纤维 Acid detergent fiber				谷氨酸 Glu	9.62	4	0.75	86	4	4.61	88	4	5.49
半纤维素 Hemicellulose				甘氨酸 Gly	2.32	4	0.08	76	4	11.21	89	4	3.79
酸性洗涤木质素 Acid detergent lignin				脯氨酸 Pro	2.73	4	0.20	73	4	19.92	112	4	24.08
总膳食纤维 Total dietary fiber				丝氨酸 Ser	2.66	4	0.25	83	4	3.89	87	4	3.30
不溶性膳食纤维 Insoluble dietary fiber				酪氨酸 Tyr	2.03	1		86	1		92	1	
可溶性膳食纤维 Soluble dietary fiber											

矿物质				维生素, mg/kg(标注单位者除外)				脂肪酸, 占乙醚提取物, %			
	$\bar{x}$	n	SD		$\bar{x}$	n	SD		$\bar{x}$	n	SD
常量元素 Macro,%				脂溶性维生素 Fat soluble				粗脂肪			
钙 Ca	0.31	3	0.04	β-胡萝卜素 β-Carotene				C-12:0			
氯 Cl				维生素 E Vitamin E				C-14:0			
钾 K				水溶性维生素 Water soluble				C-16:0			
镁 Mg				维生素 B_6 Vitamin B_6				C-16:1			
钠 Na				维生素 B_{12} Vitamin B_{12}, μg/kg				C-18:0			
磷 P	0.75	3	0.02	生物素 Biotin				C-18:1			
硫 S				叶酸 Folacin				C-18:2			
微量元素 Micro,ppm				烟酸 Niacin				C-18:3			
铬 Cr				泛酸 Pantothenic acid				C-18:4			
铜 Cu				核黄素 Riboflavin				C-20:0			
铁 Fe				硫胺素 Thiamin				C-20:1			
碘 I				胆碱 Choline				C-20:4			
锰 Mn								C-20:5			
硒 Se								C-22:0			
锌 Zn				能量, kcal/kg				C-22:1			
								C-22:5			
植酸磷 Phytate P,%				总能 GE	4451	3	20	C-22:6			
磷全消化道表观消化率 ATTD of P,%	60	1		消化能 DE	3914	2	37	C-24:0			
磷全消化道标准消化率 STTD of P,%	66	1		代谢能 ME	3536			饱和脂肪酸			
				净能 NE				单不饱和脂肪酸			
								多不饱和脂肪酸			
								碘价			
								碘价物			

续表

原料名称：压榨豆粕 (Soybean Meal, Expelled)
AAFCO#: 84.6, AAFCO 2010, p.391
IFN#: 5-04-600

常规成分, %				氨基酸, %								
					总氨基酸			可消化氨基酸				
								回肠表观消化率			回肠标准消化率		
	$\bar{x}$	n	SD	必需氨基酸 Essential	$\bar{x}$	n	SD	$\bar{x}$	n	SD	$\bar{x}$	n	SD
干物质 Dry matter	93.85	6	3.56										
粗蛋白 Crude protein	44.56	7	2.15	粗蛋白 CP	44.56	7	2.15	84	2	3.39	89	2	0.38
粗纤维 Crude fiber	5.60	2	1.13	精氨酸 Arg	3.13	6	0.46	93	2	0.78	96	2	0.82
乙醚浸出物/粗脂肪 Ether extract	5.69	5	1.3	组氨酸 His	1.17	6	0.12	88	2	1.63	91	2	0.29
酸乙醚浸出物 Acid ether extract	9.87	2	0.51	异亮氨酸 Ile	1.97	6	0.29	88	2	1.91	91	2	0.62
粗灰分 Ash	5.70	3	0.28	亮氨酸 Leu	3.29	6	0.39	88	2	0.64	89		
碳水化合物成分, %				赖氨酸 Lys	2.85	6	0.35	89	2	3.18	90		
				蛋氨酸 Met	0.56	6	0.11	88	2	0.71	91	2	0.26
乳糖 Lactose				苯丙氨酸 Phe	2.19	6	0.22	89	2	1.27	90		
蔗糖 Sucrose	7.10	1		苏氨酸 Thr	1.73	6	0.07	79	2	0.49	85		
棉子糖 Raffinose	0.77	1		色氨酸 Trp	0.67	3	0.07	88	2	0.92	89		
水苏糖 Stachyose	4.88	1		缬氨酸 Val	2.06	6	0.29	86	2	2.69	88		
毛蕊花糖 Verbascose				非必需氨基酸 Nonessential									
低聚糖 Oligosaccharides				丙氨酸 Ala	1.89	6	0.16	83	2	2.76	88	2	0.03
淀粉 Starch	1.89			天冬氨酸 Asp	4.84	6	0.47	86	2	4.31	88	2	2.93
中性洗涤纤维 Neutral detergent fiber	13.84	3	1.4	半胱氨酸 Cys	0.70	5	0.07	78	2	6.22	83		
酸性洗涤纤维 Acid detergent fiber	7.35	3	0.74	谷氨酸 Glu	7.56	6	0.77	88	2	5.80	90	2	4.66
半纤维素 Hemicellulose				甘氨酸 Gly	1.89	6	0.18	71	2	3.54	84	2	3.64
酸性洗涤木质素 Acid detergent lignin				脯氨酸 Pro	2.16	5	0.18	81	2	1.56	111	2	12.57
总膳食纤维 Total dietary fiber				丝氨酸 Ser	2.11	6	0.05	85	2	0.14	89		1.92
不溶性膳食纤维 Insoluble dietary fiber				酪氨酸 Tyr	1.50	6	0.17	87	2	0.64	89	2	1.19
可溶性膳食纤维 Soluble dietary fiber													

矿物质				维生素, mg/kg (标注单位者除外)				脂肪酸, 占乙醚提取物, %			
	$\bar{x}$	n	SD		$\bar{x}$	n	SD		$\bar{x}$	n	SD
常量元素 Macro,%				脂溶性维生素 Fat soluble				粗脂肪	1.90		
钙 Ca	0.28	1		β-胡萝卜素 β-Carotene				C-12:0	0.00		
氯 Cl				维生素 E Vitamin E				C-14:0	0.08		
钾 K				水溶性维生素 Water soluble				C-16:0	7.88		
镁 Mg				维生素 B_6 Vitamin B_6				C-16:1	0.15		
钠 Na				维生素 B_{12} Vitamin B_{12}, μg/kg				C-18:0	2.85		
磷 P	0.66	1		生物素 Biotin				C-18:1	16.28		
硫 S				叶酸 Folacin				C-18:2	39.83		
微量元素 Micro,ppm				烟酸 Niacin				C-18:3	5.55		
铬 Cr				泛酸 Pantothenic acid				C-18:4	0.00		
铜 Cu				核黄素 Riboflavin				C-20:0	0.00		
铁 Fe				硫胺素 Thiamin				C-20:1	0.00		
碘 I				胆碱 Choline				C-20:4			
锰 Mn								C-20:5	0.00		
硒 Se								C-22:0	0.00		
锌 Zn				能量, kcal/kg				C-22:1	0.00		
								C-22:5	0.00		
植酸磷 Phytate P,%				总能 GE	4692	3	29	C-22:6	0.00		
磷全消化道表观消化率 ATTD of P, %	39			消化能 DE	3876	2	345	C-24:0	0.00		
磷全消化道标准消化率 STTD of P, %	48			代谢能 ME	3573			饱和脂肪酸	10.80		
				净能 NE	2344			单不饱和脂肪酸	16.43		
								多不饱和脂肪酸	45.38		
								碘价	102.03		
								碘价物	19.39		

续表

原料名称：发酵豆粕(Soybean Meal, Fermented)
AAFCO#: 无官方定义

常规成分, %				氨基酸, %								
				总氨基酸			可消化氨基酸					
							回肠表观消化率			回肠标准消化率		
	$\bar{x}$	n	SD		$\bar{x}$	n	SD	$\bar{x}$	n	SD	$\bar{x}$	n	SD
干物质 Dry matter	92.88	3	2.80	必需氨基酸 Essential									
粗蛋白 Crude protein	54.07	4	2.67	粗蛋白 CP	54.07	4	2.67	72	2	2.76	79	2	3.46
粗纤维 Crude fiber	3.46	2	0.21	精氨酸 Arg	3.70	3	0.21	87	2	1.06	90	2	4.18
乙醚浸出物/粗脂肪 Ether extract	2.30	2	2.12	组氨酸 His	1.37	3	0.08	79	2	2.19	81	2	4.32
酸乙醚浸出物 Acid ether extract				异亮氨酸 Ile	2.55	3	0.12	79	2	3.46	82	2	5.17
粗灰分 Ash	6.98	1		亮氨酸 Leu	4.25	3	0.26	79	2	3.11	82	2	4.96
碳水化合物成分, %				赖氨酸 Lys	3.14	4	0.22	72	2	1.20	75	2	3.30
				蛋氨酸 Met	0.75	4	0.04	85	2	1.63	88	2	0.42
乳糖 Lactose				苯丙氨酸 Phe	2.87	3	0.22	77	2	7.42	80	2	9.82
蔗糖 Sucrose				苏氨酸 Thr	2.09	4	0.10	68	2	2.12	73	2	7.01
棉子糖 Raffinose				色氨酸 Trp	0.69	4	0.04	75	2	5.73	78	2	7.71
水苏糖 Stachyose				缬氨酸 Val	2.67	3	0.17	75	2	1.98	80	2	5.79
毛蕊花糖 Verbascose				非必需氨基酸 Nonessential									
低聚糖 Oligosaccharides				丙氨酸 Ala	2.45	3	0.14	74	2	1.48	79	2	2.35
淀粉 Starch				天冬氨酸 Asp	5.98	2	0.44	75	2	3.11	78	2	5.17
中性洗涤纤维 Neutral detergent fiber				半胱氨酸 Cys	0.77	3	0.02	58	2	4.60	64	2	8.57
酸性洗涤纤维 Acid detergent fiber				谷氨酸 Glu	9.12	2	0.80	76	2	7.00	78	2	8.43
半纤维素 Hemicellulose				甘氨酸 Gly	2.34	3	0.12	60	2	13.86	75	2	1.78
酸性洗涤木质素 Acid detergent lignin				脯氨酸 Pro	2.74	3	0.27	63	2	10.04	109	2	31.33
总膳食纤维 Total dietary fiber				丝氨酸 Ser	2.51	3	0.25	75	2	0.92	80	2	2.95
不溶性膳食纤维 Insoluble dietary fiber				酪氨酸 Tyr	2.08	2	0.15	84	1		88	1	
可溶性膳食纤维 Soluble dietary fiber													

矿物质				维生素, mg/kg(标注单位者除外)				脂肪酸, 占乙醚提取物, %			
	$\bar{x}$	n	SD		$\bar{x}$	n	SD		$\bar{x}$	n	SD
常量元素 Macro,%				脂溶性维生素 Fat soluble				粗脂肪			
钙 Ca	0.29	2	0.00	β-胡萝卜素 β-Carotene				C-12:0			
氯 Cl				维生素 E Vitamin E				C-14:0			
钾 K				水溶性维生素 Water soluble				C-16:0			
镁 Mg				维生素 B_6 Vitamin B_6				C-16:1			
钠 Na				维生素 B_{12} Vitamin B_{12}, μg/kg				C-18:0			
磷 P	0.80	2	0.03	生物素 Biotin				C-18:1			
硫 S				叶酸 Folacin				C-18:2			
微量元素 Micro,ppm				烟酸 Niacin				C-18:3			
铬 Cr				泛酸 Pantothenic acid				C-18:4			
铜 Cu				核黄素 Riboflavin				C-20:0			
铁 Fe				硫胺素 Thiamin				C-20:1			
碘 I				胆碱 Choline				C-20:4			
锰 Mn								C-20:5			
硒 Se								C-22:0			
锌 Zn				能量, kcal/kg				C-22:1			
								C-22:5			
植酸磷 Phytate P,%				总能 GE	4533	1		C-22:6			
磷全消化道表观消化率 ATTD of P,%	60			消化能 DE	3975	1		C-24:0			
磷全消化道标准消化率 STTD of P,%	66	1		代谢能 ME	3607			饱和脂肪酸			
				净能 NE				单不饱和脂肪酸			
								多不饱和脂肪酸			
								碘价			
								碘价物			

续表

原料名称：溶剂浸提去皮高蛋白豆粕 (Soybean Meal, High Protein, Dehulled, Solvent Extracted)
AAFCO#: 84.7, AAFCO 2010, p.391
IFN#: 5-04-612

常规成分, %				氨基酸, %									
					总氨基酸			可消化氨基酸					
								回肠表观消化率			回肠标准消化率		
	$\bar{x}$	n	SD		$\bar{x}$	n	SD	$\bar{x}$	n	SD	$\bar{x}$	n	SD
干物质 Dry matter	88.66	2	0.77	必需氨基酸 Essential									
粗蛋白 Crude protein	51.17	3	3.88	粗蛋白 CP	51.17	3	3.88	80	3	0.58	85	3	2.97
粗纤维 Crude fiber	3.62	1		精氨酸 Arg	3.78	3	0.45	90	3	1.21	92	3	2.43
乙醚浸出物/粗脂肪 Ether extract	1.10	2	1.13	组氨酸 His	1.31	3	0.23	86	3	0.23	88	3	1.71
酸乙醚浸出物 Acid ether extract				异亮氨酸 Ile	2.36	3	0.18	82	3	2.31	84	3	3.37
粗灰分 Ash	6.19	1		亮氨酸 Leu	3.87	3	0.38	83	3	1.73	85	3	2.83
碳水化合物成分, %				赖氨酸 Lys	3.11	3	0.35	85	3	1.27	87	3	2.76
				蛋氨酸 Met	0.68	3	0.09	87	3	0.87	89	3	0.06
乳糖 Lactose				苯丙氨酸 Phe	2.59	3	0.22	82	3	3.29	84	3	4.27
蔗糖 Sucrose	4.28	1		苏氨酸 Thr	1.92	3	0.15	76	3	1.42	81	3	4.09
棉子糖 Raffinose	0.68	1		色氨酸 Trp	0.68	3	0.06	83	3	1.80	85	3	3.75
水苏糖 Stachyose	3.12	1		缬氨酸 Val	2.48	3	0.23	81	3	1.10	84	3	2.50
毛蕊花糖 Verbascose				非必需氨基酸 Nonessential									
低聚糖 Oligosaccharides				丙氨酸 Ala	2.16	3	0.16	80	3	1.79	84	3	0.61
淀粉 Starch	1.89			天冬氨酸 Asp	5.81	3	0.57	82	3	1.44	84	3	2.35
中性洗涤纤维 Neutral detergent fiber	5.50	1		半胱氨酸 Cys	0.79	3	0.10	74	3	2.70	78	3	4.63
酸性洗涤纤维 Acid detergent fiber	2.95	1		谷氨酸 Glu	9.18	3	1.04	85	3	0.24	87	3	0.77
半纤维素 Hemicellulose				甘氨酸 Gly	2.13	3	0.19	77	3	5.83	88	3	0.50
酸性洗涤木质素 Acid detergent lignin				脯氨酸 Pro	2.84	3	0.01	81	3	2.60	104	3	10.96
总膳食纤维 Total dietary fiber				丝氨酸 Ser	2.42	3	0.19	81	3	2.48	84	3	4.37
不溶性膳食纤维 Insoluble dietary fiber				酪氨酸 Tyr	1.98	1		84	1		88	1	
可溶性膳食纤维 Soluble dietary fiber													

矿物质				维生素, mg/kg (标注单位者除外)				脂肪酸, 占乙醚提取物, %			
	$\bar{x}$	n	SD		$\bar{x}$	n	SD		$\bar{x}$	n	SD
常量元素 Macro, %				脂溶性维生素 Fat soluble				粗脂肪			
钙 Ca	0.56	1		β-胡萝卜素 β-Carotene				C-12:0			
氯 Cl				维生素 E Vitamin E				C-14:0			
钾 K				水溶性维生素 Water soluble				C-16:0			
镁 Mg				维生素 B_6 Vitamin B_6				C-16:1			
钠 Na				维生素 B_{12} Vitamin B_{12}, μg/kg				C-18:0			
磷 P	0.77	1		生物素 Biotin				C-18:1			
硫 S				叶酸 Folacin				C-18:2			
微量元素 Micro, ppm				烟酸 Niacin				C-18:3			
铬 Cr				泛酸 Pantothenic acid				C-18:4			
铜 Cu				核黄素 Riboflavin				C-20:0			
铁 Fe				硫胺素 Thiamin				C-20:1			
碘 I				胆碱 Choline				C-20:4			
锰 Mn								C-20:5			
硒 Se								C-22:0			
锌 Zn								C-22:1			
				能量, kcal/kg				C-22:5			
植酸磷 Phytate P, %				总能 GE	4378	2	177	C-22:6			
磷全消化道表观消化率 ATTD of P, %	39			消化能 DE	3717	1		C-24:0			
磷全消化道标准消化率 STTD of P, %	48			代谢能 ME	3369			饱和脂肪酸			
				净能 NE	2137			单不饱和脂肪酸			
								多不饱和脂肪酸			
								碘价			
								碘价物			

续表

原料名称：压榨高蛋白豆粕(Soybean Meal, High Protein, Expelled)
AAFCO#: 84.6, AAFCO 2010, p.391
IFN#: 5-04-600

常规成分, %				氨基酸, %								
				总氨基酸				可消化氨基酸				
								回肠表观消化率			回肠标准消化率		
	$\bar{x}$	n	SD		$\bar{x}$	n	SD	$\bar{x}$	n	SD	$\bar{x}$	n	SD
干物质 Dry matter	94.50	1		必需氨基酸 Essential									
粗蛋白 Crude protein	55.97	1		粗蛋白 CP	55.97	1		83	1		91	1	
粗纤维 Crude fiber				精氨酸 Arg	4.13	1		93	1		97	1	
乙醚浸出物/粗脂肪 Ether extract	5.13	1		组氨酸 His	1.39	1		89	1		93	1	
酸乙醚浸出物 Acid ether extract				异亮氨酸 Ile	2.42	1		89	1		92	1	
粗灰分 Ash				亮氨酸 Leu	4.09	1		89	1		92	1	
碳水化合物成分, %				赖氨酸 Lys	3.33	1		89	1		93	1	
				蛋氨酸 Met	0.72	1		89	1		92	1	
乳糖 Lactose				苯丙氨酸 Phe	2.71	1		90	1		93	1	
蔗糖 Sucrose	4.91	1		苏氨酸 Thr	1.96	1		81	1		89	1	
棉子糖 Raffinose	0.67	1		色氨酸 Trp	0.71	1		87	1		92	1	
水苏糖 Stachyose	4.58	1		缬氨酸 Val	2.59	1		86	1		91	1	
毛蕊花糖 Verbascose				非必需氨基酸 Nonessential									
低聚糖 Oligosaccharides				丙氨酸 Ala	2.21	1		83	1		91	1	
淀粉 Starch	1.89			天冬氨酸 Asp	6.10	1		86	1		89	1	
中性洗涤纤维 Neutral detergent fiber	9.99	1		半胱氨酸 Cys	0.80	1		77	1		84	1	
酸性洗涤纤维 Acid detergent fiber	6.30	1		谷氨酸 Glu	9.82	1		87	1		89	1	
半纤维素 Hemicellulose				甘氨酸 Gly	2.27	1		73	1		82	1	
酸性洗涤木质素 Acid detergent lignin				脯氨酸 Pro	2.74	1		80	1		121	1	
总膳食纤维 Total dietary fiber				丝氨酸 Ser	2.50	1		86	1		92	1	
不溶性膳食纤维 Insoluble dietary fiber				酪氨酸 Tyr	1.88	1		87	1		91	1	
可溶性膳食纤维 Soluble dietary fiber													

矿物质				维生素, mg/kg(标注单位者除外)				脂肪酸, 占乙醚提取物, %			
	$\bar{x}$	n	SD		$\bar{x}$	n	SD		$\bar{x}$	n	SD
常量元素 Macro,%				脂溶性维生素 Fat soluble				粗脂肪			
钙 Ca	0.29	1		β-胡萝卜素 β-Carotene				C-12:0			
氯 Cl				维生素 E Vitamin E				C-14:0			
钾 K				水溶性维生素 Water soluble				C-16:0			
镁 Mg				维生素 B_6 Vitamin B_6				C-16:1			
钠 Na				维生素 B_{12} Vitamin B_{12}, μg/kg				C-18:0			
磷 P	0.63	1		生物素 Biotin				C-18:1			
硫 S				叶酸 Folacin				C-18:2			
微量元素 Micro,ppm				烟酸 Niacin				C-18:3			
铬 Cr				泛酸 Pantothenic acid				C-18:4			
铜 Cu				核黄素 Riboflavin				C-20:0			
铁 Fe				硫胺素 Thiamin				C-20:1			
碘 I				胆碱 Choline				C-20:4			
锰 Mn								C-20:5			
硒 Se								C-22:0			
锌 Zn				能量, kcal/kg				C-22:1			
								C-22:5			
植酸磷 Phytate P,%				总能 GE	4784	1		C-22:6			
磷全消化道表观消化率 ATTD of P,%	39			消化能 DE	3717	1		C-24:0			
磷全消化道标准消化率 STTD of P,%	48			代谢能 ME	3336			饱和脂肪酸			
				净能 NE	2129			单不饱和脂肪酸			
								多不饱和脂肪酸			
								碘价			
								碘价物			

续表

原料名称：溶剂浸提去皮低寡聚糖豆粕(Soybean Meal, Low Oligosaccharide, Dehulled, Solvent Extracted)
AAFCO#: 84.7, AAFCO 2010, p.391
IFN#: 5-04-612

常规成分, %				氨基酸, %								
				总氨基酸				可消化氨基酸				
	$\bar{x}$	n	SD		$\bar{x}$	n	SD	回肠表观消化率			回肠标准消化率	
干物质 Dry matter	88.64	7	1.84	必需氨基酸 Essential				$\bar{x}$	n	SD	$\bar{x}$	n	SD
粗蛋白 Crude protein	51.84	7	1.96	粗蛋白 CP	51.84	7	1.96	82	69	5.03	87	68	4.48
粗纤维 Crude fiber	3.10	7	0.3	精氨酸 Arg									
乙醚浸出物/粗脂肪 Ether extract	1.14	7	0.23	组氨酸 His									
酸乙醚浸出物 Acid ether extract				异亮氨酸 Ile									
粗灰分 Ash	6.70	7	0.27	亮氨酸 Leu									
碳水化合物成分, %				赖氨酸 Lys									
				蛋氨酸 Met									
乳糖 Lactose				苯丙氨酸 Phe									
蔗糖 Sucrose	6.38	6	1.25	苏氨酸 Thr									
棉子糖 Raffinose	0.13	7	0.1	色氨酸 Trp									
水苏糖 Stachyose	0.43	7	0.36	缬氨酸 Val									
毛蕊花糖 Verbascose				非必需氨基酸 Nonessential									
低聚糖 Oligosaccharides	0.50	5	0.44	丙氨酸 Ala									
淀粉 Starch				天冬氨酸 Asp									
中性洗涤纤维 Neutral detergent fiber	6.30	2	0.71	半胱氨酸 Cys									
酸性洗涤纤维 Acid detergent fiber	2.55	2	0.21	谷氨酸 Glu									
半纤维素 Hemicellulose	3.75	2	0.92	甘氨酸 Gly									
酸性洗涤木质素 Acid detergent lignin				脯氨酸 Pro									
总膳食纤维 Total dietary fiber				丝氨酸 Ser									
不溶性膳食纤维 Insoluble dietary fiber				酪氨酸 Tyr									
可溶性膳食纤维 Soluble dietary fiber													

矿物质				维生素, mg/kg(标注单位者除外)				脂肪酸, 占乙醚提取物, %			
	$\bar{x}$	n	SD		$\bar{x}$	n	SD		$\bar{x}$	n	SD
常量元素 Macro,%				脂溶性维生素 Fat soluble				粗脂肪			
钙 Ca				β-胡萝卜素 β-Carotene				C-12:0			
氯 Cl				维生素 E Vitamin E				C-14:0			
钾 K				水溶性维生素 Water soluble				C-16:0			
镁 Mg				维生素 B_6 Vitamin B_6				C-16:1			
钠 Na				维生素 B_{12} Vitamin B_{12}, μg/kg				C-18:0			
磷 P				生物素 Biotin				C-18:1			
硫 S				叶酸 Folacin				C-18:2			
微量元素 Micro,ppm				烟酸 Niacin				C-18:3			
铬 Cr				泛酸 Pantothenic acid				C-18:4			
铜 Cu				核黄素 Riboflavin				C-20:0			
铁 Fe				硫胺素 Thiamin				C-20:1			
碘 I				胆碱 Choline				C-20:4			
锰 Mn								C-20:5			
硒 Se								C-22:0			
锌 Zn				能量, kcal/kg				C-22:1			
								C-22:5			
植酸磷 Phytate P,%	0.29	2	0.19	总能 GE	3985	3	233	C-22:6			
磷全消化道表观消化率 ATTD of P, %				消化能 DE				C-24:0			
磷全消化道标准消化率 STTD of P, %				代谢能 ME				饱和脂肪酸			
				净能 NE				单不饱和脂肪酸			
								多不饱和脂肪酸			
								碘价			
								碘价物			

续表

原料名称：压榨低寡聚糖豆粕 (Soybean Meal, Low Oligosaccharide, Expelled)
AAFCO#: 84.6, AAFCO 2010, p.391
IFN#: 5-04-600

常规成分, %				氨基酸, %							
				总氨基酸				可消化氨基酸			
								回肠表观消化率		回肠标准消化率	
	$\bar{x}$	n	SD		$\bar{x}$	n	SD	$\bar{x}$	n	SD	$\bar{x}$	n	SD
干物质 Dry matter	94.60	1		必需氨基酸 Essential								
粗蛋白 Crude protein	49.33	1		粗蛋白 CP	49.33	1		84	1		92	1
粗纤维 Crude fiber				精氨酸 Arg	3.77	1		94	1		98	1
乙醚浸出物/粗脂肪 Ether extract	4.62	1		组氨酸 His	1.29	1		90	1		93	1
酸乙醚浸出物 Acid ether extract				异亮氨酸 Ile	2.24	1		89	1		93	1
粗灰分 Ash				亮氨酸 Leu	3.75	1		89	1		93	1
碳水化合物成分, %				赖氨酸 Lys	3.12	1		89	1		93	1
				蛋氨酸 Met	0.68	1		89	1		92	1
乳糖 Lactose				苯丙氨酸 Phe	2.47	1		90	1		93	1
蔗糖 Sucrose	7.10	1		苏氨酸 Thr	1.81	1		81	1		88	1
棉子糖 Raffinose	0.18	1		色氨酸 Trp	0.66	1		88	1		93	1
水苏糖 Stachyose	1.55	1		缬氨酸 Val	2.43	1		86	1		91	1
毛蕊花糖 Verbascose				非必需氨基酸 Nonessential								
低聚糖 Oligosaccharides				丙氨酸 Ala	2.07	1		83	1		90	1
淀粉 Starch	1.89			天冬氨酸 Asp	5.66	1		86	1		90	1
中性洗涤纤维 Neutral detergent fiber	9.98	1		半胱氨酸 Cys	0.78	1		79	1		85	1
酸性洗涤纤维 Acid detergent fiber	6.81	1		谷氨酸 Glu	8.94	1		87	1		90	1
半纤维素 Hemicellulose				甘氨酸 Gly	2.11	1		72	1		90	1
酸性洗涤木质素 Acid detergent lignin				脯氨酸 Pro	2.47	1		82	1		124	1
总膳食纤维 Total dietary fiber				丝氨酸 Ser	2.24	1		86	1		92	1
不溶性膳食纤维 Insoluble dietary fiber				酪氨酸 Tyr	1.71	1		87	1		91	1
可溶性膳食纤维 Soluble dietary fiber												

矿物质				维生素, mg/kg (标注单位者除外)				脂肪酸, 占乙醚提取物, %			
	$\bar{x}$	n	SD		$\bar{x}$	n	SD		$\bar{x}$	n	SD
常量元素 Macro,%				脂溶性维生素 Fat soluble				粗脂肪			
钙 Ca	0.29	1		β-胡萝卜素 β-Carotene				C-12:0			
氯 Cl				维生素 E Vitamin E				C-14:0			
钾 K				水溶性维生素 Water soluble				C-16:0			
镁 Mg				维生素 B_6 Vitamin B_6				C-16:1			
钠 Na				维生素 B_{12} Vitamin B_{12}, μg/kg				C-18:0			
磷 P	0.63	1		生物素 Biotin				C-18:1			
硫 S				叶酸 Folacin				C-18:2			
微量元素 Micro,ppm				烟酸 Niacin				C-18:3			
铬 Cr				泛酸 Pantothenic acid				C-18:4			
铜 Cu				核黄素 Riboflavin				C-20:0			
铁 Fe				硫胺素 Thiamin				C-20:1			
碘 I				胆碱 Choline				C-20:4			
锰 Mn								C-20:5			
硒 Se								C-22:0			
锌 Zn				能量, kcal/kg				C-22:1			
								C-22:5			
植酸磷 Phytate P,%				总能 GE	4737	1		C-22:6			
磷全消化道表观消化率 ATTD of P,%	39			消化能 DE	3679	1		C-24:0			
磷全消化道标准消化率 STTD of P,%	48			代谢能 ME	3344			饱和脂肪酸			
				净能 NE	2151			单不饱和脂肪酸			
								多不饱和脂肪酸			
								碘价			
								碘价物			

续表

原料名称：溶剂浸提豆粕(Soybean Meal, Solvent Extracted)
AAFCO#: 84.61, AAFCO 2010, p.391
IFN#: 5-04-604

常规成分, %				氨基酸, %								
					总氨基酸			可消化氨基酸				
								回肠表观消化率			回肠标准消化率		
	$\bar{x}$	n	SD		$\bar{x}$	n	SD	$\bar{x}$	n	SD	$\bar{x}$	n	SD
干物质 Dry matter	88.79	12	0.70	必需氨基酸 Essential									
粗蛋白 Crude protein	43.90	29	1.97	粗蛋白 CP	43.90	29	1.97	80	13	5.08	85	12	2.95
粗纤维 Crude fiber	6.60	1		精氨酸 Arg	3.17	27	0.19	90	23	4.03	92	22	4.09
乙醚浸出物/粗脂肪 Ether extract	1.24	6	0.25	组氨酸 His	1.26	29	0.14	84	24	5.28	86	23	5.81
酸乙醚浸出物 Acid ether extract				异亮氨酸 Ile	1.96	29	0.19	84	24	4.15	88	23	5.08
粗灰分 Ash	6.38	3	0.24	亮氨酸 Leu	3.43	29	0.26	83	24	3.87	86	23	4.28
碳水化合物成分, %				赖氨酸 Lys	2.76	28	0.24	85	24	2.54	88	23	3.12
				蛋氨酸 Met	0.60	27	0.06	85	21	4.71	89	20	5.21
乳糖 Lactose				苯丙氨酸 Phe	2.26	29	0.16	85	24	3.43	87	23	3.50
蔗糖 Sucrose	7.63	2	0.72	苏氨酸 Thr	1.76	28	0.13	78	24	4.34	83	23	5.62
棉子糖 Raffinose	0.90	2	0.13	色氨酸 Trp	0.59	23	0.26	85	14	4.81	90	14	4.04
水苏糖 Stachyose	4.32	2	0.28	缬氨酸 Val	1.93	29	0.35	79	24	4.08	84	23	4.05
毛蕊花糖 Verbascose	0.12	1		非必需氨基酸 Nonessential									
低聚糖 Oligosaccharides				丙氨酸 Ala	1.92	25	0.18	79	19	4.55	86	19	5.04
淀粉 Starch	1.89			天冬氨酸 Asp	4.88	25	0.73	83	19	3.91	86	19	3.68
中性洗涤纤维 Neutral detergent fiber	9.82	7	1.5	半胱氨酸 Cys	0.68	23	0.20	76	13	6.81	84	13	4.98
酸性洗涤纤维 Acid detergent fiber	6.66	5	1.75	谷氨酸 Glu	7.87	25	1.15	86	19	3.59	88	19	3.38
半纤维素 Hemicellulose				甘氨酸 Gly	1.89	25	0.20	70	19	9.31	83	19	5.92
酸性洗涤木质素 Acid detergent lignin				脯氨酸 Pro	2.43	24	0.46	74	16	18.14	98	16	11.49
总膳食纤维 Total dietary fiber	17.48	1		丝氨酸 Ser	2.14	25	0.28	81	19	4.19	89	19	6.17
不溶性膳食纤维 Insoluble dietary fiber				酪氨酸 Tyr	1.55	25	0.21	83	20	10.09	86	20	10.33
可溶性膳食纤维 Soluble dietary fiber													

矿物质				维生素, mg/kg(标注单位者除外)				脂肪酸, 占乙醚提取物, %			
	$\bar{x}$	n	SD		$\bar{x}$	n	SD		$\bar{x}$	n	SD
常量元素 Macro,%				脂溶性维生素 Fat soluble				粗脂肪	1.22		
钙 Ca	0.35	12	0.09	β-胡萝卜素 β-Carotene	0.2			C-12:0	0.00		
氯 Cl	0.05			维生素 E Vitamin E	2.3			C-14:0	0.25		
钾 K	1.96			水溶性维生素 Water soluble				C-16:0	8.20		
镁 Mg	0.29	2	0.00	维生素 B_6 Vitamin B_6	6.0			C-16:1	0.25		
钠 Na	0.01	2	0.00	维生素 B_{12} Vitamin B_{12}, μg/kg	0			C-18:0	2.79		
磷 P	0.64	14	0.07	生物素 Biotin	0.27			C-18:1	16.89		
硫 S	0.39	2	0.03	叶酸 Folacin	1.37			C-18:2	38.52		
微量元素 Micro,ppm				烟酸 Niacin	34			C-18:3	5.16		
铬 Cr				泛酸 Pantothenic acid	16.0			C-18:4			
铜 Cu	17.38	2	0.62	核黄素 Riboflavin	2.9			C-20:0	0.00		
铁 Fe	235	2	75.38	硫胺素 Thiamin	4.5			C-20:1	0.00		
碘 I				胆碱 Choline	2794			C-20:4			
锰 Mn	40.64	2	9.29					C-20:5			
硒 Se	0.32							C-22:0			
锌 Zn	50.00			能量, kcal/kg				C-22:1			
								C-22:5			
植酸磷 Phytate P,%	0.36	4	0.03	总能 GE	4257	3	168	C-22:6			
磷全消化道表观消化率 ATTD of P, %	39	10	4.23	消化能 DE	3681	1		C-24:0			
磷全消化道标准消化率 STTD of P, %	48	10	4.85	代谢能 ME	3382			饱和脂肪酸	11.23		
				净能 NE	2148			单不饱和脂肪酸	17.13		
								多不饱和脂肪酸	43.69		
								碘价	99.26		
								碘价物	12.11		

续表

原料名称：全脂大豆(Soybeans, Full Fat)
AAFCO#: 84.1, AAFCO 2010, p.390
IFN#: 5-04-596

常规成分, %				氨基酸, %									
					总氨基酸			可消化氨基酸					
								回肠表观消化率			回肠标准消化率		
	$\bar{x}$	n	SD		$\bar{x}$	n	SD	$\bar{x}$	n	SD	$\bar{x}$	n	SD
干物质 Dry matter	92.36	8	1.98	必需氨基酸 Essential									
粗蛋白 Crude protein	37.56	23	1.99	粗蛋白 CP	37.56	23	1.99	74	22	8.44	79	22	9.88
粗纤维 Crude fiber	4.07	1		精氨酸 Arg	2.45	22	0.51	84	22	7.93	87	22	8.36
乙醚浸出物/粗脂肪 Ether extract	20.18	6	1.47	组氨酸 His	0.88	22	0.16	78	22	7.78	81	22	8.06
酸乙醚浸出物 Acid ether extract	15.03	2	0.66	异亮氨酸 Ile	1.60	22	0.21	75	22	8.34	78	22	9.26
粗灰分 Ash	4.89	3	0.09	亮氨酸 Leu	2.67	22	0.47	75	22	9.80	78	22	10.36
碳水化合物成分, %				赖氨酸 Lys	2.23	22	0.29	79	22	9.13	81	22	9.72
				蛋氨酸 Met	0.55	18	0.17	75	17	9.23	80	17	9.39
乳糖 Lactose				苯丙氨酸 Phe	1.74	22	0.27	77	22	9.72	79	22	10.49
蔗糖 Sucrose	6.42	3	1.02	苏氨酸 Thr	1.42	22	0.20	71	22	8.49	76	22	9.64
棉子糖 Raffinose	0.77	3	0.21	色氨酸 Trp	0.49	11	0.13	79	6	8.67	82	6	9.89
水苏糖 Stachyose	3.89	3	0.19	缬氨酸 Val	1.73	22	0.17	73	22	8.19	77	22	9.47
毛蕊花糖 Verbascose	0.03	1		非必需氨基酸 Nonessential									
低聚糖 Oligosaccharides				丙氨酸 Ala	1.59	18	0.19	74	19	8.05	79	19	9.89
淀粉 Starch	1.89			天冬氨酸 Asp	3.89	18	0.78	78	19	9.77	80	19	10.22
中性洗涤纤维 Neutral detergent fiber	10.00	4	2.16	半胱氨酸 Cys	0.59	12	0.04	70	7	10.72	76	7	13.74
酸性洗涤纤维 Acid detergent fiber	6.17	4	0.71	谷氨酸 Glu	6.05	18	1.29	81	19	7.50	84	19	7.85
半纤维素 Hemicellulose				甘氨酸 Gly	1.52	18	0.17	69	19	9.34	81	19	8.50
酸性洗涤木质素 Acid detergent lignin				脯氨酸 Pro	1.65	17	0.39	70	16	15.09	100	16	24.06
总膳食纤维 Total dietary fiber	31.45	2	4.17	丝氨酸 Ser	1.67	18	0.29	75	19	9.43	79	19	10.28
不溶性膳食纤维 Insoluble dietary fiber				酪氨酸 Tyr	1.20	15	0.30	77	15	10.07	81	12	10.16
可溶性膳食纤维 Soluble dietary fiber													

矿物质				维生素, mg/kg(标注单位者除外)				脂肪酸, 占乙醚提取物, %			
	$\bar{x}$	n	SD		$\bar{x}$	n	SD		$\bar{x}$	n	SD
常量元素 Macro,%				脂溶性维生素 Fat soluble				粗脂肪	21.62		
钙 Ca	0.31	9	0.06	β-胡萝卜素 β-Carotene	1.9			C-12:0	0.00		
氯 Cl	0.03			维生素 E Vitamin E	18.1			C-14:0	0.28		
钾 K	1.64	2	0.01	水溶性维生素 Water soluble				C-16:0	10.62		
镁 Mg	0.28			维生素 B_6 Vitamin B_6	10.8			C-16:1	0.28		
钠 Na	0.03			维生素 B_{12} Vitamin B_{12}, μg/kg	0			C-18:0	3.57		
磷 P	0.53	9	0.04	生物素 Biotin	0.24			C-18:1	21.81		
硫 S	0.30			叶酸 Folacin	3.60			C-18:2	49.79		
微量元素 Micro,ppm				烟酸 Niacin	22			C-18:3	6.67		
铬 Cr				泛酸 Pantothenic acid	15.0			C-18:4			
铜 Cu	16.00			核黄素 Riboflavin	2.6			C-20:0	0.00		
铁 Fe	80			硫胺素 Thiamin	11.0			C-20:1	0.00		
碘 I				胆碱 Choline	2307			C-20:4			
锰 Mn	30.00							C-20:5			
硒 Se	0.11							C-22:0			
锌 Zn	39.00			能量, kcal/kg				C-22:1			
								C-22:5			
植酸磷 Phytate P,%	0.33			总能 GE	5227	5	283	C-22:6			
磷全消化道表观消化率 ATTD of P,%	39			消化能 DE	4193	1		C-24:0			
磷全消化道标准消化率 STTD of P,%	48			代谢能 ME	3938			饱和脂肪酸	14.46		
				净能 NE	2874			单不饱和脂肪酸	22.09		
								多不饱和脂肪酸	56.46		
								碘价	128.24		
								碘价物	277.25		

续表

原料名称：高蛋白全脂大豆(Soybeans, High Protein, Full fat)
AAFCO#: 84.1, AAFCO 2010, p.390
IFN#: 5-04-596

常规成分, %				氨基酸, %								
				总氨基酸				可消化氨基酸				
								回肠表观消化率			回肠标准消化率		
	$\bar{x}$	n	SD		$\bar{x}$	n	SD	$\bar{x}$	n	SD	$\bar{x}$	n	SD
干物质 Dry matter	92.38	5	3.09	必需氨基酸 Essential									
粗蛋白 Crude protein	42.77	5	4.18	粗蛋白 CP	42.77	5	4.18	82	2	1.27	92	2	2.83
粗纤维 Crude fiber				精氨酸 Arg	3.16	5	0.52	93	2	0.64	97	2	2.62
乙醚浸出物/粗脂肪 Ether extract	15.59	5	1.5	组氨酸 His	1.07	5	0.14	88	2	0.14	92	2	1.27
酸乙醚浸出物 Acid ether extract				异亮氨酸 Ile	1.51	5	0.5	88	2	0.21	92	2	1.77
粗灰分 Ash				亮氨酸 Leu	3.34	5	0.79	87	2	0.07	91	2	1.63
碳水化合物成分, %				赖氨酸 Lys	2.50	5	0.33	88	2	0.14	92	2	0.78
				蛋氨酸 Met	0.57	5	0.08	88	2	0.28	92	2	2.62
乳糖 Lactose				苯丙氨酸 Phe	2.25	5	0.06	89	2	0.49	93	2	1.70
蔗糖 Sucrose	4.75	2	0.08	苏氨酸 Thr	1.57	5	0.08	78	2	2.33	87	2	0.42
棉子糖 Raffinose	0.85	2	0.49	色氨酸 Trp	0.48	2	0.21	85	2	1.06	89	2	0.99
水苏糖 Stachyose	4.01	2	0.15	缬氨酸 Val	1.76	5	0.39	84	2	0.35	90	2	2.26
毛蕊花糖 Verbascose				非必需氨基酸 Nonessential									
低聚糖 Oligosaccharides				丙氨酸 Ala	1.88	2	0.02	82	2	0.21	90	2	3.25
淀粉 Starch				天冬氨酸 Asp	5.15	2	0.14	87	2	0.99	91	2	0.28
中性洗涤纤维 Neutral detergent fiber	8.24	2	0.62	半胱氨酸 Cys	0.61	5	0.05	75	2	0.35	83	2	2.62
酸性洗涤纤维 Acid detergent fiber	5.40	1		谷氨酸 Glu	8.12	2	0.27	88	2	1.13	91	2	0.21
半纤维素 Hemicellulose				甘氨酸 Gly	1.89	2	0	68	2	8.06	91	2	3.54
酸性洗涤木质素 Acid detergent lignin				脯氨酸 Pro	2.11	2	0.09	61	2	3.11	124	2	41.51
总膳食纤维 Total dietary fiber				丝氨酸 Ser	2.04	2	0.24	84	2	1.06	91	2	0.78
不溶性膳食纤维 Insoluble dietary fiber				酪氨酸 Tyr	1.51	5	0.1	88	2	1.20	92	2	2.12
可溶性膳食纤维 Soluble dietary fiber													

矿物质				维生素, mg/kg(标注单位者除外)				脂肪酸，占乙醚提取物, %			
	$\bar{x}$	n	SD		$\bar{x}$	n	SD		$\bar{x}$	n	SD
常量元素 Macro,%				脂溶性维生素 Fat soluble				粗脂肪			
钙 Ca	0.28	1		β-胡萝卜素 β-Carotene				C-12:0			
氯 Cl				维生素 E Vitamin E				C-14:0			
钾 K				水溶性维生素 Water soluble				C-16:0			
镁 Mg				维生素 B_6 Vitamin B_6				C-16:1			
钠 Na				维生素 B_{12} Vitamin B_{12}, μg/kg				C-18:0			
磷 P	0.65	1		生物素 Biotin				C-18:1			
硫 S				叶酸 Folacin				C-18:2			
微量元素 Micro,ppm				烟酸 Niacin				C-18:3			
铬 Cr				泛酸 Pantothenic acid				C-18:4			
铜 Cu				核黄素 Riboflavin				C-20:0			
铁 Fe				硫胺素 Thiamin				C-20:1			
碘 I				胆碱 Choline				C-20:4			
锰 Mn								C-20:5			
硒 Se								C-22:0			
锌 Zn				能量, kcal/kg				C-22:1			
								C-22:5			
植酸磷 Phytate P,%				总能 GE	5306	1		C-22:6			
磷全消化道表观消化率 ATTD of P,%	39			消化能 DE				C-24:0			
磷全消化道标准消化率 STTD of P,%	48			代谢能 ME				饱和脂肪酸			
				净能 NE				单不饱和脂肪酸			
								多不饱和脂肪酸			
								碘价			
								碘价物			

续表

原料名称：低寡聚糖全脂大豆(Soybeans, Low Oligosaccharide, Full Fat)
AAFCO#: 84.1, AAFCO 2010, p.390
IFN#: 5-04-596

常规成分, %				氨基酸, %								
				总氨基酸				可消化氨基酸				
								回肠表观消化率			回肠标准消化率		
	$\bar{x}$	n	SD		$\bar{x}$	n	SD	$\bar{x}$	n	SD	$\bar{x}$	n	SD
干物质 Dry matter	94.40	1		必需氨基酸 Essential									
粗蛋白 Crude protein	39.30	1		粗蛋白 CP	39.30	1		82	1		89	1	
粗纤维 Crude fiber				精氨酸 Arg	2.79	1		93	1		96	1	
乙醚浸出物/粗脂肪 Ether extract	17.70	1		组氨酸 His	1.02	1		90	1		92	1	
酸乙醚浸出物 Acid ether extract				异亮氨酸 Ile	1.88	1		88	1		91	1	
粗灰分 Ash				亮氨酸 Leu	3.01	1		88	1		91	1	
碳水化合物成分, %				赖氨酸 Lys	2.56	1		90	1		93	1	
				蛋氨酸 Met	0.56	1		90	1		92	1	
乳糖 Lactose				苯丙氨酸 Phe	1.96	1		89	1		92	1	
蔗糖 Sucrose	5.80	1		苏氨酸 Thr	1.44	1		83	1		88	1	
棉子糖 Raffinose	0.10	1		色氨酸 Trp	0.61	1		84	1		87	1	
水苏糖 Stachyose	1.40	1		缬氨酸 Val	1.96	1		85	1		90	1	
毛蕊花糖 Verbascose				非必需氨基酸 Nonessential									
低聚糖 Oligosaccharides				丙氨酸 Ala	1.66	1		85	1		90	1	
淀粉 Starch				天冬氨酸 Asp	4.45	1		89	1		92	1	
中性洗涤纤维 Neutral detergent fiber	10.30	1		半胱氨酸 Cys	0.65	1		81	1		85	1	
酸性洗涤纤维 Acid detergent fiber	7.50	1		谷氨酸 Glu	6.83	1		90	1		92	1	
半纤维素 Hemicellulose				甘氨酸 Gly	1.67	1		77	1		90	1	
酸性洗涤木质素 Acid detergent lignin				脯氨酸 Pro	1.92	1		70	1		102	1	
总膳食纤维 Total dietary fiber				丝氨酸 Ser	1.67	1		87	1		91	1	
不溶性膳食纤维 Insoluble dietary fiber				酪氨酸 Tyr	1.40	1		88	1		91	1	
可溶性膳食纤维 Soluble dietary fiber													

矿物质				维生素, mg/kg(标注单位者除外)				脂肪酸, 占乙醚提取物, %			
	$\bar{x}$	n	SD		$\bar{x}$	n	SD		$\bar{x}$	n	SD
常量元素 Macro,%				脂溶性维生素 Fat soluble				粗脂肪			
钙 Ca	0.36	1		β-胡萝卜素 β-Carotene				C-12:0			
氯 Cl				维生素 E Vitamin E				C-14:0			
钾 K				水溶性维生素 Water soluble				C-16:0			
镁 Mg				维生素 B_6 Vitamin B_6				C-16:1			
钠 Na				维生素 B_{12} Vitamin B_{12}, μg/kg				C-18:0			
磷 P	0.60	1		生物素 Biotin				C-18:1			
硫 S				叶酸 Folacin				C-18:2			
微量元素 Micro,ppm				烟酸 Niacin				C-18:3			
铬 Cr				泛酸 Pantothenic acid				C-18:4			
铜 Cu				核黄素 Riboflavin				C-20:0			
铁 Fe				硫胺素 Thiamin				C-20:1			
碘 I				胆碱 Choline				C-20:4			
锰 Mn								C-20:5			
硒 Se								C-22:0			
锌 Zn				能量, kcal/kg				C-22:1			
								C-22:5			
植酸磷 Phytate P,%				总能 GE	5282	1		C-22:6			
磷全消化道表观消化率 ATTD of P, %	39			消化能 DE				C-24:0			
磷全消化道标准消化率 STTD of P, %	48			代谢能 ME				饱和脂肪酸			
				净能 NE				单不饱和脂肪酸			
								多不饱和脂肪酸			
								碘价			
								碘价物			

续表

原料名称：大豆浓缩蛋白(Soy Protein Concentrate)
AAFCO#: 84.12, AAFCO 2010, p.390
IFN#: 5-32-183

常规成分, %				氨基酸, %								
				总氨基酸				可消化氨基酸				
								回肠表观消化率			回肠标准消化率	
	$\bar{x}$	n	SD	必需氨基酸 Essential	$\bar{x}$	n	SD	$\bar{x}$	n	SD	$\bar{x}$	SD
干物质 Dry matter	92.64	12	1.87									
粗蛋白 Crude protein	65.20	21	4.08	粗蛋白 CP	65.20	21	4.08	85	10	3.65	89	10	1.32
粗纤维 Crude fiber	3.42	7	0.65	精氨酸 Arg	4.75	18	0.20	93	12	2.43	95	12	1.92
乙醚浸出物/粗脂肪 Ether extract	1.05	6	0.61	组氨酸 His	1.70	18	0.08	89	12	3.18	91	12	2.82
酸乙醚浸出物 Acid ether extract	0.65	5	0.41	异亮氨酸 Ile	2.99	18	0.15	89	12	3.13	91	12	2.65
粗灰分 Ash	6.11	10	0.58	亮氨酸 Leu	5.16	18	0.20	89	12	3.19	91	12	2.52
碳水化合物成分, %				赖氨酸 Lys	4.09	19	0.31	89	12	3.35	91	12	2.84
乳糖 Lactose				蛋氨酸 Met	0.87	16	0.08	90	11	2.63	92	11	2.29
蔗糖 Sucrose	0.67	3	0.35	苯丙氨酸 Phe	3.38	18	0.16	88	12	3.62	90	12	3.30
棉子糖 Raffinose	0.46	2	0.44	苏氨酸 Thr	2.52	19	0.15	82	12	4.89	86	12	3.99
水苏糖 Stachyose	0.91	2	0.06	色氨酸 Trp	0.81	13	0.27	85	10	3.91	88	10	3.29
毛蕊花糖 Verbascose				缬氨酸 Val	3.14	18	0.17	87	12	3.52	90	12	2.76
低聚糖 Oligosaccharides	2.46	1		非必需氨基酸 Nonessential								
淀粉 Starch	1.89			丙氨酸 Ala	2.82	15	0.11	85	11	4.77	89	11	2.79
中性洗涤纤维 Neutral detergent fiber	8.10	3	1.15	天冬氨酸 Asp	7.58	15	0.36	86	11	3.99	88	11	3.64
酸性洗涤纤维 Acid detergent fiber	4.42	1		半胱氨酸 Cys	0.90	16	0.14	75	11	4.35	79	11	5.36
半纤维素 Hemicellulose				谷氨酸 Glu	12.02	15	0.65	90	11	3.72	91	11	3.07
酸性洗涤木质素 Acid detergent lignin				甘氨酸 Gly	2.75	15	0.11	79	11	9.83	88	11	2.40
总膳食纤维 Total dietary fiber	18.87	3	2.06	脯氨酸 Pro	3.58	14	0.36	77	10	22.44	102	10	8.13
不溶性膳食纤维 Insoluble dietary fiber				丝氨酸 Ser	3.33	15	0.26	88	11	4.58	91	11	3.04
可溶性膳食纤维 Soluble dietary fiber				酪氨酸 Tyr	2.26	12	0.10	89	6	3.94	93	6	3.49

矿物质				维生素, mg/kg(标注单位者除外)				脂肪酸, 占乙醚提取物, %			
	$\bar{x}$	n	SD		$\bar{x}$	n	SD		$\bar{x}$	n	SD
常量元素 Macro,%				脂溶性维生素 Fat soluble				粗脂肪	0.46		
钙 Ca	0.32	5	0.05	β-胡萝卜素 β-Carotene				C-12:0	0.00		
氯 Cl				维生素 E Vitamin E				C-14:0	0.22		
钾 K				水溶性维生素 Water soluble				C-16:0	8.26		
镁 Mg				维生素 B_6 Vitamin B_6				C-16:1	0.22		
钠 Na				维生素 B_{12} Vitamin B_{12}, μg/kg				C-18:0	2.83		
磷 P	0.82	5	0.07	生物素 Biotin				C-18:1	16.96		
硫 S				叶酸 Folacin				C-18:2	38.48		
微量元素 Micro,ppm				烟酸 Niacin				C-18:3	5.22		
铬 Cr				泛酸 Pantothenic acid				C-18:4			
铜 Cu				核黄素 Riboflavin				C-20:0	0.00		
铁 Fe				硫胺素 Thiamin				C-20:1	0.00		
碘 I				胆碱 Choline				C-20:4			
锰 Mn								C-20:5			
硒 Se								C-22:0			
锌 Zn								C-22:1			
				能量, kcal/kg				C-22:5			
植酸磷 Phytate P,%								C-22:6			
磷全消化道表观消化率 ATTD of P,%	39			总能 GE	4605	2	148	C-24:0			
磷全消化道标准消化率 STTD of P,%	48			消化能 DE	4260	1		饱和脂肪酸	11.30		
				代谢能 ME	3817			单不饱和脂肪酸	17.17		
				净能 NE	2376			多不饱和脂肪酸	43.70		
								碘价	99.36		
								碘价物	4.57		

续表

原料名称：大豆分离蛋白(Soy Protein Isolate)
AAFCO#: 84.62, AAFCO 2010, p.392
IFN#: 5-24-811

常规成分, %				氨基酸, %								
				总氨基酸			可消化氨基酸					
	$\bar{x}$	n	SD		$\bar{x}$	n	SD	回肠表观消化率			回肠标准消化率		
干物质 Dry matter	93.71	3	1.75	必需氨基酸 Essential				$\bar{x}$	n	SD	$\bar{x}$	n	SD
粗蛋白 Crude protein	84.78	7	4.12	粗蛋白 CP	84.78	7	4.12	84	4	4.15	89	4	5.86
粗纤维 Crude fiber	0.17	3	0.11	精氨酸 Arg	6.14	9	0.38	93	6	3.33	94	6	4.06
乙醚浸出物/粗脂肪 Ether extract	2.76	3	1.84	组氨酸 His	2.19	9	0.14	86	6	6.21	88	6	6.65
酸乙醚浸出物 Acid ether extract				异亮氨酸 Ile	3.83	9	0.32	86	6	9.41	88	6	9.62
粗灰分 Ash	4.17	2	0.69	亮氨酸 Leu	6.76	6	0.48	88	6	6.06	89	6	6.07
碳水化合物成分, %				赖氨酸 Lys	5.19	8	0.27	90	6	3.83	91	6	3.93
				蛋氨酸 Met	1.11	9	0.20	84	5	11.49	86	5	12.11
乳糖 Lactose				苯丙氨酸 Phe	4.40	9	0.25	87	6	5.53	88	6	6.09
蔗糖 Sucrose	0.13	1		苏氨酸 Thr	3.09	9	0.27	79	6	8.20	83	6	8.42
棉子糖 Raffinose				色氨酸 Trp	1.13	5	0.07	84	2	0.21	87	2	2.89
水苏糖 Stachyose				缬氨酸 Val	4.02	9	0.20	83	6	9.91	86	6	10.21
毛蕊花糖 Verbascose				非必需氨基酸 Nonessential									
低聚糖 Oligosaccharides	0.37	1		丙氨酸 Ala	3.54	5	0.26	86	5	5.39	90	5	4.22
淀粉 Starch	1.89			天冬氨酸 Asp	9.64	5	0.67	90	5	3.74	92	5	2.98
中性洗涤纤维 Neutral detergent fiber	0.19	1		半胱氨酸 Cys	0.98	7	0.06	74	3	10.21	79	3	12.18
酸性洗涤纤维 Acid detergent fiber	0.00	1		谷氨酸 Glu	16.00	5	1.45	93	5	3.66	94	5	3.53
半纤维素 Hemicellulose				甘氨酸 Gly	3.54	5	0.27	80	5	8.98	89	5	3.12
酸性洗涤木质素 Acid detergent lignin				脯氨酸 Pro	4.45	5	0.62	83	4	9.65	113	4	29.34
总膳食纤维 Total dietary fiber				丝氨酸 Ser	4.37	5	0.66	90	5	4.67	93	5	3.28
不溶性膳食纤维 Insoluble dietary fiber				酪氨酸 Tyr	3.08	4	0.21	86	4	11.04	88	4	11.56
可溶性膳食纤维 Soluble dietary fiber													

矿物质				维生素, mg/kg(标注单位者除外)				脂肪酸, 占乙醚提取物, %		
	$\bar{x}$	n	SD		$\bar{x}$	n	SD		$\bar{x}$	n	SD
常量元素 Macro,%				脂溶性维生素 Fat soluble				粗脂肪	3.39		
钙 Ca	0.17	4	0.03	β-胡萝卜素 β-Carotene				C-12:0	0.00		
氯 Cl	0.02			维生素E Vitamin E				C-14:0	0.24		
钾 K	0.16	3	0.03	水溶性维生素 Water soluble				C-16:0	9.14		
镁 Mg	0.05	3	0.01	维生素B_6 Vitamin B_6	5.4			C-16:1	0.24		
钠 Na	1.14	2	0.01	维生素B_{12} Vitamin B_{12}, μg/kg	0			C-18:0	3.07		
磷 P	0.75	4	0.02	生物素 Biotin	0.30			C-18:1	18.79		
硫 S				叶酸 Folacin	2.5			C-18:2	42.86		
微量元素 Micro,ppm				烟酸 Niacin	6			C-18:3	5.75		
铬 Cr				泛酸 Pantothenic acid	4.2			C-18:4			
铜 Cu	12.90	3	0.45	核黄素 Riboflavin	1.7			C-20:0	0.00		
铁 Fe	15.61	3	4.00	硫胺素 Thiamin	0.3			C-20:1	0.00		
碘 I				胆碱 Choline	2			C-20:4			
锰 Mn	11.90	3	1.40					C-20:5			
硒 Se	0.14							C-22:0			
锌 Zn	40.26	3	3.84	能量, kcal/kg				C-22:1			
								C-22:5			
植酸磷 Phytate P,%				总能 GE	5386			C-22:6			
磷全消化道表观消化率 ATTD of P,%	39			消化能 DE	4150			C-24:0			
磷全消化道标准消化率 STTD of P,%	48			代谢能 ME	3573			饱和脂肪酸	12.45		
				净能 NE	2187			单不饱和脂肪酸	19.03		
								多不饱和脂肪酸	48.61		
								碘价	110.42		
								碘价物	37.43		

续表

原料名称：甜菜渣(Sugar Beet Pulp)
AAFCO#: 63.36, AAFCO 2010, p.380
IFN#: 4-00-669

常规成分, %				氨基酸, %								
				总氨基酸				可消化氨基酸				
	$\bar{x}$	n	SD		$\bar{x}$	n	SD	回肠表观消化率			回肠标准消化率	
								$\bar{x}$	n	SD	$\bar{x}$	n	SD
干物质 Dry matter	87.60	1		必需氨基酸 Essential								
粗蛋白 Crude protein	9.10	1		粗蛋白 CP	9.10	1		34				
粗纤维 Crude fiber				精氨酸 Arg	0.32			44			54	
乙醚浸出物/粗脂肪 Ether extract	0.97			组氨酸 His	0.23			46			56	
酸乙醚浸出物 Acid ether extract				异亮氨酸 Ile	0.31			41			55	
粗灰分 Ash	6.70	1		亮氨酸 Leu	0.53			44			54	
碳水化合物成分, %				赖氨酸 Lys	0.52			48			54	
				蛋氨酸 Met	0.07			52			61	
乳糖 Lactose				苯丙氨酸 Phe	0.30			38			49	
蔗糖 Sucrose				苏氨酸 Thr	0.38			16			29	
棉子糖 Raffinose				色氨酸 Trp	0.10			36			47	
水苏糖 Stachyose				缬氨酸 Val	0.45			32			42	
毛蕊花糖 Verbascose				非必需氨基酸 Nonessential								
低聚糖 Oligosaccharides				丙氨酸 Ala	0.43			36			47	
淀粉 Starch	0.00			天冬氨酸 Asp	0.73			16			26	
中性洗涤纤维 Neutral detergent fiber	44.90	1		半胱氨酸 Cys	0.06			31			46	
酸性洗涤纤维 Acid detergent fiber	23.50	1		谷氨酸 Glu	0.89			46			59	
半纤维素 Hemicellulose				甘氨酸 Gly	0.38			24			46	
酸性洗涤木质素 Acid detergent lignin				脯氨酸 Pro	0.41			21			46	
总膳食纤维 Total dietary fiber				丝氨酸 Ser	0.44			20			34	
不溶性膳食纤维 Insoluble dietary fiber				酪氨酸 Tyr	0.40			46			52	
可溶性膳食纤维 Soluble dietary fiber												

矿物质				维生素, mg/kg(标注单位者除外)				脂肪酸, 占乙醚提取物, %			
	$\bar{x}$	n	SD		$\bar{x}$	n	SD		$\bar{x}$	n	SD
常量元素 Macro,%				脂溶性维生素 Fat soluble				粗脂肪			
钙 Ca	0.81	2	0.27	β-胡萝卜素 β-Carotene	10.6			C-12:0			
氯 Cl	0.10			维生素 E Vitamin E	13.2			C-14:0			
钾 K	0.61			水溶性维生素 Water soluble				C-16:0			
镁 Mg	0.22			维生素 B_6 Vitamin B_6	1.9			C-16:1			
钠 Na	0.20			维生素 B_{12} Vitamin B_{12}, µg/kg	0			C-18:0			
磷 P	0.09	1		生物素 Biotin				C-18:1			
硫 S	0.31			叶酸 Folacin				C-18:2			
微量元素 Micro,ppm				烟酸 Niacin	18			C-18:3			
铬 Cr				泛酸 Pantothenic acid	1.3			C-18:4			
铜 Cu	11.00			核黄素 Riboflavin	0.7			C-20:0			
铁 Fe	411			硫胺素 Thiamin	0.4			C-20:1			
碘 I				胆碱 Choline	1734			C-20:4			
锰 Mn	46.00							C-20:5			
硒 Se	0.09							C-22:0			
锌 Zn	12.00			能量, kcal/kg				C-22:1			
								C-22:5			
植酸磷 Phytate P,%				总能 GE	4039			C-22:6			
磷全消化道表观消化率 ATTD of P, %	50			消化能 DE	2865			C-24:0			
磷全消化道标准消化率 STTD of P, %	63			代谢能 ME	2803			饱和脂肪酸			
				净能 NE	1734			单不饱和脂肪酸			
								多不饱和脂肪酸			
								碘价			
								碘价物			

续表

原料名称：全脂葵花籽(Sunflower, Full Fat)
AAFCO#: 71.221, AAFCO 2010, p.386
IFN#: 5-30-032

常规成分, %				氨基酸, %							
				总氨基酸				可消化氨基酸			
								回肠表观消化率		回肠标准消化率	
	$\bar{x}$	n	SD		$\bar{x}$	n	SD	$\bar{x}$	n	SD	$\bar{x}$	n	SD
干物质 Dry matter	96.83	2	1.09	必需氨基酸 Essential									
粗蛋白 Crude protein	16.60	4	1.16	粗蛋白 CP	16.60	4	1.16						
粗纤维 Crude fiber	13.10	3	0.43	精氨酸 Arg	1.72	1		89					
乙醚浸出物/粗脂肪 Ether extract	42.69	3	2.02	组氨酸 His	0.55	1		84					
酸乙醚浸出物 Acid ether extract				异亮氨酸 Ile	0.90	1		81					
粗灰分 Ash	3.25	3	0.25	亮氨酸 Leu	1.36	1		83					
碳水化合物成分, %				赖氨酸 Lys	0.54	2	0.07	77					
				蛋氨酸 Met	0.39	2	0.03	85					
乳糖 Lactose				苯丙氨酸 Phe	1.02	1		84					
蔗糖 Sucrose				苏氨酸 Thr	0.85	1		76					
棉子糖 Raffinose				色氨酸 Trp				77					
水苏糖 Stachyose				缬氨酸 Val	0.94	1		78					
毛蕊花糖 Verbascose				非必需氨基酸 Nonessential									
低聚糖 Oligosaccharides				丙氨酸 Ala	0.95	1							
淀粉 Starch	2.04			天冬氨酸 Asp	2.13	1							
中性洗涤纤维 Neutral detergent fiber	23.23	4	2.43	半胱氨酸 Cys	0.24			73					
酸性洗涤纤维 Acid detergent fiber	16.93	4	2.10	谷氨酸 Glu	4.54	1							
半纤维素 Hemicellulose				甘氨酸 Gly	1.24	1							
酸性洗涤木质素 Acid detergent lignin	4.52	2	0.17	脯氨酸 Pro									
总膳食纤维 Total dietary fiber				丝氨酸 Ser	1.00	1							
不溶性膳食纤维 Insoluble dietary fiber				酪氨酸 Tyr	0.55	1		86					
可溶性膳食纤维 Soluble dietary fiber													

矿物质				维生素, mg/kg(标注单位者除外)				脂肪酸, 占乙醚提取物, %			
	$\bar{x}$	n	SD		$\bar{x}$	n	SD		$\bar{x}$	n	SD
常量元素 Macro,%				脂溶性维生素 Fat soluble				粗脂肪	49.57		
钙 Ca	0.30	1		β-胡萝卜素 β-Carotene				C-12:0	0.00		
氯 Cl				维生素 E Vitamin E				C-14:0	0.10		
钾 K				水溶性维生素 Water soluble				C-16:0	5.64		
镁 Mg				维生素 B_6 Vitamin B_6				C-16:1	0.10		
钠 Na				维生素 B_{12} Vitamin B_{12}, μg/kg				C-18:0	4.44		
磷 P	0.20	1		生物素 Biotin				C-18:1	18.87		
硫 S				叶酸 Folacin				C-18:2	65.83		
微量元素 Micro,ppm				烟酸 Niacin				C-18:3	0.14		
铬 Cr				泛酸 Pantothenic acid				C-18:4			
铜 Cu				核黄素 Riboflavin				C-20:0	0.00		
铁 Fe				硫胺素 Thiamin				C-20:1	0.10		
碘 I				胆碱 Choline				C-20:4			
锰 Mn								C-20:5			
硒 Se								C-22:0			
锌 Zn				能量, kcal/kg				C-22:1			
								C-22:5			
植酸磷 Phytate P,%				总能 GE	6163	2	473	C-22:6			
磷全消化道表观消化率 ATTD of P,%	20			消化能 DE	4517			C-24:0			
磷全消化道标准消化率 STTD of P,%	29			代谢能 ME	4404			饱和脂肪酸	10.18		
				净能 NE	3561			单不饱和脂肪酸	19.07		
								多不饱和脂肪酸	65.97		
								碘价	136.66		
								碘价物	677.44		

续表

原料名称：溶剂浸提去壳葵花粕(Sunflower Meal, Dehulled, Solvent Extracted)
AAFCO#: 71.211, AAFCO 2010, p.386
IFN#: 5-30-034

常规成分, %	$\bar{x}$	n	SD	氨基酸, %				可消化氨基酸				
				总氨基酸				回肠表观消化率			回肠标准消化率	
					$\bar{x}$	n	SD	$\bar{x}$	n	SD	$\bar{x}$	n	SD
干物质 Dry matter	90.40	2	0.14	必需氨基酸 Essential									
粗蛋白 Crude protein	39.86	8	4.78	粗蛋白 CP	39.86	8	4.78	76	4	6.04	81	4	5.31
粗纤维 Crude fiber	18.44	2	2.53	精氨酸 Arg	3.32	6	0.27	91	5	3.43	93	5	3.35
乙醚浸出物/粗脂肪 Ether extract	2.90			组氨酸 His	0.93	6	0.10	82	5	6.48	85	5	6.28
酸乙醚浸出物 Acid ether extract				异亮氨酸 Ile	1.54	6	0.18	78	5	6.19	80	5	6.15
粗灰分 Ash	6.06	2	0.89	亮氨酸 Leu	2.47	6	0.11	77	5	5.36	80	5	5.27
碳水化合物成分, %				赖氨酸 Lys	1.45	6	0.10	75	5	4.25	78	5	5.13
				蛋氨酸 Met	0.78	5	0.17	84	4	3.84	89		
乳糖 Lactose	0.00	2	0.00	苯丙氨酸 Phe	1.63	6	0.23	79	5	7.47	81	5	7.11
蔗糖 Sucrose	0.00	2	0.00	苏氨酸 Thr	1.37	6	0.06	72	5	8.48	77	5	8.54
棉子糖 Raffinose	0.00	2	0.00	色氨酸 Trp	0.48	2	0.04	73	2	3.39	80		
水苏糖 Stachyose	0.00	2	0.00	缬氨酸 Val	1.76	6	0.21	76	5	8.37	79	5	8.06
毛蕊花糖 Verbascose	0.00	2	0.00	非必需氨基酸 Nonessential									
低聚糖 Oligosaccharides				丙氨酸 Ala	1.63	3	0.09	68	3	2.17	72	3	3.62
淀粉 Starch	2.08	2	1.03	天冬氨酸 Asp	3.55	3	0.21	74	3	1.22	77	3	1.51
中性洗涤纤维 Neutral detergent fiber	30.24	2	0.27	半胱氨酸 Cys	0.48	4	0.21	77	3	3.89	82	3	3.42
酸性洗涤纤维 Acid detergent fiber	23.00	2	2.97	谷氨酸 Glu	8.25	3	0.74	84	3	0.70	86	3	1.05
半纤维素 Hemicellulose				甘氨酸 Gly	2.09	3	0.13	63	3	4.30	70	3	4.95
酸性洗涤木质素 Acid detergent lignin				脯氨酸 Pro	2.01	3	0.61	63	3	16.86	81	3	10.91
总膳食纤维 Total dietary fiber				丝氨酸 Ser	1.66	3	0.10	72	3	3.19	76	3	4.86
不溶性膳食纤维 Insoluble dietary fiber				酪氨酸 Tyr	0.81	3	0.17	72	3	6.07	84		
可溶性膳食纤维 Soluble dietary fiber													

矿物质	$\bar{x}$	n	SD	维生素, mg/kg(标注单位者除外)	$\bar{x}$	n	SD	脂肪酸, 占乙醚提取物, %	$\bar{x}$	n	SD
常量元素 Macro,%				脂溶性维生素 Fat soluble				粗脂肪	1.7		
钙 Ca	0.39	1		β-胡萝卜素 β-Carotene				C-12:0	0.00		
氯 Cl	0.04			维生素 E Vitamin E	9.1			C-14:0	0.15		
钾 K	1.27			水溶性维生素 Water soluble				C-16:0	4.73		
镁 Mg	0.75			维生素 B_6 Vitamin B_6	13.7			C-16:1	0.30		
钠 Na	0.04			维生素 B_{12} Vitamin B_{12}, μg/kg	0			C-18:0	3.23		
磷 P	1.16	1		生物素 Biotin	1.45			C-18:1	15.23		
硫 S	0.38			叶酸 Folacin	1.14			C-18:2	48.68		
微量元素 Micro,ppm				烟酸 Niacin	220			C-18:3	0.23		
铬 Cr				泛酸 Pantothenic acid	24.0			C-18:4	0.00		
铜 Cu	25			核黄素 Riboflavin	3.6			C-20:0	0.00		
铁 Fe	200			硫胺素 Thiamin	3.5			C-20:1	0.00		
碘 I				胆碱 Choline	3150			C-20:4	0.00		
锰 Mn	35							C-20:5	0.00		
硒 Se	0.32							C-22:0	0.00		
锌 Zn	98							C-22:1	0.00		
				能量, kcal/kg				C-22:5	0.00		
植酸磷 Phytate P,%	0.89	1		总能 GE	4415	2	54	C-22:6	0.00		
磷全消化道表观消化率 ATTD of P, %	20			消化能 DE	2840			C-24:0	0.00		
磷全消化道标准消化率 STTD of P, %	29	1		代谢能 ME	2569			饱和脂肪酸	8.10		
				净能 NE	1482			单不饱和脂肪酸	15.53		
								多不饱和脂肪酸	48.90		
								碘价	102.69		
								碘价物	17.46		

续表

原料名称：溶剂浸提葵花粕(Sunflower Meal, Solvent Extracted)
AAFCO#: 71.221, AAFCO 2010, p.386
IFN#: 5-30-032

常规成分, %				氨基酸, %								
				总氨基酸				可消化氨基酸				
								回肠表观消化率		回肠标准消化率		
	$\bar{x}$	n	SD		$\bar{x}$	n	SD	$\bar{x}$	n	SD	$\bar{x}$	n	SD
干物质 Dry matter	87.93	3	0.55	必需氨基酸 Essential									
粗蛋白 Crude protein	30.70	12	2.63	粗蛋白 CP	30.70	12	2.63	77	6	5.06	83	6	4.64
粗纤维 Crude fiber	23.40	4	2.90	精氨酸 Arg	2.53	10	0.22	91	6	2.94	93	6	2.80
乙醚浸出物/粗脂肪 Ether extract	3.06	4	0.43	组氨酸 His	0.78	10	0.06	80	6	4.97	83	6	5.14
酸乙醚浸出物 Acid ether extract				异亮氨酸 Ile	1.29	10	0.06	79	6	2.96	82	6	2.62
粗灰分 Ash	5.97	4	0.26	亮氨酸 Leu	1.96	10	0.12	79	6	3.05	82	6	2.79
碳水化合物成分, %				赖氨酸 Lys	1.13	10	0.07	76	6	3.33	80	6	3.71
				蛋氨酸 Met	0.74	9	0.04	88	5	2.66	90	5	2.56
乳糖 Lactose	0.00	2	0.00	苯丙氨酸 Phe	1.39	10	0.08	83	6	4.39	86	6	3.95
蔗糖 Sucrose	0.00	2	0.00	苏氨酸 Thr	1.17	10	0.06	75	6	5.50	80	6	4.53
棉子糖 Raffinose	0.00	2	0.00	色氨酸 Trp	0.39	8	0.04	80	3	4.33	84		
水苏糖 Stachyose	0.00	2	0.00	缬氨酸 Val	1.51	10	0.09	76	6	5.00	79	6	4.37
毛蕊花糖 Verbascose	0.00	2	0.00	非必需氨基酸 Nonessential									
低聚糖 Oligosaccharides				丙氨酸 Ala	1.32	8	0.07	74	3	6.31	80	3	4.92
淀粉 Starch	2.03	1		天冬氨酸 Asp	2.68	8	0.41	80	3	3.75	84	3	3.17
中性洗涤纤维 Neutral detergent fiber	36.82	3	2.73	半胱氨酸 Cys	0.53	9	0.06	76	3	4.88	80	3	4.09
酸性洗涤纤维 Acid detergent fiber	28.67	3	2.85	谷氨酸 Glu	6.12	8	0.47	86	3	2.53	88	3	2.07
半纤维素 Hemicellulose				甘氨酸 Gly	1.76	8	0.08	65	3	5.96	74	3	5.47
酸性洗涤木质素 Acid detergent lignin	7.54	1		脯氨酸 Pro	1.29	4	0.12	79			87		
总膳食纤维 Total dietary fiber				丝氨酸 Ser	1.36	8	0.06	76	3	5.09	81	3	3.56
不溶性膳食纤维 Insoluble dietary fiber				酪氨酸 Tyr	0.70	9	0.14	83	4	5.23	88	4	5.04
可溶性膳食纤维 Soluble dietary fiber													

矿物质				维生素, mg/kg(标注单位者除外)				脂肪酸, 占乙醚提取物, %		
	$\bar{x}$	n	SD		$\bar{x}$	n	SD		$\bar{x}$	
常量元素 Macro,%				脂溶性维生素 Fat soluble				粗脂肪	1.61	
钙 Ca	0.38	3	0.04	β-胡萝卜素 β-Carotene				C-12:0	0.00	
氯 Cl	0.10			维生素 E Vitamin E	9.1			C-14:0	0.06	
钾 K	1.07			水溶性维生素 Water soluble				C-16:0	4.60	
镁 Mg	0.68			维生素 B_6 Vitamin B_6	11.1			C-16:1	0.06	
钠 Na	0.02			维生素 B_{12} Vitamin B_{12}, μg/kg	0			C-18:0	3.66	
磷 P	0.95	3	0.09	生物素 Biotin	1.40			C-18:1	15.47	
硫 S	0.30			叶酸 Folacin	1.14			C-18:2	53.91	
微量元素 Micro,ppm				烟酸 Niacin	264			C-18:3	0.12	
铬 Cr				泛酸 Pantothenic acid	29.9			C-18:4		
铜 Cu	26.00			核黄素 Riboflavin	3.0			C-20:0	0.00	
铁 Fe	254			硫胺素 Thiamin	3.0			C-20:1	0.06	
碘 I				胆碱 Choline	3791			C-20:4		
锰 Mn	41.00							C-20:5		
硒 Se	0.50							C-22:0		
锌 Zn	66.00			能量, kcal/kg				C-22:1		
								C-22:5		
植酸磷 Phytate P,%	0.84	1		总能 GE	4086	1		C-22:6		
磷全消化道表观消化率 ATTD of P,%	20			消化能 DE	2010			C-24:0		
磷全消化道标准消化率 STTD of P,%	29	2	7.39	代谢能 ME	1801			饱和脂肪酸	8.32	
				净能 NE	937			单不饱和脂肪酸	15.59	
								多不饱和脂肪酸	54.04	
								碘价	111.93	
								碘价物	18.02	

续表

原料名称：黑小麦(Triticale)
AAFCO#：无官方定义
IFN#：4-20-362

常规成分, %				氨基酸, %								
				总氨基酸				可消化氨基酸				
								回肠表观消化率			回肠标准消化率		
	$\bar{x}$	n	SD		$\bar{x}$	n	SD	$\bar{x}$	n	SD	$\bar{x}$	n	SD
干物质 Dry matter	88.48	5	1.69	必需氨基酸 Essential									
粗蛋白 Crude protein	13.60	8	1.89	粗蛋白 CP	13.60	8	1.89	79	6	4.45	87	5	3.27
粗纤维 Crude fiber	2.54	2	0.22	精氨酸 Arg	0.73	4	0.20	81	8	5.26	85	6	6.61
乙醚浸出物/粗脂肪 Ether extract	1.77	2	0.47	组氨酸 His	0.31	4	0.05	80	7	6.48	82	7	7.18
酸乙醚浸出物 Acid ether extract				异亮氨酸 Ile	0.45	4	0.09	79	8	5.66	83	8	6.83
粗灰分 Ash	2.95	2	1.49	亮氨酸 Leu	0.86	4	0.20	81	8	4.42	85	8	5.50
				赖氨酸 Lys	0.46	4	0.05	74	8	7.13	78	8	9.33
碳水化合物成分, %				蛋氨酸 Met	0.24	4	0.05	83	8	4.19	89		
乳糖 Lactose	0.00	1		苯丙氨酸 Phe	0.52	4	0.19	81	7	6.51	85	7	7.75
蔗糖 Sucrose	0.00	1		苏氨酸 Thr	0.41	4	0.09	64	8	11.62	70	8	14.66
棉子糖 Raffinose	0.00	1		色氨酸 Trp	0.16	3	0.03	76	3	9.43	82		
水苏糖 Stachyose	0.00	1		缬氨酸 Val	0.59	4	0.13	77	8	5.68	82	8	6.98
毛蕊花糖 Verbascose	0.00	1		非必需氨基酸 Nonessential									
低聚糖 Oligosaccharides				丙氨酸 Ala	0.54	4	0.10	72	7	4.70	78	7	6.47
淀粉 Starch	64.31	2	3.80	天冬氨酸 Asp	0.80	4	0.13	75	7	5.21	80	7	4.45
中性洗涤纤维 Neutral detergent fiber	10.28	5	0.96	半胱氨酸 Cys	0.29	4	0.09	80	7	7.67	83	7	5.45
酸性洗涤纤维 Acid detergent fiber	3.45	5	0.39	谷氨酸 Glu	3.75	4	0.82	89	4	4.89	91	7	4.53
半纤维素 Hemicellulose				甘氨酸 Gly	0.56	4	0.11	67	9	9.53	83	7	15.61
酸性洗涤木质素 Acid detergent lignin	0.77	1		脯氨酸 Pro	1.06	1		82	4	4.34	104	5	22.43
总膳食纤维 Total dietary fiber				丝氨酸 Ser	0.64	4	0.12	77	7	6.47	82	7	7.52
不溶性膳食纤维 Insoluble dietary fiber				酪氨酸 Tyr	0.39	4	0.11	79	6	6.39	82	6	7.00
可溶性膳食纤维 Soluble dietary fiber													

矿物质				维生素, mg/kg(标注单位者除外)				脂肪酸，占乙醚提取物, %			
	$\bar{x}$	n	SD		$\bar{x}$	n	SD		$\bar{x}$	n	SD
常量元素 Macro, %				脂溶性维生素 Fat soluble				粗脂肪	2.09		
钙 Ca	0.04	9	0.01	β-胡萝卜素 β-Carotene				C-12:0	0.67		
氯 Cl	0.03			维生素 E Vitamin E	1.7			C-14:0	0.43		
钾 K	0.46			水溶性维生素 Water soluble				C-16:0	13.11		
镁 Mg	0.10			维生素 B_6 Vitamin B_6				C-16:1	0.86		
钠 Na	0.03			维生素 B_{12} Vitamin B_{12}, μg/kg				C-18:0	1.48		
磷 P	0.33	10	0.05	生物素 Biotin				C-18:1	8.52		
硫 S	0.15			叶酸 Folacin				C-18:2	40.81		
微量元素 Micro, ppm				烟酸 Niacin				C-18:3	2.92		
铬 Cr				泛酸 Pantothenic acid				C-18:4			
铜 Cu	8.00			核黄素 Riboflavin	0.4			C-20:0	0.00		
铁 Fe	31.00			硫胺素 Thiamin				C-20:1	0.72		
碘 I				胆碱 Choline	462			C-20:4			
锰 Mn	43.00							C-20:5			
硒 Se								C-22:0			
锌 Zn	32.00			能量, kcal/kg				C-22:1			
								C-22:5			
植酸磷 Phytate P, %	0.21	5	0.02	总能 GE	4316			C-22:6			
磷全消化道表观消化率 ATTD of P, %	50	6	3.52	消化能 DE	3320			C-24:0			
磷全消化道标准消化率 STTD of P, %	56	6	3.50	代谢能 ME	3228			饱和脂肪酸	15.69		
				净能 NE	2507			单不饱和脂肪酸	10.10		
								多不饱和脂肪酸	43.73		
								碘价	90.95		
								碘价物	19.01		

续表

原料名称：黑小麦酒精糟及可溶物(Triticale DDGS)
AAFCO#：无官方定义

常规成分, %				氨基酸, %								
				总氨基酸				可消化氨基酸				
								回肠表观消化率			回肠标准消化率	
	$\bar{x}$	n	SD		$\bar{x}$	n	SD	$\bar{x}$	n	SD	$\bar{x}$	n	SD
干物质 Dry matter	89.30	1		必需氨基酸 Essential									
粗蛋白 Crude protein	27.42	1		粗蛋白 CP	27.42	1							
粗纤维 Crude fiber				精氨酸 Arg									
乙醚浸出物/粗脂肪 Ether extract	4.82	1		组氨酸 His									
酸乙醚浸出物 Acid ether extract				异亮氨酸 Ile									
粗灰分 Ash	3.93	1		亮氨酸 Leu									
碳水化合物成分, %				赖氨酸 Lys									
				蛋氨酸 Met									
乳糖 Lactose				苯丙氨酸 Phe									
蔗糖 Sucrose				苏氨酸 Thr									
棉子糖 Raffinose				色氨酸 Trp									
水苏糖 Stachyose				缬氨酸 Val									
毛蕊花糖 Verbascose				非必需氨基酸 Nonessential									
低聚糖 Oligosaccharides				丙氨酸 Ala									
淀粉 Starch				天冬氨酸 Asp									
中性洗涤纤维 Neutral detergent fiber	26.43	1		半胱氨酸 Cys									
酸性洗涤纤维 Acid detergent fiber	12.23	1		谷氨酸 Glu									
半纤维素 Hemicellulose				甘氨酸 Gly									
酸性洗涤木质素 Acid detergent lignin				脯氨酸 Pro									
总膳食纤维 Total dietary fiber				丝氨酸 Ser									
不溶性膳食纤维 Insoluble dietary fiber				酪氨酸 Tyr									
可溶性膳食纤维 Soluble dietary fiber													

矿物质				维生素, mg/kg(标注单位者除外)				脂肪酸, 占乙醚提取物, %			
	$\bar{x}$	n	SD		$\bar{x}$	n	SD		$\bar{x}$	n	SD
常量元素 Macro,%				脂溶性维生素 Fat soluble				粗脂肪			
钙 Ca	0.06	1		β-胡萝卜素 β-Carotene				C-12:0			
氯 Cl				维生素 E Vitamin E				C-14:0			
钾 K	0.88	1		水溶性维生素 Water soluble				C-16:0			
镁 Mg	0.29	1		维生素 B_6 Vitamin B_6				C-16:1			
钠 Na	0.01	1		维生素 B_{12} Vitamin B_{12}, μg/kg				C-18:0			
磷 P	0.70	1		生物素 Biotin				C-18:1			
硫 S	0.29	1		叶酸 Folacin				C-18:2			
微量元素 Micro,ppm				烟酸 Niacin				C-18:3			
铬 Cr				泛酸 Pantothenic acid				C-18:4			
铜 Cu				核黄素 Riboflavin				C-20:0			
铁 Fe				硫胺素 Thiamin				C-20:1			
碘 I				胆碱 Choline				C-20:4			
锰 Mn								C-20:5			
硒 Se								C-22:0			
锌 Zn				能量, kcal/kg				C-22:1			
								C-22:5			
植酸磷 Phytate P,%				总能 GE				C-22:6			
磷全消化道表观消化率 ATTD of P,%	56			消化能 DE				C-24:0			
磷全消化道标准消化率 STTD of P,%	61			代谢能 ME				饱和脂肪酸			
				净能 NE				单不饱和脂肪酸			
								多不饱和脂肪酸			
								碘价			
								碘价物			

续表

原料名称：硬红小麦(Wheat, Hard Red)
很多文献没有区分小麦的类型，我们把蛋白含量≥11%的小麦划分为硬质小麦。
AAFCO#：无官方定义
IFN#：4-05-258

常规成分，%				氨基酸，%									
				总氨基酸				可消化氨基酸					
								回肠表观消化率			回肠标准消化率		
	$\bar{x}$	n	SD		$\bar{x}$	n	SD	$\bar{x}$	n	SD	$\bar{x}$	n	SD
干物质 Dry matter	88.67	46	3.22	必需氨基酸 Essential									
粗蛋白 Crude protein	14.46	64	2.51	粗蛋白 CP	14.46	64	2.51	77	13	9.54	88	12	9.12
粗纤维 Crude fiber	2.57	6	0.80	精氨酸 Arg	0.60	30	0.14	83	15	5.04	91	15	5.27
乙醚浸出物/粗脂肪 Ether extract	1.82	36	0.37	组氨酸 His	0.34	31	0.10	83	15	7.46	88	15	6.30
酸乙醚浸出物 Acid ether extract	2.51	3	1.16	异亮氨酸 Ile	0.47	31	0.10	82	15	5.97	89	15	5.69
粗灰分 Ash	1.98	25	0.37	亮氨酸 Leu	0.91	31	0.15	83	15	5.24	89	15	4.98
碳水化合物成分，%				赖氨酸 Lys	0.39	34	0.08	72	15	11.73	82	15	11.31
				蛋氨酸 Met	0.22	29	0.04	83	13	6.49	88	13	6.42
乳糖 Lactose	0.00	1		苯丙氨酸 Phe	0.64	31	0.13	85	15	4.09	90	15	4.31
蔗糖 Sucrose	0.00	1		苏氨酸 Thr	0.40	32	0.07	71	15	10.61	84	15	9.30
棉子糖 Raffinose	0.00	1		色氨酸 Trp	0.17	19	0.05	82	6	5.65	88	6	4.23
水苏糖 Stachyose	0.00	1		缬氨酸 Val	0.58	31	0.10	79	15	6.07	88	15	5.91
毛蕊花糖 Verbascose	0.00	1		非必需氨基酸 Nonessential									
低聚糖 Oligosaccharides				丙氨酸 Ala	0.47	27	0.11	72	14	10.44	83	14	9.33
淀粉 Starch	59.50	26	4.32	天冬氨酸 Asp	0.71	26	0.16	73	14	9.80	84	14	9.02
中性洗涤纤维 Neutral detergent fiber	10.60	26	2.87	半胱氨酸 Cys	0.33	26	0.11	83	11	6.87	89	11	6.74
酸性洗涤纤维 Acid detergent fiber	3.55	21	0.97	谷氨酸 Glu	3.88	26	1.03	88	14	8.44	93	14	5.43
半纤维素 Hemicellulose				甘氨酸 Gly	0.57	27	0.14	70	14	13.89	92	14	13.85
酸性洗涤木质素 Acid detergent lignin	0.97	2	0.23	脯氨酸 Pro	1.36	22	0.39	78	10	18.00	105	10	27.75
总膳食纤维 Total dietary fiber	9.83	10	2.37	丝氨酸 Ser	0.60	27	0.11	81	14	8.67	89	14	7.87
不溶性膳食纤维 Insoluble dietary fiber	6.81	9	0.41	酪氨酸 Tyr	0.36	26	0.11	80	15	8.22	88	14	8.16
可溶性膳食纤维 Soluble dietary fiber	2.34	9	0.86										

矿物质				维生素，mg/kg(标注单位者除外)				脂肪酸，占乙醚提取物，%			
	$\bar{x}$	n	SD		$\bar{x}$	n	SD		$\bar{x}$	n	SD
常量元素 Macro,%				脂溶性维生素 Fat soluble				粗脂肪	1.54		
钙 Ca	0.06	25	0.05	β-胡萝卜素 β-Carotene	0.4			C-12:0	0.00		
氯 Cl	0.06			维生素 E Vitamin E	11.6			C-14:0	0.06		
钾 K	0.49	10	0.06	水溶性维生素 Water soluble				C-16:0	15.19		
镁 Mg	0.16	10	0.01	维生素 B_6 Vitamin B_6	3.4			C-16:1	0.52		
钠 Na	0.01	10	0.00	维生素 B_{12} Vitamin B_{12}, μg/kg	0			C-18:0	0.84		
磷 P	0.39	37	0.10	生物素 Biotin	0.11			C-18:1	12.47		
硫 S	0.16	10	0.01	叶酸 Folacin	0.22			C-18:2	38.96		
微量元素 Micro,ppm				烟酸 Niacin	48			C-18:3	1.75		
铬 Cr				泛酸 Pantothenic acid	9.9			C-18:4			
铜 Cu	3.00	10	1.15	核黄素 Riboflavin	1.4			C-20:0	0.00		
铁 Fe	71	10	33.88	硫胺素 Thiamin	4.5			C-20:1	0.00		
碘 I				胆碱 Choline	778			C-20:4			
锰 Mn	33.30	10	6.43					C-20:5			
硒 Se	0.33							C-22:0			
锌 Zn	31.00	9	5.61					C-22:1			
				能量，kcal/kg				C-22:5			
植酸磷 Phytate P,%	0.22	14	0.07	总能 GE	3788	25	145	C-22:6			
磷全消化道表观消化率 ATTD of P,%	46			消化能 DE	3313			C-24:0			
磷全消化道标准消化率 STTD of P,%	56			代谢能 ME	3215			饱和脂肪酸	16.10		
				净能 NE	2472			单不饱和脂肪酸	12.99		
								多不饱和脂肪酸	40.71		
								碘价	87.03		
								碘价物	13.40		

续表

原料名称：软红小麦 (Wheat Soft Red)
很多文献没有区分小麦的类型，我们把蛋白含量<11%的小麦划分为软质小麦。
AAFCO#：无官方定义
IFN#：4-05-294

常规成分, %				氨基酸, %								
				总氨基酸				可消化氨基酸				
								回肠表观消化率			回肠标准消化率		
	$\bar{x}$	n	SD		$\bar{x}$	n	SD	$\bar{x}$	n	SD	$\bar{x}$	n	SD
干物质 Dry matter	86.38	5	1.69	必需氨基酸 Essential									
粗蛋白 Crude protein	10.92	5	0.48	粗蛋白 CP	10.92	5	0.48						
粗纤维 Crude fiber				精氨酸 Arg	0.52	2	0.08	83			89		
乙醚浸出物/粗脂肪 Ether extract	1.36	3	0.06	组氨酸 His	0.28	2	0.01	84			90		
酸乙醚浸出物 Acid ether extract				异亮氨酸 Ile	0.34	2	0.04	84			90		
粗灰分 Ash	1.99	1		亮氨酸 Leu	0.68	2	0.09	85			87		
碳水化合物成分, %				赖氨酸 Lys	0.35	2	0	73			82		
				蛋氨酸 Met	0.22	2	0.01	85			90		
乳糖 Lactose				苯丙氨酸 Phe	0.52	2	0.04	87			91		
蔗糖 Sucrose				苏氨酸 Thr	0.35	2	0.02	72			85		
棉子糖 Raffinose				色氨酸 Trp	0.14	2	0.02	81			88		
水苏糖 Stachyose				缬氨酸 Val	0.47	2	0.08	80			87		
毛蕊花糖 Verbascose				非必需氨基酸 Nonessential									
低聚糖 Oligosaccharides				丙氨酸 Ala	0.42	1							
淀粉 Starch	60.04	3	1.91	天冬氨酸 Asp	0.58	1							
中性洗涤纤维 Neutral detergent fiber				半胱氨酸 Cys	0.30	2	0	84			90		
酸性洗涤纤维 Acid detergent fiber	3.55			谷氨酸 Glu	2.92	1							
半纤维素 Hemicellulose				甘氨酸 Gly	0.49	1							
酸性洗涤木质素 Acid detergent lignin				脯氨酸 Pro	1.04	1							
总膳食纤维 Total dietary fiber	9.90	3	1.07	丝氨酸 Ser	0.44	1							
不溶性膳食纤维 Insoluble dietary fiber	6.63	3	0.4	酪氨酸 Tyr	0.30	2	0.04	84			88		
可溶性膳食纤维 Soluble dietary fiber	3.27	3	0.82										
矿物质				维生素, mg/kg(标注单位者除外)				脂肪酸, 占乙醚提取物, %					
	$\bar{x}$	n	SD		$\bar{x}$	n	SD	$\bar{x}$	n	SD			
常量元素 Macro,%				脂溶性维生素 Fat soluble				粗脂肪	1.56				
钙 Ca	0.03	4	0.00	β-胡萝卜素 β-Carotene				C-12:0	0.00				
氯 Cl	0.08			维生素 E Vitamin E				C-14:0	0.13				
钾 K	0.46			水溶性维生素 Water soluble				C-16:0	17.37				
镁 Mg	0.11			维生素 B_6 Vitamin B_6	2.2			C-16:1	0.51				
钠 Na	0.01			维生素 B_{12} Vitamin B_{12}, μg/kg	0			C-18:0	0.90				
磷 P	0.30	5	0.03	生物素 Biotin	0.11			C-18:1	10.90				
硫 S	0.16			叶酸 Folacin	0.35			C-18:2	40.26				
微量元素 Micro,ppm				烟酸 Niacin	48			C-18:3	1.79				
铬 Cr				泛酸 Pantothenic acid	9.9			C-18:4					
铜 Cu	8.00			核黄素 Riboflavin	1.4			C-20:0	0.00				
铁 Fe	32			硫胺素 Thiamin	4.5			C-20:1	0.00				
碘 I				胆碱 Choline	1092			C-20:4					
锰 Mn	38.00							C-20:5					
硒 Se	0.28							C-22:0					
锌 Zn	47.00			能量, kcal/kg				C-22:1					
								C-22:5					
植酸磷 Phytate P,%	0.20	4	0.03	总能 GE	4295			C-22:6					
磷全消化道表观消化率 ATTD of P,%	46			消化能 DE	3450			C-24:0					
磷全消化道标准消化率 STTD of P,%	56	4	4.71	代谢能 ME	3376			饱和脂肪酸	18.40				
				净能 NE	2595			单不饱和脂肪酸	11.41				
								多不饱和脂肪酸	42.05				
								碘价	88.07				
								碘价物	13.74				

续表

原料名称：小麦麸(Wheat Bran)
AAFCO#: 93.1, AAFCO 2010, p.407
IFN#: 4-05-190

常规成分, %				氨基酸, %								
				总氨基酸				可消化氨基酸				
								回肠表观消化率			回肠标准消化率	
	$\bar{x}$	n	SD	必需氨基酸 Essential	$\bar{x}$	n	SD	$\bar{x}$	n	SD	$\bar{x}$	n	SD
干物质 Dry matter	87.38	8	0.55										
粗蛋白 Crude protein	15.08	10	1.08	粗蛋白 CP	15.08	10	1.08	69	2	10.57	78	2	4.96
粗纤维 Crude fiber	7.77	7	1.40	精氨酸 Arg	0.77	2	0.44	78	2	3.71	90	2	7.04
乙醚浸出物/粗脂肪 Ether extract	4.72	7	0.58	组氨酸 His	0.39	2	0.07	68	2	7.75	76	2	2.19
酸乙醚浸出物 Acid ether extract				异亮氨酸 Ile	0.47	2	0.08	72			75	2	3.90
粗灰分 Ash	4.16	7	0.59	亮氨酸 Leu	0.80	2	0.25	61	2	17.76	73	2	8.27
碳水化合物成分, %				赖氨酸 Lys	0.52	2	0.05	61	2	25.05	73	2	17.68
				蛋氨酸 Met	0.22	2	0.07	67	1		72	1	
乳糖 Lactose	0.00	7	0.00	苯丙氨酸 Phe	0.49	2	0.21	74	2	9.68	83	2	6.26
蔗糖 Sucrose	0.00	7	0.00	苏氨酸 Thr	0.60	2	0.13	48	2	20.94	64	2	6.68
棉子糖 Raffinose	0.00	7	0.00	色氨酸 Trp	0.22			59	1		73	1	
水苏糖 Stachyose	0.00	7	0.00	缬氨酸 Val	0.66	2	0.14	70	2	14.19	79	2	9.41
毛蕊花糖 Verbascose	0.00	7	0.00	非必需氨基酸 Nonessential									
低聚糖 Oligosaccharides				丙氨酸 Ala	1.79	2	1.11	52			58		
淀粉 Starch	22.56	4	7.44	天冬氨酸 Asp	3.38	2	3.07	63	2	15.60	66		
中性洗涤纤维 Neutral detergent fiber	32.28	5	6.77	半胱氨酸 Cys	0.74	1		70			77		
酸性洗涤纤维 Acid detergent fiber	11.00	6	1.61	谷氨酸 Glu	5.03	2	5.42	84	2	6.79	84		
半纤维素 Hemicellulose				甘氨酸 Gly	1.44	2	0.83	57	2	31.54	67		
酸性洗涤木质素 Acid detergent lignin				脯氨酸 Pro	0.00	1		80	2	10.78	87		
总膳食纤维 Total dietary fiber				丝氨酸 Ser	1.52	2	1.18	67	2	16.51	73		
不溶性膳食纤维 Insoluble dietary fiber				酪氨酸 Tyr	0.69	2	0.55	51	2	32.92	56	1	
可溶性膳食纤维 Soluble dietary fiber													

矿物质				维生素, mg/kg(标注单位者除外)				脂肪酸，占乙醚提取物, %			
	$\bar{x}$	n	SD		$\bar{x}$	n	SD		$\bar{x}$	n	SD
常量元素 Macro,%				脂溶性维生素 Fat soluble				粗脂肪	4.25		
钙 Ca	0.10	3	0.02	β-胡萝卜素 β-Carotene	1.0			C-12:0	0.05		
氯 Cl	0.07			维生素 E Vitamin E	16.5			C-14:0	0.16		
钾 K	1.26			水溶性维生素 Water soluble				C-16:0	13.08		
镁 Mg	0.52			维生素 B_6 Vitamin B_6	12.0			C-16:1	0.40		
钠 Na	0.04			维生素 B_{12} Vitamin B_{12}, μg/kg	0			C-18:0	0.87		
磷 P	0.99	3	0.15	生物素 Biotin	0.36			C-18:1	14.56		
硫 S	0.22			叶酸 Folacin	0.63			C-18:2	47.98		
微量元素 Micro,ppm				烟酸 Niacin	186			C-18:3	3.93		
铬 Cr				泛酸 Pantothenic acid	31.0			C-18:4			
铜 Cu	14.00			核黄素 Riboflavin	4.6			C-20:0	0.00		
铁 Fe	170			硫胺素 Thiamin	8.0			C-20:1	0.00		
碘 I				胆碱 Choline	1232			C-20:4	0.12		
锰 Mn	113							C-20:5			
硒 Se	0.51							C-22:0			
锌 Zn	100			能量, kcal/kg				C-22:1			
								C-22:5			
植酸磷 Phytate P,%	0.88	1		总能 GE	4010	7	66	C-22:6			
磷全消化道表观消化率 ATTD of P, %	46			消化能 DE	2420			C-24:0			
磷全消化道标准消化率 STTD of P, %	56			代谢能 ME	2318			饱和脂肪酸	14.16		
				净能 NE	1646			单不饱和脂肪酸	14.96		
								多不饱和脂肪酸	52.02		
								碘价	111.46		
								碘价物	47.37		

续表

原料名称：小麦酒精糟及可溶物(Wheat DDGS)
AAFCO#: 27.6, AAFCO 2010, p. 343
IFN#: 5-05-194

常规成分, %				氨基酸, %								
				总氨基酸				可消化氨基酸				
								回肠表观消化率			回肠标准消化率		
	$\bar{x}$	n	SD		$\bar{x}$	n	SD	$\bar{x}$	n	SD	$\bar{x}$	n	SD
干物质 Dry matter	92.59	20	1.77	必需氨基酸 Essential									
粗蛋白 Crude protein	36.61	23	2.78	粗蛋白 CP	36.61	23	2.78	69	10	4.82	75	10	4.96
粗纤维 Crude fiber	6.75	4	1.12	精氨酸 Arg	1.41	13	0.20	76	9	5.36	82	9	4.32
乙醚浸出/粗脂肪 Ether extract	5.34	18	1.56	组氨酸 His	0.76	13	0.09	72	10	5.33	75	10	5.37
酸乙醚浸出物 Acid ether extract	5.09	1		异亮氨酸 Ile	1.25	13	0.10	69	10	4.95	73	10	6.34
粗灰分 Ash	4.57	11	0.38	亮氨酸 Leu	2.45	13	0.23	77	10	3.84	80	10	4.02
碳水化合物成分, %				赖氨酸 Lys	0.73	15	0.17	44	11	13.66	51	10	11.14
				蛋氨酸 Met	0.52	11	0.10	70	8	7.34	78		
乳糖 Lactose				苯丙氨酸 Phe	1.67	13	0.17	82	10	3.08	84	10	2.97
蔗糖 Sucrose				苏氨酸 Thr	1.13	15	0.13	64	11	6.05	71	10	5.45
棉子糖 Raffinose				色氨酸 Trp	0.37	7	0.04	72	5	5.85	77	5	5.68
水苏糖 Stachyose				缬氨酸 Val	1.60	13	0.12	69	10	4.63	73	10	5.21
毛蕊花糖 Verbascose				非必需氨基酸 Nonessential									
低聚糖 Oligosaccharides				丙氨酸 Ala	1.35	9	0.13	64	6	2.73	70	6	2.11
淀粉 Starch	1.78	6	1.00	天冬氨酸 Asp	1.85	9	0.23	52	6	5.72	59	6	5.59
中性洗涤纤维 Neutral detergent fiber	34.7	16	8	半胱氨酸 Cys	0.61	8	0.15	69	5	11.31	76	5	8.75
酸性洗涤纤维 Acid detergent fiber	13.81	17	3.12	谷氨酸 Glu	9.59	9	1.65	79	6	13.34	87	6	1.52
半纤维素 Hemicellulose				甘氨酸 Gly	1.48	9	0.18	59	6	8.05	72	6	4.24
酸性洗涤木质素 Acid detergent lignin	4.45	1		脯氨酸 Pro	3.34	9	0.53	68	6	12.32	90	6	7.86
总膳食纤维 Total dietary fiber				丝氨酸 Ser	1.69	9	0.26	71	6	2.68	77	6	2.96
不溶性膳食纤维 Insoluble dietary fiber				酪氨酸 Tyr	1.06	7	0.05	77	5	4.27	81	5	3.82
可溶性膳食纤维 Soluble dietary fiber													

矿物质				维生素, mg/kg(标注单位者除外)				脂肪酸, 占乙醚提取物, %			
	$\bar{x}$	n	SD		$\bar{x}$	n	SD		$\bar{x}$	n	SD
常量元素 Macro,%				脂溶性维生素 Fat soluble				粗脂肪	6.50		
钙 Ca	0.16	7	0.04	β-胡萝卜素 β-Carotene				C-12:0	0.00		
氯 Cl				维生素 E Vitamin E				C-14:0	0.07		
钾 K	1.06	1		水溶性维生素 Water soluble				C-16:0	11.57		
镁 Mg	0.39	1		维生素 B_6 Vitamin B_6				C-16:1	0.26		
钠 Na	0.28	1		维生素 B_{12} Vitamin B_{12}, μg/kg				C-18:0	0.52		
磷 P	0.92	9	0.05	生物素 Biotin				C-18:1	9.88		
硫 S	0.44	1		叶酸 Folacin				C-18:2	36.66		
微量元素 Micro,ppm				烟酸 Niacin				C-18:3	3.84		
铬 Cr				泛酸 Pantothenic acid				C-18:4	0.00		
铜 Cu				核黄素 Riboflavin				C-20:0	0.00		
铁 Fe				硫胺素 Thiamin				C-20:1	0.85		
碘 I				胆碱 Choline				C-20:4	0.00		
锰 Mn								C-20:5	0.00		
硒 Se								C-22:0	0.00		
锌 Zn								C-22:1	0.00		
				能量, kcal/kg				C-22:5	0.00		
植酸磷 Phytate P,%	0.21	2	0.04	总能 GE	4650	12	165	C-22:6	0.00		
磷全消化道表观消化率 ATTD of P,%	56	3	5.98	消化能 DE	3151	6	321	C-24:0	0.00		
磷全消化道标准消化率 STTD of P,%	61	3	5.86	代谢能 ME	2902			饱和脂肪酸	12.16		
				净能 NE	1847			单不饱和脂肪酸	10.99		
								多不饱和脂肪酸	40.50		
								碘价	86.66		
								碘价物	56.33		

续表

原料名称：小麦蛋白(Wheat Gluten)
AAFCO#：无官方定义

常规成分, %				氨基酸, %								
				总氨基酸				可消化氨基酸				
								回肠表观消化率			回肠标准消化率	
	$\bar{x}$	n	SD		$\bar{x}$	n	SD	$\bar{x}$	n	SD	$\bar{x}$	n	SD
干物质 Dry matter				必需氨基酸 Essential									
粗蛋白 Crude protein	72.11	9	3.94	粗蛋白 CP	72.11	9	3.94	89	1		91	1	
粗纤维 Crude fiber				精氨酸 Arg	2.67	9	0.28	83	1		85	1	
乙醚浸出物/粗脂肪 Ether extract				组氨酸 His	1.66	9	0.36	86	1		87	1	
酸乙醚浸出物 Acid ether extract				异亮氨酸 Ile	2.66	9	0.18	86	1		87	1	
粗灰分 Ash				亮氨酸 Leu	5.06	9	0.21	90	1		91	1	
碳水化合物成分, %				赖氨酸 Lys	1.27	9	0.22	78	1		80	1	
				蛋氨酸 Met	1.08	9	0.14	83	1		85	1	
乳糖 Lactose				苯丙氨酸 Phe	3.91	9	0.32	88	1		89	1	
蔗糖 Sucrose				苏氨酸 Thr	2.42	8	0.68	68	1		72	1	
棉子糖 Raffinose				色氨酸 Trp	1.03	8	0.46	76			83		
水苏糖 Stachyose				缬氨酸 Val	2.88	9	0.24	83	1		85	1	
毛蕊花糖 Verbascose				非必需氨基酸 Nonessential									
低聚糖 Oligosaccharides				丙氨酸 Ala	2.12	1		72			79		
淀粉 Starch				天冬氨酸 Asp	3.08	1		71			79		
中性洗涤纤维 Neutral detergent fiber				半胱氨酸 Cys	1.48	1		70			76		
酸性洗涤纤维 Acid detergent fiber				谷氨酸 Glu	23.87	1		75			79		
半纤维素 Hemicellulose				甘氨酸 Gly	2.74	1		67			79		
酸性洗涤木质素 Acid detergent lignin				脯氨酸 Pro	9.67	1		68			79		
总膳食纤维 Total dietary fiber				丝氨酸 Ser	4.07	1		69			79		
不溶性膳食纤维 Insoluble dietary fiber				酪氨酸 Tyr	2.42	8	0.12	72			79		
可溶性膳食纤维 Soluble dietary fiber													

矿物质				维生素, mg/kg(标注单位者除外)				脂肪酸, 占乙醚提取物, %			
	$\bar{x}$	n	SD		$\bar{x}$	n	SD		$\bar{x}$	n	SD
常量元素 Macro,%				脂溶性维生素 Fat soluble				粗脂肪	4.00		
钙 Ca				β-胡萝卜素 β-Carotene				C-12:0	0.00		
氯 Cl				维生素 E Vitamin E				C-14:0	0.07		
钾 K				水溶性维生素 Water soluble				C-16:0	11.57		
镁 Mg				维生素 B_6 Vitamin B_6				C-16:1	0.26		
钠 Na				维生素 B_{12} Vitamin B_{12}, μg/kg				C-18:0	0.52		
磷 P				生物素 Biotin				C-18:1	9.88		
硫 S				叶酸 Folacin				C-18:2	36.66		
微量元素 Micro,ppm				烟酸 Niacin				C-18:3	3.84		
铬 Cr				泛酸 Pantothenic acid				C-18:4	0.00		
铜 Cu				核黄素 Riboflavin				C-20:0	0.00		
铁 Fe				硫胺素 Thiamin				C-20:1	0.85		
碘 I				胆碱 Choline				C-20:4	0.00		
锰 Mn								C-20:5	0.00		
硒 Se								C-22:0	0.00		
锌 Zn				能量, kcal/kg				C-22:1	0.00		
								C-22:5	0.00		
植酸磷 Phytate P,%				总能 GE				C-22:6	0.00		
磷全消化道表观消化率 ATTD of P, %				消化能 DE				C-24:0	0.00		
磷全消化道标准消化率 STTD of P, %				代谢能 ME				饱和脂肪酸	12.16		
				净能 NE				单不饱和脂肪酸	10.99		
								多不饱和脂肪酸	40.50		
								碘价	86.66		
								碘价物	34.67		

续表

原料名称：小麦细麸(Wheat Middlings)
AAFCO#: 93.5, AAFCO 2010, p.407
IFN#: 4-05-205

常规成分, %	$\bar{x}$	n	SD	氨基酸, %							
				总氨基酸				可消化氨基酸			
								回肠表观消化率		回肠标准消化率	
					$\bar{x}$	n	SD	$\bar{x}$	n	SD	$\bar{x}$	n	SD

常规成分, %	$\bar{x}$	n	SD	氨基酸	$\bar{x}$	n	SD	回肠表观消化率 $\bar{x}$	回肠标准消化率 $\bar{x}$
干物质 Dry matter	89.10	22	1.51	必需氨基酸 Essential					
粗蛋白 Crude protein	15.76	22	1.36	粗蛋白 CP	15.76	22	1.36		
粗纤维 Crude fiber	5.15	3	3.90	精氨酸 Arg	1.10	17	0.13	87	91
乙醚浸出物/粗脂肪 Ether extract	3.15	6	1.01	组氨酸 His	0.44	17	0.04	80	84
酸乙醚浸出物 Acid ether extract	2.35	1		异亮氨酸 Ile	0.51	18	0.04	77	79
粗灰分 Ash	2.05	4	0.85	亮氨酸 Leu	1.03	17	0.07	75	80
碳水化合物成分, %				赖氨酸 Lys	0.65	18	0.05	73	78
				蛋氨酸 Met	0.25	18	0.02	78	82
乳糖 Lactose	0.00	2	0.00	苯丙氨酸 Phe	0.64	17	0.06	79	84
蔗糖 Sucrose	0.00	2	0.00	苏氨酸 Thr	0.53	18	0.03	62	73
棉子糖 Raffinose	0.00	2	0.00	色氨酸 Trp	0.19	16	0.01	76	81
水苏糖 Stachyose	0.00	2	0.00	缬氨酸 Val	0.72	18	0.06	74	81
毛蕊花糖 Verbascose	0.00	2	0.00	非必需氨基酸 Nonessential					
低聚糖 Oligosaccharides				丙氨酸 Ala	0.60	2	0.03	71	77
淀粉 Starch	21.83			天冬氨酸 Asp	1.04			73	79
中性洗涤纤维 Neutral detergent fiber	34.97	17	8.52	半胱氨酸 Cys	0.35	17	0.03	71	76
酸性洗涤纤维 Acid detergent fiber	5.98	4	2.91	谷氨酸 Glu	3.10			87	91
半纤维素 Hemicellulose				甘氨酸 Gly	0.69	2	0.03	65	75
酸性洗涤木质素 Acid detergent lignin				脯氨酸 Pro	1.72	2	0.21	79	89
总膳食纤维 Total dietary fiber				丝氨酸 Ser	0.81	2	0.05	75	84
不溶性膳食纤维 Insoluble dietary fiber				酪氨酸 Tyr	0.29			77	83
可溶性膳食纤维 Soluble dietary fiber									

矿物质	$\bar{x}$	n	SD	维生素, mg/kg(标注单位者除外)	$\bar{x}$	n	SD	脂肪酸, 占乙醚提取物, %	$\bar{x}$	n	SD
常量元素 Macro,%				脂溶性维生素 Fat soluble				粗脂肪	3.60		
钙 Ca	0.11	19	0.02	β-胡萝卜素 β-Carotene	3.0			C-12:0	0.00		
氯 Cl	0.04			维生素 E Vitamin E	20.1			C-14:0	0.08		
钾 K	1.06			水溶性维生素 Water soluble				C-16:0	14.24		
镁 Mg	0.41			维生素 B_6 Vitamin B_6	9.0			C-16:1	0.32		
钠 Na	0.05			维生素 B_{12} Vitamin B_{12}, μg/kg	0			C-18:0	0.64		
磷 P	0.98	20	0.17	生物素 Biotin	0.33			C-18:1	12.16		
硫 S	0.17			叶酸 Folacin	0.76			C-18:2	45.12		
微量元素 Micro,ppm				烟酸 Niacin	72			C-18:3	4.72		
铬 Cr				泛酸 Pantothenic acid	15.6			C-18:4	0.00		
铜 Cu	10.00			核黄素 Riboflavin	1.8			C-20:0			
铁 Fe	84			硫胺素 Thiamin	16.5			C-20:1	1.04		
碘 I				胆碱 Choline	1187			C-20:4	0.00		
锰 Mn	100							C-20:5	0.00		
硒 Se	0.53	12	0.25					C-22:0	0.00		
锌 Zn	92.00			能量, kcal/kg				C-22:1	0.00		
								C-22:5	0.00		
植酸磷 Phytate P,%	0.61	1		总能 GE	3901	2	106	C-22:6	0.00		
磷全消化道表观消化率 ATTD of P,%	46			消化能 DE	3075			C-24:0	0.00		
磷全消化道标准消化率 STTD of P,%	56			代谢能 ME	2968			饱和脂肪酸	14.96		
				净能 NE	2113			单不饱和脂肪酸	13.52		
								多不饱和脂肪酸	49.84		
								碘价	106.66		
								碘价物	38.40		

续表

原料名称：小麦加工筛下物 (Wheat Screenings)
AAFCO#: 81.1, AAFCO 2010, p.389
IFN#: 4-05-216

常规成分, %				氨基酸, %								
				总氨基酸				可消化氨基酸				
								回肠表观消化率			回肠标准消化率	
	$\bar{x}$	n	SD		$\bar{x}$	n	SD	$\bar{x}$	n	SD	$\bar{x}$	n	SD
干物质 Dry matter	89.88	15	1.05	必需氨基酸 Essential									
粗蛋白 Crude protein	14.91	15	0.70	粗蛋白 CP	14.91	15	0.70						
粗纤维 Crude fiber				精氨酸 Arg									
乙醚浸出物/粗脂肪 Ether extract	5.73	15	1.63	组氨酸 His									
酸乙醚浸出物 Acid ether extract				异亮氨酸 Ile									
粗灰分 Ash				亮氨酸 Leu									
碳水化合物成分, %				赖氨酸 Lys									
				蛋氨酸 Met									
乳糖 Lactose				苯丙氨酸 Phe									
蔗糖 Sucrose	1.69	15	0.43	苏氨酸 Thr									
棉子糖 Raffinose				色氨酸 Trp									
水苏糖 Stachyose				缬氨酸 Val									
毛蕊花糖 Verbascose				非必需氨基酸 Nonessential									
低聚糖 Oligosaccharides				丙氨酸 Ala									
淀粉 Starch	46.91	15	5.12	天冬氨酸 Asp									
中性洗涤纤维 Neutral detergent fiber				半胱氨酸 Cys									
酸性洗涤纤维 Acid detergent fiber				谷氨酸 Glu									
半纤维素 Hemicellulose				甘氨酸 Gly									
酸性洗涤木质素 Acid detergent lignin				脯氨酸 Pro									
总膳食纤维 Total dietary fiber	19.22	5	1.27	丝氨酸 Ser									
不溶性膳食纤维 Insoluble dietary fiber				酪氨酸 Tyr									
可溶性膳食纤维 Soluble dietary fiber													

矿物质				维生素, mg/kg (标注单位者除外)				脂肪酸, 占乙醚提取物, %			
	$\bar{x}$	n	SD		$\bar{x}$	n	SD		$\bar{x}$	n	SD
常量元素 Macro,%				脂溶性维生素 Fat soluble				粗脂肪			
钙 Ca				β-胡萝卜素 β-Carotene				C-12:0			
氯 Cl				维生素 E Vitamin E				C-14:0			
钾 K				水溶性维生素 Water soluble				C-16:0			
镁 Mg				维生素 B_6 Vitamin B_6				C-16:1			
钠 Na				维生素 B_{12} Vitamin B_{12}, μg/kg				C-18:0			
磷 P				生物素 Biotin				C-18:1			
硫 S				叶酸 Folacin				C-18:2			
微量元素 Micro,ppm				烟酸 Niacin				C-18:3			
铬 Cr				泛酸 Pantothenic acid				C-18:4			
铜 Cu				核黄素 Riboflavin				C-20:0			
铁 Fe				硫胺素 Thiamin				C-20:1			
碘 I				胆碱 Choline				C-20:4			
锰 Mn								C-20:5			
硒 Se								C-22:0			
锌 Zn				**能量, kcal/kg**				C-22:1			
								C-22:5			
植酸磷 Phytate P,%				总能 GE				C-22:6			
磷全消化道表观消化率 ATTD of P, %				消化能 DE				C-24:0			
磷全消化道标准消化率 STTD of P, %				代谢能 ME				饱和脂肪酸			
				净能 NE				单不饱和脂肪酸			
								多不饱和脂肪酸			
								碘价			
								碘价物			

续表

原料名称：小麦次粉 (Wheat Shorts)
AAFCO#: 93.6, AAFCO 2010, p.408
IFN#: 4-05-201

常规成分, %				氨基酸, %							
				总氨基酸				可消化氨基酸			
								回肠表观消化率		回肠标准消化率	
	$\bar{x}$	n	SD		$\bar{x}$	n	SD	$\bar{x}$	n	SD	$\bar{x}$	n	SD
干物质 Dry matter	87.90			必需氨基酸 Essential							
粗蛋白 Crude protein	16.76	1		粗蛋白 CP	16.76	1		53	1		62	1
粗纤维 Crude fiber				精氨酸 Arg	1.07	1		86			88	
乙醚浸出物/粗脂肪 Ether extract	4.60			组氨酸 His	0.42	1		82			84	
酸乙醚浸出物 Acid ether extract				异亮氨酸 Ile	0.53	1		77			81	
粗灰分 Ash				亮氨酸 Leu	0.97	1		72	1		83	
碳水化合物成分, %				赖氨酸 Lys	0.59	1		62	1		76	
乳糖 Lactose				蛋氨酸 Met	0.27	1		81			84	
蔗糖 Sucrose				苯丙氨酸 Phe	0.62	1		82			84	
棉子糖 Raffinose				苏氨酸 Thr	0.51	1		72			76	
水苏糖 Stachyose				色氨酸 Trp	0.22	1		77			84	
毛蕊花糖 Verbascose				缬氨酸 Val	0.76	1		76			81	
低聚糖 Oligosaccharides				非必需氨基酸 Nonessential							
淀粉 Starch	28.60			丙氨酸 Ala	0.91	1		67	1		74	1
中性洗涤纤维 Neutral detergent fiber	29.50	1		天冬氨酸 Asp	1.11	1		66	1		73	1
酸性洗涤纤维 Acid detergent fiber	8.60			半胱氨酸 Cys	0.43	1		60	1		82	
半纤维素 Hemicellulose				谷氨酸 Glu	3.07	1		85	1		89	1
酸性洗涤木质素 Acid detergent lignin				甘氨酸 Gly	0.83	1		62	1		80	1
总膳食纤维 Total dietary fiber				脯氨酸 Pro							
不溶性膳食纤维 Insoluble dietary fiber				丝氨酸 Ser	0.63	1		67	1		75	1
可溶性膳食纤维 Soluble dietary fiber				酪氨酸 Tyr	0.26	1		78			84	

矿物质				维生素, mg/kg(标注单位者除外)				脂肪酸, 占乙醚提取物, %			
	$\bar{x}$	n	SD		$\bar{x}$	n	SD		$\bar{x}$	n	SD
常量元素 Macro,%				脂溶性维生素 Fat soluble				粗脂肪	3.50		
钙 Ca	0.08	1		β-胡萝卜素 β-Carotene				C-12:0	0.00		
氯 Cl	0.04			维生素 E Vitamin E				C-14:0	0.08		
钾 K	1.06			水溶性维生素 Water soluble				C-16:0	14.24		
镁 Mg	0.25			维生素 B_6 Vitamin B_6	7.2			C-16:1	0.32		
钠 Na	0.02			维生素 B_{12} Vitamin B_{12}, μg/kg	0			C-18:0	0.64		
磷 P	0.93	1		生物素 Biotin	0.24			C-18:1	12.16		
硫 S	0.20			叶酸 Folacin	1.40			C-18:2	45.12		
微量元素 Micro,ppm				烟酸 Niacin	107			C-18:3	4.72		
铬 Cr				泛酸 Pantothenic acid	22.3			C-18:4	0.00		
铜 Cu	12.00			核黄素 Riboflavin	3.3			C-20:0	0.00		
铁 Fe	100			硫胺素 Thiamin	18.1			C-20:1	1.04		
碘 I				胆碱 Choline	1170			C-20:4	0.00		
锰 Mn	89.00							C-20:5	0.00		
硒 Se	0.75							C-22:0	0.00		
锌 Zn	100			能量, kcal/kg				C-22:1	0.00		
								C-22:5	0.00		
植酸磷 Phytate P,%				总能 GE	4505			C-22:6	0.00		
磷全消化道表观消化率 ATTD of P,%	46			消化能 DE	2985			C-24:0	0.00		
磷全消化道标准消化率 STTD of P,%	56			代谢能 ME	2871			饱和脂肪酸	14.96		
				净能 NE	2074			单不饱和脂肪酸	13.52		
								多不饱和脂肪酸	49.84		
								碘价	106.66		
								碘价物	37.33		

续表

原料名称：啤酒酵母(Yeast, Brewers')
AAFCO#: 96.4, AAFCO 2010, p.408
IFN#: 7-05-527

常规成分, %				氨基酸, %							
				总氨基酸				可消化氨基酸			
	$\bar{x}$	n	SD		$\bar{x}$	n	SD	回肠表观消化率		回肠标准消化率	
								$\bar{x}$	n	SD	$\bar{x}$	n	SD
干物质 Dry matter	93.30			必需氨基酸 Essential							
粗蛋白 Crude protein	46.52			粗蛋白 CP	46.52						
粗纤维 Crude fiber				精氨酸 Arg	2.20			79		79	
乙醚浸出物/粗脂肪 Ether extract	2.05			组氨酸 His	1.09			77		77	
酸乙醚浸出物 Acid ether extract				异亮氨酸 Ile	2.15			74		74	
粗灰分 Ash				亮氨酸 Leu	3.13			73		73	
碳水化合物成分, %				赖氨酸 Lys	3.22			76		76	
				蛋氨酸 Met	0.74			72		72	
乳糖 Lactose				苯丙氨酸 Phe	1.83			72		72	
蔗糖 Sucrose				苏氨酸 Thr	2.20			63		66	
棉子糖 Raffinose				色氨酸 Trp	0.56			60		60	
水苏糖 Stachyose				缬氨酸 Val	2.39			70		70	
毛蕊花糖 Verbascose				非必需氨基酸 Nonessential							
低聚糖 Oligosaccharides				丙氨酸 Ala							
淀粉 Starch	4.20			天冬氨酸 Asp							
中性洗涤纤维 Neutral detergent fiber	4.00			半胱氨酸 Cys	0.50			38		48	
酸性洗涤纤维 Acid detergent fiber	3.00			谷氨酸 Glu							
半纤维素 Hemicellulose				甘氨酸 Gly							
酸性洗涤木质素 Acid detergent lignin				脯氨酸 Pro							
总膳食纤维 Total dietary fiber				丝氨酸 Ser							
不溶性膳食纤维 Insoluble dietary fiber				酪氨酸 Tyr	1.55			61		64	
可溶性膳食纤维 Soluble dietary fiber											

矿物质				维生素, mg/kg (标注单位者除外)				脂肪酸, 占乙醚提取物, %			
	$\bar{x}$	n	SD		$\bar{x}$	n	SD		$\bar{x}$	n	SD
常量元素 Macro,%				脂溶性维生素 Fat soluble				粗脂肪			
钙 Ca	0.16			β-胡萝卜素 β-Carotene				C-12:0			
氯 Cl	0.12			维生素 E Vitamin E	10.0			C-14:0			
钾 K	1.80			水溶性维生素 Water soluble				C-16:0			
镁 Mg	0.23			维生素 B_6 Vitamin B_6	42.8			C-16:1			
钠 Na	0.10			维生素 B_{12} Vitamin B_{12}, μg/kg	1			C-18:0			
磷 P	1.40	1		生物素 Biotin	0.63			C-18:1			
硫 S	0.40			叶酸 Folacin	9.90			C-18:2			
微量元素 Micro,ppm				烟酸 Niacin	448			C-18:3			
铬 Cr				泛酸 Pantothenic acid	109			C-18:4			
铜 Cu	2.70	1		核黄素 Riboflavin	37.0			C-20:0			
铁 Fe	38	1		硫胺素 Thiamin	91.8			C-20:1			
碘 I				胆碱 Choline	3984			C-20:4			
锰 Mn	8.80	1						C-20:5			
硒 Se	1.00							C-22:0			
锌 Zn	76.60	1		能量, kcal/kg				C-22:1			
								C-22:5			
植酸磷 Phytate P,%				总能 GE	4416	1		C-22:6			
磷全消化道表观消化率 ATTD of P, %	80	1		消化能 DE	4015	1		C-24:0			
磷全消化道标准消化率 STTD of P,%	85	1		代谢能 ME	3699			饱和脂肪酸			
				净能 NE	2414			单不饱和脂肪酸			
								多不饱和脂肪酸			
								碘价			
								碘价物			

续表

原料名称：白酒酵母(Yeast, Ethanol)
　　AAFCO#：无官方定义

常规成分, %				氨基酸, %								
				总氨基酸			可消化氨基酸					
							回肠表观消化率			回肠标准消化率		
	$\bar{x}$	n	SD		$\bar{x}$	n	SD	$\bar{x}$	n	SD	$\bar{x}$	n	SD
干物质 Dry matter	93.30			必需氨基酸 Essential									
粗蛋白 Crude protein	46.52			粗蛋白 CP	46.52								
粗纤维 Crude fiber				精氨酸 Arg									
乙醚浸出物/粗脂肪 Ether extract	2.05			组氨酸 His									
酸乙醚浸出物 Acid ether extract				异亮氨酸 Ile									
粗灰分 Ash				亮氨酸 Leu									
碳水化合物成分, %				赖氨酸 Lys									
				蛋氨酸 Met									
乳糖 Lactose				苯丙氨酸 Phe									
蔗糖 Sucrose				苏氨酸 Thr									
棉子糖 Raffinose				色氨酸 Trp									
水苏糖 Stachyose				缬氨酸 Val									
毛蕊花糖 Verbascose				非必需氨基酸 Nonessential									
低聚糖 Oligosaccharides				丙氨酸 Ala									
淀粉 Starch	0.00			天冬氨酸 Asp									
中性洗涤纤维 Neutral detergent fiber				半胱氨酸 Cys									
酸性洗涤纤维 Acid detergent fiber	3.00			谷氨酸 Glu									
半纤维素 Hemicellulose				甘氨酸 Gly									
酸性洗涤木质素 Acid detergent lignin				脯氨酸 Pro									
总膳食纤维 Total dietary fiber				丝氨酸 Ser									
不溶性膳食纤维 Insoluble dietary fiber				酪氨酸 Tyr									
可溶性膳食纤维 Soluble dietary fiber													
矿物质				维生素, mg/kg(标注单位者除外)				脂肪酸, 占乙醚提取物, %					
	$\bar{x}$	n	SD		$\bar{x}$	n	SD		$\bar{x}$	n	SD		
常量元素 Macro,%				脂溶性维生素 Fat soluble				粗脂肪					
钙 Ca	0.29	2	0.00	β-胡萝卜素 β-Carotene				C-12:0					
氯 Cl				维生素 E Vitamin E				C-14:0					
钾 K				水溶性维生素 Water soluble				C-16:0					
镁 Mg				维生素 B_6 Vitamin B_6				C-16:1					
钠 Na				维生素 B_{12} Vitamin B_{12}, μg/kg				C-18:0					
磷 P	0.68	2	0.01	生物素 Biotin				C-18:1					
硫 S				叶酸 Folacin				C-18:2					
微量元素 Micro,ppm				烟酸 Niacin				C-18:3					
铬 Cr				泛酸 Pantothenic acid				C-18:4					
铜 Cu				核黄素 Riboflavin				C-20:0					
铁 Fe				硫胺素 Thiamin				C-20:1					
碘 I				胆碱 Choline				C-20:4					
锰 Mn								C-20:5					
硒 Se								C-22:0					
锌 Zn				能量, kcal/kg				C-22:1					
								C-22:5					
植酸磷 Phytate P,%				总能 GE	4648			C-22:6					
磷全消化道表观消化率 ATTD of P,%	57	2	4.10	消化能 DE	4015			C-24:0					
磷全消化道标准消化率 STTD of P,%	70	2	4.10	代谢能 ME	3699			饱和脂肪酸					
				净能 NE	2394			单不饱和脂肪酸					
								多不饱和脂肪酸					
								碘价					
								碘价物					

续表

原料名称：酵母单细胞蛋白(Yeast, Single Cell Protein)
AAFCO#：无官方定义

常规成分, %				氨基酸, %								
				总氨基酸				可消化氨基酸				
								回肠表观消化率			回肠标准消化率		
	$\bar{x}$	n	SD		$\bar{x}$	n	SD	$\bar{x}$	n	SD	$\bar{x}$	n	SD
干物质 Dry matter	93.30			必需氨基酸 Essential									
粗蛋白 Crude protein	36.25	1		粗蛋白 CP	36.25	1		66	1		69	1	
粗纤维 Crude fiber				精氨酸 Arg	1.45	1		73	1		75	1	
乙醚浸出物/粗脂肪 Ether extract	2.05			组氨酸 His	0.71	1		64	1		66	1	
酸乙醚浸出物 Acid ether extract				异亮氨酸 Ile	1.36	1		57	1		59	1	
粗灰分 Ash				亮氨酸 Leu	1.81	1		59	1		61	1	
碳水化合物成分, %				赖氨酸 Lys	2.58	1		73	1		74	1	
				蛋氨酸 Met	0.84	1		87	1		88	1	
乳糖 Lactose				苯丙氨酸 Phe	1.18	1		51	1		53	1	
蔗糖 Sucrose				苏氨酸 Thr	1.42	1		51	1		54	1	
棉子糖 Raffinose				色氨酸 Trp									
水苏糖 Stachyose				缬氨酸 Val	1.53	1		55	1		58	1	
毛蕊花糖 Verbascose				非必需氨基酸 Nonessential									
低聚糖 Oligosaccharides				丙氨酸 Ala	1.45	1		51	1		52	1	
淀粉 Starch	0.00			天冬氨酸 Asp	2.30	1		52	1		55	1	
中性洗涤纤维 Neutral detergent fiber				半胱氨酸 Cys									
酸性洗涤纤维 Acid detergent fiber	3.00			谷氨酸 Glu	3.56	1		60	1		62	1	
半纤维素 Hemicellulose				甘氨酸 Gly	1.31	1		48	1		56	1	
酸性洗涤木质素 Acid detergent lignin				脯氨酸 Pro	1.10	1		55	1		65	1	
总膳食纤维 Total dietary fiber				丝氨酸 Ser	1.26	1		56	1		60	1	
不溶性膳食纤维 Insoluble dietary fiber				酪氨酸 Tyr	0.61	1		60	1				
可溶性膳食纤维 Soluble dietary fiber													

矿物质				维生素, mg/kg(标注单位者除外)				脂肪酸, 占乙醚提取物, %			
	$\bar{x}$	n	SD		$\bar{x}$	n	SD		$\bar{x}$	n	SD
常量元素 Macro,%				脂溶性维生素 Fat soluble				粗脂肪			
钙 Ca				β-胡萝卜素 β-Carotene				C-12:0			
氯 Cl				维生素 E Vitamin E				C-14:0			
钾 K				水溶性维生素 Water soluble				C-16:0			
镁 Mg				维生素 B_6 Vitamin B_6				C-16:1			
钠 Na				维生素 B_{12} Vitamin B_{12}, μg/kg				C-18:0			
磷 P	1.54	2	0.67	生物素 Biotin				C-18:1			
硫 S				叶酸 Folacin				C-18:2			
微量元素 Micro,ppm				烟酸 Niacin				C-18:3			
铬 Cr				泛酸 Pantothenic acid				C-18:4			
铜 Cu				核黄素 Riboflavin				C-20:0			
铁 Fe				硫胺素 Thiamin				C-20:1			
碘 I				胆碱 Choline				C-20:4			
锰 Mn								C-20:5			
硒 Se								C-22:0			
锌 Zn				能量, kcal/kg				C-22:1			
								C-22:5			
植酸磷 Phytate P,%				总能 GE	3725	2	1698	C-22:6			
磷全消化道表观消化率 ATTD of P, %	70	2	3.25	消化能 DE	4166	2	128	C-24:0			
磷全消化道标准消化率 STTD of P, %	75	2	1.31	代谢能 ME	3920			饱和脂肪酸			
				净能 NE	2593			单不饱和脂肪酸			
								多不饱和脂肪酸			
								碘价			
								碘价物			

续表

原料名称：圆酵母(Yeast, Torula)
AAFCO#: 96.7, AAFCO 2010, p.408
IFN#: 7-05-534

常规成分, %				氨基酸, %								
				总氨基酸				可消化氨基酸				
								回肠表观消化率			回肠标准消化率		
	$\bar{x}$	n	SD		$\bar{x}$	n	SD	$\bar{x}$	n	SD	$\bar{x}$	n	SD
干物质 Dry matter	93.30			必需氨基酸 Essential									
粗蛋白 Crude protein	51.17	1		粗蛋白 CP	51.17	1							
粗纤维 Crude fiber				精氨酸 Arg	2.99	1							
乙醚浸出物/粗脂肪 Ether extract	2.05			组氨酸 His	1.02	1							
酸乙醚浸出物 Acid ether extract				异亮氨酸 Ile	2.26	1							
粗灰分 Ash				亮氨酸 Leu	3.41	1							
碳水化合物成分, %				赖氨酸 Lys	3.39	1							
				蛋氨酸 Met	0.64	1							
乳糖 Lactose				苯丙氨酸 Phe	2	1							
蔗糖 Sucrose				苏氨酸 Thr	2.28	1							
棉子糖 Raffinose				色氨酸 Trp	0.59	1							
水苏糖 Stachyose				缬氨酸 Val	2.72	1							
毛蕊花糖 Verbascose				非必需氨基酸 Nonessential									
低聚糖 Oligosaccharides				丙氨酸 Ala									
淀粉 Starch	0.00			天冬氨酸 Asp									
中性洗涤纤维 Neutral detergent fiber				半胱氨酸 Cys	0.52	1							
酸性洗涤纤维 Acid detergent fiber	3.00			谷氨酸 Glu									
半纤维素 Hemicellulose				甘氨酸 Gly									
酸性洗涤木质素 Acid detergent lignin				脯氨酸 Pro									
总膳食纤维 Total dietary fiber				丝氨酸 Ser									
不溶性膳食纤维 Insoluble dietary fiber				酪氨酸 Tyr	1.65								
可溶性膳食纤维 Soluble dietary fiber													

矿物质				维生素, mg/kg(标注单位者除外)				脂肪酸, 占乙醚提取物, %			
	$\bar{x}$	n	SD		$\bar{x}$	n	SD		$\bar{x}$	n	SD
常量元素 Macro,%				脂溶性维生素 Fat soluble				粗脂肪			
钙 Ca	0.58			β-胡萝卜素 β-Carotene				C-12:0			
氯 Cl	0.12			维生素 E Vitamin E				C-14:0			
钾 K	1.94			水溶性维生素 Water soluble				C-16:0			
镁 Mg	0.20			维生素 B_6 Vitamin B_6	36.3			C-16:1			
钠 Na	0.07			维生素 B_{12} Vitamin B_{12}, µg/kg				C-18:0			
磷 P	1.52			生物素 Biotin	0.58			C-18:1			
硫 S	0.55			叶酸 Folacin	22.4			C-18:2			
微量元素 Micro,ppm				烟酸 Niacin	492			C-18:3			
铬 Cr				泛酸 Pantothenic acid	84.2			C-18:4			
铜 Cu	17.00			核黄素 Riboflavin	49.9			C-20:0			
铁 Fe	222			硫胺素 Thiamin	6.2			C-20:1			
碘 I				胆碱 Choline	2881			C-20:4			
锰 Mn	13.00							C-20:5			
硒 Se	0.01	1						C-22:0			
锌 Zn	99			能量, kcal/kg				C-22:1			
								C-22:5			
植酸磷 Phytate P,%				总能 GE	4718			C-22:6			
磷全消化道表观消化率 ATTD of P,%				消化能 DE	4015			C-24:0			
磷全消化道标准消化率 STTD of P,%				代谢能 ME	3667			饱和脂肪酸			
				净能 NE	2351			单不饱和脂肪酸			
								多不饱和脂肪酸			
								碘价			
								碘价物			

表 17-2 常量矿物质原料中矿物质元素含量（数据以饲喂状态为基础）[a]

项目编号	名称	国际标准饲料编号[b]	钙 /%	磷 总量 /%	磷 ATTD /%	磷 STTD /%	钠 /%	氯 /%	钾 /%	镁 /%	硫 /%	铁 /%	锰 /%
1	蒸汽骨粉 Bone meal, steamed	6-00-400	29.8	12.5	—	—	0.04	—	0.2	0.3	2.4	—	0.03
2	碳酸钙 Calcium carbonate	6-01-069	38.5	0.02	—	—	0.08	0.02	0.08	1.61	0.08	0.06	0.02
3	磷酸氢钙 Calcium phosphate (dicalcium)	6-01-080	24.8(25)	18.8(26)	73.9(16)	81.4(16)	0.20(4)	0.47	0.15	0.5(4)	0.1(4)	0.80(4)	0.14
4	磷酸一钙 Calcium phosphate (monocalcium)	6-26-334	16.9(14)	21.5(15)	82.8(14)	88.3(14)	0.2	—	0.16	0.9	0.8	0.75	0.01
5	磷酸三钙 Calcium phosphate (tricalcium)	6-01-084	34.2(3)	17.7(3)	48.0(2)	53.4(2)	6.0(1)	—	—	0.4(1)	0.01(1)	—	—
6	二水硫酸钙 Calcium sulfate, dihydrate	6-01-090	21.85	—	—	—	—	—	—	0.48	16.19	—	—
7	石粉 Limestone, ground[c]	6-02-632	35.84	0.01	—	—	0.06	0.02	0.11	2.06	0.04	0.35	0.02
8	碳酸镁 Magnesium carbonate	6-02-754	0.02	—	—	—	—	—	—	30.2	—	—	0.01
9	氧化镁 Magnesium oxide	6-02-756	1.69	—	—	—	—	—	0.02	55	0.1	1.06	—
10	磷酸镁 Magnesium phosphate	6-23-294	10.1(1)	19.7(1)	83.9(1)	98.2(1)	—	—	—	—	—	—	—
11	七水硫酸镁 Magnesium sulfate, heptahydrate	6-02-758	0.02	—	—	—	—	0.01	0	9.6	13.04	—	—
12	脱氟磷酸三钙 Phosphate, defluorinated	6-01-780	32	18	—	—	3.27	—	0.1	0.29	0.13	0.84[d]	0.05
13	磷酸一氢铵 Phosphate, monoammonium	6-09-338	0.35	24.2	—	—	0.2	—	0.16	0.75	1.5	0.41	0.01
14	磷酸盐矿石粉 Phosphate, rock curacao, ground	6-05-586	35.09	14.23	—	—	0.2	—	—	0.8	—	0.35	—
15	软质磷矿石 Phosphate, rock, soft	6-03-947	16.09	9.05	—	—	0.1	—	—	0.38	—	1.92	0.1
16	氯化钾 Potassium chloride	6-03-755	0.05	—	—	—	1	46.93	51.37	0.23	0.32	0.06	0.001
17	硫酸镁钾 Potassium and magnesium sulfate	6-06-177	0.06	—	—	—	0.76	1.25	18.45	11.58	21.97	0.01	0.002
18	硫酸钾 Potassium sulfate	6-08-098	0.15	—	—	—	0.09	1.5	43.04	0.6	17.64	0.07	0.001
19	碳酸钠 Sodium carbonate	6-12-316	—	—	—	—	43.3	—	—	—	—	—	—
20	碳酸氢钠 Sodium bicarbonate	6-04-272	0.01	—	—	—	27	—	0.01	—	—	—	—
21	氯化钠 Sodium chloride	6-04-152	0.3	—	—	—	39.5	59	0	0.005	0.2	0.01	—
22	磷酸氢二钠 Sodium phosphate, dibasic	6-04-286	—	21.15	—	—	31.04	—	—	—	—	—	—
23	磷酸二氢钠 Sodium phosphate, monobasic	6-04-288	0.09	24.7(4)	86.7(4)	93.8(4)	19.1(1)	0.02	0.01	0.01	—	0.01	—
24	十水硫酸钠 Sodium sulfate, decahydrate	6-04-291	—	—	—	—	13.8	—	—	—	9.7	—	—

注："—"表示没有可用数据。

a 括号中数字是每个观测值的平均值，如果在现有文献中没有找到相应数据，则采用 NRC（1998）中数值。

b 评估中数字表明，大多数来源的磷酸氢钙、磷酸二钙、磷酸三钙、硫酸钙、碳酸盐、脱氟磷酸盐、碳酸钙和方解石灰石的钙的生物利用率为 90%～100%，来自高铁质石灰石或白云质石灰石的钙的生物利用率非常低（50%～80%）。

c 与给出数据相比，大部分钙质石灰岩含有 38%或更多的钙和更少的镁。

d 脱氟磷酸盐中铁利用率约为 65%，与硫酸亚铁中铁相当。

表 17-3　无机微量元素的来源和生物利用率[a]

矿物元素和来源[b]	化学式	矿物元素含量/%	相对生物利用率/%
铜 Copper			
五水硫酸铜 Cupric sulfate(pentahydrate)	$CuSO_4 \cdot 5H_2O$	25.2	100
碱式氯化铜 Cupric chloride, tribasic	$Cu_2(OH)_3Cl$	58	100
氧化铜 Cupric oxide	CuO	75	0~10
碱式碳酸铜 Cupric carbonate (monohydrate)	$CuCO_3 \cdot Cu(OH)_2 \cdot H_2O$	50~55	60~100
无水硫酸铜 Cupric sulfate (anhydrous)	$CuSO_4$	39.9	100
铁 Iron			
一水硫酸亚铁 Ferrous sulfate (monohydrate)	$FeSO_4 \cdot H_2O$	30	100
七水硫酸亚铁 Ferrous sulfate (heptahydrate)	$FeSO_4 \cdot 7H_2O$	20	100
碳酸铁 Ferrous carbonate	$FeCO_3$	38	15~80
三氧化二铁 Ferric oxide	Fe_2O_3	69.9	0
六水三氯化铁 Ferric chloride (hexahydrate)	$FeCl_3 \cdot 6H_2O$	20.7	40~100
氧化亚铁 Ferrous oxide	FeO	77.8	—[c]
碘 Iodine			
二氢碘酸乙二胺 Ethylenediamine dihydroiodide(EDDI)	$C_2H_8N_2 \cdot 2HI$	79.5	100
碘酸钙 Calcium iodate	$Ca(IO_3)_2$	63.5	100
碘化钾 Potassium iodide	KI	68.8	100
碘酸钾 Potassium iodate	KIO_3	59.3	—[c]
碘化铜 Cupric iodite	CuI	66.6	100
锰 Manganese			
一水硫酸锰 Manganese sulfate(monohydrate)	$MnSO_4 \cdot H_2O$	29.5	100
氧化锰 Manganese oxide	MnO	60	70
二氧化锰 Manganese dioxide	MnO_2	63.1	35~95
碳酸锰 Manganous carbonate	$MnCO_3$	46.4	30~100
四水氯化锰 Manganous chloride (tetrahydrate)	$MnCl_2 \cdot 4H_2O$	27.5	100
硒 Selenium			
亚硒酸钠 Sodium selenite	Na_2SeO_3	45	100
十水硒酸钠 Sodium selenate (decahydrate)	$Na_2SeO_4 \cdot 10H_2O$	21.4	100
锌 Zinc			
一水硫酸锌 Zinc sulfate (monohydrate)	$ZnSO_4 \cdot H_2O$	35.5	100
氧化锌 Zinc oxide	ZnO	72	50~80
七水硫酸锌 Zinc sulfate (heptahydrate)	$ZnSO_4 \cdot 7H_2O$	22.3	100
碳酸锌 Zinc carbonate	$ZnCO_3$	56	100
氯化锌 Zinc chloride	$ZnCl_2$	48	100

a 在各矿物元素的来源中，一般来讲，以列在第一位的生物利用率为标准，其他来源的利用率是与第一位的相对值。
b 斜体为较不常用的矿物来源。
c "—"表示没有可用数据。

表17-4 不同来源油脂的特性与能值的差异（数据以饲喂状态为基础）[a]

油脂类型	IFN	≤C10	C12:0	C14:0	C16:0	C16:1	C18:0	C18:1	C18:2	C18:3	C20:1	C20:4	C20:5	C22:1	C22:5	C22:6	总饱和脂肪酸	总不饱和脂肪酸	U:S比例	IV[b]	DE[c]	ME[d]	NE[e]
动物脂肪 Animal fats																							
牛油 Beef tallow	4-08-127	0	0.9	3.7	24.9	4.2	18.9	36.0	3.1	0.6	0.3	0	0	0	0	0	48.4	44.2	0.91	44	7995	7835	6895
精选白色动物油脂 Choice white grease	—	0.2	0.2	1.9	21.5	5.7	14.9	41.1	11.6	0.4	1.8	0	0	0	0	0	40.8	59.2	1.45	60	8290	8124	7149
禽油 Poultry	4-09-319	0	0.1	0.9	21.6	5.7	6.0	37.4	19.5	1.0	1.1	0.1	0	0	0	0	28.7	64.8	2.26	79	8535	8364	7361
猪油 Lard	4-04-790	0.1	0.2	1.3	23.8	2.7	13.5	41.2	10.2	1.0	1.0	0	0	0	0	0	38.9	56.1	1.44	62	8288	8123	7148
潲水油 Restaurant grease	—	—	—	1.9	16.2	2.5	10.5	47.5	17.5	1.9	1.0	0	0	0	0	0	29.9	70.1	2.34	75	8550	8379	7374
鱼油 Fish oils																							
鲱鱼油 Herring	7-08-048	0	0.2	7.2	11.7	9.6	0.8	12.0	1.2	0.8	13.6	0.3	6.3	20.6	0.6	4.2	19.9	71.4	3.6	109	8629	8519	7496
油鲱油 Menhaden	7-08-049	0	0	8.0	15.2	10.5	3.8	14.5	2.2	1.5	1.3	1.2	13.2	0.4	4.9	8.6	26.9	60.9	2.27	161	8535	8365	7361
鲑鱼油 Salmon	—	0	0	3.3	9.8	4.8	4.3	17.0	1.5	1.1	3.9	0.7	13.0	3.4	3.0	18.2	17.4	69.4	3.99	195	8713	8538	7514
沙丁鱼油 Sardine	—	0	0.1	6.5	16.7	7.5	3.9	14.8	2.0	1.3	6.0	1.8	10.1	5.6	2.0	10.7	27.2	64.7	2.38	154	8558	8387	7381
植物油 Vegetable oils																							
油菜籽油 Canola	4-06-144	0	0	0	4.0	0.2	1.8	56.1	20.3	9.3	1.7	0	0	0.6	0	0	7.1	88.2	12.42	115	8759	8384	7554
椰子油 Coconut	—	5.6	43.8	16.8	8.4	0	2.5	5.9	1.7	0	0	0	0	0	0	0	77.0	7.59	0.11	8	7169[f]	7025	6182
玉米油 Corn	4-07-882	0	0	0	10.6	0.1	1.9	27.3	53.5	1.16	0.1	0	0	0	0	0	12.9	82.3	6.39	125	8754	8579	7549
棉籽油 Cottonseed	4-20-836	0	0	0.8	22.7	0.8	2.3	17.0	51.5	0.2	0	0.1	0	0	0	0	25.8	69.6	2.70	110	8608	8436	7424
亚麻籽油 Flaxseed	—	0	0	0.2	5.3	0	4.1	20.2	12.7	53.3	0	0	0	0	0	0	9.4	86.2	9.17	187	8759	8583	7553
燕麦油 Oat	—	0	0	0.2	16.7	0.2	1.1	34.9	39.1	1.8	0	0	0	0	0	0	18.4	76.0	4.14	107	8718	8544	7519
橄榄油 Olive	—	0	0	0	11.3	1.3	2.0	71.3	9.8	0.8	0.3	0	0	0	0	0	13.79	83.36	6.05	85	8752	8577	7548
棕榈核油 Palm kernel	—	3.7	47.0	16.4	8.1	0	2.8	11.4	1.6	0	0	0	0	0	0	0	78.0	13.0	0.17	13	7265	7119	6265
花生油 Peanut	4-03-658	0	0	0.1	9.5	0.1	2.2	44.8	32.0	0	1.3	0	0	0	0	0	16.9	78.2	4.63	99	8733	8558	7531
红花油 Safflower	—	0	0	0	4.3	0	1.9	14.4	74.6	0	0	0	0	0	0	0	6.2	89.0	14.34	148	8759	8584	7554
芝麻油 Sesame	—	0	0	0	8.9	0.2	4.8	39.3	41.3	0.3	0.2	0	0	0	0	0	13.7	81.3	5.93	111	8751	8576	7547
大豆油 Soybean	4-07-983	0	0	0.1	10.3	0	3.8	22.8	51.0	6.8	0.2	0	0	0	0	0	14.2	81.0	5.70	132	8749	8574	7545
大豆磷脂 Soybean lecithin	—	0	0	0	12.0	0.4	2.9	10.6	40.2	5.1	0	0	0	0	0	0	15.0	56.3	3.75	97	8701	8527	7504
葵花籽油 Sunflower	4-20-833	0	0	0	5.4	0.2	3.5	45.3	39.8	0.2	0	0	0	0	0	0	8.9	85.5	9.61	114	8760	8585	7555
混合油 Blends																							
动物-植物油混合油[g] Animal-vegetable blend	—	0	0.3	1.5	20.2	3.2	10.1	35.5	21.6	0.9	0.6	0.03	0	0	0	0	32.2	61.8	2.75	77	8393	8225	7238

a 脂肪酸数据按精选白色动物油脂和淡水油脂来自油脂和蛋白质质量研究基金会第23次公告（http://www.fprf.org/）外，其余全部来自美国农业部食品成分资料库（http://www.nal.usda.gov/fnic/foodcomp/search/）。
b 按脂肪酸组成计算（见第1章）。
c 按以下公式计算（Powles et al., 1995；见第3章）DE(kcal/kg)=[36.898−(0.005×FFA)−(7.330×e^(−0.906×U:S)]/0.004 184。其中，FFA为游离脂肪酸浓度假设为50 g/kg（或5%）。
d ME=DE×0.98（见第1章）。
e NE=ME×0.88（van Milgen et al., 2001; 见第1章）。
f 椰子油含量是按照Cera等（1989）的报道，3周龄仔猪断奶后2~4周的消化率（GE的89.42%）计算得到。
g 动物-植物油混合油=25%猪油、25%禽油、25%牛油和25%玉米油。

附录 A 模型使用指南

概述

此模型主要用于对仔猪、生长育肥猪、妊娠母猪和泌乳母猪四种不同类别猪的营养需要量的预测。模型将考虑每个类别营养需要量的关键决定因素（如生产水平和生产阶段），对生长性能、营养利用、营养需要各方面的内容将以图表形式呈现，并总结成报告以供打印。

日粮养分可用几种系统表示。①能量：消化能、代谢能或净能；②氨基酸和氮：总量、回肠表观可消化、回肠标准可消化；③磷：总量、全消化道表观可消化、全消化道标准可消化。在使用模型预测需要量之前，须选择养分系统。

模型还可以评估特定饲喂方案：①根据营养平衡计算流失到环境中的营养素；②将饲喂方案中模型估算的营养需要量与日粮营养水平相比较。饲喂方案是需要特定的日粮、特定时期、特定的体重范围的阶段饲喂计划。模型可以生成饲喂方案并可以存储在数据库中便于将来使用，同时生成一个饲料原料营养成分表和一个简单的饲料配方程序，并提供了日粮和饲喂方案的例子。

模型也可直接比较预测生长性能和实际观测生长性能。当模型预测生长性能和实际生长性能相近，模型预测的营养需要可信度一般都比较高。为了更加准确地预测生长育肥猪生长性能，本地的胴体评价方案就需要更加的详细。

关于模型计算的细节信息收录在 NRC（2012）第 8 章。

一系列的模型学习案例将归纳到一个 PDF 文件中，这些案例阐述了模型中每一个部分并展示出模型的特征和局限性。

使用模型

开始

至少需要微软 Excel 2002（XP）或者以上版本来支持模型的运算。模型设计基于微软 Windows 或者苹果（Apple）操作系统，不支持 Mac Excel 2008 [因为 Mac Excel 2008 无 Visual Basic 宏（macro）支持功能]。建议将模型的原始版本和个人版本以不同名字保存。在大幅度修改日粮配方和饲喂方案后，应以额外的版本形式保存。**模型含有"宏"，Excel 支持的宏可以在模型中使用。模型中的"宏"有国家科学院（National Academy of Science）的数字签名，在大多数情况下，接受该数字签名为可靠来源（trusted）即可运行"宏"。如果不能，请使用人工加载**[1]。一旦打开程序，请点击接受（Accept）按钮，表示接受模型使用风险，随后将显示主菜单（Main Menu）。将光标到有小红色三角形标记的格子，就会显示相关的备注，整个模型都是如此。

---

1 备注：如果在微软 Excel 2007 及以上版本中启用宏，先打开 Excel，点击窗口左上角的图标，然后在新窗口底部选择"Excel 选项（Excel Options）"，接着依次选"信任中心（Trust Center）"、"信任中心设置（Trust Center Settings）"、"宏设置（Macro Settings）"、"启用所有宏（Enable all macros）"。在使用模型后，"宏设置"会恢复到以前的设置。

主菜单

主菜单（图 A-1）用于选择有关能量、氨基酸和磷的不同营养系统。点击白色数据输入区的下拉菜单进入自选项。如果评估饲喂方案，则需要在主菜单指定。最初使用时，不建议使用评估饲喂方案。下面将讨论如何生成和储存饲料方案。在主菜单中，可以选择不同类别的猪。

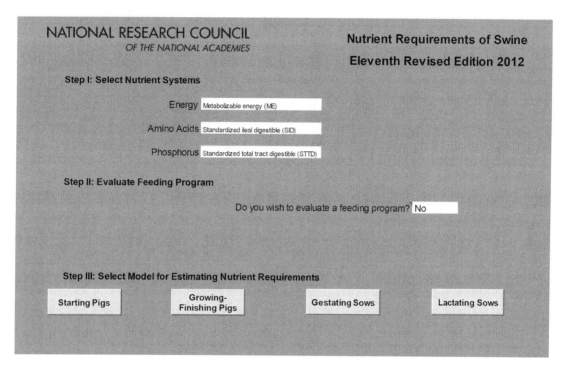

图 A-1　主菜单

模型：仔猪、生长育肥猪、妊娠母猪、泌乳母猪

对于每个模型来说（图 A-2~图 A-5），可以直接在白色数据输入区输入或者选择下拉菜单输入。当选用某些选项时，数据输入区域将显示或者隐藏。例如，当选用定义采食量的方式或者实测与模型预测性能匹配的方式时，新的数据输入区域就会出现。如果输入值改变，则需点击屏幕上方的计算（Calculate）按钮，执行模型计算。对仔猪模型而言，如果输入值改变，模型会自动运行计算。

在图 A-2~图 A-5 的结果（Results）页面，改变初始和最终体重或者时间，可观测到体重范围（仔猪、生长育肥猪）或者时间范围（妊娠母猪、泌乳母猪）对营养需要量的影响。改变体重或时间后，不用重新运行模型，因为每次模型运行时会产生表格，结果可从表格中提取。借助屏幕上端的按钮，可回到主菜单，或将输入值重新设为默认值，或观看图表、打印报告。

生长育肥猪、妊娠母猪、泌乳母猪模型中的生产水平可以改变，使预测值与实际观察值相一致。这三类猪的维持能量需要可以调整。对妊娠母猪和泌乳母猪来说，可改变母体体重变化的组成（如体蛋白和体脂肪的比例）。对生长育肥猪而言，很多操作选项既可以改变体蛋白沉积曲线，也可以改变能量摄入与体蛋白沉积的关系。这些复杂的情况要小心应对。点击胴体评估选项（Carcass Evaluation Options），可改变胴体评估参数。使预测值与实际观察值相一致是一个反复的过程，需要不断调整，反复运行，不断对比，最后得到一个合理的结果。

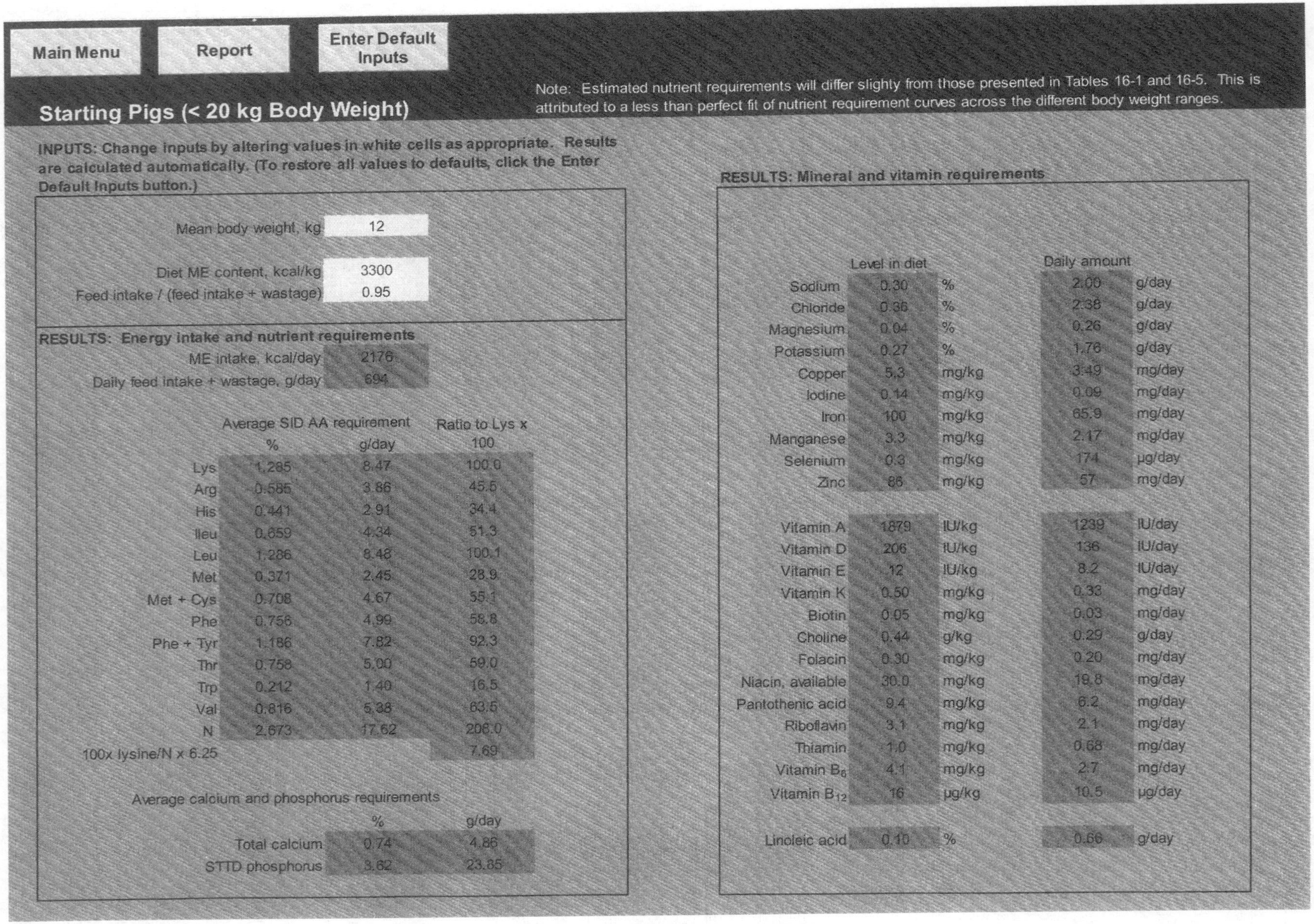

Starting Pigs (< 20 kg Body Weight)

| Main Menu | Report | Enter Default Inputs |

INPUTS: Change inputs by altering values in white cells as appropriate. Results are calculated automatically. (To restore all values to defaults, click the Enter Default Inputs button.)

Note: Estimated nutrient requirements will differ slightly from those presented in Tables 16-1 and 16-5. This is attributed to a less than perfect fit of nutrient requirement curves across the different body weight ranges.

Mean body weight, kg	12
Diet ME content, kcal/kg	3300
Feed intake / (feed intake + wastage)	0.95

RESULTS: Energy intake and nutrient requirements

ME intake, kcal/day	2176
Daily feed intake + wastage, g/day	694

Average SID AA requirement

	%	g/day	Ratio to Lys × 100
Lys	1.295	8.47	100.0
Arg	0.585	3.86	45.5
His	0.441	2.91	34.4
Ileu	0.659	4.34	51.3
Leu	1.286	8.48	100.1
Met	0.371	2.45	28.9
Met + Cys	0.708	4.67	55.1
Phe	0.756	4.99	58.8
Phe + Tyr	1.186	7.82	92.3
Thr	0.758	5.00	59.0
Trp	0.212	1.40	16.5
Val	0.816	5.38	63.5
N	2.673	17.62	206.0
100× lysine/N × 6.25			7.69

Average calcium and phosphorus requirements

	%	g/day
Total calcium	0.74	4.86
STTD phosphorus	3.62	23.85

RESULTS: Mineral and vitamin requirements

	Level in diet		Daily amount	
Sodium	0.30	%	2.00	g/day
Chloride	0.36	%	2.38	g/day
Magnesium	0.04	%	0.26	g/day
Potassium	0.27	%	1.76	g/day
Copper	5.3	mg/kg	3.49	mg/day
Iodine	0.14	mg/kg	0.09	mg/day
Iron	100	mg/kg	65.9	mg/day
Manganese	3.3	mg/kg	2.17	mg/day
Selenium	0.3	mg/kg	174	μg/day
Zinc	85	mg/kg	57	mg/day
Vitamin A	1879	IU/kg	1239	IU/day
Vitamin D	206	IU/kg	136	IU/day
Vitamin E	12	IU/kg	8.2	IU/day
Vitamin K	0.50	mg/kg	0.33	mg/day
Biotin	0.05	mg/kg	0.03	mg/day
Choline	0.44	g/kg	0.29	g/day
Folacin	0.30	mg/kg	0.20	mg/day
Niacin, available	30.0	mg/kg	19.8	mg/day
Pantothenic acid	9.4	mg/kg	6.2	mg/day
Riboflavin	3.1	mg/kg	2.1	mg/day
Thiamin	1.0	mg/kg	0.68	mg/day
Vitamin B_6	4.1	mg/kg	2.7	mg/day
Vitamin B_{12}	16	μg/kg	10.5	μg/day
Linoleic acid	0.10	%	0.66	g/day

图 A-2 仔猪模型的输入和结果界面

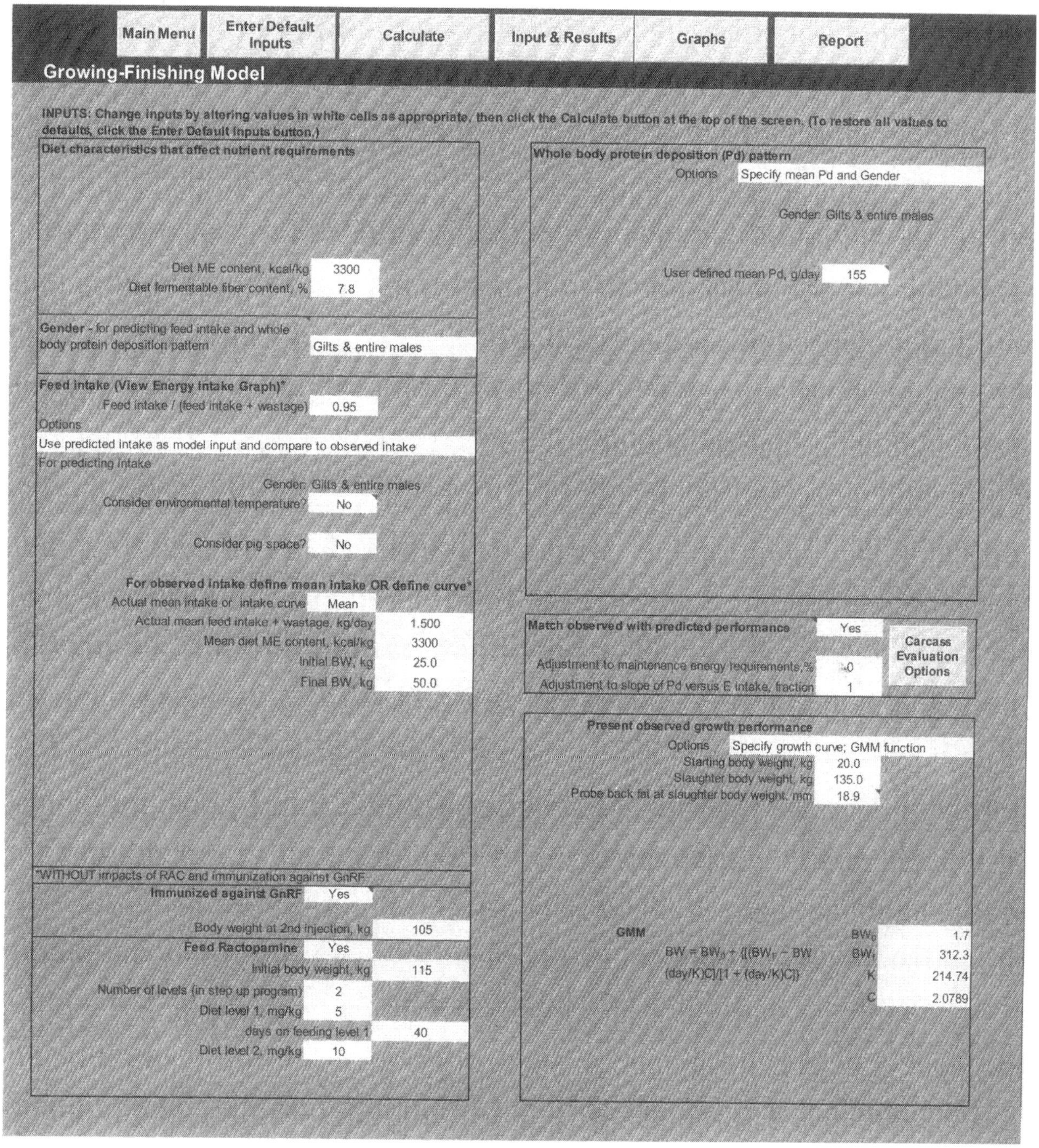

图 A-3a 生长-育肥猪模型的输入界面

图 A-3b 生长-育肥猪模型的结果界面

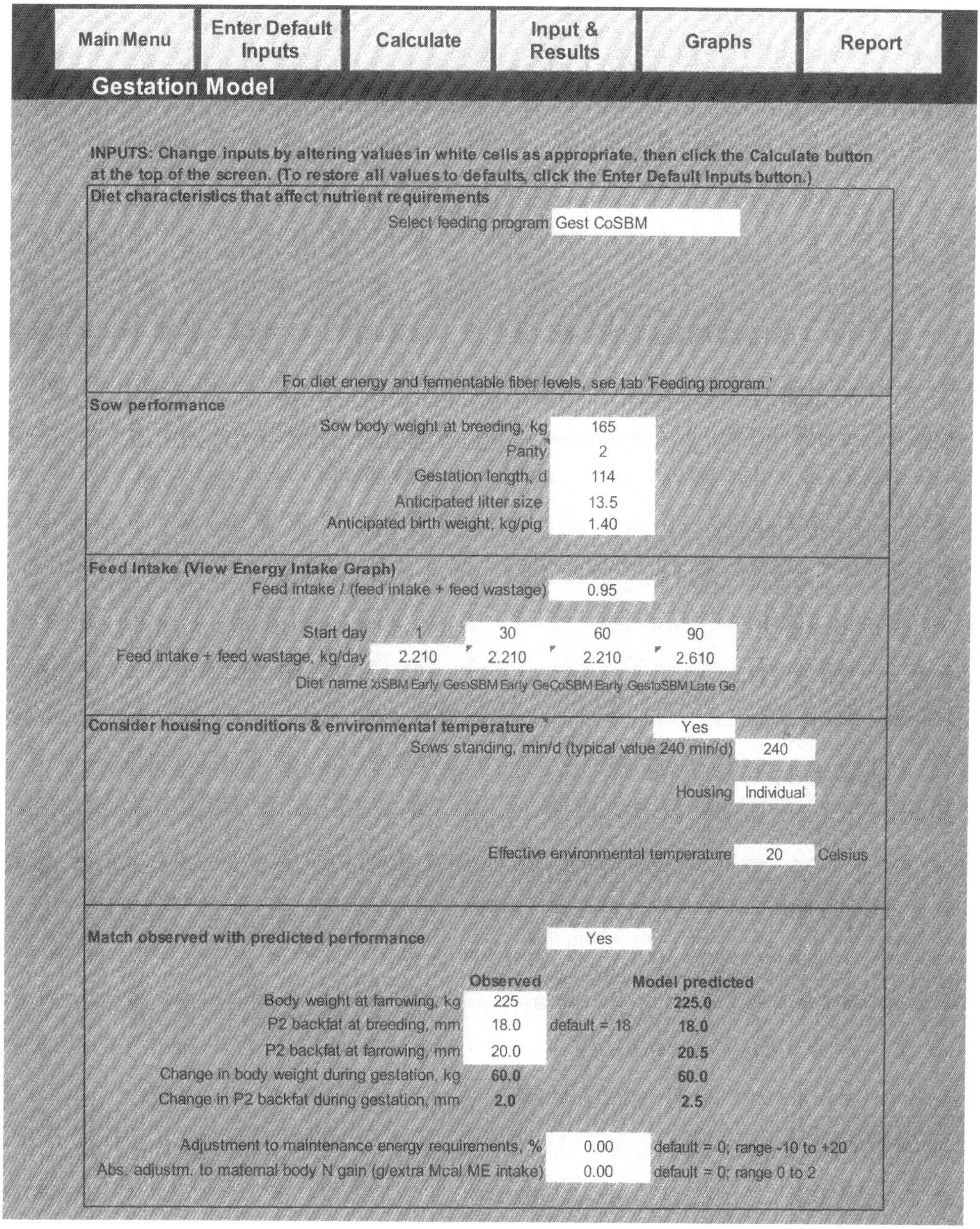

附录 A 模型使用指南

| Main Menu | Enter Default Inputs | Calculate | Input & Results | Graphs | Report |

Gestation Model

INPUTS: Change inputs by altering values in white cells as appropriate, then click the Calculate button at the top of the screen. (To restore all values to defaults, click the Enter Default Inputs button.)

Diet characteristics that affect nutrient requirements

Select feeding program: Gest CoSBM

For diet energy and fermentable fiber levels, see tab 'Feeding program.'

Sow performance

Sow body weight at breeding, kg	165
Parity	2
Gestation length, d	114
Anticipated litter size	13.5
Anticipated birth weight, kg/pig	1.40

Feed Intake (View Energy Intake Graph)

Feed intake / (feed intake + feed wastage)   0.95

Start day	1	30	60	90
Feed intake + feed wastage, kg/day	2.210	2.210	2.210	2.610
Diet name	oSBM Early Ges	SBM Early Ge	CoSBM Early Ges	toSBM Late Ge

Consider housing conditions & environmental temperature   Yes

Sows standing, min/d (typical value 240 min/d)	240
Housing	Individual
Effective environmental temperature	20 Celsius

Match observed with predicted performance   Yes

	Observed		Model predicted
Body weight at farrowing, kg	225		225.0
P2 backfat at breeding, mm	18.0	default = 18	18.0
P2 backfat at farrowing, mm	20.0		20.5
Change in body weight during gestation, kg	60.0		60.0
Change in P2 backfat during gestation, mm	2.0		2.5

Adjustment to maintenance energy requirements, %	0.00	default = 0; range -10 to +20
Abs. adjustm. to maternal body N gain (g/extra Mcal ME intake)	0.00	default = 0; range 0 to 2

图 A-4a 妊娠母猪模型的输入界面

猪营养需要

图 A-4b 妊娠母猪模型的结果界面

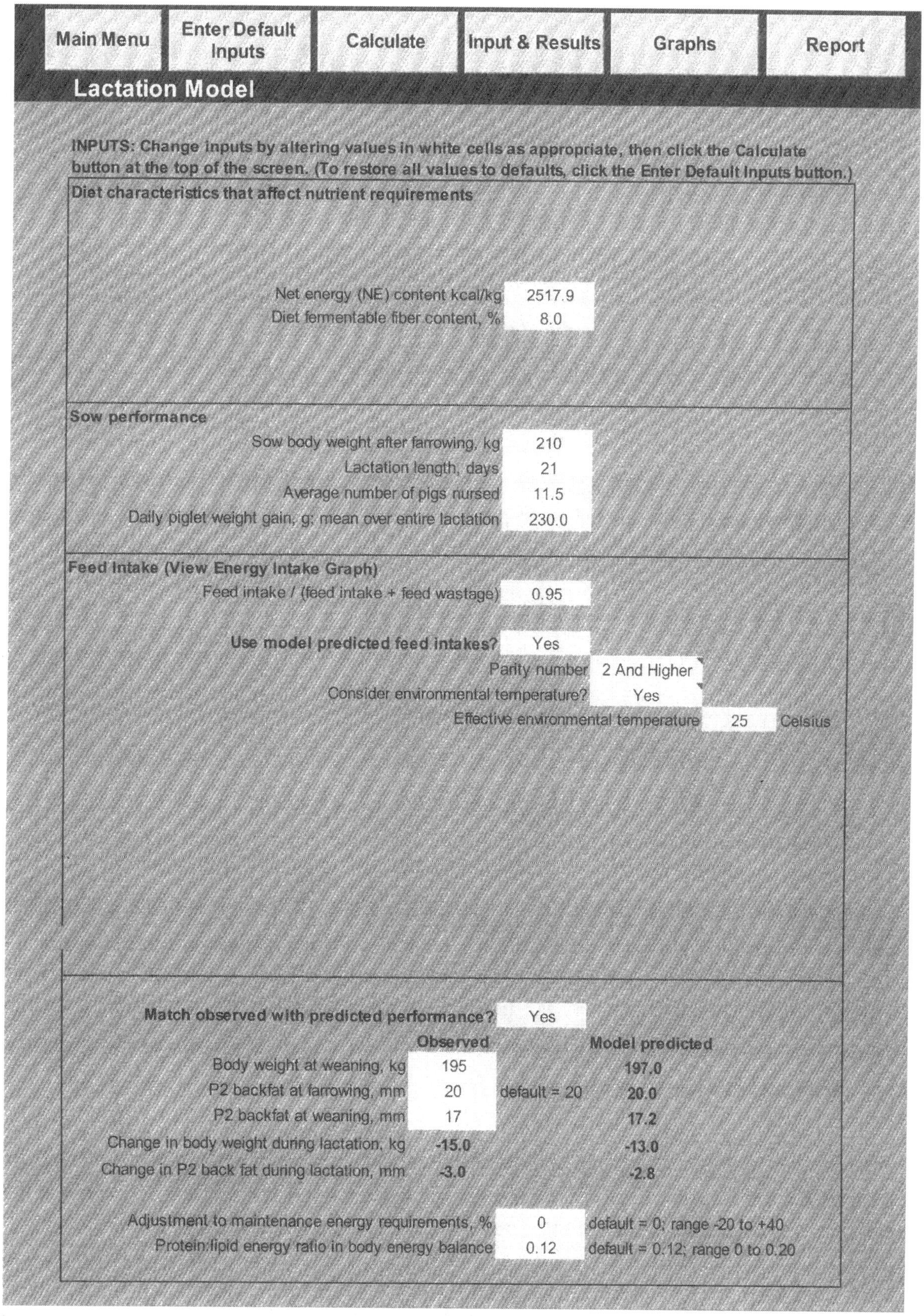

| Main Menu | Enter Default Inputs | Calculate | Input & Results | Graphs | Report |

Lactation Model

INPUTS: Change inputs by altering values in white cells as appropriate, then click the Calculate button at the top of the screen. (To restore all values to defaults, click the Enter Default Inputs button.)

Diet characteristics that affect nutrient requirements

Net energy (NE) content kcal/kg	2517.9
Diet fermentable fiber content, %	8.0

Sow performance

Sow body weight after farrowing, kg	210
Lactation length, days	21
Average number of pigs nursed	11.5
Daily piglet weight gain, g: mean over entire lactation	230.0

Feed Intake (View Energy Intake Graph)

Feed intake / (feed intake + feed wastage)	0.95
Use model predicted feed intakes?	Yes
Parity number	2 And Higher
Consider environmental temperature?	Yes
Effective environmental temperature	25 Celsius

Match observed with predicted performance?	Yes		
	Observed	Model predicted	
Body weight at weaning, kg	195		197.0
P2 backfat at farrowing, mm	20	default = 20	20.0
P2 backfat at weaning, mm	17		17.2
Change in body weight during lactation, kg	-15.0		-13.0
Change in P2 back fat during lactation, mm	-3.0		-2.8
Adjustment to maintenance energy requirements, %	0	default = 0; range -20 to +40	
Protein:lipid energy ratio in body energy balance	0.12	default = 0.12; range 0 to 0.20	

图 A-5a　泌乳母猪模型的输入界面

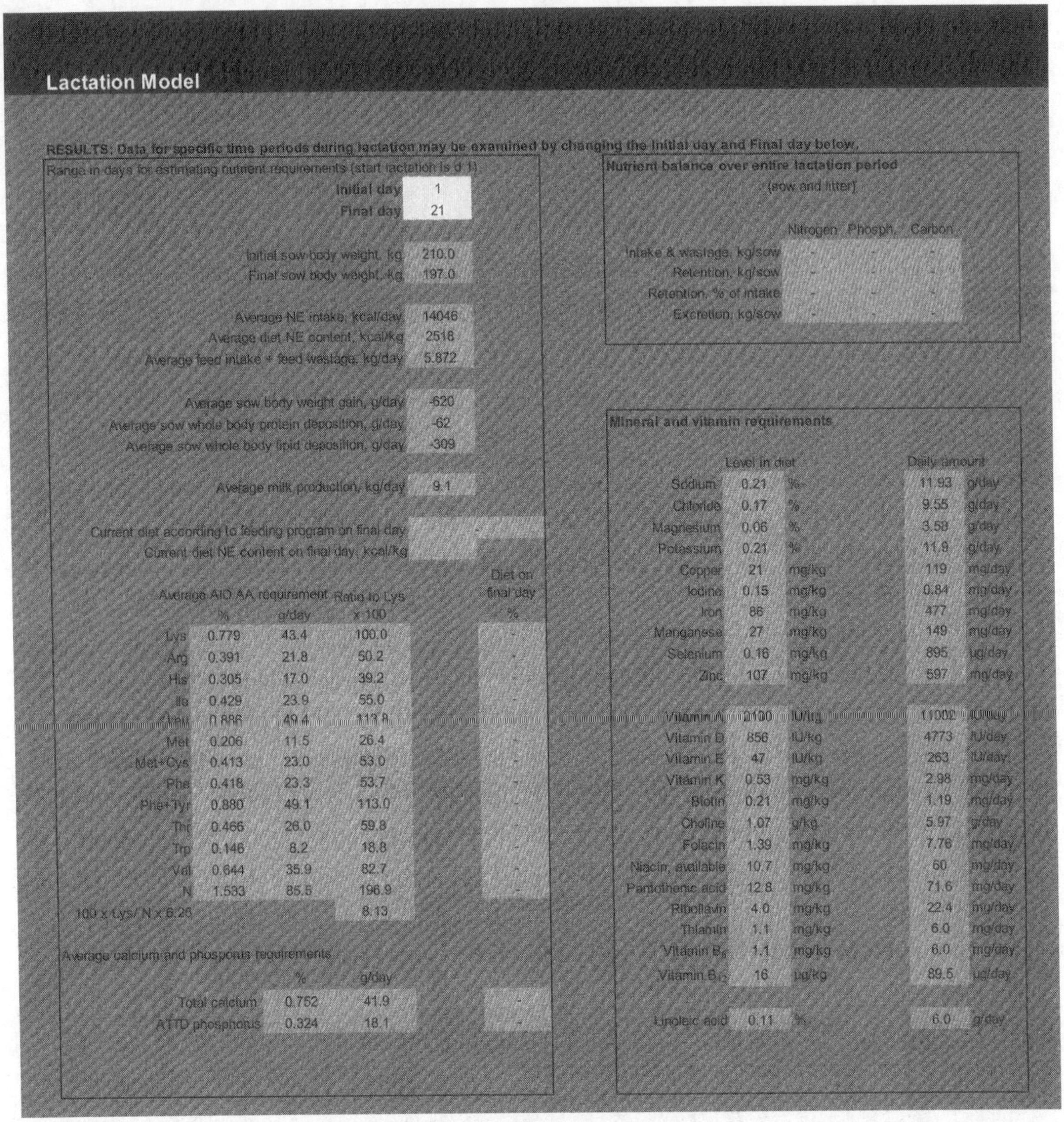

图 A-5b 泌乳母猪模型的结果界面

饲喂方案

饲喂方案和日粮产生（Feeding Program & Diet Generation）模块可从主菜单进入。对"你是否希望评估饲喂方案？（Do you wish to evaluate a feeding program?）"选择"是"，然后点击"检查饲喂方案（Review Feeding Programs）"。这个模块（图 A-6）主要是三个表格（饲料原料、日粮、饲喂方案）和四个子模块（饲料原料选择、日粮配方、检查和编辑日粮、制订饲喂方案）。选择屏幕上端的按钮，可进入各个子模块。

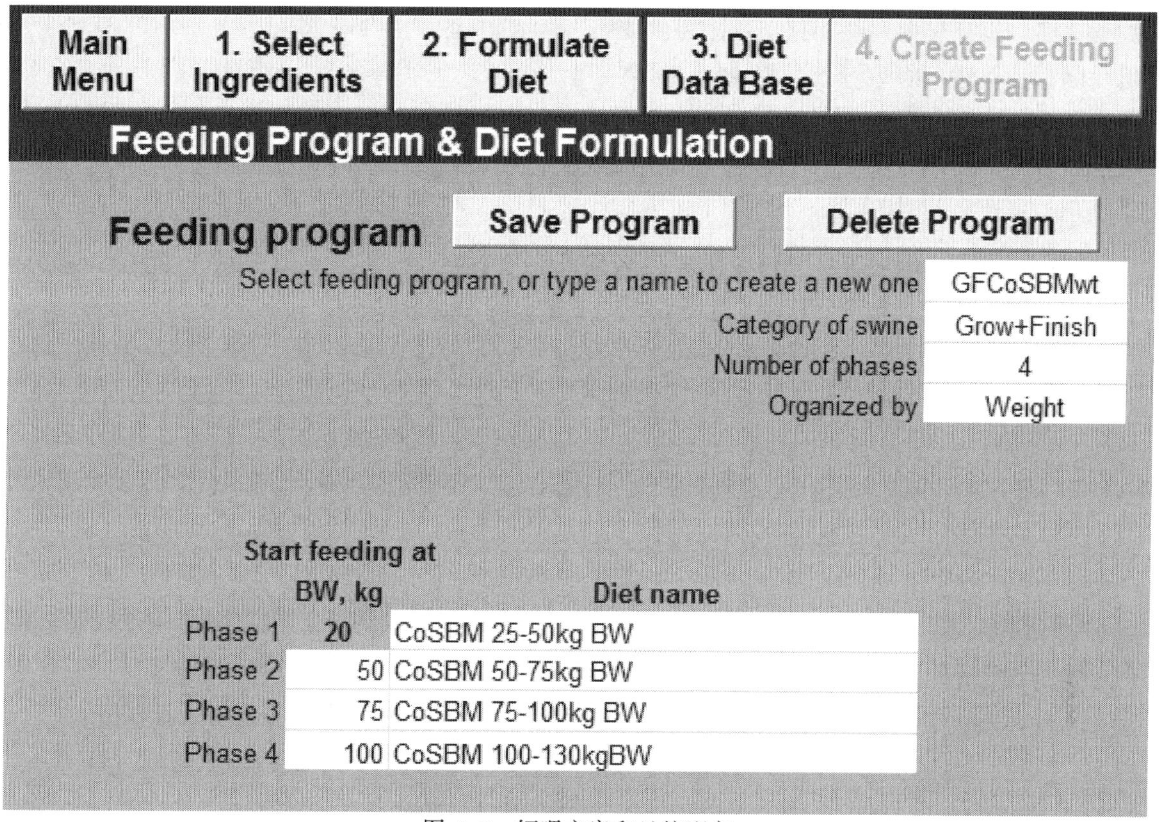

| Main Menu | 1. Select Ingredients | 2. Formulate Diet | 3. Diet Data Base | 4. Create Feeding Program |

Feeding Program & Diet Formulation

Feeding program　　Save Program　　Delete Program

Select feeding program, or type a name to create a new one　GFCoSBMwt
Category of swine　Grow+Finish
Number of phases　4
Organized by　Weight

	Start feeding at BW, kg	Diet name
Phase 1	20	CoSBM 25-50kg BW
Phase 2	50	CoSBM 50-75kg BW
Phase 3	75	CoSBM 75-100kg BW
Phase 4	100	CoSBM 100-130kgBW

图 A-6　饲喂方案和日粮配方

当确定饲喂方案时，指定生长育肥猪、妊娠母猪、泌乳母猪日粮中的能量和可发酵纤维水平，用于估测营养需要量。饲喂方案可以从输入（Inputs）页面选定（图 A-4a）。

1.饲料原料选择

在此子模块页面，在标题为原料（Ingredient）的数据输入区域有一下拉式菜单，列有原料库中的饲料原料。该原料库取自本标准的第 17 章（NRC，2012），点击原料库（Ingredient Library）按钮可查看其内容[2]。原料被选定和加载后，改变 U 至 BT 列中的数值便可改变该原料的营养成分。改变后的数值会用另一种不同的颜色标记。特别注意蓝色标记的数值，因其与主菜单指定的养分系统一致。可在该数据库中添加新的原料，即在 D 列输入原料名称，并在相应的列输入适宜的营养成分水平。原料表中所列的第一个原料作为剩余饲料原料（residual feed ingredient），用于所有的日粮中，确保日粮所有原料的百分含量之和为 100%。一种原料一旦用于日粮中，就不能被数据库中的其他原料取代。本模型原始版本将黄玉米作为剩余饲料原料（residual feed ingredient），如果要取代黄玉米而用其他原料作为剩余饲料原料（residual feed ingredient），则必须删除所有的日粮和饲喂方案。在下拉式菜单选择位于底端的"清除（Clear）"选项，可将原料从数据库中删除。数据库最多可收录 50 种饲料原料。

2.日粮配方

当"选择日粮（Select Diet）"下的数据输入区域被选中后，有一下拉菜单可以显示数据库中所有的日粮配方。在数据输入区域输入新的日粮名称，用于产生新的日粮。在"原料（Ingredient）"下数据

---

[2] 备注：饲料原料库添加了可发酵（即粪表观可消化）纤维一项指标，该值来源于荷兰动物饲料产品委员会（CVB，2004）。虽然该指标未在 NRC（2012）中列出，但本书第 8 章指出，估测可发酵苏氨酸损失和苏氨酸需要量时，需要知道可发酵纤维水平。

输入区域，用下拉式菜单选择原料。除了剩余原料（residual ingredient）之外的所有原料，必须指定其在日粮中的使用水平。剩余原料（residual ingredient）列为第一种原料，用于所有的日粮中，它在日粮中的使用水平会自动计算。如果原料使用水平发生变化，模型会自动计算和显示新日粮的营养成分。点击适宜的按钮，可保存或删除日粮，最多可保存25个配方。

3.日粮数据库

日粮数据库可显示日粮及其营养水平。在D列输入日粮名称，在U~BU列输入相应的营养成分，就可跳过"日粮配方"子模块，并添加新的日粮配方（日粮25~60）。

4.创建饲喂方案

如图A-6所示，可选择并检查饲喂方案。点击"选择饲喂方案或输入名字建立新的饲喂方案（Select a feeding program or type a name to create a new one）"，下拉式菜单会列出数据库中所有的饲喂方案。在数据输入的第一行输入名字，第二行输入猪的类别，就可建立新的饲喂方案。然后将初始体重或者开始日期输入第一列，相应的日粮名称输入第二列。点击相应的选项可以保存或删除饲喂方案，最多能保存30个饲喂方案。

附录 B 委员会的工作声明

委员会将总结有关猪营养的文献并更新能量及营养需要标准，其中包括猪的各个不同的生长阶段与生产类型。针对现代基因型猪的瘦肉率的提高，我们将重点关注氨基酸的影响。我们除了关注净能体系及其价值利用、生物燃料行业副产物、新的饲料原料（如新型大豆产品）之外，还更新可消化磷需要量和饲料原料中可消化磷含量。除此之外，日粮中常用饲料添加剂（如抗生素、酶、酸化剂、β-兴奋剂）和饲料加工（如制粒、膨化、颗粒大小）对猪的影响，以及如何提高营养物质沉积效率并且减少粪尿排泄物对环境的污染等问题也是我们重点介绍的。改进计算营养标准的计算机模型，依据最新文献更新饲料成分表信息。这篇报告将会总结上述所有方面并指明未来研究方向和热点。

附录 C 缩略语和缩写词

AA	氨基酸
AA$_{diet}$	日粮干物质中氨基酸含量
AA$_{digesta}$	回肠食糜中氨基酸含量
AAFCO	美国饲料管理协会
ADF	酸性洗涤纤维
ADFI	平均饲料日采食量
ADG	平均日增重
AFIA	美国饲料工业协会
AFSS	美国动物饲料安全体系
AID	回肠表观消化率/回肠表观可消化
Ala	丙氨酸
AOAC	美国官方分析化学家协会
AOM	活性氧法
APHIS	美国农业部动植物卫生检疫局
ARA	花生四烯酸
ARC	英国农业研究理事会
Arg	精氨酸
ASABE	美国农业与生物工程师协会
Asp	天冬氨酸
ATTD	全消化道表观消化率/全消化道表观可消化
AV	对位甲氧基苯胺值
BHA	丁基羟基茴香醚
BHT	二丁基羟基甲苯
BL	体脂肪量
BP	体蛋白量
BSAS	英国动物科学学会
BSE	牛海绵状脑病
BV	联苯胺值
BW	体重
cal	卡路里
CAST	美国农业科技委员会
CDS	浓缩酒精糟可溶物
CF	粗纤维
CFR	联邦法规代码

CLA	共轭亚油酸
CP	粗蛋白
CVB	荷兰动物饲料产品委员会
CWD	慢性消耗性疾病
Cys	半胱氨酸
d	天
Da	道尔顿
DADF	可消化酸性洗涤纤维
DCP	可消化的粗蛋白
DDE	二氯二苯二氯乙烯
DDG	干酒精糟
DDGS	干酒精糟及可溶物
DDT	二氯二苯三氯乙烷
DE	消化能
DEE	可消化乙醚浸出物
DHA	二十二碳六烯酸
DM	干物质
DMI	干物质采食量
DNA	脱氧核糖核酸
DNSP	可消化非淀粉多糖
DOM	可消化有机物
DON	脱氧雪腐镰刀菌烯醇/呕吐毒素
DP	可消化蛋白
DRES	可消化残留物
EAP	估计有效磷
EBW	空腹体重
EDTA	乙二胺四乙酸
EE	乙醚浸出物
EFA	必需脂肪酸
EPA	二十碳五烯酸
EPL	内源性磷损失
Eq	公式
EU	欧盟
FAD	黄素腺嘌呤二核苷酸
FAME	脂肪酸甲酯
FAO	联合国粮食与农业组织
FCH	可发酵碳水化合物

FDA	美国食品和药物管理局
FFA	游离脂肪酸
FH_4	四氢叶酸
FHP	绝食产热
FMN	黄素单核苷酸
FSIS	美国食品安全及检验局
FTU	植酸酶活性单位
G:F	饲料转化率
GC	气相色谱法
GE	总能
GfE	德国营养生理协会
GIT	胃肠道
Glu	谷氨酸
Gly	甘氨酸
GM	转基因
GnRH	促性腺激素释放激素
H_cE	维持体温相关的产热
HCH	六氯环己烷
H_dE	消化和同化产热
HE	产热
H_eE	维持产热
H_fE	发酵产热
H_iE	热增耗
His	组氨酸
H_jE	活动产热
HP-DDG	高蛋白干酒精糟
HP-DDGS	高蛋白干酒精糟及可溶物
HPLC	高效液相色谱法
H_rE	组织生产产热
HSCAS	水合硅铝酸钠钙盐
H_wE	废物生产产热
IFN	国际饲料编号
Ig	免疫球蛋白
IgA	免疫球蛋白 A
IgG	免疫球蛋白 G
Ile	异亮氨酸
IOM	美国医学研究所

IPCC	联合国政府间气候变化专门委员会
IU	国际单位
IV	碘值/碘价
IVGTT	静脉注射葡萄糖耐量试验
IVICT	静脉注射胰岛素耐量试验
IVP	碘值乘积
J	焦耳
k_f	动用的母体蛋白和脂肪用于胎儿和组织发育的效率
k_m	代谢能转化成奶能的效率
k_{mr}	体组织用于产奶所需能量的效率
k_p	代谢能用于蛋白质的效率
k_r	动员蛋白质和脂肪支持发育中的胎儿和组织
LA	亚油酸
LCT	低临界温度/下限临界温度
Ld	脂质沉积
L_{diet}	日粮干物质中脂质浓度
LEG	用于脂肪沉积的代谢能
Leu	亮氨酸
LN	亚麻酸
LS,ls	窝产仔数
Lys	赖氨酸
$Marker_{diet}$	日粮中指示剂的浓度
$Marker_{digesta}$	回肠食糜中指示剂的浓度
MDH	明尼苏达州卫生部
ME	代谢能
MEI	代谢能摄入量
MEIR	代谢能摄入量的减少
ME_m	维持代谢能
Met	蛋氨酸
MMA	乳房炎-子宫炎-无乳综合征
MPB	二甲基嘧啶醇亚硫酸氢钠甲萘醌
mRNA	信使核糖核酸
MSB	亚硫酸氢钠甲萘醌
MSBC	亚硫酸氢钠甲萘醌复合物
MUFA	单不饱和脂肪酸
NAD	烟酰胺腺嘌呤二核苷酸

NADP	烟酰胺腺嘌呤二核苷酸磷酸
NAS	美国国家科学院
ND	尚未确定
NDF	中性洗涤纤维
NDL	营养数据实验室
NDSC	中性洗涤剂可溶性碳水化合物
NE	净能
NE_m	维持净能
NE_p	生产净能
NFC	非纤维碳水化合物
NPB	美国国家猪肉委员会
NPPC	美国全国猪肉生产商理事会
NRC	美国科学研究委员会
NSC	非结构性碳水化合物
OIS	氧化稳定指数
PABA	对氨基苯甲酸
par	胎次
PCBs	多氯联苯
Pd	蛋白质沉积
Pd_{max}	最大蛋白质沉积速率
PEG	用于蛋白质的代谢能
PG	没食子酸丙酯
Phe	苯丙氨酸
P_{intake}	磷的日摄入量
P_{output}	磷的日排泄量
ppb	十亿分之一
ppm	百万分之一
Pro	脯氨酸
PUFA	多不饱和脂肪酸
PV	过氧化值
PVPP	交联聚乙烯基吡咯烷酮
RAC	莱克多巴胺
RE	视黄醇当量
SDF	可溶性日粮纤维
Ser	丝氨酸
SFA	饱和脂肪酸
SID	回肠标准消化率/回肠标准可消化

SOD	超氧化物歧化酶
STTD	全消化道标准消化率/全消化道标准可消化
t	时间
T	温度
TBA	硫代巴比妥酸
TBARS	硫代巴比妥酸反应物
TBHQ	叔丁基对苯二酚
TDE	四氯二苯乙烷
TDF	总日粮纤维
TDS	溶解性总固体
TFWQG	美国水质监控专家组
Thr	苏氨酸
TID	回肠末端真消化率
Trp	色氨酸
TSE	传染性海绵状脑病
Tyr	酪氨酸
U:S	不饱和:饱和（脂肪酸）比例
UCT	高临界温度/上限临界温度
USDA	美国农业部
Val	缬氨酸
VFI	自由采食量
WSC	水溶性碳水化合物

附录 D 委员会成员简介

L.Lee Southern，编委会主席，Doyle Chambers 杰出教授，现就职于路易斯安那州立大学（Louisiana State University）农业研究中心动物科技学院，单胃动物营养专家，研究领域为家禽和猪的氨基酸和矿物质利用。Southern 博士曾担任《家禽科学》（Poultry Science）和《职业动物科学家》（Professional Animal Scientist）的编辑委员会委员、《动物科学杂志》（Journal of Animal Science）的副主编和分区编委，现任家禽科学（Poultry Science）的分区编委。他从 1998~2002 年担任 NRC 动物营养的委员。Southern 博士获得的荣誉包括美国动物科学学会（American Society of Animal Science）饲料工业协会的非反刍动物营养奖（American Feed Industry Association's Nonruminant Nutrition Award）、Gamma Sigma Delta 研究奖、路易斯安那州立大学优秀教学奖（LSU Teaching Merit Honor Role）。在北卡罗来纳州立大学（North Carolina State University）获得动物科学学士和硕士学位，在伊利诺伊大学（University of Illinois）获得动物科学博士学位。

Olayiwola Adeola，普渡大学（Purdue University）动物科学教授，担任单胃动物营养教学工作，主要是氨基酸和植物来源矿物元素的利用方面。如何提高生产效益、改善机体健康和环境是 Adeola 博士的研究目标，其中提高瘦肉生产效率和降低动物排泄物对环境的影响是他的一个主要研究方向。Adeola 博士曾担任《家禽科学》（Poultry Science）的编辑委员会委员、《动物科学杂志》（Journal of Animal Science）的副主编、《加拿大动物科学杂志》（Canadian Journal of Animal Science）的分区编委。曾获得的荣誉包括美国饲料工业协会的家禽营养研究奖（Poultry Nutrition Research Award）、家禽科学学会（Poultry Science Association）的枫叶农场鸭研究奖（Maple Leaf Farms Duck Research Award）、美国动物科学学会饲料工业协会的非反刍动物营养奖（American Feed Industry Association's Nonruminant Nutrition Award）。在尼日利亚的 Ife 大学（University of Ife）获得动物科学学士学位，在加拿大圭尔夫大学（University of Guelph）获得动物科学硕士和博士学位。

Cornelis F.M. de Lange，加拿大安大略省圭尔夫大学（University of Guelph）动物科学系教授，家畜研究项目负责人。大学任教前，在饲料工业从事猪营养应用研究。他任职于圭尔夫大学期间，主要从事生长育肥猪的营养利用、减少对环境的污染、提高肉质、改善仔猪的肠道健康与发育的研究工作，以达到养猪生产可持续发展的目标。他曾获圭尔夫大学安大略农学院杰出科研奖和推广奖。分别在荷兰瓦格宁根大学（Agricultural University in Wageningen）获得动物科学学士和硕士学位，加拿大阿尔伯特大学（Alberta University）获得动物营养博士学位。

Gretchen M. Hill，密歇根州立大学（Michigan State University）动物科学系教授。她从基础的营养利用和分子水平去研究微量元素对家畜的作用，并与饲料工业紧密结合，根据现代基因型品种修正微量元素的使用量。她曾担任《营养生化杂志》（Journal of Nutritional Biochemistry）的编辑委员会委员、《动物科学杂志》（Journal of Animal Science）的副主编。她获得的荣誉包括密歇根州立大学农业和自然资源学院的杰出导师奖、美国动物科学学会饲料工业协会的非反刍动物营养奖（American Feed Industry Association's Nonruminant Nutrition Award）。她分别在肯塔基大学（University of Kentucky）获得学士学位、普渡大学（Purdue University）获得硕士学位、密歇根州立大学（Michigan State University）获得动物营养博士学位。

Brian J. Kerr，动物科学家和美国农业部农业研究服务首席科学家。任职于艾奥瓦州（Iowa）艾姆斯的一个农业部研究单位，主要研究如何提高动物生产体系利用自然资源和降低对环境的影响。在养猪生产中如何降低营养物质排出和减少恶臭气体与致病菌的排放是Kerr博士的研究重点。他还曾经在饲料公司担任研究主管和技术经理职位。Kerr博士曾担任《动物科学杂志》（*Journal of Animal Science*）的副主编和其他杂志的编辑委员会委员。他在伊利诺伊大学（University of Illinois）获得动物科学学士和硕士学位、非反刍动物营养博士学位。

Merlin D. Lindemann，肯塔基大学（University of Kentucky）动物与食品科学系猪营养与管理教授。研究领域包括饲粮中氮和磷含量的调整与动物生长性能和排污管理之间的关系、新型副产物原料的饲养价值评定、微量元素在养猪上的应用评定、添加剂对猪繁殖性能的影响。Lindemann博士曾担任《动物科学杂志》（*Journal of Animal Science*）的副主编和《职业动物科学家》（*Professional Animal Scientist*）的编辑委员会委员。他获得的荣誉包括美国动物科学学会饲料工业协会的非反刍动物营养奖（American Feed Industry Association's Nonruminant Nutrition Award）和肯塔基大学George E. Mitchell Jr.杰出导师奖（George E. Mitchell Jr. Award for Outstanding Faculty Service to Graduate Students）。他在明尼苏达大学（University of Minnesota）获得动物科学学士和博士。

Phillip S. Miller，内布拉斯加大学（University of Nebraska）动物科学系猪营养教授，主要研究生长肥育猪肝脏代谢、营养摄入、生长水平三者间的相互作用以及能量与体组成的关系。Miller博士曾担任《动物科学杂志》（*Journal of Animal Science*）的副主编和分区编委。他获得的荣誉包括Gamma Sigma Delta优秀教学奖（Gamma Sigma Delta Teaching Award of Merit）和L. K. Crowe杰出本科生导师奖（L. K. Crowe Outstanding Undergraduate Advisor Award）。他是加利福尼亚州立大学戴维斯分校（University of California, Davis）营养学学士、硕士、博士。

Jack Odle，北卡罗来纳州立大学（North Carolina State University）营养生化教授，William Neal Reynolds杰出教授。他主要研究新生幼畜的营养和代谢，尤其是从分子水平、细胞水平和动物整体水平研究油脂消化、吸收和代谢的发育变化。他的研究领域还包括肉毒碱和中链甘油三酯对新生幼畜生长的影响、生物活性肽和多不饱和脂肪酸对新生幼畜肠道发育的影响。他曾担任《营养学杂志》（*Journal of Nutrition*）副主编，并担任美国营养学会顾问。他曾荣获美国动物科学学会饲料工业协会的非反刍动物营养奖（American Feed Industry Association's Nonruminant Nutrition Award）。他是普渡大学（Purdue University）动物科学学士和威斯康星大学（University of Wisconsin）动物营养学硕士、营养学和动物科学博士。

Hans H. Stein，伊利诺伊大学厄本那香槟分校（University of Illinois, Urbana-Champaign）动物科学系教授。主要研究方向是饲料原料中能量和宏量营养物质的消化吸收和利用、消化生理、饲料原料的评估、营养管理。他曾担任动物科学杂志（*Journal of Animal Science*）的副主编，获得的荣誉包括Gamma Sigma Delta 研究奖（Gamma Sigma Delta Research Award）、养猪信息伙伴奖（Pork Information Partner Award）、2010年的美国动物科学学会饲料工业协会的非反刍动物营养奖（American Feed Industry Association's Nonruminant Nutrition Award）。他是很多国家级协会的成员，其中包括国家猪肉委员会（National Pork Board）下属的非抗生素工作小组（Noantimicrobials Working Group）和动物科学委员会（Animal Science Committee）。他从丹麦Graasten农民农业学校（Farmer's Agriculture School）获得农学文凭，在丹麦皇家兽医和农业大学（Royal Veterinary and Agriculture University）取得动物科学硕士学位，从伊利诺伊大学（University of Illinois）获得动物科学博士学位。

Nathalie L. Trottier，密歇根州立大学（Michigan State University）动物科学系副教授。她的主要研

究方向是猪生长和哺乳期间的氨基酸代谢，包括乳腺和肠道利用氨基酸的机理。Trottier博士曾担任《动物科学杂志》（*Journal of Animal Science*）的编辑委员会委员，是密歇根州立大学所办杂志《马项目通讯》（*Equine Program Newsletter*）的学术编辑，现为国际性刊物《动物》（*ANIMAL*）非反刍动物营养部分的美洲编辑。她在加拿大麦吉尔大学（McGill University）获得农学学士学位和动物营养学硕士学位，并在伊利诺伊大学（University of Illinois）获得动物营养学博士学位。

附录 E 美国国家科学院农业和自然资源委员会最近的出版物

政策和资源方面的书籍

Achievements of the National Plant Genome Initiative and New Horizons in Plant Biology (2008)

Achieving Sustainable Global Capacity for Surveillance and Response to Emerging Diseases of Zoonotic Origin: Workshop Report (2008)

Agricultural Biotechnology and the Poor: Proceedings of an International Conference (2000)

Agriculture, Forestry, and Fishing Research at NIOSH (2008)

Agriculture's Role in K-12 Education (1998)

Air Emissions from Animal Feeding Operations: Current Knowledge, Future Needs (2003)

An Evaluation of the Food Safety Requirements of the Federal Purchase Ground Beef Program (2010)

Animal Biotechnology: Science-Based Concerns (2002)

Animal Care and Management at the National Zoo: Final Report (2005)

Animal Care and Management at the National Zoo: Interim Report (2004)

Animal Health at the Crossroads: Preventing, Detecting, and Diagnosing Animal Diseases (2005)

Biological Confinement of Genetically Engineered Organisms (2004)

California Agricultural Research Priorities: Pierce's Disease (2004)

Changes in the Sheep Industry in the United States: Making the Transition from Tradition (2008)

Countering Agricultural Bioterrorism (2003)

Critical Needs for Research in Veterinary Science (2005)

Designing an Agricultural Genome Program (1998)

Diagnosis and Control of Johne's Disease (2003)

Direct and Indirect Human Contributions to Terrestrial Carbon Fluxes (2004)

Ecological Monitoring of Genetically Modified Crops (2001)

Emerging Animal Diseases: Global Markets, Global Safety: Workshop Summary (2002)

Emerging Technologies to Benefit Farmers in Sub-Saharan Africa and South Asia (2008)

Enhancing Food Safety: The Role of the Food and Drug Administration (2010)

Ensuring Safe Food: From Production to Consumption (1998)

Environmental Effects of Transgenic Plants: The Scope and Adequacy of Regulation (2002)

Evaluation of a Site-Specific Risk Assessment for the Department of Homeland Security's Planned National Bio- and Agro-Defense Facility in Manhattan, Kansas (2010)

Exploring a Vision: Integrating Knowledge for Food and Health (2004)

Exploring Horizons for Domestic Animal Genomics (2002)

Frontiers in Agricultural Research: Food, Health, Environment, and Communities (2003)

Future Role of Pesticides for U.S. Agriculture (2000)

Genetically Engineered Organisms, Wildlife, and Habitat: A Workshop Summary (2008)

Genetically Modified Pest-Protected Plants: Science and Regulation (2000)

Global Challenges and Directions for Agricultural Biotechnology (2008)

The Impact of Genetically Engineered Crops on Farm Sustainability in the United States (2010)

Incorporating Science, Economics, and Sociology in Developing Sanitary and Phytosanitary Standards in International Trade (2000)

Letter Report to the Florida Department of Citrus on the Review of Research Proposals on Citrus Greening (2008)

National Capacity in Forestry Research (2002)

The National Plant Genome Initiative (2002)

National Research Initiative: A Vital Competitive Grants Program in Food, Fiber, and Natural-Resources Research (2000)

Predicting Invasions of Nonindigenous Plants and Plant Pests (2002)

Professional Societies and Ecologically Based Pest Management (2000)

The Public Health Effects of Food Deserts: Workshop Summary—joint study with Institute of Medicine (2009)

Publicly Funded Agricultural Research and the Changing Structure of U.S. Agriculture (2002)

Review of the Methodology Proposed by the Food Safety and Inspection Service for Risk-Based Surveillance of In-Commerce Activities: A Letter Report (2009)

Review of the Methodology Proposed by the Food Safety and Inspection Service for Followup Surveillance of In Commerce Businesses: A Letter Report (2009)

Review of the U.S. Department of Agriculture's Animal and Plant Health Inspection Service Response to Petitions to Reclassify the Light Brown Apple Moth as a Non-Actionable Pest: A Letter Report (2009)

Safety of Genetically Engineered Foods: Approaches to Assessing Unintended Health Effects (2004)

Scientific Advances in Animal Nutrition: Promise for a New Century (2001)

The Scientific Basis for Estimating Emissions from Animal Feeding Operations: Interim Report (2002)

The Scientific Basis for Predicting the Invasive Potential of Nonindigenous Plants and Plant Pests in the United States (2002)

Scientific Criteria to Ensure Safe Food (2003)

Status of Pollinators in North America (2007)

Strategic Planning for the Florida Citrus Industry: Addressing Citrus Greening (2010)

Sustaining Global Surveillance and Response to Emerging Zoonotic Disease (2009)

Toward Sustainable Agricultural Systems in the 21st Century (2010)

Transforming Agricultural Education for a Changing World (2009)

The Use of Drugs in Food Animals: Benefits and Risks (2000)

家畜营养需要量的书籍及相关书籍

Mineral Tolerance of Animals: Second Revised Edition (2005)

Nutrient Requirements of Beef Cattle, Seventh Revised Edition, Update (2000)

Nutrient Requirements of Dairy Cattle, Seventh Revised Edition (2001)

Nutrient Requirements of Dogs and Cats (2006)

Nutrient Requirements of Fish and Shrimp (2011)

Nutrient Requirements of Horses: Sixth Revised Edition (2007)

Nutrient Requirements of Nonhuman Primates, Second Revised Edition (2002)

Nutrient Requirements of Small Ruminants: Sheep, Goats, Cervids, and New World Camelids (2007)

Nutrient Requirements of Swine, Tenth Revised Edition (1998)

Safety of Dietary Supplements for Horses, Dogs, and Cats (2009)

Scientific Advances in Animal Nutrition: Promise for a New Century (2001)

The First Seventy Years 1928-1998: Committee on Animal Nutrition (1998)

The Scientific Basis for Estimating Emissions from Animal Feeding Operations: Interim Report (2002)

Further information and prices are available from the National Academies Press website at http://www.nap.edu/. To order any of the titles above, go to http://www.nap.edu/order.html or contact the Customer Service Department at (888) 624-8373 or (202) 334-3313. Inquiries and orders may also be sent to the National Academies Press, 500 Fifth Street, NW, Lockbox 285, Washington, DC 20055.

索 引

DDGS 175
D 型氨基酸 19
T-2 毒素 188
α-亚麻酸 53
β-甘露聚糖酶 186
β-胡萝卜素 118,248,267
β-葡聚糖 68,70,71,185,186
β-葡聚糖酶 186

A
阿拉伯半乳聚糖 177
阿拉伯木聚糖 70,186
阿拉伯树胶 70
阿拉伯糖 67,70
安全裕量 199,231
氨基酸 1,2,6,17,21,29,37,44,116,149,150~153,155,158,159,163,171,208,226,231,260
氨基酸比例失调 21
氨基酸不平衡 19
氨基酸的利用效率 149
氨基酸的需要量 44
氨基酸分析测定 19
氨基酸合成 17
氨基酸拮抗 21,22
氨基酸利用效率 30,37,41,42,158
氨基酸模型 23,30,32,35~37,44
氨基酸缺乏 21
氨基酸失衡 21,22
氨基酸消化率 20,176,177,179
氨基酸需要量 1,2,15,17~23,26~29,36,37,41,42,149,155,159,167,168,170,190,226,231,232,234,250,254,256
氨基酸需要量的表示方法 20
氨基酸需要量的测定 28
氨基酸需要量的经验估计 23
氨基酸需要量估测值 24,159
氨基酸与赖氨酸比值 30
氨基酸与赖氨酸之比 22
氨基酸中毒 21,22
氨基酸组成 19,22,29,31,32,35,36,142,148,149,162,168,174,176,212,226
氨基酸组成模式 19,31,36
氨气 216

B
白酒酵母 386
半胱氨酸 17~19,22,27,93,170,171,267
半乳寡糖 67,185
半乳糖 67
半纤维素 70,71,205,267
饱和脂肪酸 51,52,267
保育猪 44,186,227
背膘 167
背膘厚度 36,143,153,158,165
苯丙氨酸 17,18,29,30~32,36,41,151,165,168,170,232,267
苯丙氨酸+酪氨酸 29~32,36,41,151,165,168,170,232
比较屠宰试验 6
必需氨基酸 1,17,19,51~54,267
扁豆 310
标准维持代谢需要 145
丙氨酸 18,127,267
丙酸 184,218
哺乳母猪 17,23,27,29,30,36,37,41,42,44,78,79,87,94,95,127,128,130,161,189,225,227
哺乳仔猪 44,75,76
不饱和脂肪酸 51,52,54,58,121,175,176,189
不可消化寡糖 66,183,216,218
不溶性膳食纤维 267

C
采食量 9,187
菜豆 273
菜粕 206
蚕豆 272
产后瘫痪 88
产奶 37,157
产奶量 11
产热 7,9
产乳酸细菌 184
产仔数 11
肠道损失 29,75
肠球菌 184
常量元素 84,267
超氧化物歧化酶 96
沉积效率 161

宠物食品的副产物 337
传染性海绵状脑病 201
粗蛋白 5,6,17,32,33,35~37,127,162,174,176~179,186,190,200,206,208,267
粗甘油 179
粗脂肪 7,59,162,174,176,226,267
促性腺激素释放激素 12

D

大肠杆菌 68,78,117,185,201
大豆分离蛋白 178,179,370
大豆浓缩蛋白 178,179,369
大豆皮 179,355
大麦 17,18,67,123,127,186,198,206,248,270
大米 17,342
大米浓缩蛋白 347
代谢能 1,4,9,10,11,13,23,26~28,34,38,52,141,146,148,154,157,176,217,226,267
代谢能的分配 52,142~145,147,148,153,154,157,162
代谢能摄入量 144
代谢能摄入量的分配 11
代谢能系统 13
丹宁 21,215
单不饱和脂肪酸 267
单糖 65
单位 20
胆碱 116,123~125,161,248,267
蛋氨酸 17~19,21~23,25,27,29~32,36,40,41,124,125,130,159,163~165,168,170,171,226,232,267
蛋氨酸+半胱氨酸 25,28~32,36,40,41,163~165,168,170,171,232
蛋氨酸硒 97,98
蛋白库 32,33,36,41,152,153,156,167,168
蛋白酶 215
蛋白质 1,2,5~8,10~13,17~22,29,31,32,35~38,41,42,44,52,54,56,59,74,75,77,84,87,89,92,96~99,126~128,130,141~143,147~156,158,159,162~164,166~171,174~176,179,206,208,209,212,215,217,218,225~227
蛋白质沉积 8,10,12,13,22,36,38,41,141,143,146~150,152~154,156,162,164,165,167,168,226
蛋白质沉积量 147
氮 216,231
低蛋白乳清粉,乳糖80% 317
低蛋白乳清粉,乳糖85% 318
低寡聚糖全脂大豆 368
低聚糖 65,66,267
低临界温度 144,145,154
低脂 DDGS 174
滴水损失 127

第一限制性氨基酸 22,174
碘 55,84,95,160,197,248
碘价 1,263,267
碘值 54,55,227
电解质平衡 90,91,215,227
淀粉 5,7,10,13,66,68~71,162,176,177,205,206,210,211,217,262,267
丁基羟基茴香醚 58,189
丁酸 184,218
动物生产效率 215
豆粕 2,91,123,127,174,206
短链脂肪酸 65,66,69,70,211,218
断奶仔猪 52,56,76,79,85,90,92,94,97~100,116~118,120,122~124,126,131,159,160,161,175,176,178,179,183~187,189,231
对位甲氧基苯胺值 57
多不饱和脂肪酸 53,57,227,267
多氯联苯 198,200
多肽 19
多糖 65,68

E

二丁基羟基甲苯 58,189
二噁英 198,200
二甲酸钾 184
二硫代巴比妥酸 57
二糖 65,66,210
二氧化碳当量 218

F

发酵豆粕 178,360
反式半乳寡糖 67,185
泛酸 116,127,128,161,227,248,267
泛酸钙 127
非必需氨基酸 1,17,20,22,226,267
非淀粉多糖 7,66,68,70~72,185,211,213,216~219
非限制性氨基酸 22
非营养性饲料添加剂 2,12,183,197
废气 218
分解代谢 21,22,29,31,37,150,208,209,225
分支杆菌 201
酚类吸附剂 189
粉碎 205
粪肠球菌 184,185
伏马毒素 197,198
辅酶 84
脯氨酸 17,18,130,267
副产物 2,174,176,177,179

富马酸 184
富硒酵母 97
腹部脂肪 175
腹泻 22,75~77,79,80,96,99,127,185
腹脂 176

G
钙 2,70,78~81,84~88,90,91,94,97,99,100,119,120,122,123,
　141,142,151,152,156,159,160,162,164,165,167~171,175,
　187,188,216,217,225,227,228,231,234,238,267
钙和磷 84
钙和磷的需要量 151
钙磷比例 215~217
干法碾磨 174
甘氨酸 17,18,125,129,267
甘露寡糖 67,68,185
甘露聚糖酶 186,187
甘油三酯 51,55,59
甘蔗糖蜜 323
柑橘渣 283
高比热 74
高蛋白 DDG 175,176
高蛋白 DDGS 175,176
高蛋白全脂大豆 367
高粱 17,18,87,126,127,205,248,353
高粱酒精糟及可溶物 354
高临界温度 157
高营养玉米 287
镉 199
铬 80,84,92,93,197,267
公猪 8,12,78
汞 199
共轭亚油酸 54,55,191
共济失调 91,129,130
佝偻病 96,120
估测值 8,11,21,30,42,44,52,85,86,89,99,129~131,141~
　143,149,151,156,158~160,162,164~166,168~171,231
估算有效代谢能摄入量 232,260,261
谷氨酸 17,18,122,125,267
谷氨酰胺 18
谷胱甘肽 91
谷胱甘肽过氧化物酶 93,97,98,122
谷胱甘肽还原酶 128
谷物-豆粕型日粮 20,85,125,130
骨骼钙化 86
骨软化 120
钴 93
寡糖 65~67,71,177,178,185,211

拐点分析 42
观测值 141,142,153,158,164,169
鲑鱼蛋白水解物 351
国家科学院 392
果寡糖 67,185
果胶 70,71
果胶酶 186
果糖 66
过氧化值 57

H
含硫氨基酸 18,19,22,42,91,121,174,217
汗腺 74
合成代谢 28,31
合成晶体氨基酸日粮法 209
核苷酸 185
核黄素 116,128,160,161,248,267
黑麦 13,348
黑小麦 375
黑小麦酒精糟及可溶物 376
红花粕 349
后备公猪 12
后备母猪 12,77,119,227,231
后躯麻痹 118
糊化 69,205,206
花生粕 331
化学性污染物 197
怀孕母猪 17,32,35,41,176
环境 215
黄曲霉毒素 188,189,197~199
黄曲霉毒素 B_1 188
灰分 174
挥发性脂肪酸 57,206,228
回肠氨基酸消化率 21,208
回肠标准氨基酸消化率 27
回肠标准可消化 158,162,163,165,167,168,170,231~247,
　250,252,254~257,260,261
回肠标准可消化氨基酸 37,41,141
回肠标准可消化赖氨酸需要量 38,43,44,233,235,237,239,
　241,243,245,247,251,253,255,257
回肠标准消化率 28,267
回肠表观可消化 25,141,231~247,251~257,260,261
回肠表观消化率 21,211,267
回肠消化率法 21
回归法 209
回生淀粉 69,206
混合鱼粉 304
活动 9

活性炭 188
活性氧法 57

J
机械压榨豆粕 178
计算机模型估测猪营养需要量 1
加工副产物 174
加性效应 94,100
甲酸 184
甲酸钙 184
甲状腺素 95
钾 84,90,91,160,188,179,199,217,248,267
假单胞菌 201
豇豆 334
焦耳 1,4
角化不全 121
教槽料 76,99,100,187
酵母 184
酵母单细胞蛋白 387
酵母硒 97,98
酵母细胞壁 67,189
结肠炎 127,186
浸提双低菜籽粕 281
晶体氨基酸 19 21
精氨酸 17,18,22,24,29,30,32,36,151,164,165,168,170,226,232,267
精选白油 6
精液分泌 12
精油 186,219
净能 1,4,6,13,217,226,232,263,267
净能系统 13
酒精工业 174
绝食产热 6~8

K
卡 4
抗坏血酸 130
抗菌剂 2,94,183,202
抗生素 77,100,123,183,186,202
抗性淀粉 69,70,72,206,211,216,218
抗氧化剂 58,121,130,189,227
抗氧化作用 97
抗营养因子 21,30,145,206,215,225
颗粒黏合剂 189
可加性 20,23,208,209,211,212
可溶性膳食纤维 267
空腹体重 74,88,142,143,152
矿物元素 215

矿物质 1,2,13,44,75,80,84,160,163,167,185,227,231,233,235,237,239,241,243,245,247~249,251,253,255,257,258~261,267

L
莱克多巴胺 1,2,12,31,32,142,148,162,190,191,231,244~247
赖氨酸 13,17~19,21~24,27~32,36~38,41~44,52,56,130,131,142,150,151,155,158,159,162~171,174,176,191,215,226,232,267
赖氨酸利用效率 38,41,166
赖氨酸需要量 13,30,37,38,41,42,44,150,151,155,158,159,163~166,167~171,190,191,226,231,233,235
酪氨酸 17,18,20,267
酪蛋白（来源于牛奶） 314
理想氨基酸模型 22
理想蛋白质 22
连续屠宰试验 150
连续屠宰试验 37
联苯胺值 57
亮氨酸 17,18,22,25,29,30,32,36,41,151,165,168,170,232,267
裂荚紫花豌豆 336
磷 1,2,84~91,96,97,119,120,141,142,151,152,156,159,160~163,165,167,169~171,175,176,186,187,191,199,206,208,211,212,216,217,219,227,228,232,234,238,262,263,267
磷酸酶 187
流散剂 190
硫 2,84,90~92,175,179,217,267
硫胺素 116,125,128,129,161,227,248,267
硫代巴比妥反应物 57
裸大麦 271
裸燕麦 326
铝 80
氯 78,80,81,84,90,91,160,179,227,248,267
氯化胆碱 119,123,124

M
马铃薯浓缩蛋白 339
麦芽糖酶 66
慢性消耗性疾病 201
酶处理豆粕 178,358
酶解酪蛋白法 209
霉菌毒素 117,123,188,189,197~199,215
霉菌毒素吸附剂 188,189
霉菌毒素中毒症 188
美国动物饲料安全体系 202
美国国家猪肉委员会 1
美国科学研究委员会 1
美国农业部 197

美国农业部动植物卫生检疫局 201
美国农业部食品安全检查署 201
美国全国猪肉生产者委员会 143
美国食品和药物管理局 1,2,54,98,183,197
美国饲料工业协会 197
美国饲料管理协会 197,262
美拉德反应 21,174,208,215
镁 78,79,81,84,90,91,119,120,160,188,217,248
锰 84,91,94,96,97,120,160,217,248,267
醚浸出物 5
米糠 343
泌乳 11
泌乳母猪 9,11,141,161,176,231,258,259,392,393,401
泌乳母猪模型 156,169~171,399
泌乳生产曲线 11
棉粕 301
免疫接种促性腺激素释放激素（GnRH） 231,244,246
免疫球蛋白 36,121
免疫去势 1,2,149,225,226
面包渣 269
明胶 307
模型 392
模型评估 162
模型使用指南 392
模型预测法 28
没食子酸丙酯 58,189
默认值 142,167,170,171,393
母体 32
母体蛋白 11
母猪 12,42
木聚糖酶 186
木薯粉 282
木质素 70,71,211
苜蓿 88,121,205
苜蓿粉 268
苜蓿干草 267

N
钠 78,84,90,91,160,175,179,227,248,267
钠和氯 90
内源氨基酸 21,208
内源氨基酸损失 209
内源赖氨酸 37
内源损失 21
内源性氨基酸损失 30
能量 1,2,4,158,169,208,212,231
能量摄入量 143,147

能量需要 8,10
尿能 6
尿素 6
尿液 75
柠檬酸 58,184
凝血 122
牛海绵状脑病 201
农药 198
农药残留 197

O
呕吐毒素 188,197~199

P
抛光大米 346
喷干全鸡蛋粉 302
膨化 2,116,205,206
膨化玉米 206
膨化芸豆 308
膨润土 188
膨胀 205,206
皮肤和毛的氨基酸损失 30
皮毛损失 29,30
啤酒酵母 385
啤酒糟 277
贫血 93~96,100,121,126,127,129,130
平均日增重 23,44,85,158,171,175,183,254~257
葡萄糖 65
葡萄糖耐受因子 92
普通黄玉米 286

Q
千卡 4
铅 199
禽副产物 340
禽肉粉 341
驱虫药 183
去壳红花粕 350
去壳压片燕麦 327
去壳燕麦 324
去皮豆粕 124,178,179,233
全消化道标准可消化 141,151,165,168~170,231,233,235,
  237,239,241,243,245,247,251,253,255,257,261
全消化道表观可消化 141,231,233,235,237,239,241,243,
  245,247,251,253,255,257,261
全消化道表观消化率 85
全脂大豆 177,366
全脂葵花籽 372

全脂棉籽 300
全脂双低菜籽 279

R
热处理 206
热增耗 6,7,13,51,52
妊娠 10,128
妊娠母猪 8,9,10,11,22,26,27,30,31,35,36,77~79,88~90,92,
94,97,125,126,130,131,141,152,154,156,158~162,168~170,
189,205,231,258,259,392,393,401
妊娠母猪模型 152,155,167,397~400
日粮 402
日粮氨基酸失衡 21
日粮配方 401
溶剂浸提豆粕 365
溶剂浸提葵花粕 374
溶剂浸提去壳葵花粕 373
溶剂浸提去皮低寡聚糖豆粕 363
溶剂浸提去皮豆粕 357
溶剂浸提去皮高蛋白豆粕 361
肉毒碱 56,191
肉粉 313
肉骨粉，磷>4% 312
肉碱 55
乳蛋白 37
乳房炎-子宫炎-无乳综合征（MMA） 121
乳化 52,209,210
乳清粉 319
乳清浓缩蛋白 320
乳酸 185
乳酸杆菌 67,184,185
乳糖 66,210
乳糖（来源于牛奶） 315
乳猪 227
软骨病 88,100,131
软红小麦 378
软化 88
软脂 54

S
三聚氰胺 2,197,200
色氨酸 17~19,21,23,26~32,36,39,41,42,126,127,129,155,
156,159,163~165,168,170,174,176,206,226,232,267
沙门氏菌 78,117,185,201
膳食纤维 66,67,70~72
砷 199
肾脏 75
生产性能 205

生物活性脂肪酸 53,54,227
生物利用率 21
生物素 116,123,124,160,248,267
生物性污染物 201
生物学利用率 2,20
生物学效价 21,94,96,97,99,123,126~129,232
生育酚 120
生芸豆 309
生长 10
生长激素 86
生长性能 52
生长育肥猪 17,18,22,23,27,29,32,37,41,44,52,55,56,76,85,
88~90,92,93,97,118,121,123,126,128,131,141,142,156,
160~162,164~166,170,175,176,177,179,184~187,189,
190,231,232,234,248,392,393,401
生长育肥猪模型 142,155,156,159,163
生长猪 9,31,231
湿法碾磨 174,226
湿磨工艺 176
实测值 41
视黄醇 117
适口性 187
书籍 413,414
黍 321
双歧杆菌 67,184,185
水 74
水的硬度 79
水的周转 74
水合硅铝酸钠钙 188
水解 19
水溶性 2
水溶性维生素 116,267
水中总可溶性固形物 78
丝氨酸 18,19,22,125,129,267
饲料加工 1,6,116,189,204~206,219,225,228
饲料链和食品安全常务委员会 202
饲料添加剂 1,189,199,203,219,228
饲料污染物 197
饲料营养成分表 1
饲料原料 401
饲料原料成分 262
饲料原料组成 227
饲喂方案 402
苏氨酸 17~19,21~23,26~32,36,39,41,42,151,159,160~166,
168,170,175~177,179,184~187,189,190,226,231,232,234,
248,267,392,393,401
酸化剂 184
酸洗洗涤木质素 267

酸性洗涤纤维 7,71,205,262,267
碎大米 345
梭状芽孢杆菌 67,201
缩略语 404
缩写词 404

T

胎次 42
胎儿 11,32
碳 217
碳水化合物 1,2,5,7,20,52,59,65,67,68,70,71,74,87,92,96,98,
　116,126~128,177,205,208,210~212,216~218,228,262,267
碳水化合物酶 67,69,186,215
糖苷键 65
糖原 68,70,210,211
体蛋白 12,31,36,143,146~148,151,152,156,158,159,171,
　393
体蛋白沉积 35
体温 8
体脂肪 52,143,146~148,152,154,156,158,171,217
体脂肪沉积 148
体组成 142,147
天冬氨酸 18,267
天冬酰胺 18
甜菜糖蜜 322
甜菜渣 219,371
甜味剂 187,188
条件性必需氨基酸 1,17
调味剂 187,188
调质 205
铁 80,84,87,94~96,120,121,127,129,160,217,248,267
铜 80,84,87,93,94,100,120,121,160,217,248,267
脱氧雪腐镰刀菌烯醇 197,198
脱脂 DDGS 174
脱脂米糠 344
脱脂奶粉 316

W

外源酶 186
豌豆浓缩蛋白 332
微粉碎 205
微量元素 2,84,93,116,117,119,121,123,129,217,219,232,
　263,267
微生物 267
维持氨基酸需要量 30,142,145~147,154,157
维持需要量 29
维生素 2,13,44,51,116,117,124,125,127,129,141,160,161,
　185,189,202,227,228,231~233,235,237,239,241,243,245,
　247~249,258,259,251,253,255,257,260,261,263
维生素 A 116,160,248
维生素 B_{12} 84,93,116,129,130,161,227,248,267
维生素 B_6 116,127,129,161,227,248,267
维生素 C 116,131
维生素 D 85,116,248
维生素 D_3 160
维生素 E 58,93,96~98,116,120~122,160,189,248,267
维生素 K 116,160,248
委员会成员 410
未阉公猪 144,146,148,151,165
窝仔数 171
无氮浸出物 226
无机酸 184
无效氨基酸 37
物理性污染物 202

X

硒 2,26,80,84,93,97,98,121,122,160,189,198,248
硒缺乏症 98
纤维 211
纤维素 70,205
纤维素酶 186
限制性氨基酸 20,22,23,174
香味剂 187
消化率 205,208
消化能 1,4~6,13,52,54,142,174~176,205,212,217,226,260,
　267
消化能系统 13
小麦 17,18,87,123,126,127,186,206,248
小麦次粉 384
小麦蛋白 381
小麦麸 379
小麦加工筛下物 383
小麦酒精糟及可溶物 380
小麦细麸 382
小母猪 8,146,148,150,151,165,166
斜率比值法 20
斜率法 208
缬氨酸 17,19,22,26~30,31,32,36,155,156,165,168,170,226,
　232,267
锌 70,80,84,87,88,93,94,98~100,120,160,217,227,248,267
性活跃期种公猪 44
血粉 275
血红蛋白 95,96
血浆 276
血浆蛋白粉 90
血糖碳水化合物 66

血细胞 274

Y

压榨低寡聚糖豆粕 364
压榨豆粕 359
压榨高蛋白豆粕 362
压榨花生粕 330
压榨去皮豆粕 356
压榨双低菜籽粕 280
压榨椰子粕 284
压榨棕榈仁 328
芽孢杆菌 184,201
亚麻荠粕 278
亚麻酸 54,58
亚麻籽 305
亚麻籽粕 306
亚油酸 53,54,160,161,231,248
烟曲霉毒素 197
烟曲霉毒素 B_1 188
烟酸 116,122,126,127,129,161,227,248,267
阉公猪 8,12,86,90,142,144,146,148,150,151,165,166,186,
　　231,236~239,244,246,247
炎症反应 53
燕麦 17,126,127,186,205,325
羊水 34
养分排泄 215
氧化速率 56
氧化酸败 51,54
氧化稳定指数 57
氧化应激 58
氧化作用 120
椰子粕 285
叶酸 116,125,126,130,161,248,267
一氧化二氮 219
胰蛋白酶抑制因子 21,177,178,206
胰岛素 92,98
乙醇 2,3,51,66,67,72
乙醚浸出物 177
乙醛 57
乙酸 218
乙氧基喹啉 58,189
异亮氨酸 17~19,22,25,27,29,30,32,36,155,156,164,165,168,
　　170,226,232,267
益生菌 184,185
益生素 219
益生元 67,68,185,219
益生作用 67
英国农业研究委员会 54

鹰嘴豆 333
营养物质 208
营养需要的评估方法 225
营养需要列表 231
硬红小麦 377
油酸 54
游离脂肪酸 51,52,58,179
有机酸 184,219
有效代谢能 4,12,13,142,232,233,260,261
有效环境温度 144,145
有效消化能 232,233,260,261
羽毛粉 303
羽扇豆 311
玉米 2,17,18,87,91,123,126,127,129,164,165,168,170,174~
　　176,226,232,267
玉米赤霉烯酮 188,189,197~199
玉米蛋白粉 176,297,298
玉米蛋白饲料 297
玉米-豆粕日粮 164
玉米-豆粕型基础日粮 141,142
玉米-豆粕型日粮 90~92,95,97,99,100,118,121,124~129,
　　131,142,174,175,186,231,233
玉米麸 288
玉米副产物 174,189
玉米高蛋白酒精糟 293
玉米酒精糟 174,186,189,190,289
玉米酒精糟及可溶物 174
玉米酒精糟及可溶物,6%<脂肪含量<9% 291
玉米酒精糟及可溶物,脂肪含量<4% 292
玉米酒精糟及可溶物,脂肪含量>10% 290
玉米酒精糟可溶物 294
玉米胚芽 175~177,295
玉米胚芽粕 176,177,296
玉米-去皮浸提豆粕基础日粮 13
玉米糁,玉米渣 299
预测值 10,38~42,142,153,155,158,164,166,167,169,171,393
原料库 401
原料列表 263
圆酵母 388

Z

兆卡 4
蔗糖 66,67,177,178,187,210
赭曲霉毒素 197
赭曲霉毒素 A 188,198,199
针对促性腺激素释放激素（GnRH）的免疫注射 12,148,149
针对促性腺急速释放激素（GnRH）进行免疫接种 162

支链氨基酸 17
支链淀粉 68~70,205
芝麻粕 352
脂肪 1,5,7,8,10~13,21,23,51~59,74,92,94,121,125,127,128,
　142,149,152,162,167,174,175,208,209,212,217,225
脂肪沉积 10~13,141,142,148
脂肪酸 1,51~55,59,123,179,211,218,226,248,249,258~260,
　262,263,267
脂类 1,2,10,51,52,54,58,59,92,96,98,116,121,126,186,209,
　226,227
脂溶性 2
脂溶性维生素 51,117,189,267
脂质 51~53,57~59,70,97,189
脂质氧化 56~58
脂质氧化速率 58
直链淀粉 68,69
植酸 88
植酸磷 267
植酸酶 87,99,152,175,176,187,215~217,227
植物提取物 58,183,186,219
植物提取物制粒 116
指示剂 208
制粒 2,51,87,190,205,206,219

中链脂肪酸 51,53
中性洗涤纤维 5,52,71,175~179,205,267
种公猪 44,231
重金属 199
猪肠膜蛋白 338
猪营养需要 1,4,262
主菜单 393
转基因 202
子宫 36
仔猪 23,231,392,393
仔猪模型 394
紫花豌豆 335
自发性骨折 94
自由基 56
棕榈仁粕 329
总氨基酸 28,142,233,235,237,239,241,243,245,247,251~
　253,255,257,260
总能 1,4,5,54,212,263,267
总膳食纤维 71,267
组氨酸 18,24,29,30,32,36,99,151,164,165,168,170,232,267
最低+必然分解代谢 38,41,42,151,159
最低+必然分解代谢量 151,155,156